ABARTH
The man, the machines

Luciano Greggio

ABARTH

The man, the machines

GIORGIO NADA EDITORE

GIORGIO NADA EDITORE s.r.l.

To Angela

Editorial Coordination
Leonardo Acerbi

Project, layout and cover design
Isabella Gianazza

English Translation
Robert Newman

Giorgio Nada Editore s.r.l.
Via Claudio Treves, 15/17
20090 VIMODRONE (Milano)
Tel: ++39 02 27301126
Fax: ++39 02 27301454
E-mail: info@giorgionadaeditore.it
http://www.giorgionadaeditore.it

The catalogue of Giorgio Nada Editore publications is available on request at the above address.

Abarth. The man, the machines
ISBN: 88-7911-263-5

The author and publisher would like to express their sincere thanks to Mrs Annelise Abarth, who made this publication possible by so kindly making available documents and family photograph albums, as well as the invaluable testimony she provided on the life and professional activities of Carlo Abarth.

Acknowledgements

Thanks to: Giancarlo Balbo, Gianni Cancellieri, Mario Colucci, Giuseppina Cortese and Massimo Castagnola - Ufficio Documentazione Centro Storico Fiat, Regina Costa, Ugo Fadini, Luciano Fochi, Paul Frère, Franco Lombardi, Antonio Meoni, Vincenzo Osella, Mario Poltronieri, Antonio Tomaini, Sergio Seccatore, Giorgio Valentini, Franco Varisco.
As well as: Fondazione Negri and Mondadori Press.

Thanks should also go to Professor Andrea Curami, whose techno-historic knowledge and illustrative material were of inestimable value to this book.

CONTENTS

AN APPRECIATION

This book recounts the work of a man who, in many years of total dedication to the world of engines, was able to greatly influence the habits and aspirations of millions of motorists. His tremendous ideas, contagious passion and unshakable perseverance were models of life to those who shared with Carlo Abarth the stimulating experience of working in one of the most exciting periods in the history of the motor car in Italy. To be at his side was sometimes arduous, due to the constant pressure he exerted on the search for perfection and the rigorous criteria that he considered inalienable in the daily management of his business, at times generating serious disagreements with his collaborators, but in the end his conviction, his credo as a technician and businessman always found unequivocal verification.

I first met Carlo Abarth early in my career in 1948, when my job at Cisitalia was to draw up the minutes of meetings in which the technical and motor sport heads of department took part. Together, we took our first step into the unknown territory of Abarth & C. and shared the euphoria of the early successes, the great victories but also the disappointments and the adversity of difficult moments. I lived the marvellous Abarth adventure from the beginning right up until the end and it was the finest, most interesting and happiest period of my life.

The direct testimonies and documentation contained in these pages contribute to defining the complex and unpredictable personality of Carlo Abarth, and help to better understand his conduct, reasoning, character and the outbursts that were, at times, unexpected and only apparently contradictory.

The passionate epic of the Corso Marche factory and its founder is unrepeatable, although some have tried in vain to take the same road. It is possible to say that times were different then: the conditions of the Fifties and Sixties motor vehicle market opened up favourable possibilities to cars considered impracticable today, cars that raced and won every Sunday of the year in different categories and classes. It was all made easier by the benevolent circumstances of the moment, and by the stature of the competitors. But the happy combination of all these elements materialised thanks to the prodigious intervention of Carlo Abarth, the catalyst, the one who knew how to choose the right way of exploiting them all with an enthusiastic operation of technology and competitive spirit at an international level.

Lorenzo Avidano

1908

1928

ENGINES IN HIS BLOOD

At the dawn of the 19th century, Vienna was still the capital of a vast empire made up of an imposing mosaic of ethnic and language groups, which extended across a landmass of almost 400,000 square miles. The reinstatement of an antique equilibrium in Europe that was enshrined at the Congress of Vienna following the overwhelming Napoleonic adventure had consolidated the Hapsburg monarchy, the most splendid and venerable of the old continent, over the years. Franz-Josef, last emperor of the Holy Roman Empire, came to the throne in 1848, rigorously gave the appearance of the authoritarian and unapproachable monarch surrounded and protected by an insuperable military class, but in the eyes of his subjects he also appeared as the first soldier and first servant of the State.

During the last three decades of the 19th century, Austria had successfully implemented a policy of outer peace to be able to pursue internal development on a grand scale. Although a career under arms remained the most coveted and respected within the empire's borders, the Austrians attached considerable importance to technical and industrial activities. Founded in 1815, the Imperial Academy of Science acted as a great training centre for the new generation of technicians, bankers and entrepreneurs that went to make up the class of a new, pragmatic and tumultuous bourgeoisie destined to work side by side over the years with the consolidated aristocracy of the military and bureaucracy. The World Exposition, inaugurated on 1 May 1873 at Vienna's Prater, with its fifty thousand exhibitors spread over a vast area never previously used for the purpose, opened up the planet to the conquest and boundless prospects of this rapid economic growth that went to change in significant measure the atmosphere of Viennese society and the rhythm of life of the capital's inhabitants.

That was made even more evident by the giant space of Ringstrasse, where the old medieval fortifications were demolished to make way for the urban and architectural restructuring of the city centre. At the end of the century, the style of the new buildings, often harmonious and always imposing, swirling music and dancing in many areas of the city had generated a sort of collective infatuation for the new turn in the country's fortunes, fired by a long period of tranquillity on the borders of the empire and the new and thrilling myth of speed. Riding this thrust of general euphoria for train journeys, the "felix Austria" of long ago found itself projected towards sustaining a pre-eminent role in the great historical process by which many would contribute to reducing distance - and not just geographical distance - between peoples. The opening up of new means of international communication as a result of the fundamentally important advent of the railway system and steam ships, as well as the fortunate and surprising sequence of primary scientific discoveries that took place between the 19th and 20th centuries, among them the internal combustion engine, the telephone, the microphone, the typewriter and the printing machine, contributed to giving a further push to the industrial revolution that had been taking place for a number of years, and opened the road to the use of new and unlimited sources of energy, petrol and electricity.

Karl Abarth's father, Karl Anton, had a restless youth and became involved in a host of different projects. The strict disciplinary environment of the military academy, from which he graduated as a second lieutenant, brought no end to his adventurous and voluble personality.

Vienna, where the rapid installation of numerous mechanical industries had attracted hoards of able and ambitious technicians, had become a meeting point, a sort of gymnasium in which everyone was called upon to try his hand at realising his dreams. But the city had also maintained its old function as a crossroads for the powerful aristocracies and reigning families of Europe. Heads of state from King Edward VII of Great Britain to Kaiser Wilhelm and the Russian grand dukes were often the guests of Vienna, participating in official banquets and ceremonies, attending military parades or passing entire weeks at the elegant thermal baths of Marienbad or Karlsbad. Around the monarchs revolved a multitude of courtesans and correspondents of the leading newspapers, which conferred on the city's life an opulent and sparkling atmosphere; it spread along the monumental buildings of Ringstrasse, the theatres, the concert halls, the sophisticated shops in Karntnerstrasse, in the grand cafés, where multicoloured cultural activities of a vast community found expression that demonstrated itself to be open to external inspiration but was, nevertheless, anchored in a number of rigorous parameters imposed by tradition.

A tradition that maintained itself coherent with the great Biedermeier tradition - even in family life - that constituted in the years between the 1815 Congress of Vienna and the revolution of 1848, a moment of equilibrium between progress and romanticism, between industry and poetry. In line with the Biedermeier way, children were brought up to obey the law and respect formality, in absolute harmony with the principals of discipline and the constituted hierarchy. At home, father was called upon to exercise authority over his children with a firm hand of command and by setting a good example by his demeanour, while mother was delegated to the firm management of the domestic economy. To children who manifested emotion or ill humour in public could be debited discredit that also reflected, like a violated code of life, on the parents.

In this composed atmosphere comprised of sweetness and harsh principals permeated by suggestions and symbols linked to the "belle époque", Karl Abarth was born in Vienna on 15 November 1908. Already, symptoms of far-

Karl's mother Dora Taussig, daughter of a Viennese textile manufacturer, shared the incessant wanderings of her husband for a number of years, until their marriage was dissolved after the First World War.

reaching unease were eating away at the edges of the Hapsburg empire and in the capital they were being nurtured which, in just a few years, was to lead to the deflagration of the First World War and the tragic fall of European equilibrium, maintained for over 60 years by Franz-Josef. In the Abarth home, Karl was the second-born for he arrived a year after his sister, Anna. The birth of the male child induced the mother, Dora Taussig, to hope her husband, whose name was also Karl, would finally put his life in order and apply the brake to the carousel of activities into which he threw himself with the enthusiasm of a novice and from which, equally precipitously, he walked away only to embrace another credo immediately afterwards, another passion suggested by his creative temperament. Karl's grandfather enjoyed a position of importance at Merano, where he owned the Hotel Sonne and managed the town's leading post office, including the horse and coach service that transported correspondence through the immense territory of the empire. He was a man who, in those days, passed for a prominent personage within the community and who could have equally pointed his son in the right direction to make his way in business or assist directly in managing the hotel. But the personality of the son, so restless and subject as much to overwhelming euphoria as he was to deep depression, was not fertile ground for firmness and planning. The bourgeois and insignificant life, easily predictable with pre-constituted plans, pushed Karl senior into trying his hand elsewhere so as to find a way of emerging from that greyness without hope that, to many of his contemporaries, appeared the crowning glory of their greatest aspirations. From his youth, Karl's father dedicated himself successively to painting, opera, fencing, boxing in which he did well, as he did in billiards and chess; finally, he seemed to find a conclusion of his wanderings during his military service, in which the inflexible atmosphere of the officers' academy seemed to win over his unstable and adventurous nature.

The marriage to Dora Taussig, daughter of a Viennese textile manufacturer, led him to another sudden change; he left the army in which he was a second lieutenant and went to Czechoslovakia to manage one of his father-in-law's factories where, however, once again the methodical life of office work, articulated by habit and the lack of great flights of emotion, translated itself into a new goad, an anxiety that made him transfer himself once again to Vienna with his young family.

At the time, the international relations situation was rapidly deteriorating and gave a glimpse of the birth of various and dangerous deployments by the different nations in the field. In spite of the "alliance of the three emperors" ratified in 1881 and in which Austria, Germany and Russia committed themselves to act in common accord in all matters of the Balkans, and regardless of the entry of Italy a year later into bi-lateral negotiations between Germany and

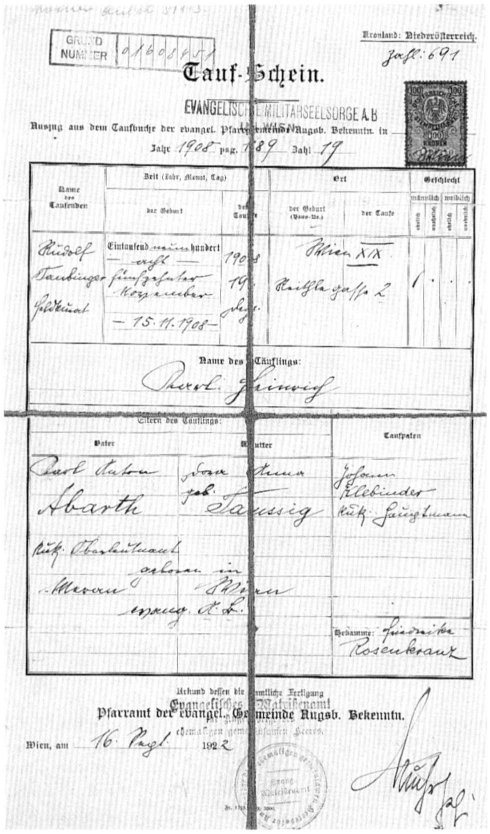
Kronland: Niederösterreich.
Tauf-Schein.
EVANGELISCHE MILITÄRSEELSORGE A.B.
Abarth

A copy of Karl Abarth's birth certificate, listing the date as 15 November 1908.

Austria that gave origin to the Double Alliance, the political and military chess game in Europe was permeated by recurring disquiet, provoked by incessant research by the great powers to construct for themselves a network of cooperation that acted as a protective curtain against the day - which was moving inexorably closer - a grave international crisis exploded in their faces.

With the soothing of the divergence between Russia and Great Britain in 1907, there developed – still in the absence of precise agreements – reconciliation between those two empires and France. From that moment, the countries of the old continent found themselves divided, even if not officially, into two opposing blocks, both of them disinclined to search for equilibrium but to erect a wall between the forces in the field. A wall made even more threatening by the huge arsenals of war that the various protagonists of this confrontation built up indiscriminately, more due to fear than a thirst for prestige and ambition. The latter pushed the military aristocracy of the different countries to assume greater authority in decision-making, to expand their intervention and their discretion in the use of arms. The historians and commentators of the period agree that it was collective fear in the first 10 years of the 20th century that took the place of reason in that desperate race towards the terrible conflict.

The first spark flew in the Balkans in 1908, the year in which little Karl Abarth was born in Vienna. Being party to the Turkish revolution, Austria annexed the province of Bosnia and Herzegovina, which it had administered since 1879, but the operation met with Russian and Italian resentment, which, while belonging to different sides, considered the move an affront by Vienna. The tension eased but that intervention rekindled, in the alliance's game that was taking place, quarrels between Germany on the one side and France and Great Britain on the other. The rapid and insane succession of incitement to war, of warlike speeches and frontier incidents, fed in all countries the psychosis of having already been sucked into an irreversible spiral that led to conflict. The unappeased ferment of the Balkan people, encouraged by an Orthodox clergy and reinvigorated by a new nationalism inspired by the west, strongly resented the influence of tribal custom, which is still not extinct today. In this ambiguous simmering of legitimate claims and easy hope came the fatal and almost expected shooting in June 1914 at Sarajevo, which caused the literal explosion of the most frightening war that has ever bloodied Europe. Four years later, in November 1918, not only did ruins and millions of dead testify to the folly of man, but the whole equilibrium of the entire old continent had been flattened, as if an enormous steamroller had passed over it and cancelled a complete chapter in the history of Europe.

The war shattered the Hapsburg Empire and dissolved the social-economic structure in which Austrians had lived and prospered for so many years, dragging into the abyss the entire collective nation. The Abarth family did not

Karl at five years old and his six-year-old sister Anna photographed at Merano in 1913, just like most families of the Hapsburg bourgeoisie.

Karl at 10 in 1918 with his sister. The defeat of Austria during the First World War did not greatly change the destiny of the family Abarth, which moved from Vienna to Merano after the conflict.

emerge too badly bruised from the tragedy of a defeated country; discharged from the army, father Karl decided to leave Vienna and return to Merano, taking with him his wife and their children Anna, 11, and Karl 10. At Merano, too, things had changed, with an Italian administration that had been installed in the place of the Austrian authorities; there was a considerable, much diffused bitterness for the adverse fortunes that marked the epilogue of Austria as a great world power. The Abarths still enjoyed an influential position in Merano and all supposed that, in that situation, the reunification of the family under the ancestral roof could bring an end to the recurring pilgrimages of father Karl. But it was not like that. After a while, he decided to move to Vienna again at the command of an umpteenth change of mood and at that point the precarious harmony in the bosom of the family was subjected to an irreversible epilogue and the unstable marriage of the Abarths broke down. Dora Taussig remained in the capital with their children, while the father returned once again to Merano where he took Italian citizenship.

Those wanderings did not disturb the economic tranquillity of the Abarths and young Karl enjoyed a serene and cushioned childhood. He showed his interest in things mechanical from infancy and he put himself to the test by taking kitchen utensils and locks to pieces and putting them back together again, developing an acute self-taught fascination for the secrets of elementary mechanisms and the arcane laws of applied electricity. Nothing in the house escaped the boy's thirst for knowledge and, once reassured by his familiarity with pinions, screws without end and solenoids, he even began to modify old mechanisms and to experiment with new devices, demonstrating an unmistakable vocation and a lively inventive spirit. At school, he was neither mediocre nor did he shine among the most gifted students, with the exception of drawing and the discipline that left him free interpretation of his inspiration and creativity. But he excelled at sport, which he played with perseverance and his robust, muscular physical constitution was of advantage to him. He particularly liked cycling and participated in club races, in which he beat his opponents in high-speed sprints with the power of his starts, clearly pointing to a scrupulous athletic preparation. He competed in his early races riding a heavy French Bastide, a bicycle he could take to pieces and re-assemble again with his eyes shut and which gave him his first victories. But his great ambition was to be able to savour the taste of a motor bicycle, that revolutionary and still abstruse machine which, at the time, was quickly modifying the rhythm of millions of lives.

In his first approach to the engine, Karl put all the ardour and passion that many of his contemporaries showed for the new civilisation of the self-moving mechanical unit. Once the heroic epoch of electrically and steam driven cars had been overtaken in just a few years, the new petrol engined vehicle established

Karl at the start of a cycle race in Vienna in 1924. He played many sports from a young age and scored his first victories with a bicycle.

Karl at 18. He had already discovered many of the secrets of the internal combustion engine and alternated school and hours spent in the Degan workshops in Vienna, where the company built chassis for motor bicycles and motorbikes.

itself in an unstoppable flood of initiatives until it found universal consecration and enormous industrial development during the 1914-1918 war. At the end of the conflict, it was the survivors of the battlefields of the first motorised war in history who became the most ardent promoters of the new conquest of motorised two and four-wheel transport. By the early Twenties, the car had already emerged from its pioneer period in the United States and had entered the family as a much-used instrument of work with great aplomb. In Europe, the euphoria of the engine was still being lived almost exclusively by the most affluent social classes, but the young generation was profoundly fascinated by it.

These, and who knows what other incitements, coursed through 15-year-old Karl Abarth's heart and mind the day he was pitch forked for the first time from a motorcycle for the first time, a machine that had been loaned to him by a mechanic near his home. What exaltation he must have felt when putting down the power of the engine onto the road's surface, and what intoxication when listening to the deafening noise of the exhaust and tasting the ecstasy of speed! But all the euphoria of that magic moment was destined to stop there. As he lost control of the bike, the young aspirant motorcyclist had his first road accident; he fell violently as the bike hit a wall. He was uninjured but the damage to the motorcycle cost him a severe dressing down by his mother, who had to pay for the machine's repair.

However, the dice had been rolled and the uncontainable demon of the engine had entered the young Abarth's blood. As a 16-year-old student at a technical-industrial institute, he was able to obtain permission to work during his free time at Castagna & C., a workshop specialised in precision mechanics for the engineering faculty of the University of Vienna. He learned to improve on the first rudiments of the construction and the experiments he carried out on his own during his infancy, but his anxiousness to engage in the assembly of the internal combustion engine led him soon afterwards to knock on the door of the Degan workshop, where they made chassis for motor bicycles and motorcycles; there, Karl could enter the fantastic world of the production of two-wheeled vehicles, powered by engines. The joy of disassembling them, repairing them and improving their performance was the realisation of a dream he had contemplated with pleasure for years; a dream that was finally within his reach. In a short space of time, he had built up valuable experience which, at just 19 years old, was soon to lead him to a stimulating new adventure, to join MT - Motor Thun, a motorcycle factory, based outside Vienna at Traischkirchen, on the pleasant road that links the capital with the Semmering Pass. This new job opportunity became a possibility the day the Austrian motorcycle racing champion Josef Opawsky met the young Abarth at the Degan workshop and a firm friendship quickly developed between the two. The considerable ability with which Karl was able to work on motorcycle chassis and engines was not lost on the champion, while to the young man it did not seem possible that the leading racing motorcyclist of the period could consider the possibility of employing him as mechanic and test rider of his racing bike.

That is how the young Abarth opened not just the door of Motor Thun, one of the best-known and most successful factories of the day, but also the prospect of being able to participate in racing as the trusted mechanic of Opawsky; within a few months, he had joined the small group of official test riders and was soon

In April 1928 Karl competed in his first motorcycle race, riding for Motor Thun in place of a regular works team member who had suddenly become ill.

able to distinguish himself with his profound technical knowledge and his untiring will to work with a spirit of devotion that bordered on exaltation.

Karl saw that MT was particularly competitive with Villiers, the British bikes, which he personally modified by lightening the moving parts to increase their output. Villiers had also begun operations in the world of motor bicycles and from the turn of the century supplied chassis and free-wheel components to a large number of bicycle factories, aspiring to a production of 80,000 units a year. In 1912, the founder of the company, John Marston, decided to compete with the well-established British motorcycle manufacturers by switching his production exclusively to engines, which were installed in the motorcycles of many different makes. Villiers became the standard bearer of the image of the mechanical parts suppliers, which play such a leading role in the automobile and motorcycle industries throughout the world today. The range of single cylinder and two-stroke Villiers engines achieved considerable success before and after the First World War and the company became highly popular in 1922 with its 247 cc Mk6A that became part of a successful triad of similar technology engines (150, 250 and 350 cc), which were bought in those days by a substantial group of European motorcycle manufacturers. The single cylinder Mk6A (bore and stroke 67x70 mm), had a large roller bearing in the piston rod head, a semi-automatic lubrication system and variable advanced ignition; it was the engine fitted to Opawsky's bike, all the darkest secrets of which Karl Abarth knew only too well. Taking it to pieces and putting it back together again with loving care, he seized on its robustness and the latitude he had with which to modify it to improve the power output and, in particular, engine torque to produce the lightning acceleration needed by racing riders.

With one of those engines, Karl was able to realise another of his dreams, perhaps the most ambitious for a lad who was not yet 20. He found the time to design and build his first motorcycle chassis, into which he invested his vast experience and his unfettered enthusiasm. Contrary to those of the period, so massive and heavy, Karl's motorcycle was of streamlined and elegant structure, yet it was extremely robust; in extensive tests, the young technician found a way to refine the manoeuvrability and reliability of the bike, its braking and stability, comparing it directly with the other bikes, the merits and defects of which he knew by heart due to his daily job of preparing and testing them. At that time of great satisfaction, Karl's life was entirely devoted to work, but what could appear to many as renunciation or even sacrifice, was, to him, a satisfactory return on his most important aspirations. During the day, he worked at Motor Thun and in the evening at Degan, spending long hours on his projects, developing early experience as a new inventor.

The magic of that period of great professionalism and maturity culminated in April 1928 when, on the eve of the motorcycle Grand Prix of Austria,

At 20 Karl already displayed the proud expression of an established champion and competed every weekend in races on sand tracks.

one of the best Motor Thun riders was suddenly taken ill and the whole team was exposed to the risk of starting the race at a disadvantage, compared to its competitor factories. Taken by surprise, the factory's technicians and Count Mathias Thun gambled and entered the youngest and most able of their test riders for the race, one Karl Abarth who had done so well in recent months and who was considered a promising talent for future racing seasons. Delighted with that recognition but aware of the danger that awaited him, Karl presented himself for practice in the saddle of the team's oldest motorcycle, determined to give the best of himself.

As always happens on practice days, the established riders risked everything to test the preparation and the competitiveness of their machines against their rivals; a few breathless laps, without holding back and without the exclu-

sion of mishaps. Even so, at the end of that infernal carousel the fastest time was Karl's, the newest entrant and an unknown young workshop mechanic. That unexpected result surprised everybody, but the quick young newcomer did not even have time to savour the joy of his success; charged with cheating by older riders, the first among them his boss Opawsky, he was accused of modifying his bike for practice with the ambitious objective of beating the other competitors, instead of simply helping to hold the head of his team high.

Karl tried to defend himself and to show his good faith but, in the end, he had to accept the demand that he exchange his motorcycle for Opawsky's for the second practice day; he should never have done it! At the beginning of his life and resolute to see himself rehabilitated from the injustice he had suffered, the young mechanic outdid himself and once again recorded the fastest time. In one fell swoop, a junior member of the team had shown more talent and courage than the champion, and that caused a great deal of fuss in the world of racing; it also caused Abarth a great disappointment that brought him into close personal contact with the malice that permeates human nature and convinced him that it is not always by means of merit and talent that some people climb the ladder of success. Lined up on the grid with the great names in the sport on race day, Karl started with cold determination to compete on equal terms, him against the rest; but after a few laps he was forced to retire with mechanical problems that he, expert technician, found mysterious, almost suspect. A flywheel could not loosen itself by chance, he decided

Karl Abarth in action on the track at Krieau, where he came third. In 1928 he also came third at Ischl, first at Salzburg and second at Baden.

Karl in 1928 wearing typical motorcycling clothes of the period.

at the end of a quick check of the engine, and suddenly he realised someone had sabotaged his motorbike to keep him out of the winner's circle.

That shattered Karl's illusions, his hopes and his reckless young belief in the supreme value of loyalty and rectitude. He suffered a sleepless night, tormented with doubt and dread that his future would be compromised by such dirty business. But the following morning he was unable to control the indignation he felt towards Count Thun and he resigned, slamming the door as he went and ready to start once more from scratch.

1928

1945

THE RACING ADVENTURE

The young Abarth wasted no time in returning to work; he now bore the scars of several years in a hard business and had scored more than a few successes, even if they were short-lived. In the first place, he was convinced that the world of racing was for him, a world in which he had already shone but from which he was expelled through no fault of his own. But he could find no other factory prepared to entrust him with a racing bike so he was reduced to buying one himself, a 250 cc British Grindlay-Peerless that he considered an adequate machine to campaign on the track. At the time, the Grindlay-Peerless was known as one of the fastest bikes in the world, having established a number of speed records ridden by racer C.W.G. Lacey over various distances in the late Twenties. Equipped with a JAP engine, almost all the models in production distinguished themselves from their contemporaries by the elegance of their streamlined tubular chassis and the use of parallelogram front forks by Brampton, which gave the bikes notable dash. Well built and with high quality components, the Grindlay-Peerless won themselves an enviable reputation, not only among British customers but also overseas enthusiasts.

A young and successful motorcyclist, in 1928 Karl achieved his ambition of competing against the greatest champions of the sport.

So it was not by chance that Karl Abarth chose the English machine with which to go racing once again. He disassembled his second hand steed piece-by-piece, lightened and balanced it in various areas and achieved surprising results with the small 250 cc. He competed in a number of races without disgrace but also without great honour, pitting himself in isolation against the toughest squads with obstinacy and fury and improving his performance each time. To be able to keep the lead from such rivals with a 'private' bike was already an achievement to him. Until 29 July 1928, when he won his first race at Salzburg riding his faithful Grindlay-Peerless. He made a brilliant return to the limelight in just a few months, overcoming difficulties and humiliations that many had thrown his way.

The Salzburg success opened the doors of new racing teams to him, for they had followed with interest the latest exploits and increasing maturity of the young champion. In 1929, after having just turned 20 years old, Abarth was engaged first by James, whose fame reached its peak in the Twenties, and then by DKW of Germany. Wrote engineer Pasquale Borracci in 'Motociclismo' magazine of 6-13 March 1926: *Whoever among motorcyclists was at the front* (in the 1915-18 war, ed.) *and had to use this make of bike (James), many of*

On 29 July 1928 Karl tasted the joy of success for the first time, winning at Salzburg on a British made Grindlay-Peerless.

which were requisitioned at the time of mobilisation, could do nothing at the end of the conflict other than immediately acquire one from war surplus, with the knowledge that they were buying eternal machines". And that is what they were, eternal machines; James motorcycles were meticulously built and the dimensions of their components infinitely correct, which enabled the British firm to climb to the most prestigious level in the international field. Among the company's technical innovations was mounting the clutch-gearbox group entirely on ball bearings (contrary to the simple use of bushes, as was the case with the majority of constructors) and in addition a clutch comprising nine pairs of bronze and steel discs in an oil bath, instead of discs of cork or other agglomerations that worked dry, as were widely adopted. For its 1926 models James also introduced a new lubrication system, in which the oil was returned to the reservoir by a pump operated by the engine drive shaft.

Much different from the technical concept point of view, but no less famous among enthusiasts for many years, was the Germany company DKW, established in 1917 by Jorge Skafte Rasmussen, a Dane who was a convinced supporter of the necessity to find an alternative system of traction to the petrol engine: he formed "Dampf Kraft Wagen" with the intention of producing steam cars. But in 1919, with Germany having been defeated in the Great War and on the edge of collapse, the new factory began to make a small 138 cc auxiliary engine for bicycles, which produced one horsepower at 2500 rpm. That was followed by others and a year later Rasmussen introduced a bicycle with double arm front forks and powered by a small single cylinder, two-stroke engine mounted horizontally behind the saddle that transmitted power to the rear wheel by belt. The young company's motorised bicycle quickly became popular, sustained by an advertising slogan in which DKW were the initial letters: "Das Kleine Wunder" (the little wonder), a catch line that soon brought notable success, contributing in no small way to the popularity of the company's products. The move from bicycles with engines to real motorcycles took place in 1923, the year in which DKW caught the attention of the market with its original "SM200" model whose tubed chassis was replaced by one of pressed steel. By the end of the Twenties, the German factory had become one of the most active in the field of technical innovation and pursued its ambitious objectives to form a powerful racing team; Karl Abarth became one of the team's highly respected standard bearers.

In 1928 and 1929 Abarth enjoyed two packed but highly successful racing seasons. In '28 he raced in the 250 cc class and came third on the dirt track

Left: Karl also competed on his 250 cc Grindlay-Peerless in 1929, his results including second places at Baden, Linz (picture), St. Polten and victory in the 175 cc class at Krieau

Right: Karl's greatest 1929 victory was in the Giesshubel hillclimb, won in record time aboard his 250 cc Grindlay-Peerless.

at Krieau, Vienna, third at Ischl, he won at Salzburg on the Grindlay-Peerless and came second at Baden. The following year saw him second again at Baden and at Linz, winner in record time at Giesshubel, second at St. Polten, first in the 175 cc class at Krieau, second on a James at the Gaisberg and Semmering hillclimbs and second again in the autumn race at Baden. By that time he had become a man dedicated to success. But his ambitions were not confined to sport and the continuous challenge of the track; his immeasurable passion and his well established ability in mechanical construction led him to design and build 'his' motorcycle bearing the Abarth brand in 1929. The bike had a light tubular chassis, heavily inclined parallelogram forks and a single cylinder, two-stroke 250 cc engine cooled by water through two radiators, a rare technique in those days. All this ferment of initiative, Karl's growing popularity on the track and his untiring commitment did not enable him to take a holiday. He always ensured he attended races, whether Austrian championship, non-title motorcycle meetings or hillclimbs; it almost seemed as if he had to prove to everyone that there was a leading role for him, Karl Abarth, in the enthusiastic world of motorcycling. And that is how it was until 4 May 1930, the day his plans and ambitions suffered a severe blow, a second and the more serious accident that seemingly tried to destroy his career and modify the course of his existence.

It happened at Wels, near Linz. Impetuous as always, Karl was leading the race and could already taste the unequalled euphoria of victory, when he

This James advertisement in the Austrian press marked Abarth's second place in an international race, in which he was up against the top riders of his day on a normal production bike.

At the 17th Semmering international race that took place on 15 September 1929, Karl came second on his James 175 cc.

lost control of his bike in an instant and came off violently. He barely saved his life, but the accident cost him multiple fractures of the right knee and injuries to the meniscus, which had been damaged in earlier falls. He was immediately operated on in Vienna and was soon able to use his leg, but his doctors told him he should never race again. For 21-year-old Karl it was a cruel but unchallengeable verdict; all at once his dreams were shattered, his intentions, plans for the future all banished, suddenly rendered illusory, unreachable. The young champion felt cheated by destiny and fell into a deep depression.

Once again he took up his life in Vienna as a mechanic and technical consultant to Degan, where at 16 years old he had begun to understand how a motorcycle was built. But his heart was not in it. With First World War military defeat and the fall of the Hapsburg dynasty, life in Vienna had changed greatly in just a few years; the ex-capital of a vast empire and effervescent heart of the "belle époque" had been transformed into an opaque and decadent city, grandiose only in its Teutonic architecture and echoes of a glorious past. That bitter and resigned atmosphere did not contribute to encouraging Karl Abarth to once again build up confidence in himself and seek other paths that would make sense of his existence. But the incessant call of engines tormented him, yet he could see no future in them that would enable him to face life with the excitement and kind of involvement that was once his. During the long hours he spent in the workshop, he experimented with a special exhaust for vehicles and with a singular intuition he developed an ingenious system

In 1929 in Europe, James went from victory to victory in the 175 cc class, winning the Grands Prix of Belgium, Hungary and Poland; they came second with Karl Abarth in the Semmering hillclimb – where he beat the previous year's fastest time by 17 seconds – and at Gaisberg.

With his usual impetuous and reckless style, Karl won in Baden in the autumn of 1929 on his James 175 cc.

The vast technical knowledge he had acquired in just a few years led Karl to design and build 'his' motorcycle, in which he installed a single cylinder water cooled 250 cc engine.

that, 20 years later, would become the beginning of his success as an entrepreneur and industrialist.

The Wels accident had marked the young Abarth both physically and morally, but as the months slowly passed a stimulating idea began to grow in his mind: an idea that was right on target and could give the fallen champion the chance of going racing again. His injured right leg may have stopped him from climbing onto a motorcycle and launching himself breathlessly into the fray again, but perhaps it would enable him to race again with a sidecar literally anchored to the bike; a sidecar, the little vehicle that was already widely in use, particularly in Britain. Karl grasped enthusiastically at the chance that signified for him a return to racing and the recovery of all those aspirations that constituted the very essence of his life.

It was in that period of great emotional stress in which Abarth went from sudden enthusiasm to the most agonising depressions, that he found a way of concentrating his technical interest on another British made motorcycle which, at the time, had forged itself a great reputation: Sunbeam. The manufacturer had already become famous in Britain towards the end of the 19th century for its hot enamelling on metal, which was first applied to kitchen utensils. Later, the technique was utilised for bicycles, which stood out from a thousand others for their shiny black paintwork and transmission by chain in an oil bath hermetically closed inside the sump. Sunbeam motorcycles first saw the light of day in 1912. John Marston, 76, owner of the factory and

Karl on his Abarth in the spring of 1929. In front of the engine are the two water radiators, a rather unusual technical feature at the time.

Karl conducted his first experiments in the development of his own exhaust system in 1930, using a British-made 500 cc Rudge bike...

...and continued with a Sunbeam, another British 500 cc machine, on which Abarth covered the 572 km from Vienna to Innsbruck over the Thurn Pass in a record time of 8h 37'.

After enforced retirement due to a serious accident at Wels on 4 May 1930, Karl decided to race again, but this time with a motorcycle and sidecar. In 1932, he competed in the 5th Prebichl Hillclimb on a Sunbeam 90, the capacity of which he had increased from 500 to 600 cc. In this photograph taken before the start, Abarth poses with his partner Joseph Holly, his regular sidecar rider for a number of years.

Karl in 1932 on an Abarth motorcycle he designed and built himself. The 496 cc engine was a two-cylinder, two-stroke Puch with split piston linked to a three-speed gearbox.

dynamic manufacturer of Villiers engines, wanted his machine designed by John E. Greenwood, well-known technician of the period who contributed to a large extent to the fortunes of Imperial Rover. The new 347 cc motorbike caught the market's attention for its engine drive shaft, which rotated on roller bearings, for its primary transmission with gears and its unusual appearance - that perfect paintwork. Shiny and rigorously black, the paint also extended to the handlebars, springs, forks and wheel rims, including the spokes.

It was in the Twenties that the British manufacturer achieved its greatest popularity by winning numerous hillclimbs and, among others, setting up the 350 cc speed record of 129.130 km/h. In 1924, Sunbeam switched from lat-

Above: The acrobatic virtuosity of Karl Abarth and Joseph Holly while cornering at the 1932 Prebichl Hillclimb, in which they clocked the fastest time of 3'54" and lowered the previous record by over a minute.

Right: Karl continually improved his Beiwagen (motorcycle and sidecar) riding technique and practiced constantly in an effort to lower his race times. The picture shows him winning the 1932 Gaisberg Hillclimb.

eral valve distribution to head valves; their "90" model, a 493 cc single cylinder, won its class in the 1928 Tourist Trophy ridden by Charlie Dodson and was the winning team, a member of which was Luigi Arcangeli of Italy. The Sunbeam 90 won the Tourist Trophy again in 1929 and its fame quickly crossed frontiers and spread through the continent; Sunbeam's 1930 racing machine ran on alcohol and had a top speed of 165 km/h.

Amid his non-stop dreams of returning to racing, Abarth heard that Sunbeam had promoted itself in the motorcycle world through its successes on the track and in endurance trials. He had already bounced into the news astride a Sunbeam in 1930, when he broke the nine hour barrier and covered the 572 kilometres from Innsbruck to Vienna in 8 hours and 37 minutes; an exceptional achievement for both rider and machine. Untiring perfectionist that he was, Karl increased the cubic capacity of his Sunbeam 90 to 600 cc, boosted the machine's engine power, sharpened the yield and next to the bike he fitted a steel tube sidecar, robust and light at the same time. He prepared to compete once more, almost bewitched by the odd-looking three-wheeled vehicle that would allow him to race again against himself and his adversaries.

Of course, sidecar racing meant competing with a courageous and experienced partner driven by the same competitive spirit as Karl, a man with a similar disregard for danger; almost Abarth's double, in the sense that he had to have an identical bent for engines and the same racing technique, completely harmonising himself with rider and machine. Karl found just such a man in Joseph Holly, a mechanic and expert turner, who had helped him more than once to modify his racing bike and admired Abarth to the point of devotion. Karl and Joseph raced together for a number of years on the turbulent tracks of many countries, competing in every event with a virtuoso commitment that gave their opponents no quarter and introduced a new way of sidecar racing.

Taken in 1933, Karl (second from the left) in a group photograph with the technicians and mechanics of the Musyl company, to which he sold his design for a robust yet light racing sidecar.

Initially sceptical, Sunbeam did not much relish the idea of sharing Karl's reckless new adventure and limited itself to providing the Austrian with a motorcycle. Abarth found a way of publicising his unusual sidecar and selling a few units, which helped finance his return to racing. Abarth's sidecar debut took place in 1932 in Austria at the fifth Prebichl hillclimb where, still only 23, young Karl immediately won his class and established a category record of 3' 54" into the bargain. The sidecar he himself designed and produced had opened the door for his return to competition and allowed him to decisively take his place among the champions of European motorcycling.

At the 19th Semmering international hillclimb in 1933, Karl clocked another record time, covering the route with his 600 cc Sunbeam motorcycle and sidecar in 8'11" and coming first in class.

Karl Abarth achieved considerable success in 1933, which also gave him a further opportunity of improving his machine and riding technique. He won for the first time outside Austria at Munich in July, and also took the Golden Helmet award with his 600 cc Sunbeam, flanked by the faithful Holly who was, more than ever, a man of high-speed acrobatics. Abarth repeated that success by winning the 600 cc class of the 19th Semmering race meeting. In the meantime, he sold the designs for his sidecar to the Muysl company, which not only competed officially during the 1933 season but also ran a considerable promotional campaign to attract public attention to the sidecar as an innovative touring vehicle. First produced as a service appendix or military device during the First World War, by the end of the Twenties the sidecar had assumed a new dignity as a valid alternative vehicle; it was considered a little snobbish and anticonformist on a steadily mounting tide of new trends among people and their conventions; the vehicle had made great strides in the areas of construction and aesthetics in more recent years and harmonised itself well with the motorcycles, for which it was built, almost becoming an integral part of some models.

XIX. Intern. Semmering-Rennen 1933

Sensationelle Erfolge

mit

Musyl-Beiwagen

Beste Zeit der Motorräder mit Beiwagen

Hermann Deimel A. J. S. 500 ccm mit Musyl-Beiwagen in 7:57

Rekord-Siege:

Neuer Klassenrekord: A. J. Schörg (Ardie) 350 ccm mit Musyl-Beiwagen in 8:48
Neuer Klassenrekord: Karl Abarth (Sunbeam) 600 ccm mit Musyl-Beiwagen in 8:11

Gesamt-Erfolge:

I. Preis: 350 ccm A. J. Schörg (Ardie) mit Musyl-Beiwagen (Klassen-Rekord)
I. Preis: 500 ccm H. Deimel (A. J. S.) mit Musyl-Beiwagen (Beste Zeit der Motorräder mit Beiwagen)
I. Preis: 600 ccm K. Abarth (Sunbeam) mit Musyl-Beiwagen (Klassen-Rekord)
II. Preis: 750 ccm M. Schmidbauer (B. M. W.) mit Musyl-Beiwagen
III. Preis: 500 ccm J. Walla (Rudge) mit Musyl-Beiwagen

The racing sidecar manufacturer Musyl celebrated its success at Semmering with an advertisement in the Austrian press showing the day's three winners, Karl Abarth (left) on his 600 cc Sunbeam, Hermann Deimel, astride his A.J.S. 500 cc and A, J. Schorg, who came first in the 350 cc class riding an Ardie.

Confirmation of the vast public interest generated by the motorbike and its new potential as a means of long distance transport came in 1934 during a historic challenge by Karl Abarth and his Sunbeam, to which an innovative new look Felber sidecar was fitted laterally and made more comfortable for the passenger. The intention was to emphasise the characteristics of the bike and sidecar and their ability so that they became better known as a new form of motorised transport. Abarth came up with an idea that sounded like a provocation; challenge the celebrated Orient Express train and the time it took to travel from Vienna to Ostend.

On paper, it was an ill-balanced challenge, considering the territory that had to be covered and the different conditions of the roads in the various countries, the customs formalities at the frontiers and the 1372 km distance route; there was also the hazard of mechanical breakdown, insidious and frequent at a time when vehicles took on long and demanding journeys. It was the latter that thrilled Karl most and attracted him to the idea; his innate talent to take almost any motorcycle apart piece by piece and put it back together again, as well as his outstanding riding ability in the worst conditions, gave him the heart to take on the challenge that, apart from its sensational publicity potential, generated much perplexity in the country.

The other protagonist of the challenge was the "king of trains and train of kings", as one of the Orient Express trains was defined; a train that, in the

Karl's run of successes continued in 1934. He won as early as the spring of that year on the Krieau sand track, where he established another record on his Sunbeam 600. Here, Abarth and Holly are shown before the start, decked out in their perfect racing gear.

Karl on his Sunbeam winning at Krieau against Walla and Vogl, his top opponents. Abarth recorded the fastest time of 43.8"on the fifth lap and set himself on the road to victory in one of the most prestigious races of the season.

last 20 years of the 19th century, had radically changed the way people travelled by rail, at least as far as the high-born bourgeoisie and the most famous personalities of the period were concerned. The idea of having the nobility and aristocracy of the day travel in a train replete with every luxury and comfort was that of engineer Georges Nagelmackers, who was the first to import the long distance travel formula of special sleepers from the United States, where George Mortimer Pullman had started it up at the end of the 1860s out of Michigan Central. In 1876, Nagelmackers founded the Compagnie Internationale des Wagon Lits and was able to convince the directors of various European countries' railways that the great conquest of steam would not realise its full potential if it were not unified with various construction regulations for rolling stock with the objective of ensuring the first international trains crossing frontiers.

The time was right because the countries of the old continent, which had remained barricaded within their own borders if not even at open war with each other, had opened up to the new era of fast and rapid communications and it was the train that assumed the vital role of the instrument by which the different cultures, experiences, languages and ways of life met. Even if originally destined to provide extravagance and amusement to the elite few, the C.I.W.L. revealed itself to be an extraordinary vehicle of social progress. The prospect of crossing Europe by rail from London to Paris, travelling on to Constantinople and then continuing as far as Peking and Vladivostock created great public enthusiasm and awakened enormous interest; King Leopold II of Belgium sup-

Karl used a small canvas-covered truck to take his bike and sidecar from race to race, where the vehicle doubled as a mobile workshop for minor repairs and tuning before each race.

ported the initiative. France followed suit, always keen to promote and safeguard its image towards the Orient, not wanting to lose the opportunity offered by Nagelmackers' enterprise of enabling its prestige to reach out far beyond its borders.

So the Compagnie Internationale des Wagon Lits inaugurated its Paris-Constantinople service in 1883 with a great official ceremony; it was 5 July when the first train, called the Orient Express, left its platform in the French capital to begin its 80-hour, 3180 km journey. The atmosphere aboard the groundbreaking train was not that of normal passenger rolling stock. The special sleeper carriages, panelled in mahogany and red velvet with furnishings in bronze and crystal, reproduced the environment of the great hotels of the period, the customers of which quickly learnt to move from one European city to another in their mobile luxury. The world of the "belle epoch", with its élan and all its intrigues of court and the bed chamber, was regularly cosseted aboard the Orient Express; a joyful ambiance yet at the same time decadent, one that died partly due to the outbreak of the First World War and the fall of the central empires, but which always constituted an element of attraction right through to the Twenties and Thirties.

The Orient Express from Ostend to Vienna was the most celebrated train in service at the time and was selected by Karl Abarth for his challenge. "To be able to cross the continent by road between the two cities with my Sunbeam and sidecar in less time..." he proclaimed more to himself than to others. And those who knew him well also knew the indomitable spirit of the young champion

In 1934, Karl Abarth visited a giant Zeppelin hangar at Friedrichshafen on Lake Constance, where the airship was being readied for commercial flights.

Karl Abarth and Martin Schneeweiss pictured outside Castrol House, London, after winning their historic challenge against the Orient Express.

In April 1934, Karl Abarth decided to challenge the Orient Express in a non-stop race against the clock from Vienna to Ostend with his Sunbeam 600 and a Felber touring sidecar, manned by Martin Schneeweiss.

would prevail on that occasion. But it was no small undertaking; the 1372 km route wound its way along the Vienna-Salzburg-Munich-Stuttgart-Cologne-Aachen-Brussells-Ostend route and meant crossing three frontiers with inevitable delays for customs procedures; all of which had to be completed in less than 24 hours, which corresponded with the duration of the train's journey.

Abarth found a first class co-rider in Martin Schneeweiss, a Viennese like him who had won the Tourist Trophy and the Grand Prix of Austria on his motorcycle. They left the Austrian capital at the same time as the Orient Express and hurled themselves at break neck speed into their challenge, alternating at the controls on the roads of half of Europe through Austria, Germany, Holland and Belgium without stopping, except to re-fuel or deal with frontier formalities. In the end, they reached Ostend and were applauded as they rode through the seaport's streets, only to find the train had arrived a few minutes earlier. Karl was incredulous and did not want to accept defeat but, in the end, he had to convince himself that customs formalities and lamentable damage to the Sunbeam's dynamo had frustrated all their efforts to win. He personally gave the Sunbeam a complete overhaul, obtained a new dynamo from Lucas and was ready to race back to Vienna. "On the return journey, everyone will see who is the fastest between me and that train"! And two weeks later, on Monday 19 April 1934, the two champions presented themselves at 8.30 am under the scrutiny of the Royal Automobile Club of Great Britain in Ostend; they raced flat out for a day and a night, Karl coaxing the impossible from his Sunbeam as he charged towards Vienna. At 8.10 am he arrived at Hutteldorf, on the outskirts of the city, where officials of the Automobile Club of Austria decreed that the two crazy motorcyclists had covered the route from Ostend in 23 hours and 40 minutes with an effective time, taking essential stops into account, of 21 hours and 50 minutes. The challenge had been won.

In 1933-34, Karl Abarth also raced outside Austria. He scored his first victory on foreign soil in Munich and also won the hillclimb on Switzerland's Klausen Pass (picture) astride his inseparable Sunbeam 90 and with Holly at his side.

For young Karl Abarth, 1934 was a year full of success. His fame as a sportsman and technician enabled him not only to consolidate his position as a champion in sidecar racing, but it also brought numerous new job opportunities. One of them was an offer by the Belgian manufacturer FN (Fabrique Nationale d'Armes de Guerre) that built its first motorcycle in 1901 and soon became well known for its multi-cylinder machines and cardan shaft transmission. The Belgian manufacturer's bike soon won an international clientele, due to its proverbial robustness and mechanical precision. In the Thirties, FN produced two motorcycle models, a 350 cc and a 500 cc; it was at the end of 1934 that Karl became a technical consultant to the Herstal, Belgium, company. He was a dynamo of activity and it seemed almost as if he wanted to recoup in one season the enforced immobility that followed the Wels accident. He moved from one circuit to another at the wheel of a small van, inside which was anchored his precious motorcycle and sidecar; the van also acted as a mobile workshop for emergency work. Karl spent the whole season at the tracks of central Europe; at Krieau, where he became Austrian champion, the hillclimbs at the Klausenpass in Switzerland and Prague, where he won his class before 60,000 spectators. He did not omit anything that would improve the performance of his machine and showed everyone that technical confines were ephemeral and passable.

In his constant anxiety to explore new roads leading to greater speed for his motorcycle and sidecar, Abarth came up with a simple and apparently intuitive device that brought a real revolution to his racing. It was called the Schwingachs (oscillating axle), and was a lever and rack system that enabled

In spite of the frenetic pace of his double life as a technician and racing motorcyclist, Karl Abarth found time to marry. In 1934, he wed the secretary of the well-known Viennese lawyer Anton Piëch, husband of Louise Porsche, daughter of the celebrated designer of the Auto Union Grand Prix racers and many other revolutionary cars, not least the Volkswagen Beetle.

Above: Two events stood out in Karl's 1935 racing year: a switch from the British Sunbeam to the Belgian FN company's machines and his invention of the Schwingachs system which, by use of a lever, enabled a sidecar's wheel to incline itself in a corner at the same angle as that of the motorcycle.

The debut of the Schwingachs took place on 25 May 1935 at Krieau, where Karl competed on a 600 cc FN bike (racing number 11) with the new adjustable axis sidecar. Right: Karl racing with the first adjustable axis sidecar at Krieau, where he won a clamorous victory in the up to 1000 cc class.

the wheel of the sidecar to incline in a corner in synthesis with the inclination the rider imposed on the motorbike. A formidable invention that made Karl feel ready to totally dominate his opponents; he tested and re-tested the mechanism at M.P. of Vienna, the factory to which he had sold the rights, and prepared himself to take on the 1935 season as champion. But first he was to marry the secretary of a well-known Viennese lawyer named Anton Piëch, himself married to Louise Porsche, daughter of the celebrated professor and the designer who conferred such prestige on German motor racing in the Thirties. Newspapers carried reports of the marriage of the young 26-year-old Abarth who, not overwhelmed by good fortune, had been able to climb his way back to the top again and achieve objectives that seemed impossible.

His home at 15 Wilhelm Exnergasse was in the same district as the M.P. sidecar factory the owner of which, Max Porges, understood that Karl's technical contribution would give the sale of his products a powerful push; so he offered Abarth a contract as technical director.

The racing debut of the Schwingachs took place on 25 May 1935 at Austria's Krieau circuit. Karl had his new sidecar anchored to the left of his 600 cc FN mono-cylinder motorcycle and took to the grid as number 11; the large crowd in the stands realised that the levered device with which to incline the sidecar's wheel enabled the rider to enter a corner at a much higher speed than before. Karl scored a decisive victory in the 1000 cc event and did the same again on 15 September at the Grand Prix of the City of Vienna, where he established a new lap record of 93.836 km/h; a few weeks later, he won again at Salzburg and then prepared to live 1936 as a second season of unstoppable success. Vienna again, Salzburg, Listenau in Voralberg and the St. Polten and Graz circuits saw Karl's FN motorcycle and its M.P. first across the finish line.

The following year his fame led him to further victories in Germany, on the Schauinsland circuit in Bremen and in June at Gmünd on his hometown

Karl and Joseph Holly being feted after the Krieau triumph on 25 May 1935. The new-concept sidecar was built by M.P. of Vienna, to whom Abarth had sold the production rights.

roads and in August in Germany once more, where he won the Grand Prix of Mecklenburg. The road to success and popularity lay open before him; at 29 years old he had already made his fortune and he set about to harvest the fruit of his brilliant career.

In the meantime, the European political panorama was causing widespread alarm in world public opinion. The political successes of Hitler in Germany, the passive wait-and-see policy of the western powers accompanied by the collapse of the League of Nations and the progressive dissolution of the precarious equilibrium created by the Treaty of Versailles severely deteriorated the international situation. The ambitious process of re-arming Germany on the one hand and the shattered chain of allies engineered by France on the other, had given birth to relationships between the nations that were just a façade; in reality a guise that masked a grave state of uncertainty. To that was added the weak expansionist policy of Italy in Africa, and its gradual alignment with the ideology that had brought the National Socialists to power in

Karl signed a lucrative contract with M.P., whose owner Max Porges could see the notable potential of Abarth as a technician, designer and rider.
The young champion began to display the logo of the Belgian motorcycle manufacturer FN on his helmet.

Germany; an alignment that became manifest political and military collaboration during the Spanish civil war and culminated with the creation of the Rome-Berlin axis, which was declared for the first time in 1936. At the end of 1937, it became inevitable that the pretentious search by the Nazis for vital lebensraum or living space to aid their expansion had to include the partition of Czechoslovakia and the reunification of Austria with the Third Reich; and it was in March 1938 that, after a precipitous popular plebiscite skilfully manipulated from Berlin, the ruling junta in Vienna sanctioned the annexation of the small state, once the hub of the Hapsburgs' power, to Hitler's Germany. The same strategy was used six months later with Czechoslovakia, whose Sudetenland territory was next to fall under the jurisdiction of Germany, in spite of strong diplomatic opposition from France and Britain.

So in the spring of 1938 Karl Abarth found himself living under the domination of the Third Reich. Not long after Nazi troops had entered Vienna, he was contacted by delegates of the city's Italian Fascist party who had learnt that his father, resident at Merano from the end of the First World War, had opted for Italian citizenship and he was, therefore, considered an Italian citi-

KARL ABARTH
WIEN IX.
Wilhelm Exnergasse 15

WIEN, den 19. Oktober 1936

Titl.
M.P.-Beiwagenwerke, Ing. Max Porges
Wien IX.
Augasse 17

Zu meinen Siegen beim Höhenstraßen-Rennen möchte ich hiemit die Erklärung abgeben, daß dieselben nicht nur meiner schnellen F/N-Maschine, sondern auch meinem herrlichen M.P.-S c h w i n g a c h s-Beiwagen zu verdanken sind, der es mir erlaubte, in den Kurven ein derart hohes Tempo einzuhalten, daß ich dadurch ein Übergewicht über alle in- und ausländischen Konkurrenten erreichte.

Hochachtungsvoll

Karl Abarth

In a letter dated 19 October 1936 – the day after his victory in the Wiener Hohenstrassenrennen – Karl told engineer Max Porges, head of M.P., how much he liked the Schwingachs sidecar, thanks to which his superiority over all riders from all nations was guaranteed.

Riding his 600 cc FN and the revolutionary Schwingachs sidecar that he invented, Karl scored a string of victories in 1935. He won the Vienna City Prize on 15 September at Krieau and scored victories at Salzburg on 6 October, Budapest on 7 November and the Three Frontiers ascent on 11 November, which he covered in record time.

This picture of the acrobatic exhibition of Karl Abarth and his partner charging through a corner at Krieau in the 600-1000 cc motorcycle and sidecar race on 15 September 1935 appeared on the front cover of the 250th edition of the Das Motorrad enthusiasts' magazine. The shot clearly shows the lever system with which to lean the wheel of the sidecar at the same angle as the inclination of the motorcycle.

zen. It would be much appreciated – they communicated – if his son would race in Italian colours in future in return for a considerable sum of money, which corresponded to the contribution that the government in Rome had budgeted to promote Italian sports activities in Austria. Between incredulity and dismay, Karl remained dumbfounded before such a proposal; still, he evaluated the advantages and possible negative factors that his decision would bring and then, fortified by the pragmatic spirit that always guided him, he accepted. All of a sudden he had a great deal of money, went by the name of Carlo and was able to afford anything he wanted; and anyway, racing a

Right: Karl Abarth (centre) with his 600 cc FN and M.P. sidecar consolidated their supremacy in 1936 in both track and road racing as well as hill trials. Their string of victories began on 19 April at Krieau and followed by others at Listenau, Voralberg (pictured), Munich on 11 July, Salzburg on the 19th, Vienna on 18 and 25 October and Graz on 8 November.

motorcycle wearing the Italian insignia could even be less unpalatable than competing with a Nazi swastika. That unexpected opportunity gave him the chance to experiment with new techno-constructional ideas on his bikes and, in particular, to re-start tests with those exhaust pipes he had conceived years before and which, due to a lack of money, he had abandoned.

But it was a short-lived euphoria. Heralded by diplomatic skirmishes that were sure to fail, war was just around the corner; the most horrendous and catastrophic conflict in the history of man. Inexorable in its arrogant process of territorial expansion, Hitler's Germany invaded Poland on 1

Karl won the Salzburg city circuit race on 19 July 1936 and established a new lap record.

At 29 years old, Karl Abarth was financially secure and had excellent professional prospects. His popularity grew month by month. In this picture (autumn 1937) he is presenting a touring motorcycle and sidecar – the bike by FN and the sidecar by M.P. – a vehicle that was coining new commercial fortunes.

Right: Karl Abarth repeated his previous year's successes in 1937, once again confirming his absolute supremacy in motorcycle and sidecar racing. Astride his 600 cc FN, he won on 18 April at Krieau, 30 May at Baden near Vienna, 20 June at Gmünd (pictured), 22 August at the Telerow hillclimb near Berlin and on 24 October at Krieau again, where he set the lap record for his kind of bike racing.

September 1939 and, with the demolition of the country's frontier barriers, inflicted a cruel blow on a world that was still living by traditions with their roots in the 19th century, even if compromised by the 1914-18 conflagration. The year 1939 changed the story of humanity in traumatic fashion, taking the world to the very edge of the abyss. For Carlo Abarth, those international events represented no more of a disastrous prospect than they did to others. But misfortune struck a month after the start of hostilities, when he had another serious accident that even made those close to him fear for his life. It happened in near Lubiana, Yugoslavia, one Sunday in October. Committed as always to give the best of himself in a race, Carlo suddenly lost control of his motorbike and sidecar and went off in disastrously; he hovered between life and death for several days and then, slowly, his strong athletic constitution helped him find his way back to recuperation.

Once he had recovered, he found himself confronted with two new and fearful realities: one was to retire from racing because of the injuries he had received, and the other, no less daunting, was the need to create a new career for himself in another land as life in Vienna under Nazi oppression had become impossible, and the offers of work were no longer as promising as they were when he was younger. At 31, Carlo suddenly felt almost a stranger in his own country and so he decided to move to Jugoslavia, far away from the winds of war, where he had found help and hospitality in the difficult weeks in hospital after his crash. Armed with his great engineering knowledge and his natural mechanical talent, he decided to try his luck in the Lubiana area, where he found work as a director of the Ignaz Vok mechanical workshop.

There was no lack of work. Wartime restrictions on petrol had induced people to find other forms of fuel suitable for the internal combustion engine; and gas made from burning coke, like that produced by the Vok plant, was widely used in those war years by many countries involved in the conflict. Abarth reverted to his well tried and tested technical knowledge to success-

Above: It was the same again in 1938, when Karl racked up another string of victories on both home tracks and road circuits in Austria, Germany, Switzerland and Hungary. The photograph shows him and his partner entertaining 40,000 people with yet another virtuoso performance at over 100 km/h at the Krieau sand circuit near Vienna.

Above, right: At that intoxicating moment in his life, Karl hand wrote a brief comment on this photograph: "From 1937 I won all the races for which I entered.
I have never been in such good form".

In some 1938 events Karl rode his ponderous 996 cc V2 machine, produced by the British company JAP and considered at the time one of the most advanced of its kind.

At the Loibl Pass hillclimb on the border of Austria and Yugoslavia, Abarth gave another demonstration of his skill and riding style. But in 1939 the winds of war began to blow precipitously across Europe.

fully convert cars and trucks by installing bulky gas generators that, while returning 30% lower performance than a petrol engine, at least allowed people to take to the roads.

It was a difficult period, made even more so by the turn the military situation was taking and the swing of power in Yugoslavia from the German and Italian occupying forces to the communist partisans who roamed the territory. Carlo had to use all his diplomatic powers to convince the various officials who came to Lubiana during those five years of war that the only allegiance he had was to engines and pure competition in the saddle of a motorcycle. He was liked by the people of the area and respected for his high degree of professionalism; to him that was another victory, achieved this time with political passion

Below, left: At a race in Yugoslavia one Sunday in October, Karl lost control of his machine, flew off the track and was severely injured. The photograph shows helpers gathered around him at the scene of the smash.

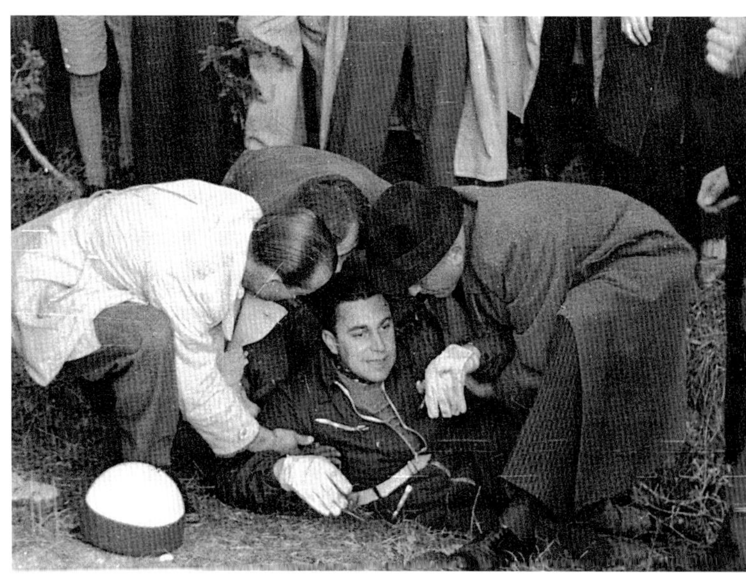

Below, right: "I can never race again" is the cold comment Abarth wrote on the back of this picture, which shows him being comforted after the accident by the first spectators who rushed to help him.

COMUNE DI MERANO
Provincia di Bolzano

Certificato di identità personale

che si rilascia a ABARTH CARLO
di Carlo e di fu Taussig Dora
nato il 15.11.1908 a Vienna
di stato civile separato legalmente di professione Direttore Tecnico
di cittadinanza Italiana
residente a Merano in Via Passeggiata Reg.Elena n.2
aventi i seguenti connotati personali:

statura: alta
capelli: neri
occhi: castani
contrassegni speciali: N.N.

Vengono inoltre autenticate la fotografia e la firma dell'interessato.
Si rilascia il presente certificato in carta bollata per tutti gli usi consentiti dalla legge.

Merano, li 20 settembre 1945

IL SINDACO

After his enforced retirement from racing and during the 1939-45 war Karl chose to live in Yugoslavia, where he found a post as technical director of the Ignaz Vok mechanical workshop near Lubiana. He fell in love with Nadina Zeriav, who later became his second wife.

On 20 September 1945, the municipality of Merano issued an identity certificate to Carlo Abarth, citizen of Italy.

and suspicion that his identity, half Austrian and half Italian, helped to fuel.

By the end of the conflict in May 1945, Europe had been razed to the ground and tens of millions of dead weighed heavily on the collective conscience of the survivors; there was not even room for compassion. Devastated by the destruction, cities and towns could offer no refuge to the great masses of terrified refugees that escaped towards the west. Carlo became involved in that chaotic migratory movement, having decided to abscond from the communist armies massing on Yugoslavia's border with Italy. Concrete and courageous as always, he began the long march on foot towards Trieste, bringing with him a cart on which he had loaded suitcases full of oriental rugs, a passion that he had harboured for years. His destination was Merano, where his 83-year-old father still lived and who was a lighthouse of salvation to him in the midst of so much misery. At 37 years old, Carlo found himself once more on the starting line, fortified only by his own vitality to face the new entrapments of life. On 20 September 1945 a certificate of personal identity was issued by the Merano council to Mr. Carlo Abarth, legally separated, profession technical director and citizen of Italy, resident in the city at Via Passeggiata Regina Elena 2.

Life laboured to return to its normal rhythm by autumn 1945, although it seemed impossible that it could be rebuilt and bring back the past. At Merano, which was saved the horrors of war, the trauma of the frightening recent conflict was still alive if latent; it had affected its two ethnic communities in equal measure, the original Tyroleans and the more recent inhabitants whose language was an Italian of the early Twenties. Carlo Abarth knew the little city well from infancy and found some serenity as well as essential sustenance in the warmth of his paternal family. Without much enthusiasm, he started a small business selling bicycles made by Bianchi, a company from which he had previously obtained the agency for Lubiana and Zagreb during the war years. He did, however, have to give up his oriental rugs, from the sale of which he obtained the money to affront those difficult moments with dignity. He was tormented by his agonising nostalgia for that period of his existence in which he raced his motorcycle and sidecar against the best riders of the day; and the imperious temptation to return to the world of engines, to the mechanical workshop where he matured his most stimulating experience led him to tempt fate once more in 1946, rich in the insatiable vocation that pulsed within him. He wrote to the daughter of Professor Ferdinand Porsche, Louise, to explore whether or not there was a chance of returning to his mechanical world, no matter what it might offer.

1946

1949

CISITALIA, THE GREAT ILLUSION

The Porsche family's war was also a harrowing one. Ferdinand, the father, was the illustrious Bohemian designer who had just turned 70 and had already given his life to the service and technical progress of the car. He had distinguished himself from the very beginning of his career when, first at Lohner of Vienna and then at Austro-Daimler in Wiener Neustadt, he completed a number of innovative projects. In 1923, his experience took him to the German Daimler company, where his creations included the Mercedes-Benz that won the Targa Florio a year later, and the exceptional series of sports cars that passed into history with the designations S, SS, SSK and SSKL. But his inspiration as an inventor culminated in the Thirties in Stuttgart with his design of the Auto Union P Wagen Grand Prix car with its V16 rear engine, and a utility prototype road car with an air-cooled engine overhanging the rear axle, a car that became a legend throughout the world in the 50 years that followed: it was the Volkswagen Beetle.

So while the Auto Unions took on the Mercedes-Benz on the circuits of Europe to exalt the feats and technical supremacy of Nazi Germany, the little professor defined the final details of the revolutionary people's car, baptised KdF from the name Kraft durch Freude or power through joy, the motto of the Nazis' recreational activities organisation. In the summer of 1938, the cornerstone was laid of a great factory in which the car was to be mass-produced, a vehicle that was to motorise the German working class; instead it motorised the Third Reich's army that wreaked fear, death and confusion across Europe between 1939 and 1945. A notable contribution to Hitler's early military success was attributed to Studio Porsche of Stuttgart, which had designed numerous vehicles of war including the Tiger and Ferdinand tanks. It was because of his support of the development of Nazi military power that the Allies hunted down the celebrated designer and imprisoned him for collaboration in 1945; he was, in fact, tracked down by the French, who kept him prisoner for a couple of years, until his family – in part through the good offices of the racing driver Raymond Sommer and Charles Faroux, the father of French motor racing journalism – was able to pay a million Franc ransom for his release. Ferdinand's daughter Louise and her brother Ferry were installed in Gmünd, Carinthia, where their father had transferred his offices and workshop in 1943 to repair small army trucks. The climate created in Austria by the Allied occu-

Reproduction of a certificate of residence issued to Carl Abarth by the Merano authorities on 10 December 1945.

1/12326

COMUNE DI MERANO
PROVINCIA DI BOLZANO

Città di Merano L. 0.25 Diritti di Segreteria

Certificato di Residenza

Il Sindaco, in base alle risultanze anagrafiche di questo Comune,

certifica

che A B A R T H CARLO

di Carlo e di Taussig Dora

nato a Vienna il 15 novembre 1908

di professione direttore tecnico di cittadinanza ITALIANA

è residente nel Comune di Merano dal 20 agosto 1945

con abitazione in Via Pass.Reg.Elena n.2

Si rilascia il presente a richiesta dell'interessato, in carta libera, per uso

ALLOGGIO=

Merano, 10 dicembre 1945

B.I.

IL SINDACO
L'assessore Delegato
Giovanni [illegible]

MERANO

Athesia, Merano - 2435

pational forces could not have filled them with easy optimism for their future; an Allied officer had already warned Ferry, saying: *If Professor Porsche had been a cobbler, he would have been able to re-start his activities; not being a cobbler, he may not go back to building cars".*

Abarth, who operated his little company called Fahrzeuge aller Art from head offices in Via Roma 168, Merano, was able to maintain continuous contact with the Porsches and obtained from them written authorisation to "represent the interests of Ferdinand Porsche Jr in the concession of licences and construction commissions in Italy. A matter, in particular, of neo-construction of various vehicles, sports and racing cars". That letter, dated 30 September 1946, played a decisive role at the foundation of Carlo Abarth's fortunes in Italy. Helped by Rudy Hruska, a young Austrian engineer who had worked with Studio Porsche, Carlo traced Tazio Nuvolari – the famous racing driver who was also intent on returning to the sport – and proposed to him a Porsche project for a small, economical tractor; he added that in the same centre for studies in Austria there were also preliminary designs for a sensational Grand Prix racing car: mid-to-rear engine, 12 cylinders opposed, 1500 cc, supercharged, four-wheel drive, lateral fuel tanks. Nuvolari immediately contacted the journalist and photographer Corrado Millanta, who organised a meeting with Piero Dusio and two German language technicians at the industrialist's Milan apartment in Via Durini 7; Piero Taruffi, another motor racing driver, was also present.

Dusio was a prominent person in Turin in the Thirties and Forties. A native of Scurzolengo in the province of Asti, he was a man full of ideas,

After the cessation of hostilities and during his first few months in Merano, Carlo Abarth lived modestly and with uncertain professional prospects. He established a small company, Fahrzeuge aller Art, for the sale of Bianchi bicycles but with the real intention of exploring every possible opportunity that could lead him back to the motoring world. These photographs, taken during the winter of 1945 and the summer of 1946, are testimony to that precarious moment in history that scarred the lives of the people of Europe.

extremely dynamic, untiring and with many interests; Dusio was an able dealer and an audacious industrialist who founded many companies, which he soon ceded to others once they had got into their stride so that he could become involved in other initiatives, other businesses. Noted in the field of textiles and clothing in general, during the war he opened the Bosco Manufacturing company in Racconigi for the production of clothes and military uniforms. From his many different activities Dusio built up the resources with which to support his great love of cars and to finance the Juventus football team, of which he was president in the uncontaminated years in which the game of soccer had not yet burst its banks and overflowed beyond the boundaries of pure sporting competition.

Dusio's great love was high performance cars, to which he dedicated much time and a huge fortune. He won the Italian Amateur Championship of speed in 1934, the year he scored a number of respectable places driving an Alfa Romeo 8C 2600 – seventh overall with Giulio Aymini in the eighth Mille Miglia on 8 April 1934, second in the Turin Hills hillclimb on 10 June, second in the Targa Vesuvio of 29 June, second in the third Ascoli Cup of 15 July and second again in the Race of the Stelvio on 26 August. The following year, Dusio successfully competed in a number of Grands Prix driving a Maserati and, among other placings, in 1936 he came seventh in the first Circuit of Milano on 28 June and sixth in the 14th Grand Prix of Italy at Monza on 13 September, while in 1937 with Ciro Basadonna he won the up to 750 cc Sports Category of the 11th Mille Miglia on 4 April in a Fiat Siata 636 cc Gran Sport

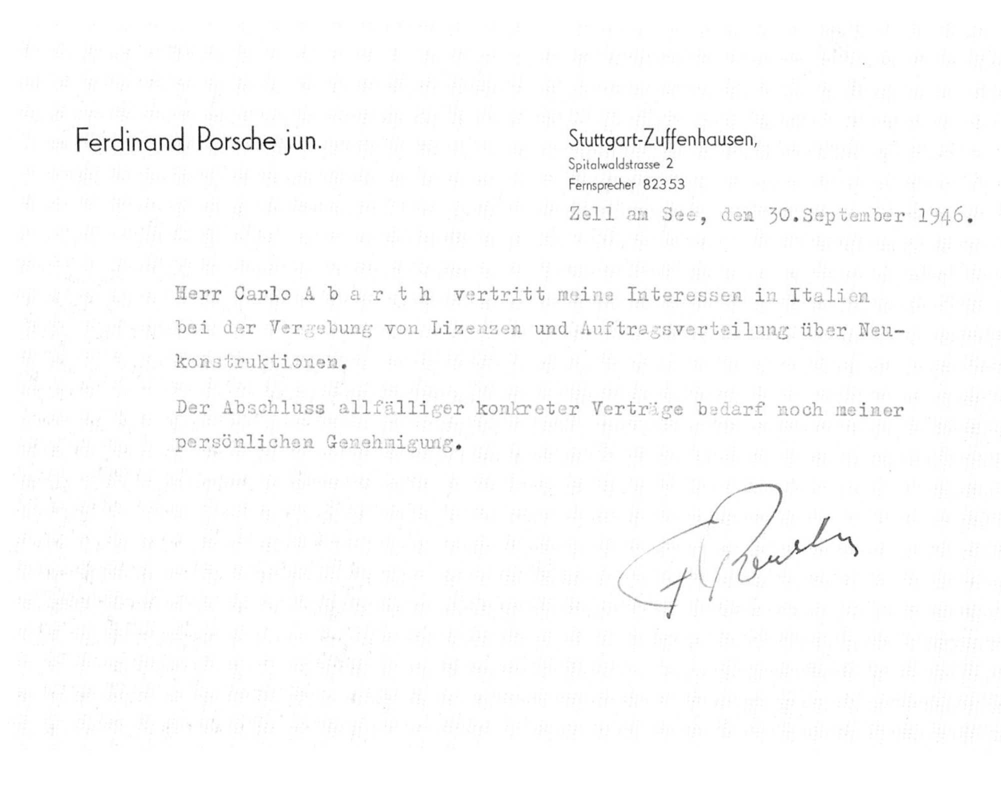

Ferdinand Porsche jun.

Stuttgart-Zuffenhausen,
Spitalwaldstrasse 2
Fernsprecher 82353

Zell am See, den 30.September 1946.

Herr Carlo A b a r t h vertritt meine Interessen in Italien bei der Vergebung von Lizenzen und Auftragsverteilung über Neukonstruktionen.

Der Abschluss allfälliger konkreter Verträge bedarf noch meiner persönlichen Genehmigung.

With these few lines from Ferdinand Porsche Jr dated 30 September 1946, Carlo Abarth was authorised to represent the interests of the Stuttgart-Zuffenhausen company in Italy for the concession of licences to construct vehicles in the Republic.

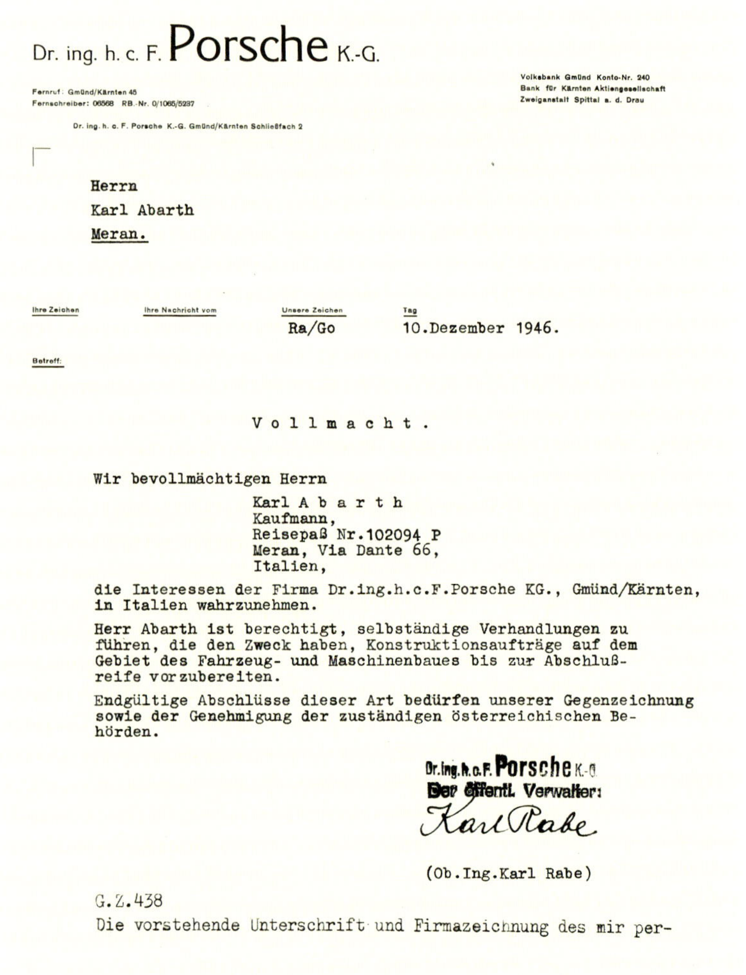

Dr. ing. h. c. F. Porsche K.-G.

Fernruf: Gmünd/Kärnten 45
Fernschreiber: 06568 RB.-Nr. 0/1065/5237

Volksbank Gmünd Konto-Nr. 240
Bank für Kärnten Aktiengesellschaft
Zweiganstalt Spittal a. d. Drau

Dr. ing. h. c. F. Porsche K.-G. Gmünd/Kärnten Schließfach 2

Herrn
Karl Abarth
Meran.

Ihre Zeichen | Ihre Nachricht vom | Unsere Zeichen Ra/Go | Tag 10.Dezember 1946.

Betreff:

V o l l m a c h t .

Wir bevollmächtigen Herrn
Karl A b a r t h
Kaufmann,
Reisepaß Nr.102094 P
Meran, Via Dante 66,
Italien,

die Interessen der Firma Dr.ing.h.c.F.Porsche KG., Gmünd/Kärnten, in Italien wahrzunehmen.

Herr Abarth ist berechtigt, selbständige Verhandlungen zu führen, die den Zweck haben, Konstruktionsaufträge auf dem Gebiet des Fahrzeug- und Maschinenbaues bis zur Abschluß-reife vorzubereiten.

Endgültige Abschlüsse dieser Art bedürfen unserer Gegenzeichnung sowie der Genehmigung der zuständigen österreichischen Behörden.

Dr.ing.h.c.F. Porsche K.-G.
Der öffentl. Verwalter:
Karl Rabe

(Ob.Ing.Karl Rabe)

G.Z.438
Die vorstehende Unterschrift und Firmazeichnung des mir per-

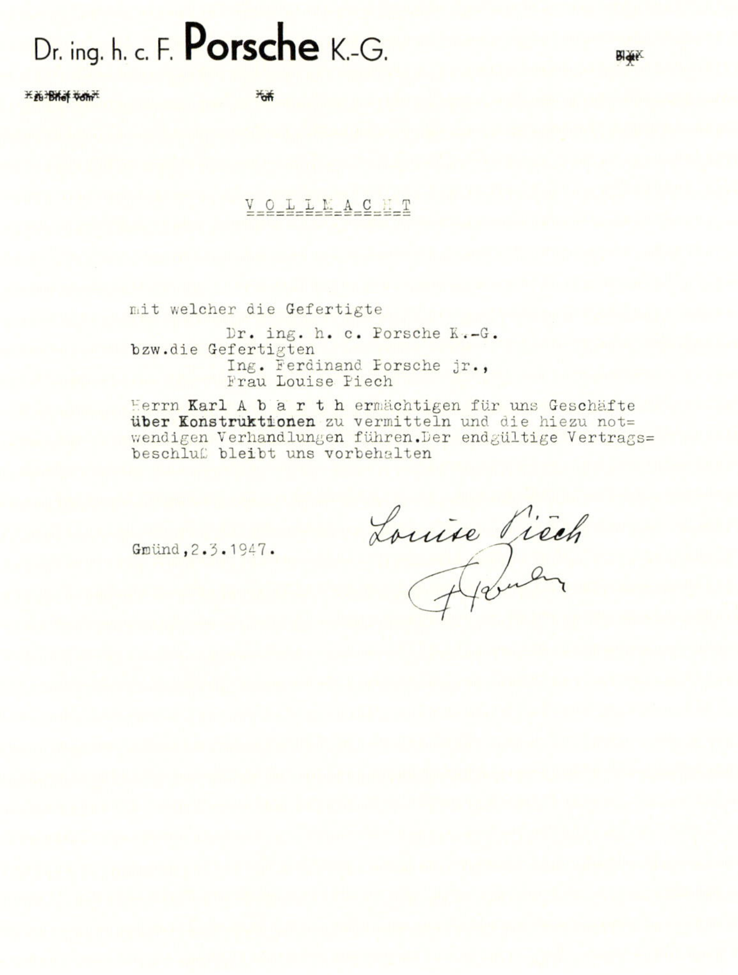

Dr. ing. h. c. F. Porsche K.-G.

zu Brief vom | an | Blatt

V O L L M A C H T

mit welcher die Gefertigte
Dr. ing. h. c. Porsche K.-G.
bzw.die Gefertigten
Ing. Ferdinand Porsche jr.,
Frau Louise Piech

Herrn Karl A b a r t h ermächtigen für uns Geschäfte über Konstruktionen zu vermitteln und die hiezu notwendigen Verhandlungen führen.Der endgültige Vertragsbeschluß bleibt uns vorbehalten

Gmünd,2.3.1947.

Louise Piëch
F. Porsche

Above: The mandate entitling Abarth to act in Italy on behalf of the Dr. ing. h.e.F. Porsche KG company, which established itself in Gmünd, Austria. Approval was sent in a letter dated 10 December 1946 signed by Karl Rabe, the engineer who was the German company's chief designer for several years.

Above, right: A reproduction of the letter dated 2 March 1947 with which brother and sister Ferdinand Jr and Louise Piech of Porsche KG authorised Karl Abarth through their then legal representatives, to conduct negotiations concerning the construction of vehicles.

and came fifth driving a Maserati in the Grand Prix of Tripoli. Another year of good results came in 1938, when Dusio and Boninsegni took third place overall in the 12th Mille Miglia on 3 April at the wheel of an Alfa Romeo 8C 2900A; he came third in the 12th Pontedecimo-Giovi on 10 July and won the Stelvio Hillclimb on 21 August; in 1939 he took fourth in the Sassi-Superga of 3 July in an Alfa Romeo 6C 2500 Super Sport.

Dusio dreamt of promoting motor sport through a racing formula in which drivers would be able to enter with technically equivalent cars or even compete in identical monotype racers of which many would be produced, so that the maximum ability of the driver would emerge in confrontation with the machine. In October 1944, Piero Dusio convinced engineer Dante Giacosa, head of Fiat's car design department, to take on the design of a single-seater racing car to be produced in a small workshop called Cisitalia (it stands for Consorzio Industriale Sportivo Italia) at Corso Peschiera 251, Turin. From that brilliant idea came a project for a small racing car with an engine derived from that of the Fiat 1100 and the front and rear suspension of the Topolino 500; christened

CISITALIA "D46" (1946)

ENGINE	front, longitudinal, 4 cylinders in line
Bore and stroke	mm 68x75
Cubic capacity	cc 1089
Valve gear	overhead valves, side camshaft
Number of valves	2 per cylinder
Compression ratio	9.5: 1
Fuel feed	1 Zenith 30 Vif/C carburettor
Ignition	magneto
Coolant	liquid
Lubrication	dry sump, oil radiator
Maximum power	62 hp at 5800 rpm
TRANSMISSION	rear wheel drive
Clutch	single dry plate
Gearbox	3 speed, pre-selector lever on the steering wheel
BODY	single-seater
CHASSIS	tubular trellis
Suspension	front independent, upper transverse leaf spring, lower arms, hydraulic dampers; rear rigid axle, longitudinal semi-elliptic springs, friction disc shock absorbers
Brakes	hydraulically operated drums front and rear
Steering	worm and sector
Fuel tank	capacity 44 litres
Tyres	front 4.00 x15; rear 5.00 x15
DIMENSIONS AND WEIGHT	
Wheelbase	mm 2050
Front track	mm 1225
Rear track	mm 1125
Length	mm 3000
Width	mm 1400
Height	mm 1020
Kerb weight	kg 370
PERFORMANCE	
Maximum speed	km/h 170

the D46, two years later the little Cisitalia became not just a symbol of a reawakened Italian motor sport, but also laid the foundation for a brilliant competitive and industrial career of which Dusio was the far-sighted advocate.

Enraptured by the idea - perhaps too ambitious – of building a Grand Prix car based on a project developed by the person who had conceived the astonishing Auto Unions of a decade earlier, Dusio did not take long to reach an agreement with Porsche. On 3 February 1947, he signed a contract for the acquisition of world patents for the single-seater, a sports car with a fuel injected V6 two-litre air cooled rear engine – project number 370 - a fast synchronised gearbox project number 380, a universal 11 hp tractor and a small hydraulic water-driven turbine for the production of energy. After the agreement was signed, Abarth and Hruska also joined the group; Carlo became sports director of Cisitalia almost immediately, while Hruska was given the task of developing the revolutionary new single-seater in record time.

In 1946, Cisitalia quickly came to the attention of the press when Piero Dusio, who had achieved his ambition of having the greatest drivers of the moment race for him at the wheel of the same type of car. He entered seven Cisitalia D46 single-seaters for the Brezzi Cup in Turin on 3 September for himself, Tazio Nuvolari, Clemente Biondetti, Piero Taruffi, Franco Cortese and the French aces Louis Chiron and Raymond Sommer. In Turin that day, they celebrated not only the victory of Dusio, the founder and driving force of Cisitalia, who came first followed by second placed Cortese; the whole city participated in the triumphant debut of the new marque. That was the beginning of a string of glorious seasons of racing on road and track.

The 1947 successes of the infant factory were clamorous. To the victories of the D46 were added those of the CMM two-seater coupé and the roadster, the latter of which made its first appearance at the Sassi-Superga **hill-**

The agreement signed on 3 February 1947 between the Porsche family and Cisitalia for the acquisition of a number of projects also included the employment of Carlo Abarth and Rudolf Hruska at the Turin factory, the former as sports director and the latter as chief designer of the Grand Prix single-seater. In Turin, Carlo began to frequent circles in which Piero Dusio was at home, including the sporting club of the Juventus football team (picture).

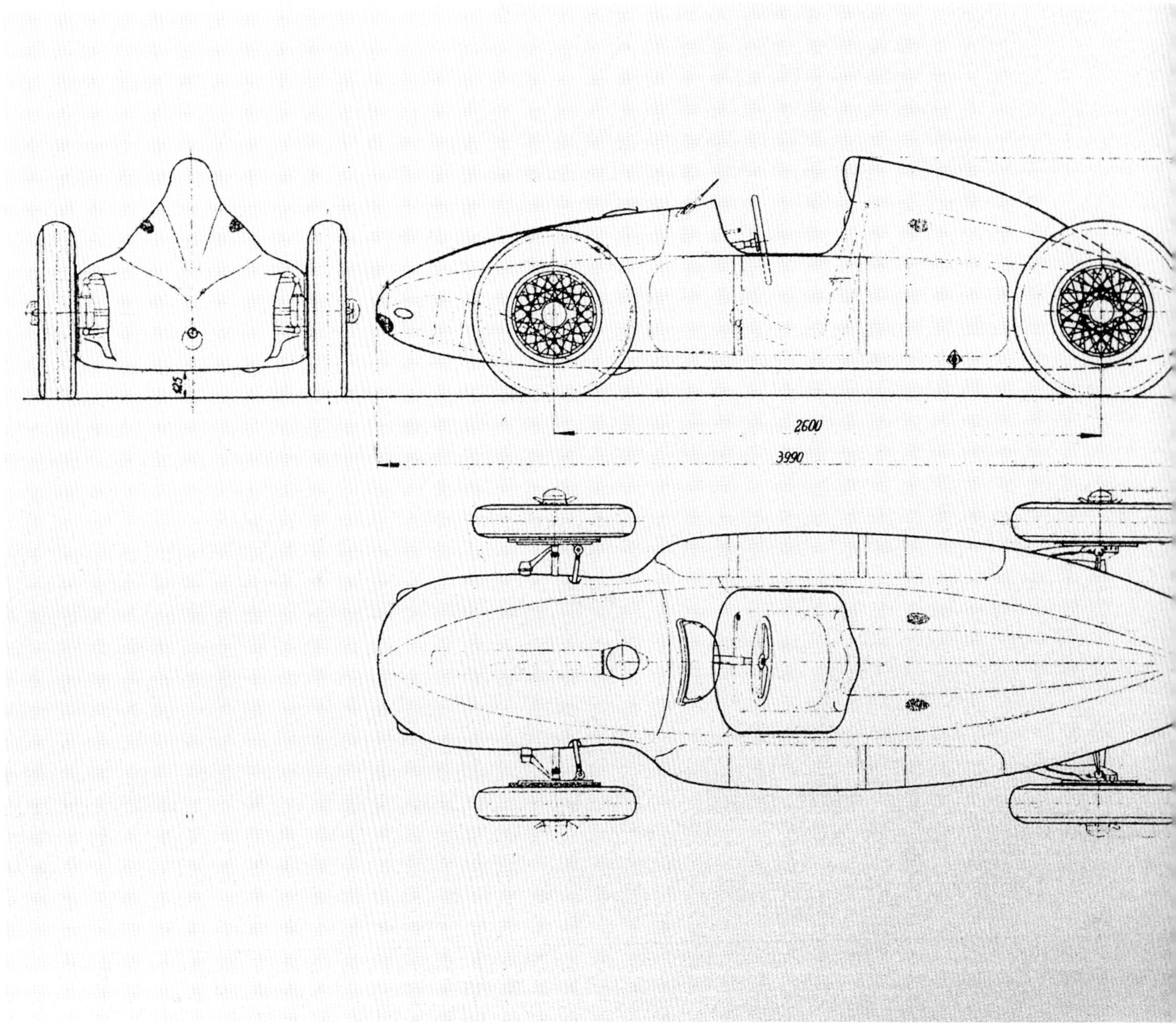

Four views of Rudy Hruska's model of the single-seater Porsche 360 Grand Prix project. Among the numerous techno-constructional innovations were the 12-cylinder opposed 1500 cc centre-rear engine, part-time (front) four-wheel drive and torsion bar suspension.

Below left: The fuel supply of the 12-cylinder four overhead camshaft engine was through two twin-choke Weber carburettors with two volumetric rotor blade superchargers.
Right: The driving position of the Grand Prix car was flanked on both sides by fuel tanks.

Below: The complete chassis of the Cisitalia single-seater, with its radiator up front.

Below: Two views of the powerful 1.5 litre engine group, accredited with a maximum of 400 hp at 12,000 rpm.

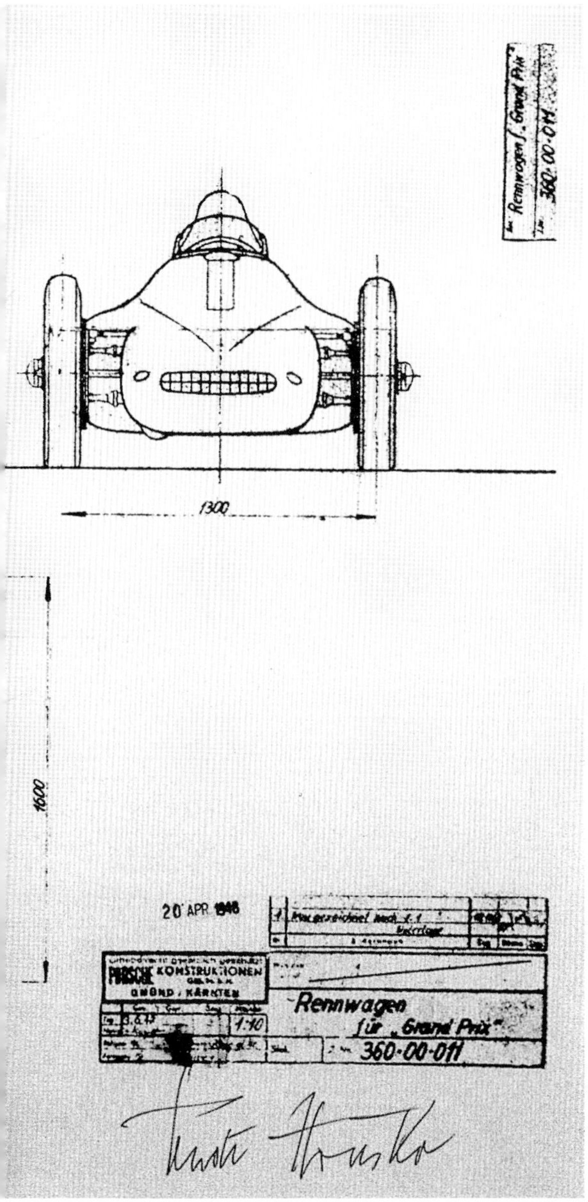

climb on 15 May and caused a stir five weeks later when Tazio Nuvolari and Giuseppe Carena won the 1100 cc Sport class and came second overall in the first post-war Mille Miglia. Cisitalia showed itself to be better than anyone had expected and their victories gave rise to enthusiastic comments in the sports press. Their moment of notoriety was further extended at the Paris Motor Show in October when Dusio unveiled the 202 Coupé, its magisterial body designed by Pininfarina; the coupé immediately became the most coveted of cars among the celebrities and sporting motorists of half the world. With the 202, which had established itself as a star after just a short motor racing career, Cisitalia also set foot in the refined world of sophisticated cars created for the elite.

At Turin, Carlo Abarth did not even have the time to get to know the city or its people. Working at Piero Dusio's side in that exalted phase in the life of Cisitalia absorbed all his energy and conceded little time for rest or entertainment, although he did agree to frequent the circles in which his boss was known and well received so as to introduce himself - with effort - into the Turin establishment. A society that was hardly penetrable at the time, almost closed in that difficult post-war period to the "Germans" in a group of technicians that surrounded a personality like Dusio at the height of his notoriety. Germans because, as well as Abarth and Rudy Hruska who were the architects of the agreement with the Porsche family and the new direction Cisitalia had taken, they were joined by Robert Eberan Eberhorst, the distinguished designer of the 1938 V12 Auto Union Type D, who had the responsibility of developing the Cisitalia Grand Prix car.

As far as his private life was concerned, when Carlo arrived in Turin he installed himself in Dusio's villa at Via Bricherasio. Abarth was accompanied by his second wife, Nadina Zeriav, with whom he had been living for several years before they were married, as he was still not divorced from his first wife. It was at the beginning of his period in Turin that, with the help of the lawyer Montel who was well known in the city, he was able to obtain a divorce and

CISITALIA GRAND PRIX (1947)

ENGINE	
	mid-rear, longitudinal, 12 cylinders opposed
Bore and stroke	mm 56 x 50,6
Cubic capacity	cc 1492
Valve gear	4 overhead camshafts, gear driven
Number of valves	2 per cylinder
Compression ratio	7.1:1
Fuel feed	2 twin-choke Weber carburettors, with two single-stage superchargers
Ignition	magneto
Coolant	liquid
Lubrication	dry sump, oil radiator
Maximum power (project)	360 hp at 10500 rpm
TRANSMISSION	
	all-wheel drive (insertable on the front wheels)
Clutch	multi-disc
Gearbox	5 speed, central lever
BODY	
	single-seater
CHASSIS	
	tubular trellis
Suspension	front independent, transverse torsion bars, hydraulic dampers; rear, independent, longitudinal torsion bars hydraulic dampers
Brakes	hydraulically operated drums front and rear
Steering	worm and sector
Fuel tank	lateral, total capacity 200 litres
Tyres	front 5.50 x 17; rear 6.00/7.50 x 18
DIMENSIONS AND WEIGHT	
Wheelbase	mm 2600
Front track	mm 1300
Rear track	mm 1300
Length	mm 3990
Width	mm 1600
Height	mm 1100
Kerb weight	kg 650
PERFORMANCE	
Maximum speed (project)	km/h 360

The tormented logotype of Cisitalia did not have time to register and become vastly popular in the public mind.

The silhouette of the Cisitalia Grand Prix car, which was several decades before its time and was a forerunner of the Formula One cars which, over the years, would develop into today's F1 racers.
The design of the Type 360, conceived by Porsche for the Cisitalia Grand Prix, was so innovative that those who had to build the car found themselves in difficulty.

Opposite. Left: The longitudinal section of the transmission group of the Cisitalia Grand Prix car, with five-speed gearbox.

Right: The first prototype of the Cisitalia Grand Prix car, which was developed and built amid many difficulties at the start of 1950 and was presented to Piero Dusio and Tazio Nuvolari (photograph). But the serious financial situation facing Cisitalia had already compromised the development of the project.

CISITALIA "202" (1947)

ENGINE	
	front, longitudinal, 4 cylinders in line
Bore and stroke	mm 68 x 75
Cubic capacity	cc 1089
Valve gear	overhead valves, side camshaft
Number of valves	2 per cylinder
Compression ratio	7,8: 1
Fuel feed	1 Weber 36DR4SP carburettor
Ignition	coil, distributor
Coolant	liquid
Lubrication	dry sump, oil radiator
Maximum power	55 hp at 5500 rpm
TRANSMISSION	
	rear wheel drive
Clutch	single dry plate
Gearbox	4 speed plus reverse, central lever
Axle ratio	3.8: 1
BODY	
	2 seat coupé by Pininfarina
CHASSIS	
	tubular trellis
Suspension	front independent, upper transverse leaf spring, lower arms, hydraulic dampers; rear rigid axle, longitudinal semi-elliptic springs, double hydraulic dampers
Brakes	hydraulically operated drums front and rear
Steering	worm and sector
Fuel tank	capacity 42 litres
Tyres	5.00 x15
DIMENSIONS AND WEIGHT	
Wheelbase	mm 2400
Front track	mm 1258
Rear track	mm 1247
Length	mm 3400
Width	mm 1450
Height	mm 1250
Kerb weight	kg 780
PERFORMANCE	
Top speed	km/h 160

Above: The Cisitalias were certainly among the first cars in the world with bodies designed as a result of extensive aerodynamic research, as can still be seen today in the 1947 202 SMM, later called the Spyder Nuvolari, a truly innovative car.

Below: The famous Cisitalia 202 created with such majesty by Pinin Farina and which became a milestone over the years in the stylistic evolution of the automobile. The car made its debut amid public and critical uproar at the Paris Motor Show in October 1947.

marry Nadina in 1950. Daughter of an important Yugoslav dignitary, Nadina met Carlo in 1940 at Lake Garda and lived with him throughout the horror of the war years. She rarely appeared with her husband in Turin and seldom travelled with him to races, living an isolated life. In those first years after the bombings and privation, existence was extremely simple if not spartan, although Carlo treated himself to a trip to the mountains or the small Piedmont lakes once in a while.

Carlo found a way of fully expressing himself in satisfying fashion at Cisitalia the following year; from the D46 he was able to develop a single-seater with a 1200 cc engine of increased power output, modified suspension and fairing at the front of the body. The most radical modifications were those

Carlo Abarth photographed with the German driver Hans Stuck and Robert Eberan Eberhorst (left), designer of the Type D Auto Union of 1938.

Right: Hans Stuck competed in several races in 1947-48 with the Cisitalia D46. The German ace drove the Auto Union to its first victory in the 1934 Grand Prix of Germany.

carried out on the Nuvolari Roadster, which was given a new tubular chassis with a central crossbar and a front end with transverse torsion bars, a development close to the heart of the Porsche design school and also used on the legendary Volkswagen. Called the 204 A, the new roadster made its debut on 9 May 1948 at the Circuit of Vercelli, where it was forced to retire due to mechanical trouble; it appeared again a month later at Mantova on 13 June, when it won the Giorgio and Alberto Nuvolari Cup having shown itself to be faster and more efficient than the original car.

A close-up of Carlo Abarth with Adolfo Macchieraldo, Piero Dusio's team test driver and racer.

But that exciting season was not to last long. The financial investment by Cisitalia in the development of the project and the construction of the legendary prototype single-seater based on the Porsche design were becoming increasingly onerous and subtracted enormous resources from the normal running of the company. As the car was being built in early 1949, the Turin company's financial crisis was dampening the enthusiasm of its founder and the promising future of the young car brand. Dusio turned to Juan Peron's Argentina in his search for a solution to the problems that gnawed away at Cisitalia. Among the various prospects was the possible transfer of the entire factory, including all the managers, technicians, workers and machines, from Italy to Peron country and a site in the Province of Santa Fé. There, the company could have reconstituted itself with its original name, or by assuming the then definitive name of Autoar-Automotor Argentina S.A.

Carlo Abarth wanted to avoid a situation in which all the hard work he had invested in the company would, once again, become frustrated by misfortune. At the beginning of 1949 he expressed his disquiet to Guido Scagliarini, the young and gifted driver who had done so well at the wheel of the 204 A; later Guido was able to convince his father Armando, a landowner from Finale Emilia, that the profound Cisitalia crisis could change into an unexpected opportunity for the Scagliarinis, who had always made their passion for the car a reason for living. Their financial resources and their skills, as well as Carlo Abarth's organisational ability were like a fundamental premise to take over the by now agonising activities of Cisitalia, so as not to lose the vast technical patrimony the company had accumulated in such a short period. On 31 March 1949 in the offices of notary Angelo Aloisi of Bologna, Carlo Abarth and Armando Scagliarini signed articles of partnership in a limited liability company called Abarth & C., with its legal headquarters at 9 Via Don Minzoni, Bologna, the home address of Scagliarini.

The company's registered stock consisted of 998 shares at Lit 1000 each,

Fourteen Cisitalias were entered for the Mille Miglia on 2 May 1948, among them the Cassone sports saloon number 86 of Bernabei-Sari, the 202 MM roadster Razzo number 23 of Porrino-Delpiano and the 202 SMM number 53 of Bonetto-Maritano. They all retired.

Other Cisitalias entered the 1948 Mille Miglia. In close-up, the 202 SMM number 22 of Piero and Carlo Dusio, who also retired. Only two Cisitalias out of the 14 starters finished the race, the 202 sports saloon of Scagliarini-Maffiodo, which came fifth overall and the 202 SMM driven by Chieregato-Volpini that came in 31st.

Above, right: The aerodynamic Cisitalia sports saloon number 7 entered for the 1948 Mille Miglia and driven by Taruffi-Rabbia, which also retired. Developed by Giovanni Savonuzzi in the wind tunnel of the Turin Polytechnic, the car sported large tail fins and was extremely fast. As a result of an increased cubic capacity from 1100 to 1200 cc, it easily exceeded 200 km/h.

which were underwritten for Lit 990,000 by Armando Scagliarini who was nominated executive director of the newly constituted venture and for Lit 8,000 by Abarth. The contract also included a clause stating Carlo would later be able to acquire 50% of the share capital. In effect, when Armando Scagliarini later assumed the post of president of the company his son Carlo – Guido's brother – joined the board as a director and reached agreement with Abarth that enabled the Viennese-born technician to secure his 50% shareholding. On 17 June, a subsidiary of the company was established in Turin with offices at 10 Via Trecate, of which Guido Scagliarini was appointed "procurer ad negotia" and Carlo Abarth was elected to the board of directors.

For Abarth, who had just turned 40 and had done all he could without reservation to refine his technical ability with engines, the moment had come to set up on his own and begin a new and exciting professional adventure.

Mantova, the Giorgio and Alberto Nuvolari Cup, 13 June 1948: Before the start, Carlo Abarth talks to Carlo Scagliarini. The 30 lap 90 kilometre Formula Two car race was won by Felice Bonetto driving a 1200 cc single-seater Cisitalia D46. The first two places in the 50-lap sports car event were taken by Cisitalia 204 A roadsters driven by Adolfo Macchieraldo and Felice Bonetto, with Guido Scagliarini fourth in a 202 SMM.

1949

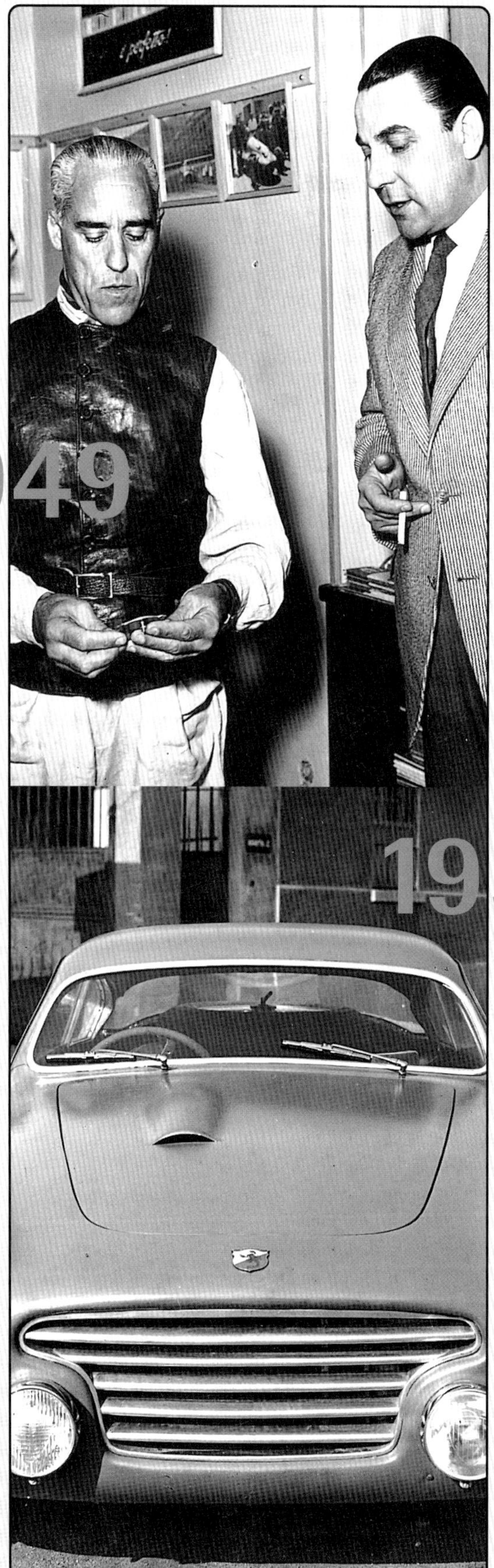

1955

EXHAUST SYSTEMS AND A LOT OF COURAGE

From the ashes of Cisitalia Abarth salvaged some good men whom he had come to appreciate in their previous two years together. Among them was the young and brilliant test driver, Cornelio Maffiodo, and Renzo Avidano, who started work for Piero Dusio in 1944 and for several years was in the sales department of Cisitalia with its manager, Levriero. Today, Avidano remembers that he was able to join Abarth because when Carlo was sports director of Cisitalia the Viennese delegated to him the logistical organisation of the team's journeys to and from the race tracks.

The new subsidiary in Via Trecate covered an area of 1815 square metres and employed 10 people at first, but by the end of 1949 that number had risen to 32. The responsibilities of the company were officially described as: "The production of vehicles and complementary equipment for sports and racing cars, as well as modification of and improvements to sports and racing cars, service, preparation of equipment for mass production, the assumption of representatives and the sale of fuel for racing cars". A range of activities like this could have sounded ambitious at the time, but the aspirations of the founder were promising and, in effect, did not go unnoticed. A scorpion, Carlo's Zodiac sign was chosen as the company's logo and was an emblem that would soon become famous throughout the world.

The first step towards establishing Abarth & C. was to quickly but fully evaluate the technical patrimony the company had inherited from Cisitalia

Below: Abarth & C. installed itself at 10 Via Trecate, Turin, where this picture was taken of the three Cisitalia 204 A Spider Sports. The cars had shone the previous year in a number of races.

Below, right: With the constitution of Abarth & C. on 15 April 1949, the Carlo Abarth Team was also introduced. The squad boasted two single-seater Cisitalia D46s, two 204 A roadsters and two cars under construction.

The Cisitalia 204 A roadster was the backbone of the newly formed Carlo Abarth Team.
Abarth himself substantially modified its chassis the year before, when he shortened and lightened it and gave it a new front end with torsion bars, which he commissioned from Porsche in Gmünd.

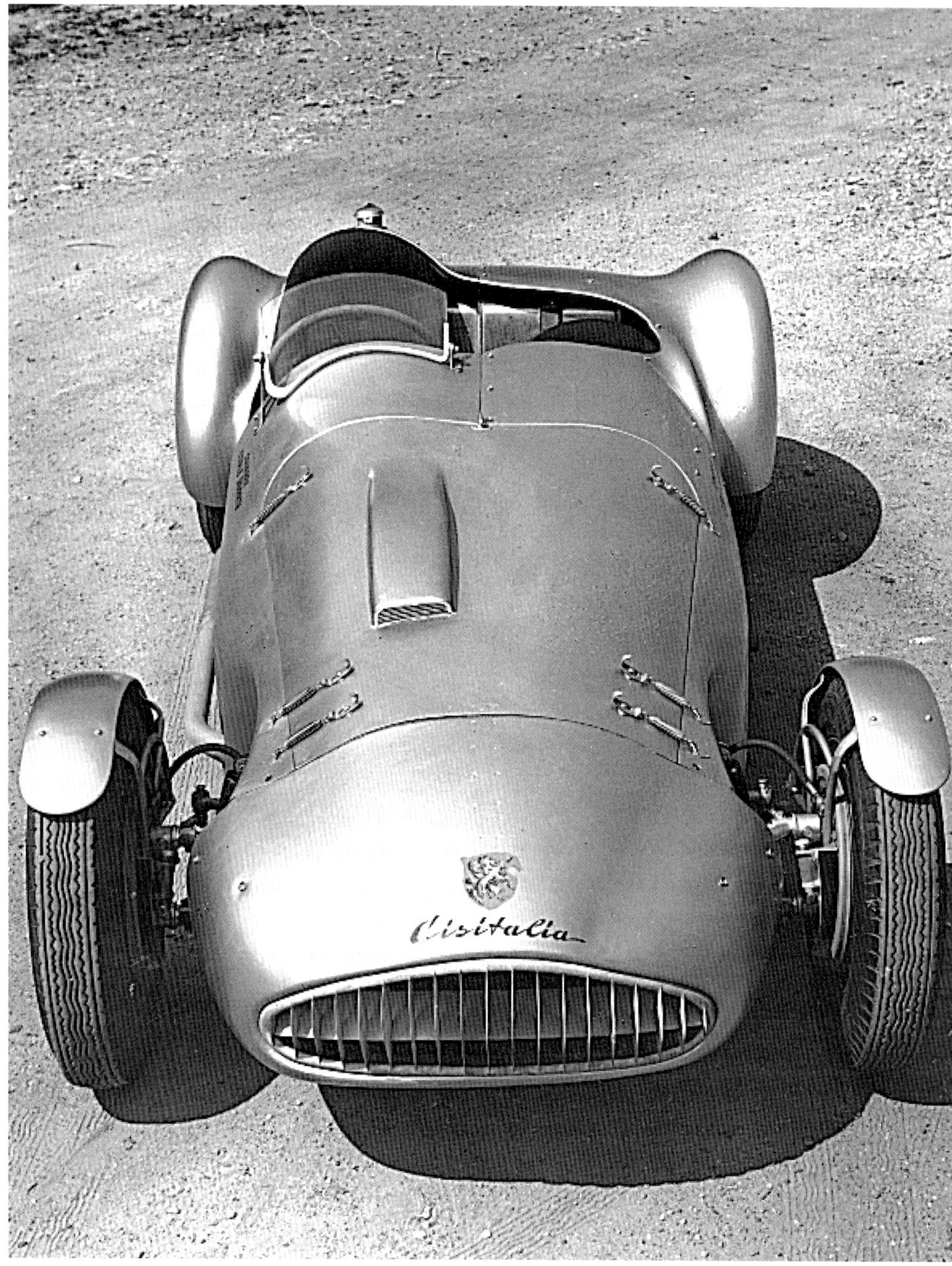

With a well-rehearsed operation that received wide coverage, Carlo Abarth introduced his racing team by recruiting the top drivers of the moment, from Franco Cortese (top picture) to Guido Scagliarini (above).

and, in more concrete terms, the material Carlo had acquired as a result of the liquidation: a D46 single-seater and two 204 A roadsters, plus two other cars under construction. The constitution of the company was officially announced on 15 April and the "Carlo Abarth Team" was presented; a team that included great names, the best-known drivers of the period including the legendary Tazio Nuvolari, Felice Bonetto, Franco Cortese, Manlio Duberti, Adolfo Macchieraldo, Emilio Romano, Franco Sardi, Guido Scagliarini, Piero Taruffi and Luigi Valenzano. It was a stroke of genius that Abarth was able to gather them together and involve all of them in an able and widely reported public-

Among the drivers who enthusiastically participated in the baptism of the Carlo Abarth Team were Luigi (Gino) Valenzano, left, and Emilio Romano, photographed in front of the company's new nameplate on the gate at the entrance to the Via Trecate premises.

ity operation. Nine days later, the new team entered four red cars in the 16th Mille Miglia, in which only Guido Scagliarini and co-driver Mario Maggio in a 204 A sports tourer did well by coming second in the up to 1100 cc class of the Sport Category and fifth overall. Scagliarini was also entered for the Circuit of Reggio Emilia on 8 May and won his class, while Piero Taruffi won the Susa-Moncenisio hillclimb on 24 July; the Lazio driver sped up through the twists and turns of the mountain in a 1200 cc single-seater, derived from the glorious D46, at an average of over 87 km/h; in the same event, Scagliarini won the 1100 Sport class again. Then Taruffi won his class at the Circuit of Senigallia on 21 August, followed a week later by him coming first overall at

The 204 A roadster to which Carlo Abarth made numerous technical and aesthetic modifications over the Cisitalia 202 SMM in 1948.
The modified car was extremely fast and reliable throughout the 1949 racing season. It was entered the first year under the banner of the Cisitalia-Carlo Abarth Team, in line with the transfer agreement between the two companies.

A little more that 40 years old and after the adversity that marked his life for years, Carlo Abarth was once again ready to take on new professional challenges.

the Aosta-Gran San Bernardo, with Scagliarini once more the winner of the up to 1100 cc Sport class.

It seemed a successful season that would lead to triumph, but on their way to Madrid at the end of October the Italian team's guardian angel momentarily turned its back on the euphoric Italian team. Present in force for the first sports car Grand Prix of Madrid and with over 100,000 spectators looking on, the 204s all dropped out of the race with mechanical problems; the writers of the period said the Cisitalia engines, which were fuelled by a special alcohol cocktail, were running on deteriorated petrol or petrol that was not suitable. The race was won by Raymond Sommer in a Simca-Gordini, but the defeat of the Abarths caused an outcry, resulting in negative repercussions hitting the management of the team and the newly formed company.

In spite of those ups and downs, the 1949 season closed with a fine collection of victories for the Scorpion; the Abarth Cisitalias won the Italian Formula 2 Championship for sports cars with Piero Taruffi, and the up to 1100 cc Sport title with Scagliarini. That induced Carlo Abarth to look for a number of alternative activities to motor sport alone, which had absorbed all the resources of the company in those first months. His feet on the ground as ever, he thought: "*We must find another option to just managing sports cars, so that the Abarth name can attract the interest of ordinary motorists and not merely a small group of enthusiasts*". He considered preparing special fuel feed kits for mass produced cars with which to improve their performance in terms of power, acceleration and maximum speed; in fact, his consummate instinct pushed him to also develop a special water pump and a column-mounted gear lever for the Fiat 500, constructive ideas which, in those years, flew in on the coat-tails of fashions introduced by the opulent post-war American cars.

That is how the production came about of a kit to transform the Topolino's gear lever, first on a small scale and later more extensively, so that the more susceptible drivers of the little Fiat could use that much more fashion-

Deeply committed to constant work, Carlo Abarth found sanctuary in a simple and unostentatious private life. Here, he is pictured with his sister Anna during one of her visits to Turin (left) and at Lake Viverone with his wife.

ABARTH "204 A" (1949)	
ENGINE	
	front, longitudinal, 4 cylinders in line
Bore and stroke	mm 68x75
Cubic capacity	cc 1089
Valve gear	overhead valves, side camshaft
Number of valves	2 per cylinder
Compression ratio	11.8: 1
Fuel feed	2 Weber 36DR4SP carburettors
Ignition	magneto
Coolant	liquid
Lubrication	dry sump
Maximum power	83 hp at 6000 rpm
TRANSMISSION	
	rear wheel drive
Clutch	dual dry plate
Gearbox	4 speed plus reverse, central lever
Axle ratio	4.87:1
BODY	
	racing roadster
CHASSIS	
	tubular with central cross
Suspension	front independent, transverse torsion bars, hydraulic dampers; rear rigid axle, longitudinal semi-elliptic springs, hydraulic dampers
Brakes	hydraulically operated drums front and rear
Steering	worm and sector
Fuel tank	capacity 48 litres
Tyres	5.00 x15
DIMENSIONS AND WEIGHT	
Wheelbase	mm 2100
Front track	mm 1250
Rear track	mm 1240
Length	mm 3900
Width	mm 1460
Height	mm 1165
Kerb weight	kg 510
PERFORMANCE	
Top speed	km/h 190

able means of changing the small car's gears. But Abarth's most successful development of the period was an exhaust system, which harked back 20 years to when he raced his motorcycle every Sunday against the toughest competitors in the sport; his bike was almost always faster because the engine, which he maintained and updated meticulously, had a special exhaust that ensured his bike performed better. Carlo was unable to take that promising idea further at the time, but at the end of 1949 it seemed to him perhaps a decisive means of creating some plain sailing for Abarth & C. He built a number of prototypes, all with a central pipe of the same diameter with lateral channels in glass wool and diaphragms to maintain the maximum back pressure; he realised that, with minor alterations, the system could be adapted to suit any car, from the little Fiat 500 to the most sophisticated high performance sports tourer. The decisive sales argument was not so much the increase in engine performance - slight but effective and made possible by a freer, more open exhaust – but the rortier sound it produced, a fuller and more harmonious bark that appealed to Topolino owners.

One of the first races in which the new team competed was the 15th Susa-Moncenisio of 24 July 1949, in which company test driver Manlio Duberti was at the wheel of the 204 A roadster (pictured).

At Madrid, however, something went wrong and the string of Abarth victories came to an abrupt halt. At the first Grand Prix for Sports cars, which took place in the Spanish capital on 30 October 1949 (pictured), the Scorpions plunged headlong into a series of mechanical problems that meant they suffered a harsh defeat.

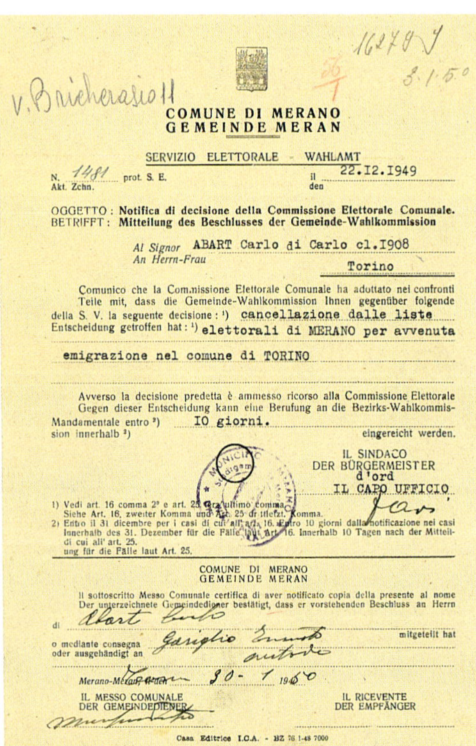

COMUNE DI MERANO
GEMEINDE MERAN

SERVIZIO ELETTORALE - WAHLAMT

N. / Akt. Zchn. 1481 prot. S. E. il / den 22.12.1949

OGGETTO: Notifica di decisione della Commissione Elettorale Comunale.
BETRIFFT: Mitteilung des Beschlusses der Gemeinde-Wahlkommission

Al Signor / An Herrn-Frau ABART Carlo di Carlo cl.1908
Torino

Comunico che la Commissione Elettorale Comunale ha adottato nei confronti della S. V. la seguente decisione: 1) cancellazione dalle liste elettorali di MERANO per avvenuta emigrazione nel comune di TORINO

Teile mit, dass die Gemeinde-Wahlkommission Ihnen gegenüber folgende Entscheidung getroffen hat: 1)

Avverso la decisione predetta è ammesso ricorso alla Commissione Elettorale Mandamentale entro 2) IO giorni.

Gegen dieser Entscheidung kann eine Berufung an die Bezirks-Wahlkommission innerhalb 2) eingereicht werden.

IL SINDACO
DER BÜRGERMEISTER
d'ord
IL CAPO UFFICIO

1) Vedi art. 16 comma 2° e art. 25 terz'ultimo comma.
Siehe Art. 16, zweiter Komma und Art. 25 dr ttlefzt. Komma.
2) Entro il 31 dicembre per i casi di cui all'art. 16. Entro 10 giorni dalla notificazione nei casi di cui all'art. 25.
Innerhalb des 31. Dezember für die Fälle laut Art. 16. Innerhalb 10 Tagen nach der Mitteilung für die Fälle laut Art. 25.

COMUNE DI MERANO
GEMEINDE MERAN

Il sottoscritto Messo Comunale certifica di aver notificato copia della presente al nome di Abart Carlo
Der unterzeichnete Gemeindediener bestätigt, dass er vorstehenden Beschluss an Herrn mitgeteilt hat

o mediante consegna / oder ausgehändigt an Gariglio Ernesto

Merano-Meran, il-den 30-1 1950

IL MESSO COMUNALE
DER GEMEINDEDIENER

IL RICEVENTE
DER EMPFÄNGER

Casa Editrice I.C.A. - BZ 76 1-48 7000

Abarth conceived an able publicity campaign for his exhaust system and backed it with the elegant presentation of the product, which was painted matt black except for the chrome terminal and included a fixed price list in his press advertisements; he assigned the creation of the campaign's graphics to the Barale agency in Turin and was able to persuade thousands of motorists to remove the normal Fiat 500 system and install the Abarth unit, creating an unusual fashion among drivers. A different exhaust was developed for specific car models, bench tested and fitted to the car, more often than not coupled with the manifold and the entire exhaust system. Renzo Avidano remembers that, in that difficult winter of 1949-1950 when dark clouds started to form over the young company, the launch of the exhaust system was a real stroke of luck. "*Abarth produced the first 50 for the Topolino and their initial impact on the market left everyone perplexed, because the selling price of Lit 4.500 seemed exorbitant against the Lit 1.600-2.000 of the competition's exhausts. But that successful advertising campaign, the involvement of Nuvolari's secretary, Nini Greggio, who was able to have Abarth's new product displayed in the windows of the most elegant shops in Verona, and the strong commercial commitment of the replaement parts and accessories distributor Autoaccessorio of Milan, managed to*

Opposite. Above: of an intense but victorious 1949 racing season, Carlo Abarth lined up for a celebration photograph with the cars that scored so many successes for the new team, two 204 A roadsters and the 1200 cc D46 single-seater.

Opposite, below: The Merano authorities certified on 30 January 1950 the cancellation of Carlo Abarth's name from the electoral role 'having moved to the municipality of Turin".

Below: The commercial launch of special Abarth inlet manifolds for mass produced cars brought a new generation of cars to life - one with modified mechanics that attracted great interest in Italy and in foreign markets. On the left are two carburettors and a manifold for the Fiat 1100S, accredited with giving the car another 9 hp. On the right the three-carburettor kit for the Fiat 1500, which produced a 10% increase in power and another 10 km/h of top speed.

create a powerful demand on the part of both the users and a number of manufacturers like Alfa Romeo, Ferrari and Maserati". Even in the United States, motorists who fitted the Abarth exhaust system to their cars stood out, whether at the wheel of a Jaguar XK120, a Porsche or a modest Chevrolet. The phenomenon, which first spread across Italy, extended to overseas markets and eventually contributed to the company's balance sheet in an unexpectedly generous manner.

Undoubtedly, the significant sales success of the Abarth exhausts should also be ascribed to the victories the Scorpion's racing cars achieved most weekends. In just a few years, the activities of the company's racing department became a formidable promotional platform for its consumer products and that was unusual, almost unrepeatable on the motoring scene of the period. The multi-form world of replacement companies began to distribute Abarth inlet manifold kits with two carburettors and an exhaust, which found many different applications among those who wanted to improve the performance of their Fiat 1100 or a six cylinder 1500; the advertising messages proclaimed an increase in maximum power of 10%, an increase in top speed of about 10 km/h,

COLLETTORI SPECIALI "ABARTH" D'ASPIRAZIONE E SCARICO

per vetture FIAT 1100 S - 1100 competizione - 1200-Cisitalia

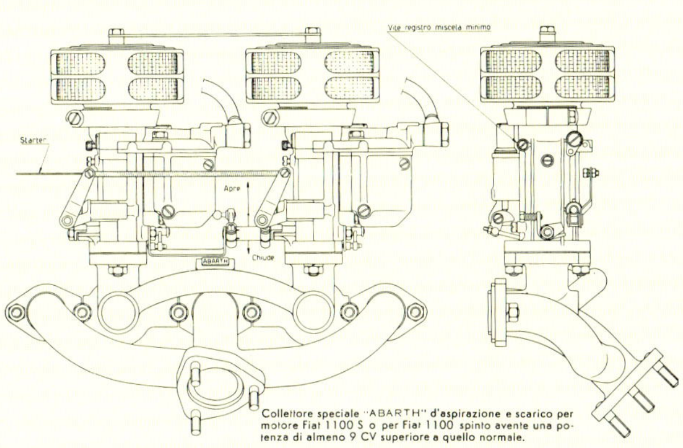

Collettore speciale "ABARTH" d'aspirazione e scarico per motore Fiat 1100 S o per Fiat 1100 spinto avente una potenza di almeno 9 CV superiore a quello normale.

Oltre ai collettori per le vetture FIAT 1100 A-B-E di serie costruiamo un tipo di collettore a due carburatori con pompa di ripresa per vetture con motore **FIAT 1100 S** destinate all'impiego sportivo ed un tipo di collettore specialissimo pure a due carburatori con pompa di ripresa, per vetture da competizione dotate di motore **FIAT 1100** modificato, 1200 e Cisitalia.

Il collettore per FIAT 1100 S. è anche adatto per motori FIAT 1100 che abbiano una potenza di almeno nove cavalli superiore a quella del motore normale. Esso è a compensazione doppia, brevettata, ed è munito di collettore di scarico pure speciale eseguito in tubi saldati e opportunamente sagomati in modo da facilitare l'espulsione dei gas. L'aumento di potenza massima conseguibile è di circa il 10 %.

Il collettore per 1100 da competizione è un diretto derivato dai collettori da noi costruiti per le nostre vetture da corsa. Esso è particolarmente adatto per motori FIAT 1100 spinti aventi una potenza considerevolmente superiore a quella del motore normale, motori quindi sui quali si sono già eseguite tutte le modifiche meccaniche volte al conseguimento di una maggior potenza attraverso all'aumento del rendimento organico e di quello termico. L'applicazione del nostro collettore assicura, in media, un aumento di potenza di circa il 10 %.

COLLETTORE SPECIALE "ABARTH" D'ASPIRAZIONE E SCARICO

PER VETTURE FIAT 1500 A-B-D-E GUIDA SINISTRA

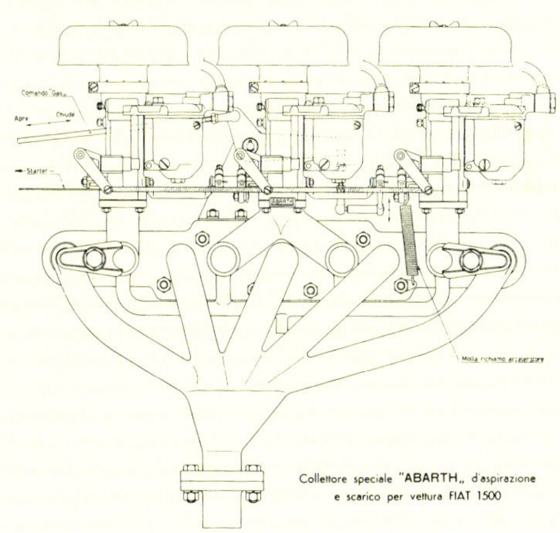

Collettore speciale "ABARTH" d'aspirazione e scarico per vettura FIAT 1500

VANTAGGI

- **AUMENTO** della potenza massima di circa il 10 %;
- **AUMENTO** della velocità massima di circa 10 Km/h.;
- **AUMENTO** considerevole della ripresa;
- **AUMENTO** dell'elasticità con conseguente possibilità di usare le marce alte anche a basso regime;
- **AUMENTO** notevole delle prestazioni in salita;
- **DIMINUZIONE** del consumo di carburante di circa il 10 % con motore funzionante agli 8/10 della potenza massima;
- **DIMINUZIONE** della sollecitazione termica del motore in seguito ad un migliore riempimento e ad una dosatura più costante. Lo speciale collettore di scarico contribuisce a tale diminuzione favorendo il deflusso dei gas combusti;
- **DIMINUZIONE** dell'usura del motore data la possibilità dell'uso prolungato delle marce alte.

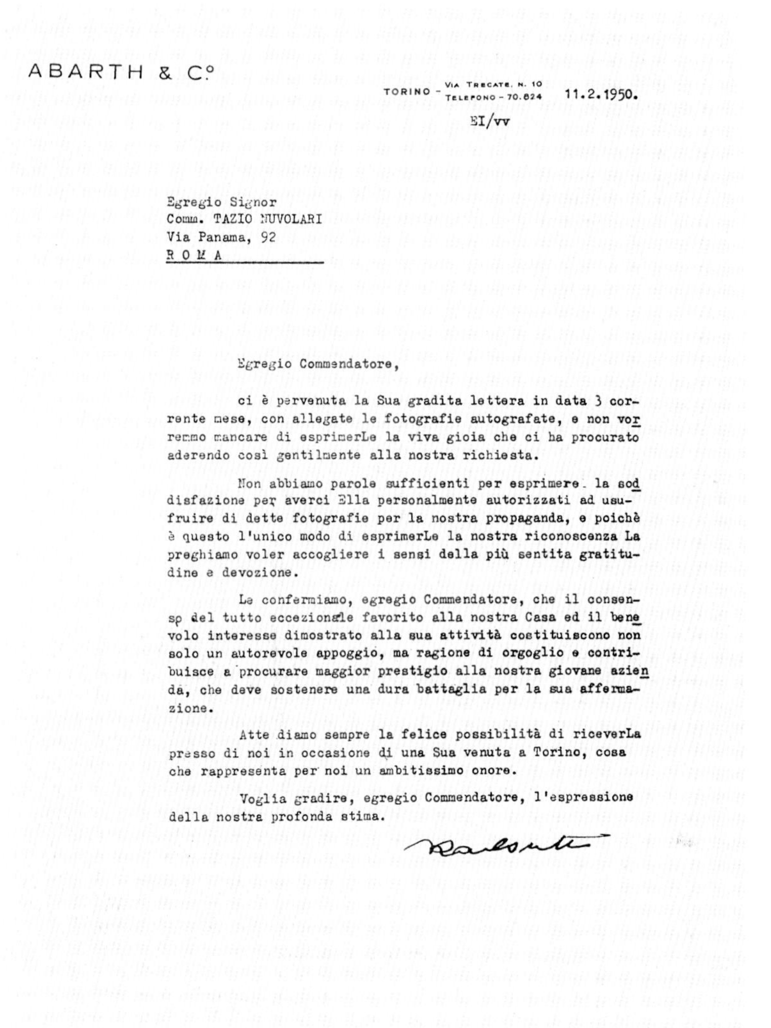

ABARTH & C°

TORINO - VIA TRECATE, N. 10 TELEFONO - 70.824 11.2.1950.

EI/vv

Egregio Signor
Comm. TAZIO NUVOLARI
Via Panama, 92
R O M A

Egregio Commendatore,

ci è pervenuta la Sua gradita lettera in data 3 corrente mese, con allegate le fotografie autografate, e non vorremmo mancare di esprimerLe la viva gioia che ci ha procurato aderendo così gentilmente alla nostra richiesta.

Non abbiamo parole sufficienti per esprimere la soddisfazione per averci Ella personalmente autorizzati ad usufruire di dette fotografie per la nostra propaganda, e poichè è questo l'unico modo di esprimerLe la nostra riconoscenza La preghiamo voler accogliere i sensi della più sentita gratitudine e devozione.

Le confermiamo, egregio Commendatore, che il consenso del tutto eccezionale favorito alla nostra Casa ed il benevolo interesse dimostrato alla sua attività costituiscono non solo un autorevole appoggio, ma ragione di orgoglio e contribuisce a procurare maggior prestigio alla nostra giovane azienda, che deve sostenere una dura battaglia per la sua affermazione.

Attediamo sempre la felice possibilità di riceverLa presso di noi in occasione di una Sua venuta a Torino, cosa che rappresenta per noi un ambitissimo onore.

Voglia gradire, egregio Commendatore, l'espressione della nostra profonda stima.

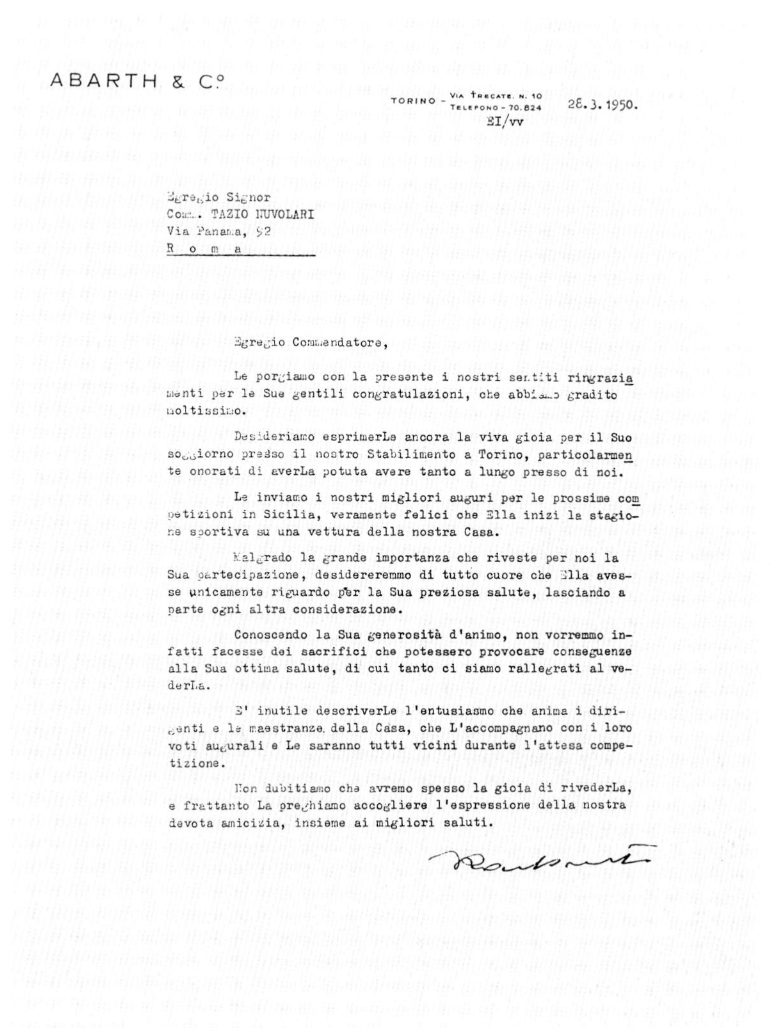

ABARTH & C°

TORINO - VIA TRECATE, N. 10 TELEFONO - 70.824 28.3.1950.

EI/vv

Egregio Signor
Comm. TAZIO NUVOLARI
Via Panama, 92
R o m a

Egregio Commendatore,

Le porgiamo con la presente i nostri sentiti ringraziamenti per le Sue gentili congratulazioni, che abbiamo gradito moltissimo.

Desideriamo esprimerLe ancora la viva gioia per il Suo soggiorno presso il nostro Stabilimento a Torino, particolarmente onorati di averLa potuta avere tanto a lungo presso di noi.

Le inviamo i nostri migliori auguri per le prossime competizioni in Sicilia, veramente felici che Ella inizi la stagione sportiva su una vettura della nostra Casa.

Malgrado la grande importanza che riveste per noi la Sua partecipazione, desidereremmo di tutto cuore che Ella avesse unicamente riguardo per la Sua preziosa salute, lasciando a parte ogni altra considerazione.

Conoscendo la Sua generosità d'animo, non vorremmo infatti facesse dei sacrifici che potessero provocare conseguenze alla Sua ottima salute, di cui tanto ci siamo rallegrati al vederLa.

E' inutile descriverLe l'entusiasmo che anima i dirigenti e le maestranze della Casa, che L'accompagnano con i loro voti augurali e Le saranno tutti vicini durante l'attesa competizione.

Non dubitiamo che avremo spesso la gioia di rivederLa, e frattanto La preghiamo accogliere l'espressione della nostra devota amicizia, insieme ai migliori saluti.

A copy of a letter dated 11 February 1950 from Abarth & C. to Commendatore Tazio Nuvolari to sincerely thank him for allowing the company to use photographs of him in its publicity material.

Another Abarth & C. letter of 28 March 1950 expressing the company's gratitude for the visit of Tazio Nuvolari to the young factory in Via Trecate, and for the decision of the great Mantuan champion to compete in a number of races in Sicily at the wheel of a Scorpion.

better pick-up and driving elasticity; in addition, they permitted a 10% fuel saving with the engine working at eight tenths of its maximum. Giovanni Savonuzzi had already studied and tested the manifold group of the Cisitalia engine at the time of his original project, but the intuition and entrepreneurial spirit of Carlo Abarth knew how to gather and exploit the product's commercial content. The modification kit proudly bearing the Scorpion logo encountered great success among drivers of all Fiat models during the Fifties - the 1400, 1900, 1100-103... - but also among those who wanted to improve the performance of the 1131 cc Volkswagen Beetle's four cylinder boxer engine.

In 1950, neo-entrepreneur Carlo Abarth came up with a wealth of ideas and projects. But the first thing to do was to consolidate the position achieved in motor racing the previous season and to convincingly repeat the validity of a number of techno-constructional choices that contributed to the company's sporting success in 1949. That string of triumphs began on 28 March with the InterEuropa Cup at Monza, where Guido Scagliarini drove the Abarth 204 A sports tourer to victory in the up to 1100 cc Sport class, followed on 10 April

The day of Tazio Nuvolari's visit to Abarth & C. during the first half of March 1950. In this photograph are, from the left: drivers Manlio Duberti and Guido Scagliarini, Nuvolari in the cockpit of the 204 A roadster and Carlo Abarth, visibly proud of the visit, which was so important to him. Note that the front of the car carries the Scorpion logo, which took the place of the Cisitalia emblem from the 1950 season.

During his Abarth visit, Tazio Nuvolari conducted extensive tests of the 204 A, with which he was to make his come back to motor racing at the 10th Giro di Sicilia on 2-3 April 1950. In the picture Nuvolari is talking to Guido Scagliarini, whose family financed the constitution of the new Via Trecate company.

For Abarth, the chance of presenting Nuvolari as a member of his team at the start of the 1950 season was a major publicity coup.
In the photograph from the left: Guido Scagliarini, chief mechanic Sambuelli, Tazio Nuvolari, Nini Greggio, Nuvolari's secretary, and Carlo Abarth during testing for the Sicilian race in April.

An unusual view of the Abarth 204 A's chassis, the car's engine derived from a Fiat 1089 cc unit. Fed by two Weber carburettors with a 50-50 mix of alcohol and petrol, the motor could generate a maximum of 70-80 hp at 6000 rpm.
Carlo substituted the original front end of the Cisitalia 202 and its transverse semi-elliptic springs with torsion bars and reduced the overall weight of the vehicle to 685 kg.
Driven by Adolfo Macchieraldo, the 204 A made its debut under the Cisitalia banner on 9 May 1948 at the opening Grand Prix of Vercelli.

by the Palermo-Monte Pellegrino hillclimb, where veteran Tazio Nuvolari scored the last emotional victory of his stunning career in a 204 A roadster. It was a very special day and race; Auto-Italiana reported the spirit of it like this: *"It is useless to say that the participation of Nuvolari was the central attraction of the Palermo event, and that most of the thousands and thousands of fans who lined the slopes from the first morning, thronging the strategic points, said that they only wanted to see the 'Flying Mantuan.'*

"Nuvolari repaid them for their wait as best he could at the time, but lacked practice - especially on that particular Monte Pellegrino route - and his Abarth did not even have the right gear ratios due to its last minute entry, but he raced

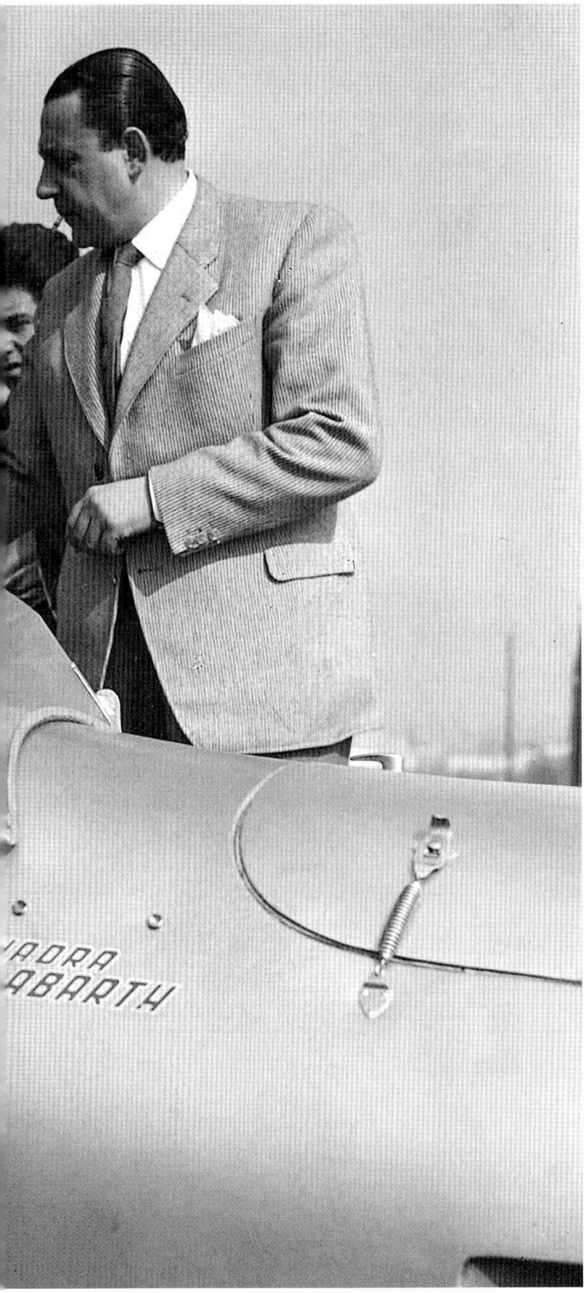

literally between two banks of an applauding crowd from start to finish.

"The class win was his and in the overall results he came immediately behind the powerful 2000 cc cars, for which his personal performance and that of the Abarth 1100 were more than gratifying".

A few days before the Sicilian race, Nuvolari – at the time 57 years old – visited the small factory in Via Trecate, Turin, and in a meeting with Carlo Abarth agreed to compete in the hillclimb that was to be the end of his amazing career as a champion of the wheel. On 28 March, Abarth & C. sent a courteous letter of thanks to Commendatore Tazio Nuvolari for his visit, in which the following passage stood out:

"We send you our best wishes for the forthcoming event in Sicily, we are extremely happy that you will begin the sports season in one of our cars.

Nuvolari coming into contact with the Abarth 204 A for the first time on the outskirts of Turin, preparing for the imminent transfer to Sicily. After retiring from the 10th Giro di Sicilia with a broken steering box, the popular champion competed in the Palermo-Monte Pellegrino hillclimb a week later and won the up to 1100 cc class of the Sport Category.

In spite of the great importance to us of your participation, we desire with all our hearts that you take complete care of your precious health, leaving aside all other considerations.

Knowing your generosity of spirit, we do not wish you to make sacrifices that could provoke consequences to your optimum health, in which we are delighted to see you..."

Nuvolari's last victory closed a heroic age of the great aces of the past. His retirement brought to an end a generation of romantic drivers, new cavaliers without blemish or fear, drivers of passion rather than profession. Tazio died of his illness three years later, on 11 August 1953.

The sports tourer version of the Abarth 204 A called the Monza made its first appearance in early 1950.
Alfredo Vignale had designed an elegant body for the car, based on initial sketches by Giovanni Michelotti.
Franco Cortese, one of the great names among the drivers of the Carlo Abarth Team, emphasised in this picture the reduced height of the new model, which was still unpainted.

Carlo Abarth had a dual-tone version of the same car built with two supplementary front lights.
Numerous styling elements of the 204 A were used by Carrozzeria Vignale to create the Orchidea panoramic saloon on the mechanics of the new Fiat 1400, which was widely acclaimed in the spring of 1950.

A special 204 A A was entered for the 17th Mille Miglia, which was run on 23 April 1950. The car's front and rear ends were considerably longer than those of the original body, with the obvious intention of improving its drag coefficient. Driven by Guido Scagliarini-Cornelio Maffiodo, the car was forced to retire.

Car number 630 was another sports saloon with an unpainted body in aluminium sheeting that competed in the historic Brescia-Rome-Brescia race. It was driven by Carlo Scagliarini and L. Garrone and came 31st overall, after having covered 1682 km in 16h 09' 46" at an average speed of over 104 km/h.

Below: In profile, the first newly designed post-war car by the Turin manufacturer, the Fiat 1400. It had an integral bodyshell, a 1395 cc 44 hp engine, independent front suspension with oscillating triangular arms and a body that tended stylistically towards the American trends of the moment and could accommodate five people. The car was introduced at the 1950 Geneva Motor Show.

The most significant sequence of Abarth victories in 1950 continued on 9 June with Franco Cortese, who won the up to 1100 cc Sport class of the first Caracalla-Roma Night Race with a 204 A sports tourer; that success was repeated by Manlio Duberti on June 10, when he won the 1100 cc Sport category of the Vues des Alpes hillclimb whit the 204 A and then again a week later at the first Circuit of Oporto, in Portugal, where Emilo Romano won. On 23 August it was Luigi Valenzano's turn; he won the 1100 Sport class of the Circuit of Pergusa (Enna); driving a long-nosed 204 A Monza Valenzano won the Sport category again, this time at the Ponente Cup, which took place at Sanremo on 12 November.

In the course of that brilliant racing season, Carlo Abarth also contemplated his dream of competing in Formula 2. Interviewed on the subject by Auto-Italiana in February 1950, he declared *"... I think Formula 2 is the most suitable compromise between the super racing cars of Formula 1 and the under-powered cars of Formula 3"*. At the time, the same authoritative Italian magazine published a cross-section picture of a V8 engine designed by Abarth to compete in single-seater two-litre races; it was an engine of extremely modern concept, with hemispheric heads and distribution by four overhead camshafts, all made in light alloy. A project on which Abarth did not follow through.

The year 1950 also included other events that marked the start of the Italian recovery, after the forced suspension of activity due to the world conflict. At the Geneva Motor Show in the spring, Fiat unveiled the 1400, its first post-war car, that was a medium-size saloon of Americanised appearance, which opened the road to the rapid process of modernisation among leading Italian manufacturers. A few weeks later, it was Lancia's turn. The company introduced the new Aurelia saloon at the Turin Motor Show; with its V6 1750 cc engine, the Aurelia was a superior class of car that was the top of the range of a long and successful series of models that remained in production until 1957. Alfa Romeo was not to be left behind and showed outside the front door of the Palazzo del Valentino in Turin the prototype of a four-door saloon that

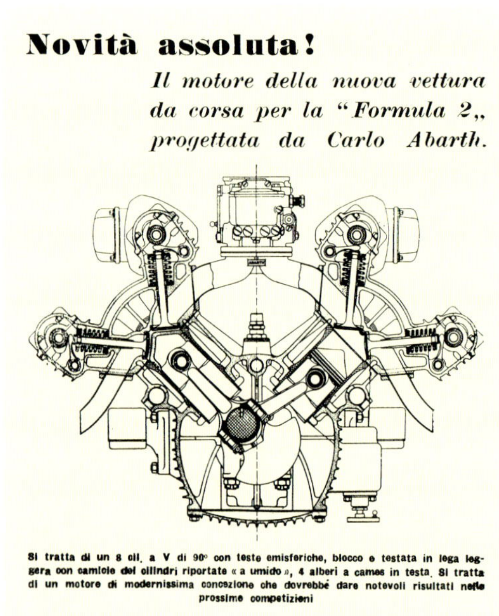

Novità assoluta!

Il motore della nuova vettura da corsa per la "Formula 2" progettata da Carlo Abarth.

Si tratta di un 8 cil. a V di 90° con teste emisferiche, blocco e testata in lega leggera con camicie dei cilindri riportate « a umido », 4 alberi a cames in testa. Si tratta di un motore di modernissima concezione che dovrebbe dare notevoli risultati nelle prossime competizioni

Edition number 3 of the magazine Auto-Italiana dated 1 February 1950 published a transverse cross-section of a two-litre eight-cylinder engine, designed by Carlo Abarth for a Formula 2 single-seater. It was a V8 with cylinder banks set at 90°, hemispheric heads, four overhead camshafts, all in light alloy. An engine which, according to Abarth, could also have been competitive in Sport category cars – but it was never put into production.

made its debut as the 1900 a year later; it was the first of the mass produced Alfas that led the company to its first large-scale commercial successes.

A definitive version of the 204 A sports tourer appeared in Abarth that year; the car was a sophistication of the model that did so well the year before, driven by Guido Scagliarini. The new version was further refined and lightened, 'dressed' in an elegant Vignale body designed by Giovanni Michelotti with three round air vents in the front wings and behind the wheel housing, with a wrap-around rear window that flowed into the tail; an experimental 204 A with a lengthened nose and rear end to improve aerodynamics. The car competed in the 17th Mille Miglia on 23 April 1950 driven by Guido Scagliarini and Cornelio Maffiodo, but it retired. A similar car without the aerodynamic modifications was driven by Guido's brother Carlo and L. Garrone who came 31st overall, two-and-a-half hours behind the winning 2340 cc Ferrari 195S of Giannino Marzotto and Marco Crosara.

Away from racing, the 204 A was shown for the first time at the International Motor Show in Turin from 4-14 May on Abarth's first stand at the show. Also on prominent display was a range of special exhaust systems, for which the factory was creating a vast and efficient sales network.

In that same year of 1950, the Italian body designers were particularly active - the old, well-established companies and the legion of newcomers who, encouraged by the new availability of more modern mechanicals, produced good workmanship that contributed much to the establishment of the reputation of the Italian school in the difficult art of body design. They generated proof of a vitality that was, however, mitigated by the modest volume of Italian car production, which had reached 100,000 units by the end of 1950, while France was already close to 200,000 cars, Britain had exceeded 520,000

Left: the 32nd Turin Motor Show that took place from 4-14 May 1950 saw the first official presence of Abarth & C., on whose stand was displayed the modified steering column gear change for the Fiat 500 as well as a series of inlet manifolds and exhaust systems for numerous cars of different makes.

Right: An inlet manifold with two carburettors sold by Abarth for the Fiat 1100 B and E. The maximum increase in power attainable according to the firm's advertising at the time was around 10%.

Carlo Abarth caught by a photographer in a rare moment of relaxation, while on holiday at Portschach am Wörther See in Carinthia. With him is his wife Nadina.

At the first Circuit of Oporto in Portugal, which took place on 18 June 1950, Emilio Romano drove the Abarth 204 A to victory in the 1100 cc Sport category.

and the United States were at over 6.5 million. Still in disarray due to the war and ruled by four different armies of occupation, car production in Germany was, nevertheless, close to 120,000 vehicles. From the underdevelopment in which the Italian automobile industry languished a vast movement in public opinion became a united spokesman, accusing the Government of not wanting to match the effective technical and social progress of Italy.

At the end of the year Abarth noted tersely, almost laconically, *"Exportation of my products (the exhausts) began: 2% of total turnover. The number of my employees has increased: now there are 41"*. With his personal commitment and iron self-discipline, Carlo was re-building a new professional life in Turin that was not without its problems. He acclimatised himself to the city fairly well and he and his wife had already moved from Via Bricherasio to 48 Via Cesare Balbo in the Vanchiglia area.

Early in 1951, it seemed evident that committed involvement in racing had brought extraordinary results in terms of promoting the Abarth brand, not only in the world of sports cars but also in mass communications. But to some it also seemed that Carlo was committing the same errors of evaluation

In the spring of 1951, the magazine Motor Italia published the layout of an inlet system designed by Carlo Abarth for the Fiat 1400 that had a special manifold and two carburettors. The advertising of the period carrying the Scorpion logo was aimed at attracting the attention of consumers to the anagrammed Abarth name in this way:
A: alto rendimento (high yield)
B: basso consumo (low consumption)
A: aumentata velocità (increased speed)
R: ripresa fulminea (lightening pick-up)
T: tranquillità di Marcia (quiet running)
H: Hp

The Turin Motor Show, which opened on 5 April 1951, included the Abarth stand debut of the 205 A saloon by Vignale.
The car was strictly derived from the racing 204 A, compared to which it was more luxuriously appointed and elegant.
The 205 A was the Via Trecate factory's first production car.

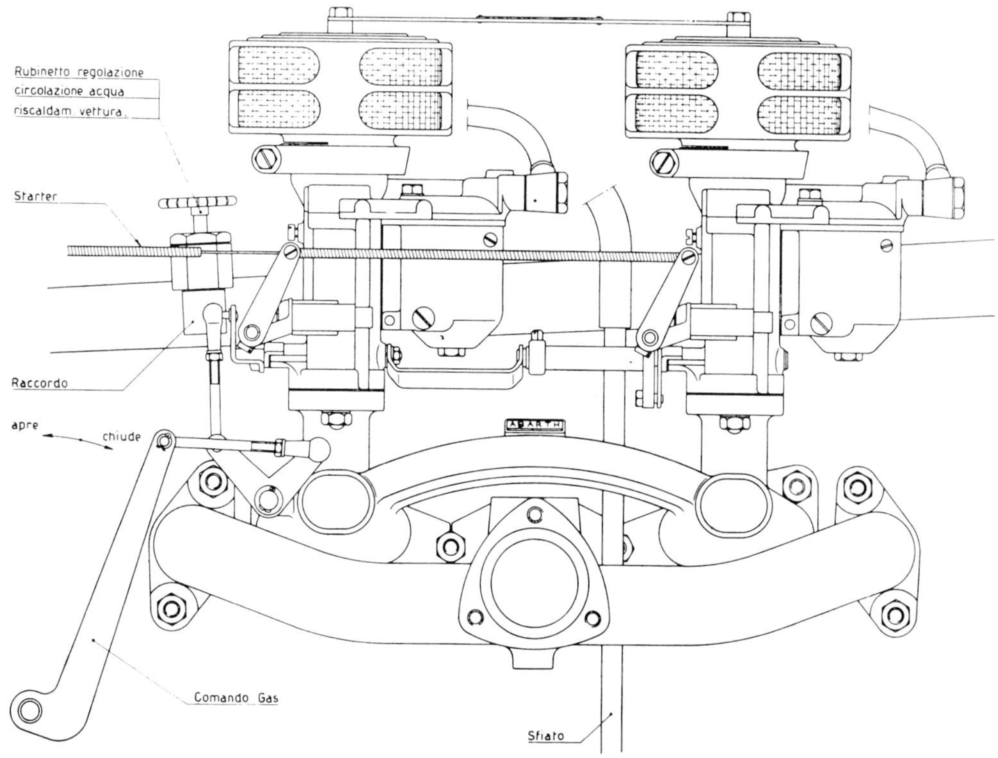

ABARTH "205 A" (1951)

ENGINE	front, longitudinal, 4 cylinders in line
Bore and stroke	mm 71,5x75
Cubic capacity	cc 1188
Valve gear	overhead valves, side camshaft
Number of valves	2 per cylinder
Compression ratio	9: 1
Fuel feed	2 Weber 32DRSP carburettors
Ignition	coil, distributor
Coolant	liquid
Lubrication	dry sump, oil radiator
Maximum power	65 hp at 5500 rpm
TRANSMISSION	rear wheel drive
Clutch	single dry plate
Gearbox	4 speed plus reverse, central lever
Axle ratio	3.7:1
BODY	two-seater coupé by Vignale
CHASSIS	boxed steel
Suspension	front independent, transverse torsion bars, hydraulic dampers; rear rigid axle, longitudinal semi-elliptic springs, hydraulic dampers
Brakes	hydraulically operated drums front and rear
Steering	worm and sector
Fuel tank	capacity 48 litres
Tyres	front 4.50 x15; rear 5.00 x15
DIMENSIONS AND WEIGHT	
Wheelbase	mm 2210
Front track	mm 1258
Rear track	mm 1248
Length	mm 3495
Width	mm 1420
Height	mm 1245
Kerb weight	kg 818
PERFORMANCE	
Top speed	km/h 170

as those attributed to Piero Dusio, who became fired up by the triumphs of Cisitalia in racing and prey to the thrill of constructing a Grand Prix car, ending up losing contact with reality and the primary needs imposed by the production of the 202 coupé for his customers. The new resources found the previous year with which to market products for mass consumption by motorists were insufficient to cover the cost of a racing department which, even if highly successful and active in all the races in the calendar, was extremely expensive to run.

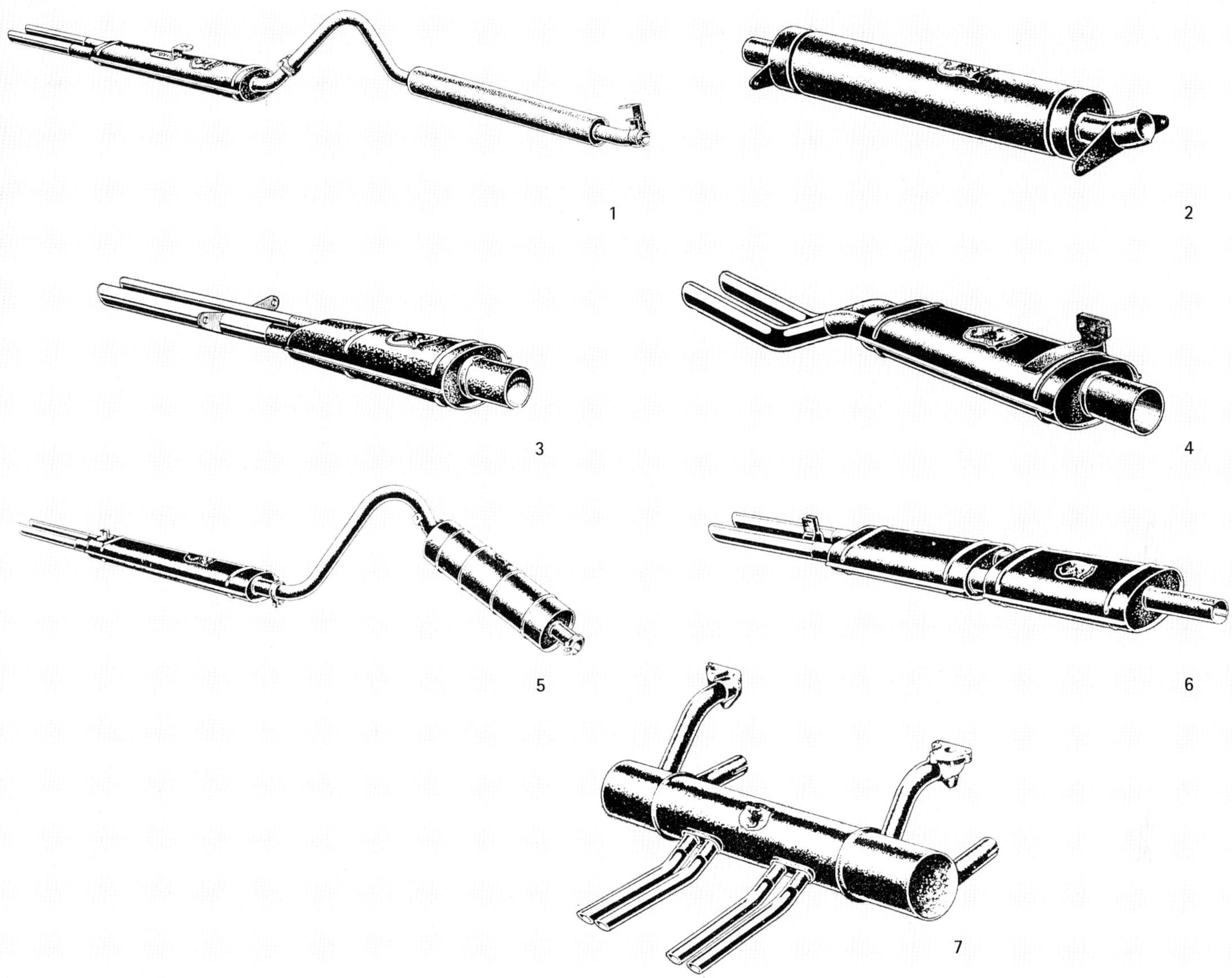

The 1951 Abarth special exhaust systems catalogue already included a good number of units for many Italian and foreign mass-produced cars. Among them were
1) the complete exhaust system for the Fiat 500C,
2) the central exhaust for the Lancia Aprilia,
3) the terminal section for the Fiat 1100S,
4) the double exhaust pipe for the Alfa Romeo 1900,
5) a complete exhaust for the Fiat 1400,
6) the two unit group for the Panhard Dyna and
7) the Volkswagen Beetle exhaust.

Abarth did not hesitate to take the necessary counter-measures and decided that, for 1951, he would concentrate all his efforts on the manufacture of exhausts and other mechanical components (steering column gear changes, water pumps, manifold groups ...) and the presentation of the coupé derived directly from the racing 204 A, aimed at his wealthy sports clientele disillusioned by the long wait for the Cisitalia 202. The car made its official debut on 5 April, opening day of the 33rd Turin Motor Show, and was called the 205 A, the lines of which resembled those of the racing sports tourer; Carrozzeria Vignale had prepared an extremely refined interior design as well as adding two narrow bumper bars in shiny metal and a pair of generous fog lamps to the vehicle's exterior. The first car to be produced by Abarth, the 205 A was well received by the critics and public; its harmonious and modern shape

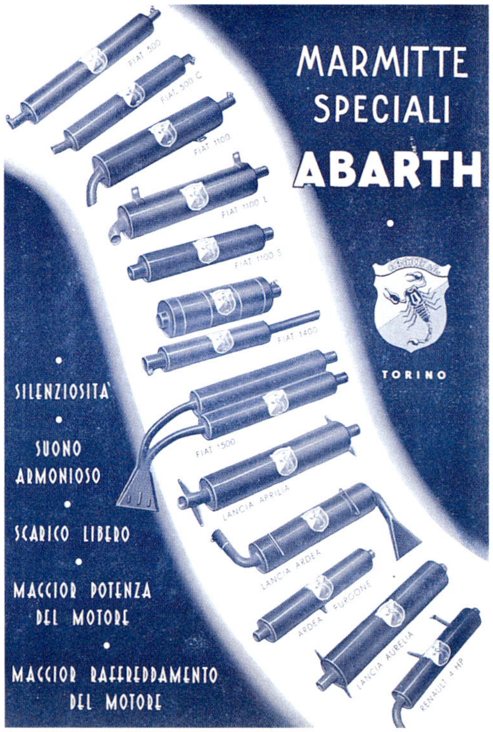

The effort made by Abarth to offer a wide range of exhausts systems was sustained by an incisive advertising campaign in the enthusiast press. The construction of special accessories aimed at replacing the mass produced components included the Abarth 51 water pump for the Fiat 500s A, B and C (illustrated on the right), promoted as being 'indispensable for winter and summer'.

looking like an ideal match for a chassis that had already become famous for the number of victories it had achieved. In the concours d'elegance held frequently in many areas of Italy at the time, the new Abarth won five first places and that recognition brought the little factory in Via Trecate further notoriety.

An extraordinary meeting of Abarth & C. board of directors was called for the beginning of April 1951, during which the substitution was discussed of Guido Scagliarini by his brother Carlo as procurator. Later, in November, Abarth himself took on that task and a few weeks later the top responsibilities of the company were redistributed, with Armando Scagliarini as president and Carlo Abarth vice-president and managing director; Carlo Scagliarini was also nominated a managing director. The new management order was intended to prepare for an extensive restructuring of the various company functions, so as to develop and manage the industrial production and sales organisation in both Italy and overseas in the best possible way. Sales success of the exhausts increased, particularly in Belgium and Germany, and at the end of 1951 exports accounted for 8% of turnover. The number of employees was also slightly up, from 41 to 44.

Carlo liked to delegate everyday activities and some functions to a small group of collaborators in which he had full confidence, and with whom he worked in direct contact. Among them was his efficient and ever-present secretary, Evangelina Isandoro, with whom he spoke only in German, creating embarrassment among the others who worked at his side; he never learned Italian well and, from the start, some suspected that the systematic use of his mother tongue by Abarth could have been interpreted as an expression of command, of power in front of the employees. He usually said, *"I don't need to speak Italian as if I were born here. What matters is to be able to make myself understood. Even more important is the fact that I love Italy. I would not be able to work and live anywhere else."*

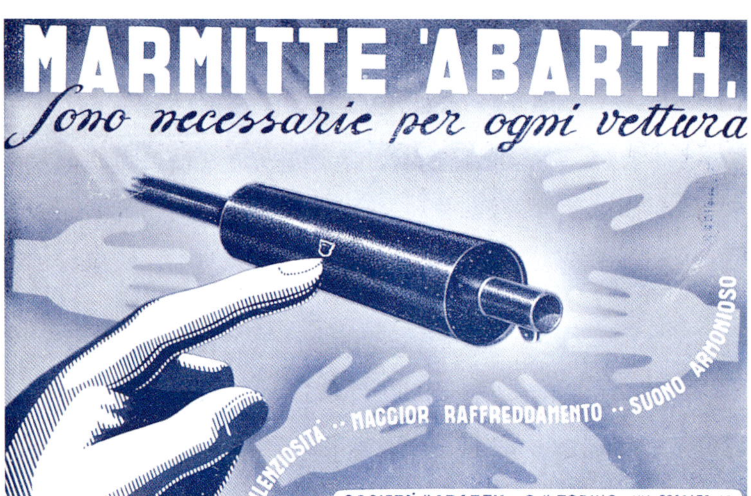

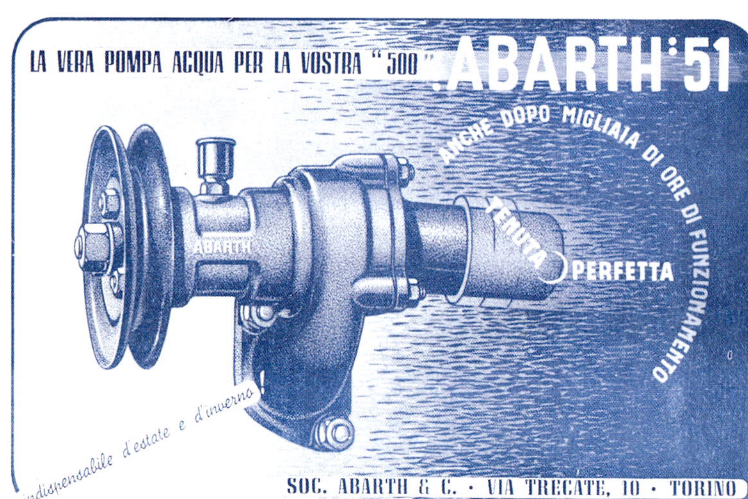

The contribution of the top drivers of the time to the advertising campaign that launched Abarth inlet manifolds had wide public appeal.

"Even the Vespa goes better with an Abarth exhaust", proclaimed the slogan aimed at the world of scooter riders.

Carlo Abarth immediately became well-known as severe and rigorous with everyone, a man who soon became irritated when his people made mistakes or, worse still, had not given the impression of being fully committed to the job he or she had been given. He was not terribly subtle in his relationship with his collaborators but he never held a grudge, on the contrary; after an initial outburst, he would speak to those with whom he had argued as if nothing had happened and wait – perhaps until Christmas – to send them a gift, sometimes expensive. He was not a diffident man, but he liked to evaluate with circumspection the people he met for the first time, and the same went for the circumstances that brought them into contact with him; his conspicuous sense of almost utilitarian pragmatism lead him to quickly conclude whether or not a new person could be of some interest to him.

True to these dictates of life, which had been drilled into him during the rigid family education he had received right from when he has a small child, Carlo made organisational changes in the company in the early years, convinced that good order and discipline were an indispensable foundation for running a very young business like his that was vulnerable to a thousand perils, and making it prosper. Past experience taught him that success could only be achieved with a solid understanding of one's ability and a profound sense of sacrifice, without being able to count in advance on propitious circumstances. Those considerations induced him to always require maximum effort from everyone, not least himself; for Abarth, the company came before anything else and justified any action, no matter how hard and unpleasant it may be.

In the summer of 1950, Carlo developed a conversion kit with double inlet manifolds and twin carburettors for the Fiat 1400, which delivered a 16% increase in the car's maximum power output. The new and long awaited Fiat

An inlet manifold and two Weber carburettors for the Fiat 1100s A, B, and E were sold to customers complete with controls, Abarth filters, tubing without carburettors, which had to be bought separately.

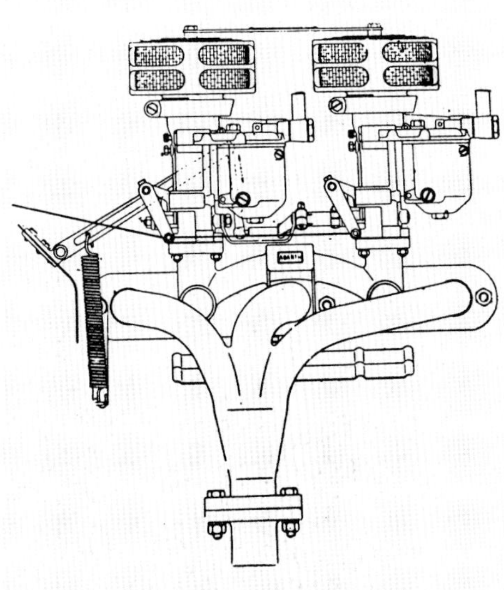

The joint name of Abarth-Weber 'the best carburettor on the best manifold' began to be advertised in 1953, following an agreement between the two factories to provide technical service to customers through their two networks.

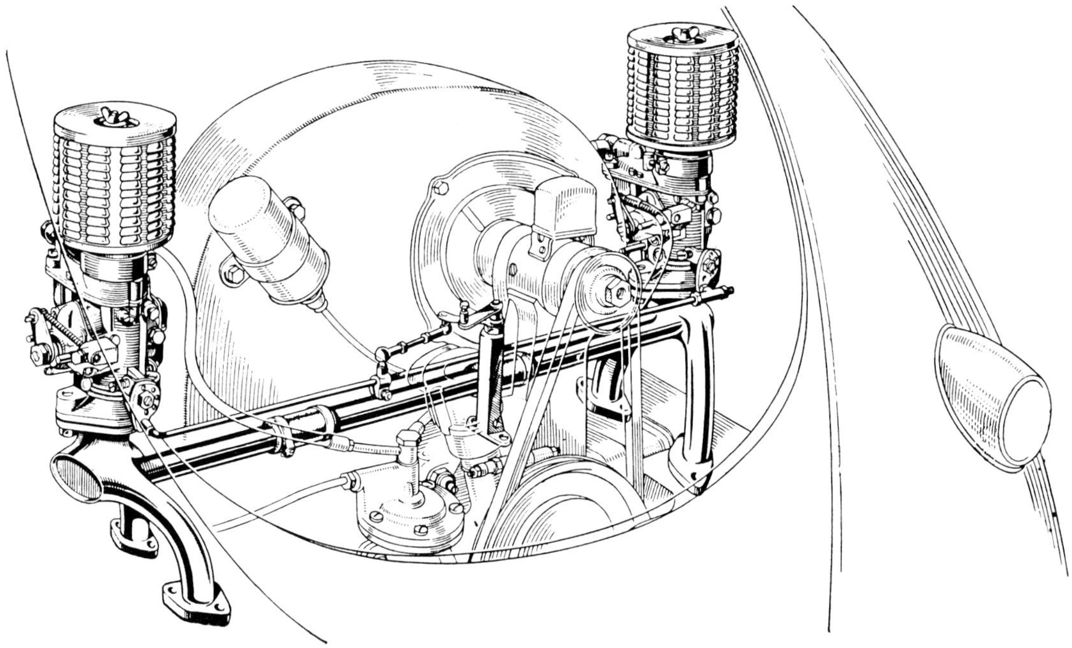

The inlet manifold and carburettor for each cylinder bank of the boxer engine of the VW 1131 cc.

Abarth's technical components stand at the 1952 Turin Motor Show was on the balcony of the Palazzo Esposizioni in the Parco del Valentino.

Carlo's wife Nadina (left) and his sister Anna pictured during a brief family holiday at Strobl, Austria.

saloon was having great success in the market due to its modern lines and innovative mechanical lay-out, which included a integral bodyshell, independent front suspension with deformable quadrilaterals and a steering column gear change; the four-cylinder 1395 cc engine produced 44 hp at 4400 rpm but, by replacing the Solex carburettor with the Abarth conversion kit of an inlet manifold and two Weber carburettors, the car's maximum power increased to over 51 hp, also improving engine torque and driving elasticity.

With the introduction of the integral Fiat bodyshell, the large family of body designers and builders in Turin found they had to confront a new problem, and that began a slow but inexorable weeding out process among those with a solid professional training and the others, who had worked for many years by trusting more to instinct and improvisation than to precise rules of mechanical construction. The abandonment of the separate chassis design formula by the major national factories caused problems for a number of body houses accustomed to building cars on the pre-existing structure of longitudinal members. They were now faced with the sorry sight of a sort of a metal cover to which the car's mechanical components were anchored, but did not present support areas on which to mount a special body in the traditional manner. So the load-bearing body imposed new technical circumstances and a company structure able to develop the engineering side of the body. In his biography, Nuccio Bertone recalled how at the initiative of its then engineering director Luigi Cajal de la Chenaye, Fiat had intervened on the 1400 production line to separate the load bearing bodyshell from the chassis so as to facilitate the work of the body designer and builder.

Untiring supporter of technical innovation that he was, Carlo Abarth was involved in research into new ideas to promote the originality of his proposals

The two-seater Abarth 1500 coupé created great interest at the 1952 Turin Motor Show. The car was based on the Fiat 1400 and its body, designed by Franco Scaglione, was by Bertone. Carlo's precise intuition when developing such a futuristic car and his well-honed sense of communication played a decisive role in creating a positive image of his young company in the minds of the public.

The Abarth 1500 had an aggressive front end and was a clear breakaway from styling convention, to which Italian and foreign designers adhered. The profile of the wheel housings and the roof of integrated glazing were elements of great interest.

Next to the production Fiat 1400 is the 1500 Abarth, which was so low and slender it seemed like a science fiction car. It was bought and shipped to Packard Motor Cars in Detroit, USA.

ABARTH "1500" (1952)

ENGINE	
	front, longitudinal, 4 cylinders in line
Bore and stroke	mm 84 x 66
Cubic capacity	cc 1480
Valve gear	overhead valves, side camshaft
Number of valves	2 per cylinder
Compression ratio	9: 1
Fuel feed	3 Weber 32DR6SP carburettors
Ignition	coil, distributor
Coolant	liquid
Lubrication	forced, oil filter on the main circuit
Maximum power	75 hp at 5700 rpm
TRANSMISSION	
	rear wheel drive
Clutch	single dry plate
Gearbox	4 speeds plus reverse, central lever
Axle ratio	4.4:1
BODY	
	2 seat coupé by Bertone
CHASSIS	
	boxed steel
Suspension	front independent, deformable quadrangles, helicoidal springs, hydraulic dampers; rear rigid axle, longitudinal half-springs, hydraulic dampers
Brakes	hydraulically operated drums front and rear
Steering	worm and roller
Fuel tank	capacity 50 litres
Tyres	5.90 x14
DIMENSIONS AND WEIGHT	
Wheelbase	mm 2650
Front track	mm 1320
Rear track	mm 1320
Length	mm 4520
Width	mm 1650
Height	mm 1380
Kerb weight	kg 870
PERFORMANCE	
Top speed	km/h 180

and did not let such an opportunity slip away. He planned and developed a chassis in boxed pressed steel for the Fiat 1400, increased the cubic capacity of the engine to 1500 cc and had Carrozzeria Bertone fit out his mechanics with a 'dress' that was equally audacious and futuristic, one that also responded to the knowledge available at the time in aerodynamic penetration. Bertone delegated the job to Franco Scaglione and the stylist came up with a masterpiece of creativity; the two-seater coupé was displayed at the 34th Turin Motor Show from 23 April–4 May 1952 and caused a sensation, with its low sleek lines, the revolutionary frontal features beginning with a large central headlight and a radiator grill split into two generous air intakes; the car seemed to have come from a science fiction book, without any probability of practical realisation. The sides also presented interestingly innovative elements, particularly in the areas

A group of manifolds for two carburettors marketed by Abarth for the Alfa Romeo 1900 Sprint, built between 1951 and 1953; the components were consigned complete with coaxial controls, Abarth filters, special cylinder head gasket, various tubes and air deflector, ready to be connected to two Weber 40DRSP or Solex 40 PAI carburettors. From 1954, Alfa themselves fitted the 1900 Sprint with two carburettors.

of the elongated wing – a feature that remained dear to Scaglione's heart – and in the roof, very composed and bright, in which an enormous panoramic windscreen with vertical pillars played a key aesthetic part; at the back end, the cockpit was closed off by a large wrap-around rear window and an small boot that flowed into the back wings by a curious and subtle fin-type profile. Stylistic intuition which, conveniently in line with mass production needs and public taste on the other side of the Atlantic Ocean, would later appear on some cars made in the USA from the 1954 model year.

It was, in fact, an American visitor to the show, the executive director of the Packard Motor Car Company of Detroit, who was enraptured as he looked at the provocative elegance of the Abarth 1500. He returned to the show on a number of days to admire the car and to try to extract in the most esoteric ideas from that extraordinary design exercise. In the end, he offered to buy that single example of the car from Abarth to freely inspire American stylists in designing Packards of the future. And Carlo, who knew how to root out a great opportunity, understood – but not without hesitation - that the splendid coupé bearing the black scorpion on the red and yellow background would transform itself into an exceptional instrument of publicity for his cars in a market that already represented Eldorado for many European car manufacturers.

Carlo's contacts among the big car manufacturers continued to increase in number. His dynamism and constant desire to become involved in new projects led him to maintain useful relations with a number of Fiat directors, in spite of the official factory attitude of not being too open with him. The abortive Cisitalia adventure in which Fiat participated – although not manifestly - was

At the XXXI Campionaria Fair of Milan from 12-29 April 1953, Abarth & C. displayed a selection of its by then well-known exhaust systems (right) and two-carburettor inlet manifolds, developed to increase the performance of production cars.

The Fiat 1100, one of the mainstays of the Italian automobile park since the late Thirties, was completely renewed and substituted in March 1953 by the 1100-103 (in profile design), which inherited from its predecessor the robust 1089 cc engine with a bore and stroke of 68x75, the same unit on which the Cisitalia and Abarth 204 A cars were based.

In just a few weeks, Carlo Abarth developed a quick version of the 1100-103: the modified 1089 cc engine produced 64 hp at 5500 rpm instead of the production car's 36 hp at 4400 rpm. The Abarth was bodied by Ghia, who designed an elegant two-seater coupé with a long bonnet and part-covered wheel housings, sober lines, headlights recessed into the radiator grill and lateral windows without frames.

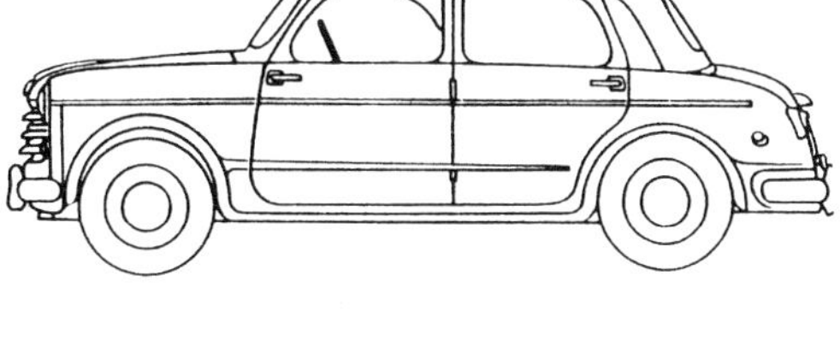

still fresh in people's minds. In those years, which were not easy for Italy's biggest manufacturer, it was thought that the Abarth sports activity with cars using basic Fiat mechanics would not constitute a disturbing factor, but could contribute to broadening the still limited horizons of the Turin factory.

During the early Fifties, Fiat ran an extensive model renewal programme and introduced in March 1953 at the Geneva Motor Show – considered the most important international event of its kind by Fiat management – the replacement for the legendary 1100, which had been the key player in the diffusion of the car in Italy since 1937. The 1100 - derived as it was from the 1932 508 Balilla – was produced in an interminable sequence of models. They included one called the 'L' that was used as a taxi in all the peninsula's cities until the Sixties, and the 'S', a 1947 aerodynamic coupé with a 51 hp engine that made its class-winning debut in the 1948 Mille Miglia and cast a shadow over Cisitalia – at least in some areas – and the fame achieved by the young upstart.

The 1953 car was the 1100-103, which retained little of the previous model except the bore and stroke of its 1089 cc engine. Almost everything else was different; the front suspension, the gear change, the simple lines of the load-bearing bodyshell, its four-door three-volume comfortable and airy body very much reflecting the styling trends of the moment. With the 1100-103, the

A few weeks after the 1953 Geneva Motor Show, where Fiat introduced the 1100-103 saloon, Abarth unveiled its one-off Ghia-bodied coupé on its stand at the 35th Turin Motor Show.
At show's end that car, too, was shipped to America.

Italians took an important step forward in their slow development of the popular car; almost 2.5 million of them were made up until 1969 in a myriad of models and derivations, including the 1200.

As far as the 1100-103 was concerned, Carlo Abarth had no intention of being second to anyone. Everybody in the business knew that all the Italian design houses would be offering special bodies on the newcomer's mechanics; anxiously awaited on the national market, the car was considered ideal, not only to initiate good number of Italian families into the world of the motoring, but also to satisfy the sporting and middle class clientele that wanted to get behind the wheel of a car that was different and a little snobbish, a special that was looked on with much favour in that far away spring of 1953. The perfect occasion on which to successfully launch the challenge was the 35th Turin Motor Show that opened on Sunday 22 April, seven weeks after the unveiling of the mass-produced Fiat. Carlo tuned the 1100 cc engine in line with techniques used with success for years and modified the other mechanical components, until his creation could accelerate to a maximum 170 km/h; he turned to Ghia, a glorious name among Turin's band of designers, for the body.

Founded in 1915 by Giacinto Ghia, the body stylist had for many years played an important role in his profession. He changed his company's corporate name twice in the following decade, first calling it Ghia & Gariglio and later Ghia & Actis. As with all specialists in the field his activities initially revolved around Fiat, for whom he stepped into the limelight by designing sleek-looking motor racing torpedoes, especially on the 501S chassis of 1924; the car won so many times the Turin car manufacturer decided to sell it directly through its sales network. In the Thirties, the craftsmanship of Giacinto

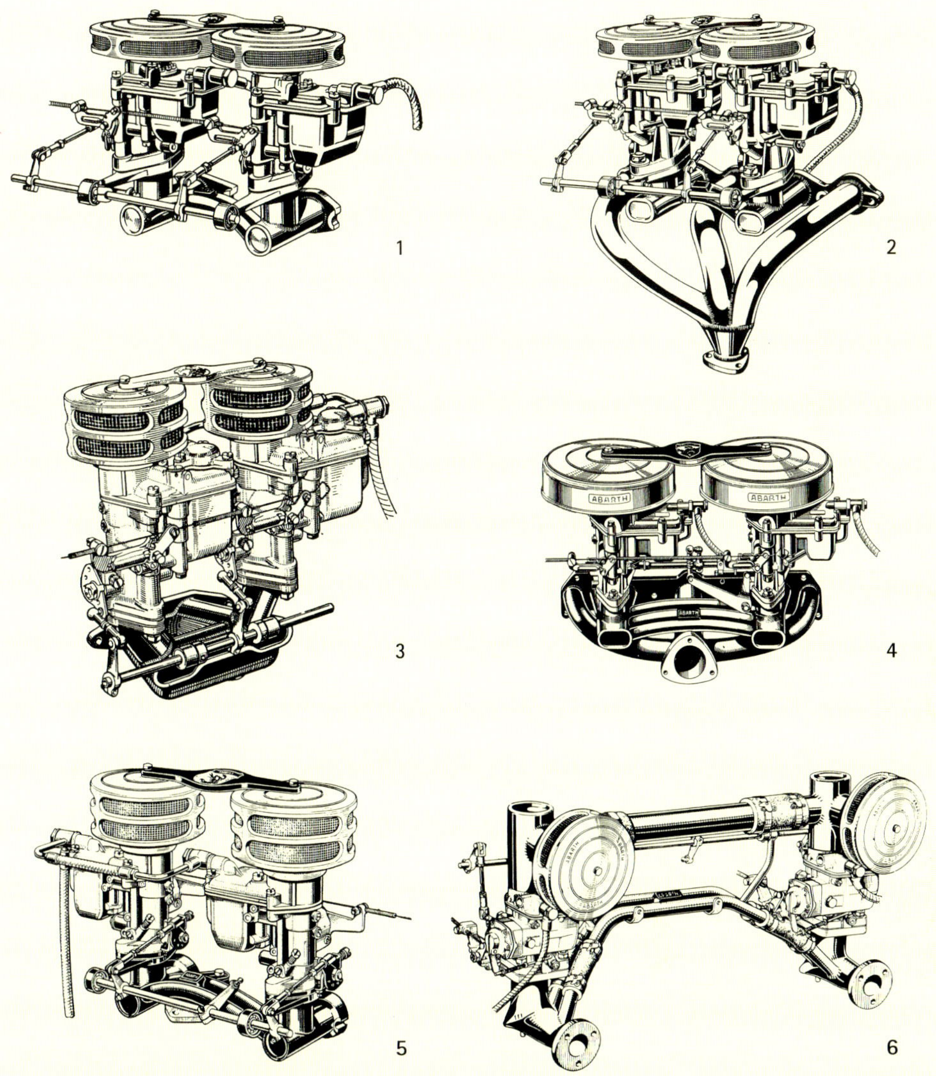

A number of inlet manifolds for mass produced cars offered in the Abarth & C. general catalogue of 1954.

1) The group for the touring version of the Fiat 1100-103, to be hooked up to two Weber 32DR10SP carburettors.
2) A group for the Fiat 1100-103 Sport, produced to link up with two Weber 32DRNP2 carburettors with internal pump.
3) The inlet manifold for the Lancia Appia, ready to be connected to two Weber 32DRNP2 carburettors.
4) A manifold for the Fiat 1900, complete with controls, ball bearings, Abarth filters, tubing, air deflector to operate with two Weber 36DR4SP carburettors.
5) The Ford Taunus manifold made to take two Weber 32DR10SP carburettors.
6) An inlet manifold for the 850 cc two-cylinder Panhard Dyna for connection to two Weber 32DR6SP or two Solex 32PBIC carburettors.

Ghia climbed every rung of the ladder leading to notoriety and entrusted a number of projects to the designer most in vogue at the time, Count Mario Revelli di Beaumont. The count became known for his innovative orientations and was the standard-bearer of the freelance stylist, who would make numerous converts after the Second World War. Another great Ghia success was the celebrated Balilla Coppa d'Oro, which Fiat launched with considerable hullabaloo in 1933, later acquiring the rights to produce it themselves; in the second half of the Thirties, Ghia's elegant creations based on the Fiat Ardita, the 1500 and the Lancia Aprilia, received numerous awards at motor shows and concours d'elegance. After the war Mario Felice Boano, who had recently become joint owner of the company, created the so-called 'flamboyant' line, a low, wide shape with closed wings and flowing sides; the Gioiello and Supergioiello (jewel and superjewel) series followed in 1949-50, built on Fiat

Another great and original idea took shape at Abarth in 1953. A car body with detachable panels of light aluminium for the Ferrari 166 MM, with which Scuderia Guastalla planned to compete in the Italian Championship. Stylistically highly attractive, the body was of a light structure on which were fixed individual panels of thin aluminium sheet, connected to each other by quick release screws; the system made the operation of replacing one or more panels after an accident both simple and quick. The large picture shows the chassis of the 166 MM, with all the body's detachable panels ready to be fitted.

and Alfa Romeo mechanics, as a result of which Ghia stepped onto the world stage to join the other masters of his craft in recognition of the innovative style and refined execution of its creations. That was the period in which Ghia's stimulating and fertile collaboration began with Chrysler and Volkswagen on a model that eventually became famous throughout the world – the Karmann-Ghia.

Carlo Abarth had a two-litre coupé in mind of really sporting appearance, but sober and elegant at the same time; a long bonnet, surmounted by a large tapered air intake to evoke great power and exciting performance. Ghia

produced one of its masterpieces and designed a car with smooth, harmonious flanks into which the wheels were set, hiding half of them from view; the front was made up of a large- mouthed air intake in which the two headlights were set with a central decorative element, a sort of metallic spinner, to give that splendid coupé an unusually aggressive personality; it was displayed on the Ghia stand accompanied by a particularly fancy plaque that described the one-off coupé as: "*The new Fiat 1100 Abarth Sport, Ghia edition*". It was given a unanimously positive reception, and after the show closed the coupé ended up in New York where it was an object of marvel and desire for many years.

The special Ferrari 166 MM with detachable panels, which weighed 54.4 kg, as it is being pushed out of the Abarth workshop by chief mechanic Garrone. In the background, Renzo Avidano (centre) and Carlo Abarth look on, pleased with their new creation.

Abarth & C. was not alien to the world of motor racing; in fact it closely followed the principal races and new technical trends. In the spring of 1953, the technicians of Via Trecate conceived a particularly light body made of removable panels for a two-litre Ferrari barchetta, with which Scuderia Guastalla intended to compete for the Italian Championship of the car's category. Not only were the car's lines extremely innovative, but the removability of various aluminium panels seemed a revolutionary concept; a light resistant structure was built on the chassis of the car from Maranello on which an aluminium bodyshell was placed, the two kept together by rapid grip screws that could be quickly removed and substituted. All of this weighed 54.4 kg, including seats and the intermediate support chassis; this solution was dictated by the need to quickly substitute the parts of the body that would be damaged in collisions or off-road excursions during a race. The extraordinary composite unit Ferrari 166 made its debut at the fifth Toscana Cup race on 31 May, driv-

In 1954 Abarth was producing a vast range of components with which to improve the performance of mass produced cars, backed by effective promotion. The picture on the left shows the company's components stand at the 54th Geneva Motor Show and on the right the Turin Show booth.

The real news at the 1954 Turin Motor Show was Abarth's high performance two-litre coupé, built on the mechanics of the Alfa Romeo 1900.

The car had a boxed pressed steel floor pan and its capacity was increased to 1975 cc (bore and stroke 84.5x88 mm) and, as a result of the adoption of a special inlet manifold and two twin-choke carburettors, its power was increased to 135 hp at 6000 rpm; the gearbox was given a centrally mounted lever. On the 2.48 metre wheelbase chassis, Carrozzeria Ghia built a coupé (left) of notable visual impact. Right: the wooden form on which the bodies' aluminium panels were beaten.

en by Giulio Musitelli. The exceptional body was designed at Abarth & C. and Carlo entrusted its construction to Giuseppe Manera, a panel beater of great experience and surprising ability; Manera used large sheets of aluminium to forge single panels with his hammer and Abarth, sceptical that the man could successfully work with such large sheets, surprisingly said publicly that in Germany nobody would have dared to beat metal panels of those dimensions.

The year 1953 was one of happy confirmation of the validity of the new direction taken by Abarth & C. The number of employees had increased from 44 to 71 in that one year and the proportion of export products on turnover had risen from 8% to 10%; 45,034 exhausts were produced, strengthening the value of Carlo's confidence as he had seen in them a valid instrument with which to help balance the company's accounts, so that he could invest in sports activities as well as special research and test projects. The following year was met with the same vehemence and spirit of initiative. The notoriety

Of modern lines and with the front end unequivocally Alfa, the Abarth 2000 received high praise from the critics and public on the Ghia stand at the 1954 Turin Motor Show. With a dry weight of 890 kg, the car had a top speed of 200 km/h. Behind the dual-tone coupé is a Chrysler Special SS, which Ghia bodied at the time for a small production run by the American manufacturer.

of the small Turin factory in Via Trecate was amply consolidated and pointed to greater success in the years to come.

At the beginning of 1954, Carlo took on a project that was even more ambitious than those of the past. He switched his attention to the mechanics of the two-litre Alfa Romeo 1900 Super saloon and created a derivation of extremely high performance, starting from a chassis designed in boxed pressed steel in which he had the four-cylinder Alfa engine installed after many of its components had been extensively modified. Two twin-choke horizontal carburettors fuelled the car and its compression ratio was taken to 8.5:1 so that the engine developed a maximum 135 hp at 6000 rpm with a specific power – exceptional for the time – of 68.5 hp/litre. The engine was hooked up to a high speed gearbox, the lever of which was shortened turret-style and moved well back compared to the original position, so that it was easier to manoeuvre. That body was also the work of Ghia, who designed a very low (1180 mm) sports tourer with a fastback tail, painted in two-tone in line with a fast-spreading fashion. Of equilibrated shape, the Abarth-Alfa 2000 had a shorter

A three quarter rear view of the Abarth Alfa 2000, photographed in Parco del Valentino.
A number of styling and constructional elements are of considerable interest, among them the long blade-like fins at the top of the massive rear wings, and the hatchback incorporating the rear window;
the roof, with its dual-tone colour scheme, was also original.

wheelbase of 2480 mm compared to the original car's 2630 mm, while the track was more or less the same. Made entirely in aluminium sheet, the car's shape and the excellent execution of the body caused quite a stir. With a weight of only 890 kg, it could easily exceed 200 km/h, in part also because of its positive power to weight ratio of 5.5 kg/hp and the state-of-the-art aerodynamic body, as far as knowledge and the type of experiments carried out at the time were able to contribute.

In 1954 Carlo Abarth also worked on the V6 2.5 litre Lancia Aurelia B20 engine, one of the most coveted cars of the period and the one that ushered in the grand touring fashion. In the European motor industry, the Aurelia was considered a car of extreme elegance and refined mechanics that was easily able to win road races, even against much more racing oriented competition. Abarth realised that work on the B20's fuel system could produce a significant increase in maximum power; he increased the compression ratio of the V6 Lancia engine from 8:1 to 8.4:1 and changed the inlet manifold and twin-choke Weber carburettor to a new configuration of four inverted Webers, as a result of which maximum power went from 112 to 138 hp at the original 5300 rpm. With other minor modifications to the calibration of the suspension, the Aurelia GT was able to exceed a 200 km/h top speed – an increase of over 30 km/h – and became difficult to beat on the road and in rallies.

In 1954, Abarth & C. had to increase its plant space to meet the needs of exhaust production and the modification of special cars.

The exhaust operation, which produced almost 58,000 units in 1954, was transferred to premises in Via Pacchiotti (picture).

The many faceted activities of Abarth & C. and the growing success of its products on the international market demanded an expansion of production and storage facilities in 1954; to meet those needs and accommodate a workforce that had risen to 90 people, the company opened new premises at 18 Via Pacchiotti, not far from its headquarters at Via Trecate, bringing its covered area to 3885 square metres. The car, engine, exhaust manifold building and other departments remained at the old premises, while exhaust production equipment was transferred to the new – a manufacturing unit that produced 57,614 systems in 1954.

The Fiat 1100-103 continued to be the car that grabbed the attention of the Italian market and attracted the efforts of a multitude of tuners, old and new. The 1100 cc capacity had consolidated itself over the years in the taste and habits of the average motorist and for the specialists of the sector like tuners and body builders, with whom it was a point of honour to be present with one or more creations to satisfy such an important category of customers. Always researching new ideas in the field of two seater sports cars, at the end of 1954 Carlo Abarth began working with the mechanics of the Fiat saloon again and designed a sensationally shaped racing roadster called the 207 A, the creation of the body shell of which was given to Carrozzeria Boano, a recently established Turin atelier.

Born in 1903, Mario Felice Boano began his career at Stabilimenti Farina in Turin and in 1930 followed one of the Farina brothers, Battista, who had gone into business on his own by founding Pinin Farina. After the Second World War Boano joined Ghia, where he designed a number of high quality bodies and actively contributed to the styling of the first series of Chrysler Corporation prototypes and the work of its leading designer, Virgil Exner. In 1953, Boano left Ghia to set up on his own and established his company in Via Collegno, Turin.

At the end of 1954, Carlo Abarth conceived and built another of his cars on the reliable mechanics of the Fiat 1100-103, in which the precision of the development and tuning of the engine harmonised with audacious formal solutions.
The 207 A roadster was born of innovation, with its boxed pressed steel chassis and modified Fiat production components.
The historic 1089 cc engine was further modified to produce a maximum 66 hp at 6000 rpm, using spares made a year before for the Abarth Ghia coupé.
The double exhaust pipe along the right side of the car looks original even today.

ABARTH "207 A" (1955)

ENGINE	
	front, longitudinal, 4 cylinders in line
Bore and stroke	mm 68x75
Cubic capacity	cc 1089
Valve gear	overhead valves, side camshaft
Number of valves	2 per cylinder
Compression ratio	9: 1
Fuel feed	2 twin-choke Weber 36DC04 carburettors
Ignition	coil, distributor
Coolant	liquid
Lubrication	dry sump
Maximum power	66 hp at 6000 rpm
TRANSMISSION	
	rear wheel drive
Clutch	single dry plate
Gearbox	4 speeds plus reverse, central lever
Axle ratio	4.3:1
BODY	
	racing roadster by Boano
CHASSIS	
	boxed steel
Suspension	front independent, trailing arms, helicoidal springs, hydraulic dampers; rear rigid axle, longitudinal struts, helicoidal springs, hydraulic dampers
Brakes	hydraulically operated drums front and rear
Steering	worm and roller
Fuel tank	capacity 32 litres
Tyres	5.20 x14
DIMENSIONS AND WEIGHT	
Wheelbase	mm 2050
Front track	mm 1229
Rear track	mm 1212
Length	mm 3620
Width	mm 1550
Height	mm 940
Kerb weight	kg 522
PERFORMANCE	
Top speed	km/h 165

The search for the sensational is well-reflected in the body of the 207 A, masterfully designed and built by Mario Felice Boano, who interpreted the racing roadster formula with a number of notable and original aesthetic features.
In the photograph above, the knowledgeable shaping of the four wings stands out as they assume a conspicuous volumetric relevance to the overall shape of the car.

Once more Abarth took the 1089 cc four-cylinder Fiat, increased its compression ratio to 9.4:1 and fitted his own special kit of an inlet manifold with two Weber 36 DO4 carburettors, as a result of which the original power output of 36 hp went up to 66 hp at 6000 rpm. Mounted on a chassis in boxed pressed steel with a wheelbase shortened to 2050 mm, the body was essential in its disposition of mass and of an extreme formal elegance; to emphasise its racing vocation, it had an slender wraparound windscreen in Plexiglas that 'surrounded' the driving position and a large rear fin that flowed into the seat's headrest, equipped for safety reasons with a tubular protection structure. The appearance of all those features was invigorated by two long stainless steel

The side and three-quarter front views of the Abarth 207 A give greater emphasis to the reduced dimensions of the car, the wheelbase of which was shortened by 290 mm compared to those of the Fiat 1100-103 saloon.
The 207 A was 3600 mm long, 1550 mm wide and 940 mm high at the upper edge of the narrow Plexiglas windscreen.
With a weight of 522 kg, according to Abarth the car could reach a maximum speed of 186 km/h.
The coloured edge along the side, with the exhaust recessed into the lower area, contributed to optically lowering the car.

exhaust pipes on the right side of the car, which sent out an eloquent message on the high level of specialisation Abarth had reached in constructing exhaust systems. The dry weight was extraordinary: 522 kg. With that mass and reduced dimensions of 3620 mm long, 1550 mm wide and 940 mm high above the edge of the windscreen plus an axle ratio of 10/43, the 207 A could accelerate to a maximum speed of 186 km/h.

The car was introduced during a brief ceremony at the Turin Exhibition in early February 1955, when Abarth and Boano announced that the new roadster was, in reality, a prototype of a small production run of cars being built at that time for the P.V. Corporation in New York, a company recently

The initial sketche of a sports saloon derived from the Abarth 207 A, retaining a number of styling elements of great visual impact. The tail is highly original and stands out for its much-tapered line, which concluded in a number plate recess, and the dual-tone livery the edge of which acted as a division between the front area containing the mechanics of the car and the purely habitable area.

The clearly racing cockpit of the Abarth 207 A was bordered on the right side by a kind of metal trapdoor that acted as a tonneau cover. At the rear there was a headrest with aerodynamic faring that diminished to a point at the end of the car.

Carrozzeria Boano created the Abarth 209 A sports saloon.
The greenhouse of the new car was extremely bright and elegant, with thin frames and large glass areas.
The 209 A also had chrome exhaust pipes under the right door sill, giving the car an even racier look, underlining its racing vocation.

established by the Italian-American Tony Pompeo and Luigi Valenzano, who was one of Abarth's early works drivers. At the same time, it was also confirmed that the car would make its debut at the New York Motor Show, scheduled for 20-27 February, and would compete in the 12 Hours of Sebring, second round in the International Sport Car Championship, in March. Elegant and incredibly low, the 207 A had an engine derived from the Fiat 1100-103, and was used as a valid sounding board for two other models that Carlo had in mind to modify within a short space of time. That is how visitors to the Turin Motor Show, which opened on 20 April, were able to witness the official intro-

On its stand at the 1955 Turin Motor Show, Abarth introduced three cars built of the one ambitious design, with mechanics derived from the Fiat 1100-103 and three splendid bodies by Boano constructed from designs by Michelotti: on the left is the 207 A competition roadster, in the centre the 208 A two-seater roadster strictly inspired by the previous car and on the right the 209 A sports tourer.

On 19 April 1955, the eve of the Turin Motor Show, Carlo Abarth had five cars lined up in Parco del Valentino, all bodied by Boano, which were to fascinate the thousands who visited the Abarth stand.

From the left: the 210 A built in a few weeks on the mechanics of the new Fiat 600, the 208 A two-seater roadster, the 207 A competition roadster, the 209 A sports tourer and the 211 A special coupé, built on the mechanics of the two-litre French Renault Fregate saloon.

duction of a triad of cars on the Abarth stand that were the centre of interest: the 207 A racing roadster, the 208 A low-slung roadster derived directly from the racer as far as its body was concerned but with the same mechanics, and the 209 A sports tourer that was stylistically based on the other two but with an enormous windscreen with vertical struts, an essential roof for interior brightness and a large wraparound rear window. Among other things, one of the first flip-up headlight systems appeared on the two road cars, enabling the stylist to design an impeccably low, sloping bonnet that was unhindered by the presence of an optical group.

The 207 A was priced at Lit 2.900.000 but the racing version was Lit 4.600.000, which corresponded to around $7,500 at a time when the '55 Ford Thunderbird, Chevrolet Corvette and the Buick Century ran out at around $ 3000 each, depending on which version, while the Cadillac Eldorado cost $

The sports successes achieved by Alfa Romeo in the early Fifties with the 1900 and 1900 T.I. saloons, the 1900 Sprint, Giulietta Sprint and Giulietta Sprint Veloce sports cars had pushed the technicians of the Milan factory into projecting a car specifically to compete in 1500 cc Sport category racing.
Construction of the prototype was assigned to Abarth & C. and in 1955 the company built a chassis of boxed pressed steel (above) in which would be installed a four-cylinder twin-cam 1488 cc engine of the Alfa Romeo school, with a bore and stroke of 76x82 mm; with a compression ratio of 10:1 and two twin-choke horizontal carburettors, the power unit generated a maximum 145 hp at 8000 rpm (specific power 97.4 hp/litre).
The car was bodied by Boano (right), who once again used the technique of housing the exhaust pipes laterally, placed on the left longitudinal member.
Called the 750 Competizione, two examples of the car were built and then abandoned because the Alfa Romeo technicians found it impossible to continue with the project.
The picture below shows the prototype at the entrance to the Monza Autodrome on its way to one of its first test drives.

6286. The car was introduced to the American market with a fitting publicity statement by Luigi Valenzano, in which he said the 207 A was faster than the Mercedes-Benz 190 SL and other road cars in vogue at the time, but his declaration was hard to confirm in the case of cars designed and built for racing. The engine of the little Abarth generated 66 hp, the factory said, but at 6000 rpm it was at its limit and even with different compression and axle ratios could hardly make the maximum declared speed of 186 km/h. In one of his numerous photograph albums, Carlo noted: "*... destined for export, the 207 A is an example of how much can be done with a four-cylinder 1100 cc front engine*".

The American Don Rosendale, who raced Abarths for several years, maintained in an article of 1962 that one of the problems with the 207 A was the excessive weight of its chassis: "*Abarth had, in fact, abandoned the light construction in steel tubes of the Cisitalia-type chassis and adopted boxed and soldered steel sheet, which was probably heavier than the Fiat 600 floor pan on which Abarth later built his noted rear engined cars; even if, in line with the usual Italian*

In the summer of 1955, a few months after the launch of the Fiat 600, Carlo Abarth was already able to offer the market three different exhaust systems to take the place the mass produced cars' units.
The slogan 'World Champion Exhaust' and the communication via the press of a price list of its products were the subject of much public discussion.

tradition, the declared weight of the 207 A of 522 kg was not optimistic, it should not be forgotten that, at the time, the Lotus 9 and 11 weighed a maximum of 450 kg and they were more powerful, had better brakes and better roadholding".

Specialist journalist John Bentley entered chassis number 001 207 A for the 12 Hours of Sebring on 13 March 1955; in testing, the car gave rear axle problems and tuner Jim McGee substituted its and its large springs made by Abarth for a normal production Fiat 1100 unit with semi-elliptics. The small Italian car had to compete against two Lotus 9s, both of which retired; the 207 A almost ran out of fuel when it was leading its class and was forced to make an extra pit stop on the 138th lap, for which Bentley and the car were disqualified; that gave victory to Paul O'Shea and Fritz Koster in a 1100 cc Porsche 356, entered by importer Hoffman. A few years later, the driver-journalist went through the episode again and attributed his retirement to a broken piston, almost certainly caused by over-revving. Other Abarths in that limited series raced in the USA – at Watkins Glen in New York State and at Torrey Pines in California - but with little success, and the model was quickly outclassed by the Lotus. Decorated with a large Scorpion on the fins of its rear wings, the 207A went down in history as one of the most elegant racing roadsters to come out of Italy in the post-war years.

1955

1958

THE HUNT FOR SUCCESS

For Carlo Abarth 1955 was a year that strongly influenced the development of his business and the consolidation of his good fortune. On 12 March Fiat introduced its 600 model at the Geneva Motor Show, a car that became not only one of the Turin company's greatest successes, but also a key contributor to the long process of popular motorisation in Italy which, in the years to come, accelerated even faster. Conceived to replace the 500 C as a leading economy car, the new utility was a four-seater, powered by a rear engine and of an entirely new concept, from its independent suspension on all four wheels to its integral bodyshell that was compact, light yet extremely roomy. The 600 represented a profound change in Fiat's car construction rules as it was designed on the basis of a rigorous 'cahier de charges', having to transport four people comfortably, achieve a top speed of 85 km/h, use less than six litres of petrol per 100 kilometres and weigh around 450 kg.

The engineer Dante Giacosa, who created the car, made this comment on the 600's engine in his fascinating memoirs published in 1979 "... *it is of exemplary simplicity and simplicity is synonymous with reliability. Any other design would have cost Fiat and its customers more. A million cars were built, but with the passing of the years the power of the engine was increased while maintaining its external dimensions; its life was further prolonged by the 850, then the A 112 and the 127...*"

In fact the new Fiat utility returned a top speed of 95 km/h, covered 100 kilometres with 5.7 litres and weighed 585 kg. With a list price of Lit 590.000, it immediately became the most coveted of objects for millions of Italians, who made the 600 an unprecedented success. With its arrival in the marketplace, the attitudes and aspirations of many citizens were profoundly transformed much as, in many ways, the entire life of the country had altered. The order of priority in the satisfaction of need in many families had changed; people's attitudes towards the hours they worked had altered and their use of free time was different - even the profile of the Italian landscape was being transformed by the first long-distance motorways as they began to snake their way across the countryside. The 600 was certainly the artifice of many aspects of life today; Fiat made almost 2.7 million of them in 15 years and in its first year the factory was able to produce more than 1000 units a day.

The Fiat 600 was a milestone in both the history of Fiat and the motorisation of the Italian population. Born of a highly advanced project and mass produced from 1955, the car set many Italian families on the road to mobility. Its small, generous 633 cc engine that developed 21.5 hp constituted the basis for a long and successful series of Abarth derivations over a 15-year period.

However, the new model did not much tempt the body designers to produce their special versions of the car. They understood the circumstances in which the notable added value represented by a special body would not easily find customers, especially considering the socio-economic profile of the average 600 buyer. But the car greatly stimulated Carlo Abarth, who developed a highly successful series of modifications for the 600's engine, which became extraordinary both for its success in motor racing and for its wide distribution on the international market. It all started with the introduction of a number of modifications to the four-cylinder mass-produced engine for an extremely trim two-seater roadster of great visual impact: the Abarth 210 A, built entirely in aluminium by Boano on a boxed pressed steel chassis that to Abarth was a much liked techno-constructional development. Introduced in that same spring of 1955, the 210 A became the first official Abarth to be developed using the mechanics of the new Fiat 600; by replacing the standard production 22 mm carburettor with a Weber 32 DRN P2, using new inlet manifolds and a special exhaust system, the car's maximum power was increased from 21.5 hp to 32 hp. Built on a 1950 mm wheelbase, the roadster only weighted 410 kg dry. Boano wanted to outdo himself in the design of the car, which was 3670 mm long, 1370 mm wide with a ground to windscreen height of 934 mm; to the front was a large central headlamp (for Carlo a happy memory of the 1952 and 1953 prototypes) the presence of which was emphasised by an attractive duo-tone paint scheme that made the little car seem even sleeker. Few examples of the Abarth 204 A were produced but they sold well, especially in export markets.

A man who by now had a deep understanding of the habits of Italian motorists, Abarth immediately understood that the mechanics of the 600, sim-

Using the floor pan of the Fiat 600, Abarth modified a number of the car's mechanical components, increasing its engine capacity to 747 cc and its power output by 20 hp. That was the first in a long line of Fiat 750 cc, 850 cc and 1000 cc cars that achieved exceptional commercial and sports success in Italy and abroad.
Magisterially bodied by Boano as a two-seater roadster, the first 750 cc Abarth derivation bowed in under the model designation of 210 A at the 1955 Turin Motor Show just a few weeks after the Fiat 600 was launched.

Compared to the prototype, in its definitive form the Abarth 210 A by Boano had a sharper front end and two normal headlights, in line with the Italian highway code.
The windscreen had also assumed a shape and dimensions compatible with homologation regulations.
The large grill on the side acted as an air intake for the engine compartment.
The lengthened and harmonious line of the car found an element of strong discord, however, in the elaborate segmented hubcaps.

ABARTH "210 A" (1955)

ENGINE	
	front, longitudinal, 4 cylinders in line
Bore and stroke	mm 60x56
Cubic capacity	cc 633
Valve gear	overhead valves, side camshaft
Number of valves	2 per cylinder
Compression ratio	9,5: 1
Fuel feed	1 Weber 32DRNP2 carburettor
Ignition	coil, distributor
Coolant	liquid
Lubrication	forced, oil filter on main circuit
Maximum power	32 hp at 5500 rpm
TRANSMISSION	
	rear wheel drive
Clutch	single dry plate
Gearbox	4 speed plus reverse, central lever
Axle ratio	4.3:1
BODY	
	2 seat roadster by Boano
CHASSIS	
	boxed steel
Suspension	front independent, upper trailing arms, lower transverse spring, hydraulic dampers; rear independent, trailing arms, helicoidal springs, hydraulic dampers
Brakes	hydraulically operated drums front and rear
Steering	worm and sector
Fuel tank	capacity 30 litres
Tyres	5.20 x12
DIMENSIONS AND WEIGHT	
Wheelbase	mm 1950
Front track	mm 1144
Rear track	mm 1154
Length	mm 3670
Width	mm 1370
Height	mm 934
Kerb weight	kg 410
PERFORMANCE	
Top speed	km/h 142

ply designed and produced as it was but a car of great flexibility, could be the object of modifications and improvements to bring its performance in line with the temperament of numerous lesser racing drivers in Italy. So he set about modifying the engine after carefully analysing its various components and their potential, the breaking strain of their materials and the dimensions of the individual parts; he became convinced that there was potential for the modification of the bore and stroke of the original Fiat power unit, taking them respectively from 60 mm to 61 mm and from 56 mm to 64 mm; but it would be necessary to change the drive-shaft, camshaft, pistons, valves and relative springs plus the sump. Carlo put these modifications into production in his own workshop and came up with an engine with a cubic capacity of 747 cc instead of the standard 633 cc. With the subsequent substitution of the inlet manifold and the original equipment carburettor with a Weber 32 IMPE as well as the adoption of an exhaust system specially designed for the Fiat utility, the new 750 engine produced 41.5 hp of maximum power at 5500 rpm and a maximum torque of 5.5 kg/m at 4000 rpm, while the original Fiat engine generated 21.5 hp at 4600 rpm and 4 kg/m of torque at 2800 rpm; increasing the compression ratio from 7.8:1 to 9:1 and further work on the moving parts contributed to giving the new engine an extraordinary ability to exploit its revolutions and made the Fiat 750 Abarth one of the most responsive and enjoyable cars of the Fifties.

Even if modified in a number of ways, the car still retained the Fiat body unchanged: in fact, lovers of 'occult speed' liked to strip the car body of all the Abarth scorpion logos to camouflage the 750 when among the legion of 600s on the roads, then accelerate away from the traffic lights and lose the rest of them in the rear view mirror, thanks to that generous Abarth engine. The

FIAT ABARTH "750" (1955)

ENGINE	
	rear, longitudinal, 4 cylinders in line
Bore and stroke	mm 61x64
Cubic capacity	cc 747
Valve gear	overhead valves, side camshaft
Number of valves	2 per cylinder
Compression ratio	9: 1
Fuel feed	1 Weber 32IMPE carburettor
Ignition	coil, distributor
Coolant	liquid
Lubrication	forced, oil filter on the main circuit
Maximum power	41.5 hp at 5500 rpm
TRANSMISSION	
	rear wheel drive
Clutch	single dry plate
Gearbox	4 speed plus reverse, central lever
Axle ratio	4,9:1
BODY	
	4 seat saloon, 2-door
CHASSIS	
	integral body
Suspension	front independent, upper trailing arms, transverse lower spring, hydraulic dampers; rear independent, trailing arms, helicoidal springs, hydraulic dampers
Brakes	hydraulically operated drums front and rear
Steering	rack and pinion
Fuel tank	capacity 27 litres
Tyres	5.20 x12
DIMENSIONS AND WEIGHT	
Wheelbase	mm 2000
Front track	mm 1144
Rear track	mm 1154
Length	mm 3215
Width	mm 1380
Height	mm 1405
Kerb weight	kg 585
PERFORMANCE	
Top speed	km/h 130

maximum speed of small four-seater jumped from the mass-produced car's 95 km/h to the 750's 130 km/h; the kit was given the name 'Abarth Derivations', which was often in full view in chrome letters on the 750's engine cover, almost as if a little nobility had touched the Mirafiori utility. Inside the car, the only elements that distinguished an Abarth from the original small Fiat was the presence on the left of the small standard dashboard of a Jaeger rev counter calibrated up to 6800 rpm, and the Abarth logo with the inevitable scorpion on the horn button.

The stir caused among young people and motor sport enthusiasts by the Abarth-modified 600 in its first months on the road immediately generated tremendous enthusiasm in a public that was being exposed for the first time to a modified mass-produced car. In a short space of time, the fame of the little Abarth 750 swept the country and did not simply remain among the most knowledgeable students of motoring but spread to professionals and elegant ladies, who saw in it not only a car of exciting road qualities but also an emergent status symbol among the rich and snobbish bourgeoisie of the moment.

After a long period of techno-scientific evaluation and exhaustive test driving of the new engine components, the advertising campaign to spread the word about the "Derivazione Abarth 750" or "Abarth 750 Derivation" of the production Fiat 600 began in the autumn of 1957. With an extensive series of well-calibrated modifications, the components substituted were the 600's drive shaft, pistons, camshaft, valves and valve springs as well as the inlet manifold, the whole exhaust system and the sump. The new 747 cc engine that once had a 633 cc capacity now boasted an additional 20 hp, jumping from a maximum 21.5 hp to 41.5 hp at 5500 rpm, while engine torque went from 4 kg/m at 2800 rpm to 5.5 kg/m at 4000 rpm. Unlike the production car, the little Fiat Abarth 750 saloon was extremely quick and had a top speed of 130 km/h.

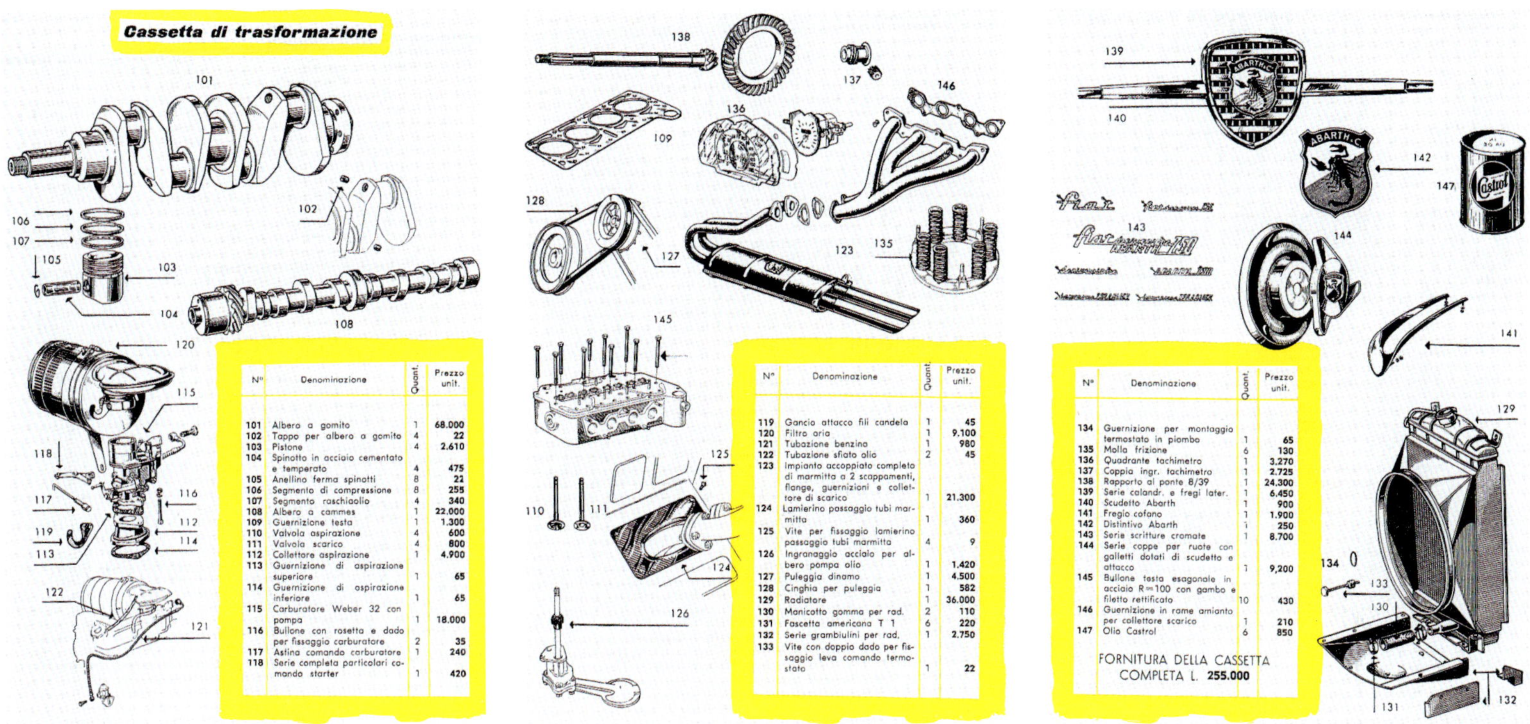

Cassetta di trasformazione

N°	Denominazione	Quant.	Prezzo unit.
101	Albero a gomito	1	68.000
102	Tappo per albero a gomito	4	22
103	Pistone	4	2.610
104	Spinotto in acciaio cementato e temperato	4	475
105	Anellino ferma spinotti	8	22
106	Segmento di compressione	8	255
107	Segmento raschiaolio	4	340
108	Albero a cammes	1	22.000
109	Guernizione testa	1	1.300
110	Valvola aspirazione	4	600
111	Valvola scarico	4	800
112	Collettore aspirazione	1	4.900
113	Guernizione di aspirazione superiore	1	65
114	Guernizione di aspirazione inferiore	1	65
115	Carburatore Weber 32 con pompa	1	18.000
116	Bullone con rosetta e dado per fissaggio carburatore	2	35
117	Astina comando carburatore	1	240
118	Serie completa particolari comando starter	1	420
119	Gancio attacco fili candela	1	45
120	Filtro aria	1	9.100
121	Tubazione benzina	1	980
122	Tubazione sfiato olio	2	45
123	Impianto accoppiato completo di marmitta a 2 scappamenti, flange, guernizioni e collettore di scarico	1	21.300
124	Lamierino passaggio tubi marmitta	1	360
125	Vite per fissaggio lamierino passaggio tubi marmitta	4	9
126	Ingranaggio acciaio per albero pompa olio	1	1.420
127	Puleggia dinamo	1	4.500
128	Cinghia per puleggia	1	582
129	Radiatore	1	36.000
130	Manicotto gomma per rad.	2	110
131	Fascetta americana T 1	6	220
132	Serie grambiulini per rad.	1	2.750
133	Vite con doppio dado per fissaggio leva comando termostato	1	22
134	Guernizione per montaggio termostato in piombo	1	65
135	Molla frizione	6	130
136	Quadrante tachimetro	1	3.270
137	Coppia ingr. tachimetro	1	2.725
138	Rapporto al ponte 8/39	1	24.300
139	Serie calandr. e fregi later.	1	6.450
140	Scudetto Abarth	1	900
141	Fregio cofano	1	1.900
142	Distintivo Abarth	1	250
143	Serie scritture cromate	1	8.700
144	Serie coppe per ruote con galletti dotati di scudetto e attacco	1	9,200
145	Bullone testa esagonale in acciaio R=100 con gambo e filetto rettificato	10	430
146	Guernizione in rame amianto per collettore scarico	1	210
147	Olio Castrol	6	850

FORNITURA DELLA CASSETTA COMPLETA L. **255.000**

The promotion piece for the 750 Abarth conversion kit for the Fiat 600 engine listed all the components that would substitute the originals, with relative retail prices.
Beside the drive shaft and the camshafts are the pistons, valves, carburettor, exhaust system, radiator, clutch spring and the rest.
The complete kit, which also included chrome identification badges for application to the body, was offered at the considerable price of Lit 255.000 but was able to ensure 600 owners enjoyed the experience of driving a car with almost sports performance.

The excitement generated by the Fiat Abarth 750 also infected foreigners and that plebiscite of comments opened the door for Carlo Abarth to a long period of activity that was distinguished by commercial success and racing victories. So for the company in Via Trecate, Turin, 1955 closed with a gratifying operating balance; exhaust production was clearly doing well and reached 92,119 units that year – 21% sent for export; and in the previous 12 months, the company's number of employees had increased from 90 to 117.

At the beginning of 1956, the Belgian racing driver and journalist Paul Frère published brief driving impressions of the Fiat Abarth 750 in Motor-Italia, which still seem eloquent today, 45 years later: " ... *Properly regulated acceleration tests were carried out with standing starts up to 80 km/h and 100 km/h. Those speeds were reached respectively in average times of 11.8 and 20.0 seconds with only the driver aboard. These results have been made possible by the extreme rapidity of manoeuvre of the central gear lever and by the excellent choice of gear ratios, which enable the car to reach real speeds of nearly 70 km/h in second and over 105 km/h in third. The brakes are those of the standard Fiat 600 without modification and on normal roads they seem sufficient for the maximum performance of the Abarth. In the same way as those of such high speed road holding – which is the proverbial level of the normal 600 – there is no need to worry; however, in this particular case more effective shock absorbers are indicated in our view.*

In our judgement, the 750 Abarth is not just great fun to drive in normal use, particularly on main roads and, certainly, in the mountains but it can

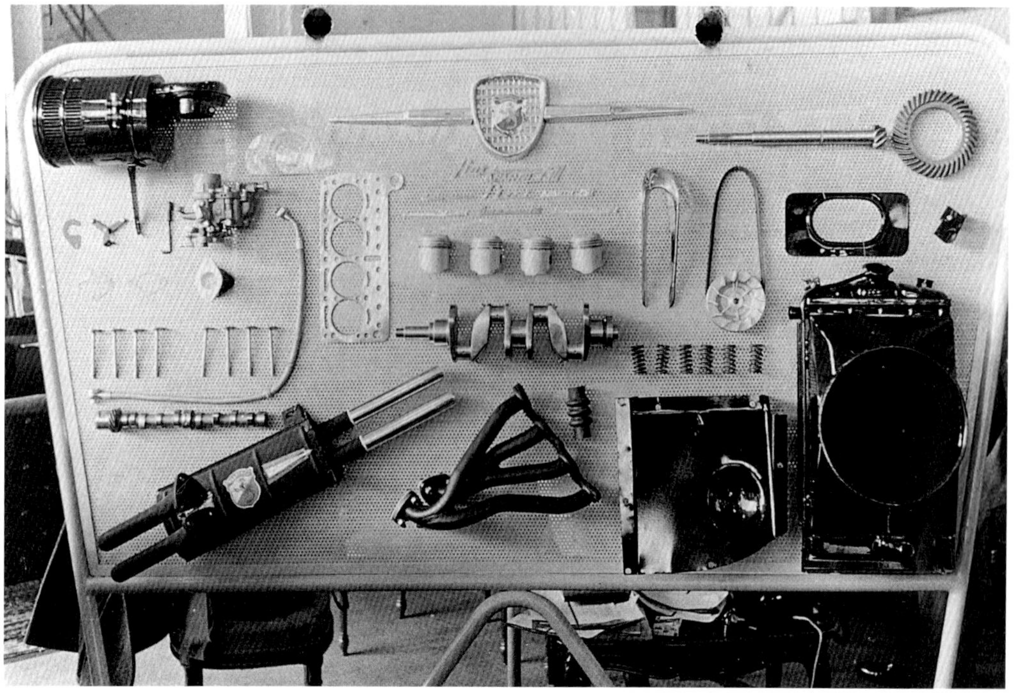

The display panel of the Abarth conversion kit for the Fiat 600 engine. Prominent among the components on the left are the Weber 32IMPE carburettor, which took the place of the 22IM, inlet valves, exhaust system and drive shafts; in the centre are the cylinder head gaskets, the four pistons, camshaft and exhaust manifold, while on the right are the pinion set with different gearbox ratios, bigger radiator, new dynamo pulley and relative belts.
As a result of that elaboration, the mass production engine's capacity was increased from 633 cc to 747 cc, bringing with it a 20 hp increase in power output from 21.5 to 41.5 hp.
Top speed of the little utilitarian, which was externally identical to the original Fiat, was almost 130 km/h.

also offer the driver an excellent possibility of a class win in important races, such as the great international rally events."

Those comments were in perfect synthesis with the impressions Carlo Abarth formed while test-driving the car. Always demanding and never content with the level of preparation achieved, he tested every car personally, criticised its handling defects or imprecise assembly; he was also capable of suspending the production of a certain model because he had diagnosed its state of readiness was still unacceptable and rectified the error himself until the origin of the defect had disappeared. On the basis of this procedure that was inalienable to him, Abarth put on sale the 600 modification kit as well as the feed manifolds, exhausts and other mechanical components; at the main motor shows he displayed a panel on which were all the new components supplied in the kit to transform the production engine into the 750 Abarth. The initiative gave rise to an unexpected avalanche of requests from half the world that the little Via Trecate factory had difficulty in satisfying for a number of months.

That orientation did not go unnoticed by some elements of the press and questions were asked in Parliament. The press asked why anyone should be allowed to 'tweak' the production mechanics of a utility car like the Fiat 600 to create a much faster car, at the wheel of which, it could be seen, that the recklessness of the young would emerge and become a serious element of miss-education, transforming itself into a severe threat to society and road safety. Abarth was forced to fight for his convictions in defence of his work and the techno-scientific validity of the modifications he had made to the 600; he defended himself vehemently, sustaining that he did not belong to the large group of car 'tweakers', but qualified himself as a real professional among the many improvisers who interfered in more or less an empirical way with the mechanics of mass produced cars. He was able to demonstrate that his work was dictated by purely technical choice, which was then evaluated in testing and based on broad scientific principles. It was a distant controversy that had the objective of being provocative and in the end, as always happens – particularly in Italy – some other news story eventually diverted attention from it and the matter returned to its natural habitat. Some still maintained, however, that increasing the performance of production cars should be banned, particularly mass-produced vehicles like the Fiat utility.

The groundswell of assent created by the Fiat Abarth 750 revealed enormous opportunities for the small Turin factory and Carlo intended to fully exploit the notoriety – almost a spell – that had been cast over his creation. The year 1956 was packed with ideas and proposals, all aimed at exploiting the magic moment the 750 had brought. As early as 3 March a Zagato coupé on 750 mechanics bowed in at Monza, while at the 23rd Mille Miglia on 28-29 April the 10 cars entered were all little 'Abarth derived' Fiat saloons, one of which won the 750 cc Grand Touring production class, driven by Domenico Ogna.

In the field of lightweight racing cars, Zagato had great credibility, strengthened by the very story of the family. Ugo, founder of the body build-

Riding the wave of its early success with the Fiat Abarth 750, the little factory concluded a deal with Carrozzeria Zagato of Milan for the production of an agile and sprightly coupé that combined good road characteristics with the ability to compete successfully in the 750 cc class of the Grand Touring category. Zagato produced a smart, harmonious coupé on the floor pan of the Fiat 600 that weighed less (535 kg) §and was able to take maximum advantage of the Abarth 747 modifications; in this case, the car's power climbed to 47 hp at 6000 rpm as a result of increasing its compression ratio to 9.8:1. The Fiat Abarth 750 Coupé acted as a potent promotional instrument, aimed at encouraging sporting Italians to compete in national and local races. A number of prototypes of the coupé were built in 1956, before the Zagato-bodied car assumed its 'double hump' roof (picture) at the end of the year.

FIAT ABARTH "215 A" (1956)

ENGINE	
	rear, longitudinal, 4 cylinders in line
Bore and stroke	mm 61x64
Cubic capacity	cc 747
Valve gear	overhead valves, side camshaft
Number of valves	2 per cylinder
Compression ratio	9: 1
Fuel feed	2 Weber 32DRNP carburettors
Ignition	coil, distributor
Coolant	liquid
Lubrication	forced, oil filter on the main circuit
Maximum power	47 hp at 6000 rpm
TRANSMISSION	
	rear wheel drive
Clutch	single dry plate
Gearbox	4 speed plus reverse, central lever
Axle ratio	4,9:1
BODY	
	2 seat coupé by Bertone
CHASSIS	
	boxed steel
Suspension	front independent, upper trailing arms, transverse lower spring, hydraulic dampers; rear independent, trailing arms, helicoidal springs, hydraulic dampers
Brakes	hydraulically operated drums front and rear
Steering	worm and sector
Fuel tank	capacity 27 litres
Tyres	5.20 x12
DIMENSIONS AND WEIGHT	
Wheelbase	mm 2000
Front track	mm 1144
Rear track	mm 1154
Length	mm 3560
Width	mm 1390
Height	mm 1180
Kerb weight	kg 595
PERFORMANCE	
Top speed	km/h 170

ing company in 1919 in Milan, was a self-taught enthusiast and untiring organiser, who created his early works on Bianchi, Chiribiri, Diatto and Fiat chassis and soon started work on Alfa Romeos, showing a particular ability with open top sports cars. His creations on the RL and RLSS running gear of the second half of the Twenties and again on Portello's 6C and 8C that won a number of Mille Miglia races in the Thirties joined the aristocracy of Italian cars holding their heads high, without being sullied by the recurring and turbulent financial problems of the company. The fame of Carrozzeria Zagato was based in large measure on building light but robust bodies with an unmistakable elegance, combining harmony of line and an extreme sobriety of execution. Ugo, who was joined by his sons Elio and later Gianni, went back to work at the end of the Second World War and there followed numerous motor sport successes with the 'Panoramica' series of sports tourers; the promoter of their cars was Elio Zagato, who raced with great energy and won three Italian championships in the Fifties, the up to 750 cc International Grand Touring in 1951, the 1955 up to 2000 cc and the 1959 up to 2600 cc. The factory concentrated even more on building coupés in aluminium and started to peak in 1956, when the production of the Fiat Abarth 750 and successive models took the company to even greater success.

In his monographic book on Zagato, Michele Marchianò described the following episode: *"At the 1955 Turin Motor Show, the Turin manufacturer Carlo Abarth took Elio (Zagato, ed) to one side and explained to him a certain*

The glorious logo of Carrozzeria Zagato of Milan, which graced hundreds of winning cars over the years in just as many races in Italy and the rest of the world.

With the arrival in the marketplace of the little coupès on Fiat Abarth mechanics, Zagato began to experience one of the most successful moments of its stormy existence.

Opposite: Besides the collaboration with Zagato, Carlo Abarth also wanted to explore other possibilities of having special bodies built on the 750's mechanics. Bertone constructed the 215 A prototype coupé based on a model by Franco Scaglione, which was noted for its considerable front overhang and two large rear fins.
In this photograph, taken on the Turin-Milan motorway at the beginning of 1956, a prototype's aerodynamic efficiency is being tested.

An open version of the Abarth 215 A was also produced and designated the 216 A.
It was, perhaps, less successful that the coupé stylistically speaking but the body had a wealth of affected and exuberant aspects to it, including the axle overhang, which today appears, frankly, over the top compared to the mechanics of the original Fiat 600.
The car was elaborated by Abarth to 750 cc using its conversion kit.

idea that had come into his mind while looking at the attractive Fiat 600 TS coupé (30 hp, 130 km/h) that Zagato presented that year on his stand. The preamble was brief and the proposal immediate: I'll give you the mechanics ... you produce the body in Milan. But no advance payment. I'll pay you only after the cars have been sold".

Finalised by Lorenzo Avidano, Abarth's motor sport director, the operation did not include the production of a real series of cars because Armando Scagliarini, president of Abarth, did not intend to become too involved. Nevertheless, the motor sport successes of the company gave the initiative sufficient credibility and Zagato accepted the proposal". The agreement with the Milan body builder stipulated that Abarth would install the modified 750 cc engine in the car and carry out final development, charging the various operations directly to the customer, and Zagato would invoice the mechanics of the Fiat production car and the body; that was to avoid double tax which would have been a burden on the various segments of work carried out before arriv-

The 23rd Mille Miglia on 28-29 April 1956 sealed the success of he Abarth 750 on the sports front. Many crews turned up to the Brescia start in the little production 600 saloon – on the left Guido Archetti in car number 37 – with their cars' mechanics modified by Abarth. Even better was the performance of the Fiat Abarth 750 Zagato, which did well not only in the Touring Production Special and 750 cc Grand Touring categories, but also in the overall results; the picture is of the A. Thiele-A. Storzini car, which came second in class and 73rd overall after having covered the 1597 kilometre route in 15 hours 5 minutes and 54 seconds, an average of almost 106 km/h.

On the Abarth stand at the Geneva Motor Show in March 1956 were four Fiat Abarth 750s with four different bodies; from the left, the Zagato coupé, the Fiat 600 saloon by Ghia, the Viotti and the Bertone 215 A coupés.

ing at the finished product. The Abarth 750 Zagato first appeared in public on 8 March at the 1956 Geneva Motor Show and a few weeks later at the Turin show; soon afterwards, its official price of Lit 1.460.000 was announced.

The Turin Show, which opened on 21 April 1956, featured another unexpected new development on the Abarth stand, this time the result of collaboration with Carrozzeria Bertone, which Carlo had involved in the development of two coupé-roadster prototypes built on Fiat Abarth 750 mechanics. Called the 215 A and the 216 A, the two cars were probably part of Abarth's plan to test the market with various bodies on those mechanics and gauge public reaction to them. Based on an outline by Franco Scaglione, Bertone produced two very streamlined bodies with long overhang on two axles and with prominent caudal fins pointed like blades. They did not meet with much success and only

An attractive picture taken in Parco del Valentino, Turin, in the spring of 1956 of the Bertone-bodied Abarth 215 A few days before the opening of the city's motor show.
The line of the body, especially at the rear, takes on number of audacious aesthetic features much liked by Franco Scaglione, who designed the three B.A.T.5, B.A.T. 7 and B.A.T.9 small saloons for the Turin body builder on the mechanics of the Alfa Romeo 1900 Sprint and Super Sprint of 1953-1955.

a small number were produced, some of which were sold in the United States at $ 4000 each.

At the time, Nuccio Bertone, the imaginative Turin designer and body builder, was increasingly taking control of his company, founded in 1912 by his father Giovanni. Of refined appearance by comparison with the men of the motor industry, he never lost an opportunity to generate interest around his creations; he always surrounded himself with gifted stylists, some of whom began their careers at his company before becoming famous and going into business for themselves. Always looking for new opportunities offered by the market, the young Bertone understood and shared the anxiety of Carlo Abarth to prepare small sports cars to be produced in limited production runs with special bodies of high design content; and he had no hesitation when Carlo Abarth asked him in the winter of 1955-1956 to design a record car with which extensive technical tests could be carried out, able to show off the

The boxed pressed steel chassis of the Fiat Abarth 215 A from which the 750 record breaking car was also derived, after design ideas by Carlo Abarth; cross and longitudinal members were integrated within the structure. The front end and rear suspension came straight from the Fiat 600, even if with different settings; the by then well-known 747 cc engine was calibrated to a maximum 47 hp at 6000 rpm, in spite of the fact that it would be placed under great stress when going for speed records over long distances.

exceptional qualities of the 'Derivazione Abarth 750' based on the Fiat 600.

Abarth intended to make a number of attempts to establish long distance records on the high speed track at the Monza Autodrome with a special car of integrated profiling and extremely light construction; the body of an aerodynamic single seater in aluminium was designed and built by Bertone; the car had a high caudal central fin, which started at the completely closed cockpit. It had a 2280 mm wheelbase while the track was 1140 mm at the front axle and 1154 mm at the rear. The engine was the classic four cylinder 747 cc derived from the Fiat 600, but its maximum power was increased to 44 hp at 6000 rpm by raising the compression ratio from 9:1 to 9.8:1. So the engine had reached its third phase of preparation after the original unit of the production Fiat saloon (41.5 hp) and the power plant for the same model but with a modified body (43 hp). Fitted with 5.20-12 tyres, the record car was 4590 mm long, 1540 mm wide, 1040 mm high and weighed 385 kg. With a final reduction ratio of 8.4:1, it was credited with a top speed of 192 km/h.

The little record car presented itself at the Monza track on 17 and 18 June 1956 and harboured all Carlo's technical hopes and aspirations; he had invested all his best experience as an established technician in that futuristic looking car, as well as considerable company resources. The record attempt included class H (from 501 to 750 cc) over 24 hours, which started at 20.17 on Sunday 17th; throughout the night the attempt was dogged by heavy rain and a thick mist that became even thicker on the track; four young test drivers at the start of their distinguished careers alternated the wheel, Umberto Maglioli, Mario Poltronieri, Alfonso Thiele and Remo Cattini. They lapped the circuit with the precision of a chronometer, penalised by persistent bad weather but helped by the efficiency of that amazing 750 cc engine, which with every lap

Seen from above, the record car designed by Franco Scaglione and built by Carrozzeria Bertone. It was a result of sophisticated research into aerodynamics and the precise execution of the various surfaces, particularly the fit of the wings and the transparent faring of the driving position, which was incorporated in the central section that ended with a small stabiliser fin.

FIAT ABARTH "750 RECORD" (1956)	
ENGINE	
	mid-rear, longitudinal, 4 cylinders in line
Bore and stroke	mm 61x64
Cubic capacity	cc 747
Valve gear	overhead valves, side camshaft
Number of valves	2 per cylinder
Compression ratio	9,8: 1
Fuel feed	1 Weber 32IMP carburettor
Ignition	coil, distributor
Coolant	liquid
Lubrication	forced, oil filter on the main circuit
Maximum power	47 hp at 6000 rpm
TRANSMISSION	
	rear wheel drive
Clutch	single dry plate
Gearbox	3 speed, central lever
Axle ratio	8.4:1
BODY	
	single-seater record car by Bertone
CHASSIS	
	caisson in boxed steel
Suspension	front independent, upper trailing arms, transverse lower spring, hydraulic dampers; rear independent, trailing arms, helicoidal springs, hydraulic dampers
Brakes	hydraulically operated front drums
Steering	worm and sector
Fuel tank	capacity 50 litres
Tyres	5.20 x12
DIMENSIONS AND WEIGHT	
Wheelbase	mm 2280
Front track	mm 1140
Rear track	mm 1154
Length	mm 4590
Width	mm 1540
Height	mm 1040
Kerb weight	kg 385
PERFORMANCE	
Maximum speed	km/h 192

was quelling the criticism that it was unreliable over long distances and did not last long under stress. For Abarth, that long distance record on the track at high speed was a better instrument with which to evaluate his theories and promote his products. At the end of the 24 hours the small, streamlined car had covered 3743.642 kilometres at an average speed of 156 km/h (155.985). A triumph for an engine with atmospheric fuel feed, rods and rocker arms!

Based on the enthusiasm of that record achieved with ease and without any mechanical problems, Abarth decided without hesitation that it was the moment to push ahead with other more demanding record attempts, in search of stupefying results on the banked corners of the Monza circuit. To publicise that event in the leading enthusiast magazines, he had the ingenious idea of inviting six noted motoring journalists to drive the car so that they could live as protagonists the exciting adventure of breaking a record on the world's most famous track; the group comprised Walter Honegger of Switzerland, Germany's H. U. Wieselmann, the Belgian Paul Frère, Gordon Wilkins from Britain, Bernard Cahier of France and prominent Italian journalist and racing driver Giovanni Lurani, at the time editor of Auto- Italiana. For this second series of attempts on 27-29 June, they used the car that did so well a few days earlier, but with a slightly less powerful engine. As always, Carlo was at the track checking everything tirelessly, a great inspiration to drivers and mechanics; Renzo Avidano, Abarth's inseparable and loyal right hand man, kept the logistical side of the event under control and worked with the press which, electrified by the previous record, wanted to personally witness the various attempts that had been announced.

The 5000 km segment started normally, but after seven hours of driving a valve follower problem forced the car to stop. Abarth did not hesitate for a minute; he loaded the car onto a truck and returned to his Turin factory where he replaced the engine with a new one and immediately returned to Monza so that everything was ready to start again the next morning. But an electrical breakdown meant long pit stops to re-charge the battery; some drivers, pressured by other urgent jobs, had to drop out and were replaced immediately by Abarth with those the company could count on in a predicament of emergency like that; Cattini, Poltronieri and Thiele returned to the track. Having overcome that problem, the attempt was a success and broke the records of 5000 km in 34h 26'. 27" at an average 145.175 km/h, the 5000 miles in 55h 26'53" (average 145.121 km/h), the 10,000 km in 71h26'53" (average 140.658), the 48 hour, having covered 6978.931 km (average 145.394 km/h) and the 72 hour in which the team drove 10,125.560 km at an average 140.632 km/h. Not everything went smoothly in those three wearying days; small recurring mechanical problems – quickly repaired – and bad weather which, between one downpour and another, even forced the car to stop for 30 minutes because of a vio-

The Abarth record car had an extremely trim body and complete faring, the work of Bertone. The cockpit of the little single-seater was protected by a dome of Plexiglas connected to the long tail fin. Total weight was kept down to 385 kg. On the Monza high speed track on 17-18 June 1956, the car broke the H class 24-hour record, covering 3743.642 kilometres at an average speed of 155.985 km/h. Driving alternated between Remo Cattini, Umberto Maglioli, Mario Poltronieri and Alfonso Thiele. The pictures show Carlo Abarth next to the car during a pit stop, and taking a few moments off with Umberto Maglioli.

A few days later, the Fiat Abarth single-seater was off to Monza again to establish other records on 27-29 June, this time driven by six motoring journalists from a number of countries.
In the small picture, from the left are Gordon Wilkins of Britain, Bernard Cahier of France, H. V. Wieselman of Germany, Paul Frère of Belgium, Walter Honegger of Switzerland, Giovanni Lurani of Italy and Carlo Abarth.
The large photograph shows Paul Frère during a driver change in the Monza pits.
In those three days, H class records were established in the 5000 km, 5000 mile, 10,000 km, 48 hour and 72 hour categories.

The exceptional record breaking drives by the Fiat Abarth were widely reported in the international press, including the German motoring magazine Das Auto Motor und Sport, which devoted its front cover number 15 of 21 July 1956 to the achievement.

lent thunder storm and pouring rain. The average of over 140 km/h achieved in the 72 hours over a distance of more than 10000 kilometres was an astonishing achievement for a small engine with little more than 40 hp.

Fired up by that mass of successes, Abarth wanted to continue the following month, so on 25 July he readied the same car, the engine of which had been reduced to 500 cc to break the international 351-500 cc Class 1 record over 2000 kilometres, 2000 miles and 24 hours; the drivers were Cattini, Poltronieri and Thiele, backed up by Della Beffa and Fraccari. They shattered all three records at an average speed of between 137.778 km/h for the 24 hours and 139.978 km/h for the 2000 kms. Not content with those results and increasingly aware of the potential of that amazing car that chewed up records so implacably, Abarth was at Monza again on 9 August in an attempt to beat the international Class G record for cars of between 751 and 1100 cc with an 800 cc engine. The same drivers were brought in again – Cattini, Poltronieri and Thiele - and the car started its attempt at 06.09 on the ninth, running for 24 hours to break six records at average speeds of between 167.722 km/h (the 24 hours) and 170.410 km/h for the 12 hours.

The 1956 season ended two months later on 12 October, when Abarth made another attempt on Class I records; the three works drivers established yet more new records for 500 miles, 1000 kilometres, 1000 miles and the six and 12 hours, returning average speeds of between 150.183 km/h (six hours) and 147.644 km/h (1000 miles). After setting up 20 international records and

During the summer of 1956 and up until 12 October, Abarth established numerous other international speed records at Monza in the I (from 351-500 cc) and G (751-1100 cc) classes. Present throughout each record attempt, Carlo Abarth coordinated technicians and time keepers and sent messages to drivers on the track using flags.

The Abarth-converted Fiat 600 dominated in the 1956 motor season. In both its Fiat production and Zagato coupé bodies, the little car stormed to the top of the results tables of national and local races, scoring 31 class wins and causing widespread alarm among independent tuners in the production car Touring and Grand Touring categories. The little Fiat family utility emerged from that successful promotional operation with a stronger image than ever.

with the enormous effect those achievements had on public opinion during the summer, the 1956 season went down in the Italian motor industry history books and the brief history of Abarth as one of the most significant and promising. For Carlo, the records also represented a brilliant return to the world of motor sport, from which he had been almost estranged for nearly a year after the exciting years of 1949-1951, the hey-day of the 204 A in roadster and sports tourer configurations.

On the industrial side the Via Trecate factory found its managerial equilibrium, with exhaust production that year reaching 113,178 units of which 21% were exported, while the great opportunity presented by the 'Derivazione 750' based on the Fiat 600 had already produced significant results in commercial terms and in motor sport. The Fiat Abarth 750 made its debut on 25 March at the Torricelle di Verona hillclimb, where Enrico Carini, Remo Cattini and Nino Crivellari took the first three places in the up to 750 cc Grand Touring Class and concluded the season after a series of victories in Italy and beyond its borders on 5 November at the Course d'Automne in Paris, where Harris won the GT Class. In total, the cars of the Scorpion scored 31 class wins; Enrico Carini won the Italian championship and the Trophy of the Mountains at the wheel of a Fiat Abarth 750 prepared by Gianni-Bosato.

At the end of the year, Abarth & C. gave the press a publicity-educational booklet in which it revealed a number of secrets and extolled the victories which the 'Derivazione 750' had accomplished in those 12 months on road and track. The publication also showed that Abarth & C. supplied five different configurations in kit form and included the price list. They were the:

- GT 750 (41.5 hp)
- GT 750 for non-standard bodies (43 hp)
- Mille Miglia GT 750 (44 hp)

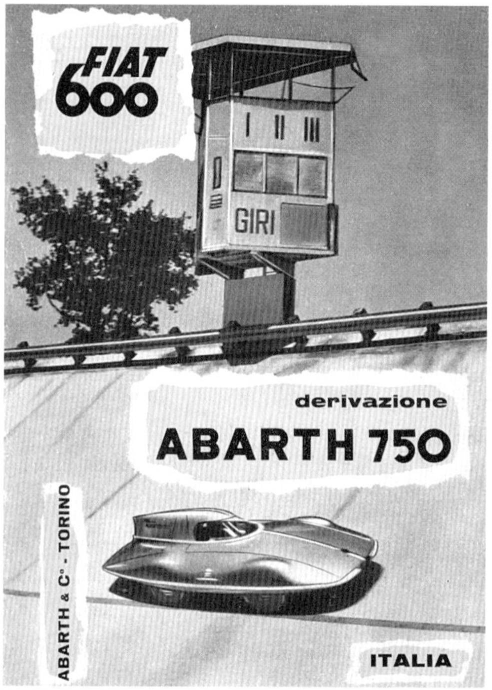

The cover of the booklet distributed by Abarth & C. at the end of 1956 for the promotion of its Fiat 600 conversion kit. The publication described the techno-constructional principles that led to the creation of the 747 cc engine, as well as giving a complete list of the modified mechanical components – with their relative retail prices – required for the substitution of production parts. The illustration showing the record car on the Monza banking is emblematic, a symbol of international motor sport success.

- Mille Miglia GT 750, prepared (46 hp)
- GT 750, prepared for hillclimbs and short tracks (47 hp)

The booklet also said that *"for those who own a 600 and want to modify its performance there is a kit containing everything necessary for the transformation, which can be carried out at our appointed garages ... ; cars modified in this way are homologated in the up to 750 Grand Touring Class"*. And: *"in the kit there is everything that we fit to the cars we modify, from the drive- shaft to our company's logo, from the radiator to the gaskets. Every modification, every technical detail is opportunely explained by communications distributed to all those who work with us"*. The price of the complete boxed kit was fixed at Lit 225.000 This was the power of the organisation wanted by Carlo Abarth and put together by Renzo Avidano during his continuous trips up and down the Italian peninsula, looking for spare parts specialists and workshops that would be suitable to represent the Scorpion brand. In Italy, still tied to the old system of car selling agents firmly bound to the factory, this new clear and simple commercial approach was a small revolution in the industry.

Fortified by that success, Carlo Abarth began 1957 rich in new experiences and driven by even more stimulating motivation. The two years of motor sport success racked up by Alfa Romeo with the Giulietta, powered by a small twin overhead camshaft 1290 cc engine, encouraged him to persevere with the exciting record adventure at Monza. He decided to play a new card using another body designer and turned to Pinin Farina to build a new record vehicle with an 1100 cc engine, derived from that of the Giulietta.

Battista Farina, nicknamed Pinin, opened his car body designing atelier in Turin on 22 May 1930 after working for a long time at his brother Giovanni's factory, where he made his name with the creative inspiration and the good taste of his work. He was already the architect of the most beautiful

The fame of Abarth and its brilliant little cars quickly reached the ears of the great and powerful. At the 1956 Brussels Motor Show in January, King Baldwin of Belgium visited the Turin firm's stand. Among the cars on display were the Fiat Abarth 750 production saloon in its characteristic dual-tone livery, and the 210 A two-seater roadster with the definitive Boano body.

In a little less than a year of motor racing, the Abarth-modified Fiat 600 widened its competitive horizons and became famous as the utility that quickly won races across Europe. The exemplary picture shows Carl Baumeister competing in the 1957 Solitude Rally in Germany, where he won the up to 1300 cc class of the Special Touring Category.

cars immediately before the war and his fame spread throughout the world due to the all-Italian style with which he 'dressed' the most prestigious cars, from Hispano-Suiza to Cadillac, from the Alfa Romeo 8C 2300 to the Lancia Astura; he formed an intensely close and fertile relationship of collaboration with Vincenzo Lancia that led to the futuristic and aerodynamic 1937 Aprilia, in which he transfused his happy intuition on the theme of aerodynamics as applied to the automobile. After the war Pinin Farina personified with majesty the design criteria of the Italian school and as early as October 1946 participated in the unofficial Caravan of Elegance that took place in Lausanne, Switzerland, in October, to move on to Paris where, heedless of the ban imposed by France as one of the winning powers of the war, parked his Alfa Romeo 2500 and Lancia Aprilia on the footpath in front of the entrance to the Motor Show at the Grand Palais, achieving enormous positive recognition. In 1947, his name was in the limelight again with the unveiling of the Cisitalia 202, which was such an innovative car with its sober lines that it became part of the history of world motor cars.

A lively and creative enthusiast, Pinin Farina designed his cars with simplicity and innovation on an architectural level, and with generous intuition not of an academic culture but the innate instinct of an artist.

He met Carlo Abarth while designing the Cisitalia 202, when the Vienna-born technician began to work with Piero Dusio. The great Pinin wrote in his memoirs, published over 30 years ago, of a number of concepts that guided him not only in the creation of that stupendous sports tourer but which were, for him, a precept of professional life: " ... *When it came out in the autumn of '47, the Cisitalia was eulogised for its design, but this also seems*

FIAT ABARTH "750 GT" ZAGATO (1956)

ENGINE	
	rear, longitudinal, 4 cylinders in line
Bore and stroke	mm 61x64
Cubic capacity	cc 747
Valve gear	overhead valves, side camshaft
Number of valves	2 per cylinder
Compression ratio	9.8: 1
Fuel feed	1 Weber 32IMPE carburettor
Ignition	coil, distributor
Coolant	liquid
Lubrication	forced, oil filter on main circuit
Maximum power	43 hp at 5800 rpm
TRANSMISSION	
	rear wheel drive
Clutch	single dry plate
Gearbox	4 speed plus reverse, central lever
Axle ratio	4.5:1
BODY	
	2 seat coupé by Zagato
CHASSIS	
	integral body
Suspension	front independent, upper trailing arms, transverse lower spring, hydraulic dampers; rear independent, trailing arms, helicoidal springs, hydraulic dampers
Brakes	hydraulically operated drums front and rear
Steering	worm and roller
Fuel tank	capacity 30 litres
Tyres	5.20 x12
DIMENSIONS AND WEIGHT	
Wheelbase	mm 2000
Front track	mm 1150
Rear track	mm 1160
Length	mm 3480
Width	mm 1340
Height	mm 1190
Kerb weight	kg 535
PERFORMANCE	
Top speed	km/h 150

The third coming of Zagato's Fiat Abarth 750 coupé went into production at the beginning of 1957. The car boasted a number of improvements to the body, particularly the rear optical group and bumpers.

The characteristic double hump roof became highly popular among sports car drivers of the period, and established a pleasant aesthetic contrast to the two upward-opening air intakes on the bonnet that also became a distinctive feature of the model.

to be a misunderstanding; the designer is, above all, a constructor and instead of concentrating too much on the cosmetics of the product, function is forgotten. ... The first Cisitalias were two-seaters; they wanted to make them 2+2 against my will, a formula which is anything but satisfactory, even arithmetically, because in this case two plus two does not make four but means the people in the rear would have to travel with their knees in their mouths. ... Anyway, I had activated my ideas without restriction; however, I did not take notice only of particular new ideas, what interested me was the renewal of the overall architecture. The bonnet, lower than the wings, represented precisely the contrary of those that were in use, especially on American cars which, in that way, did not offer great visibility ... No small number of experts say to me that the sports cars of our time follow the design of the Cisitalia. Clumsy transverse and vertical measurements, the design of the seats, the relationship between the amount of habitability respond to touring needs. And that means to offer civil treatment to passengers that, on the contrary and not rarely, are introduced into a car as common packaging merchandise".

Carlo Abarth was most impressed with that first meeting with Pinin Farina. He, so pragmatic, so meticulous, noticed that anxiousness to achieve perfection also drove the celebrated Turin designer, who had also established such a system in his own daily work. The two men agreed to build a completely new record car, to use the Giulietta engine with a re-dimensioned cubic capacity and to display the car for promotional purposes at the Geneva Motor Show in March 1957. But when Pinin mentioned to Professor Valletta, the

Right from the start, the 1957 racing season belonged to the Scorpion in the Touring and Grand Touring categories. The fourth University Championship at the Modena Aeroautodrome on 24 March saw Abarth win the Grand Touring 750 cc sub-class with a Fiat Abarth 750 Zagato driven by Guido Guidotti, who became university champion that day.

This logo identifies the most celebrated body designer and builder of all time, Pinin Farina, the creator of the most innovative and coherent car designs of the Thirties and the post-war period. Even today his custom-built cars and those deigned for numerous vehicle manufacturers still transmit messages of extraordinary Italian good taste throughout the world.

The Pininfarina stand at the 1957 Geneva Motor Show exhibited two record cars – one with a 1059 cc engine derived from the Alfa Romeo Giulietta's, and the other with a 747 cc power unit converted from the Fiat 600's motor – which seemed almost identical as far as aerodynamics and body design were concerned. Carlo Abarth made available a special chassis in boxed pressed steel available for both versions.

great leader of Fiat, that he was designing with Abarth a record car with Alfa Romeo mechanics, he was invited with a scowl to design another with a Fiat engine. That kind of imposition obliged the two constructors to work day and night for two weeks to build a second record car on the same basis, which was fitted with the classic 747 cc Abarth engine, derived from the Fiat 600. That is why two almost identical cars appeared on the Pinin Farina stand at Geneva; they were extremely attractive in their sharp lines with two large stabilising tail fins; the longest was identified as an Alfa Romeo and the other a "Fiat 600, Derivation Abarth 750".

The bodies of both cars were the result of long aerodynamic development and wind tunnel tests carried out at the Turin Polytechnic, and were made in light alloy with completely smooth surfaces to reduce parasitical tur-

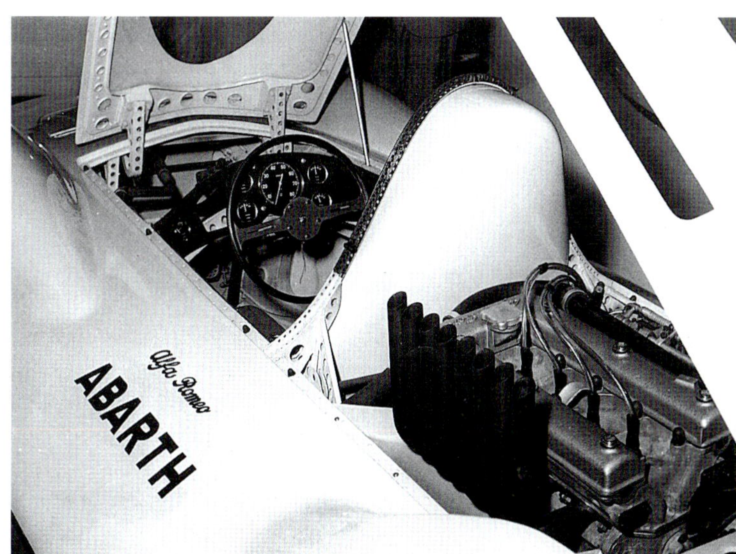

At the 24th Mille Miglia on 11-12 May 1957 there was an impressive presence of Fiat Abarths in the 750 cc class of the Grand Touring category; 20 Scorpions were entered for the race and 16 of them finished, among them a class winner driven by Alfonso Thiele.
In the three photographs above are some of the starters: the Fiat Abarth 750 saloon number 51 of Roberto Rosazza Ferraris who came 16th in class and 160th overall, the Fiat Abarth 750 Coupé Zagato number 53 driven by Lino Zanotti who climbed to fourth in class and 97th overall, the Fiat Abarth 750 Coupé Zagato number 45 driven by N. Crivellari-N.A. Nardari that made seventh in class and 114th overall.
Below: The Fiat Abarth 750 Zagato Coupé of 'Gri-Gri' that came 10th in class and 127th overall.

bulence to a minimum; the lower area was fully fared. The Alfa Romeo was an entirely new project for the small Via Trecate factory, which had to design a chassis from the beginning in boxed pressed steel with a special cradle in which to fit the twin overhead camshaft engine derived from the Giulietta, the cubic capacity of which was reduced from 1290 to 1059 cc; that was achieved by reducing the bore and stroke, while the compression ratio was increased from the basic 7.5:1 to 8.7:1; fed by two Weber 40 DCO3 carburettors, the car delivered 70 hp at 6200 rpm. The original Giulietta brakes and suspension were retained, but were suitably adapted for use on a high-speed track. The single-seater Alfa Abarth Pininfarina record car had a 2400 mm wheelbase, was 5400 mm long, 1760 mm wide and 1040 mm high; its dry weight was confirmed at 480 kg and its maximum speed 226 km/h.

Compared to the prototype shown at Geneva, the car was slightly modified at the front by the addition of two suitably fared headlights on the bumper bar and was painted in duo-tone; it was to attempt to establish a number of international records in Class G for cars from 751-1100 cc on 18 May 1957 on the Monza high speed track. The drivers were Belgian journalist Paul

Marino Guarnieri was at the start of the 1957 Mille Miglia at the wheel of Fiat Abarth 750 Coupé Zagato number 40, which came second in the 750 Grand Touring class and 68th overall. He also won the Nuvolari Grand Prix for being the fastest over the 132 kilometre Cremona-Mantua-Brescia section of the route, on which he clocked 54'31.1/5", an average of 147.277 km/h.

The winner of the 750 cc Grand Touring class of the 1957 Mille Miglia who came 63rd overall, Alfonso Thiele (no. 52) at the wheel of his Fiat Abarth 750 Coupé Zagato. He completed the 1597 kilometres in 13h32'32.33" at the formidable average speed of 117.900 km/h.

ALFA ROMEO ABARTH "1100 RECORD" (1957)

ENGINE	
	mid-rear, longitudinal, 4 cylinders in line
Bore and stroke	mm 68x75
Cubic capacity	cc 1088
Valve gear	2 chain driven overhead camshafts
Number of valves	2 per cylinder
Compression ratio	8,8: 1
Fuel feed	2 twin-choke Weber 40DC03 carburettors
Ignition	coil, distributor
Coolant	liquid
Lubrication	forced, oil filter on main circuit
Maximum power	78 hp at 6800 rpm
TRANSMISSION	
	rear wheel drive
Clutch	single dry plate
Gearbox	4 speed, central lever
Axle ratio	4.3:1
BODY	
	single-seater record car by Pininfarina
CHASSIS	
	boxed steel with cross and longitudinal members
Suspension	front independent, trailing arms, helicoidal springs, stabilizer bars, hydraulic dampers; rear independent, trailing arms, helicoidal springs, hydraulic dampers
Brakes	hydraulically operated drums on the front
Steering	worm and roller
Fuel tank	capacity 80 litres
Tyres	5.25 x15
DIMENSIONS AND WEIGHT	
Wheelbase	mm 2410
Front track	mm 1220
Rear track	mm 1178
Length	mm 5680
Width	mm 1690
Height	mm 1100
Kerb weight	kg 620
PERFORMANCE	
Top speed	km/h 260

Frère, who had been actively involved in the development and construction of the prototype, and Alfonso Thiele. That day, they set up new records for 500 kilometres in 2h 31'02" at an average speed of 198.618 km/h, the 500 miles in 4h 10'05" (193.046 km/h), the 1000 kilometres in 5h 12'. 20" (192.098 km/h), the three hours covering 580.116 kms at an average 193.372 km/h and the six hours (1,153.338 kms at an average 192.223 km/h). During the first attempt the Alfa Romeo was involved in a night accident caused by a burst tyre and was seriously damaged, but driver Thiele was uninjured. Another car was built, with which the 18 May international records were established.

The other record car did not require so much design work, as its basic structure was derived directly from its predecessor that had established 20

The first series of record attempts with the Alfa Romeo Abarth Pininfarina, launched at the 1957 Geneva Motor Show, took place in May of that year at Monza. On the 18th, Paul Frère and Alfonso Thiele broke five international class G (from 751 cc to 1100 cc) records over 500 and 1000 km, 5000 miles, three and six hours at an average speed of between 192.098 and 198.618 km/h. The pictures show the record car on a banked corner, Paul Frère as the car is being tuned and Piero Taruffi in the Monza pits with Carlo Abarth.

Powered by a 1088 cc engine derived from the Giulietta that put out 78 hp at 6800 rpm, the Alfa Romeo Abarth Pininfarina record car could exceed 250 km/h, in part due to its reduced weight of 620 kg and the sectional body in aluminium. The chassis was a sort of parallelepipedon with cross and long members in boxed pressed steel, amply perforated for maximum lightness. Built on a wheelbase of 2400 mm, a front track of 1220 mm and a rear of 1178 mm, the car was 5680 mm long, 1690 mm wide and 1100 mm high.

records the previous year with its Bertone body. However its development required numerous body and other modifications, because the Pininfarina design did not allow sufficient vertical excursion of the front wheels and was severely stressed on the cement slabs of the Monza banking; so the wings were modified. The car was longer than the previous year's version, measuring 5110 mm, its width 1615 mm and height 1100 mm; it weighed in at 510 kg and had a maximum speed of 218 km/h generated by the well-known Fiat 600-derived engine which produced 45 hp at 6000 rpm.

To complete the series of records established in 1956 in Class H (from 501 to 750 cc), the new version of the Fiat Abarth 750 took to the track at Monza from 24-27 July 1957, driven by Giulio Cabianca, Remo Cattini, Marino Guarnieri, Corrado Manfredini and Mario Poltronieri. They set up 15 international records between them in four triumphant days: they were the 500 kilometres in 2h 43'.15" at an average speed of 183.752 km/h, the three hours covering 551.719 kilometres at an average of 183.906 km/h, the 500 miles (4h 25'.09" at 182.079 km/h), the 1000 kms (5h 28'29", average 182.652 km/h), the

ABARTH "750 RECORD" (1957)

ENGINE	
	mid-rear, longitudinal, 4 cylinders in line
Bore and stroke	mm 61x64
Cubic capacity	cc 747
Valve gear	overhead valves, side camshaft
Number of valves	2 per cylinder
Compression ratio	9.0: 1
Fuel feed	1 Weber 32DRNP carburettor
Ignition	coil, distributor
Coolant	liquid
Lubrication	forced, oil filter on the main circuit
Maximum power	43 hp at 5900 rpm
TRANSMISSION	
	rear wheel drive
Clutch	single dry plate
Gearbox	4 speed, central lever
Axle ratio	4.3:1
BODY	
	single-seater record car by Pininfarina
CHASSIS	
	boxed steel with longitudinal and cross members
Suspension	front independent, trailing arms, helicoidal springs, hydraulic dampers; rear independent, trailing arms, helicoidal springs, hydraulic dampers
Brakes	hydraulically operated front drums
Steering	worm and sector
Fuel tank	capacity 73 litres
Tyres	5.00 x 15
DIMENSIONS AND WEIGHT	
Wheelbase	mm 2280
Front track	mm 1140
Rear track	mm 1154
Length	mm 5110
Width	mm 1615
Height	mm 1100
Kerb weight	kg 510
PERFORMANCE	
Maximum speed	km/h 200

six hour (1086.022 km, average 181.003 km/h), the 1000 miles (8h 59'.25", average 179.007 km/h), the 2000 kms (11h 07'01", average 179.904 km/h), the 12 hours (2162.498 kms, average 180.124 km/h), the 2000 miles (18h 9'43", average 177.220 km/h), the 24 hours (4258.909 kms, average 177.454 km/h), the 5000 kms (28h 43'29", average 174.065 km/h), the 5000 miles (47h 14'00", average 170.345 km/h), the 48 hours (8174.476 kms, average 170.301 km/h), the 10000 kms (59h 03'02", average 169.346 km/h) and the 72 hours (11906.473 kms, average 165.367 km/h; the last seven broke the same number of records established the previous year by the Fiat Abarth 750 with its Bertone body.

During those four memorable days the intrepid march of the car was unstoppable; its engine turned regularly and unerringly under the constant and attentive care of Carlo Abarth, who attributed a decisive importance to the attempt and launched himself once and for all onto the international scene with his modification of production engines; he gave precise instructions to the drivers not to force the pace so as not to compromise the safety and reliability of the car, also shortening the travel of the accelerator to avoid one of them pushing the pedal further down than he should. And apart from a couple of collisions with hares that ventured onto the track, everything went smoothly, even if Cattini complained when he noticed that, while refuelling, the tank was not filled completely. The same car returned to the track on 25 October, driven by Frère and Guarnieri, and established another three records for the 200 miles (1h 35'25", average 200.347 km/h), the 500 kms (2h 30'36", average 199.189 km/h) and the three hours (592.746 kms, average 197.582 km/h). That car had an experimental 747 cc engine with twin overhead camshafts.

July 1957 was not only a significant month in the history of Abarth & C., it was also the month of the launch of an important new Fiat that became a key player with the 600 in the process of mobilising Italy: the Nuova 500.

It was the turn of the Fiat Abarth 750 record car to take to the track at Monza from 24-27 July 1957 to make an attempt on records in class H (from 501 to 750 cc) with a carefully selected group of drivers. From the left: Corrado Manfredini, Giulio Cabianca, Marino Guarnieri and Remo Cattini.
The new Pininfarina body was built on a 2280 mm wheelbase and was 520 mm longer and 125 kg heavier than the Bertone bodied car of the previous year.
The high speed carousel on the Monza circuit continued non-stop day and night (photograph) until the car and its drivers had shattered the international records for 500, 1000 and 5000 mile, 1000, 2000, 5000 and 10,000 kilometres plus the 3, 6, 12, 24, 48 and 72 hours.

The Fiat Abarth 750 record car's high speed marathon was a major international press event, attended by many journalists.
The car charged on undisturbed for 72 consecutive hours, further confirmation of how rigorous and scientific were the modifications made by Abarth to the mass produced Fiat 600 engine. After running the first 10,000 kilometres and before going for the 72-hour record a series of meticulous checks were carried out on the car's various components under the critical eye of Carlo Abarth.
The 750 went back to the track on 25 October and set up another three records, after which it was despatched to the Paris Motor Show where it stirred up a great deal of press interest on the Pininfarina stand.

Conceived as a vehicle of initiation for millions of Italians who had never driven a car, the new super-utility was born of a project by Dante Giacosa, who intended to privilege maximum construction simplicity and low cost production. This is the principle testimony from his book "My 40 Years of Designing at Fiat": *"As far as the engine was concerned, I had no doubts: it had to be a four stroke with two cylinders in line and air cooled. Among air-cooled engines, those with cylinders in line, in other words side-by-side, are much more simple and economical. The engine-transmission group could be arranged transversally without problem, a solution that offers the advantage of simplicity and a high mechanical return because it does not need helicoidal teeth bevel gears. And if the group is placed in the rear of the car, it can transmit the drive to the wheels simply with just one universal joint for each drive shaft, being an economical car of modest performance. All our studies, including those of engines with two cylinders opposed, brought me to the same conclusion.*

At that time, we were right to consider air-cooling simpler, more practical and economical than water, also because the greater fuel consumption caused by the more powerful ventilator did not look worrying. So I had the designs developed along those lines. We decided to call the new car the 110".

That was the basis of the 'Nuova 500' project, a small two-seater powered by a two-cylinder 479 cc engine that produced 13 hp maximum and gave

The untiring coordinator of those summer days at the Monza Autodrome was Renzo Avidano, Abarth's right hand man and sports director of the Scorpion. The fifth driver, who was the most familiar with the car, was Mario Poltronieri, on the right of the picture, next to Corrado Manfredini. Carlo Abarth never left the record attempt operation for a second, conscious of how the event could be of enormous importance to the future of his company. He did everything, from pit signals to re-fuelling and rapid service, all of which was vital to ensuring the devilish little car completed its task as it constantly clocked 200 km/h.

the car an 85 km/h top speed. The Nuova 500 was not given the kind of public reception the 600 had received two years earlier; perhaps it was not exactly in line with people's expectations, or perhaps even more because the price of Lit 465.000, was too close to that of the 600, the new Fiat utility found it difficult to carve out a place for itself in the market. However, sales took off in the early Sixties and the Nuova 500 soon became a best seller as a second car until 1975, when production ceased after a run of 3,700,000 units. One of its marks of distinction was that it concluded Fiat's long crusade to spread the car across all of Italy's social classes, a process confirmed by official statistics of annual car production in the country; in 1956 279,909 cars were made by Italian manufacturers, a figure that increased to 470,674 in 1959, 693,695 in 1961, 1,105,291 in 1963, 1,719,715 in 1970; in the course of those years, the incidence of the little Fiat was determinate and the number of cars in circulation grew from 1,231,082 in 1957 to 12,484,313 in 1972.

At Fiat, they knew from the first few weeks that the Nuova 500 would find it difficult to break into the market; its spartan construction, unsynchronised gearbox, excessive hesitation of the engine at minimum, low power, the unusual and unpleasant noise of the small two-cylinder did not find favour with the public at which the car had been aimed. So it did not take much to convince Carlo Abarth that a useful injection of horsepower for a special edition of the car would have contributed to also consolidating the fortunes of the mass-produced model. He left original cubic capacity of the car unchanged but increased the compression ratio from 6.55:1 to 8.7:1 and adopted a Weber 26 IMB carburettor as well as a special exhaust system; that is how he increased the car's power output by seven horse power, generated at 5000 rpm instead of 4000 rpm, to make the car much more lively and fun to drive, able to reach almost 100 km/h. As was always the case with Abarth & C by now,

Two moments in the 9th InterEurope Cup that took place at Monza on 8 September 1957, in which a flock of Fiat Abarth 750 Coupé Zagatos competed.
On the corner kerb is car number 18 driven by Mario Poltronieri, with Alfonso Thiele's sister car number 6 at its side.
The two drivers crossed the finish line in that order at an average of 136 km/h.

With the intention of expanding the process of popular motorisation that took place after the launch of the 600, Fiat introduced the Nuova 500 two years later. A car of spartan mechanics and body, it was to enter Italian family life as a step up from the scooter.
But the disappointing list price and the modest techno-constructional characteristics of the first series did not bring the small Fiat sales success.

Launched in July 1957, the tepid market reaction to the Fiat Nuova 500 induced Carlo Abarth to step in with a number of modifications to the two-cylinder engine, which was so rough and lacking in horsepower. He did not increase the cubic capacity, which stayed at 479 cc, but upped the compression ratio from 6.55 to 8.7:1, fitted a Weber 26IMB carburettor in place of the original 24IMB and a new exhaust system; maximum power went up by 7 hp and beyond, reaching 21 hp and even 23 hp at 5000 rpm. Some Nuova 500s modified in this way were used at Monza on 15 October as demonstrators to convey to public opinion how the 'Abarth cure' could also produce extraordinary results on a car that was not a Fiat 600.

various stages of modification of the small, irksome two-cylinder Fiat were prepared, producing maximum power increases of up to 21 and also 23 hp. Some cars prepared in that way but with normal production bodies were sent to Monza on 15 October 1957 for speed trials, in which Abarth demonstrated how, with little latitude for modification, the performance of the new Fiat could be radically improved.

The Turin Motor Show opened on 30 October, when a number of Nuova 500s with non-standard bodies made their appearance, among them a sober, smooth two-seater coupé by Pininfarina and a light sports tourer by Zagato inspired by the body the company had designed for the Fiat Abarth 750, which was achieving recognition everywhere. On the Abarth stand with its strong sporting emphasis was the latest Zagato coupé on Fiat Abarth 750 mechanics,

The Zagato coupé based on the modified mechanics of the Fiat Nuova 500 made its debut at the Turin Motor Show two weeks later and was initially well received, but it did not have what it took to repeat the commercial success of the Fiat Abarth 750 Coupé. Ovidio Cappelli became the 500 cc class Italian Grand Touring champion at the wheel of the 500 Coupé Zagato.

Zagato was not alone. Other designers produced bodies for the Fiat Nuova 500 and exhibited them at the 1957 Turin Motor Show. The Pininfarina contribution looked highly equilibrated, even with exuberant overall dimensions and its large front overhang; the roof was of good design and the car had well deployed glass surfaces, but it never went into production.

In just a few months, the Fiat Abarth Coupé Zagato became extremely popular in both road and track racing in the 750 cc Grand Touring category, scoring 18 victories during the 1957 season.

Remo Cattini came second overall in the Brescia Championship for the Lumezzane Trophy, which took place on 13 September, at the wheel of the successful little car.

which took on the successful general appearance of the earlier models and introduced in definitive form the 'double hump' roof that was typical of bodies by the Milan designer in that happy period of its existence. The car was a veteran of an outstanding motor sport season, particularly Alfonso Thiele's victory at an average of over 117 km/h in the up to 750 cc Grand Touring Class of the last Mille Miglia on 11-12 May. It won the same class another 28 times during the 1957 season right from its first event, the Rally of Sestriere on 27 February driven by Colantuono-De Sanctis to Mario Poltronieri's victory in the InterEurope Cup of 7 July, Cussini's win at the Trieste Opicina of 7 October and the Sassi-Superga of 26 October by Ernesto Prinoth. The year closed with the

At the inauguration of the 39th Turin Motor Show on 30 October, the President of Italy, Giovanni Gronchi, was welcomed by Carlo Abarth to his company's stand where, among the production vehicles, the Fiat Abarth record car stood out on its dais of honour, having broken 18 class H (501-750 cc) international speed records a few months earlier.

Fiat Abarth 750 winning the Italian Speed Championship and the Trophy of the Mountains, the latter car prepared by Aldo Faccioli, and by Lanzo Cussini in the up to 750 cc Grand Touring Class. By this time, the names Abarth-Zagato featured at the top of the results list of almost all the races in which their cars competed, and that profusion of successes did not just concern Italy but also other countries, where motor sport enthusiasts were able to appreciate the extraordinary performance and reliability of the Fiat derived 750. Carlo began to note with disappointment, however, that merit for all of those victories was being attributed by the press to Zagato and not Abarth and he became so concerned about it that he suspected there was some kind of plot against him.

That highly successful season closed on 25 October with the final records of the year being established with the Fiat Abarth Pininfarina, driven by Paul Frère and Marino Guarnieri on the high speed track of Monza. They shattered three international records in Class H for cars from 501 to 750 cc; the 200 miles in 1h 35'25" at an average 202.347 km/h, the 500 kilometres in 2h 30'36" average 199.189 km/h and the three hours covering 592.746 kms at an average speed of 197.582 km/h. As well as 1957 motor sport events in which Abarth & C. earned the kind of prestige and popularity often denied the big car manufacturers, a situation of great importance had developed, one that would make a significant mark on the company. The expansion of Abarth & C.'s various activities surged ahead following the introduction to the marketplace of the Fiat Abarth 750 in its various forms. To this is added the study, management and achievement of speed records, all operations that required not only enormous financial resources but also a substantial commitment of men and equipment. The little factory in Via Trecate was no longer able to absorb this growing volume of work and could not provide for the correct conduct and rigorous organisation of the different activities; in 1956 the number

Among the more spectacular new products on Fiat Abarth 750 mechanics unveiled by a number of body atelier at the 1957 Turin Motor Show was the 'Goccia' or drop, which certainly stood out.
It was a futuristic mono-volume coupé by Carrozzeria Vignale from a design by Giovanni Michelotti.
The ample panoramic windscreen that led to an extremely short bonnet without continuity was decades ahead of its time and anticipated one of today's widely used and accepted styling formulae.
Judged extravagant and ambitious, the Vignale design created a great deal of curiosity but was not much appreciated.

The eagerness of Abarth and Zagato to produce a roadster version of the successful 750 Coupé led to the prototype introduced at the 1957 Turin Motor Show; it incorporated many of the concepts of shape already seen in the closed version, but the rear end of the car did not harmonise well with the rest of the body. The roadster was given a lukewarm reception despite anxious public anticipation, and it seems that not even Carlo liked the car.

of employees had reached 134 and it increased further to 140 in 1957, a year in which the number of exhaust systems produced rose to 118,282 units of which 30% were exported. At the end of the year Fiat ordered from Abarth a first consignment of 100 modified Nuova 500 cars for sale by their agents under the name Fiat Abarth 500.

The year 1958 began with good fortune. An increasing number of the Fiat 600s powered by the Abarth 750 engine were appearing on the roads of Italy, easily distinguishable at the beginning by their part-open engine bay – held in place by special supports – the purpose of which was to improve the circulation

FIAT ABARTH "500" (1958)

ENGINE	
	rear, longitudinal, 2 cylinders in line
Bore and stroke	mm 66x70
Cubic capacity	cc 479
Valve gear	overhead valves, side camshaft
Number of valves	2 per cylinder
Compression ratio	10.5:1
Fuel feed	1 Weber 26IMB carburettor
Ignition	coil, distributor
Coolant	air
Lubrication	forced, oil filter on the main circuit
Maximum power	26 hp at 5000 rpm
TRANSMISSION	
	rear wheel drive
Clutch	single dry plate
Gearbox	4 speed plus reverse, central lever
Axle ratio	4.56:1
BODY	
	2 door, 2-seater saloon
CHASSIS	
	integral body
Suspension	front independent, upper trailing arms, transverse lower leaf spring, hydraulic dampers; rear independent, trailing arms, helicoidal springs, hydraulic dampers
Brakes	hydraulically operated drums front and rear
Steering	worm and sector
Fuel tank	capacity 21 litres
Tyres	125x12
DIMENSIONS AND WEIGHT	
Wheelbase	mm 1840
Front track	mm 1121
Rear track	mm 1135
Length	mm 2970
Width	mm 1320
Height	mm 1300
Kerb weight	kg 440
PERFORMANCE	
Top speed	km/h 118

of air in the compartment, a connotation that was more aesthetic than functional. But it soon became a widely diffused fashion, to the point that many adopted the system their way on any model – even normal production cars – of the popular utility. Carlo Abarth liked to show that the little two-cylinder air-cooled engine of the Nuova 500 – suitably modified to produce 20 hp – was able to offer sparkling performance and could be subjected to any load. In fact, he wanted to do more; he increased the compression ratio to 10.5:1, developed a new hemispheric cylinder head, balanced every moving part and shone all the tubing, optimising the manifold and exhaust; maximum power jumped to 26 hp, double that of the production car, and the top speed rose to 118 km/h.

Adorned on the front by a large central scorpion badge, the Fiat Abarth 500 played the leading role in a whole week – from 13 to 20 February – of exceptional demonstration tests at the Monza Autodromo. Driven by Remo Cattini, Armando Giuberti, Marino Guarnieri, Corrado Manfredini, Mario Poltronieri and Elio Zagato, the little Fiat with its standard body completed a historic marathon, covering in seven days and seven nights the distance of 18,186.440 kilometres at an average speed of 108.252 km/h with a peak in the first four days of almost 109 km/h. Initially, the news of the small car's achievement passed by almost unnoticed and only mid-way through the marathon did the press start to become interested in it, with a crescendo that grew hour by hour. Many Fiat directors attended the end of the event, which Renzo Avidano carried off majestically on the organisational and coordination of time keeping fronts. Suddenly, before the eyes of Fiat's top management, the Nuova 500 appeared in a new and different and almost unexpected light, thanks to the clamorous test of efficiency passed with flying colours by the modified Abarth engine. In subsequent months the little star of those seven days and nights was displayed in Fiat showrooms all over Europe to confirm just what that popularly priced car for the masses could do.

The extraordinary exploit at Monza finally caught the imagination of Fiat's top management, which had not demonstrated a clear understanding of the successes achieved by Abarth. Faced with that achievement and the enormous coverage given by the international press to the Fiat Abarths' success on both road and track, the top management of the great Turin factory found it difficult to participate in those successes; certainly, they did not obstruct them but demonstrated clear reluctance to be involved, at least until a couple of weeks after the Monza event of February 1958. Professor Valletta asked to meet the architect of the operation, and did so in his Corso Marconi office in the presence of the factory management. At the end of the extremely cordial meeting, permeated by congratulations and veiled prospects for the future, the great Fiat boss pre-announced that Abarth would receive a suitable indication of the company's gratitude and, two weeks later, it was delivered as addressed

Carlo Abarth's masterstroke came about during the week 13 to 20 February 1959, when the Fiat Nuova 500 with a production body and mechanics modified by the Scorpion starred in a sensational exploit on Monza's fast track. With a compression ratio of 10.5:1 and a new hemispheric cylinder head, the engine was able to produce a maximum power of 26 hp at 5000 rpm and push the little utilitarian to almost 120 km/h. In a non-stop marathon on the northern Italian circuit, the car covered an astounding 18,186.440 kilometres in seven days at an average speed of 108.252 km/h. As a result of that extraordinary but unexpected performance, the Fiat 500 Abarth raced its way into the magazines and newspapers with flying colours and even managed to unfetter Fiat top management, who had followed the epilogue of that amazing record attempt with great interest.

to Engineer Carlo Abarth, although in reality, he held no such academic title. It was a commemorative plaque that said: "In recognition of six progressive international records established in February 1958 at the Monza Autodrome with a modified Fiat 500". Valletta added with a paternalistic air that Abarth could proceed without problem with his activity.

Returning from that meeting, Carlo confided to his closest colleagues that Valetta had also added: *"If you need anything, contact me directly; you know that Fiat is behind me ... I hope you are pleased with the way Fiat is treating you ..."*. And Abarth, who was not moved by that unexpected comment, shot back, almost involuntarily, in front of the many Fiat directors present, *"I am not at all pleased"*, with words pronounced abruptly and without ceremony, which was his way. The effect of that conversation made itself felt a week later, when the commercial director of Fiat, engineer Lorenzo Brinatti,

Vittorio Valletta, the supreme commander of Fiat for over 25 years and untiring architect of the company's reconstruction after the destruction of the Second World War, fought hard for the mass production of cars and the availability of the utility automobile to the widest possible public.
It was under his guidance that the 600 (1955), the Nuova 500 (1957), the 850 (1964) and the 127 (1972) were produced – all cars with which many millions of Italian families took to the road.
It was Valetta's personal intervention in early 1958 that opened up new development opportunities for Carlo Abarth and his young company.

had a further meeting with Abarth and proposed to him a draft contract that would greatly change the destiny of his guest and the company that bore his name. In general, the document proposed that for every victory or record achieved in any part of the world with a Fiat, the company would pay Abarth a certain sum of money. A list had been compiled of the various successes achieved in order of importance, on the basis of which Fiat would have paid a more or less large figure. But there was one condition - the agreement would only concerned first places: anything less would not be taken into consideration in any way. At that moment in that office at the great Fiat Carlo did not hesitate, but evaluated immediately the enormous opportunities that his signature on that contract would bring. He thought for a moment: he would be paid only for coming first and coming second was worth nothing, so that signified he would have had to concentrate on those races in which he had every probability of winning or was certain to do so.

In reality, that agreement did not contain as many original elements as it appeared to do, at least not in Abarth's eyes. Fiat wanted to base it on a similar pact of collaboration agreed at the end of the Thirties with Enrico Teodoro Pigozzi, the great 'patron' of Simca in France, and the Italian-born techno-preparer Amedeo Gordini. Those same interests, which associated Fiat with the French factory at the time, were also expressed in the Italian group's relations with a number of other external collaborator-suppliers.

Carlo signed the contract and was forced, within certain limits, to create a factory of victories, competing not only in 750 cc class races but also the 850 cc, the 1000 cc, any race in which Abarth could start with almost a certainty of winning. In subsequent years, he developed the 595 and 695 models, with which he was certain to win and so finance his business with contributions from Fiat. In a short time, Renzo Avidano recalls, Carlo established a special office linked to the whole world that had the job of documenting for Fiat all the victories achieved by Abarth with their cars every single weekend. And since many hundreds of first places were won in a season in the years that followed, the contributions received by Abarth amounted to large sums of money, on occasions up to hundreds of millions of lire of the period That agreement allowed Abarth to run his business with a breadth of view and programmes that, up until that point, he had never been able to enjoy; it also guaranteed Fiat the Viennese technician's maximum commitment in ensuring the cars with basic Fiat mechanics would win. On the other hand, Carlo found himself limited in being able to carry out tests and participating in races with cars and classes distant from the interests which that contract protected.

But from the spring of 1958, many changes took place in the company. Fortified by a sort of insurance policy constituted by the agreement with Fiat, Abarth concentrated even more on the promotion of his own cars, in particu lar in the world of the young, where his name had become a legend. Entire generations of racing drivers as well as experienced average motorists earned their spurs at the wheel of Fiat Abarths, which were rich in performance and reasonably accessible to the pockets of many people. It was an effective school that has been denied the opportunity of continuing in recent years, a negligence from which the entire world of Italian motor sport has suffered.

With the contract signed between Fiat and Abarth & C. in the spring of 1958, the 600 'alleggerita' or lightened began to take shape: it was a normal production car, but without all the components that were not needed for modification by Abarth. Until that moment, the conditions of acquisition of cars had taken into account the complete vehicle, supplied under the same conditions as those enjoyed by any of Fiat's selling agents; at Abarth they proceeded to dismantle the components (crankshaft, drive-shaft, pistons, manifolds, carburettor ...) which were substituted so that the mechanics could be modified and then sold on the replacement market, an undoubtedly onerous operation for Abarth and one that reflected negatively on the list price of the modified car. With the new opportunity conceded by Fiat to be able to operate on the basis of the '600 alleggerita' the name 600 Derivazione Abarth disappeared and the car acquired its definitive new identity - the Fiat Abarth 750.

1958

1961

DEDICATED TO WINNING

The positive reception given to the Zagato 750 GT sports saloon, a car principally conceived for racing, induced Abarth to press the body designers until they proposed a touring version of the by now renowned car, perhaps open topped and aimed at roadster drivers. In previous years the gratifying comments made about the sleek 208 A and 215 A opened a niche in the marketplace for that kind of car. Carlo raced cars with designer bodies and pushed the men who built them into innovative inspiration; he tried to make them part of his apprehension towards research into new styling solutions. A man of refined taste in dress and the objects with which he loved to surround himself, Abarth was happy to work as a 'consultant' to the body designers. But that put him in conflict with himself, because he would not tolerate customers complaining about the finish defects or squeaks his cars revealed. In the final analysis, Abarths were built to race and, as far as their mechanics were concerned they had no equal.

The 750 GT Spider was experimentally bodied by Zagato in the autumn of 1957 and the prototype went on display at the year's Turin Motor Show,

Versatile and robust in its various and successful states of modification, the Fiat 600 engine constituted the basis of an even greater step forward.
On that mono-block Carlo Abarth had engineer Gioachino Colombo design a twin overhead camshaft cylinder head to increase the performance of the small four-cylinder 747 cc power unit.
With valves set at 40° to each other and the two camshafts driven by a double chain, the new distribution opened up the engine's enormous potential, particularly in racing; the first version generated 57 hp at 7000 rpm, but the derivative installed on the record car for the attempt of 25 October 1957 returned 61 hp, a specific power output of 81.7 hp per litre. Fuel was fed into the twin-cam engine by two twin-choke Weber carburettors.

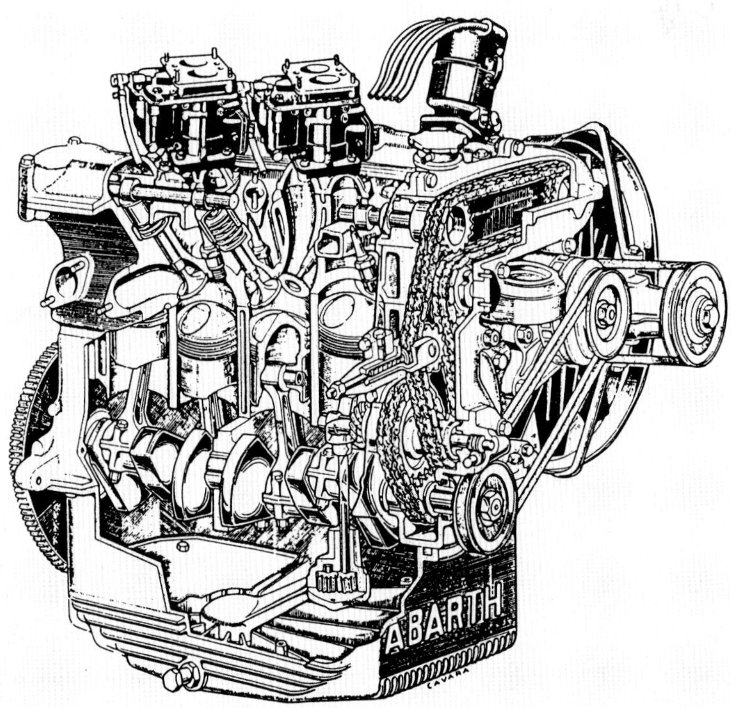

Carlo Abarth decided the mechanics revitalised by the twin-cam engine would only be used in the Zagato-bodied 750 GT, the aesthetics of which were extensively modified in the front and, in particular, the rear end, where the ample three-sectioned windows became an unmistakable feature of the car. The little coupé's dry weight was 540 kg and it had a top speed of 180 km/h. The 750 GT was officially introduced at the Paris Motor Show in October 1958 as the Fiat Abarth 750 Record Monza Zagato, price tagged Lit 2.075.000.

which opened on 30 October. The roadster by Allemano based on a design by Michelotti appeared the following year, but a front that sloped too much plus a rear end that was too high and square did not, it seems, find favour with Carlo and neither car was progressed.

Abarth knew that to transform the normal production Fiat 600 engine into the 747 cc he would need distribution by twin overhead camshafts, a solution that proved particularly effective and had evinced considerable power from the Alfa Romeo 1900's engine for the Giulietta. To fashion such a vital part of the engine Abarth contacted Gioachino Colombo, the illustrious engineer who had designed the Alfa Romeo 312, 158 and 316 in the Thirties and who had worked at Ferrari, Maserati and Bugatti after the Second World War; in 1957 he joined MV, where he designed a series of motorcycle and car engines as well as a number of turbine power units for air force use. Taking his cue from the tried and tested 747 cc block, Colombo created a twin- cam head with valves set at 40° and the distributor drive actioned by a double chain. Lorenzo Avidano still remembers the evening car journeys he made with Carlo Abarth from Turin to Milan to discuss the techno-constructional characteristics of the new head with Colombo at his home. With a compression ratio increased to 9.7:1 and a fuel feed system of two Weber 32 DCL3 carburettors, the new engine developed 57 hp at 7000 rpm and had a maximum torque of 7 kg/m at 5000 rpm; another version with two Weber 37 DCL3s produced 61 hp with a specific power of 81.7 hp per litre; that was fitted to a

The driving position, interior and ample engine compartment of the 750 Record Monza.
The car's 747 cc engine had a new twin overhead camshaft cylinder head and a new-design roof, Carrozzeria Zagato having dropped the single camshaft 750 GT's 'double hump'.

FIAT ABARTH "750 RECORD MONZA" ZAGATO (1958)

ENGINE	
	rear, longitudinal, 4 cylinders in line
Bore and stroke	mm 61x64
Cubic capacity	cc 747
Valve gear	2 overhead camshafts, chain driven
Number of valves	2 per cylinder
Compression ratio	9.7:1
Fuel feed	2 twin-choke Weber 36DCL4 carburettors
Ignition	coil, distributor
Coolant	liquid
Lubrication	forced, oil filter on main circuit
Maximum power	57 hp at 7000 rpm
TRANSMISSION	
	rear wheel drive
Clutch	single dry plate
Gearbox	4 speed plus reverse, central lever
Axle ratio	4.3:1
BODY	
	2 seat coupé by Zagato
CHASSIS	
	integral body
Suspension	front independent, upper trailing arms, transverse lower leaf spring, hydraulic dampers; rear independent, trailing arms, helicoidal springs, hydraulic dampers
Brakes	hydraulically operated drums front and rear
Steering	worm and sector
Fuel tank	capacity 45 litres
Tyres	5.20x12
DIMENSIONS AND WEIGHT	
Wheelbase	mm 2000
Front track	mm 1150
Rear track	mm 1160
Length	mm 3470
Width	mm 1350
Height	mm 1140
Kerb weight	kg 540
PERFORMANCE	
Top speed	km/h 180

record car which, driven by Paul Frère and Marino Guarnieri, set up three Class H records at an average speed of 200 km/h on 25 October 1957 at Monza.

Carlo decided that engine should only be installed in the Zagato Record Monza production car, expressly conceived for that engine and with a body that had a smooth roof instead of the double-hump; not many cars of that kind were produced with twin overhead camshafts and that meant Zagato Record Monza clients had to wait for long periods before receiving their new vehicles. The styling of the new car was improved in a number of areas compared to the Zagato 750 GT, both in the front and particularly the rear; it had an original three-part window lay-out, an ample central air intake on the bonnet, a rounded rear end and no 'little tails' at the far ends of the wings. Of identical dimensions to the 750 GT, the car's dry weight was 540 kg and it had a top speed of 180 km/h. Officially unveiled at the Paris Motor Show in October 1958 and priced at Lit 2.075.000, a semi-production run of 100 cars went ahead immediately to obtain homologation in the Grand Touring category.

Abarth was fascinated by the exciting atmosphere that went with the previous two years' successful record attempts and wanted to try again with the Fiat Abarth 500, for which Pininfarina designed a new body that was extremely well profiled and strongly inspired, some said, by the car Bertone built in 1956 for the first records established in Class H. A trellis of tubes with cross members were used for the chassis, while the two-cylinder 479 cc Fiat engine was profoundly modified (among other things, it was given a compression ratio of 10.5:1 and a Weber 36 IMP carburettor) to take maximum power up to 36 hp at 6800 rpm; the result was an absolute sensation, especially because the original Fiat engine was considered recalcitrant in increasing its revs and was only able to produce 13 hp at 4000 rpm. Suspension, steering box, wheels, wheelbase and track were derived from those of the 600, while a front air intake was built into the body, from which a long duct ran through the inside of the car to direct air into the engine compartment; a sort of aerodynamic streamlined snorkel was fitted to the rear of the body to channel air to the carburettor.

According to the regulations that governed records of that type at the time, repairs were not allowed during the attempt itself unless with spare parts carried aboard the car; the only work permitted was the substitution of wheels, tyres and electrical components – excluding the battery – as were, of course, re-fuelling and topping up with oil. The new record car was much more compact than the previous one, in that it had a new aerodynamic short tail instead of a long rear overhang with large caudal fins, which were supposed to have a directional function at high speed. The car was 4130 mm long, 1410 mm wide, 1200 mm high and weighed 368 kg, 17 kg less than the 1956 Bertone. The officially declared top speed was 178 km/h.

A cutaway of the Fiat 500 Record, designed by Pininfarina in the summer of 1958.
The car had a tubular trellis-type chassis with cross members especially designed to attempt to break speed records in Class I (from 351 to 500 cc), while the two-cylinder Fiat 479 cc engine was inclined over the rear axle.
With a compression ratio of 10.5:1 and its Weber 36 IMP carburettor, the unit developed a maximum of 36 hp at 6800 rpm.
Suspension, steering, rims and other mechanical components came from the Fiat 600.
Carburettor induction was by a snorkel-type intake on the roof, while engine-cooling air was fed in through a front intake via a long tube inside the body.
On Monza's high speed track, the 500 broke 17 international long distance records, some of which were set up at an average of over 130 km/h.

Auto-Italiana was the first to comment on the various Class I (from 351 to 500 cc) record attempts in its edition of 30 September 1958: "*On the afternoon of 22 September, record attempts were begun by Abarth with their 500 Record on the Monza high speed track. The drive started normally but a series of problems during the first few days forced a suspension of the attempts on two occasions. The first time an insufficiently dimensioned engine mount broke and on the second the halt was caused by ... rabbits, which were dazzled by the car's headlights in the middle of the track during the night.*

Continually hitting rabbits deformed the body in a number of places, so that the damage eventually interfered with several of the car's vital components.

During the second halt, the car was fitted with a stout bumper bar to protect the bodywork from such collisions.

The 500 Record is being driven by a strengthened Abarth team of about 20 drivers, among them Poltronieri, Cattini, Giuberti, Guarnieri, Toselli and Baghetti. They drove four hour stints".

The timekeepers officially started their stopwatches at Monza on 27 September and continued to log their data non-stop until 7 October in a marathon that was as punishing for the drivers as it was for the car. A string

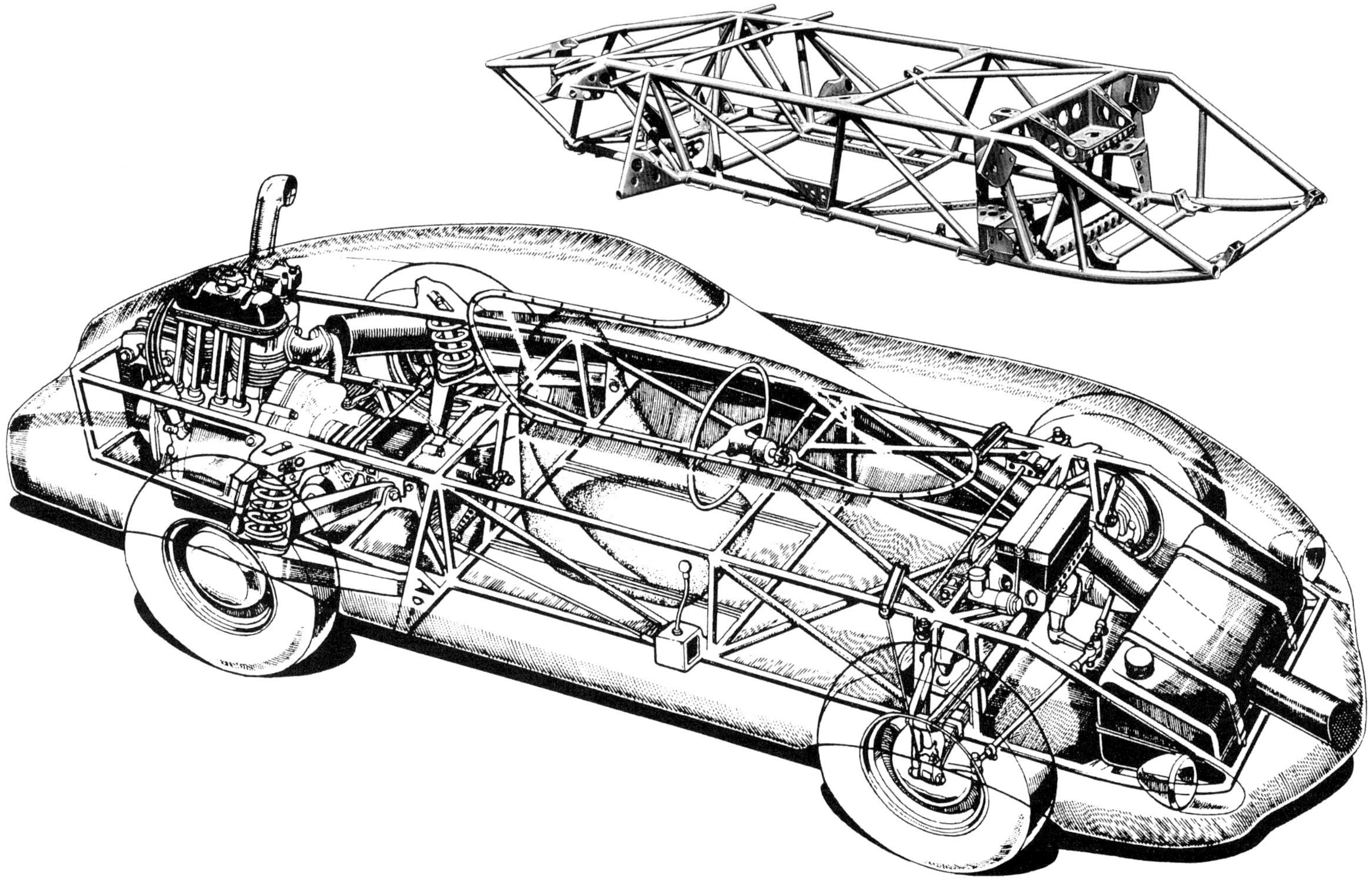

27 September-7 October 1958, Monza Autodrome: the Fiat Abarth 500 Record is about to begin its non-stop marathon in an attempt to establish new long distance records in Class I.
At the wheel in the picture below is Mario Poltronieri, one of the many drivers who did their share of the driving during the 10-day session.

FIAT ABARTH "500 RECORD" (1958)

ENGINE	
	mid-rear, longitudinal, 2 cylinders in line
Bore and stroke	mm 66x70
Cubic capacity	cc 479
Valve gear	overhead valves, side camshaft
Number of valves	2 per cylinder
Compression ratio	10.5:1
Fuel feed	1 Weber 36IMP carburettor
Ignition	coil, distributor
Coolant	air
Lubrication	forced, oil filter on the main circuit
Maximum power	36 hp at 6800 rpm
TRANSMISSION	
	rear wheel drive
Clutch	single dry plate
Gearbox	4 speed plus reverse, central lever
Axle ratio	4:1
BODY	
	single-seater record car by Pininfarina
CHASSIS	
	tubular trellis with cross members
Suspension	front independent, upper trailing arms, transverse lower leaf spring, hydraulic dampers; rear independent, trailing arms, helicoidal springs, hydraulic dampers
Brakes	hydraulically operated front drums
Steering	worm and sector
Fuel tank	capacity 36 litres
Tyres	135x13
DIMENSIONS AND WEIGHT	
Wheelbase	mm 2000
Front track	mm 1140
Rear track	mm 1150
Length	mm 4130
Width	mm 1410
Height	mm 1200
Kerb weight	kg 370
PERFORMANCE	
Top speed	km/h 180

The 23 international records established by the 500 at Record Monza were promoted in November at the 40th Turin Motor Show, where the little car with its Fiat Abarth mechanics and special Pininfarina body became a major public success.

of new international records were set up by the little Fiat, some of which were established earlier by Lloyd of Germany, others by the Abarth 500. During those 11 breathless days, the little Fiat broke the record for the 5000 kms by covering that distance in 38h 13'32" at an average speed of 130.804 km/h, the 48 hour of 6334.985 km at an average 131.978 km/h, the 5000 miles in 61h 31'52" average 130.752 km/h, the 72 hour covering 9450.787 km, average 131.286 km/h, the 10,000 kms in 76h 10'01", average 131.290 km/h, the 96 hour achieving 12,079.130 km, average 125.824 km/h, the 15,000 kms in 118h 50'35" average 126.216 km/h, the 120 hour with 15,154.958 km average 126.291 km/h, the 10,000 miles in 135h 11'28" average 119.041 km/h. In the days that followed, the plucky little car achieved the six day at 17,258.447 km average 119.850 km/h, the 20,000 km in 165h 39'39" averaging 120.728 km/h, the seven day with 20,303.508 km average 120.854 km/h, the eight day at 23,151.487 km averaging 120.580 km/h, the 15,000 miles in 119h 45'37" average 120.845 km/h, the 25,000 km in 208h 45'40' km/h averaging 119.754 km/h, the nine day covering 25,336.969 km average 117.300 km/h and the 10 day of 27,931.439 km at an average of 116.380 km/h. In all, the team set 17 long distance international records with a car that weighed less than 400 kg and was powered by a 479 cc two-cylinder engine that put out 36 hp!

A month later on 4 October a trio of drivers, Giancarlo Baghetti, Remo Cattini and Mario Poltronieri, were at it again, driving the same little car to another six international records, all previously established by the same Abarth in Class I; over shorter distances – maximum 2000 km and 12 hours – they achieved even more impressive results than during the previous month, reaching average speeds of 153.319 km/h for the 2000 km and 155.303 km/h for the 1000 miles.

Carlo Abarth welcomed the President of Italy, Giovanni Gronchi, accompanied by the president of ANFIA (National Association of Automobile Manufacturers) Rodolfo Biscaretti di Ruffia, to his stand at the 1958 Turin Motor Show, which took place from 5-16 November. Among the prominent cars in the area reserved for the new young Turin company were the Fiat Abarth 500 GT by Zagato (foreground), the Fiat Abarth 750 Spider Zagato, the 750 GT by Zagato and the Fiat Abarth 750 Record by Pininfarina (background), which had established another five international records on 25 October.

ALFA ABARTH "1000" (1958)

ENGINE	
	front, longitudinal, 4 cylinders in line
Bore and stroke	mm 74x58
Cubic capacity	cc 998
Valve gear	2 overhead camshafts, chain driven
Number of valves	2 per cylinder
Compression ratio	10:1
Fuel feed	2 twin-choke Weber 40DCO3 carburettors
Ignition	coil, distributor
Coolant	liquid
Lubrication	forced, oil filter on the main circuit
Maximum power	88 hp at 8000 rpm
TRANSMISSION	
	rear wheel drive
Clutch	single dry plate
Gearbox	5 speed plus reverse, central lever
Axle ratio	4.56:1
BODY	
	2 seat coupé by Bertone
CHASSIS	
	tubular trellis
Suspension	front independent, trailing arms, helicoidal springs, stabiliser bar, hydraulic dampers; rear rigid axle, push rod links, helicoidal springs, triangular stabiliser, hydraulic dampers
Brakes	hydraulically operated drums front and rear
Steering	worm and roller
Fuel tank	capacity 40 litres
Tyres	155x15
DIMENSIONS AND WEIGHT	
Wheelbase	mm 2160
Front track	mm 1268
Rear track	mm 1270
Length	mm 3690
Width	mm 1550
Height	mm 1200
Kerb weight	kg 640
PERFORMANCE	
Top speed	km/h 190-200

The Abarth people were all galvanised into action by those successful events. After a few days, Carlo wanted to try again, this time in Class H (501 to 750 cc) for which he had prepared a special version of the new four-cylinder twin overhead camshaft 747 cc engine that put out 75 hp at 7500 rpm and generated 100 hp per litre! Pininfarina had already designed and built a special body on a 2280 mm wheelbase, which was even sleeker than the car built the year before at a similar weight of 520 kg. Driven only by Cattini, the new version of the Fiat Abarth 750 record car took to the track on 25 October 1958 and established new international records for the 100 miles in 47'07" at an average of 204 A.924 km/h, the 200 km in 58'21" averaging 205.655 km/h, the one hour at an average 205.760 km/h, the 200 mile in 1h 33'22' averaging 206.811 km/h and the 500 km in 2h 25'04 in an average of 206.796 km/h.

On the competition front, 1958 was a triumph for the Scorpion just about everywhere. On top of the 34 international records the company established with a determination and regularity without precedent, Fiat Abarth scored 64 class wins with the 750 GT, among them at the 24 Hours of Leopoldville in the Congo and at the Asmara, Eritrea circuit on 6 January. At the 12 Hours of Sebring in Florida on 22 March, a little Fiat Abarth driven by Brumby-Aldhaus-Martinez came third in class and 30th overall in the Sports Category, while another won its class in the Grand Touring Category driven by Kessinger-West-Thiele. Success was also achieved in the Giro di Sicilia on 20 April, the Rimini-San Marino of 27 April, the second Annual Hawaiian Island Speed Week Races in Hawaii on 4 May, the Palermo-Monte Pellegrino hillclimb of 15 May, the

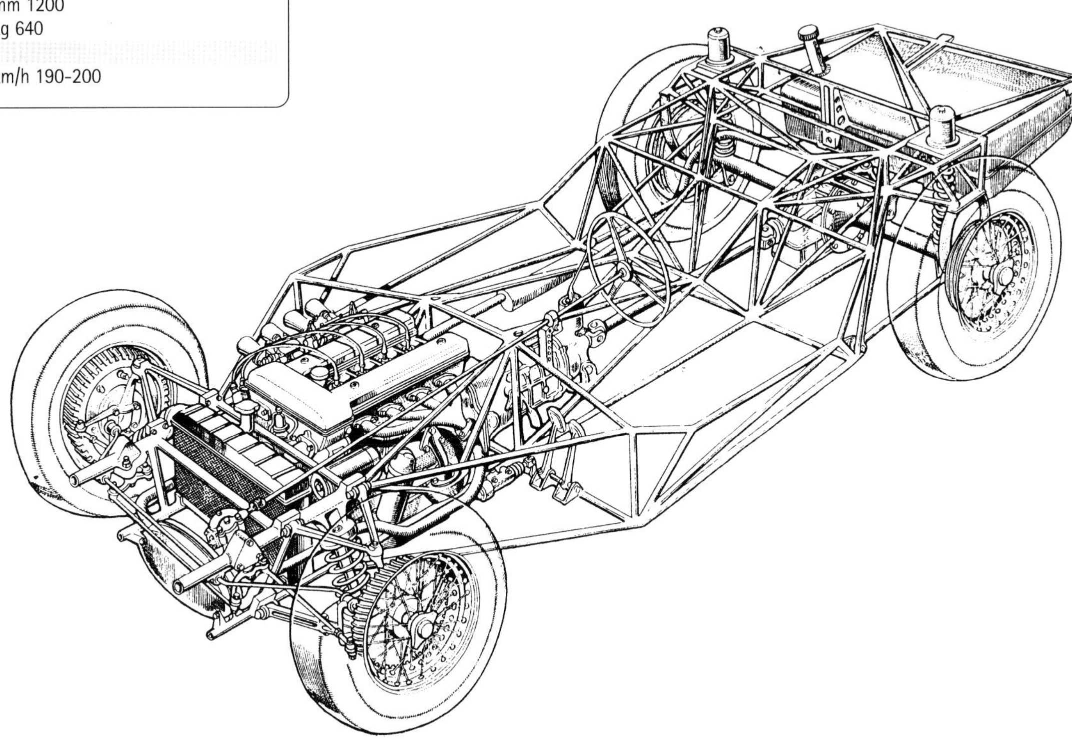

Another 1958 Turin Motor Show sensation was the Alfa Abarth 1000, which Carlo decided to build on a tubular trellis chassis instead one of boxed pressed steel. The project was assigned to Mario Colucci of Alfa Romeo. By means of a considerable reduction in stroke from 75 to 58 mm and the maintenance of the car's 74 mm bore, the capacity of the Giulietta's engine was reduced to 998 cc; with a compression ratio of 10:1 and two twin-choke Weber 40DCO3 carburettors, the engine put out 88 hp at 8000 rpm.

Bertone designed the Alfa Abarth 1000's body and the car received critical and public acclaim on the atelier's stand. It was the first time such an elegant, compact 1000 cc coupé had been seen, its length 3690 mm, height 1200 mm, weight 640 kg – and its top speed almost 200 km/h. But the car was not heard of again after the '58 Turin Show.

Consuma Cup and the Rally of Sardinia on 1 June, the Six Hours of Rome on 22 June, the InterEurope Cup on 7 September and even in the Rally of Angola and the Grand Prix of Venezuela of 9 November. The 1958 motor sport season concluded with four overall victories and three Italian Championship titles won by Ovidio Cappelli with the Fiat Abarth 500 GT prepared by Carlo Facetti, Vittorio Feroldi De Rosa in a Fiat Abarth 750 GT prepared by Gino De Sanctis and Ernesto Prinoth, who won the Trophy of the Mountains in the up to 750 cc class of the Grand Touring Category driving a Fiat Abarth 750 GT prepared by himself.

The year came to an end with a final conquest at the Turin Motor Show from 5-16 November, when Abarth unveiled another new car that had been eagerly awaited and surprised with its project layout and harmonious body created by Bertone. Carlo approached Alfa Romeo determined build a one-litre engine derived from that of the Giulietta Sprint Veloce by a generous reduction of bore from 75mm to 58 mm, while its stroke would stay at 74 mm. The new over-square 998 cc engine was modified to include a number of changes inside the block and twin-cam head; the compression ratio was increased from 9.1:1 to 10:1 and the fuel feed was by two horizontal Weber 40 DCO3 carburettors. Maximum power obtained in bench tests was 88 hp at 8000 rpm, an extremely high value for the period that was made possible by the modern and sophisticated design of the basic engine. Mario Colucci of Alfa Romeo produced the robust yet light 50 kg chassis with a trellis of steel tubes, which enabled Abarth to abandon the boxed steel formula that he had been using for a number of years since the days of the 204 A sports saloon. The other mechanical components – transmission, suspension, brakes – were strict derivations of the Giulietta, suitably adapted to the new demands that would be placed on them. Built on a wheelbase reduced from 2386 mm to 2160 mm, the Alfa Abarth 1000 GT was 'dressed' by Franco Scaglione and its body built by Carrozzeria Bertone; the project resulted in an exceedingly compact car 3690 mm long, from its lengthy sloping bonnet that blended well into the two front wings and almost without rear overhang, to the tail; it was only 1200 mm high and built entirely in aluminium to cut weight to the bone at 640 kg – 255 kg less that the Giulietta Sprint Veloce production car, which was also designed by Bertone. It was an attractive car that needed a long and precise gestation period, yet even having received much critical and public acclaim and with no other one-litre road car able to reach 200 km/h at the time, it still just slipped into oblivion immediately after the show. But no one ever knew why, after having closely followed the development and construction of such a promising prototype, Carlo Abarth abandoned it to its destiny. Some thought the attempt to come up with a 1000 cc sports coupé derived from an Alfa Romeo was not smiled upon by Fiat's top management, who held that the 'protection contract' agreed with Abarth a few months earlier enabled Mirafiori to enjoy

Carlo Abarth always maintained good relations with the motoring press, which made a substantial contribution to publicising and sustaining his initiatives in the technical and sports fields.
In these two pictures, Abarth is illustrating the technical characteristics of the 750 cc twin-cam engine on the dynamometric test bench to Gordon Wilkins of Britain (left) and Paul Frère of Belgium.
Every single Abarth engine was tested on the bench before being installed in a car.

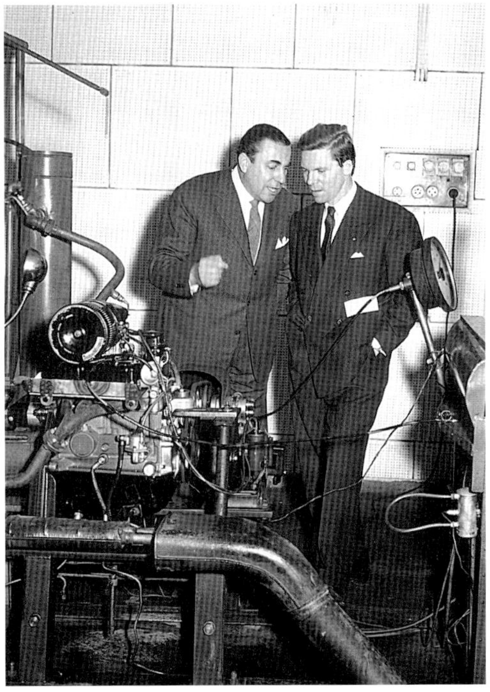

Contact with the foreign press was aimed at promoting to enthusiasts outside Italy the brilliant performance of the little Abarth 750, derived directly from a utility car par excellence which, at the time, was the Fiat 600.

In this picture, well-known British journalist Gordon Wilkins is at the wheel of a Fiat Abarth 750 Record Monza, while exchanging points of view with Abarth works driver Mario Poltronieri.

The feats of the little Zagato-bodied Abarth filled the sporting press of the period.
During the course of 1958, the 750 class of the Grand Touring category witnessed the undisputed supremacy of the little 'double hump' coupé, with its single-cam 43 hp engine.

Vittorio Feroldi De Rosa became Italian speed champion and Ernesto Prinoth won the Hillclimb Trophy at the wheel of the tiny car.
But in 1959 the Zagato was unable to stand up to the more potent Record Monza, which began its long and glorious career that year.

the exclusive services of the Viennese technician. Others speculated that perhaps Alfa Romeo themselves preferred not to open a breach in the marketplace that would see the new one-litre coupé competing against the sports versions of the Giulietta, which was beginning to achieve full commercial and sporting success in many countries.

At the end of that important year for Abarth & C. the production of special components was transferred to a new factory at 38 Corso Marche, increasing the company's total manufacturing area to 7735 square metres and the number of employees to 156. That was not only a physical injection of vital

At first, coming up with an open car derived from the 750 GT Zagato, which was enjoying such success among sporting customers, seemed quite a simple task. But neither the first example in 1957 nor the subsequent version shown by the Milan carrozzeria at the Turin Motor Show a year later (above) met with public approval, nor did they illicit favourable comments from Carlo Abarth.

So the Viennese-born technician gave the job of building the coupé to Serafino Allemano, who did his best to interpret a design by Giovanni Michelotti, but without much success (below); the car had a high and bulky bonnet, while research to visibly lighten the front using bumper bars of a different shape did not lead to a convincing result. Neither car was adequately commercial.

space but also proof of faith in the business that the company carried out with success and profit. In 1958, exhaust system production reached 138,901 units, of which 31% were exported.

Abarth company life was frenetic, without pause, driven by an unstoppable string of sports successes and the growing demand for Abarth cars, in particular the Zagato-bodied Fiat Abarth 750 Record Monza. Strengthened but not over-assured by such results, Carlo ran the factory with a hard scowl and sometimes an iron fist. He was surrounded by a few trusted colleagues, with whom he passed long hours discussing in his no-nonsense direct style the various problems that daily management brought to his desk. By his side was the ever-present Evangelina Isandoro, who had worked for him since the Cisitalia days and who proved herself to be both efficient and devoted in exemplary fashion over the years. Renzo Avidano was another man from the early days, who did everything in those heroic years, from the logistical organisation of

A group shot taken in the autumn of 1958 at the opening of the new Abarth & C. headquarters at 38 Corso Marche in Turin. Lined up against the wall from right to left are two Fiat Abarth 750 Spiders by Allemano plus a number of 750 Record Monzas and 750 GTs by Zagato. At the time, Abarth employed 156 people.

race and record attempts to car and exhaust sales and press relations at several prototype test sessions; it was normal for him to spend most of his time in Carlo's office rather than his own; the two men continually exchanged views and came up with alternative solutions to the various problems of the business. The relative language barrier sometimes acted as a damper to blunt the harshness of Abarth's uneasy temperament. He spoke to his colleagues in formal Italian since he, with no deep knowledge of the language, found it easier and closer to the German form of address. But he found it hard to tolerate the fact that his employees were not punctual and difficult to understand why they went home in the evening, leaving unfinished work instead of remaining to finish the job at hand. In that area, he was the first to give an example, often continuing to work with alacrity late into the night and enjoying being joined in those marathons by his closest colleagues. He liked to dress well and was con-

During the course of the 1958 Turin Motor Show, Carlo was visited in his new offices by Franklin Delano Roosevelt Jr, third child of the celebrated Thirties/Forties American president.
The meeting between the two entrepreneurs led to a commercial agreement stipulating that the Roosevelt Automobile Company Inc. was to be a distributor of Fiat Abarth cars on the east coast of the United States and the Roosevelt Team was to be formed to compete in a number of races on the other side of the Atlantic with the 750 Record Monza.
The large picture, autographed by Roosevelt on 12 November 1958, shows a symbolic shaking of hands between the two men. In the small picture, Roosevelt is shown talking to Mario Poltronieri (centre) and Remo Cattini.

sidered an 'arbiter elegantarium', always wearing a jacket and tie, even when he went down to the shop floor; his shirts, which he would change up to three times a day, were always crisp and immaculate and carried the initials K.A. instead of C.A., revealing a sort of coquettish nostalgia for his Viennese origins; he chose his accessories with care, good taste and able coordination and the top pocket of his jacket always sported a spotless white linen handkerchief.

Even to his closest colleagues, Carlo normally dictated brief and laconic internal communications as he subscribed to the simple principles of life and often said: *"He who does not know how to dictate does not know how to command"*. The gift of command was undoubtedly part of his DNA, even if it was never ostentatious or instrumented. Many of those who worked with him remember that, in spite of his gruff attitude and occasional unpleasantness, his employees admired him because they considered him right in what he said, impartial in his evaluations and gifted with great talent both as a technician and businessman. He knew how to transmit energy to encourage others to do better, to continually create means by which to achieve ever better results. He was driven by an unyielding work ethic that he manifested with great strength of character, unworried whether or not it was Saturday, Sunday, Christmas or New Year's Eve. And each one of his employees was infected with this daily approach, his formula for pursuing success being to insist, always insist.

Interest in small, very fast Italian cars was about to spread across the Atlantic in a big way, especially following an agreement between Carlo Abarth and Franklin Delano Roosevelt Jr., who at the time was a Fiat dealer in 17

Opposite: Three views of the record car model designed by the German engineer Leopold Schmid from an idea by Carlo Abarth, conceived to break the 1000 km/h barrier on the Utah Salt Lake Flats in America. Powered by a Bristol Orpheus BOr3 jet engine that would produce 2200 kg of thrust, the car comprised a central fuselage in which there was the cockpit and engine bay, plus two thin lateral housings containing the four enormous 1800 mm diameter wheels (section below) fitted with solid tyres, each of which had a long rubber damping element running inside the rim's Elektron structure.

The fundamental characteristics of the record car were:
Wheelbase: 4000 mm
Track: 2800 mm
Travel of the damping elements inside the wheels: +/- 80 mm
Turbine engine: Bristol Orpheus BOr3
Thrust: 2200 kg
Power at 1000 km/h: 8150 hp
Maximum rotating revolutions: 9600 rpm
Engine's dry weight: 368 kg
Hydraulically operated brakes on all four wheels
Fuel tank capacity: 800 litres
Frontal area: 1.7 square metres
Running weight (with driver): 1950 kg
Maximum speed projected: 1000 km/h.

states on the east coast of the United States. Roosevelt, third child of the celebrated U.S. president who, with his New Deal, had saved the country from its deep crisis after the 1929 Wall Street crash, was also a politician and sportsman as well as the owner of a racing car team. He wanted to be convinced of the potential of Abarth & C. and visited Turin in the second half of 1958 to meet Carlo Abarth and see his factory. The two men liked each other immediately, each fascinated by the other's personality. They prepared a plan for the sale of Fiat Abarths on the east coast of the States and a programme on the basis of which to compete in a number of races. They also discussed the possibility of developing a car capable of making an attempt on a world record in the 10,000-mile category at Daytona. With that agreement, Carlo was able to set up an American export operation of adequate business structure and would no longer become involved in sporadic initiatives, which happened and then died in a few months on the emotional machinations of one or two ephemeral salesmen. As for motor sport, the Roosevelt Team would shoulder the cost of competing in races; the cars, which would be sent by Abarth from Italy and would be sold locally at the end of the season or returned to Turin. It was also agreed that the cars would compete in Italian racing red, while cars bought by either private entrants or teams such as Cunningham would race in the American white with a central blue line.

The unconfessed interest of Carlo in the United States was also dictated by the extraordinary fascination he had for absolute speed records, which he hoped to achieve one day by following in the footsteps of Malcolm Campbell, his son Donald and John Cobb. Abarth sent Avidano to the U.S. so that he could choose the most suitable location at which to make their record attempts with a car capable of reaching and perhaps exceeding 1000 km/h; the salt lake flats at Bonneville, Utah, already the scene of similar attempts, were selected. Preliminary study of the car was entrusted to the German engineer Leopold Schmid of Stuttgart, who conceived a vehicle of three volumes, the central one of which was inspired by the shape of an aircraft's fuselage that hosted the driving position and an enormous Bristol Orpheus BOr3 turbine engine with a 2200 kg boost, able to accelerate the car to 1000 km/h with its monstrous 8150 hp. In the two side compartments, which were completely fared and had low frontal sections, were the four imposing 1800 mm diameter wheels without tyres and with suspension components incorporated into the hub; between the front and rear wheels were two thin fuel tanks with a total capacity of 800 litres. Among the fundamental characteristics of this unusual and innovative vehicle were a wheelbase of 4000 mm, a track of 2800 mm, a frontal section of 1.7 square metres and a total ready to roll weight – including the driver – of 1950 kg. The preliminary design of the Abarth record car was concluded at the end of March 1959, but the operation's huge financial cost – far greater

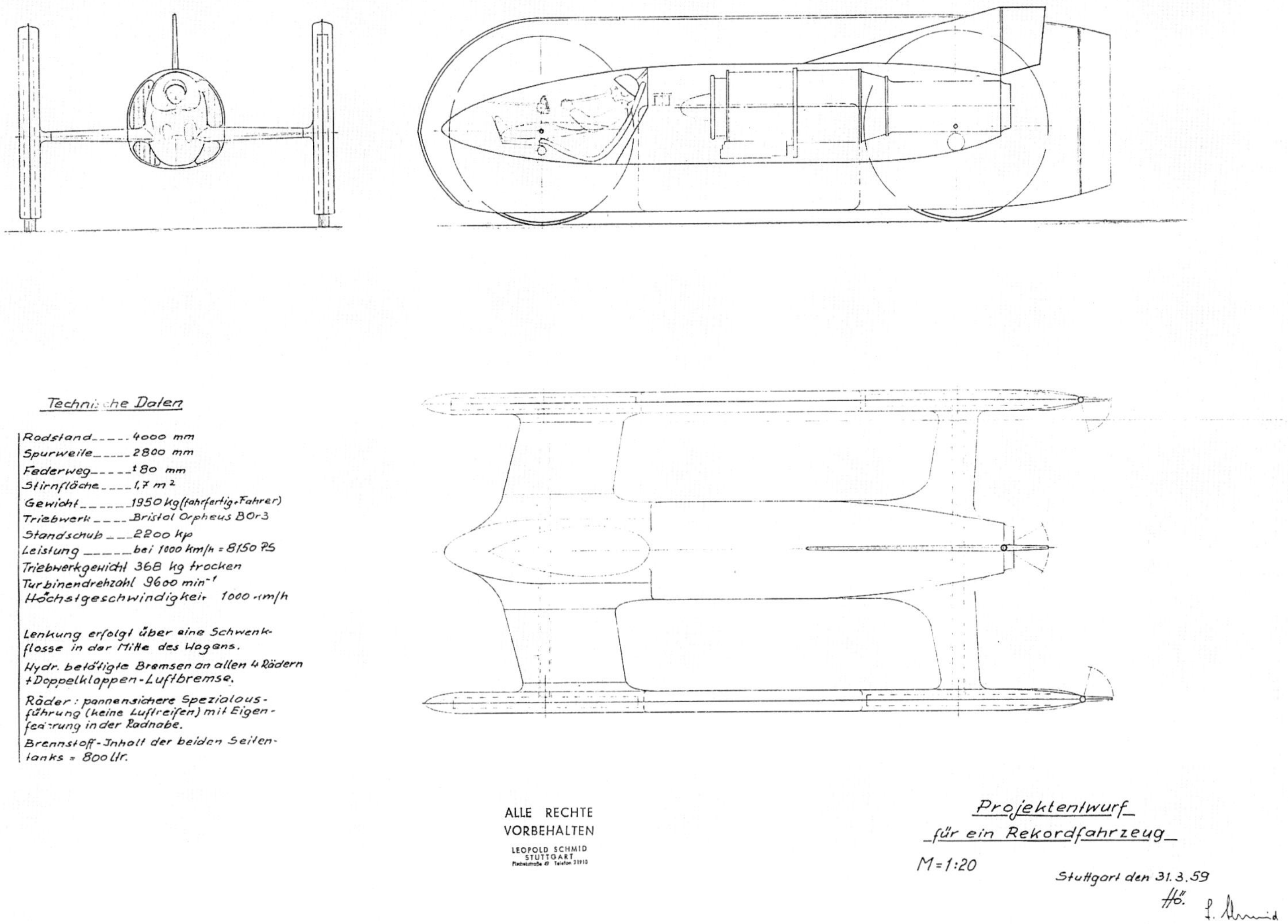
Technische Daten
Radstand ..... 4000 mm
Spurweite ..... 2800 mm
Federweg ..... ±80 mm
Stirnfläche ..... 1,7 m²
Gewicht ..... 1950 kg (fahrfertig + Fahrer)
Triebwerk ..... Bristol Orpheus BOr3
Standschub ..... 2200 kp
Leistung ..... bei 1000 km/h = 8150 PS
Triebwerkgewicht 368 kg trocken
Turbinendrehzahl 9600 min⁻¹
Höchstgeschwindigkeit 1000 km/h
Lenkung erfolgt über eine Schwenkflosse in der Mitte des Wagens.
Hydr. betätigte Bremsen an allen 4 Rädern + Doppelklappen-Luftbremse.
Räder: pannensichere Spezialausführung (keine Luftreifen) mit Eigenfederung in der Radnabe.
Brennstoff-Inhalt der beiden Seitentanks = 800 Ltr.
ALLE RECHTE VORBEHALTEN
LEOPOLD SCHMID STUTTGART
Projektentwurf für ein Rekordfahrzeug
M = 1:20
Stuttgart den 31.3.59

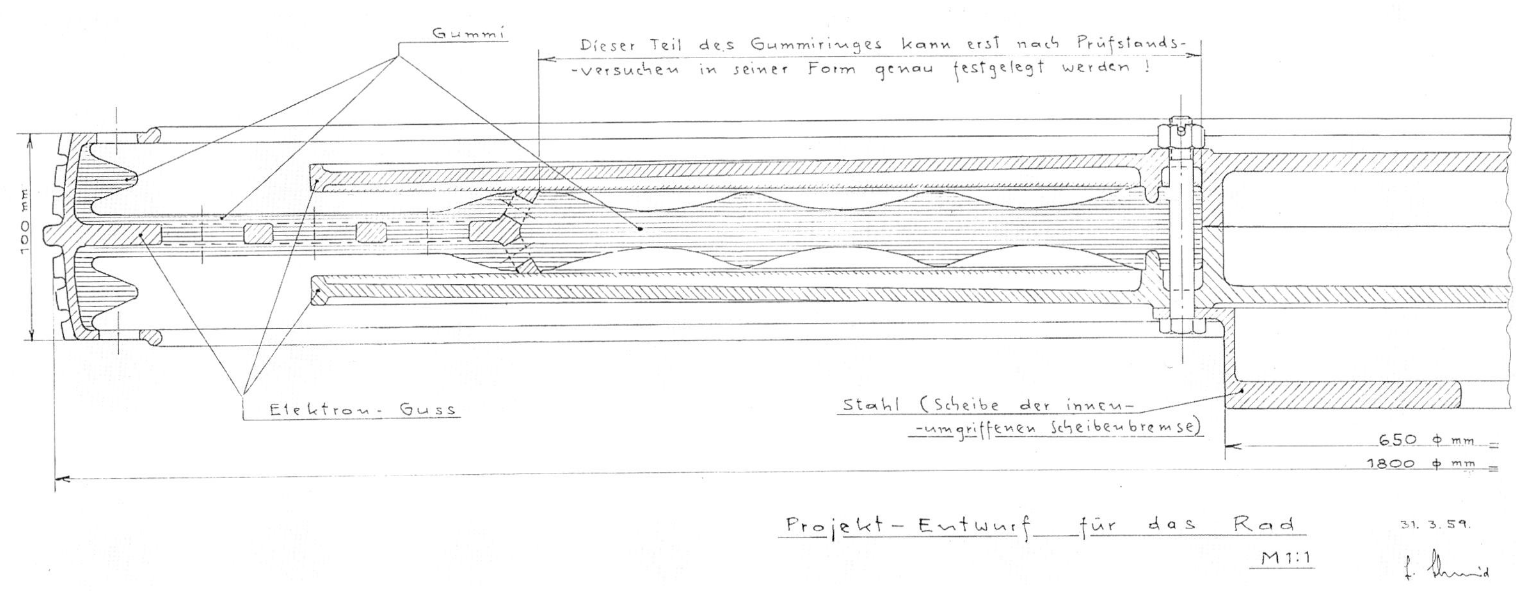
Gummi
Dieser Teil des Gummiringes kann erst nach Prüfstandsversuchen in seiner Form genau festgelegt werden!
Elektron-Guss
Stahl (Scheibe der innenumgriffenen Scheibenbremse)
650 ϕ mm
1800 ϕ mm
Projekt-Entwurf für das Rad
M 1:1
31. 3. 59.

Partial view of the Roosevelt Automobile Company Inc. headquarters, already in operation in the mid-Fifties as a Fiat distributor in 17 states on the east coast of America.
For promotional purposes, the picture shows a 'double hump' Fiat Abarth 750 GT Zagato in the centre, plus two 750 Spiders with bodies by Allemano.
Under the agreement of November 1958 between Carlo Abarth and Franklin Delano Roosevelt Jr, Abarth & C. secured for itself a suitable commercial presence in the U.S.A. and the techno-logistical contribution of a racing team already active in the States.

than the available resources of the company – suggested Abarth should await more favourable times before going ahead and building the car.

The Scorpion stung its opponents every which way during the 1959 racing season thanks to the Fiat Abarth 750 GT and Corso Marche rapidly became that factory of victories. Starting with the Rally of Sestrière on 26 January, won outright by Giancarlo Castellina and Piero Frescobaldi, who also took the up to 750 cc class of the Grand Touring Category; then came the Belmonte Cup on 15 March, when the Corso Marche factory won both the 500 GT and 750 GT classes. A week later came the great 12 Hours of Sebring in which the Roosevelt Team, not long established under the direction of John Norwood, made its debut; among the drivers Norwood engaged immediately were Paul Richards and Ray Cuomo and mechanic Jim McGee. On that occasion Carlo played a trump card by sending four 750 Record Monzas to the States plus a complete team comprising Renzo Avidano as sports director, and drivers Remo Cattini, Lanzo Cussini, Mario Poltronieri and Alfonso Thiele as well as mechanics Cornelio Maffiodo and Alberto Guerrato. Much was expected of the

One of the first events of the 1959 season was the exciting debut in America of four Fiat Abarth 750 Record Monzas in the 12 Hours of Sebring, entered by the Roosevelt Team. The picture shows the cars ready for shipment to the United States.

Commitments at Sebring, Florida, demanded a full team of seven people, shown on the steps of an aircraft bound for the United States. They are, from left to right: Lanzo Cussini, Alfonso Thiele, the two mechanics Cornelio Maffiodo and Alberto Guerrato, Remo Cattini, Mario Poltronieri and sports director Renzo Avidano. Four Fiat Abarth 750 Record Monzas were entered for the 21-22 March race, two for American crews Bill Rutan/Ray Cuomo/Paul Richards and Robert Kuhn/Jim Jeffords/George Schrafft. The all-Italian car was driven by Lanzo Cussini/Remo Cattini and the fourth Italo-American racer by Mario Poltronieri/Alfonso Thiele/John Norwood.

A converted military airfield, in 1959 the Sebring circuit was 5.2 miles or 8.369 kilometres long but now runs for 3.7 miles.
In spite of bad weather that raged for half the race, the four Fiat Abarth 750 Record Monzas scored an amazing team success, taking the first four places in the results list by category in this order:
1. Car number 64, driven by Rutan/Cuomo/Richards;
2. Number 62 with Cussini/Cattini (picture);
3. Number 65 driven by Kuhn/Jeffords/Schrafft;
4. Number 63 with Poltronieri/Thiele/Norwood.
Class winner number 64 came 28th overall after having covered 152 laps or 1416.088 kms in the 12 hours at an average of 105.976 km/h.
The American debut of Fiat Abarth and the Roosevelt Team was a great one, way beyond their most ambitious expectations, and created great opportunities for the little Italian cars in the U.S. market.

little Italian cars, which arrived almost unknown but determined to achieve success in the U.S.A. The golden prize was not only a place of honour in the final race results, but also a return to notoriety that the success of the Fiat Abarth would have generated in a market like the United States, rich and open to all types of exotica.

The race took place on 22 March in heavy rain that flooded the track, causing problems for all competitors - particularly the low-slung little Abarth Zagatos, which were swamped by walls of water while overtaking on the long straights of the circuit. From the clear memory of Avidano one learns today that he, too, had to roll up his trousers to get to the cars during pit stops and load the drivers on his shoulders to take them to their rest area. At the end of the race, the American team Rutan-Richards-Cuomo won its class; second came Cussini-Cattini, third another Fiat Abarth driven by Kuhn- Jeffords-Schrafft and fourth Poltronieri-Thiele-Norwood. It was an absolute triumph and the whole Abarth team hit the American media's headlines. It was impossible to neglect such an outstanding opportunity with which to begin a promotional operation in support of the cars from Turin, the local organisers maintained; they warmly supported and insisted on the team entering the Daytona Beach races on 5 April. Avidano, who obtained Carlo Abarth's agreement to stay on in the States until that date, sent Poltronieri back to Italy because he was due to compete in the 12 Hours of Monza a few days later; then the sports director readied himself to play his second hand on American soil. Remo Cattini came first in his category and ninth overall on the celebrated Daytona banked track, repeating the team's earlier clamorous success.

The 750 Record Monza was a real runaway success in those months of 1959. Mario Poltronieri-Elio Zagato won the first Ascari Trophy – the 12 Hours of Monza - on 11-12 April, Massimo Leto di Priolo the Rimini-San Marino on 25 April, Fischer the Wallberg, Germany hillclimb on 10 May, Carini-Ernesto Prinoth the Trento-Bondone on 12 June, Prinoth again the Westwood, Canada, circuit on 15 August, the 8 Hours of Lime Rock, U.S.A. two days later, Thiele the InterEurope Cup on 13 September, Carlo Fabi the Paris Cup on 27 September, the Circuit of Innsbruck, Austria on 4 October and, finally, the Vaduz-Triesenberg hillclimb on 11 October. Abarths scored 10 overall victories and won their class a remarkable 96 times; Prinoth won both the Italian Speed Championship and the up to 750 cc class of the Mountain Trophy's Grand Touring Category.

In the United States the Roosevelt Team's Paul Richards scored a series of 17 victories in regional and national competitions at the wheel of a Fiat Abarth 750 Record Monza. The team's Ray Cuomo recalled how the Bialbero was a little heavier than the 750 GT single-cam, but incredibly fast: *"The factory confirmed there was a 10 hp difference of power between the two engines, but at the wheel it seemed the difference was at least double; the rods and rocker arm engine could reach a maximum of 100 mph, while the twin camshaft went up to 125 mph. But the real weak point was the brakes, because they overheated so much they melted the linings on the drums making liquid leak from the hydraulic system; we got to the end of the race without a drop in the brake pump".*

In the spring of 1959, Abarth & C. decided to increase its registered stock from Lit 998.000 to Lit 1.996.000 and four months later another increase took the total to Lit 50 million. During an extraordinary board meeting in

After just a few races, the new Fiat Abarth 750 Record Monza became the 1959 season's car to beat. At the St. Ambroeus Cup on the full Monza circuit – including the high speed section – on 3 May, Massimo Leto di Priolo won in car number 124, followed respectively by Ernesto Prinoth, Carlo Fabi, Aristodemo Sala (car number 128), all driving the same model. Leto di Priolo's winning time was 37'34.3" and his average speed 159.694 km/h. The picture shows the competitors before being flagged away.

FIAT ABARTH "850 SCORPIONE" (1959)

ENGINE	
	rear, longitudinal, 4 cylinders in line
Bore and stroke	mm 62x69
Cubic capacity	cc 883
Valve gear	overhead valves, side camshaft
Number of valves	2 per cylinder
Compression ratio	9:0
Fuel feed	1 Weber 32IMPE carburettor
Ignition	coil, distributor
Coolant	liquid
Lubrication	forced, oil filter on the main circuit
Maximum power	52 hp at 6000 rpm
TRANSMISSION	
	rear wheel drive
Clutch	single dry plate
Gearbox	4 speed plus reverse, central lever
Axle ratio	4.56:1
BODY	
	2 seat coupé by Allemano
CHASSIS	
	integral body
Suspension	front independent, upper trailing arms, transverse lower leaf spring, hydraulic dampers; rear independent, trailing arms, helicoidal springs, hydraulic dampers
Brakes	hydraulically operated drums front and rear
Steering	worm and sector
Fuel tank	capacity 30 litres
Tyres	5.20x12
DIMENSIONS AND WEIGHT	
Wheelbase	mm 2000
Front track	mm 1150
Rear track	mm 1160
Length	mm 3600
Width	mm 1420
Height	mm 1190
Kerb weight	kg 600
PERFORMANCE	
Top speed	km/h 160

Turin on 11 June, it was agreed to transform Abarth limited liability to a limited company; Armando Scagliarini was nominated chairman, while Carlo Abarth and Carlo Scagliarini were appointed joint managing directors. It was also decided that year to close the Via Pacchiotti factory and open a new department at the Corso Marche plant. The manufacturing units of the company were finally united in the one purpose-built complex covering a total of 9313 square metres. Meanwhile, the number of employees had risen to 190.

The spring 1958 agreement between Abarth and Fiat was revealing itself to be most gratifying for both parties but Carlo, far from being satisfied in terms of prestige and prize money, had thought of just about nothing other than to intensify the motor sport activity. He returned to Monza on 3 August with the Fiat Abarth 500 record car, which had been updated over the previous year. A few days earlier, the car had overturned in a frightening accident at the track due to a problem with the left rear wheel; it was destroyed but, miraculously, driver Marino Guarnieri emerged unhurt. So Tony Sailer, the famous world champion skier engaged by Carlo to drive that record attempt as a publicity ploy, was unable to make his debut as a racing driver. With Giancarlo Baghetti and Mario Poltronieri at the wheel, the car returned to Monza to establish five new Class I (351 to 500 cc) records: the 500 miles in 4h 51'18" at an average speed of 165.740 km/h, the 6 hour covering 993.953 km at an average 165.658 km/h, the 1000 km in 6h 02'02" average 165.652 km/h, the 1000 miles in 10h 04'43" averaging 159.677 km/h and the 12 hour

Carlo Abarth's need to have an elegant small coupé in the marketplace, together with the decidedly racing-type 750 GT by Zagato of Milan, was finally met by a Giovanni Michelotti's design, admirably interpreted by Allemano, which was perhaps the finest designer body ever created on the floor pan of the Fiat 600.
Abarth first devised a bigger capacity 833 cc (52 hp at 6000 rpm) engine for the car, but later came up with an 847 cc unit that put out 57 hp at 6500 rpm and named it the name 850 Scorpion.
It was given a sneak preview at the Frankfurt Motor Show in September 1959, and was officially launched a few weeks later at the Turin Show.

of 1.923,330 km average 160.277 km/h. All the records they smashed had been previously established by Abarth on 21 October 1957. The achievement that surprised the technical and specialised press world alike was the astounding 165 km/h average over 1000 km in six hours. In the meantime, a new situation was taking shape in the medium-superior class of the small car market. The Fiat Abarth 750's sports successes had only a limited effect on demand for normal use, which attributed to the Zagato coupé characteristics that were too racing oriented in both its mechanics and body. The car was undoubtedly basic in its assembly, of spartan interior and with a special vocation due to its parasitic noises, squeaks, draughts and modest cockpit insulation. The needs of the customer who wanted a compact car that was nervous and fun to drive in a package that reflected a modicum of standard but essential driving comfort, accessories and build quality remained unsatisfied. Always sensitive to such indications and quick to decide, Carlo made those points to the Allemano body builders who, based on a sketch by Giovanni Michelotti, built a prototype coupé on a Fiat 600 floor-pan, which immediately appealed to the good taste of Carlo Abarth; he liked the car at first sight, considering it the most equilibrated and elegant ever devised for the popular Fiat utility. In effect Michelotti outdid himself when he designed that body, which is still current today as far as style is concerned. And Serafino Allemano, who in those days stood out among the many emergent craftsmen in that passionate business that bristled with ideas, interpreted the Michelotti design and Abarth's suggestions marvellously. Carlo concluded that to confer greater dignity on the coupé, it would be useful to give it a bigger engine than the 750 cc and began work on increasing the four-cylinder 600's power output to 833 cc, with a single camshaft able to generate a maximum of 52 horse power at 6000 rpm; as production was drawing to a close an 847 cc version also went on offer that produced 57 hp at 6500 rpm.

Out of the project came a highly spirited and pleasant car that was extremely elastic and easy to drive, particularly suited to exacting clients who were not accustomed to demanding racing cars in terms of performance. With a dry weight of 600 kg and a top speed of 160 km/h, the Fiat Abarth 850 Scorpione by Allemano made its preview debut at the Frankfurt Motor Show in September 1959 and had its official coming out at the Turin Show two months later, when the car was priced for the home market at Lit 1.550.000. By comparison, at the time the standard Fiat 600 saloon cost Lit 625.000 and the new 2100 saloon with its 95 hp six-cylinder engine cost Lit 1.560.000. A few months later the 850 Scorpione roadster appeared with the name Riviera.

In 1959, collaboration with Allemano also gave birth to more ambitious cars than the successful Fiat Abarth 500 and 600 models. The Italian market was timidly opening up to cars with a cubic capacity of around two litres and

FIAT ABARTH "2200" (1959)	
ENGINE	
	front, longitudinal, 6 cylinders in line
Bore and Stroke	mm 79x73.5
Cubic capacity	cc 2162
Valve gear	overhead valves, side camshaft
Number of valves	2 per cylinder
Compression ratio	9.5:1
Fuel feed	3 twin-choke Weber 40DCOE carburettors
Ignition	coil, distributor
Coolant	liquid
Lubrication	forced, oil filter on the main circuit
Maximum power	135 hp at 6000 rpm
TRANSMISSION	
	rear wheel drive
Clutch	single dry plate
Gearbox	4 speed plus reverse, central lever
Axle ratio	3.6:1
BODY	
	2+2 coupé by Allemano
CHASSIS	
	integral body
Suspension	front independent, deformable quadrangle, torsion bars, hydraulic dampers; rear rigid axle, push rod links with longitudinal semi-elliptic springs, helicoidal springs, hydraulic dampers
Brakes	hydraulically operated discs front and rear
Steering	worm and roller
Fuel tank	capacity 59 litres
Tyres	6.40x14
DIMENSIONS AND WEIGHT	
Wheelbase	mm 2450
Front track	mm 1340
Rear track	mm 1370
Length	mm 4610
Width	mm 1620
Height	mm 1280
Kerb weight	kg 1050
PERFORMANCE	
Maximum speed	km/h 197

beyond; to compete with Alfa Romeo and Lancia, who had just introduced the 1900 and 2000 saloons and the Aurelia and Flaminia, Fiat decided to join the fray. They did so with a modern and bright 1800/2100 saloon of angular shape and a little unprejudiced, powered by a six-cylinder in line overhead valve engine available in 1795 and 2054 cc versions with a maximum power output of 75 and 82 hp respectively. In two years, around 30,000 units were produced.

Carlo Abarth worked on those unfamiliar mechanics to create a car able to compete against the best sports cars of the period. He gave the engine a slight increase in cubic capacity to 2162 cc, by increasing the bore from 77 to 79 mm and worked considerably on the inlet and exhaust manifolds; he raised the compression ratio from 7.8:1 to 9.5:1 and fitted three twin-choke Weber 40 DCOE carburettors, so that the car's maximum power output jumped to 135 hp at 6000 rpm and torque to 18 kg/m at 3800 rpm. The gear lever was moved to the floor and the braking system was given discs on all four wheels. The normal production saloon's wheelbase was reduced by 200 mm to 2,450 mm to take a 2+2 coupé body, which was built by Allemano to Michelotti's design. The noted Turin stylist designed an extremely sober and slender car with a low, wide radiator grill at the centre of which was the large Scorpion badge; at the time, few other cars could display such formal equilibrium and moderation in its details as the Fiat Abarth 2200. It was 4,610 mm long, 1,620 mm wide, 1,280 mm high and weighed 1050 kg. According to the company, the car's maximum speed was 197 km/h – and there were not many coupés around then that could come close to 200 km/h. The car's debut was fixed for the Turin Motor Show, which opened on 31 October, and it was accompanied by the roadster which, however, was spoiled by double front headlights that had only just been permitted by Italian law. The coupé was priced at Lit

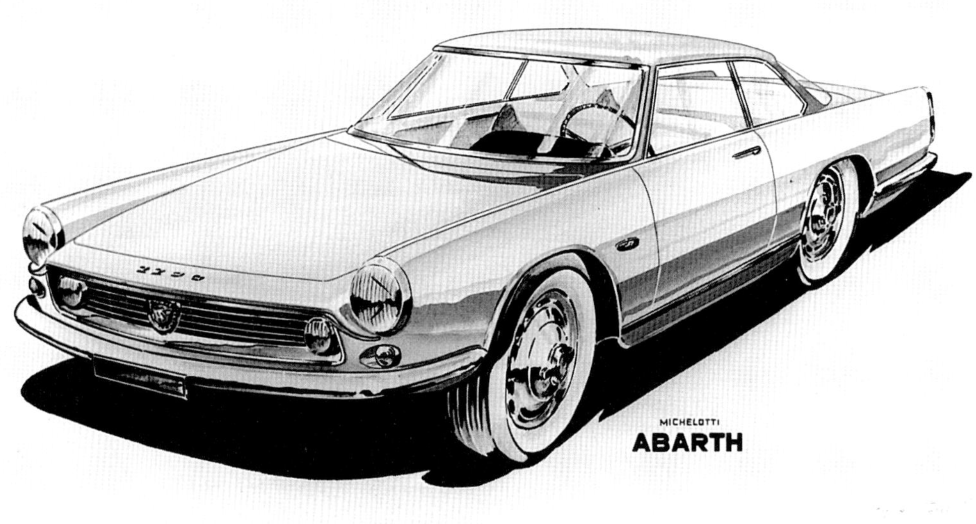

In 1959, Carlo Abarth moved into the difficult superior sports car market segment with the 2200. The car was powered by Fiat's 2100 six-cylinder engine, a model launched in March at the Geneva Motor Show. Abarth increased the Fiat's capacity from 2054 to 2162 cc and carried out numerous modifications, including the adoption of three Weber 40DCOE carburettors; maximum power went from 82 to 135 hp at 6000 rpm. Giovanni Michelotti designed a model for an attractive 2+2 coupé, which was built by Carrozzeria Allemano.

Carlo Abarth played a highly ambitious card when he introduced his 2200, which was to take to the rarefied air of the over two-litre sports car category.
The elegant body designed by Michelotti and built by Allemano in aluminium was well received on the car's introduction at the 1959 Turin Motor Show. The front, with its large lights and subtle radiator grill with the scorpion logo in the centre, the long bonnet that gave away the presence of a large and powerful engine, the essential bright cockpit, all contributed to give the car a notable formal equilibrium and sobriety of line that was not common at the time.
With leather seats and limited use of chrome, the interior was also in full synthesis with the refined personality of the car, which was 4610 mm long and had a dry weight of a little over a ton (1050 kg).
The Fiat Abarth 2200 reached a maximum speed of 197 km/h and cost Lit 2.995.000.

2.995.000 and the roadster at Lit 3.095.000, in open competition with the Touring and Zagato versions of the Lancia Flaminia coupé, both of which cost Lit 3.200.000.

The Fiat Abarth 1600 was also presented – unexpectedly – at the Turin show, its engine derived from the 1500 cc four-cylinder O.S.C.A. twin overhead camshaft by increasing the bore from 78 mm to 80.5 mm; maximum power was upped to 95 hp. The other mechanical components and basic dimensions of the car were unchanged from those of the Fiat 1500 Cabriolet, which was introduced in the summer of 1959. The Fiat Abarth 1600 was also

As was the custom in those days, a cabriolet that adopted the same slender and harmonious body line was derived from the Fiat Abarth 2200 coupé.
But it was shown at the Turin Motor Show with a different front, which frankly caused bewilderment if not open discord: in compliance with the new road traffic regulations that came into force during the summer of '59 and allowed the adoption of double headlights, the 2200 cabrio also boasted the double lamp configuration that required further modification of the radiator grill and the bonnet line, with dubious overall results.

made in both coupé and roadster form by Carrozzeria Allemano which, on that occasion, produced a less attractive car than the 2200.

Siata suddenly began to play an exceedingly ephemeral part in the destiny of Abarth & C. Founded in 1922 by Giorgio Ambrosini, the Società Italiana Trasformazioni Automobilistiche initially modified the mechanics of numerous Fiats, from the 512 to the 504 and 508; they were radical modifications to the cylinder head, even going as far as to use supercharging to obtain the most perceptible increases in power. Ambrosini had his moment in the sun with the launch of the Fiat 508 Balilla, because the different forms of trans-

FIAT ABARTH "1600" (1959)

ENGINE	
	front, longitudinal, 4 cylinders in line
Bore and stroke	mm 80.5x78
Cubic capacity	cc 1587
Valve gear	2 overhead camshafts, chain driven
Number of valves	2 per cylinder
Compression ratio	9:1
Fuel feed	1 twin-choke Weber DCL03 carburettor
Ignition	coil, distributor
Coolant	liquid
Lubrication	forced, oil filter on the main circuit
Maximum power	95 hp at 6000 rpm
TRANSMISSION	
	rear wheel drive
Clutch	single dry plate
Gearbox	4 speed plus reverse, central lever
Axle ratio	4.3:1
BODY	
	two-seater coupé/cabriolet by Allemano
CHASSIS	
	integral body
Suspension	front independent, deformable quadrangle, helicoidal springs, stabiliser bar, hydraulic dampers; rear rigid axle, semi-elliptic spring, stabiliser bar, hydraulic dampers
Brakes	hydraulically operated front and rear discs
Steering	worm and roller
Fuel tank	capacity 38 litres
Tyres	155x15
DIMENSIONS AND WEIGHT	
Wheelbase	mm 2340
Front track	mm 1287
Rear track	mm 1215
Length	mm 4230
Width	mm 1540
Height	mm 1250
Kerb weight	kg 960
PERFORMANCE	
Maximum speed	km/h 180

Bent on promoting the Scorpion brand to a diverse clientele, Abarth did not overlook the 1.5 litre sports car sector. He carried out numerous improvements to the mechanics and came up with a completely new body for the Fiat 1500, which had a twin-cam O.S.C.A. engine.

The original cubic capacity of 1491 cc was increased to 1587 cc and the power output went from 90 to 95 hp at 6000 rpm. The creation of a new body was assigned to Allemano, who came up with both coupé and open topped variants, but with less convincing results than the 2200.

formation that he came up with for the little one-litre car, which had a maximum power output of 48 hp against the production car's 20 hp, were immensely popular among enthusiasts and sports car lovers of the period. One of those specially modified Balillas achieved the remarkable speed of 154 km/h. Siata's overhead valve cylinder head for the Topolino was widely used towards the end of the Thirties and meant the little Fiat could easily exceed 100 km/h. A special cabriolet version of this sort of 'ante litteram' Abarth was the Amica, bodied by Bertone and produced in small numbers in 1939-40 using 500A and 500C components and reintroduced with a completely different shape in the early Fifties, this time with the components of the 500B and 500C. Those years also saw the Daina, the pleasant little roadster built on Fiat

The body of the Fiat Abarth 1600 seemed penalised by excessive overhang on the axles, created with the intention of giving an imposing appearance to a car that should instead have had a svelte and compact shape. Under the long bonnet was a relatively small four-cylinder engine, yet in the apparently large boot there was not even enough space for a good capacity fuel tank, which contained just 38 litres instead of the 45 litres of the Pininfarina-designed production car.

1400 mechanics and bodied by Stabilimenti Farina, as well as the 208 S, a difficult saloon to drive that was built on the Fiat V8 chassis of 1952.

In spite of that flow of vitality and initiatives and the obvious success of some of the cars, Ambrosini faced mounting difficulty in squaring the accounts of his small company; Fiat's top management went to work on Carlo Scagliarini and Carlo Abarth until they reached an agreement for the companies' joint cooption in an attempt to reinvigorate Siata's fortunes. So in 1959 Abarth & C. entered Siata with 49% of the share capital and the hope that the arrival of a new partner might resolve the situation. Siata Abarth went to the 1959 Turin Motor Show with four cars, all built on Fiat 600 mechanicals with the 31 hp 735 cc engine: a normal production saloon, a four-seater coupé, a

The one-off transient business links between Abarth & C. and Siata were on public display at the 1959 Turin Motor Show, where the cars exhibited by the two companies included a four-seater coupé and a roadster built on Fiat 600 mechanics, the engines of which were uprated to 735 cc to produce 31 hp at 5000 rpm. A sort of model duplication for Abarth and little unanimity led to the dissolution of the two partners' cooperation agreement in 1961.

A few months after the introduction of the Fiat Abarth 850 Scorpion coupé the Corso Marche factory came out with the Riviera, an open top version of the car built by Allemano to a Michelotti design. Carlo Abarth was finally able to market an attractive roadster to customers who liked small, exclusive cars with engines of less than a litre.

The British market proved particularly responsive to the Abarth brand's promotional activities overseas. In October 1959, Carlo Abarth visited his U.K. distributor, Anthony Crook Motors at Hersham, Surrey.

roadster in the same style and a special version of the 600 Multipla. The association of the two companies was not destined to last, because Carlo Abarth's spirit of initiative and activity was not in synthesis with the management criteria of his partners and that he had no intention of investing his time in running a company that, in the final analysis, was nothing more than a small scale duplication of his own. The agreement was rescinded after two years and each of the partners returned to working on his own. Siata trudged on until it went out of business in 1970 and Abarth & C. generated new energy that would lead to the golden age of its relatively brief existence.

The arrival of new models fitted with the 847 cc engine, derived from the glorious 747 cc unit, led to the same cubic capacity increase for the Record

In the early months of 1960, Abarth put a lot of productive and commercial muscle into their Scorpione and Riviera models, which were well received by the market from the start.
The picture on the left shows a number of 850 Scorpions lined up, while in the foreground is a Fiat 600 with basic bodyshell and mechanics, as delivered by Fiat to Abarth before being modified at Corso Marche.
On the right is the Abarth stand at the 1960 Geneva Motor Show, where the cars on display included the 850 Scorpione and its Riviera open top derivative.

The longitudinal section of the Porsche 356 B, which was produced in Stuttgart from 1959 to 1963.
The agreement reached in the autumn of 1959 between Carlo Abarth and the German factory called for the design and construction in Turin of a special light and compact racing coupé, which was to benefit from Abarth's experience in racing small and essential sports saloons built by Zagato on mechanics derived from the Fiat 600.
The supply contract involved the construction of an initial 20 bodies, the design of which was assigned to stylist Franco Scaglione who had distinguished himself by designing a number of prestigious cars, not least the Alfa Abarth 1000 shown by Carrozzeria Bertone at the 1958 Turin Motor Show.

Monza coupé, with its twin overhead camshafts and Zagato body; maximum power went up to 70 hp at 6800 rpm, but there were versions of 72 hp and more depending on their varying states of modification. The production of exhausts was expanding month by month with the help of strong demand from export markets; of the total 164,607 units produced in 1959, 52% were sold outside Italy. In Britain, an agreement was reached with Crook Motors of Hersham, Surrey, for distribution in the United Kingdom that included cars, with the 750 GT by Zagato enjoying an encouraging success.

At the end of Abarth's exciting 1959 season, an imminent and important event was taking shape that would turn one of Carlo's cherished dreams into reality and become a significant precedent in the history of the company. It all began on 18 September in the bar of the FrankfurterHof Hotel in Frankfurt, Germany, where a top management meeting took place between Ferry Porsche, Walter Schmidt and Klaus von Rucker on the one side, and Carlo Abarth on the other. The Stuttgart-Zuffenhausen factory had verified that the 356 model, made more powerful, comfortable and luxurious than the first generation, had also become heavier; so it was necessary to remedy the situation, perhaps by working with that specialist in Italy who was making sparks fly with several coupés of Fiat derivation, the weight of which was substantially below 600 kg. Agreement was reached at the Frankfurt meeting under which Porsche would assign to Abarth the creation of a two-seater GT body on the 356B production floor-pan.

On 22 September 1959, von Rucker took the following minutes at the Frankfurt meeting: "*the Abarth company should construct for the Porsche company a number of two-seater GT bodies on our production floorpan. The body should be as light as possible. If it is inferior to the minimum limits permitted for GT homologation as hoped, the missing weight must be made up with a ballast of steel sheets. Over 20 bodies were discussed which, according to Mr. Abarth, will cost Lit 1.000.000 each. In this initial order of Lit 20 million the wooden matrix should be included. The price would decrease to Lit 800.000 for each body in the case of subsequent orders. Mr. Porsche gave Mr. Abarth the go-ahead to begin to build the bodies. By 21 October, Mr. Abarth will deliver to Porsche the fundamental documentation concerning the future appearance of the car. By that date, the sales department of Porsche must have decided whether to order one or 20 body units on the same date. Porsche must also have calculated the sales price of the new models in relation to those acquired as indicated above. On 1 October 1959 Porsche will send a complete Carrera chassis with engine; not a working engine but a dummy. As soon as possible, Porsche will send Abarth a design of the chassis so that Abarth can determine the points of the body's union with the lower part of the chassis. Porsche will also send designs of the front end of the car as well as photographs*".

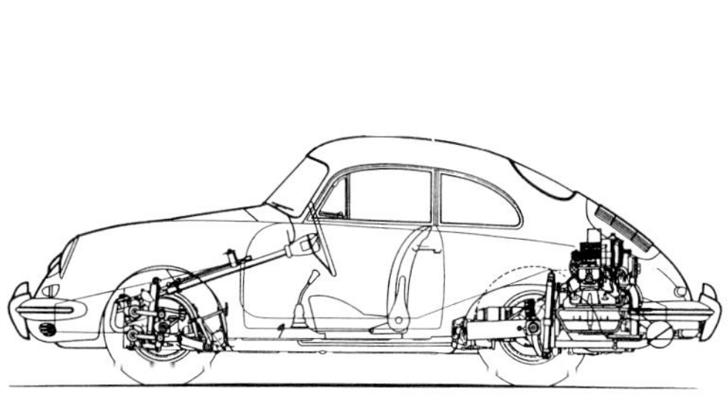

Three examples of the Porsche Carrera GTL photographed in front of the Abarth factory. The body line looks lighter and more agile than the production 356 B's. The execution of the bodyshells in aluminium and the level of finish of the Turin-produced bodies did not fully satisfy Porsche, who also complained of poorly waterproofed window seals and insufficient steering radius. The agreement lapsed after the delivery of the first 21 bodyshells.

PORSCHE "CARRERA GTL" (1960)

ENGINE	
	rear, longitudinal, 4 cylinders opposed
Bore and stroke	mm 87.5x66
Cubic capacity	cc 1587
Valve gear	4 overhead camshafts, operated by drive shafts
Number of valves	2 per cylinder
Compression ratio	9.8:1
Fuel feed	2 twin-choke Weber 40DCM2 carburettors
Ignition	double, coil, distributor
Coolant	air
Lubrication	dry sump
Maximum power	115 hp at 6500 rpm
TRANSMISSION	
	rear wheel drive
Clutch	single dry plate
Gearbox	4 speed plus reverse, central lever
BODY	
	2 seat coupé
CHASSIS	
	floor pan
Suspension	front independent, longitudinal arms, tranverse torsion bars, stabiliser bar, hydraulic dampers; rear independent, trailing half-axles, torsion bar, hydraulic dampers
Brakes	hydraulically operated drums front and rear
Steering	worm and finger
Fuel tank	capacity 80 litres
Tyres	5.90x15
DIMENSIONS AND WEIGHT	
Wheelbase	mm 2100
Front track	mm 1306
Rear track	mm 1272
Length	mm 3980
Width	mm 1672
Height	mm 1228
Kerb weight	kg 778
PERFORMANCE	
Top speed	km/h 220

The weight of the car had to be as low as possible, allowing Porsche to use steel ingot ballast in case the procedure meant the car's weight went below the minimum allowed by Grand Touring homologation. It was also agreed that the number of bodies would not exceed 20 and the unit price was estimated by Carlo at a million Lire each; for orders beyond the first 20 units the price would be reduced to Lit 800.000. The parties agreed that Porsche would send the mock-up of a complete chassis to Turin on 1 October, without the encumbrance of the 1587 cc 115 hp dry sump Carrera engine and its four overhead cams, and that by 21 October Abarth would have drawn up preliminary plan of the future car's body shape.

Franz Xaver Reimspiess arrived in Turin during the first week of October to confer with Abarth, but also to evaluate other body designers and builders to whom he could eventually entrust the construction of the lightweight Carrera. Carlo told him he would have Zagato build the 20 bodies and he himself would supervise the progress of the work with the expectation of being able to complete the consignment by the end of March 1960; he said he hoped that this first contact with Porsche could subsequently give rise to an on-going relationship and the possibility on him being involved in the direct sale of the Carreras produced in Turin. Abarth entrusted the design of the new coupé for the 356B chassis to Franco Scaglione, who had already shone in previous years by interpreting to the letter his briefings for aerodynamic sports saloons; and the able stylist, who was no longer tied to Carrozzeria Bertone by contract, designed a very low and slender coupé with a long bonnet and compact cockpit in one piece, with a heavily sloping rear end, ample back window and an engine compartment door with an intense series of air vents.

The commission for the first 20 bodyshells was given to Rocco Motto, an able Turin specialist in aluminium panel beating and a highly creative

A long prominent nose and extremely short tail were distinctive characteristics of the Porsche Carrera GTL designed by Franco Scaglione, whose first prototype was produced by the noted panel beater Rocco Motto.
The inclination of the roof's front uprights and the extreme wrap-around shape of the windscreen contributed to slightly improving aerodynamics over those of the 356 B production car; data from wind tunnel tests some years later produced a Cx of 0.376, a rather disappointing figure as the production car's rating recorded by Porsche was 0.398.

Porsche had to work on the Italian-sourced Carrera GTL to improve longitudinal habitability for the taller German driver, many of whom were unable to assume the correct driving position.
After its debut in the 1960 Targa Florio on 8 May, when the Carrera GTL driven by Paul Ernst Strahle, Herbert Linge and Dieter Lissmann won its class, the car and same crew competed in the 1000 Kilometres of the Nürburgrirg on 22 May and scored another class win.
With Linge-Walter at the wheel, the GTL won its class and came 10th overall in the 24 Hours of Le Mans from 24-25 June, having completed 269 laps or 3,619.759 kms at an average 150.811 km/h.
It was a race in which the car was particularly fast and was clocked at 222 km/h on the famous Hunaudières straight.
The picture shows the Herbert Linge and Heini Walter car number 35 exiting the Mulsanne corner.

interpreter of the drawings of numerous Italian designers. Porsche assigned the test-driving and the knockdown body to the young engineer Klaus Steinmetz, who had recently joined the German factory. When the first bodyshell was delivered on 1 February 1960, Steinmetz pointed out that there was insufficient space for the engine, radiator and oil reservoir; it was evident that the measurements of the mechanical components sent to Abarth in October were not respected and it emerged – to Carlo's consternation - that the engine had been abandoned in a corner of the workshop and had not been taken into consideration when the prototype was designed and built. *"We widened it with blows of a hammer"*, Steinmetz remembers today, *"to try to create the right fit to the exact dimensions of the engine compartment for subsequent bodies"*. Once the third bodyshell had been completed, production suddenly ground to a halt because Motto, who had just received part payment for body supply from Abarth, decided to take an unplanned holiday and disappeared, throwing everyone into confusion. They even had to turn to another Turin craftsman company, Viarengo & Filipponi, who agreed to beat the aluminium and build the bodies, overcoming no small difficulty due to the short production time available and the different techno-constructional standards of Italian body builders compared to those in Germany. On 12 March 1960, pictures of the Porsche Carrera GTL prototype – L for Leicht or Light – were published for the first time in the authoritative periodical Auto Motor und Sport; the factory announced a price of DM 25.000. After being delivered to customers, GTLs were entered for the first possible race; the new car won its class at the Targa Florio on 8 May driven by Herbert Linge and Paul Ernest Strähle, competed in the 1000 Km of the Nürburgrirg on 22 May as a car fitted with experimental disc brakes, won its class on 25-26 June in the 24 Hours of Le Mans during which a Carrera GTL was timed on the Hunaudières straight at 222 km/h. The frontal section of the car, which was lower than that of the production Carrera

FIAT ABARTH "850 RECORD MONZA" (1959)

ENGINE	
	rear, longitudinal, 4 cylinders in line
Bore and stroke	mm 62.5x69
Cubic capacity	cc 847
Valve gear	2 overhead camshafts, chain driven
Number of valves	2 per cylinder
Compression ratio	9.5:1
Fuel feed	2 twin-choke Weber 36DCL4 carburettors
Ignition	coil, distributor
Coolant	liquid
Lubrication	forced, oil filter on the main circuit
Maximum power	72 hp at 6800 rpm
TRANSMISSION	
	rear wheel drive
Clutch	single dry plate
Gearbox	4 speed plus reverse, central lever
Axle ratio	4.3:1
BODY	
	2 seat coupé
CHASSIS	
	integral body
Suspension	front independent, upper trailing arms, transverse lower leaf spring, hydraulic dampers; rear independent, trailing arms, helicoidal springs, hydraulic dampers
Brakes	hydraulically operated drums front and rear
Steering	worm and sector
Fuel tank	capacity 45 litres
Tyres	5.20x12
DIMENSIONS AND WEIGHT	
Wheelbase	mm 2000
Front track	mm 1150
Rear track	mm 1160
Length	mm 3470
Width	mm 1350
Height	mm 1140
Kerb weight	kg 570
PERFORMANCE	
Top speed	km/h 190

1600 GS, had a better Cx of 0.365 against the older car's 0.398 and weighed almost 20 kg less, which made the new model difficult to beat in both circuit and road races.

But in spite of promising motor racing results, Stuttgart management was not happy with the quality of the first examples of the car, saying the windscreen and door seals leaked, there was insufficient steering radius and finish was poor. All together, just 21 Carrera GLTs were made with chassis numbers from 1001 to 1021; and there the disappointing collaboration between Abarth and Porsche ended.

At the beginning of 1960, new motor sport regulations ushered in the FIA Cup for the Grand Touring Category in these six divisions: Division 1 up to 500 cc class; Division 2 up to 700 cc that took the place of the previous up to 750 cc class; Division 3 up to 1000 cc; Division 4 up to 1300 cc; Division 5 up to 2000 cc and Division 6 over 2000 cc. An up to 850 cc class was introduced for national competitions.

Carlo was ready with a new twin overhead camshaft 695 cc engine with two Weber 36 DCL4 carburettors that produced a maximum of 64 hp at 7900 rpm. The Fiat Abarth 700 Bialbero made its debut on 5 March at Pferdsfeld, Germany, where it won the up to 1000 cc Grand Touring class, driven by Wolfgang Seidel. But the great success early that year was when Charles Callanan and Paul Richards won the 4 Hours of Sebring in the U.S.A. on 25 March in a Fiat Abarth 750 Record Monza at an average 119.014 km/h., beating a charging Stirling Moss in an Austin Healey Sprite in the process.

The 700 Bialbero was back in the limelight again two days later at the ninth Agnano-Cappella dei Cangiani, in which Andreotti came first and then went on to win four more races in Italy, Switzerland and France in April: on 1 May Rigamonti-Sala-Zagato won the second Alberto Ascari Trophy at the 12 Hours of Monza, covering 1837.555 km at an average 153.129 km/h! At the 6 Hours of Hockenheim on 6 June, the Fiat Abarths swept the field in the up to 500 cc Grand Touring class with Raffaello Ciarpaglini and Vincenzo Antolini Ossi, the up to 700 cc class (Aristodemo and Giancarlo Sala) and the up to 1000 cc class (Wolfgang Seidel-Mario Poltronieri) with a fastest lap set up by Seidel-Poltronieri in the Fiat Abarth 750 Record Monza of 2'51.6" at an average 162.600 km/h. At Monza on 29 June Giancarlo Castellina came first overall in the 2 Hour Grand Touring Grand Prix in a Fiat Abarth 700 Bialbero, followed by Seidel in a Fiat Abarth 750 Record Monza and Ciarpaglini in a Fiat Abarth 500 Coupé Zagato. July was the month of the Turin cars' triumphs in hillclimbs: they won the Bolzano-Mendola on 3 July, the Trento-Bondone on the 10th, Garessio-San Bernardo on the 17th and Trieste-Opicina on the 24th. In August, another constant flow of victories saw Fiat Abarth name up in lights again, the cars' successes including the Sila Cup on 6 August, the Montana-Crans on the

FIAT ABARTH "700 BIALBERO" (1960)

ENGINE	
	rear, longitudinal, 4 cylinders in line
Bore and stroke	mm 61x59.5
Cubic capacity	cc 695
Valve gear	2 overhead camshafts, chain driven
Number of valves	2 per cylinder
Compression ratio	10:1
Fuel Feed	2 twin-choke Weber DCL4 carburettors
Ignition	coil, distributor
Coolant	liquid
Lubrication	forced, oil filter on the main circuit
Maximum power	64 hp at 7900 rpm
TRANSMISSION	
	rear wheel drive
Clutch	single dry plate
Gearbox	4 speed plus reverse, central lever
BODY	
	2 seat coupé
CHASSIS	
	integral body
Suspension	front independent, upper trailing arms, transverse lower leaf spring, hydraulic dampers; rear independent, trailing arms, helicoidal springs, hydraulic dampers
Brakes	hydraulically operated drums front and rear
Steering	worm and sector
Fuel tank	capacity 45 litres
Tyres	5.20x12
DIMENSIONS AND WEIGHT	
Wheelbase	mm 2000
Front track	mm 1150
Rear track	mm 1160
Length	mm 3470
Width	mm 1350
Height	mm 1140
Kerb weight	kg 570
PERFORMANCE	
Top speed	km/h 180

Fiat Abarths also competed in the 1960 24 Hours of Le Mans with their twin-cam engines and the first front radiators. Of the three cars entered, 850 cc number 49 in the hands of Tommy Spychiger and Jacques Ferret, had to retire during the 13th hour with clutch trouble; sister car number 50 driven by Paul Condriller and Jean Guichet also dropped out with clutch problems during the 16th hour. The Remo Cattini-Giancarlo Rigamonti 700 cc number 60 car had already retired during the fourth hour with engine trouble. The aluminium bodies of those Le Mans cars no longer carried the 'Z' logo of Carrozzeria Zagato of Milan, as their panels were beaten by specialist Corna of Turin and fitted by Abarth themselves.

FIAT ABARTH "750 SPIDER TUBOLARE" (1960)

ENGINE	
	rear, longitudinal, 4 cylinders in line
Bore and stroke	mm 61x64
Cubic capacity	cc 747
Valve gear	2 overhead camshafts, chain driven
Number of valves	2 per cylinder
Compression ratio	10.5:1
Fuel feed	2 twin choke Weber 36DCL4 carburettors
Ignition	coil, distributor
Coolant	liquid
Lubrication	forced, oil filter on the main circuit
Maximum power	75 hp at 7300 rpm
TRANSMISSION	
	rear wheel drive
Clutch	single dry plate
Gearbox	4 or 5 speed plus reverse, central lever
Axle ratio	4.3:1
BODY	
	2 seat racing roadster
CHASSIS	
	tubular trellis
Suspension	front independent, trailing arms, helicoidal springs, hydraulic dampers; rear independent, trailing arms, helicoidal springs, hydraulic dampers
Brakes	hydraulically operated discs front and rear
Steering	worm and roller
Fuel tank	total capacity front and lateral 70 litres
Tyres	front 4.00x13; rear 4.50x13
DIMENSIONS AND WEIGHT	
Wheelbase	mm 2040
Front track	mm 1153
Rear track	mm 1168
Length	mm 3300
Width	mm 1356
Height	mm 1082
Kerb weight	kg 410
PERFORMANCE	
Top speed	km/h 200

13th, the Giro dei due Mari on the 21st and the Sassi-Superga on the 29th. On 4 September, Giancarlo Castellina and Jean Vinatier won the 500 Kilometres of the Nürburgrirg outright and their class in a 750 Record Monza, followed by three other identical cars. On 23 October Bernard Consten and Paul Condriller scored another success in the 1000 Kilometres of Paris driving the same car in which they won their class and the performance index.

The only shadow that fell across such a successful season and its wealth of victories was the unfortunate debut of the Fiat Abarth twin-cams at the 24 Hours of Le Mans on 25-26 June, when all three cars had to retire with clutch problems: the 700 Bialbero of Remo Cattini-Giancarlo Rigamonti went out during the fourth hour, the 850 Bialbero of Tommy Spychiger-Jacques Ferret during the 16th as did the sister car of Paul Condriller-Jean Guichet, which was timed on the Hunaudières straight at 192 km/h. That was also the race in which a front radiator appeared on the little Abarths for the first time and successfully produced better engine cooling. Between March and December 1960, Fiat Abarths scored 85 class wins, seven overall victories and produced two new Italian champions in Giancarlo Castellina in the up to 700 cc Grand Touring class driving a car prepared by Bosato, and Lanzo Cussini in the up to 850 cc class, with preparation by Faccioli. The Roosevelt Team in the U.S.A. also put up a brilliant performance and by the end of the season Chandler Lawrence had become the national H Production champion at the wheel of a 750 Record Monza.

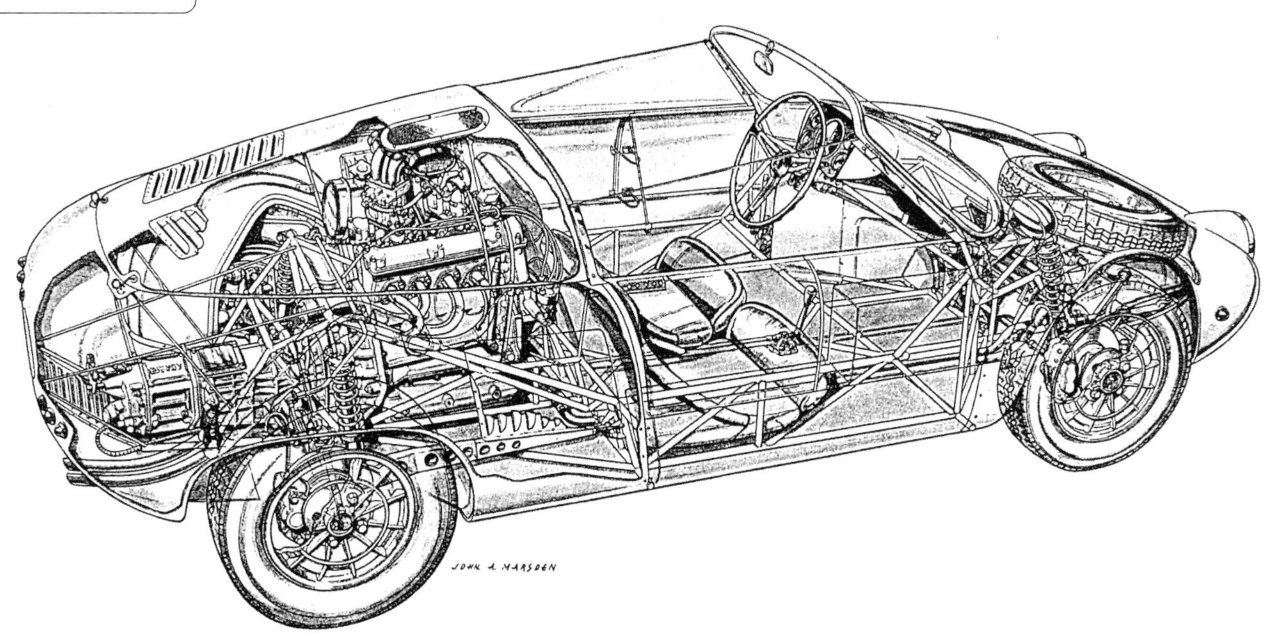

Opposite: With the 1960 arrival of Mario Colucci at Abarth from Alfa Romeo, where he was 4 the driving force behind the Giulia TZ, a new constructional philosophy for racing cars got off the ground, introducing tubular trellis chassis and engines at the centre of the car.
Among other things, that method of construction brought greater operational flexibility on the track because it permitted the use of twin-cam four-cylinder engines of varying capacities, depending on the demands of the different races.
Developed and tested during the course of 1960, the Fiat Abarth Spider Tubulare first raced the following season and became a genuine milestone in the techno-constructional evolution of Abarth racing cars.
The whole front end of the new Fiat Abarth Spider Tubolare's body was hinged at the front and tipped forwards.

The ample lid shape of the bonnet flowed laterally into the windows on both sides of the cockpit, so as to give the driving position a kind of aerodynamic faring without continuity features.
The car was also used to try out a new five-speed gearbox, developed from that of the Fiat 600 production car.
The primary sump contained the first, second and third speed ratios and reverse.
An intermediate sump bolted to the primary unit housed the cogs for fourth and fifth gears, while the mechanism actioned by the gear lever was in a small external box separate from the two sumps.
The entire engine group was mounted on three flexible supports, one on the front of the crankcase and the other two at the sides of the gearbox.
The design of the Elektron wheels produced by Amadori was the work of Carlo Abarth himself.

Mario Colucci joined the Corso Marche factory at the beginning of the year as technical director, specialising in chassis that arrived from Alfa Romeo; he had previously been involved in the Alfa Abarth 1000 project and in particular its reticular chassis with thin tubes. He remembers today how Carlo Abarth had decided the time had come to recruit an engineer able to correctly interpret designs, and offered him the opportunity of joining the company as technical boss. *"But Alfa Romeo did not want to let me go"*, Colucci recalled, *"and they even threatened Abarth, who was a supplier of exhausts for the Giulietta, that they would cancel the deal if he persisted with his intention of poaching their design staff. But Abarth did not give a damn for their deal and I joined him as technical director"*.

FIAT ABARTH "1000 BIALBERO" (1960)

ENGINE	
	rear, longitudinal, 4 cylinders in line
Bore and stroke	mm 65x74
Cubic capacity	cc 982
Valve gear	2 overhead camshafts, chain driven
Number of valves	2 per cylinder
Compression ratio	9.3:1
Fuel Feed	2 twin-choke Weber 36DCL4 carburettors
Ignition	coil, distributor
Coolant	liquid
Lubrication	forced, oil filter on the principal circuit
Maximum power	91 hp at 7100 rpm
TRANSMISSION	
	rear wheel drive
Clutch	single dry plate
Gearbox	4 speed plus reverse, central lever
Axle ratio	4.3:1
BODY	
	2 seat coupé
CHASSIS	
	integral body
Suspension	front independent, upper trailing arms, transverse lower leaf spring, hydraulic dampers; rear independent, trailing arms, helicoidal springs, hydraulic dampers
Brakes	hydraulically operated front discs and rear drums
Steering	worm and sector
Fuel tank	capacity 45 litres
Tyres	135x13
DIMENSIONS AND WEIGHT	
Wheelbase	mm 2000
Front track	mm 1150
Rear track	mm 1160
Length	mm 3470
Width	mm 1350
Height	mm 1140
Kerb weight	kg 570
PERFORMANCE	
Maximum speed	km/h 205

Rich in experience gained on the Giulia TZ project – the car was designed in 1959 and went into production in 1963 – Colucci conceived the first racing barchetta with tubular chassis in Abarth's history, apart from the 1949 204 A. In contrast with Abarth practice, which privileged the Porsche school of an outboard engine inclined behind the rear axle, he place the power unit in the centre of the car, immediately in front of the rear wheels. Soon, the Abarth 750 Sport Spider Tubolare became the watershed of a long line of racing barchettas with twin-cam engines of 750, 850 and 1000 cc. Stylistically attractive, the body of the new little car was built at Abarth in aluminium sheet; it had a bonnet that opened completely, an ample windscreen – in conformity with the special French regulation for Sport Category competitors – and a long rear engine hood that gave full accessibility to the mechanics. The car was fitted with Elektron wheels produced by Amadori to Carlo Abarth's own design.

Originally conceived to compete in the 24 Hours of Le Mans, the car was as important a stage in the Abarth process of technical evolution as it was to the company's motor sport pedigree. It was built on a 2040 mm wheelbase and had disc brakes on all four wheels. The engine was the well-known twin-cam 747 cc, with a compression ratio of 10.5:1 and fuelled by two Weber 36 DCL4 carburettors; its maximum power rose to 75 hp at 7300 rpm, which gave the Spider Tubolare a top speed of 200 km/h.

The year 1960 was also noted for the debut of a new Abarth engine derived – as usual – from the Fiat 600, the 1000 twin-cam. Bore and stroke

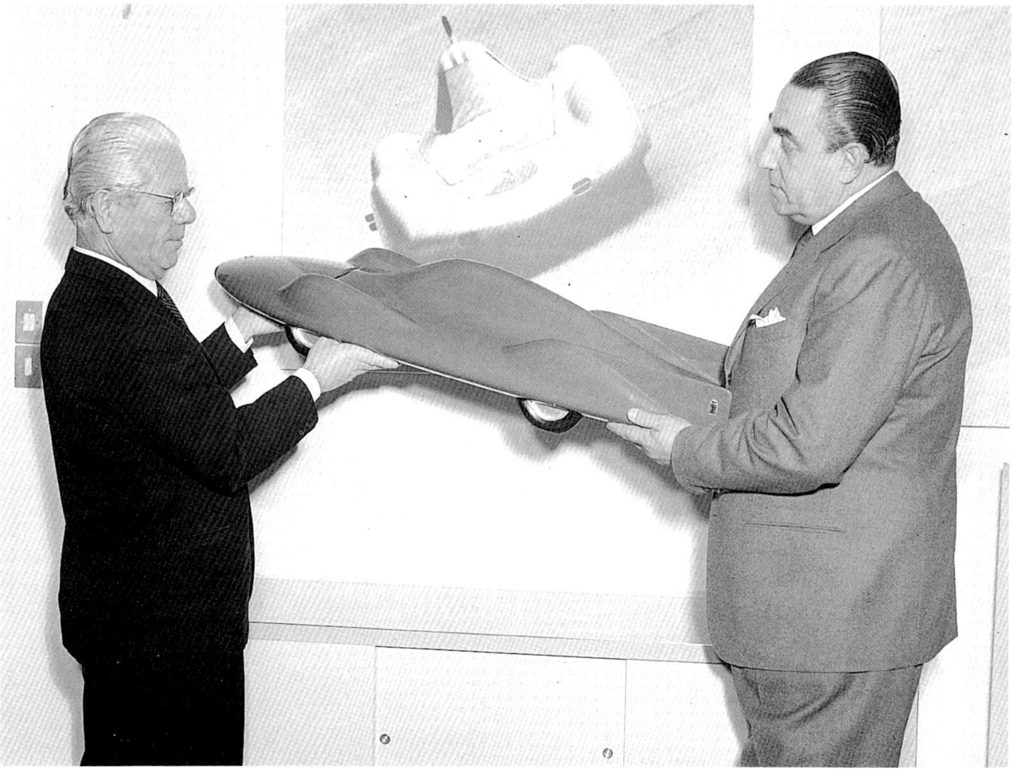

The professional relationship between Pinin Farina, the master of Italian body designers and builders, and Carlo Abarth was an intense and extremely useful one in those years, which were so rich in audacious entrepreneurial initiatives and a long series of Italian car successes on the international stage. Both men were able to come up with special cars that would establish new speed records, an outstanding promotional tool for their respective companies.

FIAT ABARTH "750 RECORD" (1960)

ENGINE	
	mid-rear, longitudinal, 4 cylinders in line
Bore and stroke	mm 61x64
Cubic capacity	cc 747
Valve gear	2 overhead camshafts, chain driven
Number of valves	2 per cylinder
Compression ratio	11:1
Fuel feed	2 twin-choke Weber 36DCL4 carburettors
Ignition	coil, distributor
Coolant	liquid
Lubrication	forced, oil filter on the principal circuit
Maximum power	75 hp at 7300 rpm
TRANSMISSION	
	rear wheel drive
Clutch	single dry plate
Gearbox	4 speed plus reverse, central lever
Axle ratio	4.3:1
BODY	
	single-seater record car by Pininfarina
CHASSIS	
	tubular trellis
Suspension	front independent, upper trailing arms, transverse lower leaf spring, hydraulic dampers; rear independent, trailing arms, helicoidal springs, hydraulic dampers
Brakes	hydraulically operated front drums
Steering	worm and sector
Fuel tank	capacity 62 litres
Tyres	5.50x15
DIMENSIONS AND WEIGHT	
Wheelbase	mm 1785
Front track	mm 1130
Rear track	mm 1160
Length	mm 4550
Width	mm 1550
Height	mm 1200
Kerb weight	kg 510
PERFORMANCE	
Maximum speed	km/h 215

Pininfarina conducted a long series of experiments in a wind tunnel to define the shape of the new record car with which Carlo Abarth wanted so badly to shatter existing records in class H (501 to 750 cc) and subject the new one litre twin-cam engine to severe testing in class G (751 to 1100 cc). The discontinuation of the large caudal fins and the long axle overhangs clearly shows how new aerodynamic dictates took the place of the beliefs and empiricism of the past in just a few years. Built on a 1785 mm wheelbase and with a length of 4555 mm, the new Abarth record car weighed 510 kg dry. Driven by Umberto Maglioli, the 750-powered version established four new international records at Monza on 22 September 1960 at average speeds of between 212 and 225 km/h.

FIAT ABARTH "1000 RECORD" (1960)

ENGINE	
	mid-rear, longitudinal, 4 cylinders in line
Bore and stroke	mm 65x74
Cubic capacity	cc 982
Valve gear	2 overhead camshafts, chain driven
Number of valves	2 per cylinder
Compression ratio	10.8:1
Fuel feed	2 twin-choke Weber 36DCL4 carburettors
Ignition	coil, distributor
Coolant	liquid
Lubrication	forced, oil filter on the main circuit
Maximum power	108 hp at 8000 rpm
TRANSMISSION	
	rear wheel drive
Clutch	single dry plate
Gearbox	4 speed plus reverse, central lever
Axle ratio	4.1:1
BODY	
	single-seater record car by Pininfarina
CHASSIS	
	tubular trellis
Suspension	front independent, upper trailing arms, transverse lower leaf spring, hydraulic dampers; rear independent, trailing arms, helicoidal springs, hydraulic dampers
Brakes	hydraulically operated front drums
Steering	worm and sector
Fuel tank	capacity 62 litres
Tyres	145x15
DIMENSIONS AND WEIGHT	
Wheelbase	mm 1785
Front track	mm 1130
Rear track	mm 1160
Length	mm 4550
Width	mm 1550
Height	mm 1200
Kerb weight	kg 510
PERFORMANCE	
Top speed	km/h 220

A few days later from 28 September-1 October, the new Fiat Abarth record car was back at Monza in an effort to shatter more records, this time in international class G (751 to 1100 cc). It was powered by a new four-cylinder 982 cc twin-cam engine, which in that special version produced 108 hp at 8000 rpm. Nine drivers took their turn at the wheel in those last three days of September to establish the new international records, which were the 12-hour, 2000 mile, 24 hour, 5000 kilometre, 5000 mile, 48 hour and the 10,000 kilometre. The attempt continued until 1 October, when Abarth broke the 72-hour record in daring fashion. The picture shows Carlo Abarth exchanging impressions during a pit stop with driver Mario Poltronieri.

were, respectively, 65x74 mm for a total capacity of 982 cc; the compression ratio was taken to 10.1:1 and a well tried and tested twin overhead camshaft cylinder head was adopted, plus two Weber 36 DCL4 carburettors, all of which developed a maximum power output of 84 hp at 7000 rpm. The engine made its first official appearance in a newly conceived single-seater record car – initially equipped with the experimental 747 cc, 75 hp twin-cam – in the hands of Umberto Maglioli on 22 September 1960 and set four new international Class H (501 – 750 cc) records at Monza – the 100 miles in 45'24.1" at an average speed of 212.672 km/h, the 200 km in 56'13.2" average 213.434 km/h, the 200 miles in 1h 29'54.1" average 214-809 km/h and the one hour in which the car covered 213.643 km. The body of the single-seater had its roots in a new aerodynamic study conducted by Pininfarina in a wind tunnel, and refined through extensive test-driving. Built of light alloy on a 1785 mm wheelbase, the car was 4555 mm long, 1,550 mm wide, 1,200 mm high and had a dry weight of 510 kg. Galvanised by the performance of the 750 cc engine, Carlo Abarth decided to get into an ambitious record programme with the 1000 cc engine and the attempts took place from 28 September – 1 October on the usual high speed track of the Monza motordrome.

Everything was carefully organised by Renzo Avidano, who was more and more becoming Carlo Abarth's right hand man in the company's activities. The best and most reliable of the Scorpion's drivers were chosen for the run: Giancarlo Baghetti, Giorgio Bassi, Giancarlo Castellina, Remo Cattini, Massimo Leto di Priolo, Corrado Manfredini, Mario Poltronieri, Giancarlo Rigamonti and Alfonso Thiele. A new four-cylinder, twin-cam 982 cc engine was fitted to the car, suitably modified with a 10.8:1 compression ratio able to develop a maximum of 108 hp at 8000 rpm, and a declared top speed of 220 km/h. Nine international Class G records were established: they were the 12 hour covering 2443.875 km at an average speed of 203.656 km/h, the 2000 miles in 16h 00'14.4" average 201.115 km/h, the 24 hour of 4771.082 km average 198.795 km/h, the 5000 km in 25h 05'43.4" average 199.238 km/h, the 5000 miles in 41h 43'8.2" average 192.878 km/h, the 48 hour covering 9132.672 km averaging 190.264 km/h and the 10,000 km in 52h 16'21.2" at an average 194.376 km/h.

The final 72 hour record attempt took place in dramatically bad weather that seemed likely to drown car and crew in a deluge of pouring rain that gushed across the track during the night of 30 September and 1 October. At the 66th hour, Carlo decided to hand over the wheel to Umberto Maglioli, his most expert bad weather driver. Maglioli leapt into the car and started to lap a flooded Monza at around 200 km/h, when suddenly the wheels on the right side of the car hit an enormous puddle at the entrance to the north curve and the car began to aquaplane, went into a terrifying skid and then rolled over and over until it stopped against the embankment inside the track. The driver, who was unhurt, climbed out and saw that the body was seriously damaged: the car could still move but not under its own power, because the crash had damaged a number of vital components. At that point, Carlo Abarth recalled intuitively that the regulations said the car could continue to pursue the record if it was able to cross the finish line without outside help. Cheered on by the

In 1960, the Fiat Abarth 750 Record Monza achieved a series of significant success in Italy and overseas. Its most prestigious victory was in Germany, where it came first overall in the 500 Kilometres of the Nürburgrirg on 4 September, driven by Giancarlo Castellina and Jean Vinatier; the car finished the race in 4 hours 32'48 3/10" at an average speed of 110.369 km/h and set up the fastest lap, averaging 119.060 km/h. Right: Carlo Abarth is taking part in a refuelling and driver change stop. Fiat Abarth and the Sant Ambroeus team of Milan triumphed that day, for they also took second overall (Sepp Liebl-Remo Cattini), third (Willy Koenig-Remo Cattini) and fourth (Siegfried Günther-Gerhard Mitter), all driving 700 Record Monzas.

While celebrating the latest speed records established at Monza with the new one-litre twin-cam engine, Carlo Abarth officially launched the new version of the 1000 coupé, to which Mario Colucci had made numerous styling modifications, in particular to the lights and roof.
After acrimoniously terminating the agreement with Zagato of Milan to supply bodies, Abarth & C. assigned the construction of its bodyshells to a handful of Turin specialists, Corna, Beccaris, Sibona & Basano, retaining the right to finish the bodies themselves at the Corso Marche factory.
The Porsche Abarth Carrera GTL bodies (in the foreground of the picture, next to the new Fiat Abarth 1000 Bialbero) were, however, built by Viarengo & Filipponi of Turin.

whole team, Maglioli pushed the Fiat Abarth along the track in an effort to cover the final two kilometres or more, at the end of which there would be the 72-hour record. The mangled car and exhausted driver slowly approached and crossed the finish line, having covered 13,441.498 km in 72 hours at an average 186.867 km/h. They had beaten the previous record of 183.910 km/h, set up by Ford. Their feat became an enormous success that once again catapulted Carlo Abarth's name and his all-conquering car factory into the headlines.

During a party in Turin attended by the best-known motor industry and racing personalities to celebrate the establishment of the new records, the

Side-by-side in the forecourt of the Corso Marche factory are the new version of the Fiat Abarth 1000 Bialbero and the Porsche Abarth Carrera GTL, the body of which was made by panel beaters Viarengo & Filipponi of Turin after the first three units produced by Motto.

The new 2+2 Fiat Abarth 1600 coupé, which was to replace its predecessor, made its debut on the Abarth stand at the 42nd Turin Motor Show from 3-13 November 1960. Modifications included a power increase from 95 hp to 100 hp at 6000 rpm as a result of the adoption of two twin-choke carburettors; maximum speed went up from 180 to 190 km/h. Built by Carrozzeria Ellena of Turin, the car was of smooth, sober lines but not very original and yet fairly equilibrated, in spite of the feasibility of the cockpit in relation to the length of the bonnet and boot.

Corso Marche's four top technicians at the end of 1960 pose behind the latest Fiat Abarth to establish new records at Monza.
From the left they are: Giorgio Valentini, head of product for large cubic capacity cars, Luigi Guerrieri, in charge of the engine production department, Mario Colucci, technical director and Luciano Fochi, chief engine designer.

new Fiat Abarth 1000 Bialbero Coupé was unveiled, the engine of which had just been given an exceptional test in efficiency and reliability at the Monza track. The new car had been subjected to numerous styling changes by Colucci, although Carrozzeria Zagato received the year's Gold Compass award for its design of the Fiat Abarth 1000, although the work had actually been carried out by Sibona & Basano of Turin. Mario Colucci recalled, *"I had changed the style making the car more modern and softer. Zagato protested, complaining that the basis of the car was his and that we could not have competed to win the Gold Compass. For political reasons, Abarth preferred to put the problem on ice and Zagato was accredited with an award they did not earn"*. In effect, the car carried the customary Z on its sides and was of softer line without the Plexiglas fared headlamps, which gave way to normal round lights with chrome surrounds, but it retained the lateral windows introduced on the first generation of the Record Monza coupé, and the large air intake on top of the bonnet.

The Fiat Abarth 1000 was introduced on 3 November at the Turin Motor Show, where it was shown with avant garde mechanics that included the 982 cc twin-cam engine that produced 91 hp at 7100 rpm; the unit had a 9.3:1 compression ratio, two twin choke Weber 36 DCL4 carburettors and was fitted with front disc brakes. The maximum declared speed was 205 km/h and the 1000's price on the road was fixed at Lit 2.600.000. Carlo also showed the new versions of the 1600 cc Grand Touring models, built by Carrozzeria Ellena with numerous variations compared to the 1959 car; the mechanics were improved, with a maximum power increase from 95 to 100 hp and a consequent increment in top speed from 180 km/h to 190 km/h. Aesthetic modifications were also made to the 2200 coupé and roadster.

At the start of the Sixties, Abarth & C. had all the characteristics of a medium-size mechanical industry, while retaining traits peculiar to a highly specialised business of extensive hand-made content.
By the end of 1960, the Corso Marche complex covered 13,000 square metres and employed 225 people, a figure that was to rise to 419 three years later.

In the picture on the left, a stockpile of crankshafts ready for rectification and static and dynamic balancing. On the right, the technical office. Carlo Abarth categorically affirmed that his modifications to normal production engines were the result of painstaking technical research, sustained in turn by experiments and testing based on rigorous scientific principles.

An event that stimulated the Italian utility car market in the late summer of 1960 was the launch of a more powerful Fiat 600 D, the capacity of which was increased from 633 to 767 cc with a consequent appreciable increase in performance, maximum power moving up from 21.5 to 29 hp. Various parties argued that the renewal of the Fiat engine drew on many aspects of Abarth's transformation of the 600 production car years earlier, with its well-known 747 cc four-cylinder; undoubtedly, the experience gathered with that engine taught many lessons. At the end of October, the Corso Marche company had already prepared its grand touring version of the new Fiat

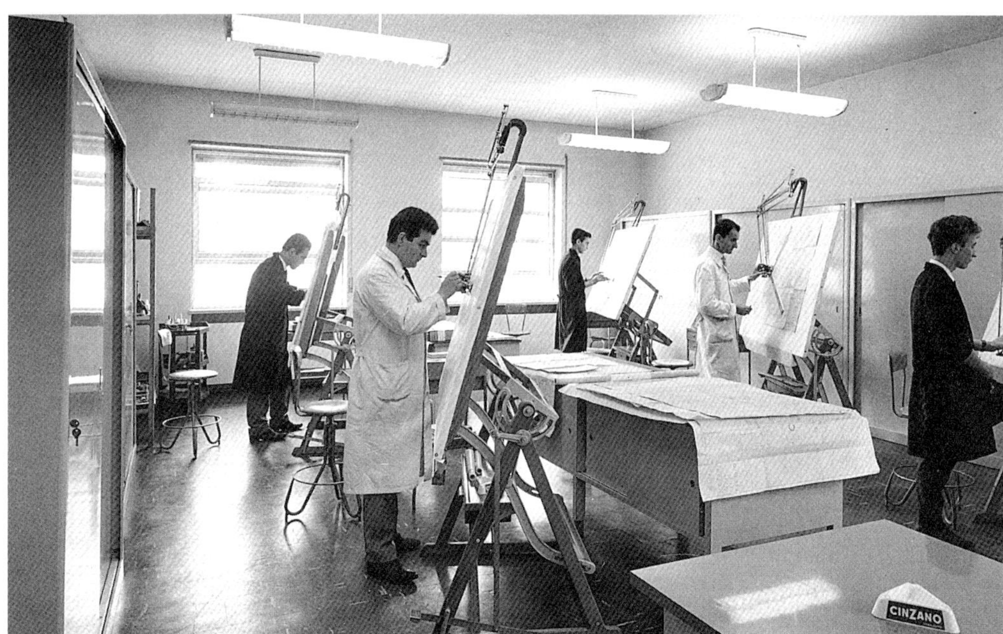

The mechanical manufacturing department enabled Abarth to respect extremely severe standards during work on individual engine components for conversion kits. With his uncanny intuition, Carlo Abarth was able to confer organisational and industrial dimensions on widespread and often empirical operations carried out by dozens of artisan workshops for the modification of mass produced engines.

engine, with its capacity increased to 847 cc; there were three different forms of modification: one produced a maximum 52 hp at 6000 rpm, the second 54 hp at 6300 rpm (the 850S) and the third – the 850 SS - 57 hp; the three differed in their compression ratios that spanned 8.8:1 to 9.9:1 and the increased diameter of the valves. The modification kit included a sump, externally identical to Fiat's normal production unit but internally the object of a series of mods, a key one of which was the increased diameter of the main bearings, the crankshaft, plus enlarged crank pins, con rods, pistons, valves, carburettor, radiator, a larger capacity sump and a completely redesigned exhaust system.

Left: A view of Abarth's engine building department and, right, the test room with dynamometric benches on which every single motor was checked for power generation and run-in briefly before being installed in its car.

Test sessions on the road were often assigned to the Abarth racing team's works drivers, as in the case of Mario Poltronieri (left) in this picture next to a 1000 Bialbero, taken at the end of 1960. Right: Giorgio Valentini, at the time the engineer responsible for Fiat Abarth 2200 product quality.

Right: New cars leaving the factory, ready for delivery to customers. The Abarths were transported by the Zust-Ambrosetti company using articulated lorries, in this case on an Autobianchi Filarete chassis.

Additional components, including perforated wheels, more powerful brakes and a supplementary radiator were available to customers interested in maximum performance. The latest generation of Fiat Abarths was test driven and the results published in the November 1960 edition of Italy's leading motoring magazine Quattroruote, a summary of whose comments were: *"It should be recognised that, especially for those who come from a normal 600, starting to drive these cars provokes a strange sensation of consternation. The rapid pick-up, the acceleration, the brio of the 850 make an impression; but one quickly recovers oneself and after a while driving an Abarth just as rapidly becomes a pleasure. To tell the truth, we would never have believed that a 600 could have such brilliant stability. One could almost say that more power had contributed to making grip more effective in a corner, and the merit we have always recognised in this kind of car, is that of being constant and in a continuous oversteer form of handling, on the 850 even more so".*

For Abarth, 1960 came to an end under the best auspices. Exhaust production had continued to expand with 218,028 units consigned to customers, but the success of the cars also forged ahead impressively. The little Abarths were in strong demand in exports markets. In the U.S.A., Franklin Delano Roosevelt Jr had established a highly active distribution centre on the Atlantic coast, Jim Parkinson of Burbank and Frank Rich of Glendale operated in California, while Motor Hoffman Co. and other concessionaires also promoted and sold the cars in America. Exhaust distribution was assigned exclusively to Fisher Products of Long Island in New York State. Renzo Avidano, who was also in charge of sales, still remembers today that the U.S. market accounted for less than 50% of total turnover and that Europe, especially Italy, France, Germany and the United Kingdom performed well. The fact that Abarth sales policy was based on somewhat unusual procedures should not be overlooked, such as demanding payment in advance for their cars, a practice that was

Carlo Abarth in his office in 1960, talking to his technical director Mario Colucci.
The walls behind him carry three large panels of perforated Masonite on which Abarth displayed the most significant photographs testifying to his activities in the industrial and sports fields. Relations with Colucci were, at times, stormy due to divergence of opinion on project technical matters, but their relationship was always a constructive one marked by maximum clarity.
Below: The great logo of the company with its popular scorpion – Carlo Abarth's zodiac sign – on the gate at the entrance to the Corso Marche factory.

hardly considered acceptable, especially in the United States. Other peculiarities included an absence of a factory guarantee for racing cars, which were delivered to customers without any cover against mechanical defects; it was also true that all Abarth engines were bench tested for a number of hours and that the cars were rigorously test driven on the road by a group of professionals before being approved for consignment.

In the case of pre-production prototypes and many racers, it was Abarth himself who tested the cars' performance with meticulous care; his ex-colleagues recall that he had a large area under his office where there was a photographic library, a table for private meetings and a cupboard containing his driving overalls and helmets for those tests. To test drive the cars, Carlo simply went down to that sort of bunker of his, as it was called in the company, put on his test gear and did his testing on the runway of Aeritalia Nord of Turin across the road from the Corso Marche factory.

The number of Abarth & C. employees had risen to 225 by the end of 1960 and a new building was being completed at 72 Corso Marche, taking the total covered area of the factory site to 13,204 square metres. The Via Trecate site was no longer used and the head office of the company was transferred to 38 Corso Marche.

The car was the biggest and richest symbol of most people's lifestyle in Italy in the early Sixties. The number of annual registrations were close to half

FIAT ABARTH "850 TC" (1961)

ENGINE	rear, longitudinal, 4 cylinders in line
Bore and stroke	mm 62.5x69
Cubic capacity	cc 847
Valve gear	overhead valves, side camshaft
Number of valves	2 per cylinder
Compression ratio	9.2:1
Fuel feed	1 Solex 32PBIC carburettor
Ignition	coil, distributor
Coolant	liquid
Lubrication	forced, oil filter on the main circuit
Maximum power	52 hp at 5800 rpm
TRANSMISSION	rear wheel drive
Clutch	single dry plate
Gearbox	4 speeds plus reverse, central lever
Axle ratio	4.55:1
BODY	2 door, 4 seat saloon
CHASSIS	integral body
Suspension	front independent, upper trailing arms, transverse lower leaf spring, hydraulic dampers; rear independent, trailing arms, helicoidal springs, hydraulic dampers
Brakes	hydraulically operated front discs and rear drums
Steering	worm and sector
Fuel tank	capacity 27 litres
Tyres	5.20x12
DIMENSIONS AND WEIGHT	
Wheelbase	mm 2000
Front track	mm 1150
Rear track	mm 1160
Length	mm 3285
Width	mm 1380
Height	mm 1405
Kerb weight	kg 610
PERFORMANCE	
Top speed	km/h 140

a million units, a significant increase on the approximately 200,000 new cars sold each year in the second half of the Fifties; in the three years 1960-62, the park increased by over a million vehicles. This powerful consumer push promoted by more effective and more widely diffused communication channels, contributed in notable measure to changing people's customs, orienting and regimenting them towards models widely available in other countries. An unfulfilled dream of many people, the car became the protagonist of this deep social change and entered into the simple fabric of daily life, closely linked to the ideals of speed and freedom of movement, both symbolised by a four-wheeled vehicle with engine. In this climate of euphoria, made even more euphoric by the presence of the latest Fiat utility models and the first timid approach to the market of a number of foreign manufacturers, the Abarth phenomenon got into top gear: it was a phenomenon that offered two extreme elements – small car, high performance - to a public at large that was particularly sensitive to them.

Carlo Abarth's fortune was not solely built on the stimulating mechanical modifications he carried out on small Fiats – in particular the 600 – but also on the willingness of Italian motorist to accept products of precise typology, in which the degree of performance assumed a priority role compared to aesthetics, even though attractiveness was also a widespread, strongly felt requirement. He often commented on the matter: *"There is probably no other country in the world with such a low number of cars driven by chauffeurs. Rich or poor, the Italians love to drive their cars themselves. Italy is also the least organised country in the world as far as cars are concerned. Where else is it*

In February 1961, Carlo Abarth played the luckiest card of his brilliant entrepreneurial career. He came up with a conversion of the new Fiat 600 D engine (bore and stroke 62x62.5 mm, 767 cc) engine that increased its power output to 847 cc, for the implementation of which he applied a number of technical improvements that were innovative in the Italian utility car world at the time. Bore and stroke went up to 62.5x69 mm, the compression ratio to 9.2:1 and the revolutions per minute by a further generous 1000 rpm over the Fiat production car. That produced a maximum power output of 52 hp at 5800 rpm, with a consequent increase in top speed to 140 km/h. With that model, which made its market debut as the 850 TC (Turismo Competizione or Touring Competition) Abarth spawned a whole new generation of high performance small cars that swiftly grabbed international public attention.

To adjust the 600 D's braking system to the much higher performance level of the 850 TC, Abarth used Girling disc brakes on the front but continued with drums on the rear; the adoption of discs was a modification of particular importance at the time, as disc brakes were considered the almost exclusive preserve of high priced sports cars. The Abarth conversion of the Fiat production saloon included the substitution of the crankshaft and camshaft, pistons and relative segments, piston pins, inlet manifold, carburettor, air filter, the whole exhaust system, sump, front semi-elliptic springs, rear helicoidal springs and existing rims for the self-ventilating type. To the right is the modified braking system, below the new front discs and above the engine cooling system, with a supplementary radiator placed transversally under the car's floor pan. Homologation in the Touring Category was requested seven to eight months ahead, after which time production had reached 1000 units.

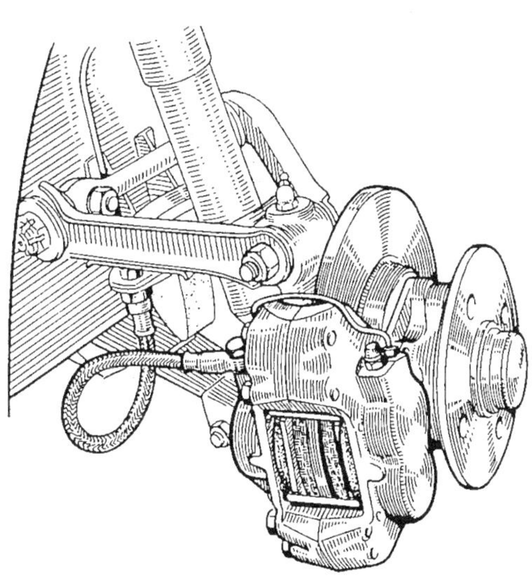

possible to built and sell a 500 cc that is faster than the majority of 1500 cc cars made elsewhere"? The name Abarth became part of the common vocabulary, not so much and not only to identify a quick and entertaining car but more to express a general concept of improvement, of excitement, of certain requisites that had, by that time, become the norm and were accepted by everyone. Abarth became a kind of synonym through which people liked to define any object or person that was out of the ordinary. And the victories the little Turin cars achieved every Sunday on the great international circuits or in national races acted like an accelerator to a sort of collective infatuation.

That is the spirit and social environment in which Carlo Abarth set out to begin the 1961 season. The opportunity of offering the market the 850 cc transformation of the Fiat 600D opened up a considerable commercial opportunity, which the Viennese born manufacturer grasped with both hands and with excellent timing. Mario Colucci explains: " ... *one of the first jobs given to me was to develop the 850 TC. We were correct and responsible in the design, in other words we did not increase engine power without improving the brakes and road holding. We adopted disc brakes. We had an excellent relationship with Fiat; they sent us the running 600 and we substituted the front wheel group with our own, which comprised wheels, hubs and disc brakes. Whatever was taken off was given back to Fiat, who sold it directly, without even test driving it".*

The Fiat Abarth 850 TC (TC stood for Turismo Competizione or Touring Competition) bowed in officially in February 1961. Abarth immediately asked that it be homologated, so that the car could also compete in the Grand Touring category, which required the construction of 1000 examples in seven to eight months. At the start of the year, Abarth's production consisted of the two 700 and 1000 cc sports models, each with a twin-cam engine and many components in common – body, floor-pan, basic architecture of the engine; the only substantial difference between the two were the brakes – disc on the

In those years, the Monza Autodrome often acted as the ideal test facility, not only for tuning Abarth single-seater record cars but also for shaking down and preparing cars that were to compete in important races. Taken in February 1961, the picture shows a team of Abarth technicians during a test session for a number of 1000 Bialberos.

1000 and drum on the 700, because when the smaller engined car was homologated it had bigger drum brakes front and rear. Next to those two cars was the production of the road cars, beginning with the new 850 TC saloon with the Fiat 600 D production car body, plus the Scorpion coupé and the Riviera roadster with an 833 cc engine, as well as the 1600 and 2200 in coupé and roadster forms. Being able to count on 'lightened' Fiats – without the components that were to be substituted by Abarth & C. – it was possible to set a list price of Lit 850.000 for the 850 TC, which also included a guarantee for the first time in the company's history. Compared to the prototypes built at the end of 1960, the production 850 TC was equipped with Girling front disc brakes as standard. To be doubly sure the modification work was conceived and carried out on the basis of rigorous criteria of efficiency and reliability, the car was fitted with a supplementary water radiator that was mounted transversely under the floor pan. Among the various additions over the original 600 D was supplementary instrumentation including a tachometer and oil thermometer, plus two spring-loaded calipers to keep the engine compartment open, which was considered useful when driving on the motorways but that became an in-crowd symbol, even among drivers of normal production Fiat 600s.

From the outset, the Scorpions began the 1961 motor sport season as they meant to continue - at full speed. After an initial victory at the Mexico City circuit with the 750 Record Monza, it was the turn of the 700 GT to win

Abarth revenge in the 850 cc class of the 1961 24 Hours of Le Mans on 10-11 June after a disappointing race the year before was taken convincingly by the New Zealand duo of Angus Hyslop and Denis Hulme, who covered 3,531.97 kms in their 850 S at an average speed of 147.166 km/h. The performance of the Fiat Abarth 700 Spider Sport number 49 was of no less impact, although negatively so. The car is shown here during practice with a frowning Carlo Abarth in the background; driven by Teodoro Zeccoli-Jean Vinatier, number 49 was at the centre of the first accident of the race, having gone off and buried itself in trackside sand after the first hour.

The three Fiat Abarth 700 Bialberos with sports saloon bodies were not much luckier than the 700 Spider Sport tubular chassis number 49. The cars with racing numbers 8 (Piero Frescobaldi-Raffaele Cammarota) and 56 (Giancarlo Sala-Giancarlo Rigamonti) were both forced to retire with mechanical problems, while number 55 (Paul Condriller-Karl Foitek-Giorgio Bassi) held the lead for much of the race, defending themselves well from attacks by D.B. of France, before having to retire a lap from the end with oil pouring from the engine.

the Routes du Nord on 10-12 February and the 1000 GT the Eisenstadter-Bergprufung during the same month; and then in Italy at the Agnano-Capella dei Cangiani on 12 March, won by a Fiat Abarth 850 GT and the St. Ambroeus Cup at Monza in which Cesare Toppetti in a 700 GT won his class in the Grand Touring Category and Mario Poltronieri came second in a sister car. March 24 was the ritual appointment with the 4 Hours of Sebring in Florida, where Harry Washburn won the up to 1000 cc class of the Grand Touring Category, followed by Robert Leiss in a similar car. From season's start to finish, the Fiat Abarth 1000 Bialbero won at every event for which it was entered: the seventh Rally du Limousin on 8-9 April, Stallavena-Boscochiesanuova on 9 April driven by Herbert Demetz, in Brussels and Vienna, the third Ascari Trophy on 7 May at Monza where the winners were Giancarlo Rigamonti and Jean Guichet, while the 700 Bialbero driven by Carlo Guandalini and Mauro Cintolesi won its class.

Twin-cam Fiat Abarth 1000s turned up in force for the third Grand Touring Grand Prix two-hour race at Monza on 29 June. It was a 'family' affair among the nine drivers competing in the one-litre class, which was won by Massimo Leto di Priolo, who also came third overall behind the Alfa Romeo Giulietta SZs of Elio Zagato and Consalvo Sanesi, who won at an average of 158.680 km/h. The picture shows Renzo Avidano (striped sweater), Abarth's sports director, congratulating the winner.

Among the most prestigious results was the 24 Hours of Le Mans on 10-11 June, when Dennis Hulme and Angus Hyslop dominated the up to 850 cc class of the Grand Touring Category in a Fiat Abarth 850 Bialbero at an average speed of 147.166 km/h. Abarth fielded five cars for the celebrated French long distance race, one 850 Biablero and four 700s with the same type of engine; one was an Abarth Sport with a tubular chassis that made its debut that year at Le Mans driven by Teodoro Zeccoli and Jean Vinatier. Avidano and Colucci both vividly remember a curious episode that could have compromised the start of that car. At the time, the regulations stipulated that it was

The 500 Kilometres of the Nürburgrirg, which took place on 3 September 1961 on the famous 22.810 km German track, confirmed Abarth's successful sporting year and the racing effectiveness of the Scorpion's latest models in the Touring and Grand Touring categories. The Corso Marche factory competed with four works 850 TCs driven respectively by Lothar Binder (car number 50), Karl Foitek (51), Ernst Furtmayr (52) and Salvatore Leto di Priolo-Aldo Nicosia (53), to which were added number 64 entered and driven by William McCowen-Heinrich Huellbuesch plus a number of 700 and 1000 Bialberos. Abarth triumphed that day, winning both categories, taking the first three places in the Touring and the first two in the Grand Touring categories.
Right: Taken at the start of the GT category race, in the foreground is the Fiat Abarth 1000 Bialbero number 111 of Eberhard Mahle-Teodoro Zeccoli, which came fifth behind winner Carlo Mario Abate (113) and second placed Massimo and Carlo Leto di Priolo (112). The Fiat Abarth 700 Bialbero of Willy Konig-Sepp Liebl, who retired, can also be seen in the background.

not permitted to locate the mouth of the petrol tank inside the cockpit but, on the incontrovertible instructions of Abarth, that is where Colucci had placed it. At scrutineering by the race marshals – the cars went through in line – the Fiat Abarth was rejected because it did not conform to the regulation covering the location of the fuel tank filler cap. *"We had to work hard and fast day and night to shift it"*, the two recall, *"but almost at the last moment, when it was necessary to empty all fuel tanks by a certain time as proscribed by the rules and re-fill them by a fixed time, one of our mechanics used a wrench incorrectly when unscrewing the fuel cap and the entire ring, which was welded onto the sheet metal, broke away from its. There were only two more hours to go before we ran out of time and could no longer fill the tank. Abarth himself, who never showed his feelings, did not know how to handle the situation. It was our general mechanic Maffiodo who saved the day with the help of a metal adhesive; to give the glue the best chance of hardening we waited until the last possible moment and then filled the tank. The little Abarth passed scrutineering, started the race, competed for the first hour and then went off, ending up in a sand bank where it remained for more than a day; and the metal plaster stayed intact".*

The Scorpions were successful again on 2 July in the Bolzano-Mendola, at the Trento-Bondone a week later, at the 6 Hours of Auvergne on 8 July where its cars came first and second in the up to 1000 cc class of the Grand Touring Category, and in another 11 races; in August Abarths won six races in three different countries and on 4 September Carlo Mario Abate won the Grand Touring class in a Fiat Abarth 1000 Bialbero at the 500 Kilometres of the Nürburgrirg, followed in second place by Massimo Leto di Priolo in a sister car. Sepp Liebl also won his up to 700 cc class at the legendary Germany track in a Fiat Abarth and in the up to 850 cc class of the Grand Touring

In the 850 cc class of the Touring category, the 850 TCs easily beat the Swedish Saab 96 and British Mini Minor. Victory went to Ernst Furtmayr in a fiercely fought 'family' battle between the little production bodied Fiat saloons.
Here as the cars race through the Eifel mountain forest, Karl Foitek in 850 TC number 51, who eventually came second in class, is shadowed by the 50 car of Lothar Binder, who came fifth.
Third place went to the 850 TC of William McCowen and Heinrich Huellbuesch and fourth to Swedes Gasta Karlsson-Ingemar Johansson with the Saab 96.

The night return of the Abarth racing team from the Nürburgrirg after its electrifying victories won on 3 September in the 500 Kilometre event. The picture shows 1000 Bialbero number 112 that Massimo and Carlo Leto di Priolo drove to victory and the 700 Bialbero number 99 of Willy Konig-Sepp Liebl, who retired. Also in picture is the 1000 Bialbero number 111 of Eberhard Mahle-Teodoro Zeccoli and a number of 850 TCs, which were a real revelation in the demanding German race.

Category victory was Ernst Furtmayr's in an 850 TC saloon, ahead of two other identical cars in the hands of Karl Foitek and B. McGowen. Six days later on 10 September 1961 at the 13th InterEurope Cup at Monza Alfonso Thiele, Herbert Demetz and Ermanno Gonella took the first three places in the up to 1000 cc class of the Grand Touring Category in Fiat Abarth 1000 Bialberos. Helmuth Zick won the up to 1000 cc class of the Grand Touring Category in at the 12 Hours of Hockenheim on 24 September, while on 2 October Richard-Vinatier came first in the performance index at the 10th Tour de France, driving a Fiat Abarth 700 Bialbero; then on 8 October at the Paris Motor Show Cup first place in the up to 1000 cc class of the Grand Touring Category went to René Richard in a Fiat Abarth 1000 Bialbero. This victorious season ended on

The Fiat Abarth 1000 Bialbero with a front radiator made its official debut at the 13th InterEurope Cup at Monza on 10 September. The car was fast from the start and proved itself able to keep the Alfa Romeo Giulietta SZ at bay. Alfonso Thiele took the overall win in car number 7 after covering 484.654 kms in the three-hour race at an average 161.551 km/h. In practice Thiele had put up the fastest time of 2'02.5" and was followed into third on the grid by Karl Foitek with 2'05.7", Massimo Dore Leto di Priolo in fourth with 2'05.8" and Herbert Demetz fifth with 2'05.9".

FIAT ABARTH "850 NÜRBURGRING" (1961)

ENGINE	
	rear, longitudinal, 4 cylinders in line
Bore and stroke	mm 62.5x69
Cubic capacity	cc 847
Valve gear	overhead valves, side camshaft
Number of valves	2 per cylinder
Compression ratio	9.8:1
Fuel feed	1 Solex 34PBIC carburettor
Ignition	coil, distributor
Coolant	liquid
Lubrication	forced, oil filter on the main circuit
Maximum power	55 hp at 6000 rpm
TRANSMISSION	
	rear wheel drive
Clutch	single dry plate
Gearbox	4 speed plus reverse, central lever
Axle ratio	4.55:1
BODY	
	2 door, 4 seat saloon
CHASSIS	
	integral body
Suspension	front independent, upper trailing arms, transverse lower leaf spring, hydraulic dampers; rear independent, trailing arms, helicoidal springs, hydraulic dampers
Brakes	hydraulically operated front discs and rear drums
Steering	worm and sector
Fuel tank	capacity 27 litres
Tyres	5.20x12
DIMENSIONS AND WEIGHT	
Wheelbase	mm 2000
Front track	mm 1150
Rear track	mm 1160
Length	mm 3285
Width	mm 1380
Height	mm 1405
Kerb weight	kg 610
PERFORMANCE	
Top speed	km/h 145

22 October at the 1000 Kilometres of Paris in which the first three places in the up to 1000 cc class of the Grand Touring Category went to Zeccoli-Sala in a Fiat Abarth 1000 Bialbero and Richard-Venatier followed by Seidel-Prinoth in 700 Bialberos.

Fiat Abarths scored seven overall victories and 116 class wins in 1961. Leo Cella won the up to 700 cc class Grand Touring Category Championship of Italy driving a 700 Bialbero prepared by Trivellato. As well as their speed, the little Turin cars demonstrated their great reliability in long distance races, in which mechanical components are submitted to great stress. That once again showed the validity of Abarth modifications to Fiat 600-based cars and the meticulous preparation with which they were readied to race. The robustness of the works racers was further spotlighted by the fact that many of them competed every Sunday and – as Avidano recalls today – there was not often enough time to check them or carry out laborious preparation work between one race and another.

At the Frankfurt Motor Show in September of that year, Fiat introduced the 2300S, a sports version of the 2300 saloon, with an elegant coupé four-seater body by Carrozzeria Ghia. Fiat turned to Carlo Abarth and his experience with the 2200 for the design of a car of such high performance – 136 hp, 200 km/h top speed – so he became a supplier to Fiat of a considerable number of mechanical items, including air intakes, exhaust manifolds, camshafts and high speed gearboxes. Abarth was also given the responsibility of checking and road testing the 2300S before delivery and Corso Marche had to form a group of 10 test drivers coordinated by engineer Klaus Steinmetz, who had

Advertising in the Italian motoring press commemorated the launch of the Fiat Abarth 850 TC (Quattroruote on the left) and the Nürburgrirg victories, from which a special version of the car was developed.

FIAT ABARTH "MONOMILLE" (1961)

ENGINE	
	rear, longitudinal, 4 cylinders in line
Bore and stroke	mm 62x74
Cubic capacity	cc 982
Valve gear	overhead valves, side camshaft
Number of valves	2 per cylinder
Compression ratio	9.8:1
Fuel feed	1 Solex 32PB1C carburettor
Ignition	coil, distributor
Coolant	liquid
Lubrication	forced, oil filter on the main circuit
Maximum power	60 hp at 6000 rpm
TRANSMISSION	
	drive
Clutch	single dry plate
Gearbox	4 speeds plus reverse, central lever
Axle ratio	4.75:1
BODY	
	2 seat coupé
CHASSIS	
	integral body
Suspension	front independent, upper trailing arms, transverse lower leaf spring, hydraulic dampers; rear independent, trailing arms, helicoidal springs, hydraulic dampers
Brakes	hydraulically operated discs all round
Steering	worm and sector
Fuel tank	capacity 30 litres
Tyres	5.50x13
DIMENSIONS AND WEIGHT	
Wheelbase	mm 2000
Front track	mm 1150
Rear track	mm 1160
Length	mm 3480
Width	mm 1410
Height	mm 1165
Kerb weight	kg 560
PERFORMANCE	
Maximum speed	km/h 175

joined the company on 1 June as a technician and test driver, to check the Fiat coupés before customers took delivery.

It was said in many quarters that such intense involvement in the production of the 2300S was decided on by Fiat after Carlo Abarth expressed his discontent. He was critical of the major national manufacturer for marketing a car that frustrated the effort and investment he had made over the two previous years to develop a prototype and build a small run of the coupé and the 2200 roadster, derived from the 2100 saloon. Compared to the 1959 model, those two cars were extensively re-styled by Carrozzeria Ellena and appeared

Carlo Abarth decided to come out with a new one-litre coupé at the 1961 Frankfurt Motor Show in September.
It was the Monomille, the body of which was designed at Corso Marche and built in aluminium by the Turin specialist Beccaris.
The commercial objective of the decision was to offer a car whose characteristics were not racy and aimed at a wide public audience.
The Monomille's name communicated the fact that distribution of the four-cylinder 982 cc engine was by a single camshaft, all of which produced 60 hp at 6000 rpm.
Sober and elegant, the car was an immediate international success, much appreciated for its sports performance, and had a top speed of 175 km/h.
Even though it differed due to a number of styling details – for example, large chrome bumper bars and front lights – the body of the Monomille was that of the Bialbero.

Simple and functional, the driving position of the Fiat Abarth Monomille clearly indicated its motor sport origins, especially the large diameter instruments that were easy to read and logically positioned, the slim, light steering wheel and the long ergonomically positioned gear lever. Less convincing, as can be seen in the photograph, was the quality of the door panels, visibly pock marked with screws, and the upper part of the dashboard. At the beginning of the Sixties, the level of finish of cars built by other special carrozzerie of the Italian school was better. The engine compartment of the Monomille houses the well-known single-cam power plant derived from the Fiat 600 D. The capacity increase from 767 cc (62x63.5 mm) to 982 cc (65x74 mm) was achieved by enlarging both the bore and stroke and modifying the compression ratio, intake, exhaust, brakes and suspension settings, using project and constructional standards that were well-established at Abarth.

for the first time at the Geneva Motor Show in March 1961. The 2200 was replaced at the Turin Show seven months later by the 2400: the new car retained more or less the same body, but its capacity was increased to over 2323 cc, power output went up to 142 hp and its maximum speed to 200 km/h. Fed by a battery of three twin choke Weber 38 DCOE carburettors, the engine was much more sophisticated in its transformed and developed form compared to the 136 hp Fiat 2300S, which had two twin choke carburettors; with a body in aluminium, the weight of the Abarth 2400 was contained at 1075 kg, while the 2300S and its all-steel Ghia body tipped the scales at a little over 1300 kg. The difference in price between the two was so great – Lit 2.600.000 for the Fiat 2300S and Lit 3.314.000 for the Abarth 2400 – that customers chose the Fiat, which was among the top sports car performers on the market.

The new unified Fiat Abarth coupés developed by Mario Colucci found success at the 1961 Turin Motor Show from 28 October-8 November, when the model also made its debut in 1000 Bialbero form. The line of the rear end, with its three-part windows and large, heavily sloped engine bay, was clearly inspired by Zagato's first Record Monza of three years earlier, although the overall shape was much smoother and softer, therefore aesthetically more pleasing.

FIAT ABARTH "2400" (1961)

ENGINE	
	front, longitudinal, 6 cylinders in line
Bore and stroke	mm 79x79
Cubic capacity	cc 2323
Valve gear	overhead valves, side camshaft
Number of valves	2 per cylinder
Compression ratio	9:1
Fuel feed	3 twin-choke Weber 38DCOE carburettors
Ignition	coil, distributor
Coolant	liquid
Lubrication	forced, oil filter on the main circuit
Maximum power	142 hp at 5800 rpm
TRANSMISSION	
	rear wheel drive
Clutch	single dry plate
Gearbox	4 speed plus reverse, central lever
Axle ratio	3.6:1
BODY	
	2 seat coupé by Ellena
CHASSIS	
	integral body
Suspension	front independent, deformable quadrangles, torsion bars, hydraulic dampers; rear rigid axle, push rod links with longitudinal semi-leaf springs, helicoidal springs, hydraulic dampers
Brakes	hydraulically operated discs front and rear
Steering	worm and roller
Fuel tank	capacity 60 litres
Tyres	165x15
DIMENSIONS AND WEIGHT	
Wheelbase	mm 2450
Front track	mm 1340
Rear track	mm 1370
Length	mm 4650
Width	mm 1624
Height	mm 1310
Kerb weight	kg 1075
PERFORMANCE	
Top speed	km/h 200

After the impressive victory of the 850 TC at the 500 Kilometres of the Nürburgrirg, Carlo Abarth wanted to introduce a higher performance version of the Fiat 600 D, which was having great sales success, using the same body as the normal production car. By increasing the compression ratio from 9.2:1 to 9.8:1, using different distribution timing and a more open exhaust system, maximum power increased to 55-57 hp at 6500 rpm and top speed rose to over 150 km/h, but among the various racing versions power output went way over 60 hp at 7,000 rpm and top speed was close on 160 km/h. Fitted with a small supplementary radiator, this version of the 850 TC was given the name

At the 1961 Turin Motor Show the Fiat Abarth 2200 and its six-cylinder engine handed over to the 2400, which was given a completely new two-seater body for the occasion. Carrozzeria Ellena created a highly stylish grand touring body on the updated mechanics of the 2300 saloon, the capacity of which was increased to 2323 cc and, by fitting three twin-choke Weber carburettors, maximum power went up to 142 hp at 5800 rpm. A class car that found few takers.

Nürburgrirg and was offered at the time of its late 1961 launch at Lit 890.000. For interest, the car's price was exactly the same as that of the Fiat 1100 Export, a medium size four-door saloon coveted by many Italian families, but which only produced 50 hp and a top speed of 130 km/h.

The new single cam, 60 hp 982 cc engine made its debut at the 1961 Frankfurt Motor Show on a model called the Monomille, a coupé designed by Mario Colucci using the 1000 Bialbero's architecture as a base and was identified by the characteristic three-part rear window and pleasant rounded line. Built on the now classical Fiat 600 D floor-pan with a wheelbase of 2000 mm, the Monomille was 3480 mm long, 1410 mm wide and 1165 mm high; it only weighed 560 kg and could easily reach 175 km/h at a declared fuel consumption of 8.2 litres per 100 km. It was a model that became extremely popular among people who did not race, but appreciated the style and sporting characteristics of the 1000 cc car. The launch of the Monomille in Europe and the United States generated a large number of contracts: Renzo Avidano recalls that the new model was a significant success in the U.S. and sold at a ratio of 10:1 against the 1000 Bialbero. By the end of a long journey to visit many of Fiat's American concessionaires, he himself had taken orders for hundreds of cars.

But the production capacity of the Corso Marche installation could not produce large numbers of the various models in the Abarth catalogue. For instance, in 1961 the plant made 1000 850 TCs with normal production Fiat bodies, and that quantity had been more than doubled to over 200 units a month in projections for the following year. Production of the 1000 Bialbero was reduced to 26 units for 1962, all for racing; a purposely small quantity to assure racing driver customers adequate service during the whole season. At the end of each racing season, the works cars were sold at prices that were sometimes higher than listed, because of the special preparation they required and the victories they had achieved on the track.

1961

1964

THE SCORPION TAKES ALL

In 1961 rumours began to circulate concerning a possible collaboration between Carlo Abarth and the French company Simca. Established in Nanterre in 1935 by Italian Enrico Teodoro Pigozzi, Simca (Société Industrielle de Mecanique et Carrosserie Automobile) began life collecting old car bodies for the Fiat foundry in Italy and later went on to assemble several of the Italian manufacturer's production models under licence. The intention was to offer the French public utility cars at a low price, taking advantage of the techno-industrial experience already acquired by Fiat in its national market. The year Simca was established, a Fiat 508 Balilla assembly line was set up at Nanterre and 12 months later production began of the Simca 5, an exact replica of the 500 Topolino launched a few months earlier by Fiat; in 1937 it was the turn of the Simca 8, the French version of the Fiat 1100. Just before the outbreak of the Second World War in 1939, Simca had assembled and marketed 70,000 cars in France during its five years of existence.

Technical cooperation continued between Simca and Fiat after the war, but the French company was to assume its own identity through the production of its own models, rather than derivations of the Italian cars. The 1949 version of the Simca 8, which closely resembled the Fiat 1100 E, was given a different radiator grill and a 1200 cc engine; the 9 Aronde, with its new and modern body shape, appeared in 1951 and was an immediate success. That was the first Simca to be designed and developed by the French company; the car's simple and robust mechanics included a four-cylinder 1221 cc engine that produced 45 hp and a top speed of 130 km/h; it descended on the French market in just the right moment. Simca's competitors did not expect the company to build a family saloon of such good performance, attractive appearance and mechanical efficiency; production stabilised at 100,000 cars a year, peaking in 1955 at 115.646 units. Produced until 1964, the company made a total of 1.425.329 Arondes.

With one of those inspired moves of a great entrepreneur that was to become typical of Pigozzi Simca incorporated Ford of Poissy in 1954, widening its range to include their V8 Vedette models and acquiring for itself an even more significant position in the industry. It was at the end of 1961 that the company achieved its greatest success with the introduction of the Simca 1000, its first rear-engined car and a four-door of square, attractive appearance, which became a direct competitor of the Renault Dauphine. The 1000 project began at Fiat and was assigned to engineer Oscar Montabone under Dante Giacosa, while the body was created in Turin by a team led by the prominent designer Mario Revelli di Beaumont. The technical side benefited from the experience gained by Fiat technicians with the 600 and resulted in a four-cylinder 944 cc engine that generated 35 hp, had a four-speed all-synchromesh gearbox, a compact and robust load-carrying structure, independent suspension all round with geometry similar to the 600's. With a 2220 mm wheelbase – over 10% longer than the little Fiat's – the new car was 3800 mm long, 1485 mm wide and 1335 mm high and had a dry weight of 698 kg. The Simca 1000 overtook the 600 D in dimensions, habitability, had six more horse power and a top speed of 120 km/h against the Fiat's 110 km/h but, of course, it had a 944 cc engine instead of the Italian car's 767 cc unit and that led to a fuel consumption of 7.5 litres per 100 kilometres against the small Fiat's 5.8 litres/100 km. By comparison, the new Simca seemed more advanced and in keeping with the needs of most European motorists than the 600, a design that was 10 years old.

The Simca 1000 elbowed its competitors to one side - and not only in the French market - to become the subject of an extraordinary commercial

Based on gossip circulating during the closing months of 1961, observers expected Abarth to work on the mechanics of the new, rear-engined Simca 1000 saloon and transform it into a sports car. But the big surprise came in February 1962, when the press were given a preview of the Simca Abarth 1300, powered by a new twin overhead camshaft engine (internal code 230) clearly inspired in its general architecture by the 1000 Bialbero motor of the by now famous Fiat Abarth. The body of the Italo-French car also felt the influence of the latest coupés designed by Mario Colucci at Corso Marche and built by Beccaris. The new model was even more attractive as a result of adopting a generously dimensioned wheelbase of 2090 mm compared to the Fiat 600's 2000 mm. Carlo Abarth invested considerable resources and enthusiasm in the 1300 coupé, which plugged a gap beyond the one-litre capacity in the range of cars he produced. Note the engine cooling radiator up front, like the more recent version of the Fiat Abarth 1000 Bialbero.

ABARTH SIMCA "1300 GT" (1962)	
ENGINE	
	rear, longitudinal, 4 cylinders in line
Bore and strok	mm 76x71
Cubic capacity	cc 1288
Valve gear	2 overhead camshafts, chain driven
Number of valves	2 per cylinder
Comprression ratio	10.4:1
Fuel feed	2 twin-choke Weber 45DCOE carburettors
Ignition	coil, distributor
Coolant	liquid
Lubrication	dry sump
Maximum power	7200 rpm
TRANSMISSION	
	rear wheel drive
Clutch	single dry plate
Gearbox	4 speed plus reverse, central lever
Axle ratio	4.1:1
BODY	
	2 seat coupé
CHASSIS	
	integral body
Suspension	front independent, upper trailing arms, transverse lower leaf spring, hydraulic dampers; rear independent, trailing arms, helicoida springs, hydraulic dampers
Brakes	hydraulically operated front and rear discs
Steering	worm and roller
Fuel tank	capacity 85 litres
Tyres	145x13
DIMENSIONS AND WEIGHT	
Wheelbase	mm 2090
Front track	mm 1260
Rear track	mm 1240
Length	mm 3555
Width	mm 1480
Height	mm 1140
Kerb weight	kg 630
PERFORMANCE	
Top speed	km/h 230

success. Unveiled at the Paris Motor Show in October 1961, it reached an annual production rate of 160.000 units as early as 1962 and lived a long life right through to 1978, overcoming the financial and management ups and downs of the company. Difficulties that meant the Chrysler Corporation acquired a 15% shareholding in the French concern in 1958. The Simca 1000 was also successful in Italy, but its diffusion was limited by its price of Lit 935.000, which was hardly in line with that of the Fiat 1100 Speciale at Lit 990.000 or the third series of the Lancia Appia costing Lit 1.215.000.

The relationship between Abarth and Simca began through the good offices of Rudy Hruska, the Viennese engineer with whom Carlo shared a brief but stimulating period at Cisitalia until 1949 and who, in the meantime, had joined the Nanterre firm as a technical consultant to Enrico Pigozzi. The Simca boss was fascinated by the Fiat Abarths' racing successes every weekend and attracted by the powerful promotional platform the victories gave the whole range of Fiat cars. He wanted Abarth to develop a number of prototypes based on his 1000 saloon with the mechanical modification of various components, which Simca would then evaluate for possible production purposes. In other words, on the basis of a specific agreement Abarth assumed the role of test centre for the French company. As part of that strictly research and test activity, Abarth was to produce a GT car using 1000 components as a base, which could contribute further to the image of the Simca brand in the eyes of motor sport enthusiast customers.

That was how the Abarth Simca 1300 came about, a car for which Carlo designed a completely new engine using the tried and tested architecture of the twin-cam 1000. With a bore and stroke of 76x71 mm for a total capacity of 1288 cc, the new four-cylinder generated 125 hp at 6000 rpm; the compression ratio was 10.4:1 and the engine was fed by two Weber 45 DCOE twin-choke carburettors. Distribution was by twin overhead camshafts with valves inclined at 80°, driven by a double chain. Lubrication was dry sump with two

Opposite: The tail of the Abarth Simca, which was longer and less invasive in appearance than that of the coupé on Fiat-derived mechanics, was much more harmonious while retaining the pleasant styling features of the rear and side windows introduced in the autumn of 1961 on the standard model.
The general appearance of the driving position was also similar to that of the Fiat Abarth coupé.
The four-cylinder, 1288 cc twin overhead cam engine – internal code 230 – had a bore and stroke of 76x71 mm and was the first power unit entirely designed by Abarth, inspired as it was in its basic architecture by the noted and well-tested Fiat-derived 982 cc four-cylinder; Carlo Abarth adopted dry sump lubrication for the new unit with two geared pumps, one to distribute and one to recoup.
With a compression ratio of 10.4:1 and fed by two twin-choke Weber carburettors, the engine generated a maximum 125 hp at 7200 rpm.

Original Simca 1000 components were used for the boxed pressed steel chassis, suspension, steering and gearbox, from which Abarth developed both five and six speed units, but not without difficulty.
Principally conceived to compete in the 1300 cc class of road and track races, the Abarth Simca was homologated in the Grand Touring category on 8 October 1962.
It was certainly no city car, but in racing it was able to take on the Alfa Romeo Giulietta effectively and win.

gear-wheel pumps – one delivery and one reciprocal – in the sump and the spark plug was in the centre of the hemispheric combustion chamber. Cooling was by front radiator, the used air being expelled through special low vents. The other suitably adapted components of the car – floor pan, transmission, steering and suspension – came from the Simca 1000, while the body reflected the successful line of the latest generation Fiat Abarth coupés. The new car was built on a 2090 mm wheelbase, was 3555 mm long, 1480 mm wide, 1480 mm high and had a dry weight of 630 kg. Accredited with a top speed of 230 km/h, it was introduced in February 1962 at a list price of Lit 3.300.000. While waiting for its Grand Touring Category homologation, which was granted on 8 October, the Abarth Simca 1300 competed in numerous Sport Category races and immediately became the car to beat.

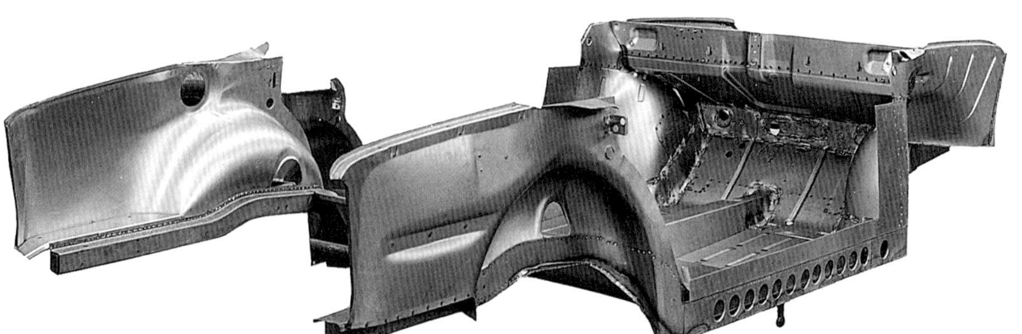

Four of the 10 Fiat Abarth 1000s that competed in the 3 Hours of Sebring on 23 March 1962, first round in the Grand Touring Constructors' World Championship. From the right, the cars of: Carlo Mario Abate (number 3) which retired, Alfonso Thiele (4) who came fourth, fifth placed Mauro Bianchi (5) and Paul Richards (6), who retired with clutch trouble. Overall and class victory went to car number 8 driven by New Zealander Bruce McLaren for the Briggs Cunningham team, followed in second place by another Fiat Abarth 1000 (7) with American Walt Hansgen at the wheel. McLaren (picture below) covered 384.254 kilometres or 46 laps at an average speed of 126.206 km/h. Third came Stirling Moss driving an Austin Healey Sprite. A lone Fiat Abarth 1000 was entered for the 12 Hours the following day for Alfonso Thiele and Jean Guichet, who won the performance index, the 1000 cc Sport category and came 10th overall.

For the 1962 season, the FIA Grand Touring Cup became the World Touring Car Constructors' Championship and was divided into a first division class of up to 1000 cc with sub-classes of up to 700 cc, 850 cc and 1000 cc; a second division for cars of up to 2,000 cc and sub-classes of up to 1300 cc, 1600 cc and 2000 cc; and a third division for over 2000 cc cars, with sub-classes of up to 2500 cc, up to 3000 cc and over 3000 cc.

Abarth's season began encouragingly with victory for the 850 TC at the eighth Circuit of Taulud at Asmara on 11 February and another 13 days later by the Fiat Abarth 1000 Bialbero in the Rally of Alicante in Spain. But it was on 23 March at the 3 Hours of Sebring in the U.S.A. that the 1000 Bialbero achieved one of its greatest successes. On the grid were many top-flight drivers, including those who were entered for the 12-hour race the following day. Stirling Moss shot into the lead in his Austin Healy Sprite on a track still awash with water from the previous night's hurricane; he kept the lead until the track began to dry out but, five minutes from the end, a Fiat Abarth driven by Bruce McLaren got the upper hand and was first across the finish line in his 1000 Bialbero, followed by Walt Hansgen in a sister car, Stirling Moss, Alfonso Thiele and Mauro Bianchi, the latter two also driving the little Italian cars. The four Abarths took four of the first five places in a flat-out race. McLaren covered 46 laps or 385 kilometres at an average 126.784 km/h. Electrified by the Saturday victory, the Abarths were on the grid again on Sunday for the 12 hour event and Alfonso Thiele and Jean Guichet won the up to 1150 cc class of the Sports Car Category driving the extraordinary 850 Bialbero. A success that was in the balance until the last moment, because before the start it was discovered that the 850's sump had a large hole, perhaps caused somehow during the voyage from Italy. There was no way to quickly solve the problem until Cornelio Maffiodo, the team's chief mechanic since its Cisitalia days, stepped in once again and plugged the hole with body shop metal filler to save the day.

On 26 April Abarth Simca 1300 GTs took the first three places in the Monte de la Lure hillclimb organised by the Automobile Club des Alpes, with Gianni Balzarini coming first overall, followed by two sister cars, one of which had an experimental roadster body. The seal was set on a positive day by Teodoro Zeccoli's class win in the Touring Category driving a Fiat Abarth 850 TC, a victory so convincing that it was protested and the car's cubic

The second round in the Grand Touring category of the World Constructors' Championship was the Garda Circuit (Angelo Maifredi Cup) at Salò on 1 May 1962 and it was another 'family' fight among the many Fiat Abarths on the grid. The winner was Ludovico Scarfiotti in 1000 Bialbero number 56, who completed the 19 laps or 311.600 kilometres in 2 hours 35'45.7" at an average 120.030 km/h, after an exciting duel with Innes Ireland driving sister car number 52. Many drivers were at the wheel of either a Fiat Abarth 1000 or 700, among them a number of well-known works team stars like Alfonso Thiele (car number 54, retired due to a mechanical breakdown), Jean Guichet (third in number 76) and Oddone Sigala (fourth in number 64).

capacity verified. Avidano still remembers with satisfaction that he was able to sell the protested car to one of its most vociferous accusers. Selling the car that had just won became a sort of commercial policy at Abarth & C. in an effort to discourage comment that the works cars were faster than those sold to customers.

Fiat Abarths added to their already impressive tally of victories every weekend in Italy, France, Switzerland, Germany and Austria, in the latter of which the Scorpions won the 700 cc, 850 cc and 1,000 cc sub-classes of the Grand Touring Division. At the Angelo Maifredi Cup on 1 May (the 12th Circuit of Garda) at Salò which was the second round in the world constructors' championship, victory went to Ludovico Scarfiotti, who just beat Innes Ireland and many other competitors driving Fiat Abarth 1000 Bialberos, and to Gianni Lado (Matich) with the 700 Bialbero in a race that was deserted the day before by the warlike Austin Healey Sprite, which made Carlo Abarth burst into one of his not infrequent fits of rage at the track. After the pre-start retirement of the Fiat Abarth's only serious rival because Stirling Moss had been involved in a serious accident a few days earlier, Abarth found itself the only competitor in a sort of family challenge. Carlo told his drivers not to engage in a head to head fight from the outset and invited them to drive to win only during the final laps. But once the race began they all rocketed away and started door handling, banging each other's cars. Furious at the drivers' insubordination at first and then disconsolate, Abarth had to control his impulse to retire the cars during the race only to eventually become excited by the enthusiasm and high competitive spirit shown by his drivers.

The City of Enna Cup at the Pergusa circuit on 15 August once again saw a flock of Fiat Abarth 1000s ready to fight for victory among themselves. The event was the fourth round in the World Constructors' Championship and was fought out over 65 laps of the 4.807 kilometre almost circular Sicilian track, a total of 312.455 kilometres.
Because the race took place two weeks before the 500 Kilometres of the Nürburgring, entries were mainly limited to many local drivers, Marsilio Pasotti ('Pam') and Oddone Sigala becoming the two key drivers of both practice and the race.
Pasotti won in Fiat Abarth 1000 number 58 at an average 163.032 km/h, followed by Claude Dubois of Belgium, who stepped into the limelight after Sigala retired with a broken axle shaft.

For Carlo Abarth, the most significant sporting event of the 1962 season was the 500 Kilometres of the Nürburgring, which took place on the historic German circuit on 2 September in perfect weather conditions.
A massive number of 14 Fiat Abarths were entered - six 850 TCs in the Touring category and a mixture of eight 1000s and 700s in Grand Touring.
The favourite, Hans Herrmann, was forced to retire his Fiat Abarth 1000 with a broken half shaft, so victory in the 1000 cc class went to Eberhard Mahle.
Right: the cars ready for the off, with the Herrmann-Ben Pon car number 85 in the foreground.
Below: There was an electrifying battle in the Touring category, won by Marsilio Pasotti in a Fiat Abarth 850 TC, between car number 48 (Paolo Renier-Gianni Marini Prada) who eventually came third in class and 11th overall, and sister car number 45 driven by Manfred Hoenig-Teodoro Zeccoli-Hubert Hahne, who took second in class and 10th overall.
Right: 1000 number 96 of Gunther Selback and Jurgen Wagner, who came second in their class and third overall.
Opposite: The Fiat Abarth 1000 of Eberhard Mahle, who won the day and, below, the Fiat Abarth 700 number 80 of Fulvio Sestili-Angela Fontana with the BMW 700 number 32 driven by Ulle Oberhammer and the DKW Junior number 60 of Wolf Dieter Mantzel-Heinrich Schuetz hot on their heels before the two German cars retired.
Right: Carlo Abarth, first time winner of the World Constructors' Championship with the three victors of the day, from the left Mahle (Grand Touring, 1000), Pasotti (Touring, 850 TC) and Spychiger (Grand Touring, 700).

At the Grand Prix of Berlin at Avus on 13 May, Robert Jenny and his Fiat Abarth 1000 Bialbero won the up to 1000 cc sub-class, a success repeated by Teodoro Zeccoli with the 700 Bialbero in his sub-class. At the City of Enna Cup on 15 August 'Pam' (Pasotti Marsilio) won with a Fiat Abarth 1000 Bialbero and Giovanni Pessina with the 700 Bialbero. To those victories were added the first convincing successes of the Abarth Simca GT. The Scorpion climbed to the pinnacle of its year's success on 2 September at the 500 Kilometres of the Nürburgring, at which Abarth won the first division world championship, having come first in five races out of five. Victory in the important German classic went to Eberhard Mahle in a Fiat Abarth 1000 Bialbero, ahead of Tommy Spychiger of Switzerland driving a Fiat Abarth 700 Bialbero

86
FIAT ABARTH

o-static
ENERGIT
BOSCH BOSCH
Autos liebe
Mobiloil

and Gunter Selbach-Jurgen Wagner in another Fiat Abarth 1000 Bialbero, while Hans Herrmann and Herbert Linge were forced to retire but Pasotti Marsilio won the Touring category with an 850 TC. On 16 September at the 400 Kilometres of Bridgehampton in the U.S.A., Bob Grossman won the up to 1000 cc Grand Touring class and came third overall, behind two Porsche Abarth GTLs. In 1962, Abarth achieved 234 class wins, seven outright victories and took three Italian championships - the up to 700 cc class of the Grand Touring Category with Matich (Gianni Lado) in a Fiat Abarth 700 Bialbero prepared by Trivellato, the up to 850 cc with Pam (Marsilio Pasotti) in a Fiat Abarth 850 TC also prepared by Trivellato and the up to 1000 cc class with Vittorio Venturi driving a Fiat Abarth 1000 Bialbero prepared by Aldo Faccioli. Abarth also won the division one up to 1000 cc World Constructors' Championship. In the new Abarth Simca 1300 GT's first year of racing the car won nine major races, among them the 15 July event at Auvergne where it took the first three places, the Freiburg-Schauinsland hillclimb won by Hans Herrmann a week later, the 1000 Kilometres of Paris won by Herrmann-Bianchi on 21 October and the Autumn Trophy at Vallelunga on 18 November when Lorenzo Bandini, Mauro Bianchi and Francesco De Leonibus took the first three places.

On the road car front, 1962 was full of new temptation for Abarth & C. Luciano Fochi, the company's engine designer, devised a 982 cc four-cylinder

The Fiat Abarth 850 TC of Mauro Bianchi and Pascal Ickx dominated the performance index class in the Touring category of the Tour de France, which took place from 15-23 September 1962 in five stages over 700 kilometres.

The Corso Marche factory entered the Abarth Simca Sport Spider Tubulare for the 1000 Kilometres of Paris, an event linked to the Paris Motor Show that took place in the French capital on 21 October. The car had a 1456 cc four-cylinder engine, derived from that of the Abarth Simca 1300, and it made a useful debut with the number 48 in the hands of Claude Bobrowski and Frank Ruata, who drove it into third place in its class and 14th overall.

Opposite: The Fiat Abarth 1000 Berlina came out in early June 1962. It had a 982 cc engine derived from the 850 TC unit that generated 60 hp at 6000 rpm, and a practically unchanged Fiat 600 D production car body.

FIAT ABARTH "1000 BERLINA" (1962)	
ENGINE	
	rear, longitudinal, 4 cylinders in line
Bore and stroke	mm 65x74
Cubic capacity	cc 982
Valve gear	overhead valves, side camshaft
Number of valves	2 per cylinder
Compression ratio	9.6:1
Fuel feed	1 Solex 32PBIC carburettor
Ignition	coil, distributor
Coolant	liquid
Lubrication	forced, oil filter in the main circuit
Maximum power	60 hp at 6000 rpm
TRANSMISSION	
	rear wheel drive
Clutch	single dry plate
Gearbox	4 speed plus reverse, central lever
Axle ratio	4.3:1
BODY	
	2 door, 4 seat saloon
CHASSIS	
	integral body
Suspension	front independent, upper trailing arms, transverse lower leaf spring, hydraulic dampers, rear independent, trailing arms, helicoidal springs, hydraulic dampers
Brakes	hydraulically operated discs front and rear
Steering	worm and sector
Fuel tank	capacity 27 litres
Tyres	5.20x12
DIMENSIONS AND WEIGHT	
Wheelbase	mm 2000
Front track	mm 1150
Rear track	mm 1160
Length	mm 3295
Width	mm 1380
Height	mm 1405
Kerb weight	kg 610
PERFORMANCE	
Top speed	km/h 158

single overhead camshaft power unit from the twin-cam 1000 cc motor, still using the same normal production Fiat 600 D block: this power unit led to Abarth winning an interminable series of victories over many years in the one-litre class of the Touring Car Category. The basic version of the Fiat Abarth 1000 Berlina developed 60 hp at 6000 rpm, but the power output of the racing version rose to 66 hp and even 68 hp at 6800 rpm, with a torque of 9.5 kg/m at 3500 rpm. The little Fiat bodied saloon could only be distinguished from the production car on the outside by the Abarth shield in the centre of the radiator grill plus Elektron rims by Campagnolo, who had acquired Amadori, and its slightly lower set-up. Dependent on which version, the 1000 Berlina weighed between 585 and 610 kg; the 60 hp derivative accelerated to a maximum 150 km/h and the racers to 170 km/h.

Abarth continued racing without respite and gave the mid-engined 1450 roadster its first outing at Innsbruck in October 1962, when it performed well in the hands of Hans Herrmann. The car was derived from the Spider Tubolare 750 cc and 1000 cc - both designed by Mario Colucci - but was subjected to a number of mechanical and body changes; the four-cylinder twin-cam engine had a 1459 cc capacity with a 76.2x80 mm bore and stroke and produced a maximum 145 hp at 7200 rpm. At first the car mounted a five-speed Colotti gearbox, but that proved too weak. So the Abarth technicians replaced it with the four-speed Simca 1000 production box and a six-speed derivation of the 'French' unit but that, too, generated numerous sealing problems. The car weighed 436 kg and its maximum speed was declared at 240 km/h.

The overwhelming success of the 1000 Bialbero in 1962 led to the introduction of a number of aerodynamic improvements to the car's body in the autumn of that year, especially at the rear end where the engine hood took on a longer, flatter appearance that improved stability at speed. The car's top

ABARTH 'SPIDER SPORT 1450" (1962)

ENGINE	
	mid-rear, longitudinal, 4 cylinders in line
Bore and stroke	mm 76.2x80
Cubic capacity	cc 1459
Valve gear	2 overhead camshafts, chain driven
Number of valves	2 per cylinder
Compression ratio	10.8:1
Fuel feed	2 twin-choke Weber 45DCOE9 carburettors
Ignition	coil, distributor
Coolant	liquid
Lubrication	dry sump
Maximum power	145 hp at 7600 rpm
TRANSMISSION	
	rear wheel drive
Clutch	single dry plate
Gearbox	4 speed plus reverse, central lever
Axle ratio	4.3:1
BODY	
	2 seat racing roadster
CHASSIS	
	tubular trellis
Suspension	front independent, trailing arms, helicoidal springs, stabiliser bar, hydraulic dampers; rear independent, trailing arms, helicoidal springs, stabiliser bar, hydraulic dampers
Brakes	hydraulically operated discs front and rear
Steering	rack and pinion
Fuel tank	capacity 56 litres
Tyres	front 4.50x13; rear 5.25x13
DIMENSIONS AND WEIGHT	
Wheelbase	mm 2090
Front track	mm 1200
Rear track	mm 1200
Length	mm 3325
Width	mm 1400
Height	mm 780
Kerb weight	kg 436
PERFORMANCE	
Top speed	km/h 250

Opposite: The designer two years earlier of the Fiat Abarth Spider Tubolari with 750 and 1000 cc mid-engines, Mario Colucci came up with a 1459 cc version in the summer of 1962. The car first appeared at the Innsbruck circuit in October, driven by Hans Herrmann, and had a four-speed Simca-derived gearbox that could not withstand the 140 hp power output of the engine and the torque the power unit generated.
But the Spider Tubolare 1450 was the watershed of a brilliant new generation of racing barchettas, which were penalised by a gearbox that revealed itself to be the real Achilles heel of the Abarth Simca.

speed was increased to 210 km/h following the adoption of a five-speed gearbox that Abarth had been testing for some time, but which had not been adopted until then because it was insufficiently reliable.

The 44th Turin Motor Show opened on 31 October and on the Abarth stand a 1000 Bialbero, which had won the first World Constructors' Championship for the Scorpion in September, was mounted on a large, elliptically shaped pedestal. That title and the brilliant sales results achieved by the company in 1962 took Carlo to a high level of notoriety and success. The number of Abarth employees had risen to 375, exhaust production was at 256,593 units of which 40% were exported and more than 100 cars a month were delivered. A special customer service department was opened at the Corso Marche headquarters and the whole Abarth complex covered around 15,000 square metres.

The name Abarth and the Scorpion logo became symbols of performance, social success and great innovative spirit, not only in the specialist environment of sports cars but also in the mind of the public; even so, Carlo still did not change his lifestyle or his entrepreneurial spirit. He almost never lost the chance to attend races or at least qualifying, where his presence imposed severity and discipline on everyone; sometimes he returned to his office on Sunday before a race so as to be ready Monday morning to begin the necessary modifications for the following event. He worked from 8 am and returned to his new home at 18 Corso Cairoli in the Valentino area or went to his club at around lunchtime; he was back at the office again at about 5 pm after his usual afternoon rest and change of suit to continue working with alacrity until late into the night. Every Monday, he habitually went to the cinema or the theatre – where he demanded a first or at least a third row seat – and did not like television except for newscasts, which he followed assiduously. He had no hobbies (*"I love my work more than anything else, more than money, more than anything. It is from work that I obtain the greatest satisfaction and all the rest"*) but he had a passion for antiques, particularly furniture, tapestries, rugs and porcelain. He said of his passion: *"My problem is that when I take interest in something, I want to become an expert and as I do not have the time I have had to give up many fine things that interested me"*.

Tall and solid, Carlo liked good cuisine in which he indulged, however, with moderation. A convinced champion of dieting as a means of preventing increased weight, in the early Sixties he started eating apples one after the other during the day, to the point of making them an affectation; there were always cases of apples around him in the office and at home and he considered them a passport to a long and healthy life. Abarth had smoked heavily – up to 60 a day – until 1952 and then he suddenly stopped, although he always kept cigarettes with him for his guests.

He had an eye for proportion and a highly developed taste; he often used the Italian expression for 'non-aesthetic' to express his disagreement when faced with formal solutions, which he considered clumsy or hardly harmonious. He liked to clarify a concept: *"We are in Italy, home of art and elegance. We love beautiful things"*.

The Fiat Abarth 1000 Bialbero was introduced with a number of mechanical and styling improvements at the 49th Paris Motor Show in mid-October 1962 and the subsequent Turin show, where the success of the model in winning the Grand Touring World Constructors' Championship, which had just concluded, was commemorated.
The car was given a five-speed box that led to an increased 210 km/h top speed. The car's rear engine cover was raised even higher and called the 'duck tail' which, at least in expectation, contributed to improving its high-speed stability.
As a result of the increase in compression ratio to 10.8:1, the installation of new twin-choke carburettors and other tuning improvements, the 982 four-cylinder twin-cam produced close on 100 hp at 7100 rpm, a power output that went up to 105 hp in the last few races of the 1962 season.

An overall view of the Abarth stand at the 1962 Turin Motor Show, where the place of honour went to the world championship-winning 1000 Bialbero. Carlo Abarth had achieved one of his most cherished ambitions by winning the world title with a well designed and built car bearing his name, which showed itself to be fast and reliable against tough international opposition.

Below: In 1962, the Fiat Abarth 1600 was re-styled and in an attempt to offer his sporting customers a car rich in personality with new design features, Carlo Abarth allocated the task of completely renewing the coupé version (left) to Carrozzeria Ellena. Allemano was given the job of coming up with a new cabriolet, which also had a hard top for the winter (right).

He never compromised in his work and defended his position with ardour. His relationship with his technical director, Mario Colucci, was built on open collaboration but it was often disturbed by violent disputes on technical matters that they saw from different angles, one from a talented designer's viewpoint of scientific preparation and the other with the rather empirical experience and exceptional intuition that Abarth had matured in over 30 years of working in a wide variety of motor vehicle plants. Mentally reliving those moments of great passion and extreme commitment again today, Colucci admits to have always received from Abarth great freedom of action in managing the design and construction of prototypes, without any constraint on expenditure; and not to be obliged to operate within limits imposed by a budget meant a designer could enjoy the maximum creative independence and exploit the inspiration of the moment in the best possible way.

The year 1963 opened in a climate of general euphoria. After overcoming the difficult period of post-war reconstruction and the state of great collective enthusiasm having passed into history as the country's 'economic miracle', the Italian economy had been consolidating itself year after year. As in all the western countries, car production was a prime parameter by which to measure the potential of modern Italian society and its level of development. In 1962 the national industry produced 877,860 cars, of which 305,429 were exported; new car registrations reached 634,706 units, while the national park broke the three million barrier for the first time at 3,030,056 cars. There were many indicators that pointed to a 1963 of record connotations; the rapid expansion of consumption as a result of an increase in family income, the relative social tranquillity and a political environment not yet pervaded by profound disagreement that upset it deeply towards the end of the decade, suggested that the country would continue to prosper for a few more years, even though it had not adequately confronted the enormous structural imbalance

The distribution of Abarth exhaust systems in international markets made a tremendous contribution to promoting the Scorpion brand beyond just the motor racing community to the public at large.

With impactful advertising for the exhausts on its sides, the Fiat Multipla was a formidable promotional tool among cars on the Italian roads of 40 years ago. In 1962, Abarth exhaust production went up to 256,593 units.

ABARTH SIMCA "1600 GT" (1963)

ENGINE	
	rear, longitudinal, 4 cylinders in line
Bore and stroke	mm 86x68.5
Cubic capacity	cc 1592
Valve gear	2 overhead camshafts, chain driven
Number of valves	2 per cylinder
Compression ratio	9.8:1
Fuel feed	2 twin-choke Weber 45DCOE9 carburettors
Ignition	double, coil, distributor
Coolant	liquid
Lubrication	dry sump
Maximum power	155 hp at 7400 rpm
TRANSMISSION	
	rear wheel drive
Clutch	single dry plate
Gearbox	6 speed plus reverse, central lever
Axle ratio	4.1:1
BODY	
	2 seat coupé
CHASSIS	
	integral body
Suspension	front independent, upper trailing arms, transverse lower leaf spring, hydraulic dampers; rear independent, trailing arms, helicoidal springs, hydraulic dampers
Brakes	hydraulically operated discs front and rear
Steering	worm and roller
Fuel tank	capacity 85 litres
Tyres	5.25x13
DIMENSIONS AND WEIGHT	
Wheelbase	mm 2090
Front track	mm 1250
Rear track	mm 1234
Length	mm 3700
Width	mm 1500
Height	mm 1120
Kerb weight	kg 740
PERFORMANCE	
Top speed	km/h 240

The sales success of the Abarth Simca 1300 GT and its early race victories - particularly in hillclimbs - induced Carlo Abarth to increase the development of the car and a number of derivatives of it for the 1963 racing season.
The Abarth Simca 1600 was given a sneak preview at the Geneva Motor Show in February; its general architecture was inherited from the 1300, although it did introduce a number of important improvements.
For example, the 1591 cc engine (internal code 235) was completely designed in-house and at 68.5 mm had a bigger bore than any other power unit produced at Abarth; its crankshaft turned on five supports and, for the first time in Abarth's history, was the car given double ignition with two dovetailed dynamos connected to two camshafts. Body changes were limited to the rear end, which had a long overhang with an engine cover shaped like a duck's tail spoiler, inspired by those of the latest generation Fiat Abarth.

that had existed from the time of the unification of Italy. In effect, 1963 revealed itself to be peppered with many positive circumstances pervaded by a climate that would favour additional phases of expansion that would be of benefit to the car.

At Abarth, the new season began with the Geneva Motor Show presentation of the Abarth Simca 1600 GT, which was of similar design to the previous year's 1300 GT except for the rear end which, in line with aerodynamic dictates warmly supported by Colucci, assumed a large inclination over the

The Abarth Simca 1600 GT did not make its debut at the Geneva Motor Show but a two-litre version did.
That car was a much more coherent choice and was highly competitive in the up to 2000 cc class of the Grand Touring Championship.
The Abarth Simca 2000 had and extremely attractive body, with all the latest developments of the period at the rear and soon became a feared racing competitor.
The road-going version developed a maximum of 177 hp at 6800 rpm with a dry weight of 725 kg.
The car was homologated at the end of 1963 and began to compete in the Grand Touring category at the start of the 1964 racing season.

rear axle at the end of the engine cover. That made for greater air pressure on the tail and, therefore, better performance of the rear end; a function that was, in part, performed on other Abarth models by the engine cover being raised and kept in position by a couple of aesthetically dubious callipers.

The more important and unexpected characteristic of the new 1600 GT was, however, its engine, which was not a derivation of the previous four-cylinder 1288 cc of the Abarth Simca 1300 GT, but of new design. The exceptional performance of the car, which beat the Alfa Romeo Giulietta many times during the 1962 racing season, was gratifying and at the last race at Vallelunga in November, it lapped at times similar to those of the two-litre cars. But it was believed the experience gained with that engine could be better exploited with a shorter racing power unit that had a crankshaft mounted on five supports. The new 1600 had a 68.5 mm bore and an 86 mm stroke for a total cubic capacity of 1591, a compression ratio of 9.8:1 and a flat combustion chamber that enabled the use of double ignition with two dovetailed

ABARTH SIMCA "2000 GT" (1963)

ENGINE	
	rear, longitudinal, 4 cylinders in line
Bore and stroke	mm 88x80
Cubic capacity	cc 1946
Valve gear	2 overhead camshafts, chain driven
Number of valves	2 per cylinder
Compression ratio	9.8:1
Fuel feed	2 twin-choke Weber 45DCOE9 carburettors
Ignition	double, coil, distributor
Coolant	liquid
Lubrication	dry sump
Maximum power	177 hp at 6800 rpm
TRANSMISSION	
	rear wheel drive
Clutch	single dry plate
Gearbox	6 speed plus reverse, central lever
Axle ratio	4.9:1
BODY	
	2 seat coupé
CHASSIS	
	integral body
Suspension	front independent, upper trailing arms, transverse lower leaf spring, hydraulic dampers; rear independent, trailing arms, helicoidal springs, hydraulic dampers
Brakes	hydraulically operated discs front and rear
Steering	worm and roller
Fuel tank	capacity 85 litres
Tyres	front 5.00x13; rear 5.50x13
DIMENSIONS AND WEIGHT	
Wheelbase	mm 2090
Front track	mm 1260
Rear track	mm 1272
Length	mm 3818
Width	mm 1500
Height	mm 1120
Kerb weight	kg 725
PERFORMANCE	
Top speed	km/h 240

The Abarth Simca 2000 Coupé was given its sneak preview in the centre of the company's stand at the 1963 Geneva Motor Show from 14-15 March. On the right in the foreground is the Abarth Simca 1300 GT, half-hidden behind which is the Fiat Abarth Monomille model. On the far left is the 1000 Bialbero, in front of which are the 850 TC and 1000 Berlina with Fiat 600 D production bodies.

coils at the extreme of the two overhead camshafts, a feature used a few months earlier on some 1300 cc engines from which a maximum power at the bench was about 138 hp at 7800 rpm. The valves were placed at 74° and fuel was fed into the engine by two twin-choke horizontal Weber 45 DCOE carburettors; as with the 1300, lubrication was dry sump and a radiator at the front of the car cooled the engine. The maximum power generated by the engine was 155 hp at 7400 rpm, with a 17 kg/m torque at 5500 rpm. The gearbox was a six speed derived from that of the normal production Simca 1000 and built at Abarth, where they added the two extra gears, which were housed in an external sump. Some mechanical components- the suspension and steering – also came from the Simca 1000, while Girling disc brakes were on all four wheels and were operated by a three-piston pump. Built on a structure of pressed steel with a 2090 mm wheelbase – the same as the 1300 coupé and 130 mm shorter than the original Simca saloon – and with a body in aluminium, the 1600 GT was 3700 mm long, 1500 mm wide, 1120 mm high and weighed 740 kg. Top speed was 240 km/h.

But the new Abarth Simca was not taken any further because with no 1600 cc racing class it would have been at a disadvantage in the up to 2000 cc class of the Grand Touring Championship. Nevertheless, it acted as a test car for new model design and in the development of the Abarth Simca 2000, which also made a low-key appearance at Geneva. During the 1963 Geneva show, Italy's CSAI confirmed the homologation of the Fiat Abarth 1000 in the Touring

Car Category and the Monomille in Grand Touring, enabling the two cars to compete in all national and international races. The Fiat Abarth 1000 Berlina Corsa (66 hp at 6800 rpm with a front radiator à la Tour de France) went on sale in March with a full guarantee, except for racing or rallying. The special versions with compression ratios of between 10.5:1 and 11.5:1 were exclusively for racing and were not guaranteed. The price for all versions was Lit 1.389.900 each. The '63 Fiat Abarth 850 TC Nürburgring Corsa was made available with four different compression ratios. The basic car's was 9.8:1, which gave it 68 hp; it was suitable for medium distance races and was supplied with a full guarantee, which immediately became null and void if the car competed in any kind of race. The 11:5:1 compression ratio engine for long distance races was sold without a guarantee, as were the 12.5:1 and the 12.8:1 derivatives. The list price of all versions, which had supplementary front radiators, was fixed at Lit 1.269.900. The 1963 Monomille coupé had a long rear end with a 'duck tail' spoiler that had already been seen on the new Fiat Abarth 1000 Bialbero and the Abarth Simca coupé.

The motor sport season began with the third Rally of the Flowers, in which Deserti-Magni won the Touring Category in a Fiat Abarth 850 TC and Marzorati-Airoldi the Prepared Touring class. The succession of victories continued, among them the 3 Hours of Daytona on 17 February in which Hans Herrmann came first in the up to 1300 cc class of the second Division and

The 3 Hours of Daytona, first race in the Grand Touring Constructors' World Championship second and third division, took place on 17 February 1963 on the famous 6.132 kilometre Florida track.
Among the 50 entries were two works Abarth Simca 1300 GTs, number 70 driven by Hans Herrmann and 71 with Mauro Bianchi at the wheel.
Herrmann is pictured on the circuit's banking overtaking the Porsche 356 B Carrera 2 number 82 of Charlie Kolb and Mike Kurkjian's 46 Porsche.
The German driver won the 1300 cc class and took ninth overall; Mauro Bianchi retired the sister car.

19
FIAT ABARTH 1000
18

Opposite: It was Sebring again for the next U.S. race five weeks later on 22 March, when the three-hour event for one-litre cars took place. Only 10 cars competed, four of them Fiat Abarth 1000 Bialberos of which three were works cars. Hans Herrmann in the 18 car and Mauro Bianchi in car number 19 dominated the event without effort, coming first and second overall. The two pictures below show the Fiat Abarths crossing the finish line (left) and works team sports director Renzo Avidano talking about the race with Carlo Abarth.

The 12 Hours of Sebring, a round in the second and third divisions of the Grand Touring World Constructors' Championship, took place a day later on 23 March. Of the four Abarth Simca 1300 GTs that started three retired, but number 61 (pictured) driven by Tommy Spychinger, Teddy Pilette, Jean Guichet and Pierre Noblet won the 1300 cc class and came 21st overall. The Hans Herrmann-Mauro Bianchi car number 60 is shown overtaking Triumph prototype number 40 with its twin cam engine prepared by Virgilio Conrero and driven by Ed Wilson, Ralph Salyer, Bob Hall, Charlie Kolb and Art Tweedale.

ninth overall in an Abarth Simca 1300 GT; he repeated that success five days later at the 3 Hours of Sebring by winning the up to 1000 cc class of the Grand Touring Category at an average 129.395 mph; then there was Armeno Degli Innocenti's overall and class victories at the wheel of an 850 TC at the 23rd Ponente Riviera Cup on 23-24th. At the 6th Stallavena-Boscochiesanuova, Europe's fastest hillclimb on 7 April, the Scorpion's cars won again, with Francesco Ghezzi in a Fiat Abarth 850 TC in Grand Touring, Maurizio Pinchetti (Fiat Abarth 700 Bialbero) and Oddone Sigala (Fiat Abarth 1000 Bialbero) in their respective classes. This avalanche of successes continued every weekend throughout the summer across Europe and in many other countries, wherever drivers were present at the wheel of those formidable little cars, which won on road and track, in long and short distance races and at hillclimbs. 'Pam'-Benedetto Guarino won the up to 1000 cc class in the

ABARTH "SPIDER SPORT 2000" (1963)

ENGINE	
	mid-rear, longitudinal, 4 cylinders in line
Bore and stroke	mm 88x80
Cubic capacity	cc 1946
Valve gear	2 overhead camshafts, chain driven
Number of valves	2 per cylinder
Compression ratio	10:1
Fuel feed	2 twin-choke Weber 58DCOE3 carburettors
Ignition	double, coil, distributor
Coolant	liquid
Lubrication	dry sump
Maximum power	188 hp at 7300 rpm
TRANSMISSION	
	rear wheel drive
Clutch	single dry plate
Gearbox	4 speed plus reverse, central lever
Axle ratio	4.9:1
BODY	
	2 seat racing roadster
CHASSIS	
	tubular trellis
Suspension	front independent, trailing arms, helicoidal springs, stabiliser bar, hydraulic dampers; rear independent, trailing arms, helicoidal springs, stabiliser bar, hydraulic dampers
Brakes	hydraulically operated discs front and rear
Steering	rack and pinion
Fuel tank	capacity 56 litres
Tyres	front 4.50x13; rear 5.25x13
DIMENSIONS AND WEIGHT	
Wheelbase	mm 2140
Front track	mm 1210
Rear track	mm 1215
Length	mm 3585
Width	mm 1540
Height	mm 1030
Kerb weight	kg 585
PERFORMANCE	
Top speed	km/h 260

Grand Touring Category at the Circuit of Garda on 12 May and Ricciardo Ricci the up to 700 cc class; Ricci did it again in his Fiat Abarth 700 Bialbero at the 7th Bologna-Raticosa of 26 May, while in the up to 1300 cc Grand Touring class Giampiero Biscaldi came first in an Abarth Simca 1300 GT and Hans Herrmann won the Sport class with an Abarth 2000; at the 25th Consuma Cup on 2 June, first place went to Ricciardo Ricci in the up to 700 cc Grand Touring class and Kurt Zuegg in the up to 1000 cc, while Mauro Bianchi came fourth overall in an Abarth Simca 1300 GT, as did Hans Herrmann in his class driving an Abarth 2000 Sport.

At the Rossfeld hillclimb on 9 June, Bianchi won the up to 1300 cc Grand Touring class in an Abarth Simca 1300 GT, while a week later Ricciardo Ricci took the up to 700 cc Grand Touring class at the 6th City of Asiago Cup in a Fiat Abarth 700 Bialbero, 'Tiger' (Romano Perdomi) the up to 1000 cc class in a Fiat Abarth 1000 Bialbero and Oddone Sigala the up to 1300 cc class driving an Abarth Simca 1300 GT. On 29 June Girolamo Capra in a Fiat Abarth 700 Bialbero won his class at the 4th Grand Touring Grand Prix at Monza, 'Tiger' did the same in a 1000 Bialbero as did Guido Rava in an Abarth Simca 1300 GT. At the Rally of Wiesbaden from 3-7 July the Fiat Abarth 1000 Bialbero won again driven by Peter Ettmuller-Xavier Perrot, on 11 August Hans Herrmann came second overall in an Abarth two-litre in the Freiburg-Schauinsland hillclimb and Tommy Spychiger won the up to 1000 cc Sport class. A week later, Giovanbattista Mainetti won the City of Enna Cup in a Fiat Abarth 700 Bialbero and 'Tiger' came first overall and won the Grand Touring up to 1000 cc class.

On 25 August at the Ollon-Villars Herrmann came second overall in an Abarth 2000 SP behind the Porsche 718 of Edgar Barth, and Spychiger won the up to 1000 cc Sport class. On 8 September Spychiger won again, this time in the Abarth Simca 1300 GT of the InterEurope Cup at Monza. Rudy Golderer and Karl von Katen won the up to 3000 cc class of the Tour de France and came 12th overall in a Fiat 2300 S during the week 14-22 September. At the 500 Kilometres of Bridgehampton on 14 September, final round in the American Grand Touring Constructors' Championship, Thomas Fleming won the 1300 cc class in an Abarth Simca 1300 GT.

Abarth became famous because its cars were unbeatable. Its 1963 racing results included 535 victories, of which 90 were scored by the Abarth Simca 1300 GT. The Scorpion won the Division 1 Grand Touring Championship for the second successive year in the three sub-classes of up to 700 cc, up to 850 cc and up to 1000 cc and came fourth in the Division 2 and second in its up to 1300 cc sub-class. And the Abarth Simca 1300 GT proved itself to be fast at the Nürburgring: it took 10 seconds a lap off the Porsche Abarth GTL.

Hans Herrmann drove an Abarth 2000 Sport Spider Tubolare number 19 in the Rossfeld hillclimb, a German round in the second Division of the World Constructors Championship, which twisted and turned for six kilometres through 31 corners in the Bavarian mountains.
He put up the fastest time in the first race but was penalised in the second by bad weather, so that the best he could do in the final results table was fourth.
Mauro Bianchi won the Grand Touring category in an Abarth Simca 1300 GT, followed by Herrmann and Piero Biscaldi, who is pictured driving number 38.

A week later on 16 June 1963, Hans Herrmann won the Grand Prix of Vienna at the 2370-metre Aspern circuit on the banks of the Danube in the outskirts of the Austrian capital at the wheel of the 2000 Sport Spider Tubolare. A beaming Carlo Abarth shared the laurels with his German driver.

Success again for Hans Herrmann at the 2.3 kilometre Norisring circuit near Nuremberg, where the German won the 1300 cc class of the Grand Touring category at the wheel of an Abarth Simca 1300 GT.

In 1963, a modified version of the Fiat 2300 S established several international speed records and won a number of circuit races. The cubic capacity of the normal production car was slightly increased to 2323 cc to equal that of the Fiat Abarth 2400 – a 'square' engine with a bore and stroke of 79 mm – with a compression ratio of 10.4:1 and a battery of three Weber 40 DCOE carburettors instead of the usual two twin-chokes, to return a power output of 172 hp at 6500 rpm; a high level, if one considers that the six cylinder Fiat had a single overhead camshaft and did not, therefore, lend itself to extensive modification. The top speed of that specially modified version was 228 km/h.

From 1-5 April Carlo and his team were at the Monza high-speed track to try and break a number of international D class (2000-3000 cc) records with the Fiat 2300 S coupé special. For that demanding undertaking, he put together a group of nine top-flight racing and test drivers whom he instructed to

The fourth round in the Grand Touring World Constructors' Championship and the fifth in the European Hillclimb Championship took place on 11 August on the edge of Germany's Black Forest over the historic 11.2 kilometre Freiburg-Schauinsland road with its 750-metre climb.
Mauro Bianchi (pictured) drove 700 cc Bialbero prototype engined Fiat Abarth Sport Spider to a class win, while Hans Herrmann came second in a 2000 Sport Spider Tubulare behind Edgar Barth and his two-litre Porsche 718WRS.
The Abarth Simca confirmed its domination of the 1300 class of the Grand Touring category in the same race, in which the protagonists were Giampiero Biscaldi and Germany's Kurt Geiss (pictured).

Luciano Bianchi won at the 11.5 kilometre Solitude circuit on the outskirts of Stuttgart on 28 July 1963 in a Fiat 1000 Bialbero. The picture shows Bianchi being pursued by Engelbert Moll, who later crashed his Simca 1300 GT.

The penultimate round in the 1963 European Hillclimb Championship was fought out on 25 August at Ollon-Villars, Switzerland. The route climbed for eight kilometres to 660 metres above sea level and attracted a huge entry for the two-heat race of 115 cars, of which 96 started and 87 finished, among them Tommy Spychiger (car 129) who campaigned an Abarth 1000 Sport Spider Tubolare with the 1000 Bialbero engine. In the foreground standing next to the car is Abarth sports director Renzo Avidano, with technician Antonio Tomaini at the rear. Spychiger won the one-litre class of the Sport/Prototype category and Hans Herrmann came second overall and in the two-litre class at the wheel of a 2000 Sport Spider Tubolare.

Opposite: Abarth confirmed its supremacy in the Touring, Grand Touring and Prototype categories on the historic Nürburgring circuit during the 500-kilometre race on 1 September 1963. Among the first to win was Fiat Abarth 1000 Berlina number 62 driven by Mauro Bianchi and Giampiero Biscaldi (photograph), then the Fiat Abarth 1000 Bialbero number 104 of Anzio Zucchi, while overall victory went to Teddy Pilette-Hans Herrmann in a Fiat Abarth 850. There were many retirements: of the 20 Abarths flagged away, 12 dropped out.

begin by maintaining an average speed of 180 km/h and progressively increase it after the first 5000 kilometres. Re-fuelling and driver change stops were scheduled for every three-and-a-half hours. The car was fitted with a supplementary fuel tank on the driver's side, as well as two batteries; the body was partially lightened to make way for selected spare parts and tools in case of emergencies, as provided for by the rules; the side chromed profiles and the radiator grill were taken off and an additional air intake on the bonnet was the only visual difference between the record car and the normal production model. This is how Mario Poltronieri recalled that umpteenth adventure at Monza in Auto-Italiana of 19 April 1963: *"Remo Cattini took the first shift at 13.20 on Monday 1 April. Then the others took their turn: Steinmetz, Gauch, Icxs, Pilette, Arena, Balzarini, Demetz, Bianchi and I alternating at the wheel. Steinmetz, the Porsche engineer on whom the production of the 2300 Abarth depended, and the 'old' drivers Cattini and I drove the most demanding night shifts; on the second night I also set the fastest lap at an average of over 196 km/h; it is curious to observe that, in all previous attempts on the record, the fastest lap was set at around 3 pm and not in the cold of the nigh). After a more than satisfactory start, a few insignificant but annoying ignition problems began to show themselves right from the first night and they forced us to frequently change the plugs, which were also standard, and that reduced our average speed. From here came the opportunity to go faster, so the average speed went up to around 190 km/h.*

"Anyway, we established our first record with Demetz at the wheel; we broke the old 5000 mile of 158.520 km/h with a taxing 179.631 km/h. Successively, Arena was driving when we broke the 48 hour with 180.208 against 158.890 km/h held, like all the others, by Austin Healey since 1958; Remo Cattini was driving when we broke the 10,000 kilometres with an average 180.778 km/h against the British car's 157.050 km/h and Balzarini was

FIAT ABARTH "2300 S" (1963)

ENGINE	front, longitudinal, 6 cylinders in line
Bore and stroke	mm 79x79
Cubic capacity	cc 2323
Valve gear	overhead valves, side camshaft
Number of valves	2 per cylinder
Compression ratio	9.8:1
Fuel feed	3 twin-choke Weber 40DCOE carburettors
Ignition	coil, distributor
Coolant	liquid
Lubrication	forced, oil filter on the main circuit
Maximum power	172 hp at 6500 rpm
TRANSMISSION	rear wheel drive
Clutch	single dry plate
Gearbox	4 speed plus reverse, central lever
Axle ratio	4.7:1
BODY	2+2 coupé by Ghia
CHASSIS	integral body
Suspension	front independent, deformable quadrangles, torsion bars, stabiliser bar, hydraulic dampers; rear rigid axle, longitudinal semi-elliptic leaf spring, stabiliser bar, hydraulic dampers
Brakes	hydraulically operated discs front and rear
Steering	worm and roller
Fuel tank	capacity 70 litres
Tyres	165x15
DIMENSIONS AND WEIGHT	
Wheelbase	mm 2650
Front track	mm 1350
Rear track	mm 1312
Length	mm 4600
Width	mm 1635
Height	mm 1380
Kerb weight	kg 1110
PERFORMANCE	
Top speed	km/h 218

in the car when were completed 72 hours at an average 178.198 km/h, beating the Healey's 156.630 km/h.

Soon afterwards the sump cracked, possibly caused by a stone, and that meant a loss of oil. As a consequence, there was the beginning of a seizure and, regardless of attempts to repair the engine, a connecting rod broke. Having repaired that, the car stopped for the first time at 1.05 am on Friday and re-started again after a final refuelling session at about 6 am. But as Carlo Abarth was not going to be satisfied with the average speeds of the records we were about to break, he felt we had achieved our objectives in part and stopped the operation, saying he would reserve the right to continue in the future to cover greater distance".

So the 2300 S was called in as it was going for the 15,000 kilometre and 96-hour records after it had covered 14,750 km at an average of 177 km/h. Slightly more modestly modified – 165 hp and a maximum of 220 km/h – the car was entered for several circuit races in 1963, when it competed with success and even beat the 3.8- litre Jaguar. A particularly significant result was brought home by Paul Frère and Luciano Bianchi driving a Fiat Abarth 2300S in the 12 Hours of the Nürburgring on 14 July, when they won their class and came second overall.

The 1963 season also saw feverish activity in the Sport category. Mario Colucci's trellis tube chassis roadster did well; it was a car in which different capacity twin-cam engines were installed, dependent on the race in which the car was to compete – 850 cc, 1000 cc, 1300 cc, 1450 cc, 1600 cc or 2000 cc. Few examples were made of those cars, which were almost entirely built in the Corso Marche workshop with the exception of a few components, like disc brakes, wheels and some engine parts bought from reliable external suppliers. Colucci re-designed the reticular chassis in 1963, using different diameter tubes in line with the power of the engines installed. He recalls: "*When I went*

ABARTH CORSE
FIAT
2300 S

SHELL
FIAT
2300 S

FIAT
2300 S
ABARTH CORSE

Abarth was back breaking records at Monza from 1-5 April 1963, this time in the international D class (2000-3000 cc). The car was the Fiat 2300 S, suitably modified to improve its performance.
The engine was given a slight capacity increase to 2323 cc, new fuel feed and exhaust systems and a voluminous supplementary fuel tank was installed next to the driving position with a quick release fuel filler cap protruding through the right window.

At 180-190 km/h, the 2300 S set up new records for the 5000 miles, 48 hours, 10,000 km, 72 hours but was stopped after 14,750 kms due to problems with the lubrication system.
The drivers who alternated at the wheel of the 2300 S are pictured below and are, from the left: Pilette, Gausch, Pascal Ickx, Mauro Bianchi, Virgilio, Poltronieri, Arena, Marin, sports director Avidano, Balzarini, Steinmetz and, on the extreme right, Carlo Abarth.

to Abarth I took with me the technology I had developed at Alfa Romeo, where I designed in 1959-60 the Giulia TZ with steel tubes of 22 millimetres. I also designed tubed chassis at Abarth. Naturally, we are talking plain tubes that became increasingly covered, eliminating cross members with panels of sheet metal and so on. At Abarth, the tendency towards rear engines inclined over the back end was the practice, because Carlo Abarth was keen on that aspect of Porsche technology and even though I did not share his keenness, the criterion had a considerable effect on the initial design of the chassis".

Identified by the internal factory code SE04, the 1963 chassis brought to life the new generation of Spider Tubolare or Tubular Roadsters, which were entered for many races during the year; they were also of a special construction that included an ample wrap-around windscreen retained at the two

"The 1300 prototype was only built for experimental purposes and, therefore, not put into production".
So said the Abarth & C. press release of 14 May 1963. Mario Colucci had redesigned the 1960-62 tubular chassis roadster, making numerous improvements; the most obvious was, without doubt, a windscreen that could be raised and had a demountable steep frame, which meant Abarth could easily transform the car from a Prototype to a Sport, depending on the regulations in force, and compete without problem in races of both categories.
As with other tubular chassis cars, the engine stayed centre-rear, with the gearbox overhanging the rear axle.
In its general structure, the 1288 cc four-cylinder engine duplicated the unit installed in the Abarth Simca 1300 GT (twin camshaft, single ignition) but power was increased to 130 hp at 7600 rpm.
The aluminium body was built with large front and rear compartments, hinged at the two extremes of the car to facilitate single opening and easier access to the mechanics.
The dry weight of the 1300 Spider Tubolare was 460 kg, 170 kg less than the Abarth Simca 1300 GT coupé.

transparent sides above the doors, dependent on the rules in force for prototypes at the time; it was a structure that could be easily removed so as to compete in Sport category races. With a 1300 cc engine producing 130 hp at 7600 cc and a dry weight of 460 kg, the Spider Tubolare made its debut in the hands of Tommy Spychiger on 14 July 1963, when it won the Trento-Bondone with an average speed of 78.094 km/h. The same car fitted with a one-litre, 105 hp twin-cam engine and driven by Vittorio Venturi won the Freiburg-Schauinsland and the Auvergne Trophy on 7 July with Mauro Bianchi driving; but it did not do so well in the 500 Kilometres of the Nürburgring, where

ABARTH "1300 SPIDER SPORT" (1963)

ENGINE	
	mid-rear, longitudinal, 4 cylinders in line
Bore and stroke	mm 76x71
Cubic capacity	cc 1288
Valve gear	2 overhead camshafts, chain driven
Number of valves	2 per cylinder
Compression ratio	9.6:1
Fuel feed	2 twin-choke Weber 45DCOE9 carburettors
Ignition	coil, distributor
Coolant	liquid
Lubrication	dry sump
Maximum power	130 hp at 7600 rpm
TRANSMISSION	
	rear wheel drive
Clutch	single dry plate
Gearbox	6 speed plus reverse, central lever
Axle ratio	4.1:1
BODY	
	2 seat racing roadster
CHASSIS	
	tubular trellis
Suspension	front independent, trailing arms, helicoidal springs, stabiliser bar, hydraulic dampers; rear independent, trailing arms, helicoidal springs, stabiliser bar, hydraulic dampers
Brakes	hydraulically operated discs front and rear
Steering	rack and pinion
Fuel tank	capacity 70 litres
Tyres	front 4.50x13; rear 5.00x13
DIMENSIONS AND WEIGHT	
Wheelbase	mm 2140
Front track	mm 1210
Rear track	mm 1215
Length	mm 3585
Width	mm 1540
Height	mm 1030
Kerb weight	kg 460
PERFORMANCE	
Top speed	km/h 230

The Abarth 1300 Sport Spider Tubolare was to have made its debut on 9 June at the Rossfeld hillclimb in Germany driven by Gianni Balzarini (car 12) and was, in fact, entered for the up to 2000 cc class of the Sport/Prototype category. The car practiced on 8 June but was not sufficiently sorted and so it did not race.

Hans Herrmann spun off at the Karusel while he was leading. Fitted with a two-litre engine, the Spider Tubolare won the Grand Prix of Vienna driven by Herrmann at Aspern on 13 June. On 20 October, the same car won the fifth San Remo of the Mountains with Luigi Taramazzo at the wheel.

Don Rosendale of America's Road and Track magazine wrote briefly about the 1450 cc version of the Spider Tubolare: *"The car seemed a little like an Italianised Lotus 19 or 23, with a more powerful engine than that of the Abarth Simca 1300. With a bore and stroke of 76.2x80 mm for an overall cubic capacity of 1459 cc, it has twin-cam distribution and two horizontal twin-choke Weber 45 DCOE9 carburettors; with a compression ratio of 10.8:1, it generates 145 hp at 7200 rpm. Gearbox, transmission and suspension are derived from the production Simca, while the brakes are Girling discs on all four wheels. Built on a wheelbase of 2120 mm, it is 3560 mm long, 1460 mm wide and 1112 mm high; its dry weight is 430 kg, less than a Porsche RS. Top speed is over 160 mph. Inside, the cockpit is simple not to say spartan; a large tachometer faces the steering wheel, as do three other smaller instruments; grouped together on the left side is a tangle of small tubes that form the reticular structure of the chassis, which make the car look like the Maserati Birdcage.*

The tail points slightly upwards for aerodynamic reasons, while the body comprises two large sections, the front and rear hinged and completely openable to facilitate maintenance; the overall appearance is not, perhaps, very attractive but it is certainly functional in terms of aerodynamic efficiency. At a track race that took place recently at Innsbruck airport, the Abarth Sport 1450 was very fast and came second behind a Lotus 2500 (which could be described as a Formula One with fenders) and ahead of all the Porsche RS 60s and RS 61s".

FIAT ABARTH "595" (1963)

ENGINE	
	rear, longitudinal, 2 cylinders in line
Bore and stroke	mm 73.5x70
Cubic capacity	cc 594
Valve gear	overhead valves, side camshaft
Number of valves	2 per cylinder
Compression ratio	9.2:1
Fuel feed	1 Solex C28PBJ carburettor
Ignition	coil, distributor
Coolant	forced air
Lubrication	forced, oil filter on the main circuit
Maximum power	27 hp at 5000 rpm
TRANSMISSION	
	rear wheel drive
Clutch	single dry plate
Gearbox	4 speed plus reverse, central lever
Axle ratio	5.13:1
BODY	
	2 door, 4 seat saloon
CHASSIS	
	integral body
Suspension	front independent, upper trailing arms, transverse lower spring, hydraulic dampers; rear independent, trailing arms, helicoidal springs, hydraulic dampers
Brakes	hydraulically operated drums front and rear
Steering	worm and sector
Fuel tank	capacity 21 litres
Tyres	125x12
DIMENSIONS AND WEIGHT	
Wheelbase	mm 1840
Front track	mm 1121
Rear track	mm 1135
Length	mm 2970
Width	mm 1320
Height	mm 1300
Kerb weight	kg 470
PERFORMANCE	
Top speed	km/h 120

In September 1963 it was the turn of the little Fiat 500 to be subjected to the fortifying medication of Carlo Abarth. Always sensitive to the aspirations of the young, to satisfying the needs of motorisation and the demand for higher performance even in utility cars, the Italo-Viennese technician believed the moment had come to also work on the mechanics of the smallest Fiat. In the meantime (1960), the car had become the Fiat 500 D, losing the word Nuova along the way and picking up a capacity increase from 479 cc to 499.5 cc, with a corresponding boost in power from 13 to 17.5 hp that pushed up its maximum speed from 85 to 95 km/h. At Corso Marche, the car's capacity was upped to 595 cc to produce a vigorous increase in power to 27 hp at 5000 rpm and maximum torque of 5 kg/m at 3500 rpm, against the 3.6 kg/m of the production car. Carlo was clearly inspired by the experience gained in February 1958, when he made news by establishing six international records with the Nuova 500 at an average speed of almost 110 km/h.

The engine modifications were extensive, to the point of joining the two cylinders in one fusion of cast iron, which made for an improved, more homogeneous distribution of heat and, therefore, also cooling; the cylinder head's inlet ducts and exhaust were polished, the valve springs substituted and a special camshaft was adopted. The increase in compression ratio from 7:1 to 9.5:1 was achieved by fitting a suitably designed piston with a crown top profile; lubrication was changed by the use of a new gear-wheel pump placed in the sump, which was also of new design and made of light alloy. The fuel feed was modified by the fitment of a Solex C28 PBJ carburettor in place of the Weber

In February 1958, Carlo Abarth made a number of modifications to a two-cylinder Nuova 500 engine, demonstrating at Monza that even the smallest Fiat could become a vetturette of great efficiency and reliability.
In September 1963, he decided to repeat that exhilarating experience with the launch of the Fiat Abarth 595, the engine capacity of which had been upped from 499.5 to 595 cc by increasing its bore from 67.4 to 73.5 mm, but leaving the stroke unchanged. Externally, the 595 differed from the mass produced Fiat by its front badging with Scorpion logo and model name on the sides and engine cover.

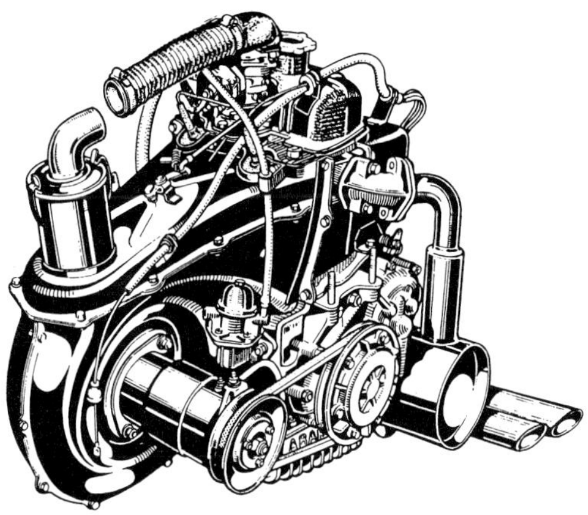

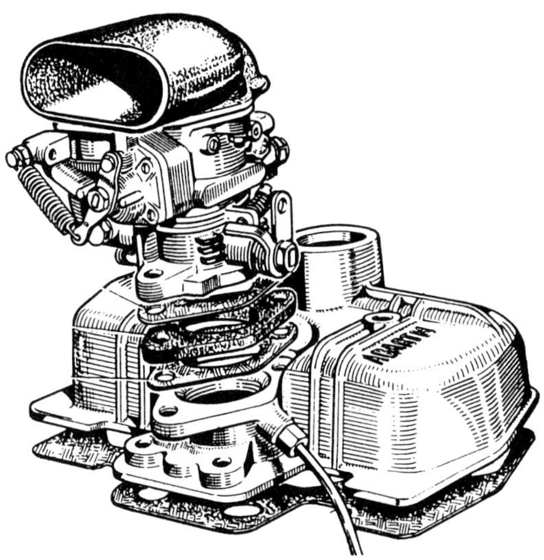

As well as increasing its capacity to 595 cc, other extensive modifications were made to the two-cylinder engine. Cast in special iron, the two cylinders were made into one to improve cooling homogeneity; a new camshaft and sump were installed, the inlet and exhaust manifolds were polished and new pistons were fitted to increase the compression ratio from 7:1 to 9.5:1.

Replacing the standard Weber carburettor with a 28 mm Solex contributed to increasing maximum power from 22 to 27 hp at 5000 rpm and maximum torque from 3.6 to 5 kg/m at 3500 rpm. Abarth gave the 595's top speed as over 120 km/h and the car's price was fixed at Lit 595.000, Lit 1000 for each cc of displacement.

26 IMB4 on the production car. With such a robust injection of new energy the little Fiat was able to exceed 120 km/h, but its price was not low for the period at Lit 595.000 or Lit 145.000 or 30% more than the basic 500 D, to which another Lit 90.000 had to be added for the special instruments that replaced Fiat's original equipment. With a vigorous acceleration and handling that would be the envy of many one-litres, the Fiat Abarth 595 became an agile car that was invincible on the various roads and in the mountains. Almost immediately, a quicker version was produced that put out 30 hp and had a top speed of 130 km/h, for which Abarth & C. applied for homologation in the Grand Touring Category.

Two months later, the Abarth technicians trained their sights on another utility car that they similarly modified. It was the Simca 1000, a small four-door saloon already known in Italy and from which the magicians at Corso Marche conjured up four different versions, all with 1137 cc engines and named the Simca Abarth 1150. Engine improvements were based on many years of long experience of the same four-cylinder Fiat production power unit: increased engine capacity was obtained by taking the bore from 68 to 69 mm, the stroke from 65 to 76 mm and strengthening other components, among them the transmission, suspension and brakes, the latter with discs on all four wheels. The basic 1150 developed 55 hp and could exceed 150 km/h; of the other three, all with front radiators, the 1150 S produced 58 hp and topped 155 km/h, the 1150 SS 65 hp and delivered 160 km/h and the 1150 Corsa generated 85 hp to reach a maximum speed of 170 km/h. The only outward difference between the production Simca and the 1150 was a number of front air vents at the front of the modified car and its designation carried on the rear wings.

That year, Abarth had a large stand at the Turin Motor Show that attracted enormous crowds, enchanted by the Scorpion's fascinating cars and

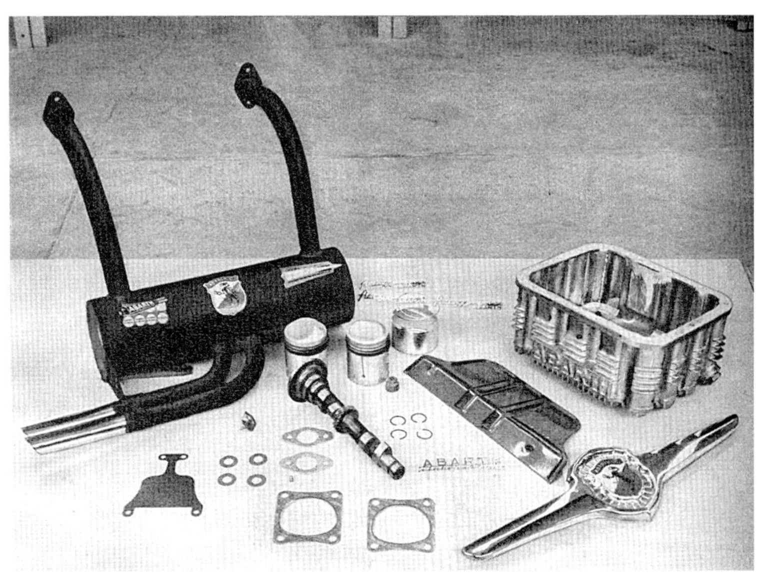

A few months after launching the 595, Abarth marketed a conversion kit for owners of the mass produced version of the little Fiat who wanted to modify their cars to 595 level. For those who did not want to increase their cars' capacity to more than half a litre, the Corso Marche factory came up with a simpler conversion kit (picture).

SIMCA ABARTH "1150" (1963)

ENGINE	
	rear, longitudinal, 4 cylinders in line
Bore and stroke	mm 69x76
Cubic capacity	cc 1137
Valve gear	overhead valves, side camshaft
Number of valves	2 per cylinder
Compression ratio	8.5:1
Fuel feed	1 Solex F32PBIC carburettor
Ignition	coil, distributor
Coolant	liquid
Lubrication	forced, oil filter on the main circuit
Maximum power	55 hp at 5600 rpm
TRANSMISSION	
	rear wheel drive
Clutch	single dry plate
Gearbox	4 speed plus reverse, central lever
Axle ratio	4.1:1
BODY	
	4 door, 4 seat saloon
CHASSIS	
	integral body
Suspension	front independent, upper trailing arms, transverse lower leaf spring, hydraulic dampers; rear independent, trailing arms, helicoidal springs, hydraulic dampers
Brakes	hydraulically operated discs front and rear
Steering	worm and roller
Fuel tank	capacity 36 litres
Tyres	145x13
DIMENSIONS AND WEIGHT	
Wheelbase	mm 2200
Front track	mm 1250
Rear track	mm 1222
Length	mm 3797
Width	mm 1485
Height	mm 1330
Kerb weight	kg 720
PERFORMANCE	
Top speed	km/h 150

excited at the prospect of experiencing the exhilaration of a demonstration lap in an Abarth on the show's test track. One of the cars on display was an updated version of the 1000 Bialbero coupé, the engine of which boasted double ignition and a power increase of 11 hp from 91 to 102 hp at 7600 rpm. That car's rear end also had a long tail with a built-in spoiler, as did the Monomille, Abarth Simca 1300 GT, 1600 GT and 2000 GT. While the first model was made entirely of aluminium by Carrozzeria Sibona & Basano, the new one would have a front end of one single pressing of glass fibre that tipped fully forward; that needed new technology for which Abarth turned to Balzaretti & Modigliani of Turin, a company with many years' experience of producing items in glass fibre and synthetic resins.

Auto-Italiana magazine of 2 December 1963 published brief driving impressions of the three most talked about Scorpions of the moment, the Fiat Abarth 595, 850 TC Nürburgring and the Simca Abarth 1150. In summary, he wrote of the 595: *"the car's acceleration is truly surprising and is enough to enthuse those who appreciate nippiness and good handling, rather than maximum power and high speed. This car would be in its element in chaotic city traffic, as it would on the tortuous roads of the mountains... Remembering the car's origin and price (Lit 595.000 excluding tax) our verdict is positive with some reserve over noise, but with much appreciation of its performance".*

Their comments on the 850 TC Nürburgring were even more flattering: "*The performance of this inspired Abarth is certainly superior to that of a car of almost double its cubic capacity... As far as its engine is concerned, we would say that, in spite of the power that permits great performance, it is elastic to the point that there is good acceleration, even at medium range revs of*

The 1963 Turin Motor Show from 30 October-10 November was the occasion on which the Simca Abarth 1150 was presented in four different stages of engine conversion with five main bearings increased to 1137 cc: the basic 1150 (pictured) of 55 hp, the 58 hp 1150 S, the 1150 SS of 65 hp and the 1150 Corsa at 85 hp. The only external modifications were to the insignia applied to the rear and – for the S, SS and Corsa – a grill for the radiator. The capacity boost was achieved by increasing the bore to 76 mm and the stroke to 69 mm. Each version had disc brakes on all four wheels.

around 3000 rpm. And that represents another positive quality of this vehicle, which is really a sports car but still relatively economical".

Finally, the clear and concise opinion on the Simca Abarth 1150: "*On a stretch of the Turin-Milan motorway, we easily overtook all the cars in front of us and among them were many of medium capacity that are normally driven at high speeds of between 140 and 150 km/h. The speedometer reached and maintained 160 km/h over long stretches, which confirms how an effective 150 km/h was exceeded.*

What seems particularly interesting is that the car's acceleration and agility in city traffic are so exciting that they can be compared to those of a Giulietta TI. In fact, this car has good pick-up from as low as 2000 rpm, sufficiently so to say that those who are not in a hurry can easily drive in the city in fourth. In addition, acceleration is really powerful from 4000 rpm onwards".

Those significant testimonies closed a 1963 full of favourable events and great sporting triumphs for the Scorpion. Employees had now increased to 419 and the workshop covered 17,583 square metres. Abarth & C. was fully mature; its reputation grew continually due to its non-stop race victories and

With the launch of the Simca Abarth 1150, the motoring press carried a page of advertising underlining Abarth's experience in transforming normal family cars into fast sporting saloons with particular emphasis on safety, as shown by the eloquent specification of disc brakes on all four wheels. The four variants of the Simca 1150 were marketed at prices ranging from Lit 1.234.440 for the basic model with a top speed of 150 km/h to Lit 1.544.340 for the 170 km/h Corsa.

On opening day of the 1963 Turin Motor Show, the Abarth stand overflowed as usual with politicians, journalists, photographers and a liberal sprinkling of enthusiasts. The cars that stood out were the 595, the updated Fiat Abarth 1000, the two Bialbero and Monomille versions plus the Simca Abarth 1150, the Abarth Simca 1600 and 2000 coupés in their definitive form.

At the end of 1963, the Fiat Abarth 1000 Bialbero was given aesthetic and functional modifications, particularly the bonnet that was longer and more slender than the previous model's; the front air intake was also lowered. The single piece front end of the bodyshell could be opened forward for better access to the car's mechanics.
The picture shows the different sizes of tyres, front 4.50x13 and rear 5.25x13, made more evident by the body's large wheel housings.
Called the 'long nose', the revised Fiat Abarth 1000 Bialbero had done well at the 500 Kilometres of the Nürburgring on 1 September by winning the Division 1 World Constructors' Grand Touring Championship for the second successive year.
The 'long nose' went into production in 1964 and was built by Sibona & Basano of Turin; its whole front end was innovative as it was made of glass fibre instead of aluminium.
The legendary 982 cc four-cylinder twin cam engine had reached a maximum power output of 104 hp at 8000 rpm and the car's top speed was up to 220 km/h.

The Fiat Abarth Monomille was also updated at the end of 1963 and adopted modifications made to the body of the 1000 Bialbero. The long rear overhang, however, emphasised by the 'duck tail' motif of the engine cover and the large screen on the car's rear view mirror, made its original line less equilibrated than its predecessor, which was considered one of the more elegant cars built on the Fiat 600 D floor pan.

important commercial successes. Over the years, it had consolidated its presence in numerous countries through an efficient network of importers; in some cases like that of Jaime Juncosa of Barcelona, Spain, they also raced in Scorpion colours and, therefore, carried out a full promotional activity as well as the institutional operations of Abarth product distribution and providing service. In France, Abarths were sold by Dimitri Mollof of Paris, in Switzerland by Edgard Schwin of Zurich, in Germany by Noeldeke of Constance, in Britain by Anthony Crook Motors of Hersham, Surrey, and later by John Anstead. Carlo became increasingly absorbed in every day activity. He wanted to witness sports events and technical testing himself and did so in transfer because he wanted to live the tension, the anxiety that competitive confrontation instilled every weekend in men who raced; it was, perhaps, a means of staying young among the young, and to constantly re-live the excitement of his youth, when he wanted to demonstrate that he was the fastest man to sit astride a motorcycle. However, such things took time away from the routine management of the company, and at a certain point Carlo began to take home his post folder; the morning afterwards, his wife Nadina read him the incoming correspondence including internal memos, and he dictated notes that were later developed in the office by Evangelina Isandoro. His every thought, every move was devoted to improving performance at work, which constituted the

real passion of his simple existence. He hid from the high life, from parties and ceremonies that he considered futile externalism, devoid of essential values that constituted the principles of his life – discipline, order, coherence with oneself, precision and scrupulousness in relations with others. His mediocre Italian, characterised by the frequent use of the formal 'you' and infinitive verbs, led to him sometimes be intransigent and aggressive, even if a strong determination in pursuing his own ideas and specific essential competence always helped him. His somewhat difficult personality, tempered by an innate sense of right and a rejection of ambiguity, created around him an aura of respect and uneasiness. He demanded much of his racing and test drivers and, in particular, did not tolerate errors committed in racing; an attentive and extremely able technician, he showed no willingness to forgive those who put the mechanical integrity of the car at risk with a wrong manoeuvre. His verbal skirmishes with Colucci were legendary, and the engineer lamented how Abarth frequently had a discriminatory, almost paternalistic attitude towards some workers who were more specialist than others. Colucci remembers he for-

Carlo Abarth at his desk, where he often worked until late and kept his closest collaborators until the small hours. Simple and almost bare, his office reflected his pragmatic spirit and reserved appearance. He was always careful about the way he looked and dressed, which was without affectation but always with elegance. His frequent dieting made him confront – not without sacrifice – the temptations of fine cuisine, which he enjoyed in moderation.

Omologazione F.I.A. N. 151
Omologazione C.S.A.I. N. HA 22

AUTOMOBILE CLUB D'ITALIA
COMMISSIONE SPORTIVA AUTOMOBILISTICA ITALIANA
FEDERATION INTERNATIONALE DE L'AUTOMOBILE

Scheda di Omologazione
secondo l'allegato J al Codice Sportivo Internazionale

CATEGORIA GRAN TURISMO

Casa costruttrice ABARTH & C.
N° di Serie Chassis 136
N° di Serie Motore 236 - S
Tipo di Carrozzeria Coupé 2 Posti
Anno inizio di Fabbricazione 1963

Modello ABARTH SIMCA 2000
Costruttore ABARTH
Costruttore ABARTH
Costruttore ABARTH
L'omologazione è valida dal 13 JANV 1964

Foto A

N° Fogli che compongono la Scheda 11
Varianti data N° Fogli aggiunti

Timbro della C.S.A.
Timbro della F.I.A.

PAGINA N. 1

The first page of a homologation form for the Abarth Simca 2000, which was given its first sneak preview at the 1963 Geneva Motor Show and of which 100 examples were later built so that it could compete in the up to 2000 cc racing class.

The twin-cam engine with five main bearings and double ignition was extremely efficient and reliable, but was not helped by inadequate transmission components, in particular the Simca derived gearbox that was of modest quality and suffered poor thermic treatment of the gears.

bade Abarth to enter the workshop to forestall such conduct, which created discontent and inopportune vindication in the working environment. Carlo's prickly character made him assume positions that were difficult to defend. The break in relations with the Roosevelt Team was, for example, provoked by Abarth's refusal to send with maximum urgency some replacement parts that were indispensable to the American racing team. Similar occurrences were not infrequent; to the American exhaust importer who had complained that an exhaust system for a Jaguar had been sent with the wrong attachments, Carlo replied: *"We don't make mistakes like that!"*; in the case of a complaint by a customer who took him to task over broken piston segments, he replied in annoyed fashion that it was not possible to give credit for such matters *"because the segments don't break"*.

The dawn of 1964 in Italy was not an encouraging one. The progressive deterioration of public funds – in particular the balance of payments, which in just one year went from a positive to a negative situation of minus Lit 783 billion in 1963 – together with a brusque and violent awakening of inflation, which jumped by +8.4%, gave unequivocal signals that the euphoric period of the 'economic miracle' was coming to an end and to make way for a trend characterised by violent economic imbalance and deep social unrest. The fiscal policy introduced in February 1964 and intended to re-dimension the acquisition of cars had poor results by the end of the year, when official production figures registered 1,028,930 units – a reduction of 7% on 1963 - while new car registrations dropped from 1963's 951.704 units to 870.175 a year later.

For Carlo Abarth, the new year began with the introduction of the Fiat Abarth 595 SS in February, a version with 5 hp more power than the 595; with a compression ratio of 10.5:1 and fed by a Solex single-choke 34 PBIC carburettor, the little twin cylinder derived from the Fiat 500 developed a maximum of 32 hp at 5000 rpm; the inlet manifold was fused into the valve cover. As a result of such a generous injection of power, the car's maximum speed went up from 120 km/h to over 130 km/h.

Abarth caused a stir at the Geneva Motor Show a month later when he presented the 695, a new car with its mechanics derived from the Fiat 500 D and scheduled, among other things, to compete in the up to 700 cc class of the Grand Touring category. By increasing bore and stroke to 76x76 mm, the two-cylinder Fiat became a 689.548 cc car; with a compression ratio of 8.5:1 and a Solex 28 PB carburettor, it generated a maximum power of 30 hp at 4900 rpm, while the SS version had a more advance compression ratio of 9.8:1 and a Solex 34 PBIC carburettor; maximum power of the SS jumped to 38 hp at 5200 rpm. As with the 595, the 695 retained the 500 D production car's braking system and that meant controversy, until then dormant, with accusations suggesting Abarth freely modified the engines of normal production cars with-

FIAT ABARTH "595 SS" (1964)

ENGINE	
	rear, longitudinal, 2 cylinders in line
Bore and stroke	mm 73.5x70
Cubic capacity	cc 594
Valve gear	overhead valves, side camshaft
Number of valves	2 per cylinder
Compression ratio	10.5:1
Fuel feed	1 Solex 34PBIC carburettor
Ignition	coil, distributor
Coolant	forced air
Lubrication	forced, oil filter on the main circuit
Maximum power	32 hp at 5000 rpm
TRANSMISSION	
	rear wheel drive
Clutch	single dry plate
Gearbox	4 speed plus reverse, central lever
Axle ratio	5.14:1
BODY	
	2 door, 4 seat saloon
CHASSIS	
	integral body
Suspension	front independent, upper trailing arms, transverse lower leaf spring, hydraulic dampers; rear independent, trailing arms, helicoidal springs, hydraulic dampers
Brakes	hydraulically operated drums front and rear
Steering	worm and sector
Fuel tank	capacity 21 litres
Tyres	125x12
DIMENSIONS AND WEIGHT	
Wheelbase	mm 1840
Front track	mm 1121
Rear track	mm 1135
Length	mm 2970
Width	mm 1320
Height	mm 1300
Kerb weight	kg 470
PERFORMANCE	
Top speed	km/h 130

out making sure they had adequate brakes to stop them. In effect, the overall 432 square centimetre braking surface of the 95 km/h Fiat 500 D's four small drums appeared inadequate for the 595 and the 695, the maximum declared speeds of which were up to 50% more than that of the basic car. The 595 was introduced at a price of Lit 595.000, while the 595 SS cost Lit 650.000; but the price of the 695 and the 695 SS was set at Lit 640.000 and Lit 695.000 respectively. By comparison, in the spring of 1964 it was possible to buy a Fiat 500 D for Lit 450.000, a 600 D for Lit 640.000 and the 1100 D for Lit 960.000.

Fiat launched the 850 in May; the car was the natural mechanical evolution of the 600 D even if more modern and rounded, but the 850's body was still only a two-door. Many believed Fiat dared a little more in France, for which they designed the Simca 1000 that appeared more functional because of its four doors and also more up to date as far as its styling was concerned. On the other hand, some prototypes being tested in 1961 and surprised by an indiscreet photographer clearly showed that Fiat was working on a much-improved saloon similar to the Simca 1000 to take the place of the mythical 600. Later, it was preferred to retain the architecture of the 600 D and design

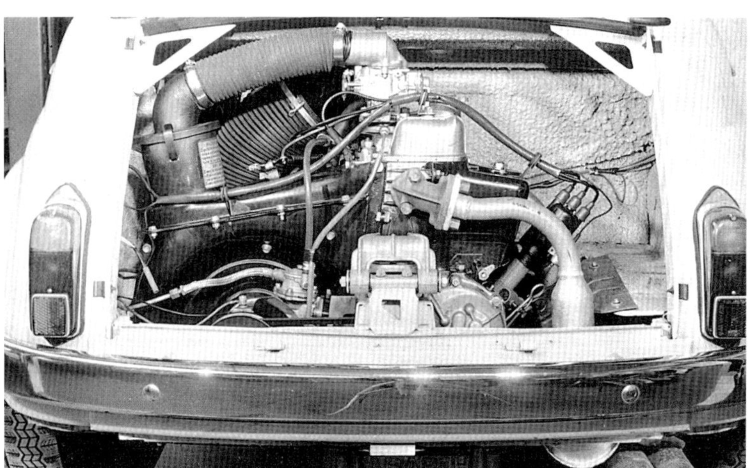

Opposite: The 595 SS, the quick version of the Fiat Abarth 595, first saw the light of day in early February 1964 and caused a stir in the sporting world for its maximum power of 32 hp at 5000 rpm, achieved in part thanks to its new light alloy inlet manifold and the adoption of a 34 mm Solex carburettor.
Externally, the SS differed from the standard 595 by its model identification badges and engine cover calipers in black rubber, fitted to the engine compartment door in place of the normal handle.
In line with its usual practice, the Corso Marche factory introduced the special conversion kit to enable Fiat 500 drivers to change their cars into the 595 SS.
The price of the kit was fixed at Lit 650.000, a Lit 55.000 increase over the normal 595 version.

Abarth unveiled its 695 and 695 SS models in March 1964 at the Geneva Motor Show.
Both were derived from the 595 launched a few months earlier and all of them from the Fiat 500 D.
The Corso Marche factory intended the car to be quick and agile in the city, but able to maintain 130 km/h on the motorway.
The faster SS version was destined almost exclusively for the racing world, due to its higher performance characteristics (38 hp at 5200 rpm).
The 695's price was fixed at Lit 640.000, which rose to Lit 695.000 for the SS.
An optional extra was an original roof in glass fibre (picture below), which would replace the canvas sunroof of the normal production Fiat 500.

the chassis ex-novo, which had habitability problems and a shape that was certainly not an evolution in the way that its predecessor had been nine years earlier. A car that was the result of such evident compromise went onto the market without receiving the same kind of welcome as the 600 in 1955; but the 850 did live on for seven years until 1971, when it was deleted from the Fiat range at the end of a production run that generated 2.203.380 units. Its 1964 launch price was Lit 750.000.

Abarth immediately began to work on 'his' version of the new 850. The July 1964 edition of the monthly motoring magazine Quattroruote said: *"Just two months after the presentation of the normal production Fiat 850, Abarth has already prepared a first quicker version of the car, the OT 850. The mod-*

FIAT ABARTH "695" (1964)

ENGINE	
	rear, longitudinal, 2 cylinders in line
Bore and stroke	mm 76x76
Cubic capacity	cc 690
Valve gear	overhead valves, side camshaft
Number of valves	2 per cylinder
Compression ratio	8.5:1
Fuel feed	1 Solex 28PB carburettor
Ignition	coil, distributor
Coolant	forced air
Lubrication	forced, oil filter on the main circuit
Maximum power	30 hp at 4900 rpm
TRANSMISSION	
	rear wheel drive
Clutch	single dry plate
Gearbox	4 speed plus reverse, central lever
Axle ratio	4.9:1
BODY	
	2 door, 4 seat saloon
CHASSIS	
	integral body
Suspension	front independent, upper trailing arms, transverse lower leaf spring, hydraulic dampers; rear independent, trailing arms, helicoidal springs, hydraulic dampers
Brakes	hydraulically operated drums front and rear
Steering	worm and sector
Fuel tank	capacity 21 litres
Tyres	125x12
DIMENSIONS AND WEIGHT	
Wheelbase	mm 1840
Front track	mm 1120
Rear track	mm 1135
Length	mm 2970
Width	mm 1320
Height	mm 1300
Kerb weight	kg 470
PERFORMANCE	
Top speed	km/h 130

ifications consist of an improvement in the engine's volumetric yield, obtained with a new Abarth manifold group and exhaust system. The carburettor is a Weber inverted by 30 mm with an acceleration pump. The other modification to the engine consists of a new sump in light alloy with cooling fins".

In effect, the new Fiat utility was offered in two different versions in its Abarth form; the OT 850/130 with 42 hp and the OT 850/150 with 50 hp, the letters OT standing for Omologato Turismo or homologated tourer; the first had a top speed of 130 km/h and the second a maximum of 150 km/h; the 842 cc capacity of both engines remained unchanged.

At the end of September the OT 850 was joined by the Fiat Abarth OT 1000, in which a four-cylinder 982 cc engine was installed, its stroke increased from 63.5 mm to 74 mm but with the same bore of 65 mm. It was a return to the by now famous 982 cc engine that had been beating everything in each season's races in its single and twin-cam versions. The OT 1000 developed 54 hp at 5200 rpm as a result of the usual modifications carried out to the inlet and exhaust manifolds and the adoption of a new carburettor. The braking system was also modified by the installation of front discs, while the normal production Fiat body remained unchanged, except at the front and the addition of the model's name on the front wings. The car's suspension was lowered by 30 mm compared to the original, and the maximum speed was 150 km/h. But the most sensational initiative undertaken by Abarth with the new Mirafiori utility car came to light at the Turin Motor Show from 31 October-11 November, when the Fiat Abarth OT 1600 was unveiled on the Scorpion's stand. The car was christened 'the monster' by many, because of such exceptionally advanced mechanics being installed in a little Fiat 850 saloon. To tell the truth, not much remained of the original car except the external panels of the body, because the entire chassis was extensively modified: suspension, brakes, wheels - everything was re-dimensioned and reinforced to match the

Inside, the 695 SS' group of instruments stood out. They replaced those of the production car and featured a large diameter rev counter calibrated to 6500 rpm on the left, and a speedometer on the right; in the centre were the thermometer and oil pressure gauge plus indicators for the directional flashers and dynamo charger.

In September 1965 the Fiat Abarth 695 SS was joined by another model that was set up for racing. The new car had a wider track – the front was increased from 1121 to 1161 mm and the rear from 1135 to 1175 mm – and wider rims plus negative camber.
In the illustrated prospectus distributed by Abarth in the autumn were the 595 SS, the 695 SS and the 695 SS 'racing set up', because by that time the normal versions of the 595 and the 695 were already being phased out of production.
Note that all three of the cars illustrated already sported the new body with doors hinged at the front, which Fiat introduced on the 500 F in March.

FIAT ABARTH "695 SS" (1964)	
ENGINE	rear, longitudinal, 2 cylinders in line
Bore and stroke	mm 76x76
Cubic capacity	cc 690
Valve gear	overhead valves, side camshaft
Number of valves	2 per cylinder
Compression ratio	9.8:1
Fuel feed	1 Solex 34PBIC carburettor
Ignition	coil, distributor
Coolant	forced air
Lubrication	forced, oil filter on the main circuit
Maximum power	38 hp at 5200 rpm
TRANSMISSION	rear wheel drive
Clutch	single dry plate
Gearbox	4 speed plus reverse, central lever
Axle ratio	4.9:1
BODY	2 door, four seat saloon
CHASSIS	integral body
Suspension	front independent, upper trailing arms, transverse lower leaf spring, hydraulic dampers; rear independent, trailing arms, helicoidal springs, hydraulic dampers
Brakes	hydraulically operated drums front and rear
Steering	worm and sector
Fuel tank	capacity 21 litres
Tyres	125x12
DIMENSIONS AND WEIGHT	
Wheelbase	mm 1840
Front track	mm 1120
Rear track	mm 1135
Length	mm 2970
Width	mm 1320
Height	mm 1300
Kerb weight	kg 470
PERFORMANCE	
Top speed	km/h 140

new engine Carlo Abarth and his colleagues decided to give that two-door utility car. It was a four-cylinder, twin cam 1592 cc power unit with double ignition and dry sump, which made its debut a year earlier in the Abarth Simca coupé 1600 GT; with a compression ratio of 9.8:1 and two twin-choke Weber 45 DCOE carburettors, it developed 154 hp at 7600 rpm. Either a four or six speed gearbox could be fitted, dependent on the circuits on which it would compete, as well as a self-locking differential. Disc brakes were on 5.00x13 wheels on the front axle and 6.00x13s at the rear. Incorporated in the new logo in the frontal area was the radiator's air intake, while in the rear had to reflect vertically the outline of the body with a special convexity that had the job of containing the bulk of the engine.

With a dry weight of 760 kg and a maximum speed of 220 km/h, the Fiat Abarth OT 1600 was built to race in the Grand Touring Category and was to compete against the Alfa Romeo Giulia T1 and the British-made Ford Lotus Cortina. But Fiat showed no interest in proceeding with the car's homologation and, in reality, only a few prototypes of the OT 1600 were built, but the car never went into production. Journalist Etienne Cornil was quick to comment on the car in an article headed "The Extraordinary Fiat Abarth OT 1600", published in the December 1964 edition of Road & Track. *"If Carlo Abarth has it in mind to build the 1000 examples of his OT 1600 required to obtain homologation, it is clear that he must have recourse to the financial support of Fiat. Regardless of its glorious motor sport pedigree, the great Turin factory no longer seems to be interested in the world of motor sport even if, in the*

ABARTH "FORMULA 2" (1964)

ENGINE	
	mid-rear, longitudinal, 4 cylinders in line
Bore and stroke	mm 76x55
Cubic capacity	cc 998
Valve gear	2 overhead camshafts, chain driven
Number of valves	2 per cylinder
Compression ratio	11:1
Fuel feed	2 twin-choke Weber 45DCOE9 carburettors
Ignition	coil, distributor
Coolant	liquid
Lubrication	dry sump
Maximum power	120 hp at 9500 rpm
TRANSMISSION	
	rear wheel drive
Clutch	single dry plate
Gearbox	4 or 6 speed plus reverse, central lever
Axle ratio	3.9:1
BODY	
	single-seater
CHASSIS	
	tubular
Suspension	front independent, triangular upper and lower trailing arms, helicoidal springs, hydraulic dampers; rear independent, reaction arms, helicoidal springs, hydraulic dampers
Brakes	hydraulically operated discs front and rear
Steering	rack and pinion
Fuel tank	capacity 35 litres
Tyres	front 5.25x13; rear 6.00x13
DIMENSIONS AND WEIGHT	
Wheelbase	mm 2300
Front track	mm 1320
Rear track	mm 1320
Length	mm 3770
Width	mm 640
Height	mm 700
Kerb weight	kg 420
PERFORMANCE	
Top speed	km/h 260

presence of the latest weapon just introduced by Carlo Abarth, it is possible that it could change its mind, with the consequent result of being able to exploit a fantastic publicity weapon throughout the world".

Even if they were changing their attitude to motor racing at Mirafiori, Abarth still went on to achieve a staggering mass of victories. After the 1 January rule change in which the first Division of the Grand Touring Category for the World Constructors' Championship was switched from 1000 cc to 1300 cc and sub-classes were excluded, the Scorpion continued on the track with the Abarth Simca, which enjoyed a brilliant season. Abarth entered five works

The debut of Abarth's single-seater open wheeler happened in April 1964 at Aspern on the outskirts of Vienna, where the Scorpion's Formula 2 car, which had appeared in the Italian enthusiast press a few days earlier, was assigned to Hans Herrmann.

With a tubular chassis designed by Mario Colucci and its 998 cc twin cam engine by Luciano Fochi that put out 120 hp at 9500 rpm, the car came fourth first time out but later the engine developed distribution problems and the entire project was dropped.

ABARTH "FORMULA 3" (1964)

ENGINE	
	mid-rear, longitudinal, 4 cylinders in line
Bore and stroke	mm 65x74
Cubic capacity	cc 982
Valve gear	overhead valves, side camshaft
Number of valves	2 per cylinder
Compression ratio	12.7:1
Fuel feed	1 twin-choke Weber 40DCD carburettor
Ignition	coil, distributor
Coolant	liquid
Lubrication	forced, oil filter on the main circuit
Maximum power	88 hp at 7900 rpm
TRANSMISSION	
	rear wheel drive
Clutch	single dry plate
Gearbox	4 speeds plus reverse, central lever
Axle ratio	5.4:1
BODY	
	single-seater
CHASSIS	
	tubular
Suspension	front independent, upper and lower triangular trailing arms, helicoidal springs, hydraulic dampers; rear independent, reaction arms, helicoidal springs, hydraulic dampers;
Brakes	hydraulically operated discs front and rear
Steering	rack and pinion
Fuel tank	capacity 35 litres
Tyres	front 5.25x13; rear 6.00x13
DIMENSIONS AND WEIGHT	
Wheelbase	mm 2300
Front track	mm 1320
Rear track	mm 1330
Length	mm 3770
Width	mm 640
Height	mm 700
Kerb weight	kg 400
PERFORMANCE	
Top speed	km/h 225

cars for the Rally of Monte Carlo, two Fiat Abarth 1000 Berlinas and three 850 TCs; two cars started from Athens, one from Paris, one from Oslo and one from Munich. But it was not an encouraging experience in spite of the careful preparation of the cars, which included the adoption of a new radiator placed between the two bars of the front bumper. Slowed by numerous problems, they were lost in the final results, with the first car home Masson-Vinatier's 1000 Berlina that started from Athens coming 32nd and the last, the other 1000 Berlina driven from Paris by Maublanc-Mosnier, 87th.

The continual technical evolution of the Abarth Simca led to the 1964 development of the 2000 GT Corsa, the mechanics of which were constantly refined as the season progressed. The five-mount engine of the 1600 GT increased to 1946 cc was readied to compete in the up to 2000 cc class of the constructors' championship's Grand Touring Category and the European Hillclimb Championship. With twin camshafts, double ignition and fed by two horizontal twin-choke Weber 58 DCOE3 carburettors, the new two-litre engine generated 185 hp at 7000 rpm, but the racing version's power output climbed to 204 hp at 7200 rpm. The four or six-speed gearbox was derived from the production unit made by Simca, which revealed itself to be the Achilles heel of that model; the transmission broke frequently in races and heavily penalised the potential of the car, which was extremely fast. The body was also updated numerous times during the season, particularly in the frontal area where Mario Colucci added two supplementary air intakes for better disc brake cooling and another wide central vent for the oil radiator. With a wheelbase of 2090 mm, the Abarth Simca 2000 GT Corsa was 3610 mm long, 1480 mm wide, 1210 mm high, weighed 665 kg and was accredited with a top speed of 250 km/h. Professional yet stubborn, Carlo Abarth never really liked the car because of the unreliability of the gearbox that stopped it exploiting its enormous potential.

The Fiat Abarth 595 won its first important race. Franco Patria, whose success was hardly ever in doubt, drove it to victory in the up to 600 cc class of the Jolly Club Trophy at Monza on 19 March. Carlo Fagioli won the up to 1000 cc class in a Fiat Abarth 1000 Berlina, while Antonio Ratti took the up to 850 cc class win in a Fiat Abarth 850 TC. At the 12 Hours of Sebring, the Fleming-Linton Abarth Simca 1300 GT won its class, while class wins in the Shell Trophy at Imola on 19 April also went to 'Mackye' alias Ricciardo Ricci in a Fiat Abarth 700 Bialbero, Anzio Zucchi in a Fiat Abarth 1000 Bialbero in the Grand Touring Category, Giuseppe Dalla Torre in an Abarth Simca 1300 GT and Franco Patria in an Abarth Simca 2000 GT.

But at the fifth Touring Car Grand Prix at Monza on 3 May, Abarth driver-customers refused to start because they felt at a disadvantage due to the participation in the same race of the works cars, three well prepared and tech-

FIAT ABARTH "OT 850" (1964)

ENGINE	
	rear, longitudinal, 4 cylinders in line
Bore and stroke	mm 63x63.5
Cubic capacity	cc 843
Valve gear	overhead valves, side camshaft
Number of valves	2 per cylinder
Compression ratio	8.8:1
Fuel feed	1 Solex 34PBIC carburettor
Ignition	coil, distributor
Coolant	liquid
Lubrication	forced, oil filter on the main circuit
Maximum power	40 hp at 5400 rpm
TRANSMISSION	
	rear wheel drive
Clutch	single dry plate
Gearbox	4 speed plus reverse, central lever
Axle ratio	4.6:1
BODY	
	2 door, 4 seat saloon
CHASSIS	
	integral body
Suspension	front independent, upper trailing arms, transverse lower leaf spring, hydraulic dampers: rear independent, trailing arms, helicoidal springs, hydraulic dampers
Brakes	hydraulically operated drums front and rear
Steering	worm and sector
Fuel tank	capacity 30 litres
Tyres	145x12
DIMENSIONS AND WEIGHT	
Wheelbase	mm 2027
Front track	mm 1146
Rear track	mm 1121
Length	mm 3575
Width	mm 1425
Height	mm 1385
Kerb weight	kg 670
PERFORMANCE	
Top speed	km/h 130

nically updated Abarth Simca 1300 GTs driven by Hans Herrmann, Franco Patria and Tommy Spychiger. The protest was a legitimate one, the private entrants felt, because they believed they were handicapped in their attempt to win the Italian Drivers' Championship by the presence of the works cars. But the position of the works team was a legitimate one, as they were exercising their right to defend their World Constructors' Championship. Of the 21 cars that qualified, only 11 started and works drivers Patria, Herrmann and Spychiger took the first three places in the race.

The season unravelled itself from one Sunday to the next with the Fiat Abarth 1000 always at the top of the up to 1300 cc Touring Category and the

Always attentive and quick to gauge the orientation of his sporting customers, Abarth believed the mechanics of the new Fiat 850 saloon, which had only just been marketed in Italy, could easily be modified to a capacity of one litre. And that was how the OT 1000 was born in September, joining the OT 850/130 and OT 850/150 that were launched in May.

Fiat's top brass also visited the Abarth stand at the 46th Turin Motor Show that took place from 31 October–11 November 1964. In the centre of the picture is Vittorio Valletta, president of the Turin carmaker, flanked by managing director Gaudenzio Bono and, further back, Gianni Agnelli.

FIAT ABARTH "OT 1000" (1964)

ENGINE	
	rear, longitudinal, 4 cylinders in line
Bore and stroke	mm 65x74
Cubic capacity	cc 982
Valve gear	overhead valves, side camshaft
Number of valves	2 per cylinder
Compression ratio	9.5:1
Fuel feed	1 Solex 34PBIC carburettor
Ignition	coil, distributor
Coolant	liquid
Lubrication	forced, oil filter on the main circuit
Maximum power	54 hp at 5200 rpm
TRANSMISSION	
	rear wheel drive
Clutch	single dry plate
Gearbox	4 speed plus reverse, central lever
Axle ratio	4:1
BODY	
	2 door, 4 seat saloon
CHASSIS	
	integral body
Suspension	front independent, upper trailing arms, transverse lower leaf spring, hydraulic dampers; rear independent, trailing arms, helicoidal springs, hydraulic dampers
Brakes	hydraulically operated discs front and drums rear
Steering	worm and sector
Fuel tank	capacity 30 litres
Tyres	145x12
DIMENSIONS AND WEIGHT	
Wheelbase	mm 2027
Front track	mm 1146
Rear track	mm 1121
Length	mm 3575
Width	mm 1425
Height	mm 1357
Kerb weight	kg 674
PERFORMANCE	
Top speed	km/h 150

On the inauguration of the Turin show the official party of personalities stopped, as was their custom, at the Abarth stand where the OT 1600 with the body of the basic Fiat 850 was making its first public appearance.
The picture above shows Carlo Abarth describing the new car to Italy's Minister of Industry and Commerce, Giuseppe Medici.
In the foreground is the powerful 1600 cc twin-cam engine of the OT 1600, derived directly from the previous year's Abarth Simca 1600 GT.

FIAT ABARTH "OT 1600" (1964)

ENGINE	
	rear, longitudinal, 4 cylinders in line
Bore and stroke	mm 86x68.5
Cubic capacity	cc 1592
Valve gear	2 overhead camshafts, chain driven
Number of valves	2 per cylinder
Compression ratio	9.8:1
Fuel feed	2 twin-choke Weber 45DCOE9 carburettors
Ignition	double, coil, distributor
Coolant	liquid
Lubrication	dry sump
Maximum power	155 hp at 7400 rpm
TRANSMISSION	
	rear wheel drive
Clutch	single dry plate
Gearbox	4 or 6 speed plus reverse, central lever
Axle ratio	3.9:1
BODY	
	2 door, 2 seat saloon
CHASSIS	
	integral body
Suspension	front independent, upper trailing arms, transverse lower leaf spring, hydraulic dampers; rear independent, trailing arms, helicoidal springs, hydraulic dampers
Brakes	hydraulically operated discs front and rear
Steering	worm and sector
Fuel tank	capacity 30 litres
Tyres	front 5.00x13; rear 6.00x13
DIMENSIONS AND WEIGHT	
Wheelbase	mm 2027
Front track	mm 1160
Rear track	mm 1320
Length	mm 3575
Width	mm 1660
Height	mm 1358
Kerb weight	kg 790
PERFORMANCE	
Top speed	km/h 220

The Fiat Abarth OT 1600 caused quite a stir at the 1964 Turin Motor Show; it retained a great deal of the 850 saloon's body but few mechanical components, which were profoundly modified to bring them into line with the exceptional performance of the car. The 1.6 litre engine was none other than that of the Abarth Simca 1600 GT, which appeared at the 1963 Geneva Motor Show with twin-cam distribution, double ignition, dry sump lubrication, 155 hp at 7400 rpm; the supplementary water radiator and the oil unit were positioned behind the wide front grill. Large wings were adopted for the rear end due to the 6.00x13 tyres with negative camber, compared to the 5.00x13 fronts.

The narrow engine compartment of the Fiat Abarth OT 1600 was barely able to contain the big four-cylinder twin-cam double ignition engine that overhung the rear axle; the rear mirror had to be given a special, much more convex shape so that the powerful engine could be accommodated in the bodyshell of the Fiat 850.

The cockpit of the popular Fiat utility was given comprehensive instruments, grouped together in front of the driver.
As well as the big-diameter rev counter and speedometer, there were thermometers for the water and oil plus a fuel gauge.
Other instruments were grouped together in front of the passenger seat, including the oil pressure gauge and voltmeter.
Bigger suspension, reinforced floor pan, self-locking differential, disc brakes and bigger tyres profoundly modified the mechanics of the original Fiat, turning it into a really fast grand tourer.
The Fiat Abarth OT 1600 was conceived to race against the then all-conquering Ford Lotus Cortina.

Abarth Simca 1300 GT heading the 1300 cc Grand Touring results table. That is how it was at the 26th Consuma Cup on 24 May, when Secondo Rodolfi won the first Division with an Abarth Simca 1300 GT. On 31 May Hans Herrmann won the first Division at the 1000 Kilometres of the Nürburgring at the wheel of a Abarth Simca 1300 GT; the eighth Bologna-Raticosa Pass hillclimb took place in Italy on the same day and that was where Fiat Abarths won the up to 700 cc Grand Touring class with Mackye, the up to 1000 cc with Enzo Buzzetti and an Abarth Simca 1300 GT won again with Angelo Gilimberti. Then came the Rossfeld race on 7 June, when Hans-Dieter Dechent won his class with an Abarth Simca 1300 GT. A week later, Abarth won again at the seventh City of Asiago Cup, at the Garessio-Colle San Bernardo on 21 June with Giuseppe Rebaudi driving a Fiat Abarth 1000 Bialbero, Vittorio Venturi in the Sport category driving a Spider Tubolare 1000 and Luigi Taramazzo, who took overall victory and a class win in an Abarth Simca 2000 GT. At the 10th Predappio-Rocca delle Caminate on 28 June the Fiat Abarth 700 Bialbero won again ('Robertino') as did the 1000 Bialbero (Cesare Sangiorgi) as well as the Spider Tubolare 1000 (Vittorio Venturi); Franco Patria also clinched the Racing Category in the first race of the Abarth Formula 3.

At the Freiburg-Schauinsland on 9 August Franco Patria at the wheel of an Abarth Sinca 2000 GT won the second Division Grand Touring class and came second overall, behind the Porsche of Edgar Barth. It was more success for the two-litre Abarth Simca on 16 August at the City of Enna Cup, where Herrmann came first overall and won the Second Division Grand Touring class; Franco Patria repeated Herrmann's result at the Sierre-Montana-Crans on 30 August in an Abarth Simca 2000 GT.

The story of the Fiat Abarth single-seater began in April, when the Formula 2 car made its debut at Vienna in the Aspern Airport race in which it came fourth with Hans Herrmann at the wheel. The car's tubular chassis,

FIAT ABARTH "1000 BERLINA CORSA" (1964)

ENGINE	
	rear, longitudinal, 4 cylinders in line
Bore and stroke	mm 65x74
Cubic capacity	cc 982
Valve gear	overhead valves, side camshaft
Number of valve	2 per cylinder
Compression ratio	12:1
Fuel feed	1 twin-choke Weber 36DCD7 carburettor
Ignition	coil, distributor
Coolant	liquid
Lubrication	forced, oil filter on the main circuit
Maximum power	76 hp at 6000 rpm
TRANSMISSION	
	rear wheel drive
Clutch	single dry plate
Gearbox	4 speeds plus reverse, central lever
Axle ratio	3.9:1
BODY	
	2 door, 4 seat saloon
CHASSIS	
	integral body
Suspension	front independent, upper trailing arms, transverse lower leaf spring, hydraulic dampers; rear independent, trailing arms, helicoidal springs, hydraulic dampers
Brakes	hydraulically controlled discs front and rear
Steering	worm and sector
Fuel tank	capacity 27 litres
Tyres	4.25x13
DIMENSIONS AND WEIGHT	
Wheelbase	mm 2000
Front track	mm 1150
Rear track	mm 1160
Length	mm 3285
Width	mm 1380
Height	mm 1400
Kerb weight	kg 590
PERFORMANCE	
Top speed	km/h 185

The Fiat Abarth 1000 Berlina celebrated its 482nd victory at the 1964 Turin Motor Show. The Corsa version shown on the stand was fitted with a supplementary front radiator placed between the bumper bars, and made its first unofficial appearance at the Rally of Monte Carlo in January.

coded SE03, was the work of Mario Colucci and the engine was designed by Luciano Fochi, who managed to squeeze 120 hp at 9500 rpm out of a four-cylinder 998 cc twin-cam engine with a bore and stroke of 76x55 derived from the 1288 cc unit of the Abarth Simca 1300 GT. The engine only produced nominal power at the dynamometric bench: once it had been installed in the car its performance suddenly deteriorated and it became unreliable, particularly in distribution. During testing at the Aeritalia Nord airport near Turin the car was driven alternately by Carlo Abarth and racing and test driver Hans Herrmann, but without being able to resolve the cause of the poor yield of the power unit, starting with the supposition that the engine from which it was derived rarely gave reliability problems. An engine under such stress was evidently not able to sustain the prolonged effort of a track race and would only do well in the Spider Tubolare on short but tough hillclimbs.

The Fiat Abarth Formula 3 car was developed directly from the Formula 2 single-seater, retaining F2's chassis and bodyshell except for the rear faring caused by a different configuration of the exhaust pipes. Expressly required by Abarth so that the company could move closer to young people in the world of pure speed and single-seaters, the F3 car adopted the well-known 982 cc single-cam engine, the maximum power of which was close to 88 hp at 7900 rpm; it was the same power plant, suitably modified and tuned, as the one in the Corsa version of the Fiat Abarth 1000 Berlina. Carlo dedicated much of his time to the development of his single-seaters, assiduously taking part in test-driving sessions and the study of prototypes. The demon of pure speed that an expert driver perceives in a special way when he is at the wheel of a single-seater was reawakened in Carlo, and made him re-live once again his intoxicating experiences of 40 years earlier in the saddle of his racing motorcycles.

Confirmation of the competitiveness of the Fiat Abarth Formula Three car came in the fifth Race of Mendola on 5 July 1964, when the cars of the

An original body design by Pininfarina on the mechanics of the Fiat Abarth 1000 Berlina with a wheelbase lengthened from 2000 to 2027 mm, received considerable public and critical acclaim at the 1964 Turin Motor Show. Two large optical groups with double headlights helped make up a single body with transparent faring that extended along the whole length of the car.
The small wraparound windscreen flowed into the lateral windows and bonnet without the impediment of pillars, moulded with the rear sides to comprise a single unit vehicle that was most unusual and elegant. The car did not go into production.

FIAT ABARTH "1000 GT PININFARINA" (1964)

ENGINE	rear, longitudinal, 4 cylinders in line
Bore and stroke	mm 65x74
Cubic capacity	cc 982
Valve gear	overhead valves, side camshaft
Number of valves	2 per cylinder
Compression ratio	11.5:1
Fuel feed	1 Weber 30DIC carburettor
Ignition	coil, distributor
Coolant	liquid
Lubrication	forced, oil filter on the main circuit
Maximum power	68 hp at 6400 rpm
TRANSMISSION	rear wheel drive
Clutch	single dry plate
Gearbox	4 speed plus reverse, central lever
Axle ratio	4.1:1
BODY	2 seat roadster
CHASSIS	integral body
Suspension	front independent, upper trailing arms, transverse lower leaf spring, hydraulic dampers; rear independent, trailing arms, helicoidal springs, hydraulic dampers
Brakes	hydraulically operated front discs and rear drums
Steering	worm and sector
Fuel tank	capacity 27 litres
Tyres	145x13
DIMENSIONS AND WEIGHT	
Wheelbase	mm 2027
Front track	mm 1158
Rear track	mm 1212
Length	mm 3780
Width	mm 1500
Height	mm 1080
Kerb weight	kg 635
PERFORMANCE	
Top speed	km/h 175

Scorpion celebrated a collective triumph in a day of rain and violent cloudbursts. The overall winner was Herbert Demetz in an Abarth Simca 2000 GT, followed by the F3 car driven by Tommy Spychiger; that day the Fiat Abarths also won the up to 850 cc class of the Touring Category with Arrigo Manavello, the 1000 cc with Arturo Merzario, the up to 700 cc class of the Grand Touring Category with Mackye and with Enzo Buzzetti, the only competitor in the up to 1000 cc Grand Touring class. Fiat Abarths won almost exactly the same races again on 12 July at the Trento-Bondone, where they took the first five places and the Abarth Simca another two in their respective categories and classes. It happened again a week later at the Trieste-Opicina, where Franco Patria came first overall in the Abarth Simca 2000 GT and seven days after that at the Cesana-Sestrière. On 2 August Herbert Demetz won the

The racing version of the Abarth Simca 2000 GT produced extremely positive results during the 1964 racing season, even though it was afflicted by gearbox problems that forced its retirement a number of times.
The car showed itself to be very fast and more than able to take on the two-litre opposition of the day. In 1963, the Abarth Simca won the European Hillclimb Championship in its original body configuration (large picture) but in the spring of 1964 Mario Colucci modified the mechanics and bodywork (small picture) and introduced large front air intakes for better front disc brake cooling, plus two small lateral vents for the engine bay.
The power of the racing version's engine, which in 1963 put out 195 hp at 7200 rpm against the 185 hp of the 2000 GT road car, was first increased to 202 hp and then 204 hp for the subsequent 1964 European Hillclimb Champion model.

third Agordo-Frassené outright in an Abarth Simca 2000 GT. But 6 September at the 15th InterEurope Cup at Monza was a black day for the Scorpion - all the Abarth Simcas had to retire with transmission trouble. Everything was back on track at the 500 Kilometres of the Nürburgring, in which Hans Herrmann and Klaus Steinmetz won the race outright in an Abarth 1300 Prototype and Abarth Simca 1300 GTs took the first two places in their class.

Other successes came at the fifth City of Paris Cup, where Franco Patria won the up to 2000 cc class of the Grand Touring Category and the up to 850 cc class of the Touring category; Claude Ballot Lena took the 1300 cc Grand Touring win in the same race on 20 September, while in the Coupe du Salon in Paris on 4 October Steinmetz won with the Fiat Abarth 850 TC and Patria with the 1000 Berlina. All of which gave the Scorpion the right to think it would close the season in triumph, until its top driver Franco Patria, who had just been appointed the company's motor sport director as a result of his victories in the preceding months, was killed on 11 October at Montlhéry by an out of control Jaguar driven by Peter Lindner. Only 21 years old, Patria joined Abarth at the start of the 1964 season and had already shown himself to be one of the team's most promising acquisitions. The 7th Trophy of the Mountains, which took place at San Remo on 29 November, was named after Patria and was won outright by Luigi Taramazzo in the Abarth Simca 2000 GT, ahead of Edoardo Lualdi's Ferrari GTO. Another four Fiat Abarth victories followed, two in Touring classes of up to 850 cc and 1150 cc, two in the up to 700 cc and 1000 cc Grand Touring classes and another two by Abarth Simcas in the up to and over 1300 cc Grand Touring classes.

There was a brusque interruption in the Scorpion's sequence of victories on 24-25 October, when the first Italian round in the European Touring

Overseas, the Fiat Abarth 850 TC scored a large number of victories in the Touring category and was often able to win against bigger engined opponents. The car also gave Abarth its greatest number of wins in 1964, and was the centre of a formidable promotional campaign for the Scorpion brand all over the world.
The two pictures show the Fiat Abarth 850 TC competing on an EE (Escursionisti Esteri or overseas excursionist) number plate in a race in Albacete, Spain, a country in which Abarth importer Jaime Juncosa was dynamic in racing and other fields.

Challenge took place at Monza; the Fiat Abarth 850 TCs were beaten by Bjorn Rothstein's three-cylinder, two-stroke Saab 96.

The season closed with the FISA Cup on 8 December at Monza, where Abarth scored six victories, and the Criterion of the Hour at Vallelunga on the 20th in which the Turin cars won five more times. Bruno Nestola's final comment on the race in Auto-Italiana of 31 December was interesting: *"A last fact to state: four cars, all derived from mass produced models but with around 100% increases in power compared to the originals, each lost a wheel. The mechanical stress that these cars must now withstand is excessive and much more superior than values established in their places of manufacture, even though they have a large margin. We are in favour of modification, provided it does not excessively compromise the safety of the car and, therefore, its driver. Also and above all in the interests of our sport"*.

The 1964 season closed after the Scorpion had achieved 741 victories – 30 of them overall wins - of which 538 were by Fiat Abarths, 177 by Abarth Simcas and the remainder by prototypes and single-seaters. Of particular significance was their geographical spread of success. The fame of the cars from Turin was radiating across Europe, from Spain to Belgium, Scandinavia and countries like Austria, France, Germany, Britain and Switzerland, in all of which they had achieved great popularity. But the unmistakable thunder of the racing Abarths was also associated with their extraordinary ability to rout their opponents in Yugoslavia, the United States, the Latin American countries, Asia and Africa, too.

Abarth won the World Constructors' Championship, Division 1 that year with 77.9 points of which 59.9 were valid, and it came third in Division 2 from 1301 to 2000 cc behind Porsche and Alfa Romeo. In Annata Automobilistica (Automobile Year) 1964/65, Cesare Facetti wrote: *"...a relevant note of merit is attributed to Abarth, the absolute protagonist and dominator virtually without rivals of the Group 1 International Makes Championship. In the up to 1300 cc class, the Turin company literally ran away with the race, inducing its adversaries to beat a precipitous retreat..."*

ABARTH "LUFTHANSA" (1964)

ENGINE	
	mid-rear, longitudinal, 4 cylinders in line
Bore and stroke	mm 86x55.5
Cubic capacity	cc 1289
Valve gear	2 overhead camshafts, chain driven
Number of valves	2 per cylinder
Compression ratio	10.5:1
Fuel feed	2 twin-choke Weber 45DCOE9 carburettors
Ignition	double, coil, distributor
Coolant	liquid
Lubrication	dry sump
Maximum power	147 hp at 8800 rpm
TRANSMISSION	
	rear wheel drive
Clutch	single dry plate
Gearbox	5 or 6 speed plus reverse, central lever
Axle ratio	3.5:1
BODY	
	2 seat coupé
CHASSIS	
	tubular
Suspension	front independent, trailing arms, helicoidal springs, stabiliser bar, hydraulic dampers; rear independent, trailing arms, helicoidal springs, stabiliser bar, hydraulic dampers
Brakes	hydraulically operated discs front and rear
Steering	rack and pinion
Fuel tank	capacity 60 litres
Tyres	front 6.00x13; rear 7.00x13
DIMENSIONS AND WEIGHT	
Wheelbase	mm 2200
Front track	mm 1310
Rear track	mm 1320
Length	mm 3550
Width	mm 1490
Height	mm 1140
Kerb weight	kg 580
PERFORMANCE	
Top speed	km/h 250

In addition, Abarth won the up to 600 cc Italian Touring Car Championship with Leo Durst in a Fiat Abarth 595 prepared by Garavello, the up to 850 class with Giampiero Raffa driving a Baggioli-prepared Fiat Abarth 850 TC and the up to 1000 cc with Achille Marzi in a Fiat Abarth 1000 Berlina prepared by Faccioli. In the Grand Touring Category, Abarth also won the up to 700 cc class of the Grand Touring Category Italian Championship with Mackye (Ricciardo Ricci) at the wheel of a Fiat Abarth 700 Bialbero prepared by Faccioli, the up to 1000 cc class with Anzio Zucchi in a Faccioli-prepared Fiat Abarth 1000 Bialbero and the up to 1300 cc with Angelo Giliberti in a self-prepared Abarth Simca 1300 GT.

An oppressive and depressed mood rippled through the stands of the 1964 Turin Motor Show, which opened on 31 October, when the Abarth – veteran of a thousand successes – polarised public curiosity. Pininfarina showed a much revered roadster prototype built on Fiat Abarth 1000 GT mechanics, the architecture of which recalled the Abarth 207 A and 208 A roadsters of a decade earlier; low-slung and long, accentuated wrap-around windscreen blended into the pillars, the car was of extreme stylistic cleanliness. It remained a prototype, but a year later it did inspire an elegant coupé, the work of Pininfarina, which became known as one of the most successful non-production cars built on a floor pan derived from the Fiat 850.

The prototypes of Corso Marche that year included a mid-engine car 'dressed' in an attractive, long-nosed body and with ample lateral air intakes immediately in front of the rear axle. Revealed by Auto-Italiana in its edition of 10 December 1964, the prototype was called the Lufthansa, because its introduction was planned the following year with sponsorship from the German national airline. The car was designed by Mario Colucci in line with project criteria he proposed some time earlier, with an engine mid-ships and not Carlo Abarth's preferred power unit inclined over the rear axle. Derived from that of the Abarth Simca 1300 GT, the engine was accredited with 147 hp at 8800 rpm and meant the unusual Scorpion accelerated to a top speed of 250 km/h. Born into a clearly controversial climate in the factory, the prototype was delivered unfinished and untested to the Lufthansa team in Germany, which proceeded to complete it under the supervision of local driver Hans-Dieter Dechent.

Carlo Abarth did not let 1964 slip away without offering the public kits of components for the modification of a number of mass produced cars, as he did years earlier for the Fiat 600. At the end of 1964, the price for the kit with which to transform the Fiat 500 into a Fiat Abarth 500 was announced at Lit 64.500, the Fiat Abarth 595 Lit 109.000 and the 595 SS Lit 152.000; a kit was also introduced to change the Fiat 850 into a Fiat Abarth OT 850 (Lit 92.000) and the Simca 1000 into a Simca Abarth 1150 (Lit 100.000). Other components

In 1964, Mario Colucci designed a new racing coupé, with its 1.3 litre engine mounted amidships and installed it in a harmonious aluminium body that was a complete breakaway from the styling traditions of Corso Marche.
The car, produced to compete in the Prototype category with sponsorship from the German airline Lufthansa, did so with mixed fortunes in a number of races.
Driven by Anton Fischhaber and Hans Dieter Dechent, they included the 500 Kilometres of the Nürburgring in September, from which the 1.3 retired with a broken clutch.

sold by Abarth & C. that further improved performance could be added to the kits, including disc brakes and alloy rims ... or motorists could limit themselves to simply embellishing their cars' appearance.

The year ended with the production of exhausts exceeding 200,000 units, of which over half were exported; that contributed to a satisfactory bottom line in spite of adverse circumstances that year severely penalising demand for cars in Italy. At year's end, Abarth & C.'s registered stock was increased to Lit 500 million and its social responsibility was renewed for the following three years 1965-67. Having expired at the end of December, the collaboration agreement between Abarth and Simca was not renewed. One main reason that brought a halt to the continuation of the profitable and stimulating relationship that had produced excellent results was that the French factory had come under the control of the Chrysler Corporation in 1963, and the American company planned to pursue totally different techno-industrial strategies. The death of Enrico Teodoro Pigozzi in November 1964 further contributed to the end of an era, which had been rich in brilliant initiatives and led the Poissy company to play an increasingly significant role in French motoring.

1965

1967

A LOT OF AMBITION AND A LITTLE BITTERNESS

At the beginning of 1965 Abarth & C. was producing small quantities of the normal and SS versions of the 595 and the 695, the 850 TC, Nürburgring and Corsa versions of the 850 TC, the 130 and 150 derivatives of the OT 850, the OT 1000, OT 1600, 1000 Berlina and the 1000 Berlina Corsa, all with bodies strictly derived from that of the Fiat 500 D, 600 D and 850. The company also built the Monomille coupé and the celebrated Bialbero coupé was fitted with either a 700 or 1000 cc engine; to these car, all different stages of modification of basic Fiat production models, should be added the 1300 and 2000 cc Abarth Simca GT coupé, both in road-going and racing formats. A range of 20 different models, all with highly sophisticated mechanicals that represented a major financial and human resources commitment by the company, especially the continual technical updates for the cars that were committed to the punishing test bed of racing every weekend.

To scrupulously refresh the cars that were to compete during the 1965 motor racing season, Abarth organised a packed test programme that took place from 1-6 February at the Vallelunga circuit near Rome, with the company's works drivers Hans Herrmann, the number one, Herbert Demetz the great hillclimb specialist, André Knorr, Leo Cella and new acquisitions Giuseppe Virgilio and Salvatore Calascibetta, candidates to compete in the 1965

From 1-6 February 1965 Carlo Abarth and his technicians went to Vallelunga for a series of shakedown tests on the cars with which the Scorpion would compete in the upcoming racing season. Important modifications had been made to the 850 TC and 1000 Berlina, in particular to the front axle that retained the Fiat 600 D production car's transverse leaf spring, which functioned only as a lower arm of the suspension. The elastic element was made up of a helicoidal spring, coaxial to the telescopic damper.

Among other improvements to the 600 D-derived Fiat Abarth was a new cooling system. A front radiator of greater capacity meant the fan, which absorbed a fair amount of power, could be eliminated from the main unit in the engine compartment. The production Fiat's water pump was replaced on the 850 TC and the 1000 Berlina by the unit fitted to the twin-cam models, which was driven by a toothed belt instead of the classical trapezoid.

European Touring Car Championship. Among the new developments extensively tested at the track was a completely new cooling system; it still comprised two radiators, one of the right of the engine compartment and the other in the front, the latter with a notable increase in capacity in an effort to eliminate the production car fan, which absorbed a fair amount of power. The original Fiat water pump was replaced with the type fitted to the Bialbero and driven by a toothed belt, instead of a normal trapezoidal unit. New front suspension was adopted by all the 850 TC and 1000 Berlina models, on which the transverse semi-elliptic springs acted only as a lower arm while the elastic ele-

Monza, 22 February: Abarth carried out another pre-season test session in readiness for the first programmed race of 1965, and in particular its participation in the European Challenge for touring cars. Among those present were, from left to right, drivers Leo Cella and Herbert Demetz and the Scorpion's sports director, Renzo Avidano.

The definitive 1965 version of the Fiat Abarth 850 TC in racing set up (left) had a large supplementary radiator located at the front that was linked to the front wings with special faring. The Corsa '65 also adopted a new four-in-one exhaust system with one pipe of higher yield, while a cardanic joint was built into the steering column to improve the angle of the steering arms.

The same technical improvements were made to the Fiat Abarth 1000 Berlina Corsa (right) that also included windscreen wipers with transverse arm linkage and, on the works cars, the quick release fuel filler cap with direct access through the bonnet. At the start of the 1965 season the Corsa version of the 850 TC and the 1000 Bialbero were being offered at Lit 1.515.000 and Lit 1.644.000.

ment function was fulfilled by two helicoidal springs, coaxial to the dampers. During Vallelunga testing, Hans Herrmann set the 3.2 kilometre lap times of 1'45.8" with the 850 TC, 1'44.3" with the 1000 Berlina and 1'37" dead with the 1000 Bialbero.

The commitment of technicians and workshops was put to a hard test in those months of the design and construction of a new 1300 cc engine, which was to replace the one derived from the Simca that Abarth had to abandon after the expiry of the agreement with the French company. Coded 237, the newborn was a four-cylinder with five main bearings; it was of 1289 cc capacity, had a bore and stroke of 86x55.5 mm, twin-cam distribution and generated a maximum of 128 hp at 7900 rpm – the prototype managed 140 hp at 8000 rpm; it was an engine that was to form the basis of the subsequent OT 1600 (type 235) and the OT 2000 (type 236), which were to be used as the Scorpion's new sports roadsters with chassis further elaborated from that of the Simca 1000 and bodyshells all in glass fibre.

The Abarth 1600 sports prototype put in its first laps on the junior track at Monza on 1 April 1965 driven by Hans Herrmann, who clocked about a minute per lap with a best time of 59.2". The new rear-engined roadster code number 136A was built from numerous production car components – parts of the floor pan, suspension, steering... - while the structural shape of the attractive body was designed at Abarth and made in glass fibre on a wheelbase of 2015 mm with a front track of 1270 mm and a rear of 1310 mm. The engine was a 1591.6 cc four-cylinder with a bore and stroke of 86x68.5 mm which, with a compression ratio of 9.5:1 and fuelled by two Weber twin-choke 45 DCOE carburettors, produced a maximum 172 hp at 8000 rpm. The car had a six-speed gearbox, but was later also offered with a five speed. The Fiat Abarth OT 1600 Sport Spider was 3760 mm long, 1570 mm wide, 810 mm high, weighed 590 kg dry and was accredited with a top speed of 240 km/h. It first raced at the Targa

FIAT ABARTH "SPORT SPIDER OT 1600" (1965)

ENGINE	
	rear, longitudinal, 4 cylinders in line
Bore and stroke	mm 86x68.5
Cubic capacity	cc 1592
Valve gear	2 overhead camshafts, chain driven
Number of valves	2 per cylinder
Compression ratio	9.5:1
Fuel feed	2 twin-choke Weber 58DCOE9 carburettors
Ignition	double, coil, distributor
Coolant	liquid
Lubrication	dry sump
Max. power (project)	172 hp at 8000 rpm
TRANSMISSION	
	rear wheel drive
Clutch	single dry plate
Gearbox	6 speed plus reverse, central lever
Axle ratio	3.5:1
BODY	
	racing roadster
CHASSIS	
	boxed steel
Suspension	front independent, upper trailing arms, transverse lower leaf spring, helicoidal springs, hydraulic dampers; rear independent, trailing arms, helicoidal springs, hydraulic dampers
Brakes	hydraulically operated discs front and rear
Steering	worm and sector
Fuel tank	capacity 60 litres
Tyres	from 5.00x13; rear 6.00x13
DIMENSIONS AND WEIGHT	
Wheelbase	mm 2015
Front track	mm 1270
Rear track	mm 1310
Length	mm 3760
Width	mm 1570
Height	mm 1010
Kerb weight	kg 590
PERFORMANCE	
Top speed	km/h 240

Florio on 9 May 1965, a round in the Italian Sports and Grand Touring Drivers Championships; driven by Hans Herrmann and Leo Cella, the Abarth 1600 came first in its class and sixth overall in 7h 17'23'', little more than 16 minutes behind the winning Vaccarella-Bandini Ferrari 275 P2.

The new generation Abarth engines were also used in the OT coupé, the first version of which was the 1300 cc that competed in 1965 Group 4 Sport events. Maximum power was increased to 147 hp at 8800 rpm in the competition version, the car having been built, like the roadster, on a floor pan of

Thursday 1 April 1965 and Abarth was at Monza to put the prototype Fiat Abarth Sport Spider OT 1600 (above) through its first test. The designation defined the car as having a chassis in boxed pressed steel with steering and suspension taken from the Fiat 850 production car, while the engine was designed and built by the Scorpion. The car covered several laps of the junior circuit driven by Hans Herrmann and recorded respectable times.

The rear engine of the Fiat Abarth Sport Spider OT 1600 was a 1592 cc four-cylinder twin-cam with double ignition, a compression ratio of 9.5:1, two twin-choke Weber carburettors and put out 172 hp at 8000 rpm. Access to the mechanics was made easier by the all-in-one aperture of the engine compartment, which was hinged at the furthest extreme of the car. Cooling was by a large front air intake, which directed the flow onto the water and oil radiators.

Conceived to race in the Sport and Prototype categories, the Fiat Abarth Sport Spider OT 1600 could be fitted with two different types of interchangeable windscreen, depending on the regulations in force at the time. Two other lateral intakes were positioned at the height of the rear wheel housings. The total weight of the prototype was 590 kg and its maximum speed was 240 km/h. Hans Herrmann and Leo Cella won their class and came sixth overall in this car in the 1965 Targa Florio.

boxed pressed steel, which was further elaborated from that of the Simca 1000, and weighed 655 kg. The rear window was inclined vertically, so that the whole rear section of the car could be tipped backwards for better access to the engine and transmission.

The final evolution of the new Fiat Abarth sport models was represented by the Spider 2000 roadster, in which the factory installed a two-litre engine that put out 215 hp at 8800 rpm. Except for the suspension settings, tyre dimensions and a number of adaptations as need dictated, the sports roadsters all had the same chassis and body but different engines; the type 237 had a 1300 cc power unit, the type 235 1600 cc and the type 236 2000 cc. This components standardisation operation was concluded in 1965, so as to replace the Abarth Simcas of the same capacity.

Fiat unveiled two sports versions of their 850 saloon at the 1965 Geneva Motor Show, the Coupé with a body by its own Styling Centre and roadster by Carrozzeria Bertone that would be manufactured at the designer's Grugliasco plant. The 843 cc engine was given a vigorous injection of power, moving on from the saloon's standard 34 hp to 47 hp for the two new sports models, giving them a top speed of over 135 km/h. A few months later, Carlo Abarth attracted attention once again with his OT 1000 Spider and OT 1000 Coupé, both of which bowed in at the Frankfurt Motor Show in September 1965. This time the transformation was fairly simple, as he used the well-known 982 cc four-cylinder engine with a 9.8:1 compression ratio that developed 62 hp at 6150 rpm; the two Fiat derivatives' top speed was 155 km/h. An additional ver-

FIAT ABARTH "SPORT SPIDER OT 2000" (1965)

ENGINE	
	rear, longitudinal, 4 cylinders in line
Bore and stroke	mm 88x80
Cubic centimetres	cc 1946
Valve gear	2 overhead camshafts, chain driven
Number of valves	2 per cylinder
Compression ratio	10.0:1
Fuel feed	2 twin-choke Weber 58DCOE carburettors
Ignition	double, coil, distributor
Coolant	liquid
Lubrication	dry sump
Max. power (project)	209 hp at 7600 rpm
TRANSMISSION	
	rear wheel drive
Clutch	single dry plate
Gearbox	6 speed plus reverse, central lever
Axle ratio	4.3:1
BODY	
	racing roadster
CHASSIS	
	boxed steel
Suspension	front independent, upper trailing arms, transverse lower leaf spring, helicoidal springs, hydraulic dampers; rear independent, trailing arms, helicoidal springs, hydraulic dampers
Brakes	hydraulically operated discs front and rear
Steering	worm and sector
Fuel tank	capacity 60 litres
Tyres	front 5.50x13; rear 6.00x13
DIMENSIONS AND WEIGHT	
Wheelbase	mm 2015
Front track	mm 1296
Rear track	mm 1340
Length	mm 3830
Width	mm 1620
Height	mm 1040
Kerb weight	kg 710
PERFORMANCE	
Top speed	km/h 260

sion of the engine that generated 68 hp at 6400 rpm was also made available for the Coupé and powered the car to a top speed of over 160 km/h. Externally, the car could be identified by its front radiator and Campagnolo wheels.

Carlo Abarth introduced yet another completely new 'radial' engine, so called, because the combustion chambers were formed by the intersection of two spherical segments of which the shorter housed a large inlet valve. The new power plant only retained the original Fiat crankcase, while the head, con rods, drive shaft and sump were all designed and made by Abarth; cooling was by a front radiator that took over the space left by the production car's spare wheel, which was mounted horizontally; initially, the engine was installed in Fiat's only coupé and called the OTR 1000 because of its front radiator, which gave the technicians a number of location problems due to the car's low, tapered front end. The radial engine, which created some uncertainty due to its doubtful efficiency on the terminological front, produced 74 hp at 6500 rpm but according to some Abarth technicians, the most advanced experimental version generated over 95 hp.

During the press presentation of these new developments at the Frankfurt Motor Show, a new version of the 695 SS – already homologated in the Touring Category - was also unveiled and had a wider track than the Fiat 500 F, which first appeared the previous March. The racing version of the little Scorpion put out 38 hp at 5350 rpm and was fitted with disc brakes, still not homologated.

The 1946 cc four-cylinder engine that delivered 209 hp at 7600 rpm was also installed in the same chassis (internal code 139) and used the same body as the Fiat Abarth Sport Spider OT 1600. That met the need to compete in races in various classes of the Sport and Prototype categories, using cars of a high degree of mechanical component standardisation and, therefore, with a correspondingly high degree of interchangeability. Driven by Hans Herrmann, the Fiat Abarth Sport Spider OT 2000 competed in the 1965 European Hillclimb Championship, but even though Herrmann won the Mont Ventoux race in France with the car on 6 June it ended up being beaten by the Porsche and Ferrari prototypes.

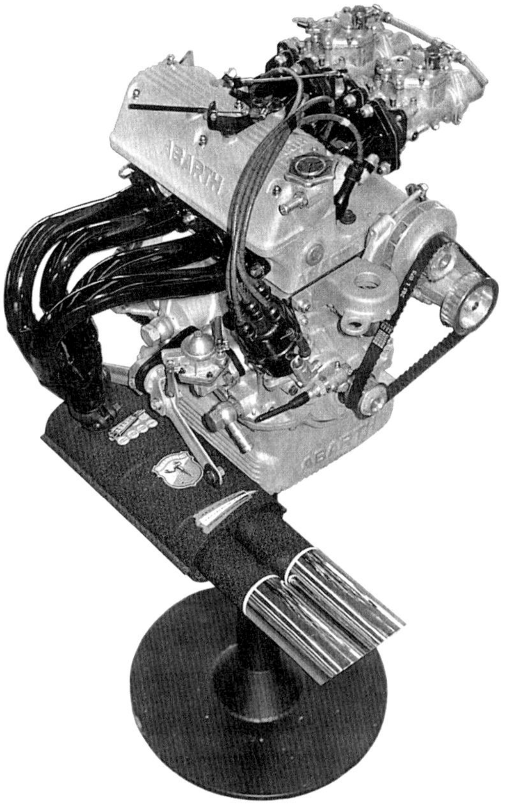

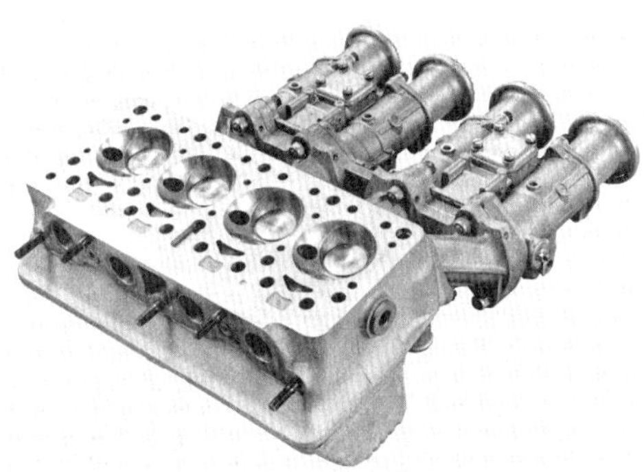

On 9 September 1965 just before the Frankfurt Motor Show, Abarth press launched the new OTR 1000, derived directly from the Fiat 850 coupé and with a new kind of engine called a 'radial'. The power unit had a new type of cylinder head in which the valves were housed and a bi-spherical combustion chamber built with two spherical domes, one containing the inlet valve and the other the exhaust. The intention was to provide fuel feed with a separate duct for each cylinder and to obtain a better filling of the combustion chamber due to the absence of counter-pressure that could obstruct the normal flow of the mixture along the combined inlet manifolds.

Abarth's 1965 motor sport programme was considerable in the Touring, Grand Touring and Sport Categories. Not even the tubular barchetta, which had produced good results years earlier, was left out. In fact the Colucci's chassis design, successively modified for competition in various cubic capacity race classes, was still extremely effective and versatile. In order to obtain homologation in the one-litre class the first 25 examples of the barchetta were built in 1965, an undertaking that demonstrated Abarth's renewed commitment to the Sport Category, which had not been given the full attention of Carlo and his technicians until a year earlier. Another 25 cars were built in subsequent years, until the model achieved Group 4 Category A homologation on 1 March 1968. That was the car that marked Abarth's definitive abandonment of aluminium bodies in favour of glass fibre; the first cars had no water radiator in the front but two rear units, one on each side of the engine compartment, which were cooled by dynamic air flow channelled through grills the shape and dimensions of which varied from one example to another. Coded 229-A, the engine was derived from the successful 982 cc four-cylinder that produced 105 hp at 8000 rpm. The car had a 2200 mm wheelbase, weighed 480 kg and was accredited with a stop speed of 230 km/h.

The new Appendix J for Touring and Grand Touring cars was published in early February 1965 and introduced these new groups:

- Group 1 for production touring cars manufactured in a minimum quantity of 5000 units a year;
- Group 2 for production touring cars with a minimum annual production of 1000 units and with great freedom of elaboration;
- Group 3 for production grand touring cars of which an annual minimum 500 are made;
- Group 4 for sports cars – ex-grand touring cars of the previous appendix – with a minimum production rate of 100 units a year; 50 from 1966;
- Group 5 for special touring cars – appertaining to Groups 1 and 2 – with ample freedom of elaboration;
- Group 6 for sports prototypes – ex-sport of the previous appendix – with greater freedom for intervention compared to the previous sports cars;
- Group 7 for single-seater racing cars – F1, F2, F3;
- Group 8 for formula libera single-seaters;
- Group 9 for two-seater cars – ex-sport in the previous appendix – without any restriction.

Contrary to previous years, in 1965 the cycle of races for the European Touring Car Challenge offered three drivers' and three constructors' titles in the 1000 cc, 2000 cc and over 2000 cc divisions respectively. The first event, the 4 Hours of the Jolly Club, took place on 19 March on Monza's full 10-kilometre track and high-speed ring circuit. Herbert Demetz won the first division

FIAT ABARTH "OTR 1000 COUPE'" (1965)

ENGINE	
	rear, longitudinal, 4 cylinders in line
Bore and stroke	mm 65x74
Cubic capacity	cc 982
Valve gear	overhead valves, side camshaft
Number of valves	2 per cylinder
Compression ratio	10.5:1
Fuel feed	2 twin-choke Solex C32PHH carburettors
Ignition	coil, distributor
Coolant	liquid
Lubrication	forced, oil filter on the main circuit
Max. power (project)	74 hp at 6500 rpm
TRANSMISSION	
	rear wheel drive
Clutch	single dry plate
Gearbox	5 speed plus reverse, central lever
Axle ratio	4.1:1
BODY	
	2 door, 2+2 coupé
CHASSIS	
	integral body
Suspension	front independent, upper trailing arms, transverse lower leaf spring, hydraulic dampers; rear independent, trailing arms, helicoidal springs, hydraulic dampers
Brakes	hydraulically operated front discs and rear drums
Steering	worm and sector
Fuel tank	capacity 30 litres
Tyres	5.50x13
DIMENSIONS AND WEIGHT	
Wheelbase	mm 2027
Front track	mm 1160
Rear track	mm 1212
Length	mm 3608
Width	mm 1500
Height	mm 1300
Kerb weight	kg 698
PERFORMANCE	
Top speed	km/h 172

of up to 1000 cc driving a Fiat Abarth 1000 Berlina Corsa, which in that state of preparation that included a 12.2:1 compression ratio and a twin-choke 36 mm Weber carburettor, had a maximum power output of 80 hp at 7400 rpm and reached a top speed of over 185 km/h. Behind Demetz came a Fiat Abarth 1000 Berlina and two 850 TCs. The victorious season continued weekend after weekend, with Abarths dominating the up to 1000 cc Touring Category and the up to 700 cc, 1000 cc and 1300 cc classes of the Grand Touring Category.

At Imola, Demetz also won the fifth Shell Grand Prix – second round in the Grand Touring Car Constructors' World Championship - on 7 April in an Abarth Simca 1300 GT, and Bruno Bonini came fourth driving a Fiat Abarth 1000 Bialbero. Klaus Steinmetz won the third round in the championship, the InterEurope Cup at Monza, third round in the Constructors' World Championship on 25 April, at the wheel of an Abarth Simca 1300 GT, while Cesare Sangiorgi came fourth in a long nosed Fiat Abarth 1000 Bialbero.

The succession of victories achieved by the Scorpion continued at the 49th Targa Florio on 9 May with two first places by Hans Herrmann and Leo Cella in the up to 1600 cc Prototype Category, driving a Fiat Abarth OT 1600 and Virgilio-Calascibetta in an Abarth Simca 1300 GT; then came the 11th 1000 Kilometres of the Nürburgring on 23 May and victory by Salvatore Calascibetta in an Abarth Simca 1300 GT, the ninth Bologna-Raticosa Pass on 30 May won outright by Hans Herrmann in a Fiat Abarth OT 2000 plus seven class wins for the Scorpion, among them Giuseppe Rebaudi's driving a Fiat Abarth 1000 Bialbero and Giuseppe Dalla Torre in an Abarth Simca 1300 GT; more success at the 24th Palermo-Monte Pellegrino (five victories), the Corsa della Mendola on 4 July in which Herbert Demetz' two-litre Abarth Simca won outright and took the class win, the Trento-Bordone of 11 July (four victories) where Hans Herrmann in a Fiat Abarth OT 2000 Spider just missed out on overall victory for a reason that was never made clear.

Adriano Ceci wrote in Auto-Italiana on 22 July: "*A few metres from the finish, Herrmann had a problem that meant he lost precious seconds; exactly how many, we shall never know. On one of the last corners his engine cut out. There are those who say the engine flooded and others that Herrmann touched the contact key when he was changing gear. Herrmann left immediately he finished the race. Having heard of the incident, we went looking for the driver to get the right version but we were not able to find him. He walked off to the numerous pine trees in the area so that he would not be disturbed at that moment of annoyance for the probable victory that went up in smoke*". A victory that went that day to Ludovico Scarfiotti, who drove a Dino 2000 in the Sports Prototype Category, behind which Herrmann and his Abarth Simca 2000 GT also had to be very happy with a second bright place in the European Hillclimb Championship.

The first cracks began to appear in Abarth's dominance of track events and hillclimbing and the company was finding the competition stiffer; even in classes in which the Scorpion did not compete, the degree of preparation and competitiveness of other entrants had reached an extremely high level. The various successes achieved by the BMW 700 in the up to 700 cc Touring class, the Morris Mini Cooper in the 1300 cc, Steyr-Puch in the up to 700 cc Grand Touring, the Alfa Romeo TZ in the 1600 cc class and the Porsche 904 in the two-litre Sport, all clearly showed how the other factories were fully committing themselves to an activity that had enormous commercial advantages. The proof of renewed interest in hillclimb championship races was confirmed in 1965 *"by the decision of Ferrari to enter a Dino for Scarfiotti, after being absent for a considerable time from that kind of competition... for two reasons: the imminent launch of the Dino, built by Fiat, and the decision to build a run of*

Among the first important events of the 1965 racing season was the 4 Hours of the Jolly Club on 19 March on Monza's full 10-kilometre circuit.
The picture above, taken on the first lap at the entrance to Lesmo, shows the Fiat Abarth 1000 Berlina number 34 of Giancarlo Baghetti being followed by John Fitzpatrick's Morris Mini Cooper S and the 1000 Berlina of Herbert Demetz, Hans Herrmann and Klaus Steinmetz. Demetz won, but Herrmann was forced to retire (below) with an overheating engine.

Carlo Abarth congratulates Herbert Demetz, who had just driven the Fiat Abarth 1000 Berlina to victory in the 1st Division of the Touring category in the 4 Hours of the Jolly Club race at Monza on 19 March '65.
Below: Ignazio Giunti won the Course de Côte du Col Bayard, France, outright and came first in class on 25 April driving a 1000 Berlina.
Opposite: the same day, Abarth won the InterEurope Cup at Monza, third round in the World Constructors' Championship, with Abarth Simca 1300 GT number 14 driven by Klaus Steinmetz. The picture shows him behind sister car 16, which Hans Herrmann had to retire with water pump trouble.

14
16
16

The Scorpion scored a brilliant victory in the Grand Touring category of the 49th Targa Florio on 19 May 1965 in the 10 laps of the 720-kilometre race. Abarth Simca 1300 GT number 26 driven by Salvatore Calascibetta-Giuseppe Virgilio took the class win and 10th place overall with a time of 7 hours 47'12.4" at an average speed of 92.463 km/h. Hans Herrmann and Leo Cella in Fiat Abarth Sport Spider OT 1600 number 152 won the 1600 cc class of the Prototype category with a time of 7 hours 17'23" and an average of 98.770 km/h.

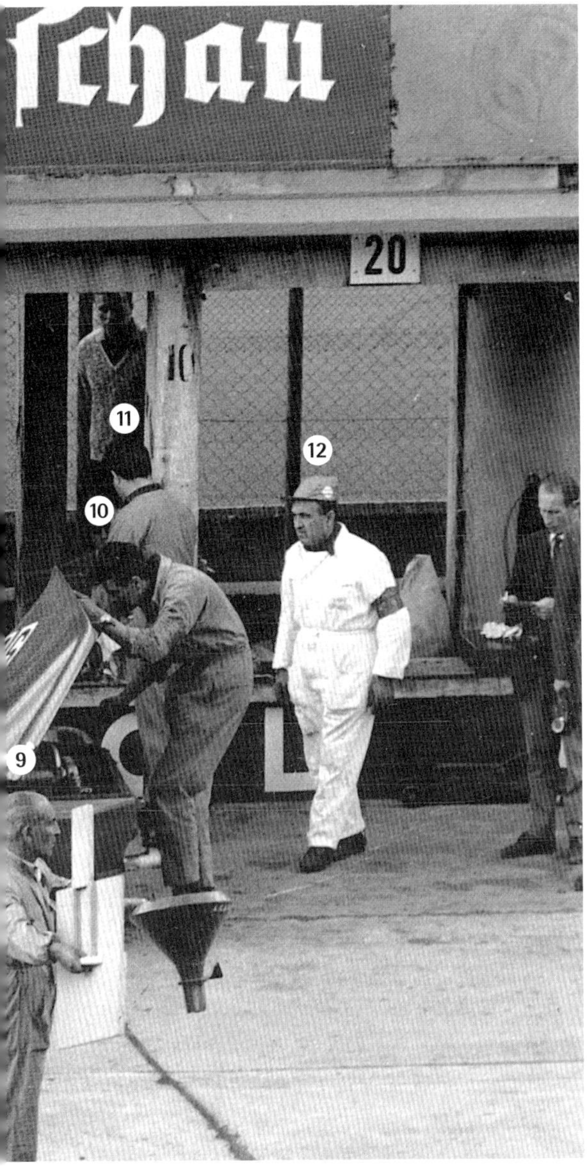

cars for homologation in the Sport category (minimum 50 examples) that will come into effect on 1 January 1966" wrote Franco Lini in Auto-Italiana of 30 September 1965.

At the fifth Aosta-Pila on 19 September, Hans Herrmann came first overall with a Fiat Abarth 2000 OT, followed by Demetz in a Sport 1000 OT but at the 500 Kilometres of the Nürburgring the same month, when the OT 1300 made its debut, the French Alpines of the Bianchi brothers got the better of the Abarths; Gerhard Mitter-Enzo Osella came second in the Group 6 Prototype in a Fiat Abarth SP 1300 followed in third place by a Fiat Abarth OT 1300 Sport Spider driven by Steinmetz-Mitter third and Kurt Ahrens' Abarth Simca 1300 GT fourth. In spite of their early failures, the Scorpion's

The 11th 1000 Kilometres of the Nürburgring, seventh counter in the International GT Constructors' Championship, took place on 23 May 1965.
After their great success six days earlier in the Targa Florio, Calascibetta-Virgilio did it again in Germany and won their class in Abarth Simca 1300 GT number 95, with which they dominated and covered the 44 laps totalling 1003.640 kilometres in 6 hours 53'45.9" at an average 128.9 km/h; the Scorpion finished the race 18th overall, five laps behind the winning Ferrari 330 P2 of John Surtees and Ludovico Scarfiotti.
The large photograph of a feverish Abarth pit stop shows (1) Renzo Avidano, (2) Friedrich Hoffmann, (3) Klaus Steinmetz, (4) Fritz Juttner (driver), (5) Antonio Tomaini, (6) Giuseppe Virgilio, (7) Esposito (mechanic), (8) Vincenzo Osella, (9) Alfonso Tirozzi, (10) Gentini, (11) Saccardi, (12) Trammonti.

cars still achieved great success in 1965: their wins included the Paris Cup (four victories), the Sauerland hillclimb (two wins), the 11th Carri Cup at Monza (3 victories), the Donau Pokal in Austria (3 firsts), the first Montjuich Trophy in Spain (2 victories), the Hockenheim Ring in Germany (2 wins), the seventh Giro di Toscana (2 firsts), the 19 December Criterion of the Hour at Vallelunga (3 victories) and the first Circuit of Jarama in Spain (2 firsts).

However there was a progressive defection of private drivers from motor sport due to the increasing cost of the cars, their preparation, tuning and maintenance necessary to compete in so many races. The list price of a Fiat Abarth 850 TC Corsa in May 1965 was Lit 1.525.000 and that went up six months later to Lit 2.340.000 after the factory's decision to make the same modifications and updates available to customer cars as were made to the works racers. With that operation, Abarth wanted to combat objections made

280949•FI

41
23

CYNAR
Pepsi-Cola
FIAT ABARTH

Opposite: The Scorpion won the 1000 cc class of the Touring category at the 6 Hours of the Nürburgring on 13 June 1965 with a Fiat Abarth 1000 Bialbero driven by Kurt Ahrens and Klaus Steinmetz (left) and on 15 August a sister car driven by John Anstead-Roy Pike (right) won their class in the 500 Kilometres of Snetterton in Britain, a round in the European Touring Championship.
Hans Herrmann in a Fiat Abarth Sport Spider OT 1600 (large photograph) won the 1600 cc class of the Ollon-Villars hillclimb in Switzerland on 29 August.

by privateers that they were being discriminated against in races compared to the works drivers, but the move also forced many private entrants to withdraw because they could no longer afford a season's motor racing.

The season ended with 900 Abarth victories, 23 of them outright. The Scorpion won the Italian Touring Championships of up to 600 cc (Leo Durst in a Fiat Abarth 595 prepared by Garavello), the up to 850 cc (Sergio Morando driving a Fiat 850 TC tuned by Varese), the up to 1000 cc class (Achille Marzi with a 1000 Berlina prepared by Faccioli). Corso Marche also won the Italian Grand Touring Championships in the up to 700 cc class (Gianni Varese at the wheel of a Fiat Abarth 700 Bialbero maintained by himself), the up to 1000 cc (Giuseppe Rebaudi in a Fiat Abarth 1000 Bialbero tuned by Faccioli) and the 1300 cc class (Roberto Bertuzzi driving an Abarth Simca 1300 GT maintained by Moroni). And the cars won the first division of the European Touring Challenge up to 1000 cc (Ed Swart driving a Fiat Abarth 1000 Berlina) as well as the World Constructors' Grand Touring Championship with the Abarth Simca 1300 GT.

Carlo Abarth did not lose the opportunity of returning to the top of the speed record pile. On 20 October 1965, he himself was at Monza driving a class G single-seater with a 982 cc twin-cam engine and double ignition to break the quarter mile standing start record, clocking an average speed of 106.344 km/h and the standing start 500 metres at an average 117.035 km/h. In order to slip into the narrow cockpit of the car, Carlo had undergone months of strict dieting at the end of which he had lost nearly 30 kilos. The following day at Monza, he climbed into the 2000 single-seater, powered by a 1946 cc four-cylinder engine with a bore and stroke of 86x80 mm that developed 216 hp at

The Fiat Abarths were victorious in the Touring category across Europe week after week.
Here, we see Ed Swart scoring touring victories at Zandvoort, Holland, on 29 August (left) and at Cannonlopett, Sweden, where he dominated the race from the first lap.

8800 rpm, and established new class E up to 2000 cc records over a quarter mile at an average 125.348 km/h and 500 metres averaging 134.629 km/h, beating the standing start records previously held by Porsche with 121.973 km/h and 125.348 km/h respectively. After his last run, Carlo was due to attempt another record, but during the re-entry manoeuvre his right foot got caught in the single-seater's pedals; the car shot forward and smashed into a pit wall under full acceleration and was seriously damaged, but Carlo got away with a slightly grazed face. That little accident far from discouraged Abarth from continuing his record programme, which had been put together that year at the Monza circuit to try to beat records in class H of up 750 cc, in class G up to 1100 cc and up to 1500 cc in class F. Driving a class H 741.9 cc twin-cam Pininfarina record car that put out 80 hp, on 27 October, Mario Poltronieri

The third round in the 1965 European Hillclimb Championship was the Trento-Bondone in Italy, for which the Scorpion entered three Fiat Abarth Sport Spider OT 2000s and three Abarth Simca two-litres.
Everyone expected a convincing victory from Hans Herrmann in his works Sport Spider OT 2000 number 54 (picture), but he had to be happy with second, behind Gerhard Mitter in a Porsche, due to a mysterious problem a few corners from the end that meant his engine cut out. But the cars of Corso Marche still won four first places in the Touring category with the Fiat Abarth 850 TC (Giuseppe Villani) and the 1000 Berlina (Achille Marzi) as well as the Grand Touring category (Giuseppe Rebaudi) and with the Abarth Simca driven by Giuseppe Dalla Torre.

Among the various culinary disciplines Carlo Abarth periodically imposed on himself was an apple diet; if eaten raw during the course of the day apples were supposed to reduce weight and revive the body's organisms. The benefit of the apple assumed an official role at Corso Marche, so much so that Carlo wanted to be immortalised with the fruit by the photographer, together with his most important cars of 1965.
That was the year the Fiat Abarth OT 1300 made its debut and went on to score the greatest victories ever achieved by a Scorpion on the leading circuits of the world.

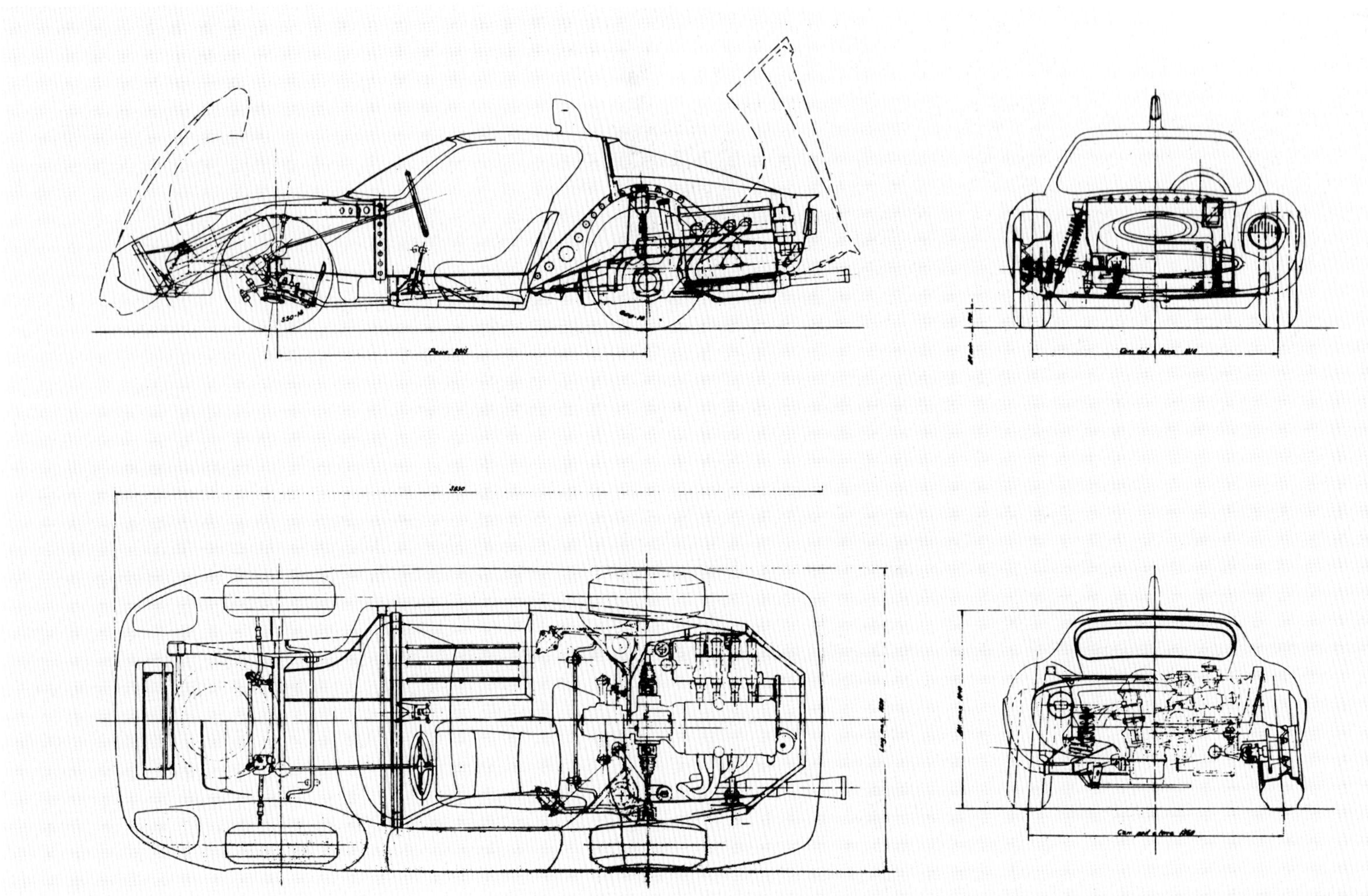

Abarth technicians developed a chassis in boxed pressed steel on the modified floor pan of the Simca 1000, to which independent suspension for all four wheels was anchored with oscillating arms and helicoidal springs derived in part from the Fiat 850. Destined to compete in the Grand Touring category, the Fiat Abarth OT 1300 raced for the first time as a prototype at the 500 Kilometres of the Nürburgring on 5 September 1965, when Klaus Steinmetz drove it into third place overall. As was now the tradition at Abarth, the low, elegant body of the OT 1300 was designed by Mario Colucci and his team and built in glass fibre by Sibona & Basano of Turin.

Mounted over the rear axle, the engine of the Fiat Abarth OT 1300 was a four-cylinder in-line twin overhead camshaft unit of 1289 cc (bore and stroke 86x55.5 mm), the early racing versions of which generated 147 hp at 8800 rpm.

FIAT ABARTH "OT 1300" (1965)

ENGINE	
	rear, longitudinal, 4 cylinders in line
Bore and stroke	mm 86x55.5
Cubic capacity	cc 1289
Valve gear	2 overhead camshafts, chain driven
Number of valves	2 per cylinder
Compression ratio	mm 10.5:1
Fuel feed	2 twin-choke Weber 45DCOE9 carburettors
Ignition	double, coil, distributor
Coolant	liquid
Lubrication	dry sump
Max. power (project)	147 hp at 8800 rpm
TRANSMISSION	
	rear wheel drive
Clutch	single dry plate
Gearbox	5 speed plus reverse, central lever
Axle ratio	5.4:1
BODY	
	2-seater coupé
CHASSIS	
	boxed steel
Suspension	front independent, trailing arms, helicoidal springs, stabiliser bar, hydraulic dampers; rear independent, trailing arms, helicoidal springs, stabiliser bar, hydraulic dampers
Brakes	hydraulically operated discs front and rear
Steering	worm and sector
Fuel tank	capacity 40 litres
Tyres	front 5.50x13; rear 6.00x13
DIMENSIONS AND WEIGHT	
Wheelbase	mm 2015
Front track	mm 1296
Rear track	mm 1340
Length	mm 3830
Width	mm 1620
Height	mm 1040
Kerb weight	kg 655
PERFORMANCE	
Top speed	km/h 245

The engine of the Fiat Abarth OT 1300 was designed from scratch at Corso Marche, helped by the experience Luciano Fochi had accrued with the four-cylinder 1600 and 2000 cc Spider Sport racing units.
The OT 1300 had five main bearings, twin-cam distribution, dry sump lubrication, double ignition and two twin-choke Weber carburettors.
Production began on 15 May 1965 and ended on 30 March 1966, by which time the minimum number of 50 cars had been built to achieve the Grand Touring homologation granted on 1 May.
The Fiat Abarth OT 1300 dominated its class for three consecutive years.

beat the standing start quarter mile record with a time of 15.25/1000" averaging 95.762 km/h and the 500 metres in 17.15/1000" at an average of 104.956; the next day, Klaus Steinmetz in an open-wheel single seater powered by a class G 982 cc, 105 hp engine established a new one kilometre standing start record of 25.2/1000", average 143.884 km/h. The previous day, Steinmetz had set up the best time over a quarter mile of 12.87/1000" – an average 112.541 km/h – and a standing start 500 metres in 14.915/1000" averaging 120.683 km/h in a class F Fiat Abarth OT 1450 with a partially fared body and a 1459 cc 148 hp engine.

FIAT ABARTH "1000 MONOPOSTO RECORD" (1965)

ENGINE	
	mid-rear, longitudinal, 4 cylinders in line
Bore and stroke	mm 65x74
Cubic capacity	cc 982
Valve gear	2 overhead camshafts, chain driven
Number of valves	2 per cylinder
Compression ratio	11:1
Fuel feed	2 twin-choke Weber 40DCOE carburettors
Ignition	coil, distributor
Coolant	liquid
Lubrication	dry sump
Max. Power (project)	105 hp at 8800 rpm
TRANSMISSION	
	rear wheel drive
Clutch	single dry plate
Gearbox	4 speed plus reverse, central lever
Axle ratio	5.3:1
BODY	
	single-seater
CHASSIS	
	tubular
Suspension	front independent, upper and lower triangular trailing arms, helicoidal springs, hydraulic dampers; rear independent, reaction arms, helicoidal springs, hydraulic dampers.
Brakes	hydraulically operated discs front and rear
Steering	rack and pinion
Fuel tank	capacity 10 litres
Tyres	front 5.25x13; rear 6.50x13
DIMENSIONS AND WEIGHT	
Wheelbase	mm 2300
Front track	mm 1320
Rear track	mm 1320
Length	mm 3770
Width	mm 1640
Height	mm 700
Kerb weight	kg 500
PERFORMANCE	
Top speed	km/h 225

On 20 October 1965, Carlo Abarth himself made a successful attempt on the quarter mile and 500 metre acceleration records at Monza in a class G twin-cam one-litre single-seater of 105 hp (centre).
The next day Abarth (side) broke the acceleration record with a 1946 cc class E single-seater that generated 216 hp at 8800 rpm.
To take on the two challenges in the narrow cockpits of those open wheel cars, Carlo had to lose 30 kg.
That day, he was involved in a banal accident, which resulted in facial grazing.

FIAT ABARTH "2000 MONOPOSTO RECORD" (1965)

ENGINE	
	mid-rear, longitudinal, 4 cylinders in line
Bore and stroke	mm 88x80
Cubic capacity	cc 1946
Valve gear	2 overhead camshafts, chain driven
Number of valves	2 per cylinder
Compression ratio	11.3:1
Fuel feed	2 twin-choke Weber 58DCOE3 carburettors
Ignition	double, coil, distributor
Coolant	liquid
Lubrication	dry sump
Max. power (project)	216 hp at 8800 rpm
TRANSMISSION	
	rear wheel drive
Clutch	single dry plate
Gearbox	4 speed plus reverse, central lever
Axle ratio	5.3:1
BODY	
	single-seater
CHASSIS	
	tubular
Suspension	front independent, upper and lower triangular trailing arms, helicoidal springs, hydraulic dampers; rear independent, reaction arms, helicoidal springs, hydraulic dampers
Brakes	hydraulically operated discs front and rear
Steering	rack and pinion
Fuel tank	capacity 20 litres
Tyres	front 4.50x13; rear 7.50x13
DIMENSIONS AND WEIGHT	
Wheelbase	mm 2300
Front track	mm 1320
Rear track	mm 1320
Length	mm 3610
Width	mm 1720
Height	mm 750
Kerb weight	kg 500
PERFORMANCE	
Top speed	km/h 260

On 11 December Steinmetz, who had shown himself to be not only an able driver but also a talented engineer right from the days of the Porsche Abarth GTL, went back to Monza with Carlo Abarth and a small group of technicians for another record attempt, this time in class E of up to 2000 cc. A special Fiat Abarth OT 2000 was prepared, with much modified aerodynamics and fitted with the engine used in October in the single-seater record breaker. Steinmetz successfully completed the kilometre from a standing start in 21.53'/1000" at an average 167.169 km/h. The car's body was in extremely smooth glass fibre and hubcaps were fitted to wheels of integral development to minimise turbulence in the wheel housing. The OT Special was 750 mm high, weighed 500 kg and was accredited with a more than 220 km/h top speed. The record, previously established a little earlier on 4 October 1965 by a Porsche driven by Huschke von Hanstein, was clocked at 22.21'/1000", an average speed of 162.074 km/h.

Frowning but determined, Carlo Abarth about to climb behind the wheel of his little single-seater on 21 October 1965 to attempt to establish new acceleration records. The Viennese-born technician was at home at Monza, where he had broken over 100 records with his Abarth specials and a number of modified production cars.

FIAT ABARTH "OT 1450 SPORT RECORD" (1965)

ENGINE	
	rear, longitudinal, 4 cylinders in line
Bore and stroke	mm 76.2x80
Cubic capacity	cc 1459
Valve gear	2 overhead camshafts, chain driven
Number of valves	2 per cylinder
Compression ratio	10.8:1
Fuel feed	2 twin-choke Weber 45DCOE9 carburettors
Ignition	double, coil, distributor
Coolant	liquid
Lubrication	dry sump
Max. power (project)	148 hp at 7600 rpm
TRANSMISSION	
rear wheel drive	
Clutch	single dry plate
Gearbox	5 speed plus reverse, central lever
Axle ratio	5.3:1
BODY	
	2 seat record car
CHASSIS	
	boxed steel
Suspension	front independent, upper trailing arms, transverse lower leaf spring, helicoidal springs, hydraulic dampers; rear independent, trailing arms, helicoidal springs, hydraulic dampers
Brakes	hydraulically operated discs front and rear
Steering	worm and sector
Fuel tank	capacity 60 litres
Tyres	front 5.00x13; rear 7.00x13
DIMENSIONS AND WEIGHT	
Wheelbase	mm 2015
Front track	mm 1296
Rear track	mm 1340
Length	mm 3830
Width	mm 1620
Height	mm 810
Kerb weight	kg 630
PERFORMANCE	
Top speed	km/h 220

And that is how the company's 1965 in motor sport ended, with eight international and two world records established by Abarth for Italy. In the production car sector, the latest Corso Marche factory move was to introduce the Fiat Abarth OT 1000 Coupé and roadster at the Frankfurt Motor Show in September, both derived from the respective 850 models but fitted with a 982 cc, four-cylinder engine producing 62 hp at 6150 rpm. On that occasion another long-awaited, much talked about engine made its debut – the 982 cc, 74 hp radial head power unit, wrongly lowered into the coupé, because the Bertone-designed roadster's body had to be modified up front to take the radiator. With a weight of 730 kg, the Fiat Abarth OT 1000 Coupé accelerated to a maximum speed of 155 km/h, while with a weight of just 698 kg the OTR 1000 Coupé had a top speed of 170 km/h. Much was expected of the new radial engine due to its advantage in terms of assimilated yield comparable to that of the twin-cam engines and the greater constructional and maintenance simplicity of the classical Fiat single camshafts.

Among the other cars introduced at the German show was the Fiat Abarth OT 1000 Coupé by Pininfarina, the general architecture of which aped the same-bodied prototype roadster that was so well received at the Turin show a year earlier by the critics and public alike. The elegance of its lines and the original feature of being able to forward-open the whole upper part of the cockpit (windscreen and roof) to facilitate access were regarded as yet more proof of the creativity and good taste of the great Turin stylist, who became gravely ill and died a few months later in Lausanne, Switzerland.

For Abarth & C. 1965 drew to a close with a long series of sporting triumphs, but motor racing was starting to become a burden on the company's balance sheet – particularly in the unfavourable economic climate that had

On 27 October it was Abarth works driver Klaus Steinmetz' turn to take to the Monza circuit and establish records in class F (1101 to 1500 cc), which had been recently instituted.
He was at the wheel of a Fiat Abarth Spider OT 1450 with its body amply modified from the racing two-seater that made its debut in the spring; there was substantial faring on the front end, which strongly tapered the transverse section, and more in the guise of a tonneau cover that extended over the majority of the cockpit, notably elevating the car's coefficient of aerodynamic penetration. Steinmetz covered the standing start quarter mile in little more than 12 seconds at an average speed of 112.541 km/h and the 500 metres in just under 15 seconds, average 120.683 km/h.

FIAT ABARTH "OT 1000 COUPE'" (1965)

ENGINE	
	rear, longitudinal, 4 cylinders in line
Bore and stroke	mm 65x74
Cubic capacity	cc 982
Valve gear	overhead valves, side camshaft
Number of valves	2 per cylinder
Compression ratio	9.5:1
Fuel feed	1 Weber 30DIC carburettor
Ignition	coil, distributor
Coolant	liquid
Lubrication	forced, oil filter on the main circuit
Max. power (project)	62 hp at 6150 rpm
TRANSMISSION	
	rear wheel drive
Clutch	single dry plate
Gearbox	4 speed plus reverse, central lever
Axle ratio	4.6:1
BODY	
	2 door, 2+2 coupé
CHASSIS	
	integral body
Suspension	front independent, upper trailing arms, transverse lower leaf spring, hydraulic dampers; rear independent, trailing arms, helicoidal springs, hydraulic dampers
Brakes	hydraulically operated front discs, rear drums
Steering	worm and sector
Fuel tank	capacity 30 litres
Tyres	5.50x13
DIMENSIONS AND WEIGHT	
Wheelbase	mm 2027
Front track	mm 1158
Rear track	mm 1212
Length	mm 3608
Width	mm 1500
Height	mm 1300
Kerb weight	kg 730
PERFORMANCE	
Top speed	km/h 155

permeated through the whole Italian manufacturing industry – due to continually having to update the cars and incessant design and development work on new models. So it became necessary to turn to a production policy and commercial measures that could convey the major thrust to associated products such as exhausts, modification kits and special parts which were in great demand in the market and produced good profits.

Prospects for Italy in 1966 were not good, even if the growth of employment and individual earnings registered in the immediate past were modifying the face of the country and the collective national approach to a new social order. The migration of great masses of people who worked in the country and lived in poverty to a few industrial centres in the north brought extensive unrest and a widespread sense of uneasiness, worsened by the primeval structural deficiency of the state system, which was not quickly able to adapt itself to the new situation. That eventually led to a period of open conflict that would soon lead

Opposite: Klaus Steinmetz set up the last record of the year on 11 December 1965 aboard a special car of low drag coefficient.
The car was based on the two-seater Fiat Abarth OT 2000.
A number of aerodynamic refinements were made to the car, including a lower, sharper front end, special faring for the wheels, an almost totally covered cockpit; with a maximum height of 750 mm, the car's weight was reduced to 500 kg and the glass fibre body was extremely smooth.
The OT Speciale, which could reach a top speed of 220 km/h, covered one kilometre in 21.535/1000", exiting at 167.253 km/h and beating the previous record established by the Huschke von Hanstein of Germany in a two-litre Porsche.

On 9 September, a few days before the opening of the 1965 Frankfurt Motor Show, Abarth unveiled the Fiat Abarth OT 1000 Coupé and OT 1000 Spider to the press.
The coupé had a Fiat production body but Abarth identification insignia, while the roadster was resplendent in a beautiful Bertone body.
Both versions had the veteran four-cylinder 982 cc single camshaft engine (internal code 202 B) with its maximum power raised to 62 hp at 6150 rpm.
Abarth declared the top speed of the two new cars of almost identical weight – the coupé 730 kg, the roadster 725 kg – at 155 km/h, with the closed bodied car priced at Lit 1.160.000 and the open top Lit 1.300.000.

to the anarchy of '68 and the loss of the social tranquillity that had become a catalytic element in favouring the economic expansion of the country.

The first racing car exhibition took place at Turin on 26 February near the Museum of the Automobile, where the Scorpion showed its Fiat Abarth OT 2000 Coupé America, a special derived from the Fiat 850 Coupé in which a two-litre twin-cam engine had been installed and produced 185 hp at 7200 rpm. The car, which boasted large wide wheel housing to take 5.50-13 tyres at the front and 6.00-13 rear, had a horizontal spare wheel that protruded from the centre front of the car and acted as a bumper bar. The OT 2000 used the same mechanical components as the Fiat production car – suspension, steering, brakes... - to which many modifications were made so that it could withstand its own stupefying performance, including a top speed of 240 km/h; the only change to the engine was the use of horizontal 45 mm Weber carburettors, instead of the 58 mm units on the Sport Spider, to make the engine more elastic and suitable for use in normal traffic. With four overhead camshafts, the new two-litre V8 engine also made its first public appearance on the stand; built entirely of light alloy, the cylinder banks were set at 90°. The engine's cubic capacity was 1987.814 cc, had double ignition and was fuelled by a battery of four twin-choke Weber carburettors located in the V of the two cam heads; with a compression ratio of 11:1, the car had a maximum power output of 230 hp at 9000 rpm.

Based on the Fiat 500 F, the various versions of the 595 and 696 were updated for the 1966 motor racing season with doors hinged at the front; maximum power of the first car was upped to 30-32 hp at 5000 rpm and the second

The mechanics of the Fiat Abarth OTR 1000 also hosted the Bertone Fiat 850 roadster body, for which the Turin designer had to slightly modify the forward line so that the supplementary radiator could fit into the car, which was unveiled at the 1965 Turin Motor Show as a closed sports saloon (picture), but which was never produced.

FIAT ABARTH "OTS 1000" (1965)

ENGINE	
	rear, longitudinal, 4 cylinders in line
Bore and stroke	mm 65x74
Cubic capacity	cc 982
Valve gear	overhead valves, side camshaft
Number of valves	2 per cylinder
Compression ratio	9.5:1
Fuel feed	2 twin-choke Solex C32PHH carburettors
Ignition	coil, distributor
Coolant	liquid
Lubrication	forced, oil filter on the main circuit
Max. power (project)	74hp at 6500 rpm
TRANSMISSION	
	rear wheel drive
Clutch	single dry plate
Gearbox	4 speed plus reverse, central lever
Axle ratio	4.1:1
BODY	
	2 door, 2+2 coupé
CHASSIS	
	integral body
Suspension	front independent, upper trailing arms, transverse lower leaf spring, hydraulic dampers; rear independent, trailing arms, helicoidal springs, hydraulic dampers
Brakes	hydraulically operated front discs, rear drums
Steering	worm and sector
Fuel tank	capacity 30 litres
Tyres	1450x13
DIMENSIONS AND WEIGHT	
Wheelbase	mm 2027
Front track	mm 1230
Rear track	mm 1248
Length	mm 3608
Width	mm 1500
Height	mm 1300
Kerb weight	kg 695
PERFORMANCE	
Top speed	km/h 180

The little 850 coupé was also clearly in evidence in the Fiat Abarth OTR 1000 leaflet, in which the car emerged much invigorated from the 'Abarth Cure' and generated 74 hp at 6500 rpm following the adoption of the new radial engine.
As was the case with other high performance Fiat Abarths, a radiator was installed in the front of the car.
The OTR 1000 also shared the body of the OTS 1000 that also boasted the single camshaft 982 cc 74 hp but not the radial engine.
The OTS was homologated to compete in the Grand Touring category on 25 March 1966 and soon started to win the following season.
The OTR 1000 entered the Abarth price list in November 1965 at Lit 1.410.000 as did the OTS 1000 a year later at Lit 1.295.000.

to 38 hp at 5350 rpm. Modifications were also made to the Fiat Abarth 1000 Berlina Corsa, which was given a 5 hp increase to 85 hp by raising the compression ratio from 10.2:1 to 13:1 and maximum revolutions per minute from 7400 to 7600. First offered with a four or five-speed, the gearbox stabilised at five-speed only. Weight was reduced to 583 kg and the maximum speed was declared at 195 km/h. Aesthetically, the new version of the 1000 Berlina Corsa boasted better aerodynamic faring on the imposing front radiator. The car was offered in apple green with a full lateral red stripe just below the waistline. Carlo Abarth also decided to personalise the Fiat-derived saloons of the works racing team with an extravagant series of coloured squares painted on the roof; they were yellow for the various 850 TCs and red for the 1000 Berlina.

Concentrating research and testing on prototypes that year resulted in the April presentation of the Fiat Abarth 1000 SP barchetta, with its tubular trellis lay-out, deformable parallelogram front and rear suspension integrated with stabiliser bars and two rear pushrod links; steering was rack and pinion

At the autumn motor shows, Pininfarina presented the OT 1000 Coupé, the overall architecture of which took its lead from the single example of the 1000 GT Spider seen the previous year at the Turin salon. Show visitors found the car highly innovative – even if not completely original, as certain features had already been seen on other prototypes – with the entire roof area – windscreen and lateral windows included – that could tip forward to facilitate access.
This was another splendid styling exercise by the celebrated Turin designer, based on the lavish floor pan of the Fiat 850.

and the alloy Campagnolo-Abarth rims were 5.00-13 on the front and 6.00-13 rear. The engine was similar to the well-known 982 cc twin-cam four-cylinder of distant Fiat 600 D derivation, mounted mid-ships in front of the rear axle group. The twin-cam cylinder head was updated by moving the inlet manifolds laterally from their original central position between the two camshafts, as was done with the 750 Record Monza in 1959. Maximum power was 105 hp at 8000 rpm; the cooling system comprised two lateral radiators at the rear with air intakes behind the driving position. The frontal change gearbox was five-speed. Made of glass fibre, the front and rear ends of the body could be tipped up for best access to the car's mechanics and their substitution in case of an accident. The large wraparound windscreen conformed

An overview of the Abarth stand at the 1965 Turin Motor Show from 3-14 November.
From the centre towards the left can be seen the Fiat Abarth OT 1000 Spider, the OTR 1000 Coupé and the OT 1000 Coupé; on the right is the new OT 2000 racing saloon – evolved from the OT 1300 – and in the background the 850 TC, 1000 Berlina and OT 1000.

FIAT ABARTH "OT 2000" (1966)	
ENGINE	
	rear, longitudinal, 4 cylinders in line
Bore and stroke	mm 88x80
Cubic capacity	cc 1946
Valve gear	2 overhead camshafts, chain driven
Number of valves	2 per cylinder
Compression ratio	10:1
Fuel feed	2 twin-choke Weber 45DCOE9 carburettors
Ignition	double, coil, distributor
Coolant	liquid
Lubrication	dry sump
Max. power (project)	185 hp at 7200 rpm
TRANSMISSION	
	rear wheel drive
Clutch	single dry plate
Gearbox	5 speed plus reverse, central lever
Axle ratio	4.3:1
BODY	
	2 door 2+2 coupé
CHASSIS	
	integral body
Suspension	front independent, upper trailing arms, transverse lower leaf spring, hydraulic dampers; rear independent, trailing arms, helicoidal springs, double hydraulic dampers, stabiliser bar
Brakes	hydraulically operated front and rear discs
Steering	worm and sector
Fuel tank	capacity 80 litres
Tyres	front 5.50x13; rear 6.00x13
DIMENSIONS AND WEIGHT	
Wheelbase	mm 2015
Front track	mm 1330
Rear track	mm 1350
Length	mm 3660
Width	mm 1500
Height	mm 1250
Kerb weight	kg 710
PERFORMANCE	
Top speed	km/h 242

to the new Group 6 Appendix J Prototype regulations. Built on a 2200 mm wheelbase, the Fiat Abarth 1000 SP was 3445 mm long, 1625 mm wide, 930 mm high and weighed 480 kg and made its first appearance in the hands of Anzio Zucchi at the 14th Coppa della Collina at Pistoia on 1 May, when it was plagued with a few teething troubles and only managed fifth in the up to 1000 cc Prototype class.

Some months later, the Fiat Abarth 1000 SP was joined by the 1300 SP, which looked like the 1000 in general architecture, but had a longer 2260 mm wheelbase. Its engine was the trusty 1289 cc four-cylinder with a bore and stroke of 86x55.5 mm that put out 145 hp at 8800 rpm. The car made its debut on 6 September at the 500 Kilometres of the Nürburgring driven by Johannes Ortner, but it retired during the second lap with a broken small bolt of the selector arm.

Opposite: The Museo dell'Automobile Carlo Biscaretti di Ruffia held its first racing car exhibition in Turin in 1966, opening on 26 February.
Centrepiece of the show was the Fiat Abarth OT 2000, with its body developed from the Fiat 850 Coupé's but with powerful mechanics comprising a 1946 cc four-cylinder twin-cam Abarth engine fed by two Weber twin-choke carburettors that generated 185 hp at 7200 rpm.
Brakes, suspension and wheels of the production car had to be reinforced and re-dimensioned to withstand the car's performance, which included a top speed of over 240 km/h.
The unusual position of the spare tyre and wheel meant the unit doubled as a bumper bar.

The same 1000 SP tubular barchetta was also made into a two-litre by fitting it with a 215 hp 1946 cc four-cylinder engine, from which both Carlo Abarth and Mario Colucci expected great race results. And that is just what they got at the Paris Cup in October, when Johannes Ortner came first overall with his Fiat Abarth 2000 Spider Tubolare, preceding the 4.4-litre works Ferrari P2 of Luciano Bianchi, the Fiat Abarth OT Spider 1600 of Jean Guichet, the Ford GT 40 driven by Jo Schlesser and a phalanx of Porsche 904s, Lotus Elans and Alpines.

The new generation of Fiat Abarth prototypes that first saw the light of day in 1966 were not only built of tubular chassis in accordance with the design philosophy that Mario Colucci brought with him from Alfa Romeo. The brilliant and versatile Abarth technical director also came up with another new version of the Fiat Abarth OT 1600 Prototype with a chassis in pressed steel and an engine inclined over the rear axle group. The power unit was the faithful 1592 cc four-cylinder unit, with a bore and stroke of 86x68.5 mm and a power output upped to 185 hp at 8200 rpm. Linked to a six-speed gearbox (code number 132) or the new internally designed five-speed (code 139), it was considered the ideal engine for fast circuits; an unusual feature was the high and carefully fared carburettor air intake, which emerged like a snorkel from the bonnet, but which was not used when the car competed in short hillclimbs. No substantial modifications were made to the chassis of the OT 1600 Spider, the wheelbase of which was 2015 mm, its track almost identical at 1370 mm front and 1380 mm rear. The body, which had long protrusions from the front and rear axle groups (3980 mm long, 1700 mm wide and 1180 mm high), displayed the typical Coke bottle shape widely used at the time, even on some U.S. production cars.

Abarth also came out with a new 1988 cc eight-cylinder engine at the Turin exhibition. Given the internal code 239, the unit had a bore and stroke of 78x52 mm, five bearing crankshaft, distribution by four overhead camshafts, inlet at the centre of the two cylinder heads (four Weber 46DCOE9 twin-choke carburettors) central exhaust between the two cylinder banks, double ignition and dry sump lubrication.

Right: The 1966 version of the Fiat Abarth 1000 Berlina Corsa differed little from the previous model with the exception of the front faring, which hosted bigger water and oil radiators, its pallid green paintwork and prominent lateral bands in red.

FIAT ABARTH "1000 SPORT PROTOTIPO" (1966)

ENGINE	
	mid-rear, longitudinal, 4 cylinders in line
Bore and stroke	mm 65x74
Cubic capacity	cc 982
Valve gear	2 overhead camshafts, chain driven
Number of valves	2 per cylinder
Compression ratio	10.5:1
Fuel feed	2 twin-choke Weber 40DCOE2 carburettors
Ignition	coil, distributor
Coolant	liquid
Lubrication	forced, oil filter on the main circuit
Max. power (project)	105 hp at 8000 rpm
TRANSMISSION	
	drive
Clutch	single dry plate
Gearbox	5 speed plus reverse, central lever
Axle ratio	4.4:1
BODY	
racing roadster	
CHASSIS	
tubular trellis	
Suspension	front independent, trailing arms, helicoidal springs, stabiliser bar, hydraulic dampers; rear independent, trailing arms, helicoidal springs, stabiliser bar, hydraulic dampers
Brakes	hydraulically operated discs front and rear
Steering	rack and pinion
Fuel tank	capacity 35 litres
Tyres	front 5.00x13; rear 6.00x13
DIMENSIONS AND WEIGHT	
Wheelbase	mm 2200
Front track	mm 1305
Rear track	mm 1320
Length	mm 3445
Width	mm 1625
Height	mm 930
Kerb weight	kg 480
PERFORMANCE	
Top speed	km/h 220

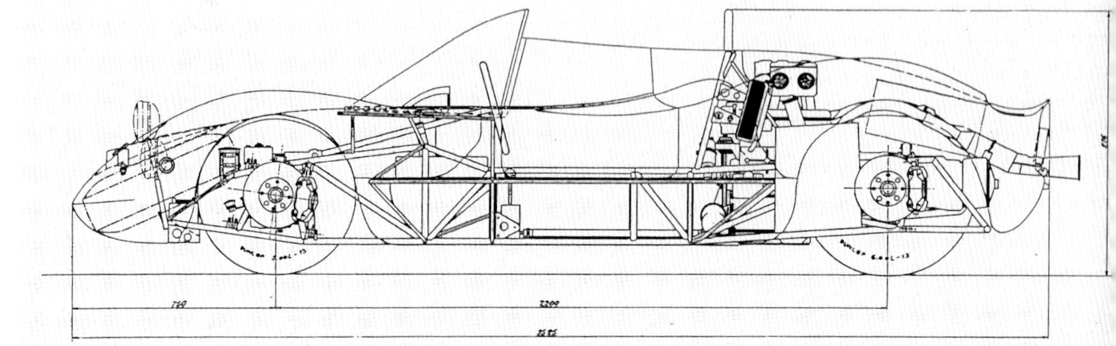

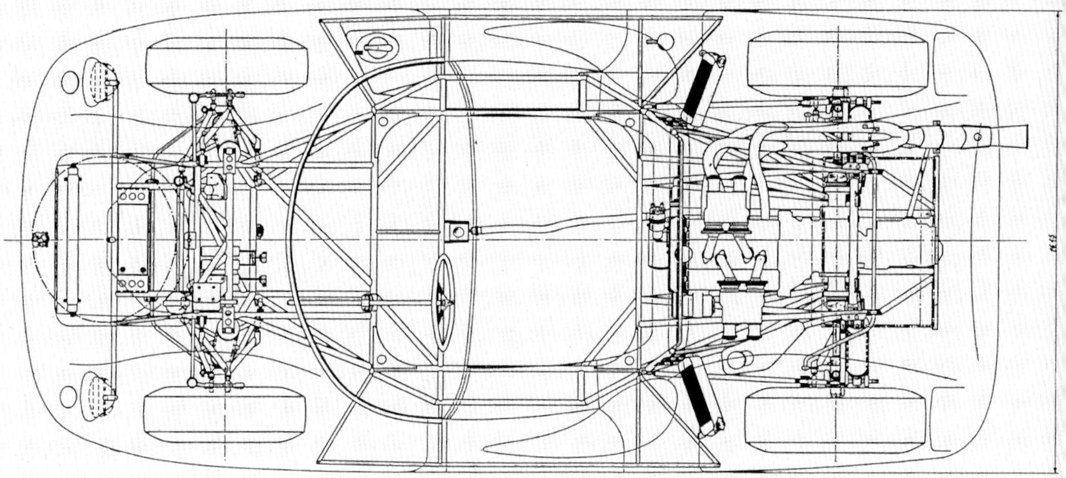

The cockpit, with its large wraparound windscreen, was bordered at the sides by transparent panels that linked with the rear roll bar, so the driver was inside a transparent cell that also contributed to the car's aerodynamics. The OT 1600 Prototipo weighed 640 kg and had a top speed of 225 km/h, but when it was initially conceived it was also designed to take other engines, among them the two-litre V8, 1.3 and 1.6 litre.

The Fiat Abarth OT 1300 Coupé that first appeared in 1965 was homologated by FIA on 1 May 1966, after the construction of the required minimum of 50 cars. Colucci also installed a two-litre four-cylinder 215 hp engine in the same boxed pressed steel chassis, retaining more or less the car's same dimensions. Built entirely of glass fibre, the body could be recognised by its integral front aperture hinged to open forwards, and the rear section hinged to open backwards, both kept in position by special two-stage external metal fasten-

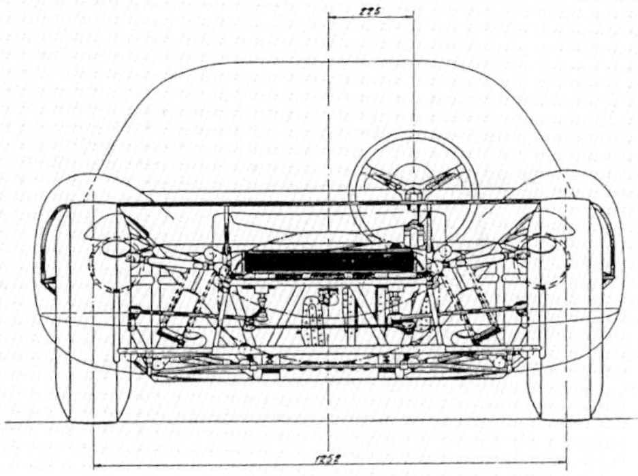

The transverse section of the Fiat Abarth 1000 SP shows the wide, flattened profile of the car. The large wraparound windscreen conformed to Appendix J Group 6 Prototype regulations.

The low, elegant body was in glass fibre and finished unusually well for a Scorpion racer. Carlo Abarth, here with British driver Jonathan Williams in front of the Corso Marche factory, had the 1000 SP make its debut at the 14th Coppa della Collina at Pistoia, where it was plagued by teething troubles and could only come fifth among the Prototypes of up to 1000 cc.

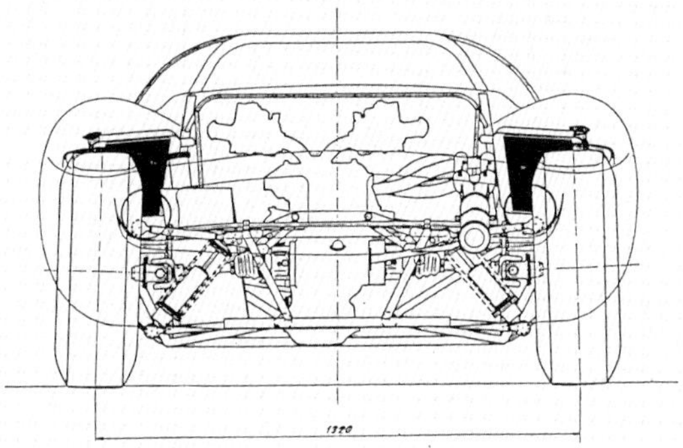

For the 1966 motor racing season, Abarth readied a barchetta with tubular chassis and engine (internal code SE04) amidships designed by Mario Colucci, fervent advocate of that kind of layout. The model with the four canonical views shows the constructional imposition of the car in which the position of the radiators stands out, one placed at the front for the lubrication liquid and the other two at the sides behind the cockpit for engine cooling. The power plant was the well-known twin-cam four with three main bearings based on the cylinder block of the Fiat 600 D, but with its capacity increased to 982 cc. With a compression ratio of 10.5:1 and fed by two twin-choke Weber carburettors, the motor produced 105 hp at 8000 rpm and with a dry weight of only 480 kg the 1000 SP had a power to weight ratio of 4.56 kg per horsepower.

ers, which allowed rapid substitution in case of necessity. With a power output of almost 108 hp and a dry weight of 655 kg, the OT 2000 Coupé had a top speed of 270 km/h.

After non-stop frenetic design and development work on cars to compete in the Touring, Sport and Prototype categories, Abarth scored a welter of victories during the 1966 season, the company's most successful: 904 wins in all, plus the Division 1 World Constructors' Championship with 50.5 points of which 37 counted towards the title (Fiat Abarth OT 1300) and Division 1 of the European Touring Car Challenge (Giancarlo Baghetti in a Fiat Abarth 1000 Berlina). Abarth also won the Touring Car Category of the Italian Championship in the up to 600 cc (Raffaele Pinto in the Fiat Abarth 595 prepared by Ferraris), the up to 700 cc (Maurizio Zanetti driving a Ferraris-prepared Fiat Abarth 695), the up to 850 class (Demetrio Martino at the wheel of

From the successful Fiat Abarth OT 1300 Mario Colucci deftly extracted the OT 2000, using the well-established 1946 cc four-cylinder engine designed by Luciano Fochi, which had done so well in other Scorpion cars.

With a power output of 215 hp at 7600 rpm, the OT 2000 had a top speed of 270 km/h and on 3 April 1966 was driven to victory by Luigi Taramazzo in the Sport Prototype category of the 1st San Remo Trophy – 2nd Franco Patria Trophy race.

FIAT ABARTH "OT 2000 (1966)

ENGINE	
	rear, longitudinal, 4 cylinders in line
Bore and stroke	mm 88x80
Cubic capacity	cc 1946
Valve gear	2 overhead camshafts, chain driven
Number of valves	2 per cylinder
Compression ratio	10.5:1
Fuel feed	2 twin-choke Weber 58DCO3 carburettors
Ignition	double, coil, distributor
Coolant	liquid
Lubrication	dry sump
Max. power (project)	215 hp at 7600 rpm
TRANSMISSION	
	rear wheel drive
Clutch	single dry plate
Gearbox	5 speed plus reverse, central lever
Axle ratio	4.3:1
BODY	
2 seat coupé	
CHASSIS	
boxed steel	
Suspension	front independent, trailing arms, helicoidal springs, stabiliser bar, hydraulic dampers; rear independent, trailing arms, helicoidal springs, stabiliser bar, hydraulic dampers
Brakes	hydraulically operated discs front and rear
Steering	worm and sector
Fuel tank	capacity 40 litres
Tyres	front 5.50x13; rear 6.00x13
DIMENSIONS AND WEIGHT	
Wheelbase	mm 2015
Front track	mm 1310
Rear track	mm 1360
Length	mm 3850
Width	mm 1620
Height	mm 1040
Kerb weight	kg 655
PERFORMANCE	
Top speed	km/h 270

a Fiat Abarth 850 TC tuned by Faccioli) and the up to 1000 cc (Mauro Nesti in a Fiat Abarth 1000 Berlina also prepared by Faccioli), as well as the Italian Championship Sport Category up to 1000 cc (Mariano Moselli in a Fiat Abarth 1000 Bialbero prepared by Faccioli and the up to 1300 cc class (Giulio Bona driving a Fiat Abarth OT 1300).

One of the season's most important races was the FISA Cup, 10 laps of the Monza road track on 12-13 March 1966, when the winners in their respective classes were Raffaele Pinto, who won the Touring Category in his Fiat Abarth 595, Franco Galli (Fiat Abarth 850 TC) and 'Matich' (Gianni Lado driving a Fiat Abarth 1000 Berlina), while in the Sport Category, which took over from Grand Touring in 1966, Maurizio Pinchetti won in a Fiat Abarth 1000 Bialbero and Giuseppe Dalla Torre with an Abarth Simca 1300 GT; in the Prototype Category Klaus Steinmetz won in a Fiat Abarth OT 1300. At the 4 Hours of the Jolly Club, a round in the European Touring Car Challenge at Monza on 19-20 March, Giancarlo Baghetti won the up to 1000 cc class of Division 1 driving a Fiat Abarth 1000 Berlina, while Johannes Ortner took the 850 cc sub-class in a Fiat Abarth 850 TC. At the eighth San Remo Trophy/second Franco Patria Trophy on 3 April, Luigi Taramazzo came first overall at the wheel of a Fiat Abarth OT 2000 and on 17 April there was the usual relentless battle between the Fiat Abarth 695s and BMW 700s in the Spring Cup at Vallelunga, with victory eventually going to the Scorpion. There was also a family fight between a swarm of Fiat Abarth 1000 Berlinas in the one-litre class, won by Vieri Finzi. That was a difficult championship, which featured protest and counter protest on the conformity of the Fiat Abarth 695 to the class' homologation parameters.

The first major races of the 1966 season included the 4 Hours of the Jolly Club, a round in the European Touring Car Challenge, which took place at Monza on 19-20 March.
The Scorpion turned up in force with 19 850 TCs and 1000 Berlinas, five of them works cars easily recognised by their new livery of a red check roof for the 1000s and yellow check for the 850s.
The two pictures show the works 1000 Berlina Corsa leading the race through a high speed corner at the entrance to the banked section of the track, with Giancarlo Baghetti (car number 30) in the lead, followed by Leo Cella (31).
Baghetti won, covering 666.615 kms in the four hours at a record average speed of 166.653 km/h. Second was Ed Swart and third Klaus Steinmetz, both driving works 1000 Berlina Corsas.
Cella's car suffered a broken alternator belt and was the only works car to retire.

A happy Giancarlo Baghetti after winning the 4 Hours of the Jolly Club at Monza in a works Fiat Abarth 1000 Berlina Coupé, having covered a longer distance than he did when he won the race in 1964, driving an Alfa Romeo Giulia TI Super.
Second came Dutchman Ed Swart in a 1000 Berlina Corsa, just four metres ahead of team-mate Klaus Steinmetz driving a sister car. Commenting on the race, Bruno Nestola said in Auto-Italiana of 31 March 1966: "Current touring cars are so light, almost ethereal, that it would not be a bad idea to oblige builders to install a robust roll-bar inside them".

The Fiat Abarth 1000 Berlina showed itself to be a brilliant hillclimb competitor.
The picture shows the Pierre Maublanc car that won the Course de Cote du Col Bayard in France on 24 April 1966 with a time of 4'29".

At the ninth Stallavena-Boscochiesanuova hillclimb also on 17 April, Abarths achieved the same series of Touring Category victories with Alberto Triboldi in a Fiat Abarth 595, Riccardo Rasia del Polo in a 695, Mario Favaretto in an 850 TC and "Matich" in a 1000 Berlina, in the Sport Category with Mariano Moselli in a 1000 Bialbero and the Sport-Prototype with Sergio Morando driving a Fiat Abarth OT 1300. A week later there was another mighty battle, this time at the 1000 Kilometres of Monza 1300 cc class between the Fiat Abarth OT 1300s and the Alpine Renaults, which had been fast in a number of earlier races. But the French cars retired with mechanical problems, and although the Fiat Abarths won the day they were plagued by a range of technical glitches. One of the OT 1300s competed with a special body with a long pointed tail, an experiment to evaluate the shape's aerodynamic effect at maximum speed on the straight.

The Abarths repeated their success again at the 10th Bologna-Raticosa Pass on 22 May, although a Fiat Abarth 2000 did come second to the Porsche Carrera 6 of Noris in the overall results.

Homologated at last, the Fiat Abarth OT 1300 of Johannes Ortner-Kurt Ahrens won its class in the 1000 Kilometres of the Nürburgring on 5 June and came 14th overall, while Klaus Steinmetz won the Garessio-Colle San Bernardo outright in a Fiat Abarth 2000 and repeated that success at the Trento-Bondone on 10 July, when the Scorpions turned in another string of class wins in the Touring Category of the up to 850 cc with Sandro Uberti in an 850 TC, the up to 1000 cc with a 1000 Berlina driven by Mauro Nesti, the Sport Category up to 1000 cc driven by Maurizio Minchetti, the up to 1300 cc with Giuseppe Rebaudi. On 17 July at Mugello in the fourth round of the European Sport Constructors' Championship, Luigi Taramazzo and Giulio

Pierre Maublanc also won the Sainte Beaune hillclimb on 8 May at the wheel of a Fiat Abarth OT 2000, taking a record 6'56.1" to complete the 8.100 kilometres at an average speed of 70.096 km/h. That car had a power to weight ratio of 3.4 kg per horsepower and a six-speed close ratio gearbox, both factors that made the OT 2000 such a feared competitor on the twisting hillclimb routes.

On 26 June, Johannes Ortner in a Fiat Abarth 1000 Berlina won the up to 1000 cc Touring category of the 1966 Mont Ventoux hillclimb, the oldest race of its kind in Europe having first been held in 1902. A 1303 metre climb of 21.700 kilometres, the Scorpion also won the Mont Ventoux 850 cc class (Mauro Cintolesi, 850 TC) and the up to 1600 cc class of the Sports Prototype category (Werner Rufenacht, OT 1600).

Abarth scored another great success at the 500 Kilometres of the Nürburgring on 4 September 1966 on the historic 22.810 kilometre German circuit.
The Fiat Abarth 1000 SP with tubular chassis driven by Herbert Muller, who took over at the last minute from an indisposed Klaus Steinmetz, won the Prototype category.

The Scorpion also won the 1300 cc class of the Sports category at the 'Ring.
It was probably Ernst Furtmayr's finest victory in a Fiat Abarth, taking 3 hours 39'17.3" to cover the distance in his works OT 1300 at an average speed of 131.1 km/h. Furtmayr, on the left with his well-deserved laurel wreath, and Abarth sports director Renzo Avidano celebrate the win immediately after the race.

Bona won the first division and came sixth overall in a Fiat Abarth OT 1300. At Snetterton, England, in the fourth round of the European Touring Car Challenge on 31 July the Fiat Abarth 1000 Berlinas of John Fitzpatrick, Giancarlo Baghetti and Johannes Ortner trounced a determined squad of Mini Cooper Ss, while the Scorpions in the Touring Category achieved similar success in the up to 850 cc class, won by John Anstead in a Fiat Abarth 850 TC. On 7 August at the City of Enna Cup, fifth event in the World Sport Constructors' Championship, Alfio Gambero won the first division and came third overall in a Fiat Abarth OT 1300. Three weeks later in Switzerland, Jean-Daniel Granjean won the Sierre Montana-Crans, sixth round in the World Grand Touring Constructors' Championship, in an Abarth Simca 1300 GT.

On 4 September at the 500 Kilometres of the Nürburgring the Alpines and Fiat Abarths crossed swords again and success went to Ernst Furtmayr at the wheel of an OT 1300 Coupé, followed in third place by Herbert Muller, who

Johannes Ortner drove the Fiat Abarth 1000 and OT 1600 to successive victories on 25 September in the Paris Cup at the 3.360 kilometre Monthléry circuit.
In the Sports category, Arturo Merzario-Jacques Vidal (picture) took third in the 1300 cc class and 10th overall behind two Alpines.

also won the 1000 cc class of the Sport-Prototype Category in his SP tubular barchetta, confirming to Abarth the value of producing another 50 of the cars to achieve homologation. A week later, Ortner won Division I and came 11th overall in the eighth Championship counter at Zeltweg, Austria.

Another significant and prestigious victory was achieved by the Scorpion at the sixth Aosta-Pila on 18 September on the classic 11.4 kilometre hillclimb that rose 786 metres up to the sky. Johannes Ortner not only won the two-litre class of the Sport-Prototype Category but also the event outright in a Fiat Abarth Tubolare 2000 Spider, finishing 3'' ahead of Edoardo Lualdi's Dino. Three Scorpions also took the next three places, with Luigi Taramazzo third in an OT 2000, Giuseppe Rebaudi fourth driving an OT 1300 and Leo Cella fifth at the wheel of the other Corso Marche barchetta, the SP 1000. A week later, during the Paris Cup at Montlhéry, Ortner cut his way through the field to take the lead in his Fiat Abarth 1600 with a dozen seconds' advantage

Driven by Pierre Maublanc-Michel Roussin, the Fiat Abarth 1000 Berlina won the 12th Rally du Var in France on 15-16 October 1966, while a sister car driven by Jaime Silvera Machado won the Targa de Oiro in Luanda, Angola, on 4 December.

Abarth went back to Monza on 27 October 1966 with a Fiat Abarth 2000 single-seater record car that had a large transparent cupola over the cockpit for better aerodynamics, and specially fared wings over the wheels to improve air penetration.
Driven by Johannes Ortner, the car established a new 500-metre record in 12.855/1000" at an average 140.023 km/h.

Right: Mounted over the rear axle, the engine was the well-known four-cylinder in-line 1946 cc power plant, tuned for the occasion to produce a maximum of 216 hp at 8800 rpm.

over Innes Ireland's Ford GT 40, but a few hundred metres from the finish the Abarth's engine suddenly died and the driver had to try everything he knew to make the finish line, losing first place to the Ford by a handful of metres. Later, it was found that the Italian car ran out of fuel due to incorrect information written on the fuel tank. But driver and car took their revenge two weeks later, beating Ford on the same Montlhéry track.

At the end of November 1966, Carlo Abarth went back to Monza in an attempt to establish five records with two single-seaters; one with a two-litre engine inclined across the rear axle group and an aerodynamic body, special rim discs and motorcycle-type mudguards for better air penetration; the other hosted 750 cc and 1000 cc engines in turn, a preferred solution to that of the Pininfarina aerodynamic body with which Mario Poltronieri broke the up to 750 cc class H record the previous year. On Thursday 27 November, Johannes Ortner covered the quarter mile with the 750 cc in 14'585/1000" at an average speed of 99.307 km/h. Immediately afterwards, he made a successful attempt on the up to 2000 cc class E 500 metre record, for which he recorded 12'.855/1000" at an average 140.023 km/h. The following Monday the two single-seaters were at Monza again, but this time transparent aerodynamic plastic domes were fitted over them after the drivers had climbed aboard; and that is how they established the new class G record of up to 1100 cc over the quarter mile in 13'.035/1000" at an average 111.116 km/h, beating the previous record of 106.344 km/h established by Carlo Abarth himself. The car with the two-litre engine broke the quarter mile record by covering the distance in 11.095/1000" averaging 130.545 km/h, while the single-seater with the 750 cc motor – which was fitted with the 1000 cc engine at Monza – took 16.955/1000" to cover 500 meters at a record average speed of 106.163 km/h. By the end of 1966, Abarth had established a total of 113 international and five world speed records.

This curious photograph, taken at Monza in October 1966, shows Carlo Abarth testing a Honda S 800 as he talks to Klaus Steinmetz. The small Japanese car, powered by a four-cylinder 791 cc engine that produced 70 hp at 8000 rpm, boasted a curious double chain transmission on the rear wheels.

That year, 686 Abarth cars were registered by Italian customers, a gratifying result, as they were almost all specials produced for racing. As Hans Herrmann had been previously, Johannes Ortner was made the number one works driver that season. Abarth favoured German language drivers, because they could more clearly understand each other and he and they had a cultural and mental affinity. While avoiding offending the sensibilities of his close Italian colleagues, Carlo was instinctively induced to evaluate men and situations by his own standards, which he had learnt as a child at home and school. In his frequent journeys to Vienna, he visited relations and friends but also doctors, dentists and various consultants, because he could express himself with greater clarity, familiarity, with more confidence and without misunderstandings. Doctors in Vienna, for example, prescribed him various diets to control his tendency to fat; and next to the apples he ate in quantity each day there appeared for a certain period hard-boiled eggs and cooked ham, which he had sliced as thick as his finger. With his half-baked and at times amusing Italian, he was able to keep a distance from those with whom he worked if necessary, depending on the approach he wanted to make to inter-personnel relationships. In the company, he could not stand loafers, unpunctuality or those who did not go into detail and did not offer solutions to problems. One day when the highly efficient Evangelina, who followed Carlo like a shadow, was absent he asked Renzo Avidano's secretary to take the following dictation, which was an internal communication to the technical director, Mario Colucci: "*Ortner tried car in flying field and if not have cross eyes could not see in the rear vision mirror*". The poor girl was in doubt for some time as to whether or not she should correct the text and turn it into good Italian – knowing Abarth had an infallible memory – or send it as it was, taking the risk of making the boss a figure of ridicule in the company. She decided to type the communication as dictated; Abarth signed it without batting an eyelid and the rear vision mirror was changed.

It was on one of his journeys to Vienna that he met the third woman of his life, with whom he would spend his remaining years and with whom he lived an intense love story far from the clamour that a personality like Abarth often generated. Those who worked with him each day are surprised even now at how Carlo, who almost never left a race track and was assiduous and attentive in the daily management of the company, could have found the time to cultivate a sentimental relationship that was so deep and demanding and to have the ability to hide it at the beginning with casualness in front of everyone. Some of his words of wisdom became famous in the company, symptomatic of a man with a prominent sense of pragmatism: "*Never give a piece of silver to an Englishman or an item of crystal to a Czech*", he maintained. He shrank from formalities, which almost seemed to irritate him; from time to time he told his closest colleagues to give themselves a salary increase, which

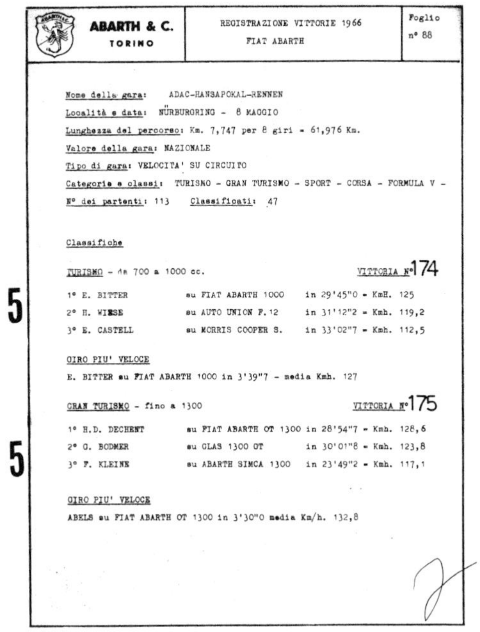

ABARTH & C.
TORINO

REGISTRAZIONE VITTORIE 1966
FIAT ABARTH

Foglio n° 88

Nome della gara: ADAC-HANSAPOKAL-RENNEN
Località e data: NÜRBURGRING - 8 MAGGIO
Lunghezza del percorso: Km. 7,747 per 8 giri - 61,976 Km.
Valore della gara: NAZIONALE
Tipo di gara: VELOCITA' SU CIRCUITO
Categorie e classi: TURISMO - GRAN TURISMO - SPORT - CORSA - FORMULA V -
N° dei partenti: 113 Classificati: 47

Classifiche

TURISMO - da 700 a 1000 cc. VITTORIA N° 174

5
1° E. BITTER su FIAT ABARTH 1000 in 29'45"0 - KmH. 125
2° H. WIESE su AUTO UNION F.12 in 31'12"2 - Kmh. 119,2
3° E. CASTELL su MORRIS COOPER S. in 33'02"7 - Kmh. 112,5

GIRO PIU' VELOCE
E. BITTER su FIAT ABARTH 1000 in 3'39"7 - media Kmh. 127

GRAN TURISMO - fino a 1300 VITTORIA N° 175

5
1° H.D. DECHENT su FIAT ABARTH OT 1300 in 28'54"7 - Kmh. 128,6
2° G. BODMER su GLAS 1300 GT in 30'01"8 - Kmh. 123,8
3° F. KLEINE su ABARTH SIMCA 1300 in 23'49"2 - Kmh. 117,1

GIRO PIU' VELOCE
ABELS su FIAT ABARTH OT 1300 in 3'30"0 media Km/h. 132,8

Carlo Abarth established an efficient information system by which Fiat could follow the successes achieved by the Scorpion week-by-week in all types of races across the world. At the beginning of each week, many forms like this were transmitted to Fiat from a special office at Corso Marche, in line with an agreement signed by the two parties in the spring of 1958.

Every single victory, indicated with progressive numbers, automatically released prize money. The only stipulation in the agreement was that the payment clause came into effect only in the case of a victory and not for any other placings. In the Sixties, the Scorpion won many hundreds of victories every season all over the world.

he later initialled without comment to signify his approval. Or he would suddenly blurt out: *"I want to give you a present ... do you want money or a gift"?* And some of them found articles in silver of notable value waiting for them when they got home.

Carlo Abarth's daily approach to work was tough, sometimes inflexible; he used a note pad on which he would explain what he wanted, the length of arms, the profile of a cam...Sometimes he had someone bring him a "sedio" – the incorrect Italian word for chair – which he would sit astride, using scissors and cardboard to indicate the shape of a car body and subsequently gave instructions to his people. His office was always home to engines and transmissions and dozens of photographs on three large wall panels: pictures of drivers, cars, races that he changed from time to time, except for some he considered fundamental to showing the progress he had made. He greatly believed in the power of the press and, right from the Thirties, carefully conserved numerous albums of newspaper and magazine cuttings telling stories of his sporting achievements.

The OT 1300 coupé gave birth to the OT 1600, which duplicated its predecessor's techno-constructional lay-out with an engine inclined over the rear axle group and a floor pan in pressed steel, a modification of the Simca 1000's. The body was directly derived from the Prototype OT 1600, with its characteristic Coke bottle shape and a periscope-style air intake on the roof for the two carburettors. The car had a 1592 cc four-cylinder 185 hp engine and a five-speed gearbox coded 139 that took the place of the six-speed, which was produced with great difficulty for the Abarth Simca GT and derived from the Simca 1000 production car. The OT 1600 went through a long programme of test-driving on the Corso Marche airstrip opposite the Abarth factory and at Monza.

At the Turin Motor Show, which opened on 3 November 1966, Carlo took everyone by surprise with another of his masterstrokes. It was the introduction of the Fiat Abarth OT 1300/124, an audacious marriage between a normal production 850 coupé body and chassis into which was lowered the engine of the 124 saloon, launched in March; the engine was given a little more power from 1197 to 1280 cc with an increase in bore and a reduction in stroke, so that maximum output went from 60 to 75 hp at 6000 rpm; other components like distribution and carburettors remained unchanged. The cooling system had a front radiator to improve performance and the engine-gearbox group was mounted on a small auxiliary tubular frame fixed to the little Fiat coupé's chassis, the principal components of which (suspension, brakes, steering) were also retained and modified in line with the car's higher performance. The OT 1300/124 generated considerable curiosity among show visitors, particularly for the original solution of that curious symbiosis between the two Fiat models that had become so popular in such a short time; but the

From 3-13 November, the Abarth stand at the 1966 Turin Motor Show displayed the company's latest production models and winning racing cars of that brilliant season, which was nearly at an end.
They included the 1000 Berlina, winner of the 1000 cc class of the European Touring Championship, and the OT 1300, the car that won the up to 1300 cc class of the 1st Division of the Grand Touring Championship.
Among the new stars of the show was another 'monster', as the production Fiats with super-powerful Abarth engines were called at the time.

car was never marketed and, regardless of its originality, was dropped.

The industrial side of the business did well in 1966, particularly in exhaust systems production, which totalled 235.000 units, 50% of which were exported. But Abarth's involvement in motor sport was so onerous and channelled into so many areas that it was clear no other constructor would take on such a commitment in a single season. The cost of all that work and the 904 victories achieved in those 12 triumphal months was about Lit 600 million. An amount that could only seemed plausible and acceptable because of the 1958 agreement with Fiat.

FIAT ABARTH "OT 1300/124" (1966)

ENGINE	rear, longitudinal, 4 cylinders in line
Bore and stroke	mm 75.5x71
Cubic capacity	cc 1280
Valve gear	overhead valves, side camshaft
Number of valves	2 per cylinder
Compression ratio	10.5:1
Fuel feed	1 twin-choke Solex C32PHH carburettor
Ignition	coil, distributor
Coolant	liquid
Lubrication	forced, oil filter on main circuit
Max. power (project)	75 hp at 6000 rpm
TRANSMISSION	rear wheel drive
Clutch	single dry plate
Gearbox	4 speed plus reverse, central lever
Axle ratio	3.9:1
BODY	2 door, 2+2 coupé
CHASSIS	integral body
Suspension	front independent, upper trailing arms, transverse lower leaf spring, hydraulic dampers; rear independent, trailing arms, helicoidal springs, hydraulic dampers
Brakes	hydraulically operated front discs, rear drums
Steering	worm and sector
Fuel tank	capacity 30 litres
Tyres	155x13
DIMENSIONS AND WEIGHT	
Wheelbase	mm 2027
Front track	mm 1220
Rear track	mm 1280
Length	mm 3608
Width	mm 1500
Height	mm 1300
Kerb weight	kg 755
PERFORMANCE	
Top speed	km/h 170

At the Turin Motor Show was also displayed the OT 1300/124, a marriage between the Fiat 850 Coupé body and the engine-transmission group derived from the Fiat 124 saloon, the capacity of which had risen from 1197 to 1280 cc by an increase in bore and a reduction in stroke. The maximum power output of the engine rose from 60 to 75 hp and top speed was confirmed as 170 km/h.

Placed on an inclined dais on the Abarth stand at the 1966 Turin Motor Show is the 1000 Berlina Corsa, celebrating its conquest of the European Touring Championship during the season just concluded.
In its most evolved form, the 982 cc Fiat 600 D- based engine could generate 85 hp at 7600 rpm.

Luciano Fochi, who was in charge of Abarth engine design, said it was possible to squeeze so much horsepower out of that production engine of rods and rocker arms because of the optimum quality of the iron and the casting of Fiat cylinder blocks, which absorbed multiple functional changes without any alteration.

In Italy, early 1967 seemed tinged with controversy. After the brief recessive phenomenon of the previous two years, the country's economy was gaining ground again, even though it was plagued by growing social unrest and widespread union disquiet. A centre left Government experiment strongly championed by Amintore Fanfani, then taken up by Aldo Moro and other Christian Democrats, decided to break with anti-communist dogma that had characterised the party since the end of the war and turned out to be more of a momentary anathema to the entrepreneurial class than an effective danger.

In 1966, car production was up 16.19% and came close to 1.300.000 units – 1.282.418 - for the first time in the country's history; new car registrations were a record at 1.014.975, after not exceeding 886.297 the previous year. In full expansion of its range, Fiat unveiled the 124 and Dino roadsters at the Turin show in November. The Dino had a six-cylinder Ferrari engine, but the car was immediately dogged by teething troubles and remained unsuccessful, even though its body designed and built by Pininfarina had projected the illusion of Fiat's return to making upper class cars. The price of the roadster, which went into production in the spring of 1967, was fixed at Lit 3.485.000 and that put the Dino in direct competition with the Lit 3.535.000 2.8 litre Lancia Flaminia 3C convertible.

But Fiat was not alone in peddling fanciful dreams in those early months of 1967. Carlo Abarth had just finalised an ambitious plan he had had in mind for years: to produce a big-engined car specifically developed to compete in races for the most powerful prototypes, against cars like the Ferrari 330 P3 and P4 and the seven-litre Ford GT 40. The new car project was agreed dur-

A meeting on the Abarth stand between two 'old-hands' who had already worked with Carlo Abarth at Cisitalia in 1947-8: Renzo Avidano, Abarth's right hand man and sports director from the inception of the company, and Rudy Hruska (left), the Viennese engineer who had been assigned the development of the Ferdinand Porsche-conceived Cisitalia Grand Prix single-seater, which came to nothing.

As technical consultant to Simca in the Sixties, Hruska promoted the agreement between Carlo Abarth and the French manufacturer for the construction of grand touring cars.
He then went to work for Fiat, until he moved to Alfa Romeo where he conceived and designed the Alfasud and the large Alfa Romeo Pomigliano D'Arco industrial complex (1967-72).

The 1:1 scale mock-up of the T 140, the ambitious V12 six-litre prototype with which Carlo Abarth was planning to enter maximum power Prototype category motor racing to compete against the likes of the Ferrari 330 P3, 330 P4 and the dreaded Ford GT40.
The project, which meant a seriously demanding commitment from the Scorpion both in terms of investment and human resources, was assigned to Mario Colucci, who took on the car's chassis and body design.
Chassis architecture was born of experience gathered by Abarth with the tubular structure of the SP barchettas; front suspension repeated the same double oscillating arm layout, while at the rear end the geometry was of inverted triangles with links, reciprocal struts, four groups of helicoidal springs with coaxial dampers and an anti-roll bar.
The car was designed on a 2500 mm wheelbase and its dimensions were 4100 long, 1850 mm wide and 1030 mm high; project weight was 1110 kg.
Considerable commitment was given not only to aerodynamic research but also the car's stylistic appearance, so that it became a rather functional and harmonious shape, with attractive, well-smoothed surfaces in all areas. But in spite of great effort, the car never saw the light of day and on 11 June 1967 the project was abandoned.

ing the previous summer; the trellis-type chassis of large section tubes in the central area where the cockpit would be located were set in plastic, the techno-constructional trend of the period and inspired, obviously, by experience obtained with the tubular SP 2000, which produced such good results under hard racing. The four-wheel independent suspension was of the usual geometry up front, with double oscillating trapezia the lower of which were hinged, attached by a link to act as push rod linkage under braking; helicoidal springs with telescopic dampers and a stabiliser bar completed the front axle architecture. The rear suspension was an inverted triangle with small connecting and other rods attributed to Len Terry and Eric Broadley (Auto-Italiana, 2 March 1967); the elastic elements were made up of four groups of spring-dampers, two per side and integrated with a transverse anti-roll bar.

The engine was of entirely new conception, with a subdivision of 12 cylinders set at a 120° V and formed by a union of two six-cylinder blocks

ABARTH "T 140" (1967)

ENGINE	
	mid-rear, longitudinal, V12
Bore and stroke	mm 92x75
Cubic capacity	cc 5982
Valve gear	overhead valves and camshafts
Number of valves	2 per cylinder
Compression ratio	12:1
Fuel feed	4 triple-choke Weber 40LDA3C1 carburettors
Ignition	double, coil, distributor
Coolant	liquid
Lubrication	forced (2 independent sumps)
Max. power (project)	610 hp at 6700 rpm
TRANSMISSION	
	rear wheel drive
Clutch	single dry plate
Gearbox	5 speed plus reverse, central lever
Axle ratio	3.3:1
BODY	
	2 seat racing coupé
CHASSIS	
	tubular trellis
Suspension	front independent, trailing arms, helicoidal springs, stabiliser bar, hydraulic dampers; rear independent, trailing arms, push rod links, double helicoidal spring groups, hydraulic dampers
Brakes	hydraulically operated double discs front and single disc rear
Steering	rack and pinion
Fuel tank	capacity 120 litres
Tyres	front 6.00x12; rear 12.00x15
DIMENSIONS AND WEIGHT	
Wheelbase	mm 2500
Front track	mm 1468
Rear track	mm 1470
Length	mm 4100
Width	mm 1850
Height	mm 1030
Kerb weight	kg 1110
PERFORMANCE	
Top speed	km/h 300

The impressive engine of the Abarth T 140 was a 120° V12 of 5982 cc, entirely designed at Corso Marche and born of the union of two six-cylinder blocks (bore and stroke 92x75 mm); there was a single camshaft for each cylinder bank, double ignition, inlet at the centre of the V, exhaust at the sides, a central forced air intake, compression ratio of 12.0:1 and fuel feed by four triple choke Weber carburettors. Lubrication was by normal wet sump, but a dry sump version was planned. Accredited with a maximum power output of 600 hp at 6800 rpm, the engine suffered a laborious gestation period and was the subject of numerous problems during its first bench tests.

The wooden frame of the Abarth T140, on which it was planned to layer a bodyshell in glass fibre, was the subject of painstaking study and development.
The front radiator was integral to the two lateral units, placed immediately behind the cockpit, attached to the long rear wings.
The enormous double-curvature wrap-around windscreen gave the car great elegance, as did the low line of the short bonnet, which was deeply set between the heavily shaped wings that surrounded the 15-inch wheels.
Carlo Abarth wanted the prototype of the T140 completed during the summer of 1967 and so that it could make its racing debut in the autumn and overcome its teething troubles in time to race competitively in 1968.
But right after the 1967 24 Hours of Le Mans in mid-June, the Fédération Internationale de l'Automobile decided to limit Prototype capacity to three litres, causing an immediate halt to the construction of the new six-litre Abarth.
The FIA decision was a severe and unexpected blow to the Scorpion and caused great financial loss.

(bore and stroke 92x75 mm) for an overall capacity of 5982 cc; such an open V between the two cylinder banks was chosen to avoid problems with the equilibration of second order forces exerted on the drive shaft and to contain the ground height of the engine; distribution was by single camshaft in each cylinder bank, the compression ratio was 12:1 and fuel was fed into the unit by four triple-choke Weber 40 IDA 3C1 carburettors. Drive was central with transmission by gears, and ignition was by two spark plugs per cylinder. The first dynamometric bench tests took place on 18 February 1967, when the V12 Abarth produced 600 hp at 6800 rpm to confirm the validity of the project, which was the work of the head of engine design at Corso Marche, Luciano Fochi. The power plant was to be mounted longitudinally in the centre of the car immediately behind the cockpit, and it was to transmit the drive to the wheels through a five-speed gearbox and a self-locking differential.

The body was another Mario Colucci design. A mock-up of it was made 1:1 scale in wood, on which it was intended to mould the final shell in glass fibre. Called the T-140, the appearance of the car was imposing and full of character; it had a 2500 mm wheelbase, a length of 4100 mm, a width of 185mm, a height of 1030 mm and large 15-inch diameter wheels. Its shape was harmonious, with an economical and extremely low front end set between the two wings and an enormous double curvature wraparound windscreen; almost half the car was occupied by the engine, with the bonnet moulded onto two long back wings that blended in with the gentle line of the rear; two large air intakes were at the base of the cockpit to ensure necessary cooling of the engine-gearbox group; among the car's other innovative characteristics was the braking system with double discs on the front wheels. The preliminary study of the car and the development of the project kept the Abarth & C. technical office busy for a long time; Carlo personally followed the evolution of

FIAT ABARTH "1000 BERLINA" (1966)

ENGINE	
	rear, longitudinal, 4 cylinders in line
Bore and stroke	mm 65x74
Cubic capacity	cc 982
Valve gear	overhead valves, side camshaft
Number of valves	2 per cylinder
Compression ratio	13:1
Fuel feed	1 twin-choke Weber 36DCD7 carburettor
Ignition	coil, distributor
Coolant	liquid
Lubrication	forced, oil filter on the main circuit
Max. power (project)	85 hp at 7600 rpm
TRANSMISSION	
	rear wheel drive
Clutch	single dry plate
Gearbox	5 speed plus reverse, central lever
Axle ratio	3.9:1
BODY	
	2 door, 4 seat saloon
CHASSIS	
	integral body
Suspension	front independent, double trailing arms, helicoidal springs, hydraulic dampers; rear independent, trailing arms, helicoidal springs, hydraulic dampers
Brakes	hydraulically operated front and rear discs
Steering	worm and sector
Fuel tank	capacity 27 litres
Tyres	4.50x13
DIMENSIONS AND WEIGHT	
Wheelbase	mm 2000
Front track	mm 1160
Rear track	mm 1170
Length	mm 3530
Width	mm 1390
Height	mm 1400
Kerb weight	kg 585
PERFORMANCE	
Top speed	km/h 195

the V12 step by step during its various building stages and savoured the idea of being able to compete with the car in the last few races of 1967 to give it a technical shakedown, so that they could refine the prototype and ready it for the 1968 season.

But they were in for a nasty shock. After the 24 Hours of Le Mans on 10-11 June, the FIA limited prototype capacity to three litres, forcing the Scorpion to immediately stop testing the new car and snuffing out Carlo Abarth's great dream. It was a harsh and unexpected letdown provoked, according to many, by a decision that seemed to be dictated by evaluations that were not exactly impartial or, at least, made on impulse. An official Abarth press release at the Turin Motor Show on 1 November censured the fact that the International Federation arrived at its decision without respecting the rule that provided for an early warning period to confirm modifications to the regulations were on the way. Abarth was forced to throw away many months of feverish work, having squandered a fortune in human resources and money through no fault of his own. The FIA decision extinguished an extraordinary charge of enthusiasm that accompanied the various phases of development of such a promising project.

Motor racing started early in 1967. From Tuesday 24 January to

The Abarth test session at Vallelunga at the end of January 1967 was aimed at tuning its cars for the imminent motor racing season.
Carlo Abarth took part with Renzo Avidano (below, left) and Friedrich Hoffman (right) specialist in high-speed gearbox design.
Among the new cars was a 1.6 litre coupé for the Prototype category (foreground, above), which had enormous driven wheels – 7.00x13 – and a large air engine compartment outlet built into the rear wings.
Four works drivers also took part in the Vallelunga tests, Antoine Hezemans, Teodoro Zeccoli, Wilhibert Kauhsen and Johannes Ortner.

Carlo Abarth photographed at the wheel of a Fiat Abarth 1000 SP during testing at Vallelunga, behind a bespectacled Antonio Tomaini.
The SE04 project included a tubular chassis, while the engine – installed amidships – was a quicker version of the well-known 982 cc four-cylinder twin-cam unit, capable of 105 hp; the inlet manifolds were placed laterally on the twin overhead camshaft cylinder head instead of in the centre, as with the original first generation 750 Record Monza engines.
Fifty 1000 SPs had been built by January 1968 so that the car could be homologated in Group 4.

Carlo Abarth and Mrs Annelise, photographed at the end of the Sixties during one of their frequent trips to Vienna. For some time they had been using the red 2400 coupé, with its aluminium body designed by Giovanni Michelotti and built by Allemano, which was easily capable of 200 km/h.

Saturday the 28th Abarth conducted a series of tests at Vallelunga in order to update its cars for the imminent season and test new tyres from Dunlop, with whom the Scorpion had renewed its contract for another year. Twelve cars were involved in the test session, including three 1000 Berlinas with new suspension being readied for Groups 2 and 5 of the Touring Category, three 1300 OTs, an OT Coupé, an OTR Coupé 2000, two SPs – a 1000 and a 2000 cc – a 1600 and a two-litre. Some of those models had been substantially modified since the previous year and others far less so in the bonnet and rear end areas; another 25 SP 1000s were to be built to obtain a Sport Category homologation. Carlo Abarth played an energetic role in the test session, with works drivers Johannes Ortner, Toine Hezemans, Teodoro Zeccoli and Wilhibert Kauhsen.

The new front suspension of the Fiat Abarth 1000 Berlina was derived from that of the OT 1300, with lower trapezia of the wheels in place of the transversal semi-elliptic springs. That was not the only modification transferred from the highly successful 1300; the 1000 engine's crank gear was also improved as were the main bearings, which became tighter-fitting to provide better flow; a new alloy inlet manifold was used instead of the pressed steel unit. A version of the car with a radial cylinder head was also tested for the numerous overseas Group 5 Special Touring Category events, especially in Britain.

New rear suspension was introduced for the OT 1300, no longer with oscillating arms but with deformable trapezium and a single spring. The two 850 Coupés at Vallelunga were an OTR with a radial engine and the OT 1300/124; Carlo Abarth was putting his money on that car, which he believed

Three Abarth 1000 Berlinas that were to compete in the Groups 2 and 5 Touring categories were also sent to Vallelunga; two of them boasted important mechanical modifications, while the third car was a 1966 model for use in comparative tests.

The front suspension had been profoundly reworked by the use of upper and lower oscillating arms in place of the previous layout, which still comprised the lower transverse semi-elliptic springs of the original Fiat 600.

was commercially valid and the right price, hoping to keep its cost at Lit 1.250.000. The two mid/rear engined SPs with gearboxes inclined over the rear axle group, chiefly designed to compete in hillclimbs or on twisting tracks, were light – they weighed less than 600 kg – rather than highly aerodynamic. But slipperiness was the object of long studies and testing for the other entirely new 1600 and 2000 cars, based on the classic lay-out so close to Carlo Abarth's heart, their engines inclined over the rear axle; with its sports tourer body and gullwing doors, the former was developed for long distance racing. Its roadster body stylistically derived from the previous season's car, the two-litre had a prominent periscope-type intake to channel air to the two twin-choke Weber 58 DCOE3 carburettors. Both roadsters had new rear suspension with tubular instead of pressed steel arms, in line with the construction technology previously introduced on one of the OT 1300. Official times clocked by the cars at the end of January on Vallelunga's 1800 metre track were:

- 1000 Berlina	1'04.6"
- 1000 Berlina radiale	1'04.0"
- 1000 SP	58.5"
- OT 1300	59.1"
- 1600	57.8"
- 2000	55.8"

Meanwhile, on the 3200 metre circuit the times were:

- 1000 Berlina	1'40.2"
- 1000 Berlina radiale	1'39.2"
- 1000 SP	1'30.7"
- OT 1300	1'29.6"
- 1600	1'28.4"
- 2000	1'25. 3"

Early that year Abarth was granted its homologation for the all-in-one five-speed gearbox and differential coded 139, designed by Friedrich Hoffman for racing. As well as having a much reduced inertia for speed of use during racing, its ratios could be changed quickly; there were two ratios for first and second, while Abarth could choose from 23 for second to fifth gear. The self-locking differential was supported by conical roller bearings, while a three-atmosphere pump performed the task of lubricating the gears, the pinion set, the transmission and the half-shafts; a supplementary oil radiator was also planned.

Before the start of the 1967 season, Carlo Abarth decided to celebrate with a party to mark Renzo Avidano's 15th year as the company's motor sport director. So on 18 February at the Hotel Ambasciatori in Turin, the gruff, at times inflexible head of the company officially expressed his appreciation of his closest colleague, who had contributed so much to the success of the Scorpion brand on the racing circuits of the world; to express his gratitude, Abarth presented Avidano with a giant silver cup 80 centimetres high and weighing five kg on which was inscribed every victory Abarth cars had achieved up to that moment: five world records, 112 international records, five world constructors' championships, two European Challenges. Carlo, who had great admiration for Enzo Ferrari but a sense of challenge, too, often liked to recall how he had seen in Avidano right from the start the ideal man to successfully represent the Scorpion in racing; on the other hand Ferrari, who understood engines and men well, frequently changed his motor sport director, sometimes after just a few months. The 18 February celebrations were interpreted at the time as a kind of indirect message from Carlo to his celebrated colleague at Maranello on the right way to manage a racing department.

The start of the 1967 season brought a change of Corso Marche's policy towards private drivers, who subsequently found a more open attitude and willingness on the part of Abarth to provide technical updates for their racers and easier terms on which to buy cars and spare parts.

Among the most important races of the year was the European Touring Cup-4 Hours of the Jolly Club, which took place at Monza on 19 March 1967. The Fiat Abarth 1000 Berlina driven by Maurizio Zanetti won Division 1 with rivals, followed by Ed Swart, Willy Kauhsen and many others in similar cars. Ed Swart recalls that 18 Fiat Abarth 1000 Berlinas were on the grid ready to start the race, but none of them were works cars. That was also the race in which the high-speed track and banking were excluded for the first time and new chicanes were introduced. A week later, Giuseppe Rebaudi won the second Fiat Cup at Vallelunga in a Fiat Abarth OT 1300. At Imola on 7 May in the Bruno Deserti Cup, Fiat Abarths steamrollered the Touring Category by winning the 600 cc class with Marco Magri in a 595, the 700 cc with Ugo Barilaro driving a 695, the 850 cc with Demetrio Martino at the wheel of an

In the five days of testing at Vallelunga, the new one-piece five-speed gearbox and self-locking differential (internal code 139) were tried out. Designed by Friedrich Hoffman, the internal gears could be quickly substituted so that the car could be easily and quickly set up for different track racing conditions.
There were two ratios for first gear and 23 different variations for second to fifth.
The axle and return group had a pinion and Gleason teeth crown with a ratio of 10/35; the internal organs were mounted on conical roller bearings.
A gear pump with a pressure of about three atmospheres assured the circulation of the sump's 2.5 litres of oil as far as the gears, the conical torque, the return and the drive shaft joints were concerned.
The gear lever selector was fitted with a system for regulating the couplings.
The whole group weighed 59.6 kg and was destined for racing engines with a power output of between 350 and 380 hp.

On 18 February 1967 at the start of the new motor racing season, Carlo Abarth decided to celebrate the 15-year career of Renzo Avidano as his company's sports director by giving a dinner attended by numerous journalists from the specialist press and motor racing personalities.
In the picture on the left is Giovanni Canestrini, doyen of Italian motoring journalists, as he gives a speech on Avedano (seated on the right of the picture), with Carlo Abarth in the centre.
Right: Abarth presenting his close collaborator with a giant silver cup on which were inscribed all the victories achieved under the sports director in the 15 year period.

850 TC and the 1000 cc with Zanetti again, this time driving a 1000 Berlina. At the 11th Bologna-Passo della Raticosa on 21 May Abarth repeated those Touring Category class wins and did well in Sport with Everardo Ostini winning the up to 1000 cc class in a Fiat Abarth 1000 and Giuseppe Villani the 1300 cc in a Fiat Abarth OT 1300, as they did in the Sport-Prototype in which the works squad won the 1000 class (Peter Schetty) and the 2000 driven by Johannes Ortner took the overall win, followed by Hezemans, Schetty and Villani; Ortner also put up the fastest time, covering the 32.700 kilometres in 18'47.8'' at an average speed of 104 km/h, 12 seconds quicker that the previous year's fastest time recorded by 'Noris' in a Porsche Carrera.

Three weeks earlier at the 3rd Cabinda Circuit race in Angola, Fiat Abarth 1000 Berlina Corsas scored a crushing 1-2 victory (picture) in class A Touring, while a Fiat Abarth 1000 OTS won the class B Grand Touring category.

Johannes Ortner at the wheel of his works Fiat Abarth 2000 Sport Spider at the 2nd Dobratsch-Villacher Alpenstrasse in Austria, an international high-speed hillclimb, which took place on 30 April 1967 over 15.800 kilometres. Victory went to Dutchman Antoine Hezemans in a two-litre Fiat Abarth Sport Prototype. Gert König won the up to 1000 class of the Touring category in a Fiat Abarth 1000 Berlina.

A round in the International Sports Car Championship took place at Mugello on 23 July 1967. It was the 14th time the track had hosted the race on 8 laps, which was over 529.600 kilometres. The 1300 cc class win went to the Fiat Abarth 1300 OT driven by Mauro Nesti, Ernst Furtmayr and Toine Hezemans. The up to 1000 cc class of the Sport Prototype category was won by the Fiat Abarth 1000 SP of Enzo Buzzetti and Secondo Ridolfi.

Originally conceived by Mario Colucci for the Spider Tubolare, a profound evolution of project SE04 took place in 1967.
The Scorpion's technical department came up with two different configurations; one rear engined (above, with Carlo Abarth standing beside the car) coded SE04/FB1 and SE04/FB2, (FB: "outboard" in Italian).

The other version was mid-engined (below) and coded SE04/L with L standing for 'lungo' or long, as the car had a 2260 mm wheelbase, longer than the original design.
The engine was the noted 1946 cc four-cylinder twin-cam with a compression ratio of 10.5:1 which developed a maximum power of 220 hp at 7800 rpm.

More success on 2 July at the Race of Mendola for Touring, Grand Touring and Sport Category cars, in which the Scorpions won the 700 cc class with a 695 SS driven by Heinz Liedl, the 850 cc with the 850 TC of Vittorio Tabucchi, the 1000 cc won by Piero Catani in a 1000 Berlina, while Maurizio Polin in a Fiat Abarth 1000 and Arduino Becchetti won their respective classes in the Sport Category.

Hockenheim was the scene of the second round in the World Sports Car Championship over 45 laps of the German circuit on 9 July; the race was won by Toine Hezemans driving a Fiat Abarth OT 1300, with Ernst Furtmayr second, Eric Bitter third and Helmut Krause, all in OT 1300s. Two weeks later, Mauro Nesti-Ernst Furtmayr-Toine Hezemans won the third round Division 1 at Mugello in another Fiat Abarth OT 1300, which also came 11th overall. On 6 August Fiat Abarth won Division 1 again, this time in the fourth round of the championship in the City of Enna Cup with Alfio Gambero, who also came second overall in his OT 1300. Then it was the turn of the works Fiat Abarth OT 1300 driven by Johannes Ortner, who won Division 1 and came seventh overall in the fifth championship counter at the 3.200 km Zeltweg circuit in Austria on 20 August.

The sixth round in the Championship was the Ollon-Villars hillclimb on 27 August, won by Toine Hezemans in another Fiat Abarth OT 1300; Peter

The elemental driving position of the mid-engined Fiat Abarth 2000 Sport Spider certainly lacked frills, but was highly essential and functional in a car destined to compete in the Prototype category and, in particular, the European Hillclimb Championship against the Dino, Porsche, Lola-BMW and Alfa Romeo 33.
With a dry weight of only 585 kg, the 2000 Sport Spider was accredited with a top speed of 270 km/h.

FIAT ABARTH "2000 SPORT SPIDER" MC (1967)

ENGINE	mid-rear, longitudinal, 4 cylinders in line
Bore and stroke	mm 88x80
Cubic capacity	cc 1946
Valve gear	2 overhead camshafts, chain driven
Number of valves	2 per cylinder
Compression ratio	10.5:1
Fuel feed	2 twin-choke Weber 58DCOE3 carburettors
Ignition	double, coil, distributor
Coolant	liquid
Lubrication	dry sump
Maximum power	220 hp at 7800 rpm
TRANSMISSION	rear wheel drive
Clutch	single dry plate
Gearbox	5 speed plus reverse, central lever
Axle ratio	3.5:1
BODY	racing roadster
CHASSIS	tubular trellis
Suspension	front independent, trailing arms, helicoidal springs, stabiliser bar, hydraulic dampers; rear independent, trailing arms, helicoidal springs, stabiliser bar, hydraulic dampers
Brakes	hydraulically operated discs front and rear
Steering	rack and pinion
Fuel tank	capacity 70 litres
Tyres	front 6.50x13; rear 8.00x13
DIMENSIONS AND WEIGHT	
Wheelbase	mm 2260
Front track	mm 1310
Rear track	mm 1360
Length	mm 3560
Width	mm 1625
Height	mm 930
Kerb weight	kg 585
PERFORMANCE	
Top speed	km/h 270

Schetty came fourth overall driving a Fiat Abarth SP 2000. The seventh World Sports Car Championship race was the 500 Kilometres of the Nürburgring on 3 September, when Ernst Furtmayr came second in an OT 1300, behind a French Alpine.

From 17-19 August, Abarth carried out three days of testing on Monza's junior track with a number of cars, including two two-litre prototypes of which the two-seater was completely new and marked the return to rear engines inclined over the rear axle group; the car also adopted another two front radiators - one for oil, the other for water – in addition to the two behind the driver's back as on the 1000 cc; the front track was also widened to 1405 mm and the rear to 1435 mm. The large windscreen of the previous model was eliminated to allow the car to compete in the Group 7 Sport Category. The car was coded SE04/FB2, meaning it had an SE04-generation tubular chassis was the second car to host an inclined engine (FB stands for fuori bordo or outboard engine), as the first with that kind of power unit had made its debut a few months earlier at the Targa Florio, driven by Jonathan Williams. Among other things, the engine was an evolution of the previous year's power unit; it was given a four valves per cylinder head that increased its maximum power from 216 hp at 7600 rpm to 240 hp at 7800 rpm. Peter Schetty was fastest in the new prototype, lapping the junior circuit in 54.3".

On the much discussed question of the mid-engine favoured by Mario Colucci and the power plant inclined over the rear axle preferred by Carlo Abarth, Colucci recalls 30 years later: *"The location of the rear engine gave vent to a diatribe without end and I found drivers were split between the two*

Peter Schetty of Switzerland at the wheel of a Fiat Abarth 2000 Sport Spider SE04/L with tubular chassis during the Trento-Bondone on 9 July 1967, run over 17.300 kilometres that climbed 1300 metres.
Of the 86 starters in the various classes and categories 72 got to the top, among them the Fiat Abarths that won the up to 850 cc (Claudio Margoni in an 850 TC) and up to 1000 cc (Johann Abt, 1000 Berlina) classes of the Touring category as well as the Grand Touring up to 1000 cc class (Paolo Brambilla, Fiat Abarth 1000) and the up to 1300 cc Sport (Franco Pilone, OT 1300) and Sport Prototype of up to 1000 cc (Enzo Buzzetti, 1000 SP). Overall, Peter Schetty's Fiat Abarth 2000 Sport Spider and Johannes Ortner's sister car came third and fifth respectively.

lay-outs. Arturo Merzario sided with Abarth, but it should also be said that he was the only one who knew how to drive the cars he test drove and helped to tune. On the other hand, Lualdi was never comfortable with the outside power unit and moved on to mid-engines. The system preferred by Abarth gave no early warning that the back end was about to break away. If a car gave early notice that it was going to spin, the drivers had time to correct but were penalised by poor lap times. If, however, the car sent a late message to the driver it was only the driver's ability that was able to correct its in extremis trajectory and that contributed to putting up a better time. It will, however, always be a more difficult car to drive".

At the seventh Aosta-Pila on 24 September over an 11.200 km route that climbed 786 metres and counted towards the Italian Championship of Speed in the Group 4 Sport Category, the Group 6 Sport-Prototype, the Group 2 Mountain Trophy-Touring and Group 3 Grand Touring, 121 cars started and 108 finished; Abarth entered two two-litre prototypes, one with a mid-engine and the other with an outboard power plant. At the wheel of the latter was Switzerland's Peter Schetty, who rose to the position of number one hillclimb driver and won the Sport-Prototype with the astonishing time of 6'40.9", chopping more than 28" off Ortner's record of the previous year. Gianni Varese won the up to 1000 cc class of the same category in a one-litre barchetta. Sergio Morando took the 1300 cc Sport in a 1300 OT, while the 1000 cc was won by Fulgenzio Marino driving an evergreen Fiat Abarth 1000 Bialbero. More of the same success for the Abarths in the Touring Category, with class wins by Ugo Barilaro in a 595, Bruno Pagliero driving an 850 TC and Sergio Dal Cason in the 1000 cc sub-class of the up to 1300 cc at the wheel of a 1000 Berlina.

The Fiat Abarth OT 1300 went from victory to victory in 1967.
The Scorpion entered four cars for the Grand Prix of Hockenheim for sports cars on 9 July, when other Fiat Abarths competed as well as a host of NSU Roadsters with Wankel rotary engines.
The OT 1300 number 3 of Dutchman Toine Hezemans won the 304.605 km 45-lap race (above) followed by the sister car of Ernst Furtmayr; car number 1 driven by Kurt Ahrens retired.
Hezemans covered the distance in 1 hour 47'50.4" at an average of 169.476 km/h.
Hezemans did it again in the Swiss Grand Prix at Ollon-Villars on 27 August, a hillclimb counter in the International Sports Car Championship in two heats over 16 kilometres. He won the event's up to 1300 cc Sport category (below) and took 13th overall with a time of 9'20.8" at an average speed of 103.078 km/h.

FIAT ABARTH "2000 SPORT SPIDER" FB (1967)

ENGINE	
	rear, longitudinal, 4 cylinders in line
Bore and stroke	mm 88x80
Cubic capacity	cc 1946
Valve gear	2 overhead camshafts, chain driven
Number of valves	4 per cylinder
Compression ratio	10.5:1
Fuel feed	2 twin-choke Weber 58DCOE3 carburettors
Ignition	double, coil, distributor
Coolant	liquid
Lubrication	dry sump
Max. power (project)	240 hp at 7800 rpm
TRANSMISSION	
	rear wheel drive
Clutch	single dry plate
Gearbox	5 speed plus reverse, central lever
Axle ratio	3.5:1
BODY	
	racing roadster
CHASSIS	
	tubular trellis
Suspension	front independent, trailing arms, helicoidal springs, stabiliser bar, hydraulic dampers; rear independent, trailing arms, helicoidal springs, stabiliser bar, hydraulic dampers
Brakes	hydraulically operated discs front and rear
Steering	rack and pinion
Fuel tank	capacity 70 litres
Tyres	front 4.75/10.00x13; rear 6.00/12.00x13
DIMENSIONS AND WEIGHT	
Wheelbase	mm 2085
Front track	mm 1405
Rear track	mm1435
Length	mm 3440
Width	mm 1750
Height	mm 910
Kerb weight	kg 595
PERFORMANCE	
Top speed	km/h 285

The 12th Carri Cup was a counter towards the Italian Touring Trophy and took place on 1 October, when Demetrio Martino won the 850 cc class driving an 850 TC, while Nicola Parenti was first among the 1000 Berlinas, but the race brought to light a situation that had been ongoing for some time and that caused no small amount of concern. Aldo Seregni put it into words in Auto-Italiana on 12 October 1967: *"It is our duty to make a rather sad observation; in a very delicate moment for Italian motor sport, the touring car sector has gone into a severe crisis!...Today, the entrants are always the same names, the same resistant enthusiasts who are not renewed year by year any more, in fact their numbers diminish each year. Costs have increased substantially; the purchase of the car, its preparation, extremely costly maintenance, accessories and hard-to-find vital components costing ever more. Certainly, the technical results have improved. Lap times today are of prestige, average speeds in general are very high, but the races lack vitality".*

It was an alarm call that became increasingly persistent, and was to be added to the more widespread unsporting conduct of the drivers, guilty of continued irregularity in the preparation of cars that were too often found by the technical marshals to be underweight or with tracks that were too wide, had rims that were not homologated and not listed as being homologated for the various cars. A state of malaise that was in contrast with the extensive participation of young drivers in the races of a few years earlier, when the enthusiasm and competitive spirit of the dozens of competitors in every class on the grid at the start were the most important ingredients of Sunday racing.

Confirmation of this profound sense of unease and reduced participation that pervaded the world of motor racing came in October, when official figures for new car registrations by brand were published in Italy; in the first eight months of 1967 the total market expanded by 8.8% compared to the

Based on the first SE04 project for the 1000 SP tubular chassis barchetta, in the summer of '67 Mario Colucci produced the Fiat Abarth 2000 Sport Spider, which had a wider track and an engine overhanging the rear axle, from which its additional FB2 code comes.
The car was powered by the 1946 cc four-cylinder twin-cam engine coded 236 and adopted for the first time the four valves per cylinder system to produce 240 hp at 7800 rpm.
Minor modifications were made to the body over the mid-engined SE04/L that appeared a few months earlier, including the replacement of the large wrap-around windscreen of the 1000 SP with a smaller screen for Group 7 two-seater car races.

The Scorpion had a particularly good day at the 500 Kilometres of the Nürburgring on 3 September 1967.
It won the up to 1000 cc class of the Touring category (Gustav Dieter Edelhof, Fiat Abarth 1000 Berlina, above) and the Sport category (Ernst Furtmayr, OT 1300)...

.... while at the Grand Prix of Austria high speed hillclimb at Gaisberg over an 8.625 kilometre ascent, the Corso Marche factory won the up to 1000 cc class of the Touring category (Johann Abt, 1000 Berlina) and the up to 1300 cc Sport (Antoine Hezemans, OT 1300).

At the opening day of the 1967 Turin Motor Show on 1 November, President of the Italian Republic Giuseppe Saragat stopped by the Abarth stand, where Carlo Abarth described the cars on display to him.
Among the show cars were the 1000 Berlina Corsa, which won that year's up to 1000 cc class of the Italian Touring Championship (Nicola Parenti) and the European Touring Challenge (Wilhibert Kausen).
Also on display were the tubular chassied 1000 SP barchetta, the OT 1300/124 and the latest version of the OT 1300, which won the Division 1 International Grand Touring Championship for the second successive year.

Under the elegant gaze of Turin's Palazzo Esposizioni, the Abarth stand exhibited the Fiat Abarth 1000 SP (foreground), which brought the Scorpion a great deal of success in the Prototype and, after 1 March 1968, in the Sport category. In the background are the Touring category cars that won the 1000 cc class of the first division of the 1967 European Touring Challenge plus the Italian Touring Trophy.

same period in 1966, while the delivery of Abarths to national customers suffered a reduction of 50.2% at 287 cars against the total 576 of January to August of the previous year. This data, in complete contrast with the general progress of demand, raised serious concerns within the company. In addition, the aggressiveness of other manufacturers and tuners had been evident for a couple of seasons in the Touring Category with the Steyr-Puch, BMW 700, NSU Prinz TTS, Renault 8 Gordini, Morris Mini Cooper S and Alfa Romeo GTA, even if in classes in which Abarth did not compete. Of the protagonists large and small at the track every Sunday, it was always Carlo who shouldered most of the financial and organisational burden imposed by such incessant activity. His personal commitment to the development of new cars and continuous research into constructive new solutions were both surprising and admirable in a man close to 60 years old. Such a vital man who contributed to both workshops in Corso Marche and the constant test sessions, which he always attended. In that period, Abarth moved house and went to live in the hills on the other side of the Po in Parco Europa at Vicinale delle Torrazze Road, a residential complex with a swimming pool that he felt was important to his physical well being.

In mid-November '67, he went with his best and most loyal colleagues to Vallelunga for another test session a few weeks after the team had tuned their cars at the track. As well as Abarth and Avidano, there were seven drivers – Goedemans, Hezemans, Kauhsen, Merzario, the two Spanish Juncosa brothers and Ortner, but Schetty did not attend because he was on honeymoon. This time the team had a test schedule involving six different models: a 1000 SP, two 1000 cc Group 5 Touring cars and three 2000 cc Sport Prototypes with different engine configurations, central on one and inclined on the other, the latter readied to compete in Group 7 for two-seaters. The V8 three-litre, which was still under development, was not involved in the test session.

The Group 7 car boasted several important innovations compared to the model that had competed in a number of hillclimbs in the preceding three months. To avoid the kind of roadholding peculiar to a car with the weight of its engine over the rear axle - an advantage on certain tracks, but which meant such cars had to be driven in a special and constant manner - its chassis was profoundly modified so that the engine, the maximum power of which had risen to 245 hp, could be installed in a much lower position, improving the car's centre of gravity and set-up on the track's surface.

On the new Group 5 up to 1000 cc Touring Category car, which was much lower than the version that had just finished the '67 season, the rear suspension was modified with the oscillating arms made of a tubular trellis instead of moulded pressed steel like the original Fiat; in addition, the wings and wheel housings were widened. The 982 cc engine was also refined with

Tested at the end of 1967, the new rear suspension layout of the Fiat Abarth 1000 Berlina Corsa comprised parallel longitudinal oscillating arms of tubular structure in place of the pressed steel triangles of the Fiat 600; the new system meant re-dimensioning the wheel housing and widening the wings. Those modifications were carried out on the Group 5 1000, on which the radial cylinder head was also extensively tested before it successfully completed its experimental phase.

the compression ratio taken to 12.5:1 and bigger carburettors being fitted; in readiness for the car's homologation for the 1968 season in both Group 2 (Touring) and Group 5 (Touring Specials) a twin-choke Weber DCD7 carburettor was adopted for G2 to take power output up to 88 hp at 7600 rpm, while two twin-choke Weber 40 DCOEs were used for G5 and they boosted maximum power to 110 hp at 8000 rpm. Dry weight for both cars was declared as 583 kg and maximum speed over 190 km/h.

Abarth also decided to build 50 one-litre Sport-Prototype cars to obtain the Group 4 Sport Category homologation for the '68 season. Of new regulations that changed the championship from a constructors' to a drivers' title chase, Carlo Abarth said he was perplexed because as far as he was concerned the participant manufacturers would have lost the publicity interest that had been so effective in previous years. And as the prices of future Group 5 Touring cars looked like they were going to be much higher, observers asked how many drivers could saddle themselves with the enormous expense of buying and running such a machine for the entire season.

The Scorpion's 1967 closed with a total of 803 victories being achieved across the world and winning the World Constructors' Championship with the Fiat Abarth OT 1300 as well as the Division One of the European Touring Car Challenge with Wilhibert Kauhsen driving a Fiat Abarth 1000 Berlina. The company also won the Italian Touring Car Championships in the up to 600 cc class (Marco Magri in a Fiat Abarth 595 prepared by Romani), the up to 700 cc (Giovanni Brancadori driving a Fiat Abarth 695), the up to 850 cc (Demetrio

A number of mechanical and body modifications had been carried out to the Fiat Abarth OT 1300 to make the car more competitive than the original version. Mario Colucci had revised the pressed steel chassis, derived from the Simca 1000's production floor pan, modified the suspension anchorage, increased the front track from mm 1295 to 1310 and the rear track from mm 1340 up to 1370. The engine output was also improved from 147 to 157 hp at 8800 rpm. Such a brilliant motor achieved extraordinary racing results which led to strong demand for Fiat Abarth cars throughout the world.

The OT 1300's second edition also underwent some radical body changes, mostly in the rear section, which displayed a completely new look thanks to a large Plexiglas rear window and an effective tail spoiler. Note the "snorkel" air scoop placed on the car's roof, which enabled to search the best air intake for the carburettors at high speed. On an experimental basis the car was also fitted with up to four different lengthened front and rear body ends, and one official Long-tailed OT 1300 was entered at the Nürburgring 500 Kilometres in 1966 showing significant top speed improvement. The Corso Marche factory found it difficult to meet daily production targets that reached three cars a day during peak working periods.

The entire bodyshell of the OT 1300 was composed of three large sections, the central unit with the whole cockpit and two large front and rear bays with one-piece opening to fully facilitate access to the car's mechanics.
Minor interior modifications were also made that emphasised the exquisitely racing pedigree of the car, including the large snorkel-like air intake that appeared on the roof to aspirate the engine.
The OT 1300 won the 1966 and 1967 1st Division Grand Touring World Championships and the 1966, 1967, 1968 Group 4 Italian Sport titles.

FIAT ABARTH "OT 1300 II SERIES" (1967)

ENGINE	
	rear, longitudinal, 4 cylinders in line
Bore and stroke	mm 86x55.5
Cubic capacity	cc 1289
Valve gear	2 overhead camshafts, chain driven
Number of valves	2 per cylinder
Compression ratio	10.5:1
Fuel feed	2 twin-choke Weber 45DCOE9 carburettors
Ignition	double, coil, distributor
Coolant	liquid
Lubrication	dry sump
Max. power (project)	157 hp at 8800 rpm
TRANSMISSION	
	rear wheel drive
Clutch	single dry plate
Gearbox	5 speed and reverse, central lever
Axle ratio	3.5:1
BODY	
	2 seat coupé
CHASSIS	
	boxed steel
Suspension	front independent, trailing arms, helicoidal springs, stabiliser bar, hydraulic dampers; rear independent, trailing arms, helicoidal springs, stabiliser bar, hydraulic dampers
Brakes	hydraulically operated discs front and rear
Steering	worm and sector
Fuel tank	capacity 40-80 litres
Tyres	front 5.50x13; rear 6.50x13
DIMENSIONS AND WEIGHT	
Wheelbase	mm 2015
Front track	mm 1310
Rear track	mm 1370
Length	mm 3830
Width	mm 1620
Height	mm 1222
Kerb weight	kg 655
PERFORMANCE	
Top speed	km/h 255

Martino at the wheel of an 850 TC tuned by Faccioli), the up to 1000 cc (Nicola Parenti in a 1000 Berlina prepared by Baggioli) plus the up to 1300 cc class of the Italian Sports Car Championship with 'Nikor' (Guido Nicolai) driving a Fiat Abarth OT 1300, a car that was the one to beat for the second successive year. The OT 1300 had reached an extremely high state of techno-constructional maturity and demand from drivers in many countries forced the factory to increase chassis production for the whole OT family to two or three racers a day. That also required maximum logistical effort from Carrozzeria Sibona & Basano, who made the glass fibre bodyshells.

Word of Carlo Abarth's biggest and most ambitious secret project began to leak out at the end of the '67 season through indiscretion: it concerned Formula One, the king of motor racing. In line with his rapid way of facing up to a situation, Carlo spoke of it himself in an interview given in mid-January 1968 to Auto-Italiana, during his brief stay in Sicily to assess drivers and cars for the Scorpion's participation in the Targa Florio, scheduled for 5 May; he took the whole motor racing world by surprise: "*...We worked until new year's eve on our F1. The design of a single-seater demands a lot of work, a lot of money and a lot of time. But I can confirm now that we have made good progress... We will be able to carry out the first tests next May. We'll try the car on the road with a three-litre engine. But in its definitive version, the car will have a new engine, which we started to design some time ago. The definitive version should be finished in February. It will be an eight-cylinder*".

Asked what made him want to create a Formula One car and enter the world of the Grands Prix, he replied without hesitation or fear of causing possible disputes: "*Essentially, I believe, for two reasons: the policy recently adopted by Ferrari in this sector – too many problems and nothing concrete – is also the reason why F1 is the only category among those planned by the CSI today that guarantees constructors a certainty of continuity in its regulations. Even today, in a way, I am paying the consequences of constructing a prototype. I monopolised the whole of my factory for that expensive car. Then, when it was 90% ready after Le Mans, the CSI decided to limit prototype capacity. That was a severe blow to my factory*".

During the interview, Carlo clarified his thoughts on a point on which there had been a great deal of discussion in the past, the criteria of evaluation that he adopted in the choice of his works drivers, almost all of whom were foreigners- "*... In Italy, the atmosphere is not the best for the development of champions. Italian drivers cannot be professional. They race because they like to, for a particular reason and that's it. They don't dedicate themselves to this activity, submitting themselves to sacrifices like, I don't know, the English, the Germans and so on. At the beginning they go very fast, always improving and then, when they have got to the top, when they should begin to reap the benefit of their sac-*

rifices, they give it up. We have Merzario in our team, who showed he was a driver of great potential last year. Until when will we have him available to us? We can't say, but we are sure that the day he gets married he'll abandon us and we will lose a valuable member of our team". Faithful to this viewpoint, Abarth works drivers for '68 included only one Italian and that was Merzario, whose team mates were Schetty of Switzerland, Ortner of Austria, Hezemans of Belgium and Goedemans of Holland and Williams of Britain.

By 31 December 1967, 369 cars had been delivered to Italian customers, against 686 registered the previous year, an alarming reduction of over 46%.

1968

1969

THE PRICE OF TRIUMPH

From the start, 1968 revealed itself to be a year dogged by controversy in which Carlo Abarth was involved, often against his will. By nature, Carlo hated disputes and half-truths even if he did face up to problems and their effective dimensions with a scowl, leaving others to the lay claim to their sterile positions of principle. His first skirmish took place at the beginning of March, when news reached Corso Marche that homologation had not been granted for the new rear suspension of the Fiat Abarth 1000 Berlina in Group 5 Special Touring events; the regulation stipulated that the modification of normal production suspension must retain unchanged the principal function of the original, and as the new rear end geometry of the 1000 cc saloon was far removed from the production Fiat's, homologation was denied. As they found to their chagrin at Abarth, that was because the ban on making changes of that nature was only put in place by an addition to the regulations and that additional clause was not circulated to competitors until much of the 1000 Berlina's modification work had been completed. It was the unexpected appendix that not only meant Abarth had to produce another type of suspension in record time, it was also late in the final stages of building the prototype and its two-litre engine for which, among other things, it was intended to drop carburettors in favour of Lucas indirect fuel injection.

On a more positive note, early that season the1000 SP (internal code SE04) was homologated to compete in the Sport Category with a tubular chassis and was able to race in Group 4 from that moment. To make the car more appetising to potential customer-drivers, a launch price of Lit 4.120.000 was established – 10% discount for teams – which was a million lire less than the car's price a few weeks later. By comparison, in March 1968 Lit 3.5-Lit 5 million bought by a sporting customer who did not race an Alfa Romeo Grand Tourer 2600 SZ (Lit 3.970.000), a Lancia Flaminia 3C Supersport 2.8 (Lit 3.615.000) or a Dino 206 GT (Lit 4.950.000).

Abarth competed in the 4 Hours of the Jolly Club at Monza on 24 March with the new Group 5 and Group 2 1000 Berlinas. The winner of Division 1 was the Peter Schetty-Ab Goedemans car, which covered 595.735 km at an average speed of 148.993 km/h. But the victory was marred by a large number of mechanical problems experienced by the 1000 Berlinas, which started on the 10th lap when Toine Hezemans was forced to retire with a broken engine

Abarth entered a squadron of Group 2 and Group 5 1000 Berlina Corsas for the 4 Hours of the Jolly Club, a counter in the European Challenge for touring cars, which took place at Monza on 24 March. Peter Schetty and Ab Goedemans won the 1st Division in car number 12, but there were many Abarth retirements due to mechanical difficulties, in particular Arturo Merzario's car 18 (above) with engine problems and Johannes Ortner's 16 car (gearbox flange).
The five works 1000 Corsas were fitted with radial cylinder heads, homologated by FIA in Group 5; in the engine bay were plastic manifolds to enable oil vapour to flow from the sump and cylinder head to a purpose-built tank (large picture) as required by the regulations to avoid losing oil on the track surface.

and continued on lap 14 with the retirement of Johannes Ortner (gearbox flange pin), the 17th when Pam stopped (cylinder head gasket) and on the 30th lap when Arturo Merzario went out with another broken engine. The tendency of engines to overheat plus the breakdown of segments and pistons were not unknown to Abarth technicians and drivers. Two days before the Monza race, a 1000 Berlina entered by the Squadra Faza team of Brewster, New York, competed in the 3 Hours of Sebring and had to retire on the third lap due to engine overheating.

April 7 was an important date for the Scorpion, with the victorious debut in the Ampus hillclimb in France of the definitive 2000 Sport Prototype, a car with a tubular chassis built from considerable previous experience with the SE04 series of barchettas. The chassis, a trellis of steel in 22 mm section chromomolybdenum tubes, was strengthened with a double lamination of glass fibre and its overall weight was kept down to 47 kg; Mario Colucci gave preference to the tubular solution, of which he had been a convert since the Fifties, not only for weight reasons but to improve access to components. Contrary to his project preference the engine was outboard and inclined over the rear axle, from which came a car with a weight distribution split of 38% on the front axle and 62% on the rear. The well-known 1946 cc four-cylinder engine of the first 1968 derivative was not fitted with fuel injection but retained the two twin-choke Weber 58 DCO3 carburettors, yet maximum power still increased by a few horsepower to almost 250 hp at 8000 rpm; cooling was by two water and two oil radiators, placed in pairs behind the front

Engine trouble forced Toine Hezemans of Holland to retire his works Fiat Abarth 1000 Berlina Corsa 15 (above) from the 4 Hours of the Jolly Club at Monza.

air intake and the driver's seat. The chassis was unchanged; the wheelbase was 2085 mm and the front and rear track 1405 mm/1435 mm. The body was a light demountable glass fibre shell that weighed 50 kg including its windscreen and various accessories, which made it possible to keep the dry weight of the car down to 575 kg. Maximum speed was declared at 270 km/h.

The body shape was unusual due to the large paired headlamps and the small front air intake that extended across the entire frontal section of the car; without doubt, one of the most attractive cars ever to take to the race track. Over 50 examples of this successful model were built and they enjoyed a long and distinguished motor racing career; on the car's debut over the 6.8 kilometre course of the fourth Course de Cote d'Ampus in France, Peter Schetty won an outstanding victory and established a record of four minutes dead.

On 12 April at the 500 Kilometres of Snetterton in England, the Fiat Abarth of Holland's works drivers Toine Hezemans and Ab Goedemans beat the challenge of the Mini Cooper S and Ford Anglias and won. A success that was repeated on 5 May at the Grand Prix of Limbourg at Belgium's Zolder circuit, where Goedemans' Fiat Abarth 1000 Berlina beat Hezemans in a sister car and John Fitzpatrick's Ford Anglia.

The great stress to which the latest generation of 982 cc engines was subjected induced the Scorpion's technicians to issue strict instructions to the drivers for testing and racing -"*watch the water and oil temperatures ... warm up the engine carefully, using warm plugs that have to be substituted with other colder ones ... and do not exceed 8200 rpm for any reason whatsoever*". A curious problem that arose in later races was the poor seal of the 1000

An attractive group shot taken at the Monza Autodrome on 24 March at the 4 Hours of the Jolly Club.
From the left: Italians Marsilio Pasotti ('Pam') and Arturo Merzario, Antoine Hezemans of Holland, Carlo Abarth, Britain's Jonathan Williams, Peter Schetty from Switzerland, Austrian Johannes Ortner, Dutchman Ab Goedemans and Abarth motor sport director Renzo Avidano.

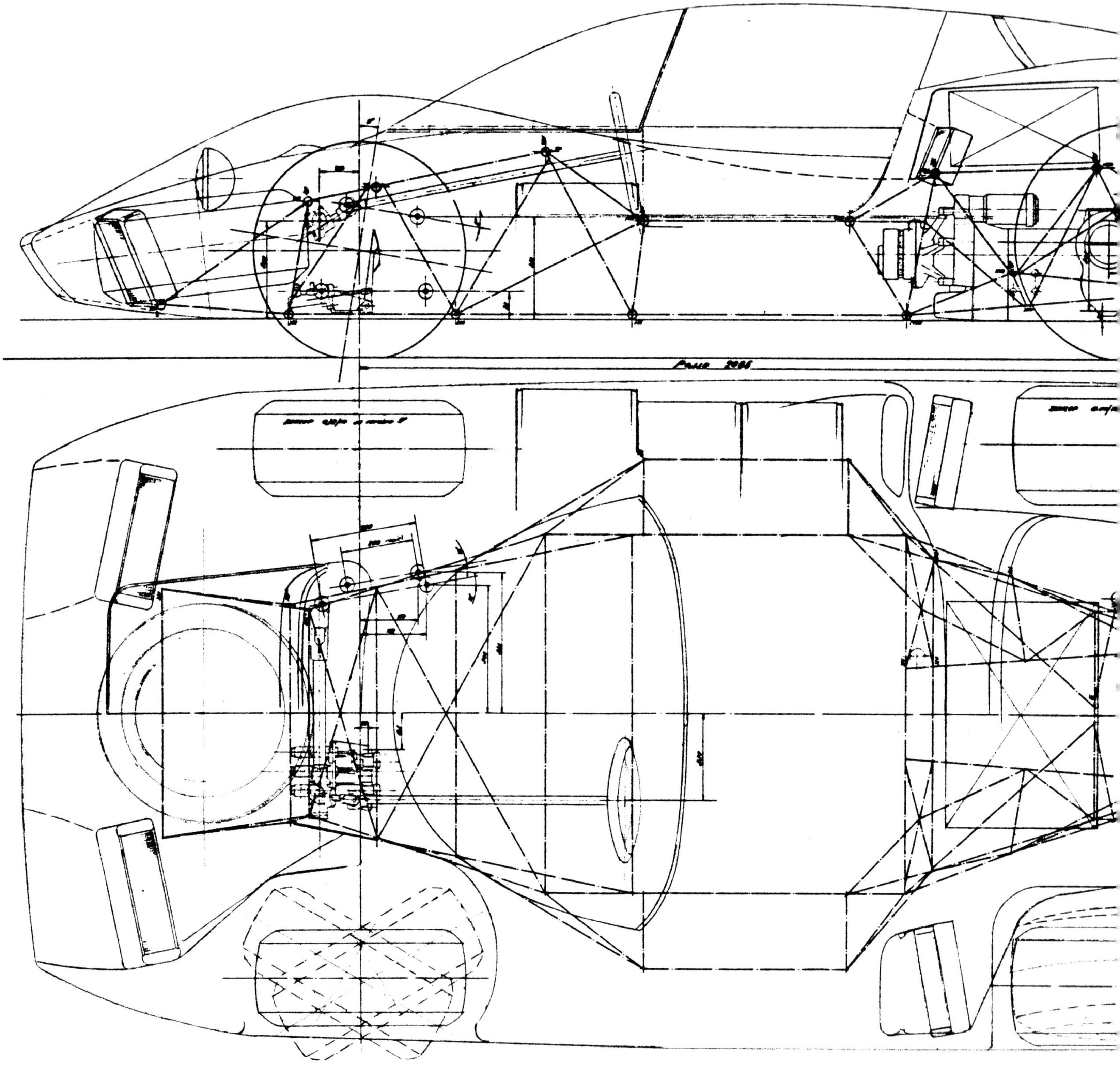

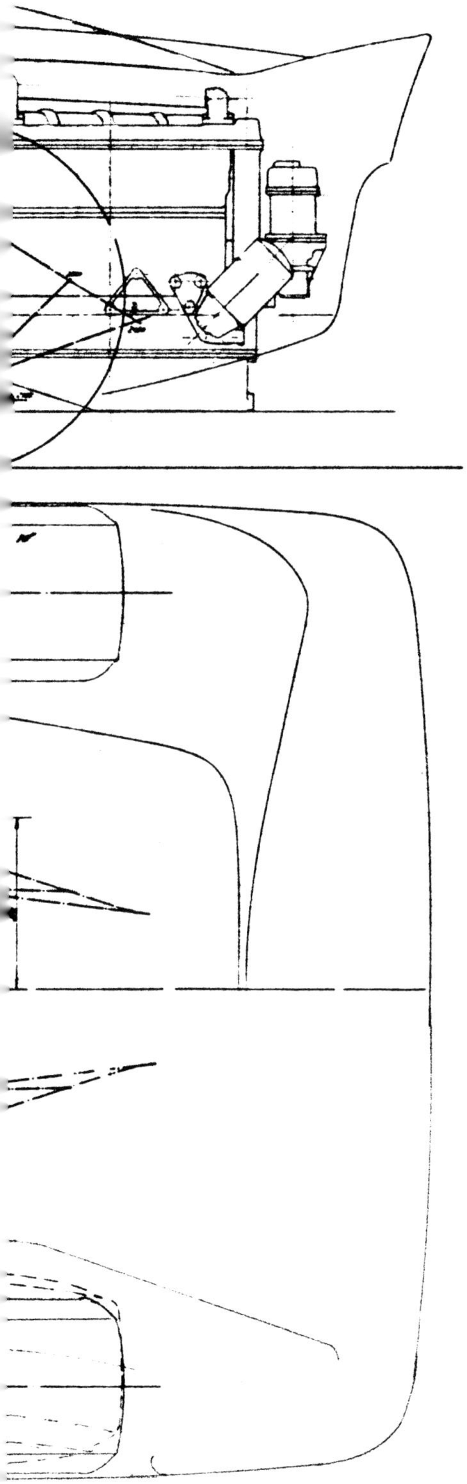

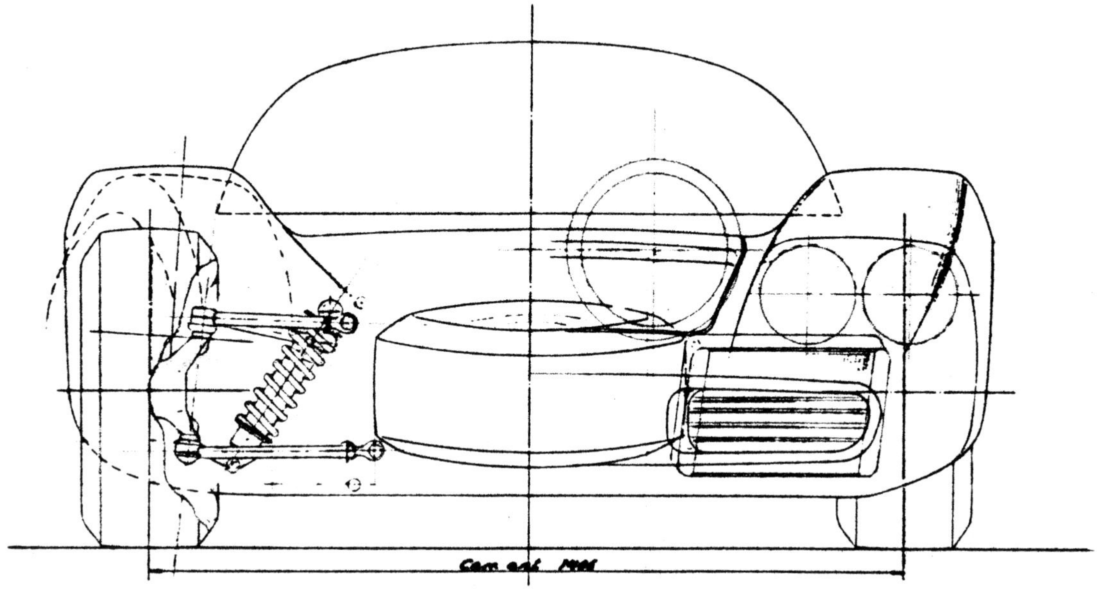

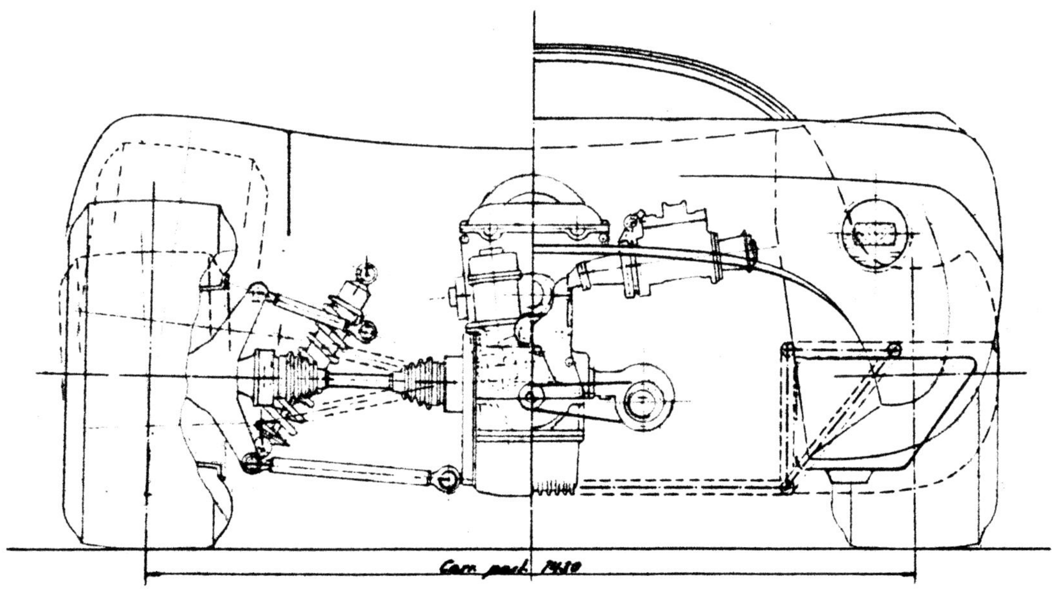

Four views of the Fiat Abarth 2000 Sport Spider model (internal code SE010) had a trellis of steel tubes in chromomolibdeno of 20 and 22 mm sections; the whole thing weighed 39 kg, to which should be added another 8 kg in glass fibre reinforcement.

The first 25 examples of the 2000 Sport Spider were produced for Group 4 homologation, which was issued on 1 April 1969.
In all, over 50 cars were built with this 1946 cc four-cylinder engine and its 16-valve cylinder head that gave the car 250 hp at 8700 rpm.

FIAT ABARTH "2000 SPORT SPIDER" (1968)

ENGINE	
	rear, longitudinal, 4 cylinders in line
Bore and stroke	mm 88x80
Cubic capacity	cc 1946
Valve gear	2 overhead camshafts, chain driven
Number of valves	4 per cylinder
Compression ratio	11.5:1
Fuel feed	2 twin-choke Weber 58DCO3 carburettors
Ignition	double, coil, distributor
Coolant	liquid
Lubrication	dry sump
Max. power (project)	250 hp at 8700 rpm
TRANSMISSION	
	rear wheel drive
Clutch	single dry plate
Gearbox	5 speed plus reverse, central lever
Axle ratio	3.5:1
BODY	
	racing roadster
CHASSIS	
	tubular trellis
Suspension	front independent, trailing arms, helicoidal springs, stabiliser bar, hydraulic dampers; rear independent, trailing arms, helicoidal springs, stabiliser bar, hydraulic dampers
Brakes	hydraulically operated discs front and rear
Steering	rack and pinion
Fuel tank	capacity 100 litres in total
Tyres	front 4.75/10.00x13; rear 6.00/12.00x13
DIMENSIONS AND WEIGHT	
Wheelbase	mm 2085
Front track	mm 1405
Rear track	mm 1430
Length	mm 3850
Width	mm 1780
Height	mm 970
Kerb weight	kg 575
PERFORMANCE	
Top speed	km/h 270

The driving position of the Fiat Abarth 2000 Sport Spider, with the car's large rev counter in front of the driver and the other three small instruments – water and oil thermometers and the oil gauge; to the left is a small console housing the ignition key and a number of operational switches.

Berlina's fuel cap; petrol slopped over onto the windscreen so that the wipers wore faster and would no longer do their job.

More controversy blew up at the end of April that pitched Abarth against the organisers of the Targa Florio and the 1000 Kilometres of Monza. Carlo maintained they had not correctly assessed the injection of fresh energy and interest the presence of his team would bring to the promotion and success of the two events and that, as a consequence, the start money they offered the Scorpion for its participation was completely inadequate to cover his expenses in sending his men and machines. So after having tried in vain to reach and agreement on money, Carlo dropped a bombshell by refusing to compete in the two events, adding in Auto-Italiana on 2 May 1968: *"It will certainly not be championship matters that will make me change my mind"*.

Abarth's refusal to compete in the Targa Florio and the Monza 1000 Kilometres plus lateness in the development of the Formula One car meant a period of enforced inactivity for British driver Jonathan Williams, and that was the cause of an unexpected divorce between the Englishman and the Scorpion. The situation inevitably led to other thinly veiled controversial claims and disregarded expectations; Abarth, who had a contract that tied Williams to the factory until the end of 1969, complained that the British driver conducted himself in too a casual manner to break the letter of the agreement.

On 21 April, two weeks after the success at Ampus, Peter Schetty was entered for the 11th Stallavena-Boscochiesanuova hillclimb in a Fiat Abarth

The Monza 1000 Kilometres (Filippo Caracciolo Trophy) race took place at the northern Italian track on 25 April 1968 and was a round in the International Constructors' Grand Touring Championship. Entries included two Fiat Abarth 1000 SP with Gianfranco Palazzoli ('Pal Joe') and Piero Botalla in the 60 car and Luciano Pasotto-Umberto Grano driving car 59. Over 100 laps of the circuit by 35 competitors, the race concluded without drama but with the Ford GT40 taking the win in the Sport category, the Porsche 910 and 907L winning the two and three-litre classes respectively of the Prototype category. The two 1000 SP tubular barchettas had no serious opposition and came first and second in the up to 1000 cc Sport category, with car 60 leading 50 over the finish line. In practice, the Palazzoli-Botalla car lapped the 10.100 kilometre circuit in 3'48.3" and Pasotto-Grano in 4'00.7".

Issued after the requisite minimum of 50 examples had been built, the first page of the FIA homologation form for the Fiat Abarth 1000 SP in the Sport category, valid from 1 March 1968.

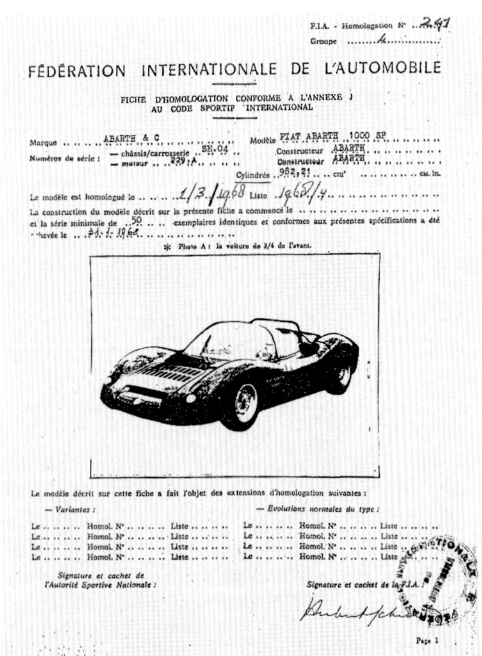

F.I.A. - Homologation N° 241
Groupe 4

FÉDÉRATION INTERNATIONALE DE L'AUTOMOBILE

FICHE D'HOMOLOGATION CONFORME A L'ANNEXE J AU CODE SPORTIF INTERNATIONAL

Marque ABARTH & C Modèle FIAT ABARTH 1000 SP
Numéros de série : — châssis/carrosserie SE.04 Constructeur ABARTH
— moteur 229.A Constructeur ABARTH
Cylindrée 982,21 cm³ ... cu. in.

Le modèle est homologué le 1/3/1968 Liste 1968/4
La construction du modèle décrit sur la présente fiche a commencé le ...
et la série minimale de 50 exemplaires identiques et conformes aux présentes spécifications a été achevée le 31-1-1968

* Photo A : la voiture de 3/4 de l'avant.

Le modèle décrit sur cette fiche a fait l'objet des extensions d'homologation suivantes :

— Variantes :
Le ... Homol. N° ... Liste ...
Le ... Homol. N° ... Liste ...
Le ... Homol. N° ... Liste ...
Le ... Homol. N° ... Liste ...

— Evolutions normales du type :
Le ... Homol. N° ... Liste ...
Le ... Homol. N° ... Liste ...
Le ... Homol. N° ... Liste ...
Le ... Homol. N° ... Liste ...

Signature et cachet de l'Autorité Sportive Nationale :

Signature et cachet de la F.I.A. :

Page 1

2000; he drove the Lucas fuel injected version during practice, but preferred to race the carburettor car. On race day, the focused Basle driver climbed up through the twists and turns of the course in record time, beating Johannes Ortner's fastest time of the previous year by just under 19 seconds. That day, the Scorpion also won the up to 1000 cc class of the Sport Category with Paolo Lado and his Fiat Abarth SP and in the up to 1300 cc class with a Fiat Abarth OT 1300 driven by Franco Pilone. Then there were Touring Category victories in their respective classes for 'Arch' in a Fiat Abarth 595, Enrico Pasolini at the wheel of a 695 and Nicola Parenti in a 1000 Berlina.

It was Ortner's turn at the fifth City of Volterra Cup on 23 May; he won the event and set a new record time. Throughout the season the much-talented Swiss and Austrian drivers, who got on well together, scored alternate victories in the demanding latest version of the Fiat Abarth 2000 Spider. At Vallelunga on 2 June Arturo Merzario won the Sport-Prototype Category in a Fiat Abarth SP 1000, while Carlo Zuccoli took the up to 1300 cc class of the Sport Category in an omnipresent OT 1300. At the sixth Coppa Acqua Cerelia hillclimb of 9 June over 12.200 kms, Ortner came first overall driving a Fiat Abarth SP 2000 and on 16 June at the 17th Vittorio Veneto-Cansiglio it was down to Peter Schetty - he took the overall win in a sister car. Fiat Abarths won both races' Group 2 Touring Category classes with the 595, 695 SS, 850 TC and the 1000 Berlina, as they did the up to one-litre Sport Category with an SP 1000.

At the 12th Bologna-Raticosa Pass on 23 June, the duel between the Fiat Abarth 2000s of Ortner and Schetty was an in-house confrontation that

Two pictures of the 1000 Kilometres of Monza, showing the two protagonists of the 1000 cc Sport class; the 1000 SP number 60 of "Pal Joe"-Botalla and the 59 car of Pasotto-Grano. Car 60 won, after covering 82 laps of the track.

The start of the 1968 motor racing season saw both versions of the 2000 Sport Spider competing, the mid-engined car of which Carlo Abarth was no fan but in which technical director Mario Colucci believed, and the model with its engine overhanging the rear axle, about which the opinion of the Scorpion's works drivers was divided.
Left: Pierre Maublanc at the wheel of the mid-engined car (SE04/L) during the Col Bayard hillclimb in France on 28 April.

Right: the last model of the Fiat Abarth 2000 Sport Spider with outboard engine (SE010) competing in the Savona-Cadibona race on 5 May, when Peter Schetty came second overall.
That model made a successful debut a month earlier at the Ampus hillclimb in France on 7 April, when Schetty covered the 6.8 kilometre route in a sensational record time of 4' dead, clipping 14.6" off the previous record.

brought the event's record tumbling down. Ortner won and also managed to beat the event's record time by over a minute; Schetty was second 1.4'' behind him. But the tables were turned at the Bolzano-Mendola on 7 July, when Schetty won and easily broke the 90 km/h average barrier, covering the distance at an average 102.248 km/h. Goedemans-Hezemans won the Touring Category of the 6 Hours of the Nürburgring on the same day in the Fiat Abarth 1000 Berlina, while Maurizio Zanetti and Edmund Swart took the up to 850 cc class victory with a Fiat Abarth 850 TC. Then came the Trento-Bondone on 14 July, won outright by privateer Karl Federhofer in a Fiat Abarth OT 1300, followed in second place by Gianni Varese in an SP 1000 and Gianni Lado, who posted the fastest time in qualifying, third driving a sister car. Edoardo Gatti won the up to 850 cc class of the Group 2 Touring Category in another Fiat Abarth 850 TC and Maurizio Campanini the up to 1000 cc driving a 1000 Berlina. The Scorpion also took the upper hand in the up to 1000 cc Grand Touring Category, won by Giancarlo Galmozzi with the 1000 OTS beating Mühlbauer's Honda.

Peter Schetty competed in the Swiss St. Ursanne-Les Ranglers hillclimb at the wheel of a formula libera single-seater, in which some thought they saw the Abarth F1 car that was under construction at the time, except for the engine which that day was the well-known 1946 cc four-cylinder with four

On the same 5 May on the Cote d'Azur in France, Arturo Merzario won the Touring class and came seventh overall in the Cannes-Tenneron hillclimb in car number 70, a Fiat Abarth 1000 Berlina.
Johannes Ortner and his works Fiat Abarth 1000 SP number 100 took the Sport Prototype and came fourth overall.

A week later on 12 May, the new Fiat Abarth 2000 Sport Spider made another race appearance, this time in the 3rd Castione Baratti-Nerviano Arduini hillclimb near Parma over a 5.500 kilometre route. Johannes Ortner won in car number 320, followed by Peter Schetty in a sister car and Edoardo Lualdi driving a Dino 2000. In those first competitive outings, the Scorpion's two-litre SE010 was also entered or tested with a Lucas indirect fuel injection system.

The 2000 Sport Spider immediately confirmed itself as being full of potential as far as its mechanical and chassis yield were concerned, as well as being appreciated as one of the most attractive competitors for its elegant and equilibrated lines. The car's reduced dimensions and unusual power to weight ratio of 2.3 kg per horsepower made it a formidable competitor in both hillclimbs and long distance track races.

valves per cylinder. With his car lacking in preparation, particularly as far as grip is concerned, the best Schetty could do was to bring it in eighth.

Next was the dramatic 500 Kilometres of the Nürburgring of 1 September, which saw the outright triumph of the mid-engined Fiat Abarth 1600 Prototype driven by Peter Schetty, but which also cost the life of works driver Ab Goedemans during practice. The Scorpion's success was completed by Johannes Ortner and Arturo Merzario, who came second and third overall driving Fiat Abarth 1600 Prototypes with outboard engines. Abarths also won the Touring Category's up to 1000 cc class with 'Pam' (Pasotti Marsilio), who covered the 19 laps of the circuit in 3h 32'35" at an average 122.5 km/h driving a Fiat Abarth 1000 Berlina. In the up to 1300 cc class of the Sport Category, Jeremy Walton came first in a Fiat Abarth OT 1300. At the 3 Hours of Jarama in Spain on 29 September, last round in the FIA Challenge for touring cars, Pam won the up to 1000 cc class of the Touring Category in his Fiat Abarth 1000 Berlina and Edouard Swart the up to 850 in a Fiat Abarth 850 TC. But Pam's success was not enough to clinch the one-litre European Championship, which went to John Handley and his BMC Mini Cooper after

This attractive picture, autographed by Dutch driver Edouard Swart, shows his works Fiat Abarth 850 TC Corsa during a race on the historic Zandvoort circuit on 12 May 1968.

Abarth had won the title every year since 1965.

Under the direction of Carlo Abarth himself, the Scorpion went to Austria with five cars for the Donau Pokal on Aspern airport on 6 October. Among them was the new three-litre V8 Group 7, the 1000 Berlina with new rear geometry and engine – the usual 982 cc single shaft, four-cylinder – capable of 106 hp at 8400 rpm, a five-speed gearbox, 4.50-13 tyres and a weight of 585 kg. The car could easily exceed 200 km/h but its body was still unequivocally the Fiat 600's, which certainly gave no indication of its surprising performance. The new deformable trapezoidal rear end permitted minimum variation of camber, allowing more suspension travel as became evident from the characteristic behavioural curve of the car; on the full 10.100 km Monza circuit, the Group 5 1000 Berlina was 12 seconds faster than the same

FIAT ABARTH "850 TC CORSA" GR. 5 (1968)

ENGINE	
	rear, longitudinal, 4 cylinders in line
Bore and stroke	mm 62.5x69
Cubic capacity	cc 847
Valve gear	overhead valves, side camshaft
Number of valves	2 per cylinder
Compression ratio	12.5:1
Fuel feed	2 twin-choke Weber 40DCOE carburettors
Ignition	coil, distributor
Coolant	liquid
Lubrication	forced, oil filter on the main circuit
Max. power (project)	93 hp at 8400 rpm
TRANSMISSION	
	rear wheel drive
Clutch	single dry plate
Gearbox	5 speed plus reverse, central lever
Axle ratio	4.55:1
BODY	
	2 door, 4 seat saloon
CHASSIS	
	Integral body
Suspension	front independent, trailing arms, helicoidal springs, hydraulic dampers; rear independent, trailing arms, helicoidal springs, hydraulic dampers
Brakes	hydraulically operated discs front and rear
Steering	worm and sector
Fuel tank	capacity 27 litres
Tyres	5.50x13
DIMENSIONS AND WEIGHT	
Wheelbase	mm 2000
Front track	mm 1160
Rear track	mm 1160
Length	mm 3530
Width	mm 1390
Height	mm 1400
Kerb weight	kg 583
PERFORMANCE	
Top speed	km/h 190

One of the greatest Scorpion 1968 Touring category victories was in the 6 Hours of the Nürburgring, which took place on the Eifel Mountain circuit on 7 July. Both the 850 TC Corsa number 113 (above) driven by Ed Swart-Maurizio Zanetti and the 1000 Berlina Corsa of Ab Goedemans-Toine Hezemans (106) won their respective classes. Note the characteristic tendency of the cars to lift their inside wheel in a corner, caused by both the unusual front and rear suspension geometry and the wealth of power generated by the engine, which contributed to lightening the front end under acceleration.

Above: A group photograph taken at the Trento-Bondone on 14 July 1968 of the 2000 Sport Spider.
From the left: Swiss driver Peter Schetty, Carlo Abarth, Mrs Annelise and Antonio Tomaini.

The 500 Kilometres of the Nürburgring saw the Scorpion win again on 1 September 1968. Peter Schetty (picture), Johannes Ortner and Arturo Merzario took the first three places in that order in the 1600 class of the Sports Prototype category.

model in Group 2 configuration. The three-litre Group 7 car (factory code SE013) was completely new: its V8 engine with a bore and stroke of 88x81 mm had a capacity of 2968 cc and was installed in the centre of the car; with a compression ratio of 12:1 and fuelled by four Weber 48 IDA carburettors, it produced 350 hp at a maximum 8200 rpm. Built on a wheelbase of 2265 mm and with a front and rear track of 1405/1403 mm respectively, it was 4031 mm long, 1780 mm wide and 1190 mm high. The body was in glass fibre and was of harmonious and smooth line and the car was generally well built, very much in line with Corso Marche tradition.

An unusual feature of the V8 was its water radiator, placed up high on two tubular supports behind the cockpit; it was cooled by airflow while racing and also acted as an aerodynamic spoiler. Peter Schetty, to whom the three-litre had been assigned for the Aspern race, confirmed the car was not quite ready as far as grip was concerned as it was much inclined to spin when accelerating out of a corner. The Swiss driver, who was accustomed to prototypes with inclined engines, was asked to drive a mid-engined car for the first time, which was evident during qualifying. Only his skill was able to dominate

FIAT ABARTH "3000 SPORT SPIDER" (1968)

ENGINE	
	mid-rear, longitudinal, 4 cylinders in line
Bore and stroke	mm 88x81
Cubic capacity	cc 2968
Valve gear	4 overhead camshafts, chain driven
Number of valves	2 per cylinder
Compression ratio	12:1
Fuel feed	4 twin-choke Weber 48IDA carburettors
Ignition	double, coil, distributor
Coolant	liquid
Lubrication	dry sump
Max. power (project)	350 hp at 8200 rpm
TRANSMISSION	
	rear wheel drive
Clutch	single dry plate
Gearbox	5 speed plus reverse, central lever
Axle ratio	3.5:1
BODY	
	racing roadster
CHASSIS	
	tubular chassis
Suspension	front independent, trailing arms, helicoidal springs, stabiliser bar, hydraulic dampers; rear independent, trailing arms, helicoidal springs, stabiliser bar, hydraulic dampers
Brakes	hydraulically operated discs front and rear
Steering	rack and pinion
Fuel tank	capacity 120 litres
Tyres	front 4.50/11.60x13; rear 5.50/13.60x13
DIMENSIONS AND WEIGHT	
Wheelbase	mm 2265
Front track	mm 1405
Rear track	mm 1435
Length	mm 4031
Width	mm 1780
Height	mm 1190
Kerb weight	kg 725
PERFORMANCE	
Top speed	km/h 300

the V8's precarious instability, to which the unusual and heavy radiator in mid-air above the carburettor trumpets also contributed. Regardless, Schetty was able to take pole position and the following day dominated the race for all of its 25 laps, which he covered in 25' 54.38". Second was Arturo Merzario in a Group 6 Fiat Abarth Sport Spider 2000, while success in the Group 5 Touring Category went to Toine Hezemans driving the radial head 1000 TCR.

Abarth & C. closed its 1968 motor racing season with 875 victories achieved throughout the world in the various categories and classes. The Scorpion won the National Touring Car Trophy's up to 600 cc class with

Mario Colucci produced the first Fiat Abarth 3000 Sport Spider with its V8 engine and tubular chassis SE04/L, it became clear that the excessive weight of the engine group overhanging the rear axle would make it necessary to ballast the front end to make the car manageable. That led to place the power unit mid-ships (internal code SE013). The motor was a 2968 cc V8 (type 245) with four Weber carburettors and a compression ratio of 12.0:1, all of which developed 350 hp at 8200 rpm to produce a specific power output of 116 hp per litre. A curious feature of the car was the position of the water radiator that was located externally, high up on two supports above the carburettor trumpets.

The Fiat Abarth 3000 Sport Spider Group 7 car made its racing debut on 6 October 1968 in the Donau Pokal, which took place on the 2.600 kilometre circuit at Aspern airport, near Vienna. Driven by Peter Schetty, the car was not sufficiently well prepared and it was only the skill of the Swiss that enabled him to take pole position. The next day, Schetty won the three-litre class of the Sports Prototype category with the car, followed by Arturo Merzario in second driving a Fiat Abarth 2000 Sport Spider. That day, the Scorpion also won the up to 1000 class of the Touring category (Toine Hezemans, 1000 Berlina) and the Sport (W. Donà, 1000 SP).
Taken just before the start, the picture above shows Schetty's Fiat Abarth 3000 Sport Spider number 3 in the foreground, behind which is Merzario's 2000 Sport Spider number 70.
Below: Carlo Abarth leans on the car to better hear Schetty's comments.

Piergiorgio Poggi in a Fiat Abarth 595 prepared by his father, the 850 cc with GiBi-Giovanbattista Guarnieri in a Ferraris-prepared Fiat Abarth 850 TC and the 1000 cc with Maurizio Campanini at the wheel of a Fiat Abarth 1000 Berlina prepared by Baistrocchi. The company also won the Italian Sports Car Championship in the classes of up to 1000 cc with Salvatore Calascibetta driving a Fiat Abarth 1000 SP and in the up to 1300 cc Romano Ramoino won the C.S.A.I. Cup in a Fiat Abarth OT 1300.

In early December it was announced that Renzo Avidano, who had served for 16 years as the Scorpion's sports director, had been appointed managing director of the company. Peter Schetty became Abarth's new sports director, continuing as a works driver.

In spite of his considerable personal commitment to prototype racing, Carlo Abarth never lost sight of the transformation of production Fiats and special cars for sports enthusiasts. In fact, 1968 signalled the start of research into a new radial cylinder head for the Group 2 695 cc car that Luciano Fochi had designed, taking the lead from the 982 cc four-cylinder's homologated head of two years earlier. In its new configuration, the valves were no longer parallel but inclined in V position. The Solex 34 PBIC carburettor was replaced by a horizontal Weber 45 DCOE, which provided a significant increase in power and engine torque, but consumed much more fuel. Because it revealed itself to be expensive and difficult to tune, the engine was later dropped.

On the road car side, Abarth offered a coupé of generous mechanics and distinguished by a custom-built body of original and sporting line; it was

The SE013 mid-engined Fiat Abarth 3000 Sport Spider repeated its Aspern success on the airport circuit at the Preis von Tirol Innsbruck on 20 October.
Other Scorpion drivers also proved themselves winners again, with victory in the Touring category (Johannes Ortner, 1000 Berlina), the Sports Prototype with Toine Hezemans (above) winning the up to 2000 cc class in a 2000 Sports Spider and Peter Schetty taking up to 3000 cc class honours in the brand new Group 7 3000 Sport Spider.

The big, heavy water radiator anchored externally above the V8 engine did not perform aerodynamically as expected.
In fact, it contributed to the precarious stability of the car in corners when accelerating and decelerating and showed a distinct liking for spinning.

inspired by the extremely attractive coupé created by the Francis Lombardi Carrozzeria of Vercelli on the mechanics of a Fiat 850 coupé, ambitiously named Grand Prix. Abarth and Colucci gave it Fiat Abarth OT 1300/124 component groups, made a number of other modifications to the body and re-baptised the car the Scorpion. A saloon of notable impact emerged from the exercise, built on a 2045 mm wheelbase and with a large front and rear track of 1255 mm and tyre/wheel dimensions of 155 SR 13; the Scorpion was 3610 mm long, 1495 mm wide and 1100 mm high, made visually sleeker by the large overhang of the front end and essential roof with a lot of glass. The engine was derived from that of the Fiat 124, with its capacity increased to 1280 cc (75 hp at 6000 rpm); the gearbox was four-speed and developed from the Fiat 850's. With a chassis built entirely of steel, the Scorpion weighed 670 kg and accelerated to a maximum speed of 185 km/h. The suspension and brakes were also modified to bring them up to the car's performance level.

Born a year earlier in the bodyshell of the Fiat 850 Coupé, the OT1300/124 was powered by an engine derived from that of the 124 saloon but was given a tepid reception because of the body, which was unified with that of the original Fiat and lacked individuality.
So Corso Marche fitted the car's mechanics to a coupé of innovative line, produced by body builder Francis Lombardi on the 850 floor pan.
From that marriage came a new kind of Scorpion, small and agile, with the brio of a real sports car.
The capacity of the original Fiat engine was increased to 1280 cc and developed 75 hp at 6000 rpm, pushing the 670 kg Scorpion to a maximum speed of 180 km/h.
Abarth technicians made significant modifications to the SS version's braking system and suspension.
The limited dimensions of the body can be clearly seen in the picture above. Carlo Abarth towers over the little coupé.
The car created much interest at the 1968 Turin Motor Show, where it made its official debut at Lit 1.510.000 and Lit 1.750.000 dependent on the version selected.

A more rigidly suspended 'S' version of the little coupé was also produced, as was an SS, in which Elektron wheels already fitted to the 1000 Berlina Corsa took the place of steel rims. The performance of the Scorpion SS engine was further increased by the adoption of two Weber 32 DCOF carburettors and the replacement of the chain-driven distribution with a cascade gear system. The car enjoyed a gratifying and immediate sales success, but its potential in the American market came up against new laws in the State of California governing safety and exhaust emission, as did all small production cars.

At the end of December, Carlo Abarth went to Spain to evaluate with

FIAT ABARTH "SCORPIONE SS" (1968)

ENGINE	rear, longitudinal, 4 cylinders in line
Bore and stroke	mm 75.5x71
Cubic capacity	cc 1280
Valve gear	overhead valves, side camshaft
Number of valves	2 per cylinder
Compression ratio	10.5:1
Fuel feed	1 twin-choke Solex C32PHH carburettor
Ignition	coil, distributor
Coolant	liquid
Lubrication	forced, oil filter on the main circuit
Max. power (project	75 hp at 6000 rpm
TRANSMISSION	rear wheel drive
Clutch	single dry plate
Gearbox	4 speeds plus reverse, central lever
Axle ratio	3.9:1
BODY	2 seat coupé by Francis Lombardi
CHASSIS	integral body
Suspension	front independent, double trailing arms, helicoidal springs, stabiliser bar, hydraulic dampers; rear independent, trailing arms, helicoidal springs, stabiliser bar, hydraulic dampers
Brakes	hydraulically operated discs front and rear
Steering	rack and pinion
Fuel tank	capacity 30 litres
Tyres	155x13
DIMENSIONS AND WEIGHT	
Wheelbase	mm 2045
Front track	mm 1255
Rear track	mm 1255
Length	mm 3610
Width	mm 1495
Height	mm 1100
Kerb weight	kg 670
PERFORMANCE	
Top speed	km/h 180

Jaime Juncosa, importer of his exhaust systems, the possibility of opening a factory to produce complete exhaust systems in the country. Accompanied by Mrs. Annelise, Carlo, who had just celebrated his 60th birthday, seemed animated and made his usual volcanic approach to the new opportunities that were taking shape in both the industrial and sports fields. On 29 December, El Mundo Deportivo published a long interview with Abarth in which he vindicated once again his qualifications, not as a modifier of cars but as a constructor who, after much research and testing, was able to extract from a mass-produced Fiat 600 a small car of extremely high performance that exceeded 200 km/h on the track: *Achieving these results is not the work of a simple craftsman. In my factory in Turin"* he added, *"I have 65 people whose job it is to do nothing but design and test. On the other hand, we ourselves have produced entirely new cars, the results of which have been equally positive. We are one of a small number of manufacturers able to survive while operating almost exclusively in the difficult field of racing cars"*. And he concluded with a witty remark that was openly controversial and critical of the restrictions of the period imposed by Spain on the import of foreign cars: *"What do the drivers of this country do if they cannot buy themselves cars for practice and racing?"*

In Italy, the events of 1968 caused a profound change in the socio-political panorama of the country. Following the example of the rebellion of contestation at the University of California at Berkeley and interpreting with intransigence the teachings of the philosopher Herbert Marcuse aimed at violently rejecting the development of civilisation, the student movement became an aggressive promoter of unrest that spread across most of Italy. Marxist ideology soon gave way to extremism, the use of iron bars and Molotov cocktails

Carlo Abarth and Mrs Annelise, having just arrived at Barcelona airport on 29 December 1968. The volcanic Viennese technician was considering manufacturing his exhaust systems in Spain and wanted to evaluate various possibilities in Catalonia for the purpose. But limitations affecting car distribution and motor sport promotion in force at the time did not favour such an operation.

that took new impetus from the May uprising in France and the anarchistic activities of the German, Daniel Cohn Bendit. With the motto 'it is prohibited to prohibit', a kind of populist revolt gained ground in Italy, much more intolerant in that it originated from the double student-worker component that found nothing better than to compress the two currents of their remonstrations to legitimise the 'proletariat protest' damage exerted on shops in city and town centres; from this to the next stage of terrorism was but a short step. Waving Mao's little red book as a symbol of a new society without rules, the improvised but fearsome revolutionaries of the Student Movement opened the bidding with the old economic demands of the workers' world, immediately finding the unions ready to engage in the battle of the 'hot autumn". The ideological matrix was not, however, completely convincing or deep-rooted. It was fulfilled and contradicted at the same time, protesting against international events of importance such as the controversial and unpopular war of the U.S.A. in Vietnam and, at the other end of the scale, the 'Prague Spring' that Russian tanks intervened to smash Alexander Dubcek's and Czechoslovakia's dreams of freedom. In this tense climate packed with uncertainty, attempts were made to also subvert the fabric of relations between the industrial world and its employees, aided and abetted by the unions, who did not seem to believe they could ride the wave of the collective revolt of everyone against everyone to radicalise their demands in the factories.

Vittorio Valletta, Fiat's illustrious strategist, died a year earlier in August 1967. He was the man who had given historic impulse to the mass production of cars for the people; often in the face of opposition, he turned himself into the architect of that extensive process of migration from the south of Italy to the northern industrial regions that indelibly marked one of the most significant of Italian socio-economic periods. With his death came the acceleration of the *"...great changes in the organisation of Fiat that would, without serious upheaval, lead to the radical decentralisation of responsibility..."* (Dante Giacosa, 'My 40 Years of Planning at Fiat). The gigantic Turin company was assuming the dimensions and characteristics of a great integrated group, which had already been hinted at a year earlier with the incorporation of Autobianchi of Desio, near Milan and OM of Brescia into the group. A year later in 1969, Fiat also entered the share capital of Ferrari (the laconic press release of 21 June recited that *"in the pre-eminent intention of assuring Ferrari Automobiles continuity and development, it has been decided that the technical collaboration relationship which exists with Fiat be transformed by the end of the year into joint participation"*). At Gianni Agnelli's press conference at the Turin Motor Show, which opened on 29 October, it was learnt that the great Turin group had also acquired Lancia for the symbolic price of one lira per share, as the company was drowning in debt and near collapse.

At the end of 1968 Carlo Abarth turned his attention to the little two-cylinder Fiat 500, which he first converted for the better 10 years earlier and which blotted out the irritations many found with the Turin manufacturer's utility. Luciano Fochi started all over again with the 595 engine, which he gave a radial cylinder head, a system well tried and tested with the 982 cc four; the original carburettor was replaced by a twin-choke horizontal Weber 45DCOE unit to give the motor more maximum power and torque. That also meant the adoption of a new light alloy inlet manifold, a different configuration of gear distribution, the elimination of the thermostat and the modification of the air-cooling conveyor. The development and tuning of the new head became more labour intensive than expected and the engine elaborated in that way was not homologated by FIA until 1 July 1971.

In the space of two years, the entire industrial panorama of the Italian automobile industry had been profoundly changed, almost traumatically so, and it is not difficult to understand that, finding himself suddenly in a position of splendid isolation in respect of the great Fiat – in which new men were steadily climbing up towards top management jobs, men with different management criteria to those of their predecessors – Carlo Abarth asked himself how much longer he could continue in business without new restrictions being placed on his entrepreneurial activities. To that must be added the fact that the Italian car market was going through a moment of rapid transformation, characterised by the offer of increasingly better and more diversified Italian and foreign production cars. That contributed to forcing the small manufacturers into a corner and to progressively restricting the activity and existence of the body designers and builders. The production of special cars became less appetising, even in the eyes of the public, not just because of their price – hand-built car production became almost prohibitively expensive and presented undeniable difficulty as far as the absence of spare parts was concerned – but also because of a new and widespread criterion, according to which the car had, to a large extent, lost its elements of ostentation and privilege to the more practical aspects of rationality and functionality. The political climate of the moment and the consequent orientation of the consumer towards more thoughtful choices, acted as an undeniable factor in this change of convention.

For Carlo Abarth, 1969 began on a sour note. Peter Schetty, the Basle driver who seemed to have agreed the terms of a new contract with the Scorpion to work side-by-side with Abarth as his number two in sports management, announced at the end of the winter holidays on 15 January that he had signed a contract with Ferrari for the '69 season. Taken by surprise and extremely annoyed, Carlo recounted what had happened in the Auto-Italiana of 30 January: *"Ferrari had one possibility to try and win against our cars at hillclimbs – our cars dominated almost all of last year's events in Italy and many outside the country – and that was to take our best driver. We employed Schetty when he was starting out and it was with us that he had the chance to become an optimum driver: he had the makings and from our side we left no stone unturned so that he could show what he could do. Now Ferrari finds itself with a good driver brought on at our expense, who will be very useful to them and at the same time create inside our company the problem of substituting him in mid-January, when the best drivers are already engaged..."*

The motor racing season looked like it was going to be full of contradiction and controversy, triggered off by restrictions imposed by the Ministry of Public Works concerning the conduct of speed racing at hillclimbs and on the roads; they were the consequence of frequent accidents that devastated some of those competitions in earlier seasons. A special technical commission approved

FIAT ABARTH "2000 PININFARINA" (1969)

ENGINE	
	rear longitudinal, 4 cylinders in line
Bore and stroke	mm 88x80
Cubic capacity	cc 1946
Valve gear	2 overhead camshafts, chain driven
Number of valves	4 per cylinder
Compression ratio	10.5:1
Fuel feed	2 twin-choke Weber 45DCOE9 carburettors
Ignition	double, coil, distributor
Coolant	liquid
Lubrication	dry sump
Max. power (project)	220 hp at 7600 rpm
TRANSMISSION	
	rear wheel drive
Clutch	single dry plate
Gearbox	5 speeds plus reverse, central lever
Axle ratio	4:1
BODY	
	2 seat coupé by Pininfarina
CHASIS	
	tubular trellis
Suspension	front independent, trailing arms, helicoidal springs, stabiliser bar, hydraulic dampers; rear independent, trailing arms, helicoidal springs, stabiliser bar, hydraulic dampers
Brakes	hydraulically operated discs front and rear
Steering	rack and pinion
Fuel tank	capacity 60 litres
Tyres	front 4.50/8.00x13; rear 5.50/10.00x13
DIMENSIONS AND WEIGHT	
Wheelbase	mm 2085
Front track	mm 1405
Rear track	mm 1435
Length	mm 3780
Width	mm 1780
Height	mm 970
Kerb weight	kg 740
PERFORMANCE	
Top speed	km/h 270

only 65 of the programmed races, leaving the others 'sub judice' while additional checks were carried out on the suitability of the routes proposed.

On its stand at the Brussels Motor Show in January 1969, Pininfarina unveiled an original stylistic interpretation of a formula prototype racing coupé on the chassis of a Fiat Abarth 2000 Sport Spider. Features introduced as new developments included the inclination of the bonnet without continuity problems with the windscreen, and a two-seater cockpit surmounted by a prismatic-shaped roof that tipped forward completely to facilitate occupant access; the roof's rear pillars hid from view large engine cooling air intakes; the engine itself was in its original position, inclined over the rear end; an enormous rear window and a fine, oblique-cut grill comprised the rear end of the car, the vertical mirror of which was out in the open, leaving the engine and exhaust system in view. The mechanics were partly modified to road rather than racing performance, with a compression ratio reduced from 11.5:1 to 10.5:1 and two twin-choke Weber 45 DCO3 carburettors instead of the 58 DCO9s so the car generated 220 hp at 7600 rpm, instead of the original 250 hp at 8700 rpm. The new Pininfarina was greeted with public and critical acclaim, but remained a one-off.

Pininfarina was the man behind a splendid styling exercise using the chassis of the Fiat Abarth 2000 Sport Spider, which was unveiled in January 1969 at the Brussels Motor Show.
Some formal elements of great originality included the extremely short bonnet that flowed with a continuing line to an ample windscreen and the central section of the body, which held the lateral cooling radiators for an engine that overhung the rear axle.
The rear end had no tail mirror so that the mechanics remained in view and allowed hot air to exit the engine compartment; an aesthetic yet functional fine-mesh grill ran along the two sides, immediately behind the cockpit.
A unique example of high-level aesthetics and construction.

The large transparent dome that enclosed the cockpit was hinged at the front and could be lifted forward for easy access. About a third of the car's body was covered in glass, which lightened its overall weight in spite of its metal panels being in aluminium.
Conceived exclusively for racing, chassis SE010 was adapted for possible road use and the exuberant engine was detuned; the compression ratio was wound down from 11.5:1 to 10.5:1 and the two Weber twin-choke 58DCO9s were replaced by a pair of 45DCOE3 carburettors, reducing maximum power from 250 to 220 hp.
Only one example of the Pininfarina Fiat 2000 Coupé was built; it was a car that embodied a number of new styling parameters in the conception of a modern high performance two-seater coupé.

The first important race of the season took place on 16 March at Vallelunga and was the first Dalla Vecchia Trophy, in which over 200 competitors competed passionately in 'family' events. In the Prepared Touring Category, Raimondo De Sanctis won the up to 600 cc class driving a Fiat Abarth 595 and outclassing Renato Mazzeschi, while Angelo Mola, Del Panta and Barberini in Fiat Abarth 695s roundly defeated the BMW 700s to win the up to 700 cc class. There was an in-house battle between Abarths in the up to 850 cc class, which was won by Piero Falorni in an 850 TC. In the up to 1000 cc class of the Grand Touring Category, the Fiat Abarth OTS of Pierpaolo Brambilla came first, with Carmignani second. In the up to 1300 cc Sport class Alessandro Fabroni (Fiat Abarth OT 1300) won, while in the Sport Prototype 1000 cc class Stefano Sebastiani drove the Fiat Abarth SP 1000 to victory and Gianpiero Biscaldi took the up to 2000 cc class in a Fiat Abarth 2000.

There were few private competitors at the 4 Hours of the Jolly Club at Monza a week later, but there was a powerful presence of manufacturers'

FIAT ABARTH "2000 SPORT PROTOTIPO CUNEO" (1969)

ENGINE	
	rear, longitudinal, 4 cylinders in line
Bore and stroke	mm 88x80
Cubic capacity	cc 1946
Valve gear	2 overhead camshafts, chain driven
Number of valves	4 per cylinder
Compression ratio	11.5:1
Fuel feed	2 twin choke Weber 58DCO3 carburettors
Ignition	double, coil, distributor
Coolant	liquid
Lubrication	dry sump
Max. power (project)	250 hp at 8700 rpm
TRANSMISSION	
	rear wheel drive
Clutch	single dry plate
Gearbox	5 speed plus reverse, central lever
Axle ratio	3.5:1
BODY	
	racing roadster
CHASSIS	
	tubular trellis
Suspension	front independent, trailing arms, helicoidal springs, stabiliser bar, hydraulic dampers; rear independent, trailing arms, helicoidal springs, stabiliser bar, hydraulic dampers
Brakes	hydraulically operated discs front and rear
Steering	rack and pinion
Fuel tank	capacity 70 litres
Tyres	front 4.75/10.00x13; rear 6.00/12.00x13
DIMENSIONS AND WEIGHT	
Wheelbase	mm 2085
Front track	mm 1405
Rear track	mm 1435
Length	mm 3680
Width	mm 1780
Height	mm 900
Kerb weight	kg 560
PERFORMANCE	
Top speed	km/h 280

teams. Adriano Ceci of Auto-Italiana wrote in the 3 April edition: "*We believe this was because of the high cost of Group 5 cars against that of Group 2 and 1 machines, which have no chance of success or even a good placing. It is thought a Fiat Abarth 850 TC costs a little more than five million lire and a 1000 TC not much more than six million. With maintenance and the cost of transfer, these figures are sky high*". At the Monza race, a round in the European Touring Car Championship, Dutchman Edouard Swart won Division 1 driving a Group 5 Fiat Abarth 1000 Berlina, followed by Johannes Ortner in a similar car.

Abarth decided to compete in the first important hillclimb – the Ampus Pass in France on 30 March – with the new Fiat Abarth 2000 Sport Prototipo Cuneo (internal code SE 014), the mechanics of which were largely derived

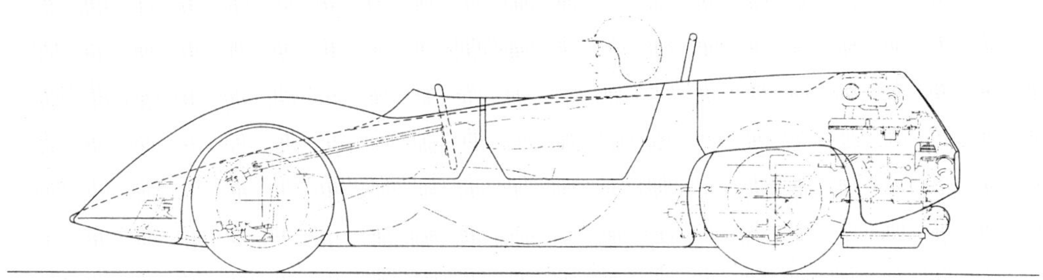

The Fiat Abarth 2000 Sport Prototipo Cuneo (code SE014) arrived almost unexpectedly on the 1969 sporting scene.
Mario Colucci designed the car on the basis of the 2000 Sport Spider SE010's tubular trellis chassis, but developed a more compact car – even if on the same 2085 mm wheelbase – and a body with an accentuated wedge shape.
The well-established 1946 cc four-cylinder, 16-valve engine that produced 250 hp at 8700 rpm was mounted outboard, inclined behind the rear axle.
The body was built in aluminium by Giuseppe Manera (picture), the veteran maestro panel beater who had worked with Carlo Abarth since the foundation of the Corso Marche company; the weight of the car was 560 kg and promised well by keeping the power to weight ratio down at 2.24 kg per horsepower. Maximum speed: 280 km/h.

Edoardo Lualdi Gaboardi took the Fiat Abarth 2000 Sport Prototipo Cuneo on its motor sport debut in the 5th Course de Côte d'Ampus in France on 30 March 1969 on a difficult and insidious 6.900 kilometre course. He had only been able to try the car a few days earlier on the Campo Volo track opposite the Corso Marche factory and raced the car unpainted.

Lualdi on one of the many difficult sections of the d'Ampus climb. He bought the first example of the Fiat Abarth 2000 Prototipo Cuneo, which was specifically designed to take on the most tortuous routes of hillclimbs; during practice he had asked the works mechanics to meticulously adjust the driving seat. He rocketed off the start line and drove with the skill of a veteran, but 600-700 metres from the finish a tyre lost its pressure and Lualdi ended up in sixth place with a time of 4'21.6".

from the 2000 Sport Spider of the previous year, but which was 'dressed' in an a completely new and compact aluminium body – even if built on an unchanged wheelbase of 2085 mm – and was aerodynamically better. The front suspension was updated, while the engine that inclined over the rear end was unchanged from its last 16-valve configuration. Assigned to Edoardo Lualdi – driving a car with an outboard engine for the first time – its baptism was a modest one for it came sixth overall behind the victorious Ferrari 212 E of Peter Schetty, who made a brilliant start to his first season at Maranello, Johannes Ortner's three-litre Abarth, the two two-litre Abarths of Biscaldi and Finkel and Rouveyran's Tecno F2.

At the first round in the European Hillclimb Championship at Montseny, Spain, on 25 May over a 16.3 kilometre route and with an incline of 884 metres, Ortner's Fiat Abarth 2000 Cuneo was unable to win and was beaten by over 28'' by Peter Schetty. The same day at the Trieste-Opicina Franco Pilone, driving only his third race in a Fiat Abarth 2000, beat the Porsche Carrera 10 of 'Noris' (Giacomo Moioli), while in the Sport Prototype up to 1000 cc class success was the prerogative of the Fiat Abarth 1000 SP driven by Maurizio

Zanetti. In the Grand Touring Category up to 1000 cc, Eugenio Rossi won in his Fiat Abarth 1000 OTS, while in the Touring Category up to 600 cc class the win went to Carlo Bettazzi in a Fiat Abarth 595, to GiBi with the Fiat Abarth 850 TC in the up to 850 cc and Achille Soria at the wheel of a Fiat Abarth 1000 Berlina in the up to one litre class.

The third round in the European Hillclimb Championship took place on Mont Ventoux in France on 22 June over 21.6 km and was won by Schetty driving a Ferrari 212 E, while in practice Arturo Merzario and Johannes Ortner alternated fastest times in works Fiat Abarth 2000 Sport Spiders. Schetty won again at the Trento-Bondone, fourth round in the championship, in which Merzario could only come second. By this time it was clear to everyone that the one to beat in the Sport Prototype Category was the 12 cylinder car from Maranello; that day, Merzario won the up to 2000 cc Sport Category, while in the Touring class of up to 850 cc and 1000 cc there was the ritual Fiat Abarth family fight from which Claudio Margoni (850 TC) and Paolo Lado (1000 Berlina) emerged victorious. The up to 1000 cc Grand Touring class went to Camillo Spica at the wheel of a Fiat Abarth 1000 OTS.

Merzario's sequence of victories in the Fiat Abarth in the two-litre Sport continued on 27 July at the fifth European Hillclimb Championship counter, the Freiburg-Schauinsland and then the Cesana-Sestrière on 3 August, when he was followed by Ortner, Pilone and Taramazzo, all driving two-litre Fiat Abarths. But Merzario also surprised everyone at the Grand Prix of Mugello on 20 July, which took place over eight laps of a 66 km road circuit. Once he had taken the lead in the works Fiat Abarth 2000 he stayed there, much against predictions that had the Alfa Romeo 33 or Porsche 908 winning. The

The first confrontation took place at Ampus between the new Ferrari 212E driven by Peter Schetty, who had recently joined Maranello, and the Abarth squad competing in the French event with a number of 2000 Sport Spiders and a 3000 Sport Spider, the latter in the hands of Johannes Ortner. Schetty, who had put up a modest time in morning practice, surprised everyone during the race by completing the climb in 3'56.7" at an average speed of 103.442 km/h, exactly 4" less than the record he had established the year before with a Fiat Abarth 2000 Sport Spider. Second came Ortner with the Abarth three-litre SE010 and third Giampiero Biscaldi in a Group 4 2000 Sport Spider (picture). It was the car's last Prototype race, as it was granted a homologation in the Sport category two days later, on 1 April 1969.

The 1000 Kilometres of Monza, fourth round in the 1969 International Constructors' Championship, took place on 25 April over the full 10,000 kilometre circuit, made up of the normal and banked tracks. Among the 52 starters were five Fiat Abarth 1000 SPs, three of which would finish and take first and second places in the Sport category (car number 58 driven by Luciano Pasotto-Umberto Grano and 60 in the hands of Eris Tondelli-Marco Crosina) and victory in the Group 6 Prototype by the 30 car of Ugo Locatelli-Maurizio Zanetti (pictured here).

same happened again in Austria at the Salzburg-Gaisberg hillclimb on 7 September, when Merzario won the up to 2000 cc class of the Sport Category; that day, though, the Fiat Abarth 850 TC's grip on the up to 850 cc class was loosened when the race was won by Willi Sommer in a Steyr-Puch; but the Scorpion did win the up to 1150 cc class with Karl Wendlinger in a Fiat Abarth 1000 Berlina. In that class too, though, the NSUs and Mini Coopers were a threat and eventually managed to finish high up in the final results table.

At the 500 Kilometres of Imola on 14 September, 100 laps of the circuit and a round in the Italian Championships, the up to 2000 cc Sport Category victory went to Toine Hezemans in a Fiat Abarth 2000 Sport Spider, a car that confirmed itself reliable and perfectly set up as far as its mechanics and aero-

On 27 April 1969, the Scorpion competed in the fourth Dobratsch Bergrennen near Villach, Austria, a hillclimb counter for the Touring, Grand Touring, Sport, Sports Prototype and Corsa categories over 15.800 kilometres, its cars including Fiat Abarth 2000 Sport Spider number 54 for Giampiero Biscaldi (below, left) and the mid-engined 3000 Sport Spider number 61 for Johannes Ortner.
The three-litre car carried large aerodynamic devices at the front, but no longer sported the curious external water radiator above the engine first seen in competition the previous year.

The Fiat Abarth 2000 Sport Prototype Cuneo was at the Montseny, Spain, first race in the European Hillclimb Championship on 25 May 1969, to give some opposition to the Ferrari 212E driven by Peter Schetty. Johannes Ortner in the two-litre Scorpion could do no better than second (picture) and a yawning 28" gap between him and Schetty, who covered the 16.300 kilometre, 884 metre climb in 9'12.46" at an average speed of 106.215 km/h. Fourth and fifth came the two Fiat Abarth 2000 Sport Spiders driven by Giampiero Biscaldi and Luigi Taramazzo respectively.

dynamics were concerned; it was a car that enabled the Turin factory to do well in long distance races but was also a feared opponent in hillclimbs.

In early September Carlo Abarth sent for testing an updated version of the three-litre prototype, introduced the previous year. No major changes had been made to the V8 engine, meaning no attempt was made to use the four valves per cylinder head or Kugelfischer fuel injection system in place of its four Weber 48 IDA carburettors. But the whole body was redesigned and based on a wheel-base that was 80 mm shorter, had large front aerodynamic appendices and a variable inclination transverse rear spoiler. The car was not often seen in races in 1969, in part due to the wave of strikes that hit Italy in the autumn causing major delays to the supply of components and the definitive set-up of the car.

The 1969 season was winding down on 12 October when the 14th Carri Cup, a counter in Group 2 of the National Touring Trophy, was held at Monza where a Fiat Abarth duel in the 1000 cc class saw Paolo Lado win in his 1000

Berlina. By the end of the racing year, the Scorpion had won the European Touring Car Challenge, Division 1 ('Pam'-Pasotti Marsilio in a Fiat Abarth 1000 Berlina) and the Sport Category of the European Hillclimb Championship (Arturo Merzario in a Group 4 Fiat Abarth 2000 Spider). The team also won the up to 600 cc class of the National Touring Trophy (Piergiorgio Poggi, Fiat Abarth 595), the up to 700 cc (Liberatore Cestra, Fiat Abarth 695 prepared by Ferraris) and the up to 850 cc (GiBi-Giovanbattista Guarnieri, Ferraris-prepared 850 TC); the up to 1000 cc was not awarded that year due to a lack of valid results. The Turin factory also won the up to 1000 cc class of the Italian Sports Car Championship (Enzo Buzzetti, 1000 SP) and Paolo Gargano took the National Hillclimb Trophy in a Fiat Abarth 1000 Bialbero. Worldwide, Abarth won about 800 different events during the season.

At the Turin Motor Show from 29 October-9 November, the Abarth stand displayed a coupé built by Italdesign, founded a year earlier by Giorgetto Giugiaro, who had previously made a name for himself at Bertone and Ghia. He designed a sober body for the rear engined Fiat Abarth 1600 chassis that was, perhaps, a little loaded at the back end where the insertion of the air intakes in the third pillar of the roof seemed like slightly forced styling. Introduced as a prototype suitable for a small production run, the Fiat Abarth 1600 GT (internal code 104 G) was powered by a four-cylinder 1591 cc twin-cam engine that put out a maximum 145 hp; it had a five-speed gearbox, independent suspension on all four wheels and front and rear disc brakes. Built on a wheelbase of 2045 mm, the car was 3800 mm long, 1565 mm wide, 1100 mm high, weighed 674 kg and was accredited with a 240 km/h top speed. But the turbulent climate in which the show took place, protected by a massive deployment of police to safeguard visitors and prevent sabotage, was not conducive to easy decisions on the future of the 1600 GT, and the interesting Italdesign body remained just that.

Contrary to all the predictions that confidently prophesied a battle for victory in the Grand Prix of Mugello on 20 July 1969 between the Alfa Romeo 33 and the Porsche 908, a two-litre Abarth won an extremely tough race that took place in temperatures of over 34°C.
An interminable sequence of corners, climbs and descents that called for brusque acceleration and braking over each 66.200 kilometre lap for eight laps severely stressed the cars' mechanics and the drivers' physical resistance.
Abarth, who had entered a 2000 Sport Prototype Cuneo with an updated chassis driven by Arturo Merzario for a definitive shakedown, and a 2000 Sport Spider for Gijs Van Lennep and Johannes Ortner, found themselves unexpectedly at the top of the classification, having taken the lead from the early stages. Merzario (left) won with car number 14, while the 81 car took second overall and won the Sport category.

On 2 September 1969, Abarth introduced the new prototype three-litre SE016, an evolution of SE013, which made its debut at Aspern the previous October for which lead ballast had to be used in the front to better equilibrate the distribution of weight over the two axles.
The new car was built on a wheelbase shortened from 2265 to 2185 mm and had its driving position on the right; it was given a sleeker, more compact and aerodynamic body, but no change was made to the four valves per cylinder engine, with its indirect injection system in place of the four Weber carburettors.
The planned 15-inch rims could not be used due to union agitation, so 13-inch wheels were on the car again.
Side: Johannes Ortner during practice for the 500 Kilometres of Imola on 14 September.
When the race was stopped on the 72nd lap due to bad weather, the Austrian was lying third, four laps behind the Gulf-Mirage of winner Jackie Ickx.

Abarth confirmed its participation in the Grand Touring category at the 51st Turin Motor Show, which opened its doors on 29 October 1969, by introducing a road car evolved from the OT 1600 with a four-cylinder, twin-cam 1592 cc engine that produced 145 hp at 7200 rpm. Giorgetto Giugiaro was given the task of designing an outboard rear engined 2+2 coupé.

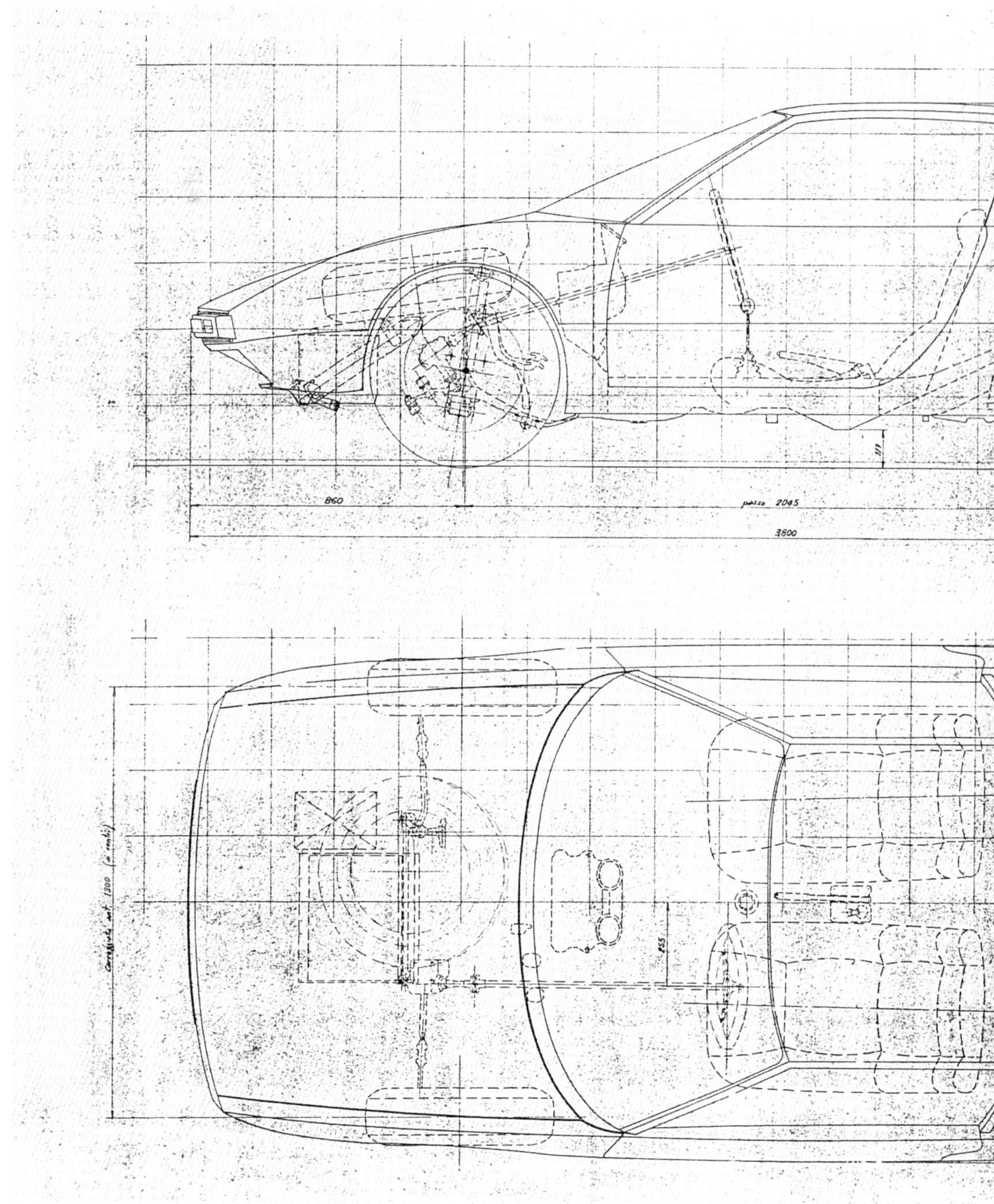

FIAT ABARTH "1600 GT ITALDESIGN" (1969)

ENGINE	
	rear, longitudinal, 4 cylinders in line
Bore and stroke	mm 86x68.5
Cubic capacity	cc 1592
Valve gear	2 overhead camshafts, chain driven
Number of valves	2 per cylinder
Compression ratio	9.8:1
Fuel feed	2 twin-choke Weber 45DCOE9 carburettors
Ignition	double, coil, distributor
Coolant	liquid
Lubrication	dry sump
Max. power (project)	145 hp at 7200 rpm
TRANSMISSION	
	rear wheel drive
Clutch	single dry plate
Gearbox	5 speed plus reverse, central lever
Axle ratio	3.1:1
BODY	
	2 door 2+2 by Italdesign
CHASSIS	
	boxed steel
Suspension	front independent, trailing arms, helicoidal springs, hydraulic dampers rear independent, trailing arms, helicoidal springs, hydraulic dampers
Brakes	hydraulically operated discs front and rear
Steering	worm and sector
Fuel tank	capacity 48 litres
Tyres	4.75/10.00x13
DIMENSIONS AND WEIGHT	
Wheelbase	mm 2045
Front track	mm 1300
Rear track	mm 1270
Length	mm 3800
Width	mm 1565
Height	mm 1100
Kerb weight	kg 674
PERFORMANCE	
Top speed	km/h 240

At a turning point after 20 years in business, Abarth & C. had an enormous patrimony of experience and technical success with which it won five world constructors' championships and almost 5000 race victories in various classes and categories across the world. Many of those successes took the company to victory in the European Challenge for touring cars in the up to 850 and 1000 cc classes in the three years 1965-67 and a long series of victories in the Italian Touring Car and in Grand Touring Championships. That extraordinary cache of results, however, confirmed a serious operational situation in the company, which was involved in non-stop expensive sports activities on

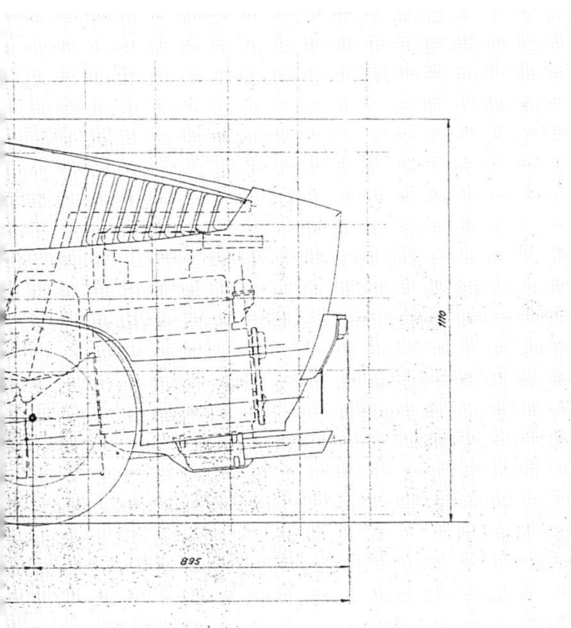

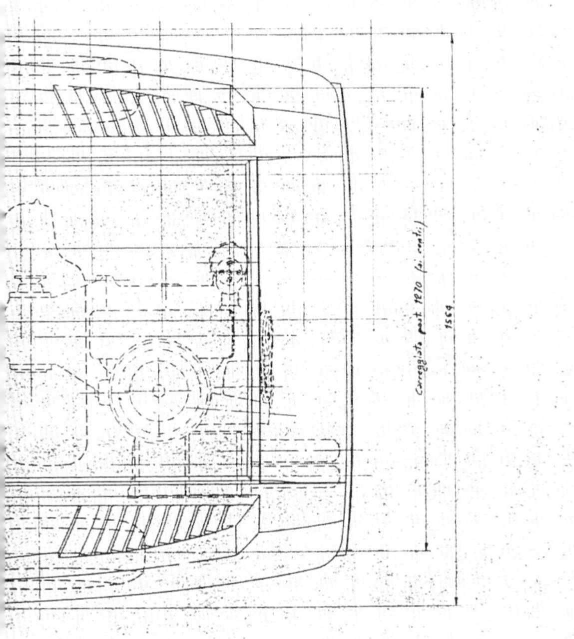

For a small production run, Abarth decided to use a modified Fiat 125 engine.
The body seemed heavy at the rear and that caused perplexity: it was also in open contrast with the harmonious lines of the front of the vehicle.
But it was the uneasy socio-political climate of the moment that stopped further development of the project.

various fronts, aimed at promoting the image of the Scorpion on road and track every Sunday. A commitment of men and equipment that no other manufacturer in the world could have sustained at such a punishing rhythm, year after year. Motor sport regulations had become increasingly restrictive; spiralling operating costs, tougher safety and anti-pollution regulations in the United States – which outlawed the export to that rich market of cars not designed to newly imposed standards – all contributed to the reorganisation of Abarth's industrial and commercial activities, in spite of the fact that the exhaust, technical components and accessories sectors for production cars continued to yield appreciable results. Prospects for the near to medium term did not look good, even if Abarth was able to deliver 315 cars to its Italian customers in 1969, against 228 the previous year.

At the end of the year the 'Merzario Case' exploded. The young Como driver accepted an offer to drive for Ferrari in 1970, even though he was bound to a contract with Abarth until the end of '71. That is how the chain of 'irritation, vexation and spite' (Autosprint, 1-8 December 1969) between Corso Marche and Maranello was rekindled, one that harked back to the Prancing Horse's entry into the Fiat orbit and which had already manifested itself resoundingly in early January with the sudden defection of Peter Schetty.

On 31 December 1969, news broke saying Mario Casoni of Modena, who had already driven for Ferrari and Alfa Romeo, was to join Abarth for the 1970 season, competing in the two and three-litre Prototype races of the International Constructors' Championship, the European Hillclimb Championship and the European Trophy for Prototypes. Casoni became a team-mate of Johannes Ortner, who had already renewed his contract for the following season.

The national automobile industry panorama had been affected by a number of new Fiat developments during the course of 1969: the launch of the 130 saloon at the Geneva Motor Show as the company's top-of-the-range car and, a few weeks later, the presentation of the 128, a people's car that took the place of the legendary evergreen 1100, as well as the launch in later months of the Autobianchi A 111 and A 112. The 130 could not have entered the market at a worse moment. It was to have introduced Fiat to the sophisticated world of superior class cars and, as such, its imposition embodied a number of innovative or at least unusual features, including a V6 twin-cam 2.8 litre engine, automatic gearbox as original equipment, independent suspension on all four wheels, light alloy rims and a wealth of accessories as standard. All of that was contained in a massive body, the stylistic management of which was born of compromise and considerable re-thinking and was inspired much more by American tendencies of the late Fifties than the sober elegance of the Mercedes-Benz 300, against which it was said the new Fiat would compete. The project was *"subjected to an uninterrupted elaboration in a succession of different perspectives, especially as regards the engine"* (wrote Dante Giacosa on page 268 of his previously mentioned autobiography). But the 130 had already been overtaken by time the day it was born for it lacked performance and that level of quality the market considered indispensable for a top-flight saloon. With a fully laden weight of almost two tons, its feeble engine of 140 hp was considered insufficient and was replaced two years later by another V6 with an increased capacity of 3.2 litres and a power output of 165 hp. That last move decreed the definitive failure of the model, the list price of which on

1 March 1970 was Lit 3.300.000 against the Lit 3.450.000 of a Mercedes-Benz 280 S, Lit 3.650.000 for a BMW 2800 and 3.750.000 of a 2.8 litre Jaguar XJ6.

The task of the 128 was a lot easier and the result simpler to predict, both as far as the typology of the car's utility character was concerned – much more congenial to the men of Fiat – and the mountain of experience the company had gained in previous years with the Autobianchi Primula, which was to open the road in 1969 to the 'all up front' solution long opposed by many in Fiat. The 128 could trace its origins back to that car and became the first front-wheel drive Fiat, with an engine layout similar to the transverse engine arrangement introduced by Alec Issigonis' Mini 10 years earlier. The car appeared in company with the homologous Autobianchi A 111 of greater capacity, power, dimensions and price and inherited the long and positive legacy in the marketplace of the 1100. Unveiled at the end of March 1969 and priced at Lit 930.000 for the four-door saloon, it took many Italian motorists from the 'all behind' of the 600 and 850 to the 'all up front', a design formula to which everyone attributed a great future. In Italy, the 128 remained in production in its different forms until 1982 during which period 2.776.000 of them were produced.

The fourth new Fiat Group happening in 1969 was the Autobianchi A 112, a model that achieved an unconditional public success right from the first few days after its launch at the Turin Motor Show; it was derived from the Fiat 128 in its general architecture but the engine – the well-known 903 cc, 44 hp four-cylinder – was inherited from the sportier 850 coupé and roadster. Destined to become an extremely popular model in the Italian park, the A 112 curried favour with a wide and diverse group of motorists in a historically important segment of the market in which practicality and rationality typical of a small city car were highly valued.

1970

1979

DRIVING INTO THE SUNSET

As a result of the new labour contract signed the previous December by Fiat, which gave an equal salary increase to everyone as well as the institution of a form of union control on the rhythm of production, list prices of Italian-made cars increased by an average of 5% from 26 January 1970. At the same time, Abarth introduced a new version of the 1000 Berlina with a radial cylinder head, updated to meet Appendix J. The new regulation stipulated that, for the new season, overall victory in the European Challenge would be decided in favour of the driver who had gained the highest number of points in all three previous divisions (up to 1000 cc, 1600 cc and over 1600 cc); a European Cup would also be put up for constructors.

The new sub-division of categories in line with the latest updates of Appendix J, which took effect on 1 January 1970, were: Group 1 Production Touring Car for homologation in which 500 examples had to be built; Group 2 Special Touring, 1000 cars; Group 3 Production Grand Touring, 5000 cars but 1000 from 1971; Group 4 Special Grand Touring, 500; Group 5 Sport, 25; Group 6 Sport Prototype. The CSI also homologated fuel feeding the Abarth 1000 Berlina by radial head and two twin-choke Weber 40 DCOE2 carburettors, its new-design front and rear wings plus lateral and rear windows in Plexiglas instead of glass. Plexiglas windows were also homologated for the 850 TC, while plastic tail ends were sanctioned for the front and rear wheel housings of the 595. Power output of the Group 2 1000 Berlina Corsa for the

The Fiat Abarth 1000 Berlina prepared by the Scorpion for the 1970 motor racing season complied with the new Appendix J international regulations, which came into effect in January. The new 1000 Berlina Group 2 car was updated with a radial cylinder head engine from the previous Group 5 and accredited with a maximum power output of 115 hp at 8200 rpm and a top speed over 190 km/h.

FIAT ABARTH "1000 BERLINA CORSA" GR. 2 (1970)

ENGINE	
	rear, longitudinal, 4 cylinders in line
Bore and stroke	mm 65x74
Cubic capacity	cc 982
Valve gear	overhead valves, side camshaft
Number of valves	2 per cylinder
Compression ratio	13:1
Fuel feed	2 twin-choke Weber 40DCOE carburettors
Ignition	coil, distributor
Coolant	liquid
Lubrication	forced, oil filter on the main circuit
Max. power (project)	112 hp at 8200 rpm
TRANSMISSION	
	rear wheel drive
Clutch	single dry plate
Gearbox	5 speed and reverse, central lever
Axle ratio	4.3:1
BODY	
	2 door, 4 seat saloon
CHASSIS	
	integral body
Suspension	front independent, double trailing arms, helicoidal springs, hydraulic dampers
	rear independent, trailing arms, helicoidal springs, hydraulic dampers
Brakes	hydraulically controlled discs front and rear
Steering	worm and sector
Fuel tank	capacity 70 litres (on request)
Tyres	4.25/9.50x13
DIMENSIONS AND WEIGHT	
Wheelbase	mm 2000
Front track	mm 1215
Rear track	mm 1250
Length	mm 3530
Width	mm 1420
Height	mm 1300
Kerb weight	kg 583
PERFORMANCE	
Top speed	km/h 190

The little saloons transformed by Abarth into quick competition cars became immensely popular in Germany, due in part to the many successes they achieved in local races.
At the Munster Motor Show (above) in February 1970, the Abarth stand displayed three Scorpions developed from Fiat production cars and one coupé with its 124 engine increased to 1280 cc and bodied by Francis Lombardi.

1970 season went up to 112 hp at 8200 rpm as a result of adopting two twin-choke Weber 45 DCOE carburettors and the new 'cow tail' exhaust system, which exited from the bonnet; the rear hood in resin was more profiled than the metal version derived from the Fiat production car.

At the bi-annual exhibition of racing cars that opened at the Museo dell'Automobile in Turin in March, Abarth showed two new cars. The first was the 1000 Biposto Corsa (internal code SE 018), a roadster derived from the SP barchetta with a tubular chassis but an outboard engine instead of its power plant mid-ships. The engine group was the veteran 982 cc twin-cam four-cylinder with its maximum power increased to 120 hp at 8200 rpm, and the

At the end of February 1970, Abarth unveiled its new car aimed mainly at youngsters moving into four-wheel competition: the Fiat Abarth 1000 Biposto Corsa (SE018) was a light and compact car with a one-litre engine. Unlike the previous 1000 SP (SE04), the engine was inclined over the rear axle, a solution dear to Carlo Abarth.

The 1000 Biposto Corsa came out at the fourth Racing Car Exhibition in Turin, which opened on 28 February 1970 at the Museo dell'Automobile Carlo Biscaretti di Ruffia (above).
In the centre of the Abarth stand, the new car looked lower and more collected than the Autobianchi A112 and the 1000 Berlina next to it.
Those characteristics soon earned the new comer the diminutive name of 'Millino'.

The chassis of the Fiat Abarth 1000 Biposto Corsa was a trellis of steel tubes, its engine the well-known 982 cc four-cylinder twin-cam which, for the occasion, developed 120 hp of maximum power at 8200 rpm.
Compact and smooth, the body was made of glass fibre; the rear end had no vertical mirror, leaving the whole engine group in plain view.
Built on a wheelbase of 1950 mm, the new little Abarth had a dry weight of 400 kg and was considered particularly suited to competing on extremely tortuous tracks.
The Millino's maximum speed was not given officially when the car was unveiled, but it was later confirmed as 220 km/h.

FIAT ABARTH "1000 BIPOSTO CORSA" (1970)

ENGINE	
	rear, longitudinal, 4 cylinders in line
Bore and stroke	mm 65x74
Cubic capacity	cc 982
Valve gear	2 overhead camshafts, chain driven
Number of valves	2 per cylinder
Compression ratio	12:1
Fuel feed	2 twin-choke Weber 40DCOE carburettors
Ignition	coil, distributor
Coolant	liquid
Lubrication	dry sump
Maximum power	120 hp at 8200 rpm
TRANSMISSION	
	rear wheel drive
Clutch	single dry plate
Gearbox	5 speed plus reverse, central lever
Axle ratio	4.4:1
BODY	
	racing roadster
CHASSIS	
	tubular trellis
Suspension	front independent trailing arms, helicoidal springs, stabiliser bar, hydraulic dampers; rear independent, trailing arms, helicoidal springs, stabiliser bar, hydraulic dampers
Brakes	hydraulically operated discs front and rear
Steering	rack and pinion
Fuel tank	capacity 35 litres
Tyres	front 4.25/9.50x13; rear 4.75/11.50x13
DIMENSIONS AND WEIGHT	
Wheelbase	mm 1950
Front track	mm 1280
Rear track	mm 1300
Length	mm 3360
Width	mm 1590
Height	mm 780
Kerb weight	kg 400
PERFORMANCE	
Top speed	km/h 220

The cockpit of the Fiat Abarth Millino was spartan; the instrumentation was clear and essential. The car was made in large demountable sections with the intention of giving maximum access to the mechanics. The engine group of the 1000 Biposto Corsa was inclined over the rear axle and was the tried and tested four-cylinder type 229, initially fed by two horizontal twin-choke Weber 40DCOE carburettors, which were replaced in the last few cars by two Weber 45 mm units. The gearbox was the usual five-speed plus reverse (type 102) that had been previously installed in the 1000 SP SE04. During the first few weeks of 1970, the prototype of the new car went through an intense test session at the Paul Ricard circuit in France.

five-speed gearbox was mounted in front of the power plant. Extremely compact on a 1950 mm wheelbase, the car was 3360 mm long, 1590 mm wide, just 780 mm high and weighed in at 400 kg. Aimed at young drivers who had already gained experience in the Touring category and who wanted to step up to a more demanding car, the new Fiat Abarth became reasonably popular in the sporting community and won numerous races; small and essential, fast – 220 km/h top speed – and agile, it was soon baptised the 'Millino' and went

AUTOBIANCHI "A112 PROTOTIPO" (1970)

ENGINE	
	front, transverse, 4 cylinders in line
Bore and stroke	mm 65x74
Cubic capacity	cc 982
Valve gear	overhead valves, side camshaft
Number of valves	2 per cylinder
Compression ratio	12.5:1
Fuel feed	2 twin-choke Weber 40DCOE carburettors
Ignition	coil, distributor
Coolant	liquid
Lubrication	forced, oil filter on the main circuit
Max. power (project)	108 hp at 8200 rpm
TRANSMISSION	
	rear wheel drive
Clutch	single dry plate
Gearbox	4 speed plus reverse, central lever
Axle ratio	4.1:1
BODY	
	3 door, 4 seat saloon
CHASSIS	
	integral body
Suspension	front independent, lower trailing arms, vertical pillars, helicoidal springs, stabiliser bar, hydraulic dampers; rear independent, lower trailing arms, transverse leaf spring, stabiliser bar, hydraulic dampers
Brakes	hydraulically operated discs front and rear
Steering	rack and pinion
Fuel tank	capacity 30 litres
Tyres	4.50/10.00x13
DIMENSIONS AND WEIGHT	
Wheelbase	mm 2038
Front track	mm 1250
Rear track	mm 1225
Length	mm 3280
Width	mm 1528
Height	mm 1280
Kerb weight	kg 680
PERFORMANCE	
Top speed	km/h 205

on sale for Lit 7.800.000. The other new car at the museum's mini show was the Abarth prototype of the new Autobianchi A 112, in which was installed a slightly lower powered 108 hp version of the 982 twin-cam four-cylinder engine of the Group 2 Fiat Abarth 1000 Berlina. Given a generous tyre size of 4.50/10.00-13 and, dependent on need, large rear plastic tail ends for the four-wheel housings, the car weighed 680 kg and was credited with a top speed of 208 km/h. But it was a prototype that did not go into production, and that constituted a clear message to Carlo Abarth on how much less Fiat now appreciated modified versions of their mass produced cars. The men at Corso Marche had their confirmation that the scant interest demonstrated by Fiat in the past in the OT 1600 and a recent proposal for a racing version of the 128 saloon reflected the new strategy of the Turin group's top management. The new guard was beginning to distance itself from motor racing, in which Abarth competed every weekend in Italy and overseas; the Scorpion also knew how difficult it would be to persevere with that expensive operation without the full support of Italy's top car maker. This new attitude induced Carlo to acquire a number of 600 Ds before the model went out of production so that he could continue to race the 850 TC and 1000 Berlina, which were still the cars to beat in the Touring category.

On 5 March at Campo Volo, the airfield opposite his Corso Marche factory, Abarth lined up the cars with which he would compete during the 1970 motor racing season. There was the revised Fiat Abarth 3000 Spider Prototipo (internal code ES 016) with a central V8 engine that reduced the wheelbase from 2260 mm of the previous prototype to 2185 mm to produce a racer in keeping with the Scorpion's other Sport and Sport Prototypes and resulted in

The Racing Car Show in Turin also saw the debut of a racing prototype built by Abarth on the basics of the new Autobianchi A112 utility car, in line with the successful formula used in the past for the Fiat 600, 500 and 850.
The engine was the 982 cc radial cylinder head unit also bolted into the Fiat Abarth 1000 Berlina Group 2 car, the maximum power of which was slightly reduced from 112 hp to 108 hp. Unfortunately the Autobianchi A112, as other projects, was given the thumbs down by Fiat top management, leaving Abarth with nothing more than the satisfaction of modifying the production engine for a special version of the car, which became a tremendous sales success.

FIAT ABARTH "595 SS" CORSA (1970)	
ENGINE	
	rear, longitudinal, 2 cylinders in line
Bore and stroke	mm 73.5x70
Cubic capacity	cc 594
Valve gear	overhead valves, side camshaft
Number of valves	2 per cylinder
Compression ratio	10.5x1
Fuel feed	1 Solex 34PBIC carburettor
Ignition	coil, distributor
Coolant	forced air
Lubrication	forced
Max. power (project)	34 hp at 5000 rpm
TRANSMISSION	
	rear wheel drive
Clutch	single dry plate
Gearbox	4 speed plus reverse, central lever
Axle ratio	4.9:1
BODY	
	2 door, 4 seat saloon
CHASSIS	
	integral body
Suspension	front independent, upper trailing arms, transverse lower leaf spring, hydraulic dampers; rear independent, trailing arms, helicoidal springs, hydraulic dampers
Brakes	hydraulically operated drums front and rear
Steering	worm and sector
Fuel tank	capacity 21 litres
Tyres	front 125x12; rear 135x12
DIMENSIONS AND WEIGHT	
Wheelbase	mm 1840
Front track	mm 1185
Rear track	mm 1190
Length	mm 2970
Width	mm 1415
Height	mm 1210
Kerb weight	kg 520
PERFORMANCE	
Top speed	km/h 130

Updated 1970 racing versions of the Fiat Abarth 595 and 695 were also exhibited at the Turin museum show. Improvements to them mainly concerned an increase in track, the adoption of wider rims and with the consequent modification of the wheel housings that were given extenders of generous dimensions, painted red to contrast with the white bodies. The quickest version of the 595 produced a maximum power of 34 hp at 5000 rpm, which gave the car a top speed of 130 km/h.

a sleeker, more essential body especially at the amply exposed rear end. In all, 84 improvements had been made to the Fiat Abarth three-litre, excluding those introduced directly by Abarth and Colucci during the construction of the first example. Carlo himself supervised the various stages of the car's construction and, as was his habit, was the first to test drive the finished car. With a weight of 585 kg and a maximum output of 350 hp, the power to weight ratio was 1.67, one of the lowest in its category and that promised formidable performance, particularly at hillclimbs and on rough, uneven tracks where acceleration, agility and braking were more important than sheer speed. It was not by chance that the Scorpion aimed to debut the new 3000 MC (motore centrale or mid-engine) prototype at the 54th Targa Florio on 3 May.

Next to the three-litre was the 2000 cc for Sport and Sport Prototype races and some European Hillclimb Championship events. Built on a wheelbase of 2085 mm, the Fiat Abarth 2000 Sport Spider Group 5 car benefited from only minor improvements over the '69 model, among them the body's rear end - the same as the SP 3000 - and safety tanks. The car was to compete officially in the European Constructors' Championship, while private entrants were more or less left to compete for the European hillclimb title. A version of the SP 1000 developed from the SE 018 was also at Campo Volo that day, with its rear engine inclined over the back axle. The Group 7 1000 Biposto Corsa was given its first test run during the usual winter shakedown at the beginning of 1970 at Le Castellet, the new Paul Ricard circuit in southern France, during one of the regular winter test sessions that Carlo was accustomed to organising at Vallelunga, in Sicily or, indeed, Ricard, where weather conditions were more suited to such events at that time of year.

The 1970 version of the Group 2 1000 Berlina one-litre also made an appearance at that time, although its destiny was a different one; the car's engine had a single camshaft radial head, a 13:1 compression ratio and two

FIAT ABARTH "3000 PROTOTIPO" (1970)

ENGINE	
	mid-rear, longitudinal, V8 (90°)
Bore and stroke	mm 88x81
Cubic capacity	cc 2968
Valve gear	4 overhead camshafts, chain driven
Number of valves	2 per cylinder
Compression ratio	12:0
Fuel feed	4 twin-choke Weber 48IDA carburettors
Ignition	double, coil, distributor
Coolant	liquid
Lubrication	dry sump
Max. power (project)	355 hp at 8200 rpm
TRANSMISSION	
	rear wheel drive
Clutch	single dry plate
Gearbox	5 speed plus reverse, central lever
Axle ratio	3.5:1
BODY	
	racing roadster
CHASSIS	
	tubular trellis
Suspension	front independent, trailing arms, helicoidal springs, stabiliser bar, hydraulic dampers; rear independent, trailing arms, helicoidal springs, stabiliser bar, hydraulic dampers
Brakes	hydraulically controlled discs front and rear
Steering	rack and pinion
Fuel tank	capacity 120 litres
Tyres	front 4.50/11.60x13; rear 5.50/13.60x13
DIMENSIONS AND WEIGHT	
Wheelbase	mm 2185
Front track	mm 1405
Rear track	mm 1435
Length	mm 4031
Width	mm 1820
Height	mm 970
Kerb weight	kg 600
PERFORMANCE	
Top speed	km/h 300

Weber 40 mm carburettors, all of which gave it a maximum power output of 115 hp. The new version of the successful model in its Fiat 600 body weighed 563 kg and accelerated to 215 km/h.

At the start of the season, Abarth & C. also unveiled a special elaboration of the Autobianchi A 112, still fresh from its debut at the Turin Motor Show of the previous year. The intention was to repeat the recent great success of the Fiat Abarth 850 TC, a car on which many young drivers had cut their teeth. But as Fiat's top management had vetoed the proposal of an A 112 with a one-litre radial head engine, the car was powered by the 903 cc, four-cylinder Fiat unit fitted to the production A 112, but with its stroke increased from 68 to 74 mm to take its capacity to 982 cc.

With a compression ratio of 10:1 and a Weber 32 DMTR carburettor, the engine developed 58 hp at 6600 rpm, a value that was considered the starting point for other types of intervention. Equipped in that way, the prototype was subjected to a long series of motorway tests – during which it easily reached a top speed of 150 km/h against the 135 km/h of the production version - before going into production at Autobianchi's Desio factory and making its debut in the marketplace in September 1971.

Other new developments were in preparation at Corso Marche. At the traditional 4 Hours of the Jolly Club, first round in the European Touring Car Championship at Monza on 15 March 1970, the Scorpion won its class four times with the Fiat Abarth 595 of Marco Magri prepared by Romani, the 695 driven by Angelo Mola tuned by Garavello, the 850 TC of Fulvio Tandoj prepared by Baistrocchi and Johannes Ortner's works 1000 Berlina. But on that day of bad weather and a cold, slippery track surface the team's new 1000 TCRs were dogged by a series of problems, principally caused by the engine's inability to

A brief presentation of cars that would compete in the 1970 motor racing season at the Corso Marche airfield included the latest evolution of the prototype Fiat Abarth 3000, with its mid-mounted engine (SE020) developed from the positive experience gained the previous year. In general, the car's technical characteristics remained the same, except for a small increase of five hp in maximum horsepower and a 40 mm widening of the glass fibre body. Weight was kept at just 600 kg.

reach its operational temperature and produce its best performance on the track.

At the end of March, another version of the 2000 Sport Spider was presented (SE 019), which had an inboard rear engine instead of a power plant inclined over the back axle; it was the well-known 1946 cc four-cylinder (bore and stroke 88x80 mm) the maximum power of which had been increased by 10 hp from 250 to 260 hp. There were slight changes to the body compared to the version of a few months earlier, including front and rear spoilers. With a dry weight of 610 kg, the new derivative of the Group 6 Fiat Abarth two-litre was to debut on 19 April in the first up to two-litre class event of the European Constructors' Championship driven by Mario Casoni at Paul Ricard, and then at the Targa Florio on 3 May.

The Scorpion's transfer to Sicily was not exactly blessed with good fortune that year; during testing on Thursday 30 April, Casoni went into a ruinous spin and broke an upright. Then Johannes Ortner snapped the drive shaft of his sister car; Carlo Abarth was not pleased by the damage, for which he blamed the drivers. And he was even more critical of them because their cars lagged behind the Alfa Romeo 33-3 and threatened to retire the team and return to Turin that night. The intervention of Arturo Merzario convinced him to stay on, but on race day the newly repaired Fiat Abarth 2000 Sport Spider had to retire after one lap with a broken gearbox, coded 139 and originally designed to be hooked up to the 1.3 and 1.6 litre engines; it seemed under-dimensioned for the power and torque developed by two and three-litre motors, which were generating 250 and 350 hp respectively by then.

But the Scorpion managed to win the up to 2000 cc Group 5 Sport Category with 'Pam' and 'GiBi', the up to 1300 cc class wit 'Black and White'-Garufi (Fiat Abarth 1000 SP) and the up to 1000 cc class of the Group 6 Sport Prototype Category with Pellegrin-Ruspa driving a Fiat Abarth 1000.

Abarth won its first 1970 race at the 11th Fleron hillclimb over a two kilometre course from Bellevue to Forêt in Belgium on 15 March. Eris Tondelli in a 2000 Sport Spider took the Group 5 and 6 Sport and Sport Prototype over 1000 cc categories, returning a time of 1'08.0"...

The chance to shine with the Fiat Abarth 2000 in the European Constructors' Championship for two-litre cars came at the Scandinavian Raceway at Anderstorp, Sweden, on 7 June, when Carlo Abarth had another slight disagreement with one of his drivers, Austrian Dieter Quester, who was guilty of having semi-destroyed one of the works cars during practice. In the Sport Prototype Category, success went to Joachim Bonnier's Lola T210, while Johannes Ortner won the Group 5 Sport in a Fiat Abarth 2000, followed by another two sister cars driven by Ed Swart and Sergio Morando.

More improvements were carried out to the mechanics and body of the Fiat Abarth 2000 Sport Spider for the upcoming Mugello race; the front air intake was modified and two slender fins were fitted as stabilisers to the engine hood. Abarth entered four cars for the event, three mid-engined machines and one outboard, the latter driven by Arturo Merzario who, unlike the other works drivers, was fully accustomed to the car, which he drove on the limit – and looked like he had his hands full doing it.

Abarth's predilection for cars with outboard engines paid off many times during hillclimbs, in which that lay-out – unloved by his technicians, starting with Colucci – limited rear wheel spin in acceleration coming out of a corner, due to the greater load placed on the back end by the engine and, consequently, the car's better pick-up and more speed when exiting hairpins. This engine disposition also accentuated the Sport Spider's oversteer, which could be exploited positively by drivers on twisting ascents that were not so fast.

That year the 17th Grand Prix of Mugello of 19 July, sixth round in the European Constructors' Trophy and the Italian Grand Touring, Special, Sport and Sport Prototype counter, was shortened from the previous year's eight laps and a total distance of 530 km to five laps covering 331 km. Arturo Merzario won for the second successive year in a Fiat Abarth 2000 Sport Prototype at

...while Fernand Neri came second in a Fiat Abarth 3000 Sport Prototype with 1'08.8". Third was William Scheeren in a 2000 Sport Spider in a time of 1'10.2".

A new episode was played out at the end of March 1970 in the technical approach to the outboard engine and the inboard mid-car power unit when the Fiat Abarth 2000 Sport Spider SE019 arrived on the scene with its engine amidships. The body minus its rear end was unified with that of the rear engined version and its racing debut was slated for 19 April 1970 at the International Paul Ricard Trophy race in France. It won first time out, with Mario Casoni at the wheel of the new car winning the up to 2000 cc Group 5 Sport category in 2h20'35" at an average speed of 135.298 km/h. Second and third were two Fiat Abarth 2000 Sport Spiders driven by Edouard Swart and 'Pogo' respectively.

FIAT ABARTH "2000 SPORT SPIDER" (1970)

ENGINE	
	mid-rear, longitudinal, 4 cylinders in line
Bore and stroke	mm 88x80
Cubic capacity	cc 1946
Valve gear	2 overhead camshafts, chain driven
Number of valves	4 per cylinder
Compression ratio	11.5:1
Fuel feed	2 twin-choke Weber 58DCO3 carburettors
Ignition	double, coil, distributor
Coolant	liquid
Lubrication	dry sump
Max. power (project)	260 hp at 8800 rpm
TRANSMISSION	
	rear wheel drive
Clutch	single dry plate
Gearbox	5 speed plus reverse, central lever
Axle ratio	3.5:1
BODY	
	racing roadster
CHASSIS	
	tubular trellis
Suspension	front independent, trailing arms, helicoidal springs, stabiliser bar, hydraulic dampers; rear independent, trailing arms, helicoidal springs, stabiliser bar, hydraulic dampers
Brakes	hydraulically operated discs front and rear
Steering	rack and pinion
Fuel tank	capacity 70 litres
Tyres	front 4.50/11.60x13; rear 5.50/13.60x13
DIMENSIONS AND WEIGHT	
Wheelbase	mm 2100
Front track	mm 1405
Rear track	mm 1435
Length	mm 3600
Width	mm 1780
Height	mm 900
Kerb weight	kg 610
PERFORMANCE	
Top speed	km/h 275

an average speed of 131.977 km/h and led another two sister cars to the flag, driven by Leo Kinnunen and Gijs Van Lennep. The Scorpion's triumph was completed by the victories of Mauro Nesti driving a 1000 SP in the up to 1000 cc class of the Sport Prototype Category, 'White' in the up to 1300 cc class of the Sport Category at the wheel of a Fiat Abarth 1000 and 'Pago' in the 2000 cc class with a Fiat Abarth 2000. After the Mugello race, Abarth was third in the 2000 cc European Constructors' Championship with 27 points, with Lola top (36 points) and Chevron second (30).

In the first round of the European Hillclimb Championship on 24 May at Montseny, Spain, Mario Casoni took the Fiat Abarth 2000 Sport Spider to victory, followed by 'Riccardone' (Carlo Benelli) and Luigi Taramazzo in sister cars; Johannes Ortner in a works Fiat Abarth 3000 Sport Prototype won the second round on Mont Ventoux, France, on 21 June when he made the climb

The start of the 1970 motor racing season was intense and highly gratifying for the cars of the Scorpion. At the seventh City of Volterra Cup 10.150 km hillclimb on 17 May, Fiat Abarths triumphed in the Touring category of the up to 2000 cc Group 5 Sport with Mario Casoni (right) and in the up to 2000 cc Sport Prototype in which Johannes Ortner not only won his category but also took outright victory with a time of 5'06.4" and an average 119.255 km/h.

Mario Casoni won his class again at the seventh Montseny, Spain, a round in the European Hillclimb Championship on 24 May. He came first overall and won the up to 2000 cc in the Group 5 Sport category at the wheel of a Fiat Abarth 2000 Sport Spider with outboard engine, covering the 16.300 kms in 9'30.28", average 102.896 km/h.

in 10'4. 8'' at an average 128.076 km/h, but without breaking Peter Schetty's record of the previous year.

The third Championship event was the 30th Trento-Bondone on 5 July, when Ortner was involved in an accident that took him out of the Group 6 Sport Prototype Category event, which was won by Antonio Zadra in a Fiat Abarth 2000 Sport Spider followed by Franco Pilone in second overall and winner of the two-litre Sport category driving a Fiat Abarth 2000 and another army of drivers in cars of the Scorpion. That day, the Corso Marche factory also won the Group 2 Special Touring Category with Claudio Margoni (Fiat Abarth 850 TC), Achille Soria (Fiat Abarth 1000 Berlina), the up to 1000 cc

Then it was the 16th Mont Ventoux, one of the top hillclimbs of the year and the second round in the European Championship, which took place in France on 21 June 1970 over a 21.600 km route. That day, Johannes Ortner won the event outright in a 3000 Sport Spider with a time of 10'04.8" averaging 128.571 km/h (picture).
Ortner also won his up to 3000 cc class and then the up to 2000 cc class in a 2000 Sport Spider, while 'Cyprien' took the Sports Prototype category win in a Fiat Abarth OT 1300.

Ortner missed out on victory at the classic Bolzano-Mendola hillclimb on 28 June. He came second (above) in the 14.400 km event driving a Fiat Abarth 2000 Sport Spider, having been beaten by 5/10 second by 'Pam' in a sister car. A week later, Ortner (below) was the favourite in the Trento-Bondone, third counter in the European Hillclimb Championship, over a 17.300 km route from Montevideo to Vason but he went of and retired. Antonio Zadra won the Group 6 Sports Prototype in a Fiat Abarth 2000 Sport Spider in a triumphant day for the Scorpion.

After its almost surprise victory in the Grand Prix of Mugello the previous year, Abarth went all out in 1970 to repeat its success.
The Scorpion was in Scarperia, Florence, on 19 July, ready to compete over five laps of the circuit for a total of 331 km in a vitally important race in the Group 5 Sport and Group 6 Sports Prototype categories.
As always, Carlo Abarth contributed untiringly to tuning the cars in all stages of practice, which took place from Monday of the previous week to ensure the cars were honed to competitive perfection.
In the small picture below, Carlo Abarth discusses the results of the first few laps of testing with Arturo Merzario, and in the large photograph Abarth directs tuning operations with a proud sternness.
Opposite, large photograph: Carlo Abarth, Annelise and Renzo Avidano walk among the 2000 Sport Spiders of Gijs Van Lennep (18), Leo Kinnunen (19), Nino Vaccarella (9) and Arturo Merzario (1).

In the two small pictures, a relaxing moment for Carlo and a lap for Merzario, who won outright and took the class win in his outboard engined 2000 Sport Spider in 2h30'28.8" at an average speed of 131.977.

Motor racing in the United States was kept alive in those years by individual teams and many enthusiasts who competed, often successfully, against tough opponents from various origins.

The picture shows the Group 2 Fiat Abarth 1000 Berlina leading a pack comprising an Austin Healey Sprite, Morris Mini Cooper, NSU TTS and other Fiat Abarth 1000 Berlinas.

class of the Group 5 Sport Category with Aldo Bersano and the up to 1300 class with Karl Federhofer with the 1300 OT. Fiat Abarth also took the one-litre Sport Prototype event with a 1000 Spider in the hands of Mauro Nesti.

The fourth round in the European Hillclimb Championship took place on the Cesana-Sestrière on 12 July and was won outright by Arturo Merzario in a Fiat Abarth 2000 Sport Prototype, with Johannes Ortner (Fiat Abarth 2000 Sport Spider) second and then Franco Pilone, 'Pogo' and Gigi Taramazzo, all in sister cars. Luigi Moreschi took the 1300 cc class and Gianni Varese the 1000 cc class of Group 5 Sport, with the customary victories in the Special Touring Category's up to 600 cc class (Renato Mazzeschi, Fiat Abarth 595), up to 700 cc (Santo Famoso, Fiat Abarth 695), 850 cc (Aldo Gunnella, Fiat Abarth 850 TC) and the 1000 cc (Maurizio Campanini, Fiat Abarth 1000 Berlina).

On 26 July the fifth round in the Hillclimb Championship took place on the classic Freiburg-Schauinsland ascent on which Johannes Ortner was rele-

At the fifth Dobratsch hillclimb, final round in the European Championship, at Villach, Austria, on 27 September, Johannes Ortner won the race and the title of hillclimb champion of Europe.
The event took place in two heats over a 6.293 km climb and the Scorpions also took the win in the up to 1000 cc Special Touring Group 2 (Karl Wendlinger, 1000 Berlina), up to 1300 cc Group 5 (Karl Federhofer, OT 1300), and up to 2000 cc (Johannes Ortner, 2000 Sport Spider).
The picture shows the new European Hillclimb Champion listening to the comments of an excited Carlo Abarth.

gated to fifth, but he did better on the sixth counter on Monte Dore, France, on 9 August when he came second. The seventh and last round was in Dobratsch, Austria, on 27 September, when Ortner won the event and clinched the European title with 64 points to second placed Franco Pilone's 58, both driving Fiat Abarth 2000 Sport Spiders.

The Scorpion ended 1970 with about 790 victories in various categories and classes. Abarth won Division 1 of the European Touring Car Championship with Johann Abt of Germany in a Group 2 Fiat Abarth 1000 Berlina) and the European Hillclimb Championship (Johannes Ortner, Fiat Abarth 2000 Sport Spider), as well as the Italian Sport Category with 'White' (Luciano Pasotto) driving a Fiat Abarth 1000 SP prepared by Crivellari, the up to 600 cc class of the Italian National Touring Car Trophy (Renato Mazzeschi, Fiat Abarth 595 tuned by Garavello), the up to 700 cc class (Santo Famoso, Fiat Abarth 695, also prepared by Garavello), the up to 850 cc (Mario Barone, Fiat Abarth 850 TC, tuned by Baistrocchi) and the up to 1000 cc class (Maurizio Campanini, Fiat Abarth 1000 Berlina, another car prepared by Baistrocchi). The Corso Marche factory also won the Italian Sport Prototype National Hillclimb Trophy ('Riccardone' – Carlo Benelli – Fiat Abarth 2000 SP tuned by Cortini) and the Sport Category title with Luigi Moreschi (Fiat Abarth 2000). Victory was also theirs in the CSAI Cup, won by Giancarlo Galmozzi (Fiat Abarth 1000 OTS) and Mauro Nesti (Fiat Abarth 1000 SP).

But the 1970 motor sport season set alarm bells ringing at Corso Marche, for it was clear that the Scorpion's cars were coming under increasing attack from other manufacturers, at times even in the categories and classes in which the cars from Turin habitually raced from triumph to triumph. Even in the Touring Category classes in which the battle for success was almost exclusively 'in family' between the Fiat Abarth 595s, 695s, 850 TCs and 1000 Berlinas, stiff competition was now coming from the Alfa Romeo GTA Junior and Giulia, the BMC Mini Cooper and Cooper S, the NSU TT, the Ford Escort and Lotus Cortina, the Lancia Fulvia HF, the BMW 2002 and 2800 CS, the Porsche 906 and 908. Abarth was also finding it increasingly difficult to keep the Lolas and Chevrons at bay in the Sport and Sport Prototype Categories. At the 500 Kilometres of the Nürburgring on 6 September, for instance, the works drivers said they were happy with the engine, but were unhappy with the chassis and suspension of both the outboard and mid-engined versions of the Fiat Abarth 2000.

The battle was becoming increasingly open on all fronts and Carlo found difficulty in meeting the challenge alone, in spite of his ceaseless and passionate personal commitment over the years. Many circumstances had changed; Fiat's invaluable and unconditional support was now more cautious and less forthcoming, making the whole working environment tenser and less

The Abarth stand at the 52nd Turin Motor Show, which opened on 28 October 1970, tended to emphasise the sports activities of the Scorpion. The large letters at the back that commemorated more than 700 Corso Marche victories achieved during the season just ended, and the cars on show, which were all racers, seemed to transmit a sort of tacit message concerning the company's future.
In the foreground are the mid and outboard-engined 2000 Sport Spiders, which had dominated the year. From January to December 1970, Abarth & C. consigned 540 new cars to Italian customers, peaking in July with a delivery that month of 77 units.

willing. Union agitation, with which Fiat had infected other Turin car manufacturers, had usurped the feeling of close team work that had bonded Abarth the entrepreneur to his workers; and for him, a man resolved and intransigent when confronted with any trouble that could frustrate his work or even the mission that he felt he had undertaken, those events humiliated his every action and forestalled his enthusiasm.

This state of affairs weighed heavily on Carlo because he felt he could not intervene, as he would have liked in favour of those who worked at his side. It had already become clear that he could not enter cars for races modified or turned out as he would have liked, because one strike or another meant it had not been possible to finish the work in time. He grappled discontentedly with this conflict within himself that made him accept compromises, not so much because he had to adjust to a situation with which he was not in agreement, but because the values on which he had always called to confront life and create success were being taken away from him.

At the ordinary and extraordinary meetings of Abarth & C. on 26 June 1970, an increase in company capital to Lit 750 million was approved, as well as a company recovery plan; the scheme's objectives were to rationalise car and component production and concentrate sports activity on sectors in which the Scorpion would have the greatest chance of winning. At the same time,

FIAT ABARTH "3000 SPORT PROTOTIPO" (1970)

ENGINE	mid-rear, longitudinal, V8 (90°)
Bore and stroke	mm 88x81
Cubic capacity	cc 2968
Valve gear	4 overhead camshafts, chain driven
Number of valves	2 per cylinder
Compression ratio	12:1
Fuel feed	Lucas indirect injection
Ignition	distributor
Coolant	liquid
Lubrication	dry sump
Max. power (project)	365 hp at 8400 rpm
TRANSMISSION	rear wheel drive
Clutch	dry multiple plate
Gearbox	5 speed plus reverse, central lever
Axle ratio	3.5:1
BODY	racing roadster
CHASSIS	tubular trellis
Suspension	front independent, trailing arms, helicoidal springs, stabiliser bar, hydraulic dampers; rear independent, trailing arms, helicoidal springs, stabiliser bar, hydraulic dampers
Brakes	hydraulically controlled discs front and rear
Steering	rack and pinion
Fuel tank	lateral, overall capacity 120 litres
Tyres	front 4.50/11.60x13; rear 5.50/15.00x13
DIMENSIONS AND WEIGHT	
Wheelbase	mm 2100
Front track	mm 1405
Rear track	mm 1470
Length	mm 3500
Width	mm 1880
Height	mm 840
Kerb weight	kg 615
PERFORMANCE	
Top speed	km/h 300

Carlo Abarth was nominated chairman and managing director of the company and Carlo Scagliarini was appointed a managing director.

In the Abarth & C. price list of 1970 there were eight versions of the 595 (from Lit 675.000 to Lit 900.000), five of the 695 (from Lit 850.000 to Lit 975.000), plus the Fiat 850 TC Corsa (Lit 2.690.000), the Fiat Abarth 1000 Berlina Corsa (3.710.000) and two Fiat Abarth OTS 1000s (from Lit 1.380.000 to Lit 2.050.000). Not included in the list were the Fiat Abarth 1000 Sport (Lit 7.800.000) and the 2000 Sports (Lit 11.500.000).

The 52nd Turin Motor Show that opened on 28 October marked the debut of the Fiat Abarth 3000 Prototipo (SE 020) with a completely re-designed chassis and body, which was immediately re-baptised the 'Red Porsche' due to its considerable similarity to the latest version of the Zuffenhausen-built Porsche 908. The 2968 cc V8 engine was, as before, mounted up front but it had a new twin-cam cylinder head and Lucas indirect fuel injection, as well as a Dinoplex ignition system with an impulse fuel feed; those modifications limited increased power to 15 hp (from 350 to 365 hp) but it appeared clear that, updated in that way, the engine would lend itself to further improvements in performance. The tubular chassis was re-designed and shortened from 2185 mm to 2100 mm and the driving position advanced significantly; the corresponding balancing of the rear end was achieved by moving the water radiators further back on either side of the cockpit and the oil radiators to the two extremes of the tail.

The show also saw the 595 and 695 two-cylinders with radial heads, almost vertical induction valves at the centre of the combustion chamber and the exhaust valve heavily inclined. Designed by engineer Stefano Jacoponi, the new radial cylinder head was submitted for homologation, which was granted in July 1971.

At the centre of the Turin show's Abarth stand was Corso Marche's newborn 3000 Sport Prototype (SE020), which overturned the techno-constructional formula developed years earlier for racing barchettas. The tubular chassis had been re-designed and the wheelbase shortened by 85 mm compared to the previous model; the cockpit was moved forward and water radiators placed on both sides of the driving position, immediately in front of the engine compartment. Oil radiators were mounted at the extreme end of the tail to even out weight distribution between the two axles. The new mechanical configuration greatly changed the shape of the body, which became more solid and essential, inspired directly by the Porsche 908/3 that was seen a few months earlier at the Targa Florio.

Rumours also circulated at the Turin show that the new 3000 Prototype might take part in a comparative test against the Ferrari 312 prototype in the hands of Maranello's two contracted drivers for 1971, Ignazio Giunti and Arturo Merzario. The leak seemed to strengthen the hypothesis that a common Scorpion-Prancing Horse motor sport programme - or at least one that was in synthesis - may be agreed, perhaps conducted and run by Fiat. However, the gossip did not turn into fact.

In 1970, the Italian car market registered a promising improvement with a total number of 1,150,000 new cars registrations – an increase of 6.53% against 1969. Fiat Abarth deliveries to Italian clients totalled 540 units, with a high point of 77 consignments in July.

In the closing moments of the motor sport season the fantastic 1000 Berlina engine, which had gone from victory to victory in the Touring Car Category since 1962, appeared again but in more up-to-date and sophisticated guise, producing a maximum of 115 hp. Several times during the '70 season, people had the impression that the generous four-cylinder with only three main bearings and rod and rocker arm distribution had really reached the limit of its possibilities; the adoption of the radial head and two twin-choke Weber 45 mm carburettors provided further increases in power, but the car was breaking down in races more and more frequently and that confirmed no more could be asked of the glorious power unit. The final versions of the Group 2 1000 Berlina had also become difficult to drive, because the new suspension was effective at high speed but fairly difficult to control. So at the start of the 1970 season it had already been decided to drop the deformable quadrilateral rear end with pushrod link and return, in a certain sense, to the old system with a triangle of large tubes but a much different geometry than that of the original Fiat.

The Sixties changed Italy profoundly. The significant increase in people's buying power created ample pockets of demand in the market and began the first euphoric race towards consumerism for a community that, except for the middle and more affluent classes, had not yet had a taste of the 'economic miracle' of a decade earlier. With an ever more dissatisfied and aggressive proletariat and the administrative disorder and indolence that afflicted the opaque governments of the centre-left, the field was wide open to union organisations which, not satisfied with the conspicuous conquests made with the renewal of 1969-70 labour contracts, aimed at sharing in all national politics with a weak executive, driven by the Christian Democrats. At Alfa Romeo alone there were 1200 strikes during the 12 months of 1970, of which 70 were proclaimed for reasons that had nothing to do with company problems; agitation was attributed to political reasons in general or a number of international events such as the war in Vietnam and Cambodia, or even to express solidarity with the workers of other manufacturing sectors. During his press con-

Carlo Abarth pictured at the presentation of his cars for the 1971 racing season, which took place in early February at the Corso Marche airfield in Turin.
At over 62 years old, the indomitable Viennese technician still showed great determination in pursuing his activities as both an entrepreneur and sportsman.
The passing years had left their mark on Carlo's physique, but his tremendous personality remained intact and expressed itself through the proverbial elegance of his dress.

ference that inaugurated the 51st Turin Motor Show on 29 October the president of Fiat, Gianni Agnelli, said in 1969 his company had lost nine million man-hours due to union agitation, equal to 174,000 cars not being produced. "That loss is equivalent to the production of a medium-size car factory; four times the production of Lancia, for example", he said.

Industry was also beginning to feel the sharp rise in production costs and a series of operational orders (economy in the use of manpower, contraction of new investments with the political power, production decentralisation plans...), which constituted new and certainly disturbing factors in company management. In that kind of climate of collective uncertainty so unfavourable to medium-long term planning many businessmen attempted to restructure their businesses, while others were forced to close, burdened by new costs that often added to internal inefficiency. Foreign car brands were becoming a growing threat in the Italian market, increasing their impact on sales during the period 1966-1970 from 10.67% to 27.80% by exploiting the free competition that followed the liberalisation of customs tariffs within the EEC.

After his 62nd birthday, Carlo Abarth no longer lived the moments of great euphoria and extraordinary success that had marked his life in previous years, although he continued to show signs of initiative and untiring creativity. At the end of February 1971 he went with his team to the Paul Ricard circuit in France, which he had selected as his official test track as Vallelunga was no longer available and Monza's bad weather could not provide enough time for testing. With Abarth was Enzo Tomaini, drivers Johannes Ortner, Arturo Merzario and new acquisitions Giovanni Salvati and Nanni Galli, plus four cars. There was the 1000 Spider barchetta, the 2000 Sport Spider and two three-litres, one in the latest Prototype Category (650 kg) configuration and the other a free weight version of the 'monotipo' for 1971, similar to the most recent version except for its large wraparound windscreen, the retention of the spare wheel and the presence of a small boot as required by the regulations. The two-litre Sport also adopted the mid-engine layout and the chassis and bodyshell were almost the same as those of the 3000 Monotipo, on a wheelbase of 2100 mm, with a length of 3500 mm, a width of 1760 mm and a height of 840 mm. Further reduced, the weight was down to 525 kg and the maximum speed was given as 300 km/h.

Testing confirmed the Fiat Abarth 2000 Sport Prototype and the one-litre Millino were the most efficient cars. This is how Franco Lini expressed his impressions of the three-litre Fiat Abarths in Autosprint of 1-8 March 1971: "*...they are very good as far as road holding and aerodynamics are concerned, but they decidedly lack power. About 360 hp has been found with the current two valves per cylinder engine, but that is really very little compared to the 450 hp and more of the Ferrari...*". And he continued: "*...we could think that*

Carlo Abarth invited Arturo Merzario to be with him that morning in February 1971 at the Aero Club of Turin airfield for the presentation of the three new cars that would compete in the imminent racing season: from the left, the Fiat Abarth 3000 Prototipo (SE020) ready for the World Constructors' Championship, the 3000 Monotipo (SE022) built especially for the European Hillclimb Championship and the 2000 Spider Prototipo (SE021) for the European Constructors' Championship.

All of them had inboard engines and, except for minor details, the same chassis and glass fibre body.

Opposite, above: Two views of the Fiat Abarth 2000 Sport Prototipo (SE021), photographed inside the Corso Marche factory before its official introduction.
Below: Arturo Merzario and Carlo Abarth pose proudly between the two 3000 Sport Spiders, the car on the right (SE022) with a large wrap-around windscreen as required by the international regulations.

Abarth will race the two cars seen at Paul Ricard in pursuit of the European and Italian Hillclimb Championships, and in the meantime work on something more powerful for the 1972 World Constructors' Championship. That is our feeling". In effect, the Fiat Abarth 3000 Monotipo duplicated precisely the external dimensions of the 2000 Sport Prototype, but weighed 175 kg more – a total of 700 kg – and, powered by a 365 hp engine, had a top speed of 306 km/h. Abarth gave the price of the Sport Prototype 2000 as Lit 12 million and the 3000 as Lit 19 million.

Two weeks before the transfer to Le Castellet, Carlo announced that the updated version of the 2000 Spider Prototype would compete in the European Constructors' Championship driven by Arturo Merzario and Giovanni Salvati, while the 3000 V8 Monotipo would go after the European Hillclimb Championship with Johannes Ortner at the wheel.

On the subject of the acquisition of drivers for the 1971 season, Autosprint of 2 February hosted a long arm debate between its editor, Marcello

Sabbatini, and Carlo Abarth concerning the constant commitment of the Turin manufacturer to hillclimbs, compared to an affirmed gradual disinterest in track and endurance racing. In an open letter to the magazine, Carlo affirmed: "*...when international regulations permitted my cars to participate with success in the World Constructors' Championship (which they won for no less than six years) we competed on the international circuits, not excluding Sebring and Daytona*".

To this and other subjects, Sabbatini replied that: "... *the European Constructors' Championship with two-litre prototypes was not under discussion, but clearly the commitments with the three-litre that Carlo Abarth built a year or two ago regularly ended up being seen in hillclimbs. Being put to the test in the Targa Florio or at Imola and one or two other races is nothing other than the exception that confirms the rule*".

FIAT ABARTH "2000 SPIDER PROTOTIPO" (1971)

ENGINE	
	mid-rear, longitudinal, 4 cylinders in line
Bore and stroke	mm 88x80
Cubic capacity	cc 1946
Valve gear	2 overhead camshafts, chain driven
Number of valves	4 per cylinder
Compression ratio	11.5:1
Fuel feed	Lucas indirect injection
Ignition	double, coil, distributor
Coolant	liquid
Lubrication	dry sump
Max. power (project)	260 hp at 8800 rpm
TRANSMISSION	
	rear wheel drive
Clutch	single dry plate
Gearbox	4 speed plus reverse, central lever
Axle ratio	3.5:1
BODY	
	racing roadster
CHASSIS	
	tubular trellis
Suspension	front independent, trailing arms, helicoidal springs, stabiliser bar, hydraulic dampers; rear independent, trailing arms, helicoidal springs, stabiliser bar, hydraulic dampers
Brakes	hydraulically operated discs front and rear
Steering	rack and pinion
Fuel tank	lateral, capacity 120 litres
Tyres	front 4.50/11.60x13; rear 5.50/13.60x13
DIMENSIONS AND WEIGHT	
Wheelbase	mm 2100
Front track	mm 1285
Rear track	mm 1350
Length	mm 3500
Width	mm 1760
Height	mm 840
Kerb weight	kg 525
PERFORMANCE	
Top speed	km/h 300

The three new mid-engined Fiat Abarths could trace their origin to the same tubular chassis and were fitted with the same body with a wheelbase of 2100 mm. Their weight varied between the 525 kg of the two-litre Prototipo and the 700 kg of the 3000 cc Monotipo for hillclimbs, which was fitted with a 20-litre fuel tank while the other two could take 120 litres apiece.

It was success again for the Fiat Abarths at the traditional 4 Hours of the Jolly Club at Monza on 14 March, a round in the Italian Touring Car Championship, in which the Scorpions won the 700 cc class with Gianni Giudici (Fiat Abarth 695), the 850 cc with Edoardo Gunnella (Fiat Abarth 850 TC) and the 1000 cc with Gino Ravinale (Fiat Abarth 1000 Berlina). But there was no 1150 cc class race for Corso Marche, even though the company wanted to enter four cars for Merzario, Ortner, Salvati and 'Pam' and compete openly against the Fiat 128s of the Swiss team Filippinetti, because three of the cars' radial head engines broke during practice. Some said such an unexpected retirement was caused by the direct intervention of Fiat, so that the Abarths would be unable to beat the manufacturer's own cars with its 1000 Berlinas.

Abarth only entered two Sport 2000s for the seventh 1000 Kilometres of Monza on 25 April, one for 'Paco'-'Riccardone' who retired on the 48th lap with a broken gearbox, and the other 'Pal Joe'-Becchetti who won their class and finished 11th overall, 35 laps down on the winning Porsche 917 K of Rodriguez-Oliver. A week later, Ortner scored an easy win at the 12.5 km Pieve S. Stefano-Passo dello Spino hillclimb in Arezzo, covering the distance in 6' 00.5" at an average 124.826 km/h driving the agile Group 6 Fiat Abarth Prototipo. At the 55th Targa Florio on 16 May the company competed low key and won the Group 6 up to 1000 cc class of the Sport Prototype Category with a 1000 SP driven by Enzo Barone and Maurizio Campanini, who finished 14th overall.

As well as the works cars, the Scorpion was able to rely on the powerful presence of privately owned Abarths, which were further elaborated, tuned and tested by a number of specialist mechanics. To better indirectly service this large number of demanding and enthusiastic privateers, at the beginning of the year Carlo Abarth set up an external organisation under the technician Vincenzo Osella, who took on the task of maintaining and technically updating the racing cars sold to private entrants. That initiative was also dictated by a need to counter the widespread opinion among privateers that the works cars against which they competed were faster and more reliable; that suspicion had created considerable discontent among those who lined up on the grid to compete every Sunday on, it was believed, equal terms against the cars of Corso Marche. Osella operated more or less as a branch of the factory at his Via Guastalla workshop in Turin, even guaranteeing private Abarth owners that their cars were mechanically identical to those of the factory. He also distributed replacement parts and other special components to other tuners, so that everyone could have available to them the same components as those used by the works team.

A large entry of 124 cars competed in the fifth La Roquette hillclimb in France on 4 April 1971 over a 6.100 km route, on which Luigi Taramazzo in a Fiat Abarth 2000 Sport Spider won outright victory, the Group 5 Sport, Group 6 Sports Prototype and even set up a new record of 4'31.6", an average speed of 80.854 km/h. Taramazzo drove the 1970 car, with which he won the previous year's European Hillclimb Championship.

The 1971 motor racing season got off to a good start for Abarth in the European and Italian national hillclimb Championships. As early as 28 March a Fiat Abarth three-litre driven by Franco Pilone had an easy time in a non-championship event at Ampus, France, covering the 6.800 km in 3'14. 8", an average speed of 110.369 km/h. The first round in the European Hillclimb Championship took place at Dobratsch, Austria, on 24 April, when Johannes Ortner won without difficulty in a Fiat Abarth 3000 Sport Spider, followed in second place by Franco Pilone in a sister car. On 30 May it was Arturo Merzario's turn to win, driving the Fiat Abarth 2000 Sport Prototype into first place in the 19th Vittorio Veneto-Cansiglio; 'Riccardone' won the Coppa della Collina at Pistoia with a sister car the same day, but on 2 June Ortner and Merzario at the wheel of Fiat Abarth 2000 Sport Prototypes were unable to do better than second and third respectively on the refurbished Vallelunga circuit, where they were beaten into first place by Carlo Facetti in a Chevron B19. Four days later, though, Ortner drove a Fiat Abarth 3000 Prototype into first place at the Rossfeld, Germany hillclimb and on the same day Domenico Scola won the 22nd Sila Cup, an Italian Championship counter, in a similar car, with other drivers of Fiat Abarths scoring a string of class wins – Enzo Cantucci (Fiat Abarth 595), Mario Casciaro (Fiat Abarth 695), Giuseppe D'Ippolito (Fiat Abarth 850 TC) and Luigi Ferragina (Fiat Abarth 1000 Berlina) in the Touring Category. The Special Grand Touring class wins went to Alfredo Torricelli (Fiat Abarth OTS), the Sport victories to Rocco Caratozzolo (Fiat Abarth 1000 Spider) and Frank McBoden (Fiat Abarth 1300 OT). In the Tolmezzo-Verzegnis hillclimb on the same day, Dante Gargan won outright in a Group 2 Fiat Abarth 1000 Berlina, covering the distance of 6.2 kilometres in 4'02"3 at an average 92.117 km/h.

At the Mont Ventoux hillclimb in France on 20 June, a round in the European Mountain Championship, a Fiat Abarth 3000 driven by Pierre Maublanc came first; 'Amphicar' took overall victory in the Monte Pellegrino event near Palermo in a Group 5 Fiat Abarth 2000 Sport Spider on 27 June. At the 31st Trento-Bondone, the fourth counter in the championship, on 4 July Carlo Facetti and his two-litre Chevron B 19 took the win and Johannes Ortner came second overall in a Fiat Abarth 3000 Sport to consolidate his lead in the title chase. The Austrian driver won the Cesana-Sestrière a week later in a Fiat Abarth 3000 Monotipo and Arturo Merzario won the Group 6 Sport Prototype Category driving a 2000 SP, which also confirmed the good results achieved with the introduction of fuel injection to the two-litre engine and the use of a new cylinder head with closer camshafts.

The fifth round in the two-litre European Constructors' Trophy was the AC of Bologna Cup at Imola on 25 July, an event that gave Abarth little chance to get back into the fight for the cup against Chevron and Lola; in effect, the Fiat Abarth SP 2000 with its 265 hp engine and driven by Arturo Merzario took the lead and kept it until the eighth lap, when the Italian had to retire with a blown cylinder head gasket, leaving the road clear for Helmut Marko and his Lola T212.

The Scorpion's 1971 motor racing year concluded early with 344 overall and class wins and Johannes Ortner as European Hillclimb Champion in his Group 5 Fiat Abarth 3000. In addition, Abarth won the up to 700 cc class of the Touring Category of the European Championship (Cesare Garrone, Fiat Abarth 695 prepared by Audisio/Benvenuto), the up to 850 cc class (Edoardo Gatti, Fiat Abarth 850 TC also prepared by Baistrocchi) and the up to 1000 cc class (Maurizio Campanini, Fiat Abarth 1000 Berlina, another Baistrocchi-prepared car). The company also won the Italian Grand Touring Car Championship (Eugenio Tinghi, Fiat Abarth 1000 SP tuned by Cambini) and the Italian Special Grand Touring Championship (Camillo Spica, Fiat Abarth 1000 OTS prepared by himself).

Abarth did not often enter the races of the last two months of the season; Ortner and Merzario competed in the last few events in cars wearing Osella team colours, which everyone knew was the Abarth subsidiary workshop for customer service. But what happened behind the scenes?

The Abarth racing department had been demobilised and, since the end of the summer, had been transferred little by little to the Osella premises; the cars that went across included two prototype 3000s, six prototype

Johannes Ortner became the outright winner of the 6th Dobratsch Bergrennen in Austria on 24 April '71 as well as winning his Group 5 Sport category, all at the wheel of a new Fiat Abarth 3000 Monotipo (SE022), covering the distance in 5'56.34". Second was Franco Pilone in a sister car with a time of 5'59.49".

The last stage in the evolution of the new generation two-litre Fiat Abarth Sport Prototipo goes back to late spring 1971, when the design department at Corso Marche created the technical layout (above) and body model of the SE023.

The chassis was no longer a trellis of steel tubes, like those of the prototypes on display at the airfield the previous February, although the new car retained the same 2100 mm wheelbase and boasted wider track of 1340 mm front and 1350 rear.

The mid-engine was the noted 1946 cc four-cylinder 16-valve with indirect fuel injection, but the cylinder head was of new design with closer camshafts and a slight valve angle to allow for a high compression ratio, while maintaining the piston heads almost flat; the four inlet manifolds were of different lengths to improve progressive power output.

According to Abarth & C. archives, project 023 dated 4 May 1971 never went ahead, perhaps due to the imminent and traumatic change in the top management of the company, which had already been defined.

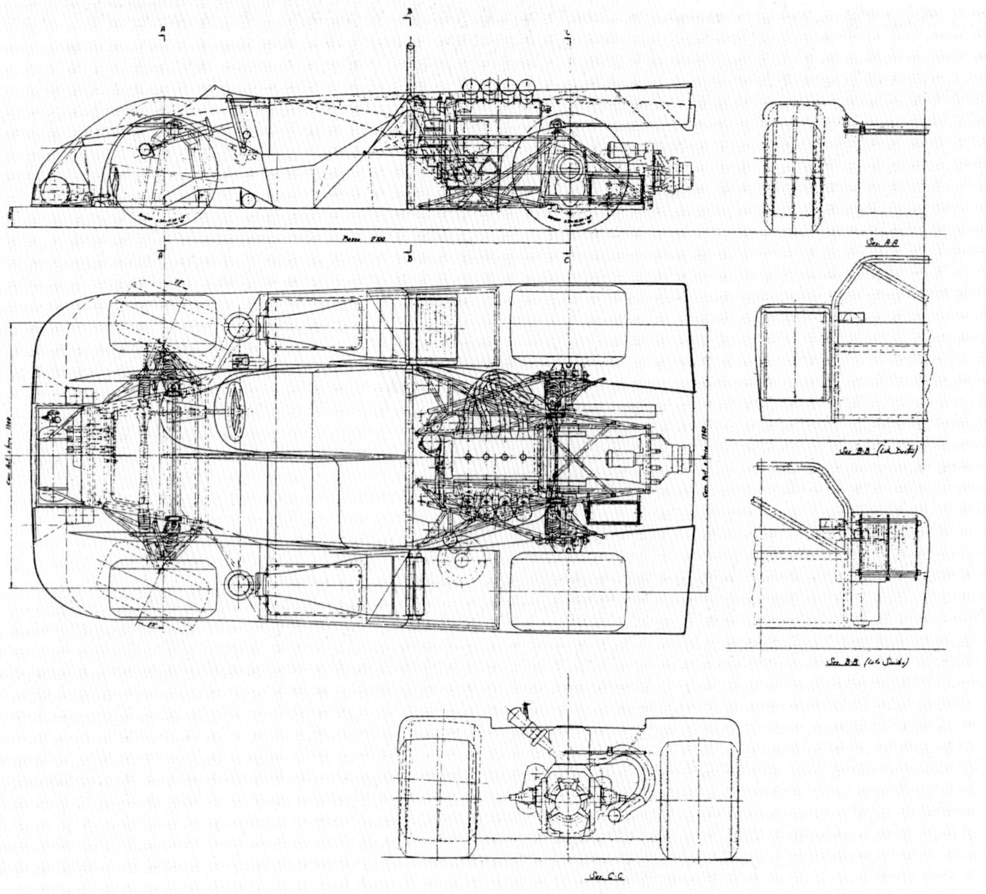

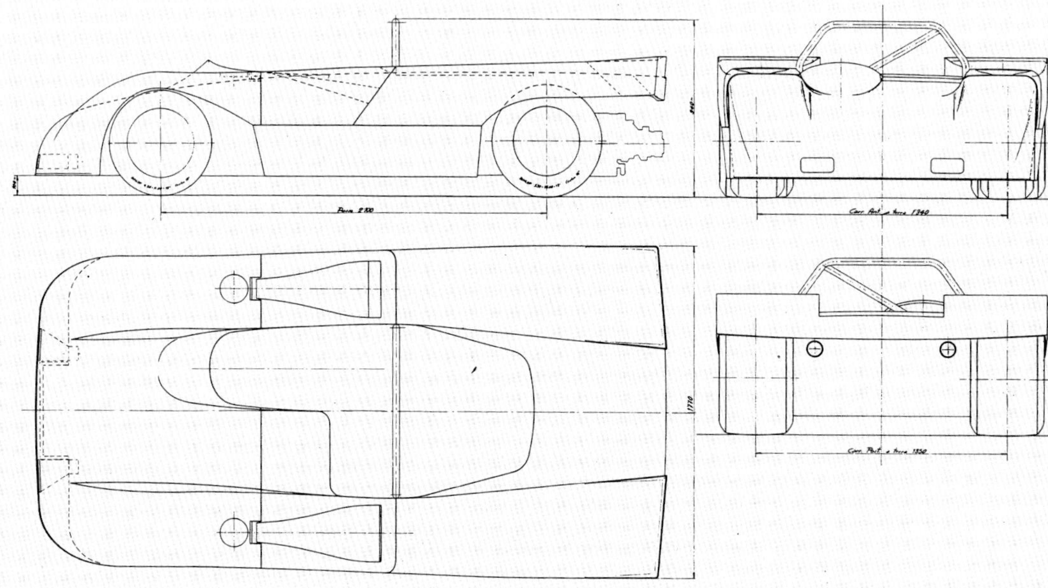

The 11th Cesana-Sestrière, a fast hillclimb over 10.400 km and a counter in the European championship, took place on 11 July 1971 and the Scorpion cleaned up. It won 10 times in the various categories and classes as well as taking outright victory with Johannes Ortner at the wheel of a Group 5 3000 Sport (picture), which covered the distance in 4'59.7", an average of 124.925 km/h.

2000s and 10 SP 1000s, plus Antonio Tomaini, a dozen technicians and mechanics. Although it was not confirmed until mid-October, the Fiat Group had taken over Abarth & C., all its employees and activities in early August.

In 1969, Fiat informed Carlo that it intended to end its 1958 agreement with Abarth & C. and he wrote directly to Umberto Agnelli, asking if he could be given an extension to enable him to reconstruct the company in readiness for its substantial reorganisation. The extension was given, but only until engineer Giovanni Sguazzini, who had quietly built up a team of technicians at Fiat with the aim of participating in rallying, suggested in 1971 that, instead of paying the usual sum of money to Abarth, would it not be more advantageous to buy the company and so end track racing, which Fiat did not seem to appreciate as much as they had previously. So one day in the summer of 1971, Fiat's sales director, engineer Righini summoned Carlo Scagliarini to his office and informed the Abarth managing director that the Group would no longer pay money under the 1958 contract but that, in consideration of the enormous investments made over the years in the Abarth brand, felt it opportune to buy the entire Corso Marche complex.

It is possible that Fiat management felt too involved in the non-stop cascade of Sunday victories in which the presence of the name Abarth with its

These paltry lines of routine dated 31 August 1971 sent by Giovanni Agnelli, president of Fiat, to Carlo Abarth in reality marked the incorporation of the Scorpion into the giant Turin group.

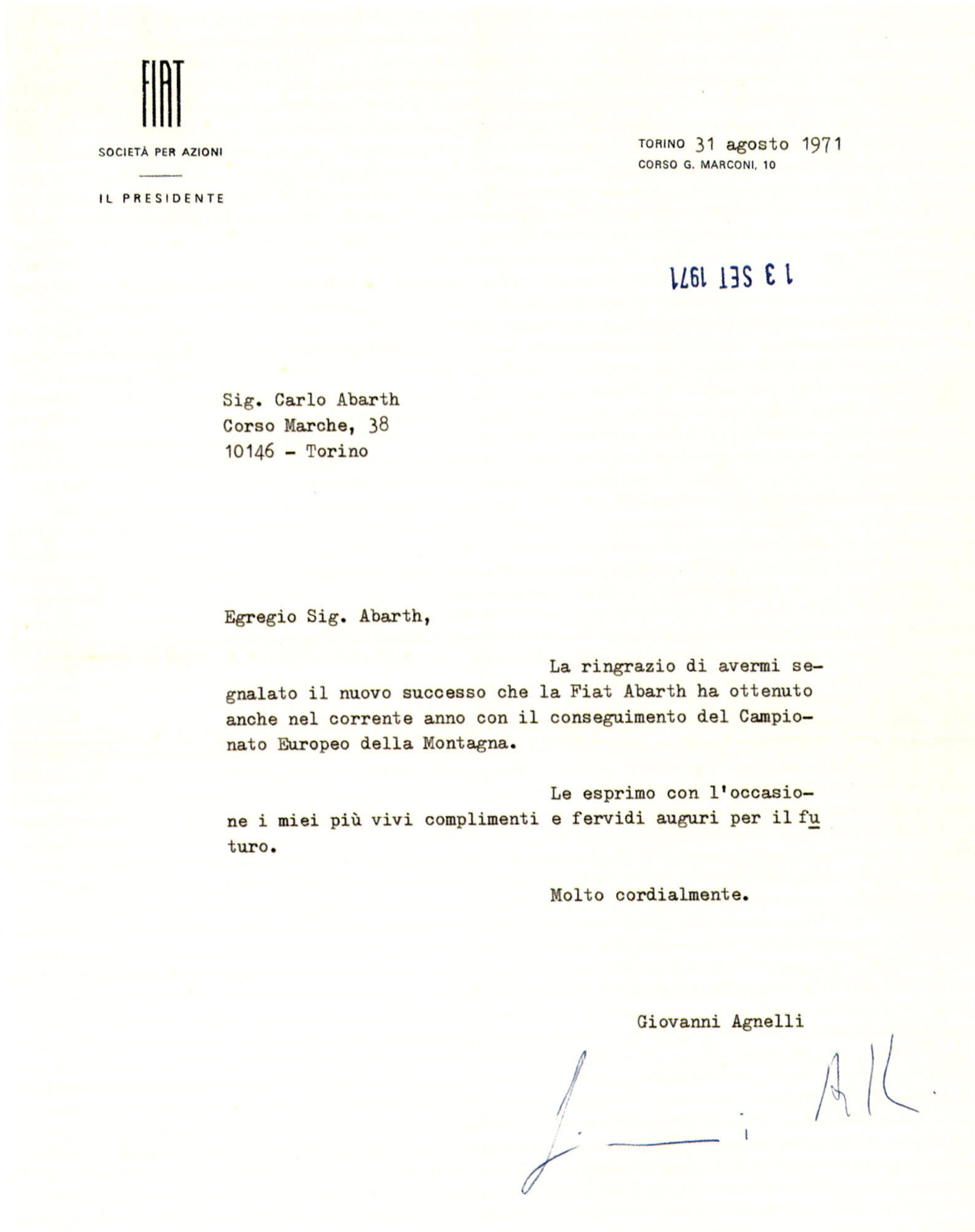

FIAT

SOCIETÀ PER AZIONI

IL PRESIDENTE

TORINO 31 agosto 1971
CORSO G. MARCONI, 10

13 SET 1971

Sig. Carlo Abarth
Corso Marche, 38
10146 - Torino

Egregio Sig. Abarth,

La ringrazio di avermi segnalato il nuovo successo che la Fiat Abarth ha ottenuto anche nel corrente anno con il conseguimento del Campionato Europeo della Montagna.

Le esprimo con l'occasione i miei più vivi complimenti e fervidi auguri per il futuro.

Molto cordialmente.

Giovanni Agnelli

own had, perhaps, become too cumbersome, almost inconvenient. It had certainly become inconvenient to have to pay that large sum of money each year in recognition of all those victories, which Professor Valletta had previously decided was a powerful promotional tool for the Fiat brand across the world.

Many things had changed during those 13 years and the first to realise that fact was Carlo Abarth. In more recent times, he had had to accept compromises that were contrary to his temperament, hardly the hallmark of a tough fighter. On conclusion of the CSAI's mid-summer meeting, he was still not completely convinced of the irksome operation that had caused problems in the national motor sport environment and went by the name Formula Italia.

After the meeting on 2 July 1971, a press release was issued announcing the organisation of the new Formula Italia, a single-seater training scheme of rigidly fixed characteristics with cars produced solely by Abarth & C., and aimed at young enthusiasts who wanted to compete in open-wheel racing. The new single-seater was to have a 1600 cc Fiat 124 S engine, a Lancia five-speed gearbox and all other components from normal production cars except its suspension, which had to be designed specifically for the car; other characteristics included a tubular chassis, a plastic body and safety devices that conformed to formula regulations in force at the time. The whole operation was underwritten by Fiat, although not officially, at an internal cost of around Lit 300 million and was guaranteed at a time when the Group – only just out of a long and onerous struggle with the unions – was certainly not ready to invest money in other complementary activities.

A promotional price of Lit 16-17 million was projected for a Formula Italia single-seater and the cars would only be sold to holders of a third category racing licence who were no older than 26-27. About 20 races were fixed for the following season for about 30 of the new cars, which had to be ready by April 1972. Carlo Abarth had already designed the car, using experience gained in that sector. He built a chassis (internal code SE025) of large steel tubes that were simple and economical, in which he installed the 1608 cc, four-cylinder twin-cam Fiat engine, fed by two Weber twin-choke 40 IDF13 carburettors that produced 110 hp at 6200 rpm. The car had a wheelbase of 2250 mm, front and rear track of 1300/1325 mm, four normal production disc brakes, 13-inch tyres and weighed 450 kg. The announcement of the project was well received and at least 50 young men signed a purchase order for the car, although their hope of taking delivery by the deadline was a feint one due to a long prototype development period and a succession of strikes that delayed car construction.

The definitive version of the Formula Italia car was presented at the 53rd Turin Motor Show from 3-14 November; the single-seater had a Fiat 124 Sport engine, a Lancia Fulvia gearbox, Autobianchi A 111 uprights, Fiat 128 steering box and radiator and Fiat 125 disc brakes. In the four months since July, the project had become much more clearly defined and ambitious; people were talking of 100 cars being built and the possibility extending the initiative to overseas markets, in which the Fiat sales network could have extracted notable advantages with suitable promotion of the cars. In the meantime, though, the criteria for the assignment of the single-seaters had not been established even though engineer Alberto Rogano, president of CSAI, saw to it that the entire programme was quickly made available. It made a number of primary points - a production rate of five cars per week, first deliveries to drivers in March 1972, a sufficient number of cars –15 - with which to run the first races and that racing was to begin in May.

FIAT ABARTH "FORMULA ITALIA" (1971)

ENGINE	
	mid-rear, longitudinal, 4 cylinders in line
Bore and stroke	mm 80x80
Cubic capacity	cc 1608
Valve gear	2 overhead camshafts, chain driven
Number of valves	2 per cylinder
Compression ratio	9.8:1
Fuel feed	2 Weber 40IDF13 carburettors
Ignition	coil, distributor
Coolant	liquid
Lubrication	forced, oil filter on the main circuit
Max. power (project)	6200 rpm
TRANSMISSION	
	rear wheel drive
Clutch	single dry plate
Gearbox	5 speed plus reverse, central lever
Axle ratio	4.3:1
BODY	
	single-seater
CHASSIS	
	tubular trellis
Suspension	front independent, trailing arms, helicoidal springs, hydraulic dampers; rear independent, trailing arms, helicoidal springs, helicoidal springs, hydraulic dampers
Brakes	hydraulically operated discs front and rear
Steering	rack and pinion
Fuel tank	capacity 30 litres
Tyres	front 4.75/8.00x13; rear 5.00/9.60x14
DIMENSIONS AND WEIGHT	
Wheelbase	mm 2250
Front track	mm 1300
Rear track	mm 1325
Length	mm 3700
Width	mm 620
Height	mm 660
Kerb weight	kg 450
PERFORMANCE	
Top speed	km/h 200

The little Formula Italia single-seater was coded SE025 at Abarth & C. and resulted from a joint initiative by Carlo and the CSAI, the Italian motor sport authority, to help set young people on the road to open-wheel motor racing. The SE025 was born as a training car with set technical characteristics and was built of mass produced car components.
For example, the 1608 cc four-cylinder twin-cam engine belonged to the Fiat 124S, the five-speed gearbox to the Lancia Fulvia 1600 HF, the water radiator was from the Fiat 128 and the drive shaft joints were borrowed from the Autobianchi A111.
The Formula Italia car, which was unveiled on 26 October 1971 at the Turin Polo Club and made its public debut a week later at the Turin Motor Show, was designed by Mario Colucci, who conceived a simple but robust trellis tubed chassis on a 2250 mm wheelbase. With a maximum power output of 110 hp and a dry weight of 450 kg, the SE025 was accredited with a top speed of over 200 km/h.
Competitors could choose to fit a maximum of three final ratios to the transmission, but no gear ratio variation was permitted.

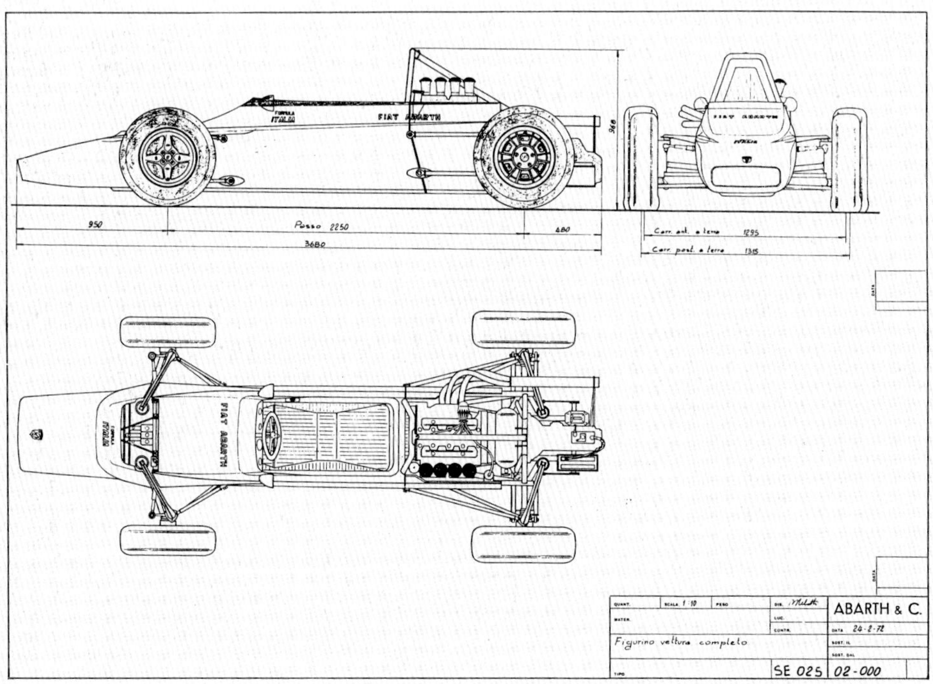

Three views of the Formula Italia car model and the building phase of the SE025 in the Corso Marche factory (picture).
The whole operation, was underwritten by Fiat as a promotional campaign for the recruitment and training of youngsters interested in competing in motor racing.
The series was to be of about 20 races on the circuits of Monza, Vallelunga, Imola, Varano Melegari as well as Turin's Cerrina track.
Thirty Formula Italia cars, which were to have been built by the end of April, were to compete. First tests of the prototype were carried out at Vallelunga in the last two weeks of November, when Arturo Merzario clocked 1'26" during his early laps, which he later lowered to 1'24.6".

Mario Colucci had a great talent for not only designing chassis but also car bodies, as he showed many times with the elegant Sports barchettas that growled around the world's race tracks during those years. The SE025 single-seater also displayed an essential but elegant shape, made even more attractive by a livery that refined its lines. The CSAI decided to assign the Formula Italia cars to young third category racing licence holders who were no more than 26 years old.

The Abarth stand at the 53rd Turin Motor Show from 3-14 November 1971 shocked Scorpion enthusiasts, who were for many years used to seeing a sparkling exhibition of cars with which the company had won world and European titles in the various categories and capacity classes.
The very formal and almost withdrawn 1971 display transmitted a clear message to everyone that Fiat Abarths would never be the same again. The presence on the stand of only the Formula Italia and Autobianchi A112 cars, the latter with an engine modified to a 58 hp output, contributed to making the new brand strategy public.
At the show, the price of the Formula Italia car previously announced as between Lit 1.6 and Lit 1.7 million was set at Lit 1.820.000.

Fiat's acquisition of Abarth hit many like a thunderbolt out of a clear blue sky, but Carlo Abarth had seen it coming, knowing as he did that there would be no acceptable alternative solution to his company's difficulties. For two years competing in races had been under constant threat from strikes, to the point that the Abarth team was forced on occasions to send its cars to locations near races tracks – where they were hosted by local Fiat concessionaires - several days in advance to avoid sudden union agitation that would stop them leaving the Corso Marche factory. Abarth and Carlo Scagliarini realised that, as well as the company's financial problems, relationship issues further complicated matters as a result of which they could no longer rely on their workers to rigorously respect the team's motor racing calendar.

It was an extremely frustrating experience for Carlo, who neither understood the situation nor found it plausible. The situation had become so awkward, so wretched that the only important matter left was the conservation of

Abarth & C.'s enormous patrimony of technical knowledge and experience built up over the last two decades, which should be transferred to those who would take over the management of 'his' company. That patrimony consisted, among other things, of a sort of insurance policy for those who would remain in their jobs, committed to continue without a break the meritorious work of technical refinement and testing. Lucid and pragmatic as always, Abarth persuaded himself that he should affront the situation without recrimination.

In his visits to Vienna, which had become more frequent in recent times, he certainly abandoned himself to memories of the past, to events in which he had been the protagonist all those years ago, from the time he took his courage in his hands and started to build Abarth & C. with Scagliarini from the ashes of Cisitalia, to when those first victories began to roll in with his modified Fiat 600s and the clamorous success he had achieved on the race tracks of the world with those little demons of cars that bore his name. In the end he reluctantly agreed to push his small but ingenious company into the waiting arms of Fiat, from which he had extracted the indispensable resources that had turned his dreams of being a technician and entrepreneur into reality.

The laconic communication released by Fiat on 15 October 1971 put it like this: "*Fiat has completely taken over the company Abarth & C. S.p.A. of 38 Corso Marche, Turin, and its capital of Lit 750 million.*

Abarth will continue to maintain its official name.

Production will be integrated with the preparation of Fiat cars for rallying. The construction of racing prototypes will cease and activity will be oriented towards the elaboration of sports versions of Fiat and Autobianchi engines and cars. The first result, the Autobianchi A112 Abarth, was presented recently.

All other activities in the field of accessories, especially exhausts, will continue normally.

The meeting of Abarth shareholders held today nominated engineer Giovanni Sguazzini executive director of the company, to which Carlo Abarth will remain a consultant.

A terse text that not only confirmed inexorably the end of the company, but also expressed between the lines the embarrassment, almost annoyance felt at Fiat for what had been done at Corso Marche up until that moment. Those closest to Carlo said that cold, almost hostile release was not even agreed with him; he had seen it a day later in a newspaper in Vienna, where he was enjoying a brief period of rest. The position of consultant attributed to him without comment in the last line sounded to many like a casual last minute afterthought, suggested more than anything else by the artificial needs of protocol. Renzo Avidano, Carlo's most loyal colleague from the outset, was appointed to the key position of managing director, while the Fiat rally depart-

AUTOBIANCHI ABARTH "A112" (1971)	
ENGINE	
	front, transverse, 4 cylinders in line
Bore and stroke	mm 65x74
Cubic capacity	cc 982
Valve gear	overhead valves, side camshaft
Number of valves	2 valves per cylinder
Compression ratio	10:1
Fuel feed	1 twin-choke Weber 32DMTR carburettor
Ignition	coil, distributor
Coolant	liquid
Lubrication	forced, oil filter on the main circuit
Max. power (project)	58 hp at 6600 rpm
TRANSMISSION	
	front wheel drive
Clutch	single dry plate
Gearbox	4 speed plus reverse, central lever
Axle ratio	4.7:1
BODY	
	3-door, 4 seat saloon
CHASSIS	
	integral body
Suspension	front independent, lower trailing arms, vertical pillars, helicoidal springs, stabiliser bar, hydraulic dampers; rear independent, lower trailing arms, transverse leaf spring, stabiliser bar, hydraulic dampers
Brakes	hydraulically operated front discs, rear drums
Steering	rack and pinion
Fuel tank	capacity 30 litres
Tyres	135SR13
DIMENSIONS AND WEIGHT	
Wheelbase	mm 2038
Front track	mm 1250
Rear track	mm 1220
Length	mm 3230
Width	mm 1480
Height	mm 1340
Kerb weight	kg 660
PERFORMANCE	
Top speed	km/h 150

ment began its transfer to the Corso Marche premises. Of the company's 60 top technicians 30 were retained, but the track racing programme was cancelled.

Not yet a consultant, Carlo organised the launch of the Autobianchi A 112 Abarth for September, which gave a clear indication of how the relationship between him and the Fiat Group had changed; the quicker version of the A 112 was not projected as a modification of a basic car by the Scorpion, but as a version technically developed by Abarth for construction at the Autobianchi's Desio factory. That successful little car proved popular in both the Italian and overseas markets in its first coming with its 982 cc 58 hp engine, and even more so in 1050 cc 70 hp garb, launched in January 1974.

The presence of Abarth in 'his' factory rapidly decreased as early as 1972. He participated sporadically in the new management's technical meetings and made rare appearances at race tracks to visit his team which, under the name of Osella, competed in the 1972 European Constructors' Championship in the 2000 cc Sports Prototype Category and won the title with Arturo Merzario. In the same year, the company won the Touring Category of the Italian Championship in classes up to 600 cc with 'York' (Maurizio Lastraioli) driving a Fiat Abarth 595 prepared by Vaccari, up to 850 cc with Edoardo Gatti in a Fiat Abarth 850 TC tuned by Baistrocchi and the up to 1000cc with Dante Gargan in a Spagnol-prepared Fiat Abarth 1000 Berlina. The Scorpion also won the Italian Grand Touring Special Championship with Camillo Spica in a Fiat Abarth 1000 OTS he prepared himself. In 1973, the Corso Marche cars won the up to 850 cc class of the European Touring Car Championship with a Baistrocchi-prepared Fiat Abarth 850 TC driven by Samuele Valerio and the 1000 cc by Rino Amighini in a Fiat Abarth 1000 TCR injection prepared by Garavello. The Scorpion also won the Italian Special Grand Touring Championship with 'Poker' (Domenico Cedrati) in a Fiat Abarth 1000 OTS.

Opposite: A month ahead of time the A112 Abarth, the quick version of the little utility launched so successfully a year earlier by Autobianchi, made its first public appearance at the 1971 Turin Motor Show.
Unlike the special versions of the Fiat 600, 500 and 850 made in limited number at Corso Marche, the A112 was a mass produced derivative of the model assembled at the Autobianchi plant in the Milan suburb of Desio. The Abarth engine – evolved from the venerable 982 cc four-cylinder unit (bore and stroke 65x74 mm) was detuned to 58 hp at 6000 rpm. That generous power plant, born under the glorious banner of the Scorpion, gave the little Fiat Group car an extraordinary vitality and helped it achieve great sales success in Italy and abroad.

Carlo Abarth's last official appearance representing 'his' company was on 8 December 1971 at the 'Casco Iridato' motor racing awards ceremony organised by the motor sport weekly, Autosprint. He was presented with the bronze 'Oscar' with the following citation: "First of all, he was able to give the young (and continues to give them with his ever-present passion) a utility sports car". The picture above shows Abarth receiving his award from the president of the CSAI, Alberto Rogano.
In 1969, the bronze 'Oscar' was presented to Enzo Ferrari and Giuseppe Luraghi, while in 1970 it went to Ferdinand Piëch (Porsche) and Luciano Pederzani (Tecno).

In 1972, Mario Colucci developed another chassis from the Formula Italia design for the Fiat Abarth Formula Libre, which had a different engine that came from the Fiat 124 Spider Rallye (1756 cc, four-cylinder, 128 hp at 6200 rpm) but with its maximum power output raised to 170 hp; as with the SE025, the speed gearbox came from the Lancia Fulvia. Built only as an experimental car, the Formula Libre was given front spoilers and the car boasted a large rear wing, which would incline depending on prevailing aerodynamic conditions.

The Abarth name and its Scorpion logo were still around at the end of 1972 on the 2000 Sport Spider that Vincenzo Osella built for racing the following year in the 10 events of the two-litre class of the European Championship.
Designed by Tomaini, the chassis was a trellis of square and round section tubes; the front end and cockpit constituted a sort of sub-chassis of tubes to which panels of aluminium were glued and riveted, while at the rear there was another auxiliary tubular chassis, the task of which was to give rigid support to the engine group.
The power unit was the noted four-cylinder twin-cam 16-valve Abarth, with a slight rise in capacity from 1946 cc to 1981 cc following an increase in bore from 88 to 88.8 mm.
With a compression ratio of 11.5:1 and Lucas indirect fuel injection – even though a Kugelfischer unit was also tested – maximum power went up to 265 hp at 8200 rpm.
The speed gearbox was a five-ratio Hewland FG400 that replaced the FT200, which had not proved efficient or reliable enough during testing the previous year.
Particular attention was paid to the aerodynamics of the new car; the front area of the body exploited the positive experience gained in 1971-72; the long, angular upper line that ran along the entire car integrated itself with two large tail wings with a negative lift function.
The new Abarth-Osella came about following an agreement reached between Osella and the new Scorpion management (engineer Sguazzini and Renzo Avidano), under which Osella would build the chassis and race the car, while Abarth would produce the engine.
The two parties also agreed on drivers who were to campaign the car in the 1973 European Championship; they were to be Arturo Merzario and Derek Bell, but they also considered Nanni Galli, Carlo Facetti and Nino Vaccarella as alternatives.

The debut of the single-seater designed by Carlo Abarth for Formula Italia took place at Monza on 29 June 1972, the day of the Formula 2 Lottery Grand Prix. It was the year the Fiat 124 Spider emerged as a dominant car in motor sport and was specially prepared at Corso Marche to compete in the European Rally Drivers' Championship, won by Raffaele Pinto and in 1975 by Maurizio Verini.

In 1973-4 Carlo Abarth contributed to the development of the new 2000 Sport Spider (internal code SE027), a completely new car based from the beginning on the previous two-seater 2000 cc Sports from Corso Marche. A mid-engined car, the power unit was a 1981 cc four-cylinder (bore and stroke 88.8x80 mm) with an 11.5:1 compression ratio and fuel injection; it developed 280 hp at 9000 rpm and had a maximum torque of 27.4 kg/m at 6900 rpm. Compact on a 2300 mm wheelbase, the SE027 had a prototype body in aluminium that had been extensively tested in the new Pininfarina air tunnel; its front end was short and sloped and from it protruded fared quadrangular headlights; the elaborated profile of the sides boasted a large NACA intake to channel air to the engine bay; the wide, flat tail without rear uplift was surmounted by a large wing on three supports that played a obvious aerodynamic role. The car, the development and construction of which took more time than anticipated, was 3800 mm long, 1870 mm wide, 940 mm high, weighed 575 kg dry and was credited with a top speed of 318 km/h.

Carlo Abarth was given the responsibility of testing the new Sport Spider, the first session of which took place at Campo Volo, Turin, and later at the Misano race track in March 1974. In the late autumn of that year the car's

A further evolution of the car that competed at the start of the 1973 season was the Abarth-Osella. The car won at Enna and the Nürburgring and assumed the designation PA1, which stood for Prototype Alberto and was named after Alberto Guerrato, a greatly talented technician who worked at Corso Marche for many years and to whom Enzo Osella wanted to pay tribute after Guerrato's death. The PA1 (picture) generally followed on from the two-litre of the previous December, but the body was much more aerodynamically effective in the tail area, with the adoption of a large transverse wing in place of the previous two spoilers at the extremes of the two rear wings. Two boxed air intakes were located down low at the sides, corresponding with two lateral radiators. A prominent 'snorkel' on the bonnet acted as a dynamic air intake for the manifold trumpets and relative fuel injectors.

designer, Mario Colucci, created a second version with a sharper chassis, suspension and body shape and the four-cylinder engine was replaced by a new two-litre six-cylinder in-line unit (bore and stroke 86x57 mm) designed by engineer Stefano Jacoponi. With a compression ratio of 11.4:1, the power plant developed a top speed of 300 hp at 11,000 rpm and maximum torque of 34.4 kg/m at 9200 rpm. With its wheelbase lengthened to 2350 mm the six-cylinder version was 4500 mm long, while other dimensions, weight and top speed remained the same as the four-cylinder. But at the end of 1974 the 2000 Sport Spider, in which a great deal of money had been invested, was cancelled and abandoned.

The motor industry and racing were hard hit in October 1973 by the Yom Kippur war and the serious oil crisis that followed it, causing a consequent economic depression that affected the countries of the globe in different ways. The conflict between Israel and the Arab nations that surrounded it provoked a series of reprisals by the Middle East oil producers, which led to petrol rationing in many western lands and an alarming contraction of industrial activity almost

The Abarth SE027 project developed entirely at Corso Marche by Mario Colucci and strongly supported by Carlo Abarth himself (small picture) had taken further steps forward, even if the technicians insisted on maintaining it was a car-laboratory to act as a test bed for numerous innovative experiments. The chassis was a monococq in riveted and glued aluminium panels; front suspension and steering component mounts were fitted, while behind the cockpit was an auxiliary tubular chassis to support the engine group, which consisted of the latest evolution of the Abarth 1986 cc four-cylinder (bore and stroke 88.9x80 mm) fed by Lucas indirect fuel injection, which developed 280 hp at 9000 rpm. The car also had a Hewland FG400 five-speed gearbox. With a wheelbase of 2300 mm, the SE027 was 3800 mm long, 1870 mm wide, 940 mm high at the roll bar, SE027 weighed 575 kg. Its maximum speed was 315 km/h. In 1974, the four-cylinder engine was replaced by a new six-cylinder in line twin-cam designed by engineer Stefano Iacoponi; with a capacity of 1987 cc (bore and stroke 86x57 mm) and a compression ratio of 11.4:1, the unit developed a maximum 300 hp at 11000 rpm. In that final version, SE027 had a 2350 mm wheelbase.

ABARTH "SE027" (1974)

ENGINE	
	mid-rear, longitudinal, 6 cylinders in line
Bore and stroke	mm 86x57
Cubic capacity	cc 1987
Valve gear	2 overhead camshafts, chain driven
Number of valves	4 per cylinder
Compression ratio	11.4:1
Fuel feed	indirect injection
Ignition	distributor
Coolant	liquid
Lubrication	dry sump
Max. power (project)	300 hp at 11000 rpm
TRANSMISSION	
	rear wheel drive
Clutch	double dry plate
Gearbox	5 speed plus reverse, central lever
Axle ratio	3.1:1
BODY	
	racing roadster
CHASSIS	
	monococq
Suspension	front independent, trailing arms, helicoidal springs, stabiliser bar, hydraulic dampers; rear independent, trailing arms, helicoidal springs, stabiliser bar, hydraulic dampers
Brakes	hydraulically operated discs front and rear
Steering	rack and pinion
Fuel tank	capacity 100 litres
Tyres	front 9.30/20x13; rear 13.00/24x15
DIMENSIONS AND WEIGHT	
Wheelbase	mm 2350
Front track	mm 1325
Rear track	mm 1405
Length	mm 4500
Width	mm 1870
Height	mm 940
Kerb weight	kg 575
PERFORMANCE	
Top speed	km/h 315

The body of the SE027 was slim and much tested in the Pininfarina wind tunnel, which had recently been completed. That testing heralded the abandonment of the car's spiky shape, which had been in vogue for a number of years, in favour of the dawn of new curved lines and a more rounded appearance overall. The car made its official debut at the 1974 Geneva Motor Show, which took place from 14-24 March, when it was displayed on the Pininfarina stand with a chassis to demonstrate the advanced state of development it had reached. In May, the new car was subjected to an intense series of tests at the Santamonica di Misano Adriatico Autodrome, attended by designer Mario Colucci, drivers Giorgio Pianta, Pino Pica and Vittorio Brambilla, the new sports director Guido Rosani, Carlo Abarth and Renzo Avidano. The first promising results of the test session were, however, rejected and the entire project was stopped.

everywhere. The dramatic increase in the cost of crude oil on the international market and the soaring inflation it spawned, particularly in Europe, was one of the causes of a long recession that had its roots in an uncontrolled spiral of raw material costs and, as far as Italy was concerned, a deepening balance of payments deficit. Forced by increased costs to defer the time of the renewal of plant needed to affront the situation through increased productivity, Italian industry had to face up to the harshest crisis that had hit the country since the Second World War. Car production, which had climbed to 1.823.333 in 1973, dropped to 1.630.686 in 1974, to 1.348.514 in 1975 and did not increase greatly until 1979-80, when it climbed to about 1.5 million cars, only to plummet again in 1981 to 1.257.340 units. International experts estimated that the ruinous effect of the 1973-74 oil crisis would last a decade, until the end of 1983.

In that difficult period the Italians' buying power ended up being heavily penalised by a devaluation of the lire, which almost hit the 25% level. The Fiat 127 cost Lit 920.000 when it was launched in 1971, but in January 1983 the economical version was price-listed at Lit 7.200.000. In the 'Story of ANFIA, 1912-1992', Bruno Bottiglieri wrote: "*Basic Fiat demand came from labourers, office workers and small business owners, who were the first to feel the severity of the crisis. It was for those reasons that Fiat suffered more than other manufacturers at the moment of the fall in buying power of the public at large, its traditional customers*". In such a climate that certainly did not provide

The Autobianchi A112 had a long and successful career in the Italian market, where 290,000 cars out of a total production of 408,000 were registered between 1969 and 1974, the remaining 27.5% being exported.
The Abarth version, with its 982 cc engine that put out 58 hp at 6600 rpm, was significantly improved in December 1974 (left), when the car's capacity was raised to 1050 cc after its bore had been increased from 65 to 67.2 mm and the power raised to 70 hp at the usual 6600 rpm.
So the quick little car went into its second generation with notable sales success, not only among younger drivers but also the more demanding motorist, who considered it an ideal second car.
The first 70 hp A112 was offered at Lit 2.016.000 , but in November 1977 after a number of aesthetic improvements had been made (right), the price went up to Lit 3.823.000.

great opportunities for expansion in racing by the car manufacturers, the name Abarth appeared on the Fiat 124 Rally, the Fiat X1/9 2000 Stradale and Corsa (Road and Racing) of 1975 and the 131 Rally of the Fiat works team, which won the World Constructors' Rally Championship in 1977, 1978 and 1980.

The following year, the Lancia HF racing team moved to Corso Marche, which later became Fiat's motor sport operations centre. Under the same roof were built 70 hp Autobianchi A 112 Abarth production engines, of which the 50.000th was produced in 1978.

In 1979 design work began on a new single-seater. It was the Formula Fiat Abarth 2000, a training car for the young designed to take over from the brilliant Formula Italia, conceived with cooperation between Fiat and 11 component manufacturers, who sponsored the car as parts suppliers. The car was designed by Mario Colucci and was a single-seater with a simple but robust tubular auxiliary chassis front and rear; it had a transverse engine-gearbox group derived from the two-litre Lancia Beta production car that developed 132 hp at 6000 rpm, specially designed suspension and brakes, a steering box from the Fiat 131, a radiator from the Ritmo and numerous other components drawn from mass production.

Carlo Abarth's trips to Turin had become even less frequent by that time, but he made one of them during the development of the new single-seater. The atmosphere in the car world was much changed, perhaps too much so for his strong principles and in addition his uncertain state of health meant he spent long periods at his Vienna residence, to which he had returned without difficulty and where he still had many friends; he had his villa built on a hill in the Grinzing area with a swimming pool, on the bottom of which was a huge scorpion, symbol of the success that had accompanied his professional life for over 20 years.

AUTOBIANCHI ABARTH "A112" (1974)

ENGINE	
	front, transverse, 4 cylinders in line
Bore and stroke	mm 67.2x74
Cubic capacity	cc 1050
Valve gear	overhead valves, side camshaft
Number of valves	2 per cylinder
Compression ratio	10.4:1
Fuel feed	1 twin-choke Weber 32DMTR carburettor
Ignition	coil, distributor
Coolant	liquid
Lubrication	forced, oil filter on the main circuit
Max. power (project)	70 hp at 6600 rpm
TRANSMISSION	
	front wheel drive
Clutch	single dry plate
Gearbox	4 speed plus reverse, central lever
Axle ratio	4.7:1
BODY	
	3 door, 4 seat saloon
CHASSIS	
	integral body
Suspension	front independent, lower trailing arms, vertical pillars, helicoidal springs, stabiliser bar, hydraulic dampers; rear independent, lower trailing arms, transverse leaf spring, stabiliser bar, hydraulic dampers
Brakes	hydraulically operated discs front, drums rear
Steering	rack and pinion
Fuel tank	capacity 30 litres
Tyres	135SR13
DIMENSIONS AND WEIGHT	
Wheelbase	mm 2038
Front track	mm 1250
Rear track	mm 1220
Length	mm 3230
Width	mm 1480
Height	mm 1360
Kerb weight	kg 700
PERFORMANCE	
Top speed	km/h 160

In July 1979, the A112 Abarth was given a five-speed all-synchromesh gearbox and the popular Autobianchi had reached a complete state of maturity.
At that time, a number of modifications were made to the interior; the three small complimentary instruments – amp meter, water and oil gauges – were housed in a little horizontal panel in the centre below the dashboard.
The characteristic two-spoke steering wheel in bled metal remained unchanged. Renewed in that way, the A112 went into the market at a price of Lit 5.717.000 and was produced for another six years until 1985.
A total of 117,351 70 HP versions of the car rolled off the Desio factory production lines.

FORMULA FIAT ABARTH (1979)

ENGINE	
	mid-rear, longitudinal, 4 cylinders in line
Bore and stroke	mm 84x90
Cubic capacity	cc 1955
Valve gear	2 overhead camshafts, chain driven
Number of valves	2 per cylinder
Compression ratio	8.9:1
Fuel feed	2 Weber carburettors
Ignition	coil, distributor
Coolant	liquid
Lubrication	forced, oil filter on the main circuit
Max. power (project)	132 hp at 6000 rpm
TRANSMISSION	
	rear wheel drive
Clutch	single dry plate
Gearbox	5 speed plus reverse, central lever
Axle ratio	3.8:1
BODY	
	single-seater
CHASSIS	
	tubular trellis
Suspension	front independent, trailing arms, helicoidal springs, hydraulic dampers; rear independent, trailing arms, helicoidal springs, helicoidal springs, hydraulic dampers
Brakes	hydraulically operated discs front and rear
Steering	rack and pinion
Fuel tank	capacity 32 litres
Tyres	front 175/50x13; rear 265/40x13
DIMENSIONS AND WEIGHT	
Wheelbase	mm 2300
Front track	mm 1426
Rear track	mm 1471
Length	mm 3450
Width	mm 1380
Height	mm 660
Kerb weight	kg 530
PERFORMANCE	
Top speed	km/h 230

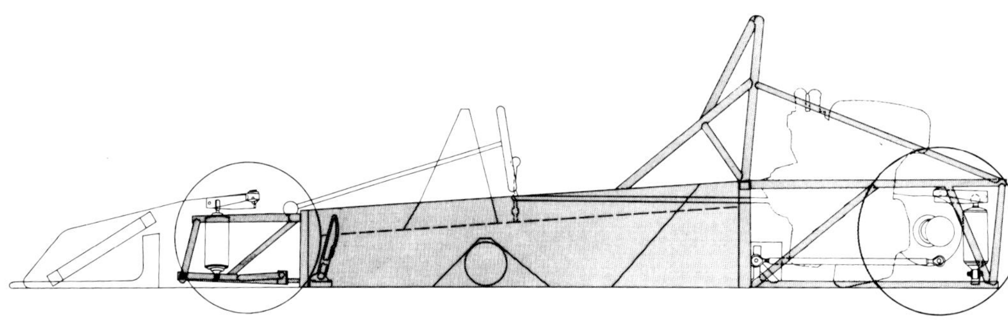

On the basis of the good results achieved with the Formula Italia single-seater, Abarth wanted to build other open-wheel training cars in 1979. The Formula Fiat Abarth design was once more assigned to Mario Colucci, while Fiat itself raised significant finance for the project and also involved 11 of its component suppliers as sponsors. The monococq body was made up of a central unit of progressive collision shock absorption, which carried the front suspension mounts and a sub-chassis in steel tubes 25CD4 and OC30, on which was installed the transverse four-cylinder twin-cam two-litre Lancia Beta production engine that generated 132 hp at 6000 rpm. The five-speed gearbox came from the Lancia Beta Montecarlo, the radiator from the Fiat Ritmo, the steering components from the Fiat 131, but the suspension and brakes were designed ad hoc for the new single-seater. The central part of the structure was rigid and light, while the glass fibre body was pleasantly sculpted and extremely attractive.

One day in the spring of 1979, Renzo Avidano received a 'phone call from Vienna in which Abarth told him he had stomach cancer and wanted to return to Turin one more time. Avidano, who had been at Abarth's side as his shrewd collaborator without interruption for almost 25 years, went to meet him in Milan and was deeply shocked to see the emaciated Abarth who was visibly suffering and wearing a suit that had become too big for him.

The following morning, Carlo had a look at the Formula Fiat Abarth single-seater under construction and afterwards, seated like any other visitor on the other side of Avidano's desk, asked for some coffee. Then he burst out unrestrainedly: "*I've never seen such an ugly car as the one you've built*". Mario Colucci, the designer with whom Abarth had shared the intoxication of suc-

The Formula Fiat/Abarth was built on a 2300 mm wheelbase with front and rear track of 1426/1471 mm. It was 3450 mm long, 1380 mm wide and weighed 530 kg. The first 60 cars were built for February 1980, after which more followed to make a construction total of 150 cars.

cess since 1960, came to see him and the embrace of the two strong-willed men aroused deep emotion and anxiety among those present; Colucci asked Carlo to give him a hand to improve the single-seater and, perhaps, Abarth had expected such an invitation. So Abarth had someone bring him a 'sedio' - his usual mispronounced Italian word for chair - a pair of scissors and some pieces of cardboard; a touch here, a cut there, he worked for an hour in the role that he loved so much because he wanted to leave his mark one last time on a car which, deep down, he still felt was his.

That was Carlo Abarth's last, almost mute contribution as a car designer and technician. He died in Vienna on 23 October 1979, leaving those who had worked with him for years in that fascinating adventure that was Abarth & C stricken with grief.

The company also ceased to exist on 30 September 1981, following an announcement during the course of an extraordinary meeting of Abarth & C. on 13 May, an announcement that decreed a 'union by incorporation' within Fiat Auto S.p.A.

Abarth, from victory to victory

The cars of Carlo Abarth won literally thousands of races in the widest range of categories and classes over many years. The exciting motor racing exploits of the cars bearing the scorpion emblem began on 15 April 1949, immediately after the constitution of the Squadra Corse Carlo Abarth, and drew to a close under the guidance of the indomitable Viennese technician at the end of the summer of 1971, although they continued under different managements with different strategies and limitations for a few more years.
The successes achieved in 1950 were of extraordinary value to the new company on the promotional front, but their organisation became a burden to the management of the young company, which had to suspend its sports activities until 1956.
From then on, though, with the creation of the first modified Fiat 600, enormous and unexpected opportunities opened up in motor racing for Abarth & C. It was from 1958, when an agreement signed with Fiat sanctioning a financial reward for every race won by the Scorpion, that the results of every single race were the subject of systematic inter-company communication.
The tables in the following pages reproduce most of that official information, listing all the Abarth victories with the exception of some of those in 1968 and 1969, for which the records are incomplete. The same goes for a limited number of other races, the results of which were not recorded by Abarth for some unknown reason and were not sent to Fiat. The tracks on which the various races were held varied from one year to the next, due to the changing local conditions or the decisions of the organisers, which sometimes makes it difficult to conduct a comparative evaluation of the various runnings of the same race.
Whether they competed in a round of a world championship, a short hillclimb or a simple small town gymkhana the cars of Carlo Abarth, derived from the mechanics of Fiats, Simcas or completely designed and built by the company itself, were first across the finish line in over 7,300 races in the seasons 1956-1971. Fifteen years of glory.

DATE	LOCATION	RACE	CATEGORY	CLASS	WINNER	TIME OR DISTANCE	CAR
1949							
8 May	Reggio Emilia, Italy	Circuito di Reggio Emilia (2,100 kms x 40 laps)	Sport	1100 cc	G. Scagliarini	50'02"0	Cisitalia 204
22 May	Erlen, Switzerland	Gran Prix Eastern Switzerland (circuit-2,820 kms x 50 laps)	F2	1100 cc	L. Noverraz	1h02'02"0	Cisitalia D46
5 June	Palermo, Italy	Corsa al Monte Pellegrino (hillclimb-8,750 kms)	Sport	1100 cc	A. Pucci	6'44"1	Cisitalia 202 SMM
3 July	Sorrento, Italy	Sorrento-Sant'Agata (hillclimb-10,500 kms)	Sport	1100 cc	S. La Motta (overall)	9'13"2	Cisitalia 202 SMM
10 July	Bolzano, Italy	Bolzano-Mendola (hillclimb-25,000 kms)	Sport	1100	P. Taruffi	16'37"5	Cisitalia 204
10 July	Develiers, Switzerland	Course de côte Develiers-Les Rangiers (hillclimb)	F2	1100 cc	A. Canonica	4'01"6	Cisitalia D46
24 July	Susa, Italy	Susa-Moncenisio (hillclimb-22,100 kms)	Sport	1100 cc	G. Scagliarini	15'41"0	Cisitalia 204
			F2	up to 1200 cc	P. Taruffi	15'5"8	Cisitalia D46
24 July	Freiburg, Germany	Freiburg-Schauinsland (hillclimb-1,300 kms)	F2	1100 cc	A. Canonica	0'56"6	Cisitalia D46
31 July	Avignon, France	Course de côte du Mont Ventoux (hillclimb-21,600 kms)	Sport	1100 cc	Grangé	19'29"0	Cisitalia 202 SMM
16 August	Messina, Italy	Messina-Colle San Rizzo (hillclimb-10,100 kms)	Sport	1100 cc	S. La Motta	8'24"4	Cisitalia 202 SMM
21 August	Senigallia, Italy	Circuito Senigallia (2,790 kms x 15 laps)	Sport	1100 cc	P. Taruffi	1h14'27"	Cisitalia 204
28 August	Aosta, Italy	Aosta-Gran San Bernardo (hillclimb-33,910 kms)	Sport	1100 cc	G. Scagliarini	26'47"1	Cisitalia 204
			F2	1200 cc	P. Taruffi (overall)	25'17"1	Cisitalia D46
11 September	Caltanissetta, Italy	Coppa Nissena (hillclimb-12,000 kms)	Sport	1100 cc	A.Pucci	8'33"0	Cisitalia 202 SMM
18 September	Genova, Italy	Pontedecimo-Giovi (hillclimb-9,560 kms)	Sport	1100 cc	G. Scagliarini	6'51"2	Cisitalia 204
25 September	Biella, Italy	Biella-Oropa (hillclimb-11,000 kms)	Sport	1100 cc	G. Scagliarini	9'19"2	Cisitalia 204
2 October	Roma, Italy	Vermicino-Rocca di Papa (hillclimb-14,400 kms)	Sport	1100 cc	G. Scagliarini	8'27"3	Cisitalia 204
23 October	Salerno, Italy	Salerno-Cava dei Tirreni (hillclimb)	Sport	1100 cc	E. Ricciardi	4'23"8	Cisitalia 204
1950							
26 March	Monza, Italy	Coppa Intereuropa (circuit-6,300 kms x 2 hours)	Sport	1100 cc	G. Scagliarini	kms 263,869	Abarth 204A
10 April	Palermo, Italy	Corsa al Monte Pellegrino (hillclimb-8,750 kms)	Sport	1100 cc	T. Nuvolari	6'21"7	Abarth 204A
9 June	Roma, Italy	Notturna di Caracalla (circuit x 3 hours)	Sport	1100 cc	F. Cortese	kms 251,170	Abarth 204A
10 June	Vue des Alpes, Switzerland	Vue des Alpes (hillclimb-10,000 kms)	Sport	1100 cc	M. Duberti	6'19"2	Abarth 204A
18 June	Oporto, Portugal	Circuito Oporto	Sport	1100 cc	E. Romano	---	Abarth 204A
25 June	Villa Real, Portugal	Circuito de Villa Real	Sport	1100 cc	E. Romano	---	Abarth 204A
23 August	Enna, Italy	Circuito di Pergusa (4,797 kms)	Sport	1100 cc	L. Valenzano	---	Abarth 204A
12 November	Sanremo, Italy	Coppa Ponente	Sport	1100 cc	L. Valenzano (overall)	---	Abarth 204A
1956							
25 March	Verona, Italy	Trofeo Torricella (hillclimb-1,880 kms x 2)	Grand Touring	750 cc	E.Carini	---	Fiat Abarth 750 GT
25 March	Liège, Belgium	Course de côte de la Roche (hillclimb-2,200 kms)	Grand Touring	750 cc	Harris	1'19"2	Fiat Abarth 750 GT
29 April	Brescia, Italy	Mille Miglia (1597 kms)	Grand Touring	750 cc	D.Ogna	16h48'06"0	Fiat Abarth 750 GT
13 May	Monza, Italy	Lombard Drivers' Championship	Grand Touring	750 cc	A.Thiele	25'00"2	Fiat Abarth 750 GT
20 May	Cagliari, Italy	Trofeo Sardo (800 kms)	Grand Touring	750 cc	E.Carini	3h28'24"5	Fiat Abarth 750 GT
10 June	Piacenza, Italy	Castell'Arquato-Vernasca (hillclimb-9,775 kms)	Grand Touring	750 cc	G.Pizzagalli	6'47"3	Fiat Abarth 750 GT
24 June	Roma, Italy	Coppa Gallenga (hillclimb-10,400 kms)	Grand Touring	750 cc	E.Carini	8'40"2	Fiat Abarth 750 GT
1 July	Bolzano, Italy	Bolzano-Mendola (hillclimb-14,500 kms)	Grand Touring	750 cc	E.Carini	12'55"0	Fiat Abarth 750 GT
15 July	Bologna, Italy	Bologna-San Luca (hillclimb-2,080 kms)	Grand Touring	750 cc	L. Cussini	---	Fiat Abarth 750 GT
29 July	Catanzaro, Italy	Giro di Calabria	Grand Touring	750 cc	E. Carini	7h53'05"0	Fiat Abarth 750 GT

DATE	LOCATION	RACE	CATEGORY	CLASS	WINNER	TIME OR DISTANCE	CAR
29 July	Avignon, France	Course de côte du Mont Ventoux (hillclimb-21,600 kms)	Grand Touring	750 cc	M. Poltronieri	16'44"3	Fiat Abarth 750 GT
18-19 August	Congo Belga	Rallye du Lac Kivu	Grand Touring	750 cc	Castellier	---	Fiat Abarth 750 GT
3 September	Monza, Italy	Coppa Intereuropa (circuit-6,300 kms x 1 hour)	Grand Touring	750 cc	E. Carini	kms 126,804	Fiat Abarth 750 GT
3 September	Cuneo, Italy	Garessio-Colle San Bernardo (hillclimb-7,100 kms)	Grand Touring	750 cc	E. Gonella	5'36"7	Fiat Abarth 750 GT
16 September	Brescia, Italy	Sarezzo-Lumezzane (hillclimb)	Grand Touring	750 cc	R. Cattini	6'02"7	Fiat Abarth 750 GT
16 September	Sorrento, Italy	Sorrento-Sant'Agata (hillclimb-10,500 kms)	Grand Touring	750 cc	Ricceri	---	Fiat Abarth 750 GT
16 September	Modena, Italy	Vignola-Guiglia (hillclimb)	Grand Touring	750 cc	L. Cussini	4'46"3	Fiat Abarth 750 GT
23 September	Trento, Italy	Trento-Bondone (hillclimb-17,300 kms)	Grand Touring	750 cc	G. Guidotti	11'46"4	Fiat Abarth 750 GT
23 September	Mitholz, Switzerland	Mitholz-Kandersteg Bergrennen (hillclimb-2,640 kms)	Grand Touring	750 cc	Gurzeler	2'23"0	Fiat Abarth 750 GT
23 September	Viterbo, Italy	Coppa del Cimino (hillclimb-10,300 kms)	Grand Touring	750 cc	A. Thiele	6'05"7	Fiat Abarth 750 GT
15 October	Padova, Italy	Treponti-Castelnuovo (hillclimb-8,020 kms)	Grand Touring	750 cc	E. Carini	5'48"5	Fiat Abarth 750 GT
20 October	Roma, Italy	Gran Premio Roma (circuit)	Grand Touring	750 cc	M. Poltronieri	---	Fiat Abarth 750 GT
28 October	Torino, Italy	Sassi-Traforo del Pino (hillclimb-3,800 kms)	Grand Touring	750 cc	E. Gonella	4'13"4	Fiat Abarth 750 GT
5 November	Rimini, Italy	Coppa San Marino (hillclimb-8,350 kms)	Grand Touring	750 cc	A. Thiele	6'35"9	Fiat Abarth 750 GT
5 November	Bologna, Italy	Campionato Regionale A.C. Emiliani	Grand Touring	750 cc	L. Cussini	14'47"7	Fiat Abarth 750 GT
1957							
27 February	Sestrière, Italy	Rallye del Sestrière (2832 kms)	Grand Touring	750 cc	Colantuomo-De Sanctis	---	Fiat Abarth 750 GT
27 March	Liège, Belgium	Course de côte de La Roche (hillclimb-2,200 kms)	Grand Touring	750 cc	Perlingeri	5'19"7	Fiat Abarth 750 GT
31 March	Napoli, Italy	Coppa Agnano-Cappella dei Cangiani (hillclimb-8,660 kms)	Grand Touring	750 cc	Harris	1'21"8	Fiat Abarth 750 GT
1 April	Verona, Italy	Trofeo Torricelle (hillclimb-1,880 kms x 2 laps)	Grand Touring	750 cc	L. Cussini	3'29"2	Fiat Abarth 750 GT
3 April	Huy, France	Douze Heures d'Huy (circuit)	Grand Touring	750 cc	Harris	---	Fiat Abarth 750 G
7 April	Frascati, Italy	Frascati-Tuscolo (hillclimb-4,500 kms)	Grand Touring	750 cc	Brocchi	4'05"3	Fiat Abarth 750 GT
12 April	Milano, Italy	Coppa Milano-Sanremo	Grand Touring	750 cc	L. Pozzoli	---	Fiat Abarth 750 GT
14 April	Palermo, Italy	Giro di Sicilia 1058 kms)	Grand Touring	750 cc	G. Minaudo	13h16'37"0	Fiat Abarth 750 GT
1 May	Bologna, Italy	Bologna-San Luca (hillclimb-2,080 kms)	Grand Touring	750 cc	L. Cussini	1'38"9	Fiat Abarth 750 GT
2 May	Monza, Italy	Trofeo Vigorelli (circuit-5,750 kms)	Grand Touring	750 cc	Megna	29'51"7	Fiat Abarth 750 GT
6 May	Firenze, Italy	Coppa della Consuma (hillclimb-15,000 kms)	Grand Touring	750 cc	V. Feroldi de Rosa	7'40"2	Fiat Abarth 750 GT
19 May	Brescia, Italy	Mille Miglia (1597 kms)	Grand Touring	750 cc	A. Thiele	13h32'33"0	Fiat Abarth 750 GT
11 August	Macerata, Italy	Sfercia-Camerino (hillclimb)	Grand Touring	750 cc	L. Cussini	5'20"8	Fiat Abarth 750 GT
18 August	Trapani, Italy	Trapani-Monte Erice (hillclimb-6,670 kms)	Grand Touring	750 cc	Bologna	10'35"0	Fiat Abarth 750 GT
4 September	Cuneo, Italy	Garessio-Colle San Bernardo (hillclimb-7,100 kms)	Grand Touring	750 cc	V. Gianni	5'05"3	Fiat Abarth 750 GT
7 September	Monza, Italy	Coppa Intereuropa (circuit-5,750 kms x 1 hour)	Grand Touring	750 cc	M. Poltronieri	kms 136,135	Fiat Abarth 750 GT
7 September	Brescia, Italy	Trofeo Lumezzane (hillclimb-8,500 kms)	Grand Touring	750 cc	R. Cattini	5'59"9	Fiat Abarth 750 GT
8 September	Martigny, Switzerland	Martigny-La Forclaz (hillclimb)	Grand Touring	750 cc	Dieter	7'17"6	Fiat Abarth 750 GT
16 September	Viterbo, Italy	Coppa del Cimino (hillclimb-10,300 kms)	Grand Touring	750 cc	V. Feroldi de Rosa	6'05"1	Fiat Abarth 750 GT
21-22 September	Mitholz, Switzerland	Mitholz-Kandersteg Bergrennen (hillclimb-2,640 kms)	Grand Touring	750 cc	W. Lier	2'20"5	Fiat Abarth 750 GT
30 September	Genova, Italy	Pontedecimo-Giovi (hillclimb-9,560 kms)	Grand Touring	750 cc	L. Cussini	6'53"2	Fiat Abarth 750 GT
7 October	Trieste, Italy	Trieste-Opicina (hillclimb-10,150 kms)	Grand Touring	750 cc	L. Cussini	5'35"6	Fiat Abarth 750 GT
7 October	Modena, Italy	Coppa Intersociale	Grand Touring	750 cc	Fogli	14'18"6	Fiat Abarth 750 GT
7 October	Sorrento, Italy	Sorrento-Sant'Agata (hillclimb-10,500 kms)	Grand Touring	750 cc	V. Feroldi de Rosa	9'11"9	Fiat Abarth 750 GT
14 October	Monza, Italy	Coppa Leopoldo Carri (circuit-2,405 kms per 15 laps)	Grand Touring	750 cc	M. Poltronieri	15'9"3	Fiat Abarth 750 GT
20 October	Sondrio, Italy	Rallye della Valtellina	Grand Touring	750 cc	Roiatti (overall)	---	Fiat Abarth 750 GT
26 October	Torino, Italy	Sassi-Superga (hillclimb-4,500 kms)	Grand Touring	750 cc	E. Prinoth	3'37"6	Fiat Abarth 750 GT
11 November	Modena, Italy	Scuderia Arena	Grand Touring	750 cc	L. Cussini	---	Fiat Abarth 750 GT
1958							
---	Leopoldville, Congo	24 Heures de Leopoldville	Grand Touring	750 cc	Grosso	---	Fiat Abarth 750 GT
6 January	Genève, Switzerland	Rallye de Genève	Grand Touring	750 cc	Hunger-Tranzal	---	Fiat Abarth 750 GT
6 January	Asmara, Eritrea	Taulud Circuit (48,225 kms)	Grand Touring	750 cc	Soldi	---	Fiat Abarth 750 GT
6 January	Melbourne, Australia	---	Grand Touring	750 cc	Grosso	---	Fiat Abarth 750 GT
2-3 February	Megève, France	Rallye Neige et Glace	Grand Touring	750 cc	Bertoglio	---	Fiat Abarth 750 GT
9 February	France	Rallye International des Routes du Nord	Grand Touring	750 cc	Harris	---	Fiat Abarth 750 GT

DATE	LOCATION	RACE	CATEGORY	CLASS	WINNER	TIME OR DISTANCE	CAR
24/28 February	Sestrière, Italy	Rallye del Sestrière (2832 kms)	Grand Touring	750 cc	L. Cussini	---	Fiat Abarth 750 GT
9 March	Napoli, Italy	Coppa Agnano-Cappella dei Cangiani (hillclimb-8,660 kms)	Grand Touring	750 cc	Mancino	---	Fiat Abarth 750 GT
16 March	Modena, Italy	University Championship	Grand Touring	750 cc	E. Crivellari	---	Fiat Abarth 750 GT
16 March	Avola, Italy	Coppa Belmonte (hillclimb-7,400 kms)	Grand Touring Grand Touring	500 cc 750 cc	Vuture V. Arena	--- ---	Fiat Abarth 500 GT Fiat Abarth 750 GT
22 March	Sebring, Florida, USA	12 Hours (circuit-5,205 miles)	Grand Touring	750 cc	Kessinger-Thiele-West	kms 1238,612	Fiat Abarth 750 GT
31 March	Genève, Switzerland	Kilomètre d'Eau Morte	Grand Touring	750 cc	Bianchi	---	Fiat Abarth 750 GT
14 April	Frascati, Italy	Coppa 2 Passi	Grand Touring	750 cc	Stefani	---	Fiat Abarth 750 GT
20 April	Palermo, Italy	Giro di Sicilia (1072 kms)	Grand Touring	750 cc	V. Arena	---	Fiat Abarth 750 GT
27 April	Napoli, Italy	Gran Premio di Napoli	Grand Touring Grand Touring	500 cc 750 cc	Argenti L. Cussini	--- ---	Fiat Abarth 500 GT Fiat Abarth 750 GT
27 April	Rimini, Italy	Coppa San Marino (hillclimb-8,350 kms)	Grand Touring Grand Touring	500 cc 750 cc	Argenti L. Cussini	--- ---	Fiat Abarth 500 GT Fiat Abarth 750 GT
4 May	Bologna, Italy	Bologna-San Luca (hillclimb-2,080 kms)	Grand Touring Grand Touring	500 cc 750 cc	O. Cappelli E. Prinoth	--- ---	Fiat Abarth 500 GT Fiat Abarth 750 GT
4 May	Haway Island	Annual Hawaiian Island Week Race	Grand Touring	750 cc	F. Aldhaus	---	Fiat Abarth 750 GT
4 May	Asmara, Eritrea	Asmara-Keren-Asmara	Grand Touring	500 cc	Nardoni	---	Fiat Abarth 500 GT
15 May	Palermo, Italy	Corsa al Monte Pellegrino (hillclimb-8,750 kms)	Grand Touring Grand Touring	500 cc 750 cc	O. Cappelli C. Avventurieri	--- ---	Fiat Abarth 500 GT Fiat Abarth 750 GT
1 June	Firenze, Italy	Coppa della Consuma (hillclimb-10,500 kms)	Grand Touring	750 cc	E. Prinoth	---	Fiat Abarth 750 GT
1 June	Cagliari, Italy	Rallye di Sardegna (772 kms)	Grand Touring	750 cc	Stefani-Tecilla	---	Fiat Abarth 750 GT
8 June	Napoli, Italy	Targa Vesuvio (9,200 kms)	Grand Touring	750 cc	S. Mandato	---	Fiat Abarth 750 GT
8 June	Livorno, Italy	Rallye dei 7 Passi	Grand Touring	750 cc	Cecchini-Ansano	---	Fiat Abarth 750 GT
15 June	Asiago, Italy	Coppa Città di Asiago (hillclimb-14,500 kms)	Grand Touring	750 cc	E. Prinoth	---	Fiat Abarth 750 GT
15 June	Piacenza, Italy	Castell'Arquato-Vernasca (hillclimb-9,775 kms)	Grand Touring	750 cc	M. Poltronieri	---	Fiat Abarth 750 GT
15 June	Varese, Italy	Varese-Campo dei Fiori (hillclimb-10,000 kms)	Grand Touring Grand Touring	500 cc 750 cc	Argenti C. Avventurieri	--- ---	Fiat Abarth 500 GT Fiat Abarth 750 GT
22 June	Vallelunga, Italy	6 Ore di Roma (circuit-1,800 kms x 25 laps)	Grand Touring Grand Touring	500 cc 750 cc	Argenti C. Avventurieri	--- ---	Fiat Abarth 500 GT Fiat Abarth 750 GT
22 June	Pavia, Italy	Coppa Lomellina	Grand Touring	750 cc	Bottacchi	---	Fiat Abarth 750 GT
22 June	Dübendorf, Switzerland	Gara nazionale di slalom	Grand Touring	750 cc	Bianchi	---	Fiat Abarth 750 GT
22 June	Brescia, Italy	Mille Miglia (1500 kms)	Grand Touring	750 cc	Poltronieri-Bassi	---	Fiat Abarth 750 GT
29 June	Forlì, Italy	Predappio-Rocca delle Caminate (hillclimb-4,000 kms)	Grand Touring GranTurismo	500 cc 750 cc	Argenti L. Cussini	--- ---	Fiat Abarth 500 GT Fiat Abarth 750 GT
29 June	Verbois, Switzerland	Course de côte de Verbois (hillclimb-1,150 kms)	Grand Touring	750 cc	Munger	---	Fiat Abarth 750 GT
29 June	Knockke, Belgium	Acceleration race (standing 1 km)	Grand Touring Grand Touring	500 cc 750 cc	Swaters Blary	--- ---	Fiat Abarth 500 GT Fiat Abarth 750 GT
6 July	Bolzano, Italy	Bolzano-Mendola (hillclimb-14,500 kms)	Grand Touring	750 cc	E. Prinoth	11'56"1	Fiat Abarth 750 GT
6 July	Ferrara, Italy	Rallye degli Estensi (110 kms)	Grand Touring	750 cc	Ansano-Cecchini	---	Fiat Abarth 750 GT
13 July	Trento, Italy	Trento-Bondone (hillclimb-17,300 kms)	Grand Touring Grand Touring	500 cc 750 cc	G. Kerschbaumer E. Prinoth	13'30"3 11'01"4	Fiat Abarth 500 GT Fiat Abarth 750 GT
20 July	Cuneo, Italy	Garessio-Colle San Bernardo (hillclimb-7,100 kms)	Grand Touring	750 cc	Castellina-Nova	5'12"4	Fiat Abarth 750 GT
3 August	Catanzaro, Italy	Giro delle Calabrie (720 kms)	Grand Touring Grand Touring	500 cc 750 cc	O. Cappelli V. Feroldi de Rosa	--- ---	Fiat Abarth 500 GT Fiat Abarth 750 GT
31 August	Torino, Italy	Sassi-Superga (hillclimb-4,500 kms)	Grand Touring Grand Touring	500 cc 750 cc	G. Kerschbaumer E. Prinoth	4'36"5 3'36"6	Fiat Abarth 500 GT Fiat Abarth 750 GT
7 September	Monza, Italy	Coppa Intereuropa (circuit-5,750 kms x 1 hour)	Grand Touring Grand Touring	500 cc 750 cc	O. Cappelli M. Poltronieri	--- ---	Fiat Abarth 500 GT Fiat Abarth 750 GT
7 September	Chieti, Italy	Coppa Città di Chieti (hillclimb-6,000 kms)	Grand Touring	750 cc	R. Recchi	5'7"3	Fiat Abarth 750 GT
7 September	Belluno, Italy	Coppa Nevegal (hillclimb)	Grand Touring	750 cc	E. Prinot	5'9"2	Fiat Abarth 750 GT
14 September	Brescia, Italy	Trofeo Lumezzane (hillclimb-10,300 kms)	Grand Touring	750 cc	R. Falotti	5'54"2	Fiat Abarth 750 GT
14 September	Asti, Italy	Mulino del Rocco-Cucconato (hillclimb-2,860 kms)	Grand Touring	750 cc	V. Gianni	2'50"5	Fiat Abarth 750 GT
14 September	Sorrento, Italy	Sorrento-Sant'Agata (hillclimb-10,500 kms)	Grand Touring	750 cc	S. Mandato	8'52"8	Fiat Abarth 750 GT
14 September	Verona, Italy	Stallavena-Boscochiesanuova (hillclimb-15,300 kms)	Grand Touring	750 cc	E. Prinoth	9'15"0	Fiat Abarth 750 GT
14 September	Angola	Angola Rally	Grand Touring	750 cc	R. de Fonseca	---	Fiat Abarth 750 GT
0 November	Venezuela	Venezuela Grand Prix	Grand Touring	750 cc	A. Manzini	6h03'11"0	Fiat Abarth 750 GT

DATE	LOCATION	RACE	CATEGORY	CLASS	WINNER	TIME OR DISTANCE	CAR
1959							
26 February	Sestrière, Italy	Rallye del Sestrière (2832 kms)	Grand Touring	750 cc	Castellina-Frescobaldi (overall)		Fiat Abarth 750 GT
8 March	Napoli, Italy	Coppa Agnano-Cappella dei Cangiani (hillclimb-8,660 kms)	Grand Touring	750 cc	M. De Luca	5'00"8	Fiat Abarth 750 GT
15 March	Avola, Italy	Coppa Belmonte (hillclimb-7,400 kms)	Grand Touring Grand Touring	500 cc 750cc	A. Tiné N. Bertozzi	7'53"1 6'48"8	Fiat Abarth 500 GT Fiat Abarth 750 GT
22 March	Sebring (Florida), USA	12 Hours (circuit-5,205 miles)	Grand Touring	750 cc	Rutan-Richards-Cuomo	kms 1272,088	Fiat Abarth 750 GT
29 March	Borne, Luxembourg	Course de côte de Borne (hillclimb)	Grand Touring	750 cc	H. Wagner	1'53"3	Fiat Abarth 750 GT
6 April	Daytona (Florida), USA	Daytona Beach 100 Miles (circuit-2,500 miles)	Grand Touring	1000 cc	R. Cattini	---	Fiat Abarth 1000 RM
12 April	Monza, Italy	12 Ore di Monza (circuit-10,000 kms for 12 hours)	Grand Touring	750 cc	Poltronieri-Zagato	kms 1841,565	Fiat Abarth 750 RM
12 April	Liège, Belgium	Course de côte de La Roche (hillclimb-2,200 kms)	Grand Touring	750 cc	A. Spiers	5'17"9	Fiat Abarth 750 GT
19 April	Treviso, Italy	Gymkana	---		A. Manavello	---	Fiat Abarth 750 RM
19 April	Frascati, Italy	Trofeo Venturi (hillclimb-4,500 kms)	Grand Touring	750 cc	S. Mandato	3'56"0	Fiat Abarth 750 GT
19 April	Genève, Switzerland	Kilomètre d'Eau Morte	Grand Touring	750 cc	Munger	0'33"6	Fiat Abarth 750 RM
25 April	Rimini, Italy	Rimini-San Marino (hillclimb-8,350 kms)	Grand Touring	750 cc	M. Leto di Priolo	6'01"8	Fiat Abarth 750 RM
30 April	Roma, Italy	Giornata dei Primati (6,617 kms)	Grand Touring Grand Touring	500 cc 750 cc	Priori 'Mariottino'	1'18"8 1'05"7	Fiat Abarth 500 GT Fiat Abarth 750 GT
3 May	Monza, Italy	Coppa Sant'Ambroeus (circuit-10,000 kms x 10 laps)	Grand Touring	750 cc	M. Leto di Priolo	37'34"3	Fiat Abarth 750 RM
3 May	Verbois, Switzerland	Course de côte de Verbois (hillclimb-1,150 kms)	Grand Touring	750 cc	F. Hussy	1'06"9	Fiat Abarth 750 GT
7 May	Palermo, Italy	Valdesi-Santa Rosalia (hillclimb-7,000 kms)	Grand Touring	500 cc	Tiné	6'16"5	Fiat Abarth 500 GT
8 May	Palermo, Italy	Corsa al Monte Pellegrino (hillclimb-8,750 kms)	Grand Touring	750 cc	Guercia	6'18"0	Fiat Abarth 750 RM
10 May	Wallberg, Germany	Wallberg Rennen (hillclimb)	Grand Touring	750 cc	L. Fischer	5'13"6	Fiat Abarth 750 RM
10 May	Mitholz, Switzerland	Mitholz-Kandersteg Bergrennen (hillclimb-2,640 kms)	Grand Touring	750 cc	'Basilius'	2'12"7	Fiat Abarth 750 RM
17 May	Vallelunga, Italy	6 Ore di Vallelunga (1,800 kms 25 laps)	Grand Touring	750 cc	E. Prinoth	31'35"8	Fiat Abarth 750 RM
24 May	Piacenza, Italy	Castell'Arquato-Vernasca (hillclimb-9,775 kms)	Grand Touring	750 cc	M. Poltronieri	6'27"8	Fiat Abarth 750 RM
24 May	Palermo, Italy	Targa Florio (circuit-72,000 kms x 10 laps)	Grand Touring	750 cc	Carini-Prinoth	12h53'39"0	Fiat Abarth 750 RM
30 May	Brescia, Italy	Mille Miglia	Grand Touring	500 cc	Giuberti-Bocchio	---	Fiat Abarth 500 GT
			Grand Touring	750 cc	Prinot-Pitscheider		Fiat Abarth 750 RM
31 May	Payerne, Switzerland	Slalom de Payerne (hillclimb-4,800 kms)	Grand Touring	750 cc	R. Egli	3'15"0	Fiat Abarth 750 RM
31 May	Pferdsfeld, Germany	Internationales Pferdsfeldrennen (circuit)	Grand Touring	750 cc	M. Poltronieri	38'31"8	Fiat Abarth 750 RM
2 June	Lecce, Italy	Rallye del Salento	Grand Touring	750 cc	Milella (overall)	---	Fiat Abarth 750 GT
7 June	Udine, Italy	Cividale-Castelmonte (hillclimb)	Grand Touring	750 cc	Zuliani	5'55"2	Fiat Abarth 750 GT
7 June	Parma, Italy	Salsomaggiore-Sant'Antonio (hillclimb)	Grand Touring Grand Touring	500 cc 750 cc	Prandoni M. Poltronieri	7'31"7 6'49"6	Fiat Abarth 500 GT Fiat Abarth 750 RM
7 June	Firenze, Italy	Coppa della Consuma (hillclimb-10,500 kms)	Grand Touring	750 cc	L. Cussini	7'05"2	Fiat Abarth 750 RM
13 June	Genève, Switzerland	Rallye de Genève	Grand Touring	750 cc	R. Egli	---	Fiat Abarth 750 RM
14 June	Bari, Italy	Rallye dei Trulli e delle Grotte	Grand Touring	750 cc	Milella (overall)	---	Fiat Abarth 750 GT
14 June	Como, Italy	Trofeo Val d'Intelvi (hillclimb-8,700 kms)	Grand Touring	750 cc	D. Leto di Priolo	7'08"8	Fiat Abarth 750 RM
14 June	Reggio Emilia, Italy	Compiano-Vetto d'Enzo (hillclimb)	Grand Touring Grand Touring	500 cc 750 cc	Giuberti Sala	4'30"3 3'57"4	Fiat Abarth 500 GT Fiat Abarth 750 RM
14 June	Napoli, Italy	Targa Vesuvio (9,200 kms)	Grand Touring	750 cc	S. Mandato	10'01"3	Fiat Abarth 750 GT
14 June	Thun, Switzerland	Autoslalom	Grand Touring	up to 1000 c	Gurzeler	5'45"8	Fiat Abarth 750 GT
21 June	Venezia, Italy	Rallye Lido Venezia	---	---	E. Prinoth	---	Fiat Abarth 750 RM
21 June	Asiago, Italy	Coppa Città di Asiago (hillclimb-14,500 kms)	Grand Touring	750 cc	M. Leto di Priolo	10'01"1	Fiat Abarth 750 RM
28 June	Dübendorf, Switzerland	Gara nazionale di slalom	Grand Touring	up to 1000 cc	'Basilius'	7'16"6	Fiat Abarth 750 RM
29 June	Cosenza, Italy	Coppa della Sila (hillclimb-14,000 kms)	Grand Touring Grand Touring	500 cc 750 cc	Tiné Lenza	25'29"1 21'01"4	Fiat Abarth 500 GT Fiat Abarth 750 GT
29 June	Novara, Italy	Vegliomosso-S. Maria (hillclimb-3,400 kms x 2)	Grand Touring	750 cc	E. Prinoth	5'56"4	Fiat Abarth 750 RM
29 June	Rovigo, Italy	Rallye di Rovigo	---	---	Cavallari (overall)	---	Fiat Abarth 750 GT
29 June	Rovigo, Italy	Rallye del Polesine	Grand Touring	750 cc	Cavallari (overall)	---	Fiat Abarth 750 GT
29 June	Forlì, Italy	Predappio-Rocca delle Caminate (hillclimb-4,000 kms)	Grand Touring	750 cc	Cavallari	4'14"6	Fiat Abarth 750 GT
5 July	Bolzano, Italy	Bolzano-Mendola (hillclimb-14,500 kms)	Grand Touring	750 cc	C. Fabi	12'12"9	Fiat Abarth 750 RM
10-11 July	Endenwale, Canada	Endenwale Receway					Fiat Abarth 750 RM
12 July	Trento, Italy	Trento-Bondone (hillclimb-17,300 kms)	Grand Touring	750 cc	E. Prinoth	15'17"0	Fiat Abarth 750 RM
12 July	Strasbourg, France	Course de côte de Turckheim Trois Epis	Grand Touring	up to 1300 cc	R. Maggiorini	4'38"0	Fiat Abarth 750 RM
19 July	Cuneo, Italy	Garessio-Colle San Bernardo (hillclimb-7,100 kms)	Grand Touring	750 cc	C. Fabi	4'58"3	Fiat Abarth 750 RM

DATE	LOCATION	RACE	CATEGORY	CLASS	WINNER	TIME OR DISTANCE	CAR
19 July	Perpignan, France	Rallye de Roussillon	---	---	---	---	Fiat Abarth 750 GT
26 July	Trieste, Italy	Trieste-Opicina (hillclimb-10,150 kms)	Grand Touring	750 cc	E. Prinoth	5'08"5	Fiat Abarth 750 RM
8 August	Trapani, Italy	Trapani-Monte Erice (hillclimb-6,670 kms)	Grand Touring Grand Touring	500 cc 750 cc	Tiné Lenza	7'07"0 5'38"1	Fiat Abarth 500 GT Fiat Abarth 750 GT
9 August	Roma, Italy	Coppa Gallenga (hillclimb-10,400 kms)	Grand Touring	750 cc	L. Cussini	4'31"2	
12 August	Fribourg, Switzerland	Slalom de la Poya	Grand Touring	750 cc	'Basilius'	---	Fiat Abarth 750 RM
14 August	Greece	Salita al Monte Prilerimos	Grand Touring	750 cc	M. Leto di Priolo	3'35"9	Fiat Abarth 750 RM
15 August	Westwood, Canada	Circuito di Westwood	Grand Touring	750 cc	---	---	Fiat Abarth 750 RM
15 August	Rodi, Greece	Circuito di Rodi	Grand Touring	750 cc	M. Leto di Priolo (overall)	---	Fiat Abarth 750 RM
16 August	Gaisberg, Austria	Grosser Bergpreis von Österreich (hillclimb-8,600 kms)	Grand Touring	750 cc	C. Fabi	12'14"6	Fiat Abarth 750 RM
17 August	Lime Rock (Connecticut), USA	8 Hours of Lime Rock (circuit)	--	--	Callahan-Penske	---	Fiat Abarth 750 RM
18 August	Brindisi, Italy	Coppa Fasano-Selva (hillclimb-7,400 kms)	Grand Touring	750 cc	L. Cussini	4'44"0	Fiat Abarth 750 RM
29 August	Davos, Switzerland	Klosters-Davos Bergrennen	Grand Touring	750 cc	R. Egli	3'56"9	Fiat Abarth 750 RM
30 August	Torino, Italy	Sassi-Superga (hillclimb-4,500 kms)	Grand Touring	750 cc	E. Prinoth	4'05"2	Fiat Abarth 750 RM
6 September	Terni, Italy	Coppa Borzacchini	Grand Touring	750 cc	Montecatini	2'54"4	Fiat Abarth 750 GT
6 September	Aosta,Italy	Aosta-Pila (hillclimb-11,200 kms)	Grand Touring	750 cc	E. Prinoth	4'11"4	Fiat Abarth 750 RM
13 September	Green Acres, Canada	---	---	750 cc	Polivka	---	Fiat Abarth 750 RM
13 September	St. Ursanne, Switzerland	Saint Ursanne-Les Rangiers (hillclimb-4,700 kms)	Grand Touring	750 cc	A. Thiele	kms 153,237	Fiat Abarth 750 RM
20 September	Viterbo, Italy	Coppa del Cimino (hillclimb-10,300 kms)	Grand Touring	750 cc	L. Cussini (overall)	5'38"4	Fiat Abarth 750 RM
20 September	Genova, Italy	Pontedecimo-Giovi (hillclimb-9,560 kms)	Grand Touring	750 cc	E. Zagato	7'01"9	Fiat Abarth 750 RM
23 September	Zeltweg, Austria	Grosser Preis von Österreich (circuit-3,200 kms x 157 laps)	Grand Touring	up to 1300 cc	Friederich	---	Fiat Abarth 750 RM
27 September	Monthléry, France	Coupes de Paris (circuit-7,784 kms)	Grand Touring	up to 1000 cc	C. Fabi	58'53"3	Fiat Abarth 750 RM
27 September	Wurzeenpass, Germany	Bergrennen Wurzenpass (hillclimb)	Grand Touring	750 cc	E. Prinoth	3'14"0	Fiat Abarth 750 RM
27 September	Wasgau, Germany	ADAC Wasgau Bergprüfung (hillclimb-5,100 kms x 2)	Grand Touring	up to 1000 cc	L. Fischer	4'19"9	Fiat Abarth 750 RM
27 September	Catania, Italy	Catania-Etna (hillclimb-28,200 kms)	Grand Touring	750 cc	Bertazzi	22'23"7	Fiat Abarth 750 RM
27 September	Genève, Switzerland	Course de côte de Marchairuz (hillclimb-4,800 kms per 2)	Grand Touring	750 cc	'Basilius'	3'50"4	Fiat Abarth 750 RM
27 September	Brescia, Italy	Trofeo Lumezzane (hillclimb-8,500 kms	Grand Touring	750 cc	Saruggia	6'53"4	Fiat Abarth 750 GT
30 September	Chieti, Italy	Coppa Città di Chieti (hillclimb-6,000 kms)	Grand Touring	750 cc	A. Rizzo	4'29"0	Fiat Abarth 750 GT
4 October	Caltanissetta, Italy	Coppa Nissena (hillclimb-10,000 kms)	Grand Touring	750 cc	Guercia	5'48"3	Fiat Abarth 750 RM
4 October	Innsbruck, Austria	Preis von Tirol (circuit-2,800 kms x 25 laps)	Grand Touring	750 cc	M. Leto di Priolo	22'53"8	Fiat Abarth 750 RM
4 October	Belluno, Italy	Coppa Alpe del Nevegal (hillclimb-7,500 kms)	Grand Touring	750 cc	Cavallari	5'20"3	Fiat Abarth 750 GT
4 October	Ravenna, Italy	Gymkana a Lugo di Romagna	Grand Touring	over 600 cc	Minardi	---	Fiat Abarth 750 GT
4 October	Wolfsfelden, Germany	Wolfsfeldener Bergprüfung (hillclimb)	Grand Touring	up to 1000 cc	L. Fischer	---	Fiat Abarth 750 RM
5 October	Montlhéry, France	Coupe du Salon (circuit-3,300 kms x 30 laps)	Grand Touring	up to 1000 cc	M. Poltronieri	42'44'4	Fiat Abarth 750 RM
10 October	Graz, Austria	---	Grand Touring	750 cc	E. Gassler	25'45"9	Fiat Abarth 750 GT

1960

DATE	LOCATION	RACE	CATEGORY	CLASS	WINNER	TIME OR DISTANCE	CAR
5 March	Pfersfeld, Germany	Internationales Pferdsfeldrennen (circuit)	Grand Touring	up to 1000 cc	W. Seidel		Fiat Abarth 700 BA
19 March	Frascati, Italy	Frascati-Tuscolo (hillclimb-4,500 kms)	Grand Touring	up to 850 cc	Montecatino	3'54"1	Fiat Abarth 750 GT
25 March	Sebring (Florida), USA	4 Hours (circuit-5,205 miles)	Grand Touring	up to 1000 cc	Richards-Callanan	kms 477,009	Fiat Abarth 750 RM
27 March	Napoli, Italy	Coppa Agnano-Cappella dei Cangiani (hillclimb-8,660 kms)	Grand Touring	700 cc	Andreotti	5'05"6	Fiat Abarth 700 BA
3 April	Verona, Italy	Stallavena-Boscochiesanuova (hillclimb-15,300 kms)	Grand Touring	700 cc	M. Leto di Priolo	8'41"8	Fiat Abarth 700 BA
3 April	Avola, Italy	Coppa Belmonte (hillclimb-7,400 kms)	Grand Touring	850 cc	V. Arena	6'27"9	Fiat Abarth 750 RM
7-9 April	Genève, Switzerland	Rallye de Genève	Grand Touring	700 cc	Richard-Chabert	---	Fiat Abarth 700 BA
10 April	Reggio Calabria, Italy	Targa del Busento	Grand Touring Grand Touring	700 cc 850 cc	Casagrande V. Arena	5'11"0 4'39"4	Fiat Abarth 700 BA Fiat Abarth 750 RM
24 April	Genova, Italy	Coppa Mare e Monti	Grand Touring Grand Touring	500 cc up to 1150 cc	Prandoni D. Leto di Priolo	21'09"1 19'54"4	Fiat Abarth 500 Fiat Abarth 700 BA
24 April	Liège, Belgium	Course de côte de La Roche (hillclimb-2,200 kms)	Grand Touring	850 cc	R. Egli	4'34"5	Fiat Abarth 750 RM
1 May	Monza, Italy	12 Ore di Monza (circuit-10,000 kms)	Grand Touring	700 cc	Sala-Rigamonti-Zagato	kms 1837,55	Fiat Abarth 700 BA
8 May	Firenze, Italy	Rallye di Toscana (250 kms)	Grand Touring	up to 850 cc	Guandalini	---	Fiat Abarth 700

DATE	LOCATION	RACE	CATEGORY	CLASS	WINNER	TIME OR DISTANCE	CAR
14-15 May	Germany	Semperit Rallye	Grand Touring	850 cc	Kohlweiss	---	Fiat Abarth 750 RM
14-15 May	Mitholz, Switzerland	Mitholz-Kandersteg Bergrennen (hillclimb-2,640 kms)	Grand Touring Grand Touring Grand Touring Sport	700 cc 850 cc up to 1000 cc 850 cc	J. Egli 'Basilius' R. Stierli R. Zehr	2'15"6 2'11"8 2'10"9 2'09"4	Fiat Abarth 700 BA Fiat Abarth 750 RM Fiat Abarth 750 RM Fiat Abarth 750
15 May	Palermo, Italy	Corsa al Monte Pellegrino (hillclimb-8,750 kms)	Grand Touring	850 cc	V. Arena	6'15"9	Fiat Abarth 750 RM
22 May	Piacenza, Italy	Castell'Arquato-Vernasca (hillclimb-9,775 kms)	Grand Touring Grand Touring	500 cc 850 cc	R. Ciarpaglini R. Fusina	7'39"0 6'54"0	Fiat Abarth 500 Fiat Abarth 750 RM
6 June	Hockenheim, Germany	Hockenheim 6 Stunde (circuit-6,768 kms)	Grand Touring Grand Touring Grand Touring	500 cc 700 cc 1000 cc	R. Ciarpaglini Sala-Sala Seidel-Poltronieri	kms 708,653 kms 916,350 kms 955,407	Fiat Abarth 500 Fiat Abarth 700 BA Fiat Abarth 750 RM
6 June	Firenze, Italy	Coppa della Consuma (hillclimb-12,500 kms)	Grand Touring Grand Touring	700 cc up to 1150 cc	G. Castellina F. Sapia	7'09"3 7'34"1	Fiat Abarth 700 BA Fiat Abarth 750 RM
12 June	Le Touquet, France	Rallye du Touquet (410 kms)	---	---	Richard-Herriau	---	Fiat Abarth 700 BA
12 June	Terni, Italy	Coppa Borzacchini (hillclimb-4,000 kms)	Grand Touring	850 cc	R. Arfé	2'53"1	Fiat Abarth 750 RM
12 June	Parma, Italy	Salsomaggiore-Monte Sant'Antonio (hillclimb)	Grand Touring	850 cc	M. Leto di Priolo	6'35"0	Fiat Abarth 700 BA
12 June	Como, Italy	Trofeo Val d'Intelvi (hillclimb-8,700 kms)	Grand Touring	850 cc	E. Gonella	7'00"1	Fiat Abarth 700 BA
12 June	Austria	Gara di Campionato Austriaco (hillclimb)	Sport	1300 cc	Kohlweiss	2'04"2	Fiat Abarth 750 RM
12-14 June	Brescia, Italy	Rallye della Stampa	---	---	M. Ventriglia (overall)	---	Fiat Abarth 750 RM
16 June	Reggio Emilia, Italy	Compiano-Vetto d'Enza (hillclimb)	Grand Touring	850 cc	G. Capra	3'46"0	Fiat Abarth 750 RM
19 June	Roma, Italy	Coppa Gallenga (hillclimb-10,400 kms)	Grand Touring	700 cc	Andreotti	8'05"3	Fiat Abarth 700 BA
26 June	Varese, Italy	Varese-Campo dei Fiori (hillclimb-10,000 kms)	Grand Touring	850 cc	F. Sabbiac	8'34"5	Fiat Abarth 750 RM
26 June	Palermo, Italy	Valdesi-Santa Rosalia (hillclimb-7,000 kms)	Grand Touring	850 cc	G. Bonomo	4'47"0	Fiat Abarth 750 RM
29 June	Monza, Italy	Gran Premio Grand Touring (circuit-5,750 kms x 2 hours)	Grand Touring Grand Touring	700 cc 850 cc	G. Castellina W. Seidel	kms 295,133 kms 294,922	Fiat Abarth 700 BA Fiat Abarth 750 RM
29 June	Forlì, Italy	Predappio-Rocca delle Caminate (hillclimb-4,000 kms)	Grand Touring Grand Touring	700 cc 850 cc	G. Capa Massotti	4'11"7 4'47"1	Fiat Abarth 700 BA Fiat Abarth 750 RM
3 July	Bolzano, Italy	Bolzano-Mendola (hillclimb-14,500 kms)	Grand Touring	850 cc	G. Capa	12'25"3	Fiat Abarth 700 BA
3 July	Napoli, Italy	Targa Vesuvio (9,200 kms)	Grand Touring Sport	700 cc 850 cc	'Mariottino' Casagrande	8'35"2 9'08"8	Fiat Abarth 700 BA Fiat Abarth 750 RM
3 July	San Benedetto del Tronto, Italy	San Benedetto-Acquaviva (hillclimb-6,500 kms)	Grand Touring	700 cc	Alba	5'53"4	Fiat Abarth 700 BA
10 July	Trento, Italy	Trento-Bondone (hillclimb-17,300 kms)	Grand Touring Grand Touring	700 cc 850 cc	G. Sala E. Gassler	16'14"3 16'16"3	Fiat Abarth 700 BA Fiat Abarth 750 RM
17 July	Cuneo, Italy	Garessio-Colle San Bernardo (hillclimb-7,100 kms)	Grand Touring Grand Touring	700 cc up to 1150 cc	E. Gonella F. Sapia	4'59"1 4'57"3	Fiat Abarth 700 BA Fiat Abarth 750 RM
24 July	Trieste, Italy	Trieste-Opicina (hillclimb-10,150 kms)	Grand Touring Grand Touring	700 cc 850 cc	G. Castellina F. Sapia	5'15"8	Fiat Abarth 750 Fiat Abarth 750 RM
6 August	Cosenza, Italy	Coppa della Sila (hillclimb-14,000 kms)	Grand Touring	850 cc	R. Arfé	23'35"1	Fiat Abarth 750 RM
7 August	Trapani, Italy	Trapani-Monte Erice (hillclimb-6,670 kms)	Grand Touring Grand Touring	700 cc 850 cc	V. Arena M. Perillo	5'36"2 5'35"3	Fiat Abarth 700 BA Fiat Abarth 750 RM
7 August	Freiburg, Germany	ADAC Bergpreis Freiburg-Schauinsland (hillclimb-11,200 kms)	Grand Touring	850 cc	R. Stierli	9'36"4	Fiat Abarth 750 RM
13 August	Sierre, Switzerland	Course de côte de Sierre-Montana (hillclimb-12,000 kms)	Grand Touring Grand Touring Sport	700 cc 850 cc 850 cc	J. Egli R. Stierli J. Oggenfuss	10'18"2 10'29"4 9'54"0	Fiat Abarth 700 BA Fiat Abarth 750 RM Fiat Abarth 750
21 August	Catanzaro, Italy	Giro Automobilistico dei Due Mari (170 kms)	Grand Touring Grand Touring	700 cc 850 cc	V. Arena Testarotta	1h53'54"9 1h55'37"5	Fiat Abarth 700 BA Fiat Abarth 750 RM
28 August	Brindisi, Italy	Cappa Fasano-Selva (hillclimb-7,400 kms)	Grand Touring	850 cc	Gambardella	5'16"4	Fiat Abarth 750 RM
28 August	Chieti, Italy	Coppa Città di Chieti (hillclimb-6,200 kms)	Grand Touring Grand Touring	700 cc 850 cc	S. Smedile A. Rizzo	3'58"4 4'22"4	Fiat Abarth 700 BA Fiat Abarth 750 RM
28 August	Villars, Switzerland	Ollon-Villars (hillclimb-8,000 kms)	Grand Touring	700 cc	J. Egli	6'02"4	Fiat Abarth 700 BA
29 August	Torino, Italy	Sassi-Superga (hillclimb-4,500 kms)	Grand Touring	850 cc	F. Sapia	3'29"1	Fiat Abarth 750 RM
4 September	Adenau, Germany	Internationales 500 Kms Nürburgring Rennen (circuit-22,810 kms x 22 laps)	Grand Touring	850 cc	Castellina-Vinatier	4h32'48"3	Fiat Abarth 750 RM
11 September	Brescia, Italy	Trofeo Lumezzane (hillclimb-8,500 kms)	Grand Touring Grand Touring	700 cc 850 cc	V. Arena F. Sapia	5'16"0 5'18"0	Fiat Abarth 700 BA Fiat Abarth 750 RM
26 September	Catania, Italy	Catania-Etna (hillclimb-28,000 kms)	Grand Touring Grand Touring	500 cc 850 cc	Spoto V. Arena	26'42"0 21'33"0	Fiat Abarth 500 Fiat Abarth 750 RM
26 September	Osimo, Italy	Coppa Luigi Fagioli (hillclimb-5,300 kms)	Grand Touring Grand Touring	700 cc 850 cc	Toppetti Vimbel	3'14"4 3'24"4	Fiat Abarth 700 BA Fiat Abarth 750 RM

DATE	LOCATION	RACE	CATEGORY	CLASS	WINNER	TIME OR DISTANCE	CAR
28 September	Caltanissetta, Italy	Coppa Nissena (hillclimb-10,000 kms)	Grand Touring	850 cc	V. Arena	5'38"2	Fiat Abarth 750 RM
2 October	Genova, Italy	Pontedecimo-Giovi (hillclimb-9,560 kms)	Grand Touring Grand Touring	700 cc 850 cc	G. Castellina F. Sapia	6'37"3 6'44"2	Fiat Abarth 700 BA Fiat Abarth 750 RM
2 October	Genève, Switzerland	Course de côte de Marchairuz (hillclimb-4,800 kms x 2 laps)	Grand Touring Grand Touring Grand Touring Special Sport	750 cc 850 cc 850 cc 850 cc	J. Egli RR. Stierli 'Basilius' Terzariol	3'56"9 3'48"9 3'42"2 4'01"5	Fiat Abarth 750 RM Fiat Abarth 750 RM Fiat Abarth 750 RM Fiat Abarth 750 RM
2 October	Belluno, Italy	Coppa Alpe del Nevegal (hillclimb-7,500 kms)	Grand Touring	850 cc	---	4'52"2	Fiat Abarth 750 RM
9 October	Padova, Italy	Tre Ponti-Castelnuovo (hillclimb-8,020 kms)	Grand Touring	850 cc	E. Carini	5'48"5	Fiat Abarth 750 RM
10 October	Innsbruck, Austria	Preis von Tirol (circuit-2,800 kms x 12 laps)	Grand Touring	up to 1000 cc	C. Fabi	17'45"1	Fiat Abarth 750 RM
15 October	Modena, Italy	Coppa d'Oro A. C. Modena	Grand Touring	850 cc	L. Cussini	31'59"3	Fiat Abarth 750 RM
17 October	Schorndorf, Germany	Schorndorfer Bergrennen (hillclimb)	Grand Touring	850 cc	Lohssträter	4'41"6	Fiat Abarth 750 RM
17 October	Adenau, Germany	Nürburgring Rennen	Grand Touring	700 cc	W. Seidel	45'15"3	Fiat Abarth 700 BA
23 October	Montlhéry, France	1000 Km de Paris (circuit-7,784 kms x 129 laps)	Grand Touring	850 cc	Consten-Condriller (performance index)	---	Fiat Abarth 750 RM
30 October	Monza, Italy	Coppa Leopoldo Carri (circuit-2,405 kms x 20 laps)	Grand Touring	850 cc	'T. R. '	12'20"3	Fiat Abarth 750 RM
15 November	Palermo, Italy	Passo di Rigano-Bellolampo (hillclimb-9,750 kms)	Grand Touring	850 cc	M. Perillo	6'00"4	Fiat Abarth 750 RM
20 November	Messina, Italy	Coppa Colle San Rizzo (hillclimb-9,400 kms)	Grand Touring	850 cc	Fichera	8'18"7	Fiat Abarth 750 RM
8 December	Vallelunga, Italy	Coppa Petrolcaltex (circuit-1,800 kms x 25 laps)	Grand Touring	up to 1150 cc	Tomassini	19'48"8	Fiat Abarth 750 RM

1961

DATE	LOCATION	RACE	CATEGORY	CLASS	WINNER	TIME OR DISTANCE	CAR
---	Ciudad de Mejico, Messico	Circuito de Ciudad de Mejico	Grand Touring	750 cc	A. Silvestre	28'54"6	Fiat Abarth 750 RM
10-12 February	France	Routes du Nord	Grand Touring	700 cc	Richard-Chopin	---	Fiat Abarth 700 BA
12 February	Eisenstadt, Germany	Eisenstadter Bergprüfung (hillclimb)	Grand Touring	1000 cc	Steinbacher	3'41"2	Fiat Abarth 1000 BA
12 March	Napoli, Italy	Coppa Agnano-Cappella dei Cangiani (hillclimb-8,660 kms)	Grand Touring	850 cc	Kirke	5'07"0	Fiat Abarth 850 GT
12 March	Monza, Italy	Coppa Sant'Ambroeus (circuit 5,750 kms 10 laps)	Grand Touring	700 cc	C. Toppetti	20'28"6	Fiat Abarth 700 BA
24-25 March	Sebring (Florida), USA	4 Hours (circuit-5,205 miles)	Grand Touring	1000 cc	H. Washburn	kms 493,771	Fiat Abarth 1000 BA
25 March	Frascati, Italy	Frascati-Tuscolo (hillclimb-4,500 kms)	Grand Touring Grand Touring	700 cc 1000 cc	C. Toppetti B. Carotenuto	3'35"6 3'45"4	Fiat Abarth 700 BA Fiat Abarth 1000 BA
3 April	Modena, Italy	Coppa Tricolore	Grand Touring	up to 1150 cc	V. Venturi	12'15"8	Fiat Abarth 1000 BA
9 April	Genova, Italy	Coppa Riviera di Ponente (270 kms)			J. Demcenko	---	Fiat Abarth 750 GT
9 April	Verona, Italy	Stallavena-Boscochiesanuova (hillclimb-15,300 kms)	Grand Touring Grand Touring	700 cc 1000 cc	Schenk H. Demetz	8'44"2 7'59"1	Fiat Abarth 700 BA Fiat Abarth 1000 BA
9 April	Bruxelles, Belgium	Coupe de Bruxelles	Grand Touring	1000 cc	L. Van Hophen	28'50"3	Fiat Abarth 1000 BA
16 April	Aspern, Austria	Preis von Wien (circuit-2,600 kms x 38 laps)	Grand Touring	1000 cc	K. Foitek	36'39"8	Fiat Abarth 1000 BA
30 April	Achum, Germany	ADAC Flugplatzrennen	Grand Touring	1000 cc	H. Zich	---	Fiat Abarth 1000 BA
30 April	Genova, Italy	Coppa Mare Monti	Grand Touring	1000 cc	H. Demetz (overall)	17'22"7	Fiat Abarth 1000 BA
7 May	Monza, Italy	Coppa Alberto Ascari (circuit-10,000 kms x 6 hours)	Grand Touring Grand Touring	700 cc 1000 cc	Guandini-Cintolesi Rigamonti-Guichet	kms 901,080 kms 989,880	Fiat Abarth 700 BA Fiat Abarth 1000 BA
7 May	Mitholz, Switzerland	Mitholz-Kandersteg Bergrennen (hillclimb-2,640 kms)	Grand Touring Grand Touring	850 cc 1000 cc	Aeberli P. Ettmüller	2'11"0 1'59"4	Fiat Abarth 750 RM Fiat Abarth 1000 BA
7 May	Pferdsfeld, Germany	Internationales Pferdsfeldrennen (circuit)	Grand Touring	1000 cc	H. Zich	30'59"5	Fiat Abarth 1000 BA
14 May	München, Germany	Flugplatzrennen München (circuit)	Grand Touring	1000 cc	M. Hönig	16'59"7	Fiat Abarth 1000 BA
14 May	Piacenza, Italy	Castell'Arquato-Vernasca (hillclimb-9,775 kms)	Grand Touring Grand Touring	850 cc up to 1150 cc	Sestilli A. Becchetti	6'38"8 6'22"0	Fiat Abarth 750 RM Fiat Abarth 1000 BA
14 May	Asiago, Italy	Coppa Città di Asiago (hillclimb-14,500 kms)	Grand Touring	700 cc	L. Cella	9'53"6	Fiat Abarth 700 BA
14 May	Montlhéry, France	Grand Prix de Paris (circuit-3,300 kms x 15 laps)	Grand Touring	700 cc	R. Richard	31'41"0	Fiat Abarth 700 BA
21 May	Udine, Italy	Cividale-Castelmonte (hillclimb)	Grand Touring Grand Touring	700 cc 1000 cc	L. Galli Sacilotto	6'14"1 7'25"4	Fiat Abarth 700 BA Fiat Abarth 1000 BA
21 May	Parma, Italy	Salsomaggiore-Sant'Antonio (hillclimb)	Grand Touring Grand Touring	700 cc 1000 cc	C. Toppetti H. Demetz	6'23"4 6'06"3	Fiat Abarth 700 BA Fiat Abarth 1000 BA
21 May	Verbois, Switzerland	Course de côte de Verbois (hillclimb-1,150 kms)	Grand Touring	up to 1300 cc	Vaglio	1'08"2	Fiat Abarth 1000 BA
21 May	Palermo, Italy	Corsa al Monte Pellegrino (hillclimb-8,750 kms)	Grand Touring	700 cc	R. Arfé	6'22"3	Fiat Abarth 700 BA
27-28 May	Brescia, Italy	Mille Miglia	Grand Touring	700 cc	Pessina-Bonini	4h7'29"4	Fiat Abarth 700 BA
27-29 May	Dübendorf, Switzerland	Gara nazionale di slalom	Grand Touring Grand Touring	850 cc 1000 cc	Aeberli P. Hettmüller	11'40"3 11'07"1	Fiat Abarth 750 RM Fiat Abarth 1000 BA

DATE	LOCATION	RACE	CATEGORY	CLASS	WINNER	TIME OR DISTANCE	CAR
28 May	Nice, France	Criterium Jean Behra	Grand Touring	up to 1000 cc	R. RIchard	23'42"	Fiat Abarth 700 BA
28 May	Regensburg, Germany	Ratisbona Bergrennen (hillclimb)	Grand Touring	1000 cc	M. Hönig	2'52"0	Fiat Abarth 1000 BA
4 June	Firenze, Italy	Coppa della Consuma (hillclimb-12,500 kms)	Grand Touring	700 cc	L. Cella	7'10"0	Fiat Abarth 700 BA
			Grand Touring	1000 cc	Redes	7'30"2	Fiat Abarth 1000 BA
4 June	Gex, France	Course de côte de la Fancille (hillclimb)	Grand Touring	1000 cc	Cassano	7'24"2	Fiat Abarth 1000 BA
10-11 June	Le Mans, France	24 Heures du Mans (circuit-13,483 kms x 24 hours)	Grand Touring	up to 850 cc	Hulme-Hyslop	kms 3531,974	Fiat Abarth 850 BA
11 June	Roma, Italy	Coppa Gallenga (hillclimb-10,400 kms)	Grand Touring	700 cc	Smedile	6'17"8	Fiat Abarth 700 BA
			Grand Touring	1000 cc	V. Venturi	5'54"0	Fiat Abarth 1000 BA
18 June	Como, italia	Trofeo Val d'Intelvi (hillclimb-8,700 kms)	Grand Touring	700 cc	Rodes	7'03"2	Fiat Abarth 700 BA
			Grand Touring	up to 1150 cc	H. Demetz	6'20"2	Fiat Abarth 1000 BA
18 June	Bobbio, Italy	Bobbio-Penice (hillclimb)	Grand Touring	850 cc	G. Capra	10'52"7	Fiat Abarth 700 BA
18 June	Berchtesgarden, Germany	Rossfeld Alpen Bergpreis (hillclimb-6,000 kms)	Grand Touring	850 cc	E. Gabler	---	Fiat Abarth 700 BA
			Grand Touring	1000 cc	R. Stierli	---	Fiat Abarth 1000 BA
18 June	Terni, Italy	Coppa Borzacchini (hillclimb)	Grand Touring	700 cc	C. Toppetti	2'44"8	Fiat Abarth 700 BA
			Grand Touring	1000 cc	B. Carotenuto	2'39"2	Fiat Abarth 1000 BA
25 June	Cuneo, Italy	Garessio-Colle San Bernardo (hillclimb-7,100 kms)	Grand Touring	1000 cc	F. Sapia	4'57"4	Fiat Abarth 700 BA
			Grand Touring	1000 cc	Rodes	4'55"0	Fiat Abarth 1000 BA
25 June	Palermo, Italy	Valdesi-Santa Rosalia (hillclimb-7,000 kms)	Grand Touring	up to 1000 cc	M. Baratore	4'57"9	Fiat Abarth 750 RM
25 June	Forlì, Italy	Predappio-Rocca delle Caminate (hillclimb-4,000 kms)	Grand Touring	700 cc	G. Capra	4'04"0	Fiat Abarth 700 BA
			Grand Touring	1000 cc	V. Venturi	3'44"4	Fiat Abarth 1000 BA
25 June	Avignon, France	Course de côte du Mont Ventoux (hillclimb-21,700 kms)	Grand Touring	1000 cc	R. Richard	13'33"7	Fiat Abarth 1000 BA
25 June	Remich, Luxembourg	Remich Autoslalom	Grand Touring	850 cc	H. Wagner	2'18"4	Fiat Abarth 850 BA
25 June	Monza, Italy	Gran Premio Grand Touring (circuit-5,750 kms x 2 hours)	Grand Touring	700 cc	L. Cella	kms 289,491	Fiat Abarth 700 BA
			Grand Touring	1000 cc	M. Leto di Priolo	kms 317,360	Fiat Abarth 1000 BA
25 June	Reggio Emilia, Italy	Compiano-Vetto d'Enza (hillclimb)	Grand Touring	700 cc	G. Capra	3'17"4	Fiat Abarth 700 BA
2 July	San Benedetto del Tronto, Italy	Coppa San Benedetto-Acquaviva (hillclimb-6,500 kms)	Grand Touring	1000 cc	B. Carotenuto	5'35"0	Fiat Abarth 1000 BA
2 July	Bolzano, Italy	Bolzano-Mendola (hillclimb-14,500 kms)	Grand Touring	700 cc	Frisori	12'03"4	Fiat Abarth 700 BA
			Grand Touring	1000 cc	H. Demetz	11'02"2	Fiat Abarth 1000 BA
2 July	Siracusa, Italy	Tre Ore Notturna (circuit-5,500 kms)	Grand Touring	up to 1150 cc	Lavaggi		Fiat Abarth 1000 BA
2 July	Saint Ursanne, Switzerland	Course de côte de Saint Ursanne-Les Rangiers (hillclimb-4,700 kms)	Grand Touring	850 cc	Aeberli	3'06"3	Fiat Abarth 750 RM
			Grand Touring Special	850 cc	Baur	3'08"7	Fiat Abarth 850 BA
			Grand Touring	1000 cc	R. Stierli	2'50"0	Fiat Abarth 1000 BA
8 July	Clermont-Ferrand, France	6 Heures d'Auvergne (circuit-8,055 kms)	Grand Touring	1000 cc	J. Guichet	kms 628,722	Fiat Abarth 1000 BA
9 July	Trento, Italy	Trento-Bondone (hillclimb-17,300 kms)	Grand Touring	700 cc	C. Toppetti	15'16"8	Fiat Abarth 700 BA
			Grand Touring	1000 cc	H. Demetz	14'42"8	Fiat Abarth 1000 BA
12 July	Monopoli, Italy	Coppa Città di Monopoli (hillclimb-4,500 kms)	Grand Touring	850 cc	Triggiani	2'53'3	Fiat Abarth 850 BA
16 July	Aosta, Italy	Aosta-Pila (hillclimb-11,200 kms)	Grand Touring	700 cc	Sestilli	8'48"3	Fiat Abarth 700 BA
			Grand Touring	up to 1000 cc	Serra di Cassano	8'08"8	Fiat Abarth 1000 BA
			Sport	up to 1000 cc	E. Prinoth (overall)	7'44"1	Fiat Abarth 1000 BA
23 July	Trieste, Italy	Trieste-Opicina (hillclimb-10,150 kms)	Grand Touring	700 cc	E. Gonella	6'52"5	Fiat Abarth 700 BA
			Grand Touring	1000 cc	H. Demetz	6'26"1	Fiat Abarth 1000 BA
30 July	Platschberg, Germany	Platschbergrennen (hillclimb)	Grand Touring	1000 cc	P. Hammerschmied	2'19"4	Fiat Abarth 1000 BA
6 August	Sestrière, Italy	Cesana-Sestrière (hillclimb-10,400 kms)	Touring	850 cc	T. Zeccoli	7'45"5	Fiat Abarth 850 TC
			Grand Touring	700 cc	Frisori	7'22"0	Fiat Abarth 700 BA
			Grand Touring	1000 cc	C. Sigala	6'56"1	Fiat Abarth 1000 BA
13 August	Cuneo, Italy	Cuneo-Colle della Maddalena (hillclimb-17,000 kms)	Touring	850 cc	P. Renier	13'06"5	Fiat Abrth 850 TC
			Grand Touring	700 cc	De Bonis	13'16"7	Fiat Abarth 700
13 August	L'Aquila, Italy	Avezzano-Salviano (hillclimb-6,000 kms)	Touring	850 cc	S. Abbandonati	4'30"5	Fiat Abarth 850 TC
			Grand Touring	1000 cc	Chalfaut	4'23"0	Fiat Abarth 1000 BA
20 August	Taunus, Germany	Taunus Bergrennen (hillclimb)	Grand Touring	1000 cc	H. Zick	5'03"3	Fiat Abarth 1000 BA
27 August	Brindisi, Italy	Coppa Fasano-Selva (hillclimb-7,400 kms)	Touring	850 cc	B. Guarini	4'55"2	Fiat Abarth 850 TC
			Grand Touring	1000 cc	V. Venturi	4'19"3	Fiat Abarth 1000 BA
27 August	Klosters, Switzerland	Klosters-Davos (hillclimb)	Touring	850 cc	R. Egli	4'04"2	Fiat Abarth 850 TC
			Grand Touring	700 cc	J. Egli	3'57"0	Fiat Abarth 700 BA
			Grand Touring	1000 cc	R. Stierli	3'25"2	Fiat Abarth 1000 BA
3 September	Brescia, Italy	Sarezzo-Lumezzane (hillclimb)	Touring	850 cc	Pasotti	5'26"0	Fiat Abarth 850 TC
3 September	Pergusa, Italy	Coppa Città di Enna (circuit-4,797 kms)	Grand Touring	1000 cc	M. Perrillo	30'15"4	Fiat Abarth 1000 BA
3 September	Adenau, Germany	Internationales 500 Km Nürburgring Rennen (circuit = 22,810 kms x 22 laps)	Touring	850 cc	E. Furtmayr	4h05'46"8	Fiat Abarth 850 TC
			Grand Touring	700 cc	S. Liebl	4h08'23"4	Fiat Abarth 700 BA
			Grand Touring	1000 cc	C. M. Abate	4h25'16"8	Fiat Abarth 1000 BA

DATE	LOCATION	RACE	CATEGORY	CLASS	WINNER	TIME OR DISTANCE	CAR
10 September	Monza, Italy	Coppa Intereuropa (circuit-5,750 kms x 3 hours)	Grand Touring	1000 cc	A. Thiele	kms 484,654	Fiat Abarth 1000 BA
10 September	Gaisberg, Austria	Grosser Bergpreis von Österreich (hillclimb-8,600 kms)	Grand Touring	1000 cc	P. Hammerschmied	11'13"6	Fiat Abarth 1000 BA
17 September	Genova, Italy	Pontedecimo-Giovi (hillclimb-9,560 kms)	Touring Grand Touring Grand Touring	850 cc 700 cc 1000 cc	'Pam' Raffi H. Demetz	6'49"9 6'48"3 6'06"6	Fiat Abarth 850 TC Fiat Abarth 700 BA Fiat Abarth 1000 BA
24 September	Catania, Italy	Catania-Etna (hillclimb-28,200 kms)	Grand Touring Grand Touring	700 cc 1000 cc	C. Toppetti Lavaggi	21'32"8 20'17"5	Fiat Abarth 700 BA Fiat Abarth 1000 BA
24 September	Osimo, Italy	Coppa Luigi Fagioli (hillclimb-5,300 kms)	Touring Grand Touring Grand Touring	850 cc 700 cc 850 cc	R. Bacci C. Castellano G. Garzone	3'23"8 3'11"0 3'32"4	Fiat Abarth 850 TC Fiat Abarth 700 BA Fiat Abarth 850 BA
24 September	Hockenheim, Germany	12 Ore di Hockenheim (circuit-6,768 kms)	Grand Touring	1000 cc	H. Zick	---	Fiat Abarth 1000 BA
2 October	Rouen-Reims	Tour de France (14 special stages + 5000 kms)	Grand Touring	700 cc	Richard-Vinatier	---	Fiat Abarth 700 BA
8 October	Montlhéry, France	Coupe du Salon (circuit-3,300 kms)	Grand Touring	1000 cc	R. Richard	---	Fiat Abarth 1000 BA
8 October	Genève, Switzerland	Course de côte de Marchairuz (hillclimb-4,800 kms)	Grand Touring Grand Touring	700 cc 1000 cc	J. Egli R. Stierli	3'53"7 3'29"0	Fiat Abarth 700 BA Fiat Abarth 1000 BA
22 October	Cognac, France	Rallye du Cognac (regolarità)	Grand Touring	1000 cc	Guichet-Ruata (overall)	---	Fiat Abarth 1000 BA
24 October	Montlhéry, France	1000 Km de Paris	Grand Touring	1000 cc	Zeccoli-Sala	6h33'06"2	Fiat Abarth 1000 BA

1962

11 February	Asmara, Eritrea	Circuito di Taulud (circuit 48,225 kms)	Grand Touring	850 cc	L. Pastacaldi	33'12"	Fiat Abarth 850 TC
11 February	Chieti, Italy	Coppa Città di Chieti (hillclimb-6,200 kms)	Touring	850 cc	De Benedictis-Amati	3'55"2	Fiat Abarth 850 TC
23-24 February	Alicante, Spagna	Rallye de Alicante (regolarità-900 kms)	Grand Touring	1000 cc	V . Lepori	---	Fiat Abarth 1000 BA
11 March	Napoli, Italy	Agnano-Cappella dei Cangiani (hillclimb-4,900 kms)	Touring Grand Touring	850 cc 700 cc	C. Avventurieri C. Castellano	4'24"3 4'11"3	Fiat Abarth 850 TC Fiat Abarth 700 BA
18 March	Barcelona, Spagna	Carrera en cuesta de Costa Brava (hillclimb)	Grand Touring	1000 cc	V. Lepori	---	Fiat Abarth 1000 BA
23 March	Sebring (Florida), USA	Sebring 3 Hours (circuit-5,205 miles)	Grand Touring	1000 cc	B. McLaren	kms 385	Fiat Abarth 1000 BA
24 March	Sebring (Florida), USA	Sebring 12 Hours (circuit-5,205 miles)	Sport	1000 cc	Thiele-Guichet	kms 1724	Fiat Abarth 1000 BA
24-25 March	Valencia, Spagna	Rallye Folias (800 kms)	Grand Touring	1000 cc	V. Lepori	---	Fiat Abarth 1000 BA
25 March	Oise, France	Course de côte du Valois (hillclimb)	Grand Touring	1000 cc	Lobry	---	Fiat Abarth 1000 BA
25 March	Avola, Italy	Coppa Belmonte (hillclimb-7,400 kms)	Touring Grand Touring Grand Touring	up to 1150 cc 700 cc 1000 cc	A. Spadaro A. Monaco G. Lavaggi	6'12"4 6'21"5 5'46"2	Fiat Abarth 850 TC Fiat Abarth 700 BA Fiat Abarth 1000 BA
25 March	Frascati, Italy	Frascati-Tuscolo (hillclimb-4,200 kms)	Grand Touring	700 cc	C. Castellano	3'42"6	Fiat Abarth 700 BA
25 March	Arezzo, Italy	Rallye Vallate Aretine (266 kms)	Touring	850 cc	'Etrusco'	---	Fiat Abarth 850TC
1 April	Saint-Lô, France	Course de côte d'Hébécrevon (hillclimb-1,200 kms)	Grand Touring	1000 cc	P. Richard	0'45"2	Fiat Abarth 1000 BA
8 April	Montserrat, Spagna	Carrera en cuesta de Montserrat (hillclimb-7,000 kms)	Grand Touring	1000 cc	V. Lepori	5'33"	Fiat Abarth 1000 BA
8 April	Verona, Italy	Stallavena-Boscochiesanuova (hillclimb-15,300 kms)	Touring Grand Touring Grand Touring	850 cc 700 cc 1000	'Pam' G. Lado F. Ruata	8'45"2 8'32"4 8'26"4	Fiat Abarth 850 TC Fiat Abarth 700 BA Fiat Abarth 1000 BA
8 April	Grosseto, Italy	Rallye della Maremma (340 kms)	Touring	850 cc	M. Ventriglia	---	Fiat Abarth 850 TC
15 April	Digne, France	Course de côte du Mont de la Lure (hillclimb 12,700 kms)	Touring Grand Touring Sport	850 cc 1300 cc 1000 cc	T. Zeccoli G. Balzarini C. Clément	9'48"2 8'30"1 9'53"3	Fiat Abarth 850 TC Abarth Simca 1300 Fiat Abarth 1000 BA
15 April	Aspern, Austria	Gare nazionali in circuit	Grand Touring	1000 cc	M. Hönig	34'59"1	Fiat Abarth 1000 BA
15 April	Monza, Italy	Trofeo Grand Touring (circuit 5,750 kms x 1 hour)	Grand Touring Grand Touring	700 cc 1000 cc	'Matich' O. Sigala	kms 137,913 kms 144,380	Fiat Abarth 700 BA Fiat Abarth 1000 BA
15 April	Trieste, Italy	Rallye del Biancospino (148 kms)	Touring	850 cc	A. Bonduri	---	Fiat Abarth 850 TC
15 April	Roma, Italy	Trofeo E.Lazzaroni (regolarità-187 kms)	Touring	850 cc	F. Tecilla	---	Fiat Abarth 850 TC
15 April	Eberbach, Germany	Internationales Bergrennen Eberbach (hillclimb-3,000 kms)	Grand Touring	1000 cc	R.Stierli	4'05"5	Fiat Abarth 1000 BA
15 April	Marlboro (Maryland), USA	Marlboro Nat'ls (circuit-72 miles)	---	---	E. Hessert	---	Fiat Abarth 750 GT
22 April	Budapest, Hungary	Corsa Internazionale di Budapest (circuit-3,380 kms x 22 laps)	Grand Touring	1000 cc	P. Hammerschmied	---	Fiat Abarth 1000 BA
23 April	Beaufort, Luxembourg	Course de côte de Kautenbach (hillclimb)	Touring Grand Touring	850 cc 1000 cc	H. Wagner R. Stierli	2'34"3 2'15"3	Fiat Abarth Scorpione Fiat Abarth 1000
28-29 April	Steckborn, Switzerland	Steckborn-Eichhölzli (hillclimb)	Touring	850 cc	Pichler	2'22"1	Fiat Abarth 850 TC

DATE	LOCATION	RACE	CATEGORY	CLASS	WINNER	TIME OR DISTANCE	CAR
28-29 April	Danville (Virginia), USA	Danville Nat'ls (circuit-176 miles)	---	---	E. Astri	---	Fiat Abarth 750 GT
29 April	Lucca, Italy	Rallye di Lucca (187,500 kms)	Grand Touring	up to 1150 cc	S. Lunatici	---	Fiat Abarth 1000 BA
29 April	Macerata, Italy	Coppa Villa Potenza (hillclimb-5,200 kms)	Touring	850 cc	R. Bacci	2'59"3	Fiat Abarth 850 TC
29 April	Monthléry, France	Coupes de Vitesse (circuit 3,340 kms x 15 laps)	Grand Touring	1000 cc	C. Bobrowski	---	Fiat Abarth 1000 BA
1 May	Salò, Italy	Circuito del Garda (16,400 kms x 19 laps)	Grand Touring Grand Touring	700 cc 100 cc	'Matich' L. Scarfiotti	2h43'38"8 2h35'45"7	Fiat Abarth 700 BA Fiat Abarth 1000 BA
5 May	Nice, France	Criterium Jean Behra (kms 424)	Touring	850 cc	Bose-Conso	---	Fiat Abarth 850 TC
5 May	Wallberg, Germany	Wallberg-Rennen (hillclimb)	Grand Touring	1000 cc	M. Hönig	4'41"6	Fiat Abarth 1000 BA
5-6 May	Bordeaux, France	Rallye de Bordeaux (1375 kms)	Grand Touring	1000 cc	'Sarayac'-Daret	---	Fiat Abarth 1000 BA
5-6 May	Mitholz, Switzerland	Mitholz-Kandersteg (hillclimb-2,640 kms x 2)	Touring Grand Touring Grand Touring	850 cc 700 cc 1000 cc	Pichler Aeberli H. Illert	2'13"8 2'14"2 1'52"7	Fiat Abarth 850 TC Fiat Abarth 700 BA Fiat Abarth 1000 BA
6 May	Firenze, Italy	Rallye della Toscana (250 kms)	Touring	850 cc	F. Tecilla	---	Fiat Abarth 850 TC
6 May	Hockenheim, Germany	Circuito di Hockenheim (6,768 kms x 10 laps)	Touring Grand Touring	850 cc 1000 cc	H. Zick Plaut	--- ---	Fiat Abarth 850 TC Fiat Abarth 1000 BA
13 May	Cagliari, Italy	Rallye di Sardegna (772 kms)	Touring	850 cc	F. Tecilla	---	Fiat Abarth 850 TC
13 May	Piacenza, Italy	Castell'Arquato-Vernasca (hillclimb-9,775 kms)	Touring Grand Touring	850 cc 700 cc	Scarambone 'Felice'	6'45"5 6'41"2	Fiat Abarth 850 TC Fiat Abarth 700 BA
13 May	Cumberland (Maryland), USA	Cumberland Nat'ls (circuit-46,4 miles)	---	---	E. Hessert	---	Fiat Abarth 750 GT
13 May	Berlin, Germany	Avus Rennen (circuit-8,300 kms x 40 laps)	Grand Touring Grand Touring	700 cc 1000 cc	T. Zeccoli R. Jenny	1h53'2"2 1h52'23"8	Fiat Abarth 700 BA Fiat Abarth 1000 BA
13 May	Asiago, Italy	Coppa Città di Asiago (hillclimb-14,500 kms)	Touring Grand Touring Grand Touring	850 cc 700 cc 1000 cc	'Pam' 'Matich' Gonella	10'38"8 10'29"7 10'37"8	Fiat Abarth 850 TC Fiat Abarth 700 BA Fiat Abarth 1000 BA
20 May	Monthléry, France	Grand Prix de Paris (circuit-3,360 kms x 15 laps)	Touring Grand Touring	1000 cc 1000 cc	Schlesser R. Richard	32'33"3 25'26"3	Fiat Abarth 850 TC Fiat Abarth 1000 BA
20 May	Villefranche-sur-Saône, France	Course de côte de Marchant-Croix Rosier (hillclimb)	Touring Grand Touring	850 cc 1000 cc	T. Zeccoli H. Demetz	11'36"1 10'25"0	Fiat Abarth 850 TC Fiat Abarth 1000 BA
20 May	Dudelange, Luxembourg	Course de côte Tele-Dudelange (hillclimb-1,400 kms)	Touring	850 cc	H. Wagner	1'09"2/5	Fiat Abarth 850 TC
20 May	Francorchamps, Belgium	Gran Prix de Spa (circuit-14,106 kms)	Grand Touring	1000 cc	C. Bobrowski	1h0'3"8/10	Fiat Abarth 1000 BA
20 May	Villefranche sur Saône, France	Course de côte de Beaujolais (hillclimb-16,000 kms)	Touring Grand Touring	850 cc 1000 cc	T. Zeccoli H. Demetz	11'36"1 10'25"0	Fiat Abarth 850 TC Fiat Abarth 1000 BA
20 May	Payerne, Switzerland	Slalom National de Payerne (hillclimb-5,200 kms x 2)	Touring Grand Touring Grand Touring	850 cc 700 cc 1000 cc	Pichler Aeberli H. Illert	4'09"7 4'16"5 3'52"6	Fiat Abarth 850 TC Fiat Abarth 700 BA Fiat Abarth 1000 BA
20 May	Wien, Austria	Martha Goldpokal (regolarità-1,200 kms)	Grand Touring	1000 cc	F. Kleine	---	Fiat Abarth 1000 BA
20 May	Metz, France	Course de côte d'Arry (hillclimb)	Touring	1000 cc	H. Wagner	50"	Fiat Abarth 850 T
20 May	Sassari, Italy	Alghero-Scala Picada (hillclimb-8,700 kms)	Touring Grand Touring	850 cc up to 1150 cc	A. Spataro Zaretti	5'44"7 5'50"5	Fiat Abarth 850 TC Fiat Abarth 700 BA
20 May	Wasgau, Germany	ADAC Wasgau Bergprüfung (hillclimb-5,100 kms x 2)	Grand Touring	1000 cc	G. Selbach	3'08	Fiat Abarth 1000 BA
26 May	Dübendorf, Switzerland	Gara nazionale di slalom	Touring Grand Touring Sport	850 cc 1000 cc 1000 cc	Buess H. Illert Egger	9'15"4 8'32"3 8'48"6	Fiat Abarth 850 TC Fiat Abarth 1000 BA Fiat Abarth 1000
26-27 May	Stuttgart (Arkansas), USA	Stuttgart Nat'ls (circuit-90 miles)	---	---	B. Stone	---	Fiat Abarth 750 GT
27 May	Palermo, Italy	Corsa al Monte Pellegrino	Touring Grand Touring Grand Touring	850 cc 700 cc 1000 cc	S. Sansica G.Bonomo W. Jenny	6'16"4 6'25"2 5'45"9	Fiat Abarth 850 TC Fiat Abarth 700 BA Fiat Abarth 1000 BA
27 May	Bologna, Italy	Bologna-Passo della Raticosa (hillclimb-32,700 kms)	Touring Grand Touring Grand Touring	850 cc 700 cc 1000 cc	A.Zucchi 'Felice' V. Venturi	22'38"2 23'20"8 22'09"5	Fiat Abarth 850 TC Fiat Abarth 700 BA Fiat Abarth 1000 BA
27 May	San Benedetto del Tronto, Italy	San Benedetto-Acquaviva (hillclimb-6,500 kms)	Touring Grand Touring Grand Touring	850 cc 700 cc 1000 cc	Rinaldi C. Castellan Valerio	5'21"1 5'09"4 5'20"0	Fiat Abarth 850 TC Fiat Abarth 700 BA Fiat Abarth 1000 BA
27 May	Marseille, France	Course de côte de la Sainte Baume (hillclimb-8,100 kms)	Touring Grand Touring	850 cc 1000 cc	Boso R. Richard	--- ---	Fiat Abarth 850 TC Fiat Abarth 1000 BA
31 May	Aix en Provence, France	Course de côte d'Aix-Saint Antonin (hillclimb-2,700 kms)	Touring Grand Touring	850 cc 1000 cc	Martin R. Richard	2'12"8 2'00"6	Fiat Abarth 850 TC Fiat Abarth 1000 BA
1-3 June	Cataluna, Spagna	Rallye de Cataluna	Grand Touring	1000 cc	V. Lepori	---	Fiat Abarth 1000 BA
2-3 June	Luxembourg	Tour de Luxembourg	Touring	850 cc	Wagner-Eschbourg	---	Fiat Abarth 850 TC

DATE	LOCATION	RACE	CATEGORY	CLASS	WINNER	TIME OR DISTANCE	CAR
3 June	München, Germany	Oberschleissheim Rennen (circuit)	Touring Grand Touring	850 cc 1000 cc	M. Hönig M. Hönig	--- 32'45"2	Fiat Abarth 850 TC Fiat Abarth 1000 BA
3 June	Touquet, France	Rallye du Touquet (410 kms)	Grand Touring	700 cc	Bichat	---	Fiat Abarth 700 BA
3 June	Piacenza, Italy	Bobbio-Penice (hillclimb-12,800 kms)	Touring Grand Touring Grand Touring	850 cc 850 cc up to 1150 cc	T.O. Bach A. Libeti 'Oscar'	10'24" 12'07" 10'14"1	Fiat Abarth 850 TC Fiat Abarth 700 BA Fiat Abarth 1000 BA
3 June	La Baule, France	Rallye de la Baule (170 kms)	Grand Touring	1000 cc	'Sarayac'	---	Fiat Abart 1000 BA
3 June	Gex, France	Course de côte de la Faucille (hillclimb-10,396 kms)	Grand Touring	1000 cc	R. Richard	6'48"6	Fiat Abarth 1000 BA
9-10 June	Udine, Italy	Rallye di Lignano (178 kms)	Touring Grand Touring	850 cc up to 1150 cc	S. Lipizer Minen	--- ---	Fiat Abarth 850 TC Fiat Abarth 700 BA
11 June	Parma, Italy	Parma-Poggio di Berceto (hillclimb-13,000 kms)	Touring Grand Touring Grand Touring	850 cc 700 cc 1000 cc	'Pam' 'Felice' E. Gonella	9'21"3 9'38"4 9'28"4	Fiat Abarth 850 TC Fiat Abarth 700 BA Fiat Abarth 1000 BA
16 June	Scheveningen, Holland	24 Uurs Rit-Scheveningen	Sport	---	H. van der Heijden Jr.	---	Fiat Abarth 850 TC
16-17 June	Lake (Wisconsin), USA	Elkart Nat'ls (circuit-152 miles)	---	---	E. Astri	---	Fiat Abarth 750 GT
17 June	Bourges, France	Grand Prix de la Châtre (circuit 1,267 kms x 20 laps)	Grand Touring	1000 cc	R. Richard	---	Fiat Abarth 1000 BA
17 June	Antwerpen, Belgium	Autoslalom Antwerpen	Grand Touring	up to 1150 cc	H.Wagner	1'51"7	Fiat Abarth 850 TC
17 June	Potenza, Italy	Coppa dei Laghi di Monticchio (regolarità-251 kms)	---	---	I. Tarsia	---	Fiat Abarth 850 TC
17 June	Firenze, Italy	Coppa della Consuma (hillclimb-12,500 kms)	Touring Grand Touring	850 cc 700 cc	'Pam' G. Orengo	6'48"1 7'07'3	Fiat Abarth 850 TC Fiat Abarth 700 BA
17 June	Avignon, France	Course de côte du Mont Ventoux (hillclimb-21,700 kms)	Grand Touring Touring Preparato Grand Touring	up to 1150 cc 850 cc up to 1150 cc	V. Venturi M. Martin 'Oscar'	6'38"2 15'25"6 10'14"1	Fiat Abarth 1000 BA Fiat Abarth 850 TC Fiat Abarth 1000 BA
21 June	Palermo, Italy	Valdesi-Santa Rosalia (hillclimb-7,000 kms)	Touring Grand Touring	850 cc 1000 cc	S. Sansica Walker	4'41"2 4'12"7	Fiat Abarth 850 TC Fiat Abarth 1000 BA
21 June	Forlì, Italy	Predappio-Rocca delle Caminate (hillclimb-4,000 kms)	Touring GranTurismo	850 cc 1000 cc	M. Nesti V. Venturi	3'53" 3'38"2	Fiat Abarth 850 TC Fiat Abarth 1000 BA
24 June	Monza, Italy	Gran Premio Touring (circuit-5,750 kms x 1 hour)	Touring	850 cc	G. Dalla Torre	kms 139,544	Fiat Abarth 850 TC
24 June	Amalfi, Italy	Coppa della Primavera (hillclimb-8,750 kms)	Touring Grand Touring Grand Touring	850 cc 700 cc 1000 cc	C. Avventurieri P. Salerno Pezzano	7'40"2 8'20"1 7'30"1	Fiat Abarth 850 TC Fiat Abarth 700 BA Fiat Abarth 1000 BA
24 June	Chianciano Terme, Italy	Rallye di Chianciano (hillclimb-8,750 kms)	Touring	850 cc	A. Degli Innocenti	---	Fiat Abarth 850 TC
29 June	Cuneo, Italy	Garessio-Colle San Bernardo (hillclimb-7,100 kms)	Touring Grand Touring Grand Touring	850 cc 700 cc 1000 cc	F. Ghezzi F. Sapia G. Acutis	5'00"7 4'57"1 4'55"7	Fiat Abarth 850 TC Fiat Abarth 700 BA Fiat Abarth 1000 BA
30 June	Torino, Italy	Rallye dei Rododendri (280 kms)	Touring	850 cc	A. Degli Innocenti	---	Fiat Abarth 850 TC
1 July	Neuchâtel, Switzerland	Course de côte de la Tourne (hillclimb-5,000 kms)	Touring Grand Touring Sport	850 cc 1000 cc 1300 cc	Laubacher H. Illert T. Spychiger	3'53"8 3'12"4 3'01"8	Fiat Abarth 850 TC Fiat Abarth 1000 BA Abarth Simca 1300
1 July	Bolzano, Italy	Bolzano-Mendola (hillclimb-14,400 kms)	Touring Grand Touring Grand Touring	850 cc 700 cc 1000 cc	'Pam' Bergamo O. Sigala	11'25"3 11'44"4 11'28"4	Fiat Abarth 850 TC Fiat Abarth 700 BA Fiat Abarth 1000 BA
7 July	Siracusa, Italy	Tre Ore Notturna (circuit-5,500 kms)	Touring Grand Touring	850 cc up to 1150 cc	Rollo A. Spatafora	--- ---	Fiat Abarth 850 TC Fiat Abarth 1000 BA
7-8 July	Garnett (Kansas), USA	Lake Garnett Nat'ls (circuit-70 miles)	---	---	B. Larson	---	Fiat Abarth 750 GT
8 July	Avellino, Italy	Rallye dell'Irpinia	---	---	F. Gomez	---	Fiat Abarth 850 TC
8 July	Trento, Italy	Trento-Bondone (hillclimb-17,300 kms)	Touring Grand Touring Grand Touring	850 cc 700 cc 1000 cc	G. Dalla Torre F. Sapia T. Spychiger	15'10"3 15'26"1 14'12"9	Fiat Abarth 850 TC Fiat Abarth 700 BA Fiat Abarth 1000 BA
8 July	Adenau, Germany	Nürburgring 12 Stunden (circuit-22,810 kms)	Touring	850 cc	Springer-Plaut	1277,360 kms	Fiat Abarth 850 T
8 July	München, Germany	ADAC Flughafenrennen (circuit)	Grand Touring	1000 cc	M. Hönig	---	Fiat Abarth 1000 BA
15 July	Maron, France	Course de côte de Macon-Solutre (hillclimb-2,650 kms)	Grand Touring	1000 cc	Egger	1'37"4	Fiat Abarth 1000 BA
15 July	Clermont Ferrand, France	Course Trophée d'Auvergne (circuit-8,055 kms x 38 laps)	Sport	1300 cc	L. Bianchi	---	Abarth Simca 1300
15 July	Monopoli, Italy	Coppa Città di Monopoli (hillclimb-4,500 kms)	Touring Grand Touring	850 cc 1000 cc	Barattolo B. Guarini	2'48" 2'31"4	Fiat Abarth 850 TC Fiat Abarth 1000 BA
15 July	Vinci, Italy	Vinci-San Baronto (hillclimb-11,000 kms)	Touring Grand Touring	850 cc 700 cc	F. Glori G. Capra	8'38" 8'55"2	Fiat Abarth 850 TC Fiat Abarth 700 BA

DATE	LOCATION	RACE	CATEGORY	CLASS	WINNER	TIME OR DISTANCE	CAR
15 July	Sestrière, Italy	Cesana-Sestrière (hillclimb-10,400 kms)	Touring	850 cc	'Pam'	7'11"0	Fiat Abarth 850 TC
			Grand Touring	700 cc	'Felice'	7'11"9	Fiat Abarth 700 BA
			Grand Touring	1000 cc	G. Acutis	6'54"8	Fiat Abarth 1000 BA
			Sport	1000 cc	Deserti	6'42"8	Fiat Abarth 1000 SP
22 July	Trieste, Italy	Trieste-Opicina (hillclimb-10,150 kms)	Touring	850 cc	F. Ghezzi	6'00"8	Fiat Abarth 850 TC
			Grand Touring	700 cc	F. Sapia	5'50"8	Fiat Abarth 700BA
			Grand Touring	1000 cc	V. Venturi	5'27"9	Fiat Abarth 1000 BA
22 July	Freiburg, Germany	ADAC Bergpreis Freiburg-Schauinsland (hillclimb-11,200 kms)	Grand Touring	1000 cc	E. Mahle	7'52"3	Fiat Abarth 1000 BA
			Sport	1300 cc	H. Herrmann	15'07"8	Abarth Simca 1300
22 July	Elbeuf, France	Course de côte de Lyons-La Forêt (hillclimb-1,250 kms)	Touring	850 cc	R. Richard	1'14"2	Fiat Abarth 850 TC
			Grand Touring	1000 cc	Bichat	1'11"2	Fiat Abarth 850 TC
25 July	Luxembourg	Slalom du Findel	Touring	850 cc	H. Wagner	---	Fiat Abarth 850 TC
			Grand Touring	1000 cc	H. Wagner	---	Fiat Abarth 850 T
25 July	Villars, Switzerland	Ollon-Villars (hillclimb-8,000 kms)	Touring	850 cc	Buess	6'16"5	Fiat Abarth 850 TC
			Grand Touring	1000 cc	H. Illert	5'19"9	Fiat Abarth 1000 BA
29 July	Belluno, Italy	Coppa Agordo-Frassenè (hillclimb-9,000 kms)	Touring	850 cc	U. Di Giorgio	7'35"2	Fiat Abarth 850 TC
			Grand Touring	700 cc	G. Capra	6'49"6	Fiat Abarth 700 BA
			Grand Touring	1000 cc	Scotti	6'45"4	Fiat Abarth 1000 BA
4-5 August	Carpentersville (Illinois), USA	Meadowdale Nat'ls (circuit-131 miles)	---	---	B. Larson	---	Fiat Abarth 750 GT
5 August	Adenau,Germany	AVD Rheinland-Pfalz-Preis Nürburgring (circuit-22,810 kms x 6 laps)	Grand Touring	1000 cc	H. Zick	1h13'45"	Fiat Abarth 1000 BA
6 August	Trapani, Italy	Trapani-Monte Erice (hillclimb-6,670 kms)	Touring	850 cc	S. Sansica	5'23"3	Fiat Abarth 850 TC
			Grand Touring	700 cc	A. Scimone	5'50"4	Fiat Abarth 700 BA
			Grand Touring	1000 cc	Riolo	5'15"4	Fiat Abarth 1000 BA
11-12 August	Mont Dore, France	Course de côte du Mont Dore (hillclimb-5,000 kms)	Touring Preparato	up to 1150 cc	Bianchi	4'02"2	Fiat Abarth 850 TC
			Grand Touring	up to 1150 cc	Bichat	3'57"4	Fiat Abarth 1000 BA
12 August	Sciacca, Italy	Coppa Città di Sciacca (hillclimb-17,500 kms)	Touring	up to 1150 cc	'Pam'	12'07"6	Fiat Abarth 850 TC
			Grand Touring	700 cc	A. Scimone	13'04"6	Fiat Abarth 700 BA
			Grand Touring	1000 cc	Riolo	11'51"4	Fiat Abarth 1000 BA
15 August	Pergusa, Italy	Coppa Città di Enna (circuit-4,797 kms x 65 laps)	Grand Touring	700 cc	Pessina	---	Fiat Abarth 700 BA
			Grand Touring	1000 cc	'Pam'	1h54'59"	Fiat Abarth 1000 BA
19 August	Chieti, Italy	Coppa Città di Chieti (hillclimb-6,200 kms)	Touring	850 cc	Amati/De Benedictis ex aequo	3'55"2	Fiat Abarth 850 TC
19 August	Reggio Calabria, Italy	Santo Stefano-Gambarie (hillclimb-7,830 kms)	Touring	850 cc	'Trinacria'	6'27"2	Fiat Abarth 850 TC
			Grand Touring	1000 cc	Pezzano	6'20"	Fiat Abarth 1000 BA
26 August	Avezzano, Italy	Trofeo Micangeli (hillclimb-6,000 kms)	Grand Touring	up to 1150 cc	Valerio	4'06"2	Fiat Abarth 1000 BA
26 August	Luxembourg	Slalom de Ettelbruck	Touring	850 cc	H. Wagner	2'41"3	Fiat Abarth 850 TC
26 August	Brindisi, Italy	Coppa Selva di Fasano (hillclimb-7,400 kms)	Touring	850 cc	A. Zucchi	4'43"7	Fiat Abarth 850 TC
			Grand Touring	850 cc	C. Marazia	5'08"2	Fiat Absrth 850 BA
			Grand Touring	1000 cc	V. Venturi	4'15"	Fiat Abarth 1000 BA
2 September	Roma, Italy	Coppa Gallenga (hillclimb-10,400 kms)	Touring	850 cc	S. Sansica	5'43"8	Fiat Abarth 850 TC
			Grand Touring	700 cc	Scaglia	6'00"9	Fiat Abarth 700 BA
			Grand Touring	1000 cc	Riolo	5'23"	Fiat Abarth 1000 BA
2 September	Grenoble, France	Course de côte Chamrousse (hillclimb-17,200 kms)	Touring	1000 cc	M. Bianchi	13'30"4	Fiat Abarth 850 TC
			Sport	up to 2000 cc	J. Guichet	12'19"3	Abarth Simca 1300
2 September	Adenau, Germany	Internationales 500 Km Nürburgring Rennen (circuit-22,810 kms x 22 laps)	Touring	850 cc	'Pam'	4h1'42"4	Fiat Abarth 850 TC
			Grand Touring	700 cc	T. Spychiger	4h16'51"6	Fiat Abarth 700 BA
			Grand Touring	1000 cc	E. Mahle	4h07'46"4	Fiat Abarth 1000 BA
2 September	Bologna, Italy	Rallye dell'Alto Appennino Tosco-Emiliano (132 kms)	---	---	Angiolini	---	Fiat Abarth 850 TC
9 September	Catania, Italy	Catania-Etna (hillclimb-28,000 kms)	Touring	up to 1150 cc	A. Spataro	20'31"8	Fiat Abarth 850 TC
			Grand Touring	1000 cc	A. Monaco	19'52"	Fiat Abarth 1000 BA
9 September	Dijon, France	Course de côte d'Urcy	Grand Touring	1000 cc	C. Bobrowski	---	Fiat Abarth 1000 BA
9 September	Gaisberg, Austria	Gaisberg Bergpreis von Österreich (hillclimb-8,625 kms)	Touring	850 cc	K. Zuegg	5'38"2	Fiat Abarth 850 TC
			Grand Touring	1000 cc	H. Demetz	5'04"2	Fiat Abarth 850 TC
14-15 September	Vaals, Holland	Heuvelklim Vaals (hillclimb-1,200 kms)	Touring	850 cc	H. Wagner	1'04"2	Fiat Abarth 850 TC
16 September	Ascoli Piceno, Italy	Corsa al Colle San Marco (hillclimb-10,200 kms)	Touring	850 cc	L. Nasalli Rocca	7'22"3	Fiat Abarth 850 TC
			Grand Touring	700 cc	'Clu'	7'08"4	Fiat Abarth 700 BA
			Grand Touring	1000 cc	Primo	7'29"4	Fiat Abarth 1000 BA
16 September	Brescia, Italy	Trofeo Lumezzane (hillclimb-8,500 kms)	Touring	850 cc	F. Ghezzi	5'09"4	Fiat Abarth 850 TC
			Grand Touring	700 cc	'Felice'	5'07"2	Fiat Abarth 700 BA
			Grand Touring	1000 cc	A. Becchetti	5'15"5	Fiat Abarth 1000 BA
16 September	Innsbruck, Austria	Timmelsjoch-Hochalpen Bergrennen (hillclimb-12,050 kms)	Grand Touring	1000 cc	H. Demetz	2'44"8	Fiat Abarth 1000 BA

DATE	LOCATION	RACE	CATEGORY	CLASS	WINNER	TIME OR DISTANCE	CAR
16 September	Long Island (NY), USA	Bridgehampton Race (circuit-2,85 miles x 87 laps)	--- ---	--- ---	Weiser Grossman	--- ---	Fiat Abarth 850 BA Fiat Abarth 1000 BA
15-23 September	Rouen-Reims	Tour de France (14 prove speciali-5000 kms)	Touring		Bianchi-Ickx	---	Fiat Abarth 850 TC
22 September	Watkins Glen (NY), USA	Glen National (circuit-101 miles)	---	---	E. Hessert	---	Fiat Abarth 750 GT
23 September	Dudelange, Luxembourg	Slalom de Dudelange	Touring	850 cc	H. Wagner	---	Fiat Abarth 850 TC
23 September	Osimo, Italy	Coppa Luigi Fagioli (hillclimb-5,300 kms)	Touring Grand Touring Grand Touring	850 cc 850 cc 1000 cc	A. Zucchi Suardi V. Venturi	3'07"4 3'26"4 3'01"	Fiat Abarth 850 TC Fiat Abarth 700 BA Fiat Abarth 1000 BA
23 September	Montlhéry, France	Coupes de Paris (circuit-3,600 kms x 18 laps)	Grand Touring	1000 cc	C. Bobrowski	---	Fiat Abarth 1000 BA
23 September	Belluno, Italy	Coppa Nevegal (hillclimb-7,500 kms)	Touring Grand Touring Grand Touring	850 cc 700 cc 1000 cc	J. Trivellato Pinchetti 'Matich'	4'41"3 4'43"4 4'31"4	Fiat Abarth 850 TC Fiat Abarth 700 BA Fiat Abarth 1000 BA
30 September	Namur, Belgium	Course de côte de Namur (hillclimb)	Touring	850 cc	H. Wagner	1'50"2	Fiat Abarth 850 TC
3-6 October	Andorra	Rallye d'Andorra			Branche	---	Fiat Abarth 850 TC
7 October	Montlhéry, France	Course du Salon (circuit-3,600 kms x 18 laps)	Touring Grand Touring	850 cc 1000 cc	Demortemart C. Bobrowski	--- ---	Fiat Abarth 850 TC Fiat Abarth 1000 BA
7 October	Innsbruck, Austria	Preis von Tirol (circuit-2,800 kms)	Touring Grand Touring Sport	1000 cc 1000 cc 1300 cc	M. Bianchi H. Herrmann T. Spychiger	15'43"6 21'40" 26'11"42	Fiat Abarth 850 TC Fiat Abarth 1000 BA Abarth Simca 1300
7 October	Monza, Italy	Coppa Leopoldo Carri (circuit 2,405 kms x 20 laps)	Touring	850 cc	'Pam'	25'21"9	Fiat Abarth 850 TC
7 October	Pescara, Italy	Cronoscalata Svolte di Popoli (hillclimb-8,000 kms)	Touring Grand Touring	850 cc 1000 cc	L. Nasalli Rocca Del Pozzo	4'10"4 4'23"7	Fiat Abarth 850 TC Fiat Abarth 1000 BA
14 October	Monza, Italy	Coppa d'Autunno (circuit-5,750 kms x 10 laps)	Touring Grand Touring Grand Touring	850 cc 700 cc 1000 cc	F. Ghezzi 'Bonarda' O.Sigala	20'52"6 20'22" 24'13"9	Fiat Abarth 850 TC Fiat Abarth 700 BA Fiat Abarth 1000 BA
14 October	Cognac, France	Rallye du Cognac	Grand Touring	1300 cc	J. Guichet	---	Abarth Simca 1300
21 October	Wien, Austria	Berglandfahrt	Touring	850 cc	K. Zuegg	---	Fiat Abarth 850 TC
21 October	Montlhéry, France	1000 Km di Parigi (circuit-7,784 kms x 129 laps)	Grand Touring	1300 cc	Herrmann-Bianchi	---	Abarth Simca 130
4 November	Messina, Italy	Coppa Colle San Rizzo (hillclimb-9,400 kms)	Touring Grand Touring	850 cc 1000 cc	A. Spataro Riolo	7'29"6 7'30"	Fiat Abarth 850 TC Fiat Abarth 1000 BA
17-18 November	Sud-Ouest, France	Rallye d'Automne	---	---	'Sarayac'	---	Fiat Abarth 850 TC
18 November	Vallelunga, Italy	Trofeo d'Autunno (circuit-1,800 kms x 20 laps) (25 laps) (25 laps)	Touring Grand Touring Grand Touring	850 cc 1000 cc 1300 cc	A. De Adamich V. Venturi L. Bandini	26'5" 30'18"2 22'06"9	Fiat Abarth 850TC Fiat Abarth 1000 BA Abarth Simca 1300
9 December	Argentina	Asociacion Argentina Automoviles Sport (circuit-2,126 kms x 22 laps)	---	---	O. Silich	28'00"8	Fiat Abarth 850 TC
1963							
17 February	Daytona (Florida), USA	Daytona 3 Hours (circuit-3,810 kms)	Grand Touring	1300 cc	H. Herrmann	kms 454	Abarth Simca 1300
22-24 February	Alicante, Spagna	Rallye de Alicante (regolarità-900 kms)	Grand Touring	1000 cc	V. Lepori	---	Fiat Abarth 1000 BA
22-24 February	Sanremo, Italy	Rallye dei Fiori (503 kms)	Touring Prepared Touring	850 cc 850 cc	Deserti-Magri Marzorati-Airoldi	--- ---	Fiat Abarth 850 TC Fiat Abarth 850 TC
3 March	Aix en Provence, France	Rallye Mistral	Grand Touring	1000 cc	Hanrioud-Arbez	---	Fiat Abarth 1000 BA
9-10 March	San Sebastian, Spagna	Rallye Internacional Vasco-Navarro	Touring	850 cc	V. Sagi	---	Fiat Abarth 850 TC
12-17 March	Algarve, Portugal	Rallye A. C. Portugal	Touring	---	Berenguer	---	Fiat Abarth 850 TC
17 March	Le Mans et Rouen, France	Rallye des Capitales de l'Ouest (hillclimb + circuit)	Touring Grand Touring	850 cc 1300 cc	R. De la Geneste F. Pasquier	--- ---	Fiat Abarth 850 TC Abarth Simca 1300
17 March	Barcelona, Spagna	Rallye Costa Brava	Touring	850 cc	J. Juncosa	---	Fiat Abarth 850 TC
19 March	Roma, Italy	Rallye del Mare	Touring	701-1150 cc	A. Lambertini	---	Fiat Abarth 1000 Berlina
22 March	Sebring, (Florida), USA	3 Hours (circuit-5,205 miles)	Grand Touring	1000 cc	H. Herrmann	kms 393	Fiat Abarth 1000 BA
23 March	Sebring (Florida), USA	12 Hours (circuit-5,205 miles)	Grand Touring	1300 cc	Spychiger- Herrmann-Pilette	kms 1453	Abarth Simca 1300
23-24 March	Genova, Italy	Coppa Riviera di Ponente (270 kms)	Touring Touring	850 cc 1000 cc	A. Degli Innocenti C. Benelli	--- ---	Fiat Abarth 850 TC Fiat Abarth 1000 Berlina
24 March	Bruxelles, Belgium	Course de côte de Alle sur Semois (hillclimb-4,200 kms)	Touring	1000 cc	L. Bianchi	2'24"5	Fiat Abarth 1000 Berlina
24 March	Königsstetten, Austria	Dopplerhütten Wertungsfahrt (hillclimb-2,700 kms)	Grand Touring	1000 cc	D. Quester	2'23"0	Fiat Abarth 1000 Berlina

DATE	LOCATION	RACE	CATEGORY	CLASS	WINNER	TIME OR DISTANCE	CAR
30 March	Genève, Switzerland	Kilomètre d'Eau Morte	Touring Touring Grand Touring Grand Touring	850 cc 1000 cc 850 cc 1000 cc	C. Pasche B. Mauris J. P. Pinoch R. Sacchi	0'35"0 0'32"4 0'37"5 0'30"2	Fiat Abarth 850 TC Fiat Abarth 1000 Berlina Fiat Abarth 700 BA Fiat Abarth 1000 BA
31 March	Frascati, Italy	Frascati-Tuscolo (hillclimb-4,200 kms)	Touring Touring Grand Touring Grand Touring Grand Touring	850 cc 1000 cc 700 cc 1000 cc 1300 cc	R. Addamieno L. Arena A. Marino S. Salerno E. Buzzetti	3'33"2 3'39"7 3'54"7 3'30"6 3'17"7	Fiat Abarth 850 TC Fiat Abarth 1000 Berlina Fiat Abarth 700 BA Fiat Abarth 1000 BA Abarth Simca 1300
31 March	Bologna, Italy	Coppa Acqua Cerelia (hillclimb-8,000 kms)	Touring Touring Grand Touring Grand Touring Grand Touring	850 cc 1000 cc 700 cc 1000 cc 1300 cc	M. Cintolesi 'Donald Duck' 'Janna Areshi' 'Tiger' L. Conti	7'03"1 7'01"1 7'36"2 7'01"7 6'47"4	Fiat Abarth 850 TC Fiat Abarth 1000 Berlina Fiat Abarth 700 BA Fiat Abarth 1000 BA Abarth Simca 1300
31 March	Saint-Lô, France	Course de côte d'Hébécrevon (hillclimb-1,200 kms)	Touring Grand Touring Grand Touring	850 cc 1000 cc 1300 cc	F. Ruata C. Marchais F. Pasquier	0'49"3 0'43"0 0'46"3	Fiat Abarth 850 TC Fiat Abarth 1000 BA Abarth Simca 1300
7 April	Verona, Italy	Stallavena-Boscochiesanuova (hillclimb-15,300 kms)	Touring Grand Touring Grand Touring Grand Touring	850 cc 700 cc 1000 cc 1300 cc	F. Ghezzi M. Pinchetti O.Sigala G. Biscaldi	8'54"7 8'44"5 8'18"2 7'46"0	Fiat Abarth 850 TC Fiat Abarth 700 B Fiat Abarth 1000 BA Abarth Simca 1300
7 April	Eifel, Germany	ADAC Bergpreis Bad Neuenahr (hillclimb-3,300 kms x 2)	Grand Touring	1000 cc	A. Schmalbach	4'47"2	Fiat Abarth 1000 BA
7 April	Napoli, Italy	Targa Vesuvio (hillclimb-9,200 kms)	Touring Touring Sport	850 cc 1300 cc 1000 cc	R. Di Dato L. Rizzo A. Capostella	8'10"3 7'58"8 9'14"5	Fiat Abarth 850 TC Fiat Abarth 1000 Berlina Fiat Abarth 1000 Sport
14 April	Cesenatico, Italy	Circuito Riviera di Cesenatico (2,750 kms x 22 laps)	Sport	1000 cc	V. Venturi (overall)	31'50"4	Fiat Abarth 1000 Sport
20 April	Verbois, Switzerland	Course de côte de Verbois (hillclimb-1,150 kms)	Touring Grand Touring Grand Touring	850 cc 850 cc 1000 cc	C. Pasche Terzariol C. Chaudet	1'04"9 1'07"0 1'02"2	Fiat Abarth 850 TC Fiat Abarth 700 BA Fiat Abarth 1000 BA
21 April	Eberbach, Germany	Eberbach Bergrennen (hillclimb-3,000 kms)	Grand Touring Grand Touring	1000 cc 1300 cc	J. Calderari K. Geiss	4'16"7 4'20"0	Fiat Abarth 1000 BA Abarth Simca 1300
21 April	Monza, Italy	Trofeo Grand Touring (circuit-5,750 kms x 15 laps)	Touring Touring	850 cc 1000 cc	U. De Giorgio D. Romanello	36'42"6 36'29"0	Fiat Abarth 850 TC Fiat Abarth 1000 Berlina
21 April	Roma, Italy	Trofeo E. Lazzaroni (244 kms)	Touring Grand Touring	850 cc 1000 cc	S. Lipizer B. Benelli	--- ---	Fiat Abrth 850 TC Fiat Abarth 1000 BA
21 April	Avola, Italy	Coppa Belmonte (hillclimb-7,400 kms)	Touring Grand Touring	850 cc 1000 cc	"Odar" G. Iacono	6'03"8 5'45"0	Fiat Abarth 850 TC Fiat Abarth 1000 BA
21 April	Barcelona, Spagna	Rallye 600 (regolarità-555 kms)	---	---	J. Juncosa	---	Fiat Abarth 850 TC
25 April	Trieste, Italy	Rallye del Biancospino (144 kms)	---	---	G. Sferrella	---	Fiat Abarth 850 TC
28 April	Danville (Virginia), USA	National Race SCCA (circuit-2,000 miles x 88 laps)	---	---	P. Richards	---	Fiat Abarth 1000 BA
28 April	Gap, France	Course de côte du Col Bayard (hillclimb-6,300 kms)	Touring Grand Touring	850 cc 1300 cc	R. Terlizzi J. Clément	4'31"1 3'54"3	Fiat Abarth 850 TC Abarth Simca 1300
28 April	Liège, Belgium	Course de côte de la Roche-Samree (hillclimb-8,000 kms)	Touring	850 cc	'Thierry'	5'14"7	Fiat Abarth 850 TC
28-29 April	Madrid, Spagna	Rallye Nacional R.A.C.E. (2217kms)	Touring	850 cc	J. Juncosa	---	Fiat Abarth 850 TC
4-5 May	Amiens, France	Rallye de Picardie (492 kms)	Touring	1000 cc	R. De la Geneste -Du Genestoux	---	Fiat Abarth 1000 Berlina
4-5 May	Nice, France	Criterium Jean Behra (527 kms)	Touring Grand Touring	850 cc up to 850 cc	Arbez-Rey Battesti-Python	--- ---	Fiat Abarth 850 TC Fiat Abarth 700 BA
5 May	Alessandria, Italy	Coppa d'Oro (180,700 kms)	---	---	'Villi' (overall)	---	Fiat Abarth 850 TC
5 May	Piacenza, Italy	Castell'Arquato-Vernasca (hillclimb-9,775 kms)	Touring Grand Touring Grand Touring Grand Touring Sport	850 cc 700 cc 1000 cc 1300 cc up to 2000 cc	L. Ogna G. Varese F. Rovetta A. Tedeschi V. Venturi	6'27"8 6'12"2 5'56"5 5'44"7 5'32"4	Fiat Abarth 850 TC Fiat Abarth 700 BA Fiat Abarth 1000 BA Abarth Simca 1300 Fiat Abarth 1000
5 May	Roma, Italy	Coppa Gallenga (hillclimb-10,400 kms)	Touring Touring Grand Touring	850 cc 1000 cc 700 cc	M. Cintolesi D. Romanello A. Pagano	5'52"7 5'42"7 6'21"7	Fiat Abarth 850 TC Fiat Abarth 1000 Berlina Fiat Abarth 700 BA
5 May	Trier, Germany	Trierer Flugplatzrennen (circuit-1,900 kms x 27 laps)	Grand Touring	1000 cc	A. Schmalbach	31'31"2	Fiat Abarth 1000 BA
5 May	Earls Colne Airfield, United Kingdom	Earls Colne Express Driving Tests (200 yards x 2)	----	----	B. Pinkerton	---	Fiat Abarth 500

DATE	LOCATION	RACE	CATEGORY	CLASS	WINNER	TIME OR DISTANCE	CAR
11 May	Newburg, United Kingdom	Goodwood Sprint (circuit-2,486 miles x 5 laps)	---	---	Buchanan-Michaelson	9'35"8	Fiat Abarth 1000 Berlina
12 May	La Baule, France	Rallye de la Baule (384 kms)	Touring	850 cc	Ruata-Gillet	---	Fiat Abarth 850 TC
			Grand Touring	1300 cc	Pasquier-Pasquier	---	Abarth Simca 1300
12 May	Valence, France	Course de côte du Pin (hillclimb-7,050 kms)	Touring	850 cc	Vulin	5'31"8	Fiat Abarth 850 TC
			Grand Touring	up to 1000 cc	Battesti	5'50"8	Fiat Abarth 700 BA
			Grand Touring	1300 cc	J. Clément	5'30"0	Abarth Simca 1300
12 May	Salò, Italy	Circuito del Garda (circuit-311,600 kms)	Grand Touring	700 cc	R. Ricci	2h 38'50"	Fiat Abarth 700 BA
			Grand Touring	1000 cc	'Pan'-Guerini	2h 32'42"	Fiat Abarth 1000 BA
12 May	Payerne, Switzerland	Slalom Nationale de Payerne (hillclimb-4,800 kms x 2)	Touring	1000 cc	S. Hauser	8'02"6	Fiat Abarth 1000 Berlina
			Prepared Touring	850 cc	R. Egli	7'55"5	Fiat Abarth 850 TC
			Sport	1000 cc	R. Stierli	7'25"1	Fiat Abarth 1000
12 May	Arezzo, Italy	Rallye delle Vallate Aretine (266 kms)	Touring	850 cc	P. Faloni	---	Fiat Abarth 850 TC
			Touring	1000 cc	R. Volpi	---	Fiat Abarth 1000 Berlina
12 May	Bayerischzell, Germany	Sudelfeld Bergrennen (hillclimb-3,500 kms)	Grand Touring	1000 cc	A. Schmalbach	4'54"2	Fiat Abarth 1000 BA
			Grand Touring	1300 cc	K. Geiss	5'00"5	Abarth Simca 1300
12 May	Barcelona, Spagna	Subida en cuesta Flor de Mayo (hillclimb)	Touring	850 cc	J. Juncosa	---	Fiat Abarth 850 TC
18-19 May	Elbeuf, France	Rallye Automobile Elbeuf (circuit-518 kms)	Touring	up to 1150 cc	R. De la Geneste -Du Genestoux	---	Fiat Abarth 1000 Berlina
19 May	Firenze, Italy	Rallye della Toscana (239,600 kms)	Touring	850 cc	P. Palai	---	Fiat Abarth 850 TC
			Touring	1000 cc	I. Minzoni	---	Fiat Abarth 1000 Berlina
19 May	Montlhéry, France	Grand Prix de Paris (circuit-3,360 kms x 15 laps)	Touring	850 cc	F. Ruata	29'25"'	Fiat Abarth 850 TC
			Grand Touring	up to 1600 cc	Lobry	23'25"8	Abarth Simca 1300
19 May	Foggia, Italy	Targa di Capitanata (regolarità-306 kms)	Grand Touring	850 cc	F. Gomez Paloma	---	Fiat Abarth 850 TC
19 May	Villefranche sur Saône, France	Course de côte du Beaujolais (hillclimb-16,000 kms)	Touring	1000 cc	Ramu-Caccia	10'25"00	Fiat Abarth 1000 BA
19 May	Barcelona, Spagna	Carrera en cuesta de Valvidrera (hillclimb-2,450 kms)	Touring	850 cc	J. Juncosa	---	Fiat Abarth 850 TC
23 May	Aix en Provence, France	Course de côte d'Aix-Saint Antonin (hillclimb-2,700 Km)	Grand Touring	700 cc	Battesti	2'13"8	Fiat Abarth 700 TC
			Grand Touring	1300 cc	J Clément	2'00"3	Abarth Simca 1300
			Sport	700 cc	M. Second	2'12"9	Fiat Abarth 700 Sport
23 May	Roulers, Belgium	Slalom de Roulers (1,650 kms)	Touring	851-1300 cc	X. Boulanger	2'42"9	Fiat Abarth 1000 Berlina
23 May	Cesena, Italy	Coppa Città di Cesena (regolarità-167 kms)	Touring	850 cc	C. Massagrande	---	Fiat Abarth 850 TC
			Touring	850 cc	C. Pasche	10'04"7	Fiat Abarth 850 TC
25 May	Dübendorf, Switzerland	Gara nazionale di Slalom	Prepared Touring	850 cc	R. Egli	9'38"5	Fiat Abarth 850 TC
			Grand Touring	1000 cc	P. Ettmüller	8'46"5	Fiat Abarth 1000 BA
25-26 May	Zaragoza, Spagna	Rallye Internacional de Zaragoza	Touring	850 cc	J. Juncosa	---	Fiat Abarth 850 TC
26 May	Nancy, France	Rallye de Lorraine (790 kms)	Prepared Touring	850 cc	Krause-Maury	---	Fiat Abarth 850 TC
			Prepared Touring	1000 cc	De la Geneste-Du Genestoux	---	Fiat Abarth 1000 Berlina
26 May	Sassari, Italy	Alghero-Scala Picada (hillclimb-8,700 kms)	Touring	850 cc	'Odar'	5'31"1	Fiat Abarth 850 TC
			Touring	up to 1150 cc	L. Rizzo	5'37"3	Fiat Abarth 1000 Berlina
			Grand Touring	1000 cc	G. Rebaudi	5'13"7	Fiat Abarth 1000 BA
26 May	Bomeree, Belgium	Course de côte de Charleroi (hillclimb-2,000 kms)	Touring	850 cc	'Thierry'	1'29"5	Fiat Abarth 850 TC
			Touring	up to 1150 cc	X. Boulanger	1'27"9	Fiat Abarth 1000 Berlina
			Prepared Touring	850 cc	Lefevre	1'30"0	Fiat Abarth 850 TC
			Grand Touring	700 cc	Therstappen	1'30"6	Fiat Abarth 700 BA
			Grand Touring	1000 cc	C. Marchaux	1'19"7	Fiat Abarth 1000 BA
26 May	Marseille, France	Course de côte de la Sainte Baume (hillclimb-8,100 kms)	Touring	850 cc	M. Martin	8'24"6	Fiat Abarth 850 TC
			Grand Touring	up to 850 cc	De Saint Michel	8'14"9	Fiat Abarth 700 BA
			Grand Touring	1300 cc	J. Clément	7'53"8	Abarth Simca 1300
26 May	Bologna, Italy	Bologna-Passo della Raticosa (hillclimb-32,700 kms)	Touring	850 cc	C. Zuccoli	22'35"6	Fiat Abarth 850 TC
			Touring	up to 1150 cc	M. Bianchi	22'06"2	Fiat Abarth 1000 Berlina
			Grand Touring	700 cc	R. Ricci	22'10"1	Fiat Abarth 700 BA
			Grand Touring	1000 cc	A. Zucchi	21'35"7	Fiat Abarth 1000 BA
			Grand Touring	1300 cc	G. Biscaldi	20'41"8	Abarth Simca 1300
			Sport	up to 2000 cc	H. Herrmann	20'01"6	Fiat Abarth 2000
26 May	Suffolk (Virginia), USA	Annual Old Dominion Jaycee Sprints (circuit-2,3 miles x 12 laps)	---	---	C. Pelouze	---	Fiat Abarth 1000 BA
27 May	Ascoli Piceno, Italy	San Benedetto del Tronto-Acquaviva Picena, (hillclimb-7,000 kms)	Touring	850 cc	F. Chiari	5'20"0	Fiat Abarth 850 TC
			Touring	up to 1150 cc	L. Verrocchio	5'10"3	Fiat Abarth 1000 Berlina
			Grand Touring	1000 cc	'Gin'	5'07"0	Fiat Abarth 1000 BA
1-2 June	Lubiana, Jugoslavia	Premio del Litorale a Portorose (circuit)	Grand Touring	850 cc	S. Matevz	---	Fiat Abarth 850 TC

DATE	LOCATION	RACE	CATEGORY	CLASS	WINNER	TIME OR DISTANCE	CAR
2 June	Firenze, Italy	Coppa della Consuma (hillclimb-12,500 kms)	Touring	850 cc	F. Ghezzi	8'03"6	Fiat Abarth 850 TC
			Touring	1000 cc	F. Frescobaldi	8'11"4	Fiat Abarth 1000 Berlina
			Grand Touring	700 cc	R. Ricci	8'24"4	Fiat Abarth 700 BA
			Grand Touring	1000 cc	K. Zuegg	7'56"0	Fiat Abarth 1000 BA
			Grand Touring	1300 cc	M. Bianchi	7'13"7	Abarth Simca 1300
			Sport	1000 cc	T. Spychiger	7'17"1	Fiat Abarth 1000
2 June	Palermo, Italy	Corsa al Monte Pellegrino (hillclimb-8,750 kms)	Touring	850 cc	C. Melilli	6'59"0	Fiat Abarth 850 TC
			Touring	up to 1150 cc	G. Spampinato	6'27"2	Fiat Abarth 1000 Berlina
			Grand Touring	700 cc	A. Rizzo	6'26"2	Fiat Abarth 700 BA
			Grand Touring	1000 cc	M. Spadafora	5'43"0	Fiat Abarth 1000 BA
			Grand Touring	1300 cc	'Gordon'	5'40"3	Abarth Simca 1300
2 June	Long Island (NY), USA	Bridgehampton Circuit National Championship Race (2,85 miles x 32 laps)	---	---	P. Richards	---	Fiat Abarth 1000 BA
2-3 June	Lerida, Spagna	Rallye Valle de Aran (regolarità-838 kms)	Touring	850 cc	J. Juncosa	---	Fiat Abarth 850 TC
8-9 June	Barcelona, Spagna	Rallye de Cataluna (811 kms)	Touring	850 cc	J. Palau Ribes	---	Fiat Abarth 850 TC
			Grand Touring	1000 cc	V. Lepori	---	Fiat Abarth 1000 BA
8-9 June	Nürburg, Germany	Wertungsfahrt Nürburg (430 kms)	Prepared Touring	850-1000 cc	H. P. Koepchen	---	Fiat Abarth 850 TC
9 June	Piacenza, Italy	Bobbio-Penice (hillclimb-12,800 kms)	Touring	850 cc	M. Crosina	10'04"0	Fiat Abarth 850 TC
			Touring	1000 cc	C. Benelli	10'14"8	Fiat Abarth 1000 Berlina
			Grand Touring	700 cc	R. Ricci	11'03"1	Fiat Abarth 700 BA
			Grand Touring	up to 1150 cc	P. Marsiaj	10'29"2	Fiat Abarth 1000 BA
			Grand Touring	1300 cc	C. Corini	10'04"2	Abarth Simca 1300
9 June	Berchtesgaden, Germany	Rossfeld Alpen Bergpres (hillclimb-5,890 kms)	Grand Touring	1000 cc	E. Mahle	7'52"7	Fiat Abarth 1000 BA
			Grand Touring	1300 cc	M. Bianchi	7'31"4	Abarth Simca 1300
			Sport	700 cc	M. Bianchi	8'45"7	Fiat Abarth 700
9 June	Napoli, Italy	Targa Vesuvio (9,200 kms)	Touring	850 cc	R. Bosurgi	8'08"0	Fiat Abarth 850 TC
			Grand Touring	up to 1300 cc	M. Moselli	7'35"8	Fiat Abarth 1000 BA
			Sport	1000 cc	F. Natella	8'22"3	Fiat Abarth 1000
9 June	Digne, France	Course de côte de la Lure (hillclimb-12,700 kms)	Touring	850 cc	R. Maggiorini	10'10"4	Fiat Abarth 850 TC
			Grand Touring	up to 850 cc	R. Terlizzi	10'22"0	Fiat Abarth 700 BA
			Grand Touring	1300 cc	J. Clément	9'11"9	Abarth Simca 1300
			Sport	up to 1150 cc	C. Bobrowski	9'38"7	Fiat Abarth 1000
9 June	Belvaux, Luxembourg	Slalom Automobile de Belvaux	Touring	1000 cc	H. Wagner	1'08"5	Fiat Abarth 850 TC
9 June	Berlin, Germany	Goldner Bär von Berlin (325 kms)	Touring	1000 cc	H. J. Mack	---	Fiat Abarth 1000 Berlina
13 June	Aspern, Austria	Preis von Wien (circuit-2,600 kms x 25 laps)	Touring	1000 cc	H. Herrmann	22'51"94	Fiat Abarth 1000 Berlina
			Grand Touring	1000 cc	H. Herrmann	28'19"52	Fiat Abarth 1000 BA
			Grand Touring	1300 cc	M. Bianchi	27'21"22	Abarth Simca 1300
			Sport	over 1600 cc	H. Herrmann	24'47"61	Fiat Abarth 2000
15-16 June	Sierre, Switzerland	Course de côte de Sierre-Montana (hillclimb-12,000 kms x 2)	Touring	850 cc	E. Kuecht	19'17"1	Fiat Abarth 850 TC
			Touring	1000 cc	J. P. Spicher	18'31"6	Fiat Abarth 1000 Berlina
			Grand Touring	1000 cc	A. Knor	16'17"4	Fiat Abarth 1000 BA
			Grand Touring	1300 cc	E. Moll	16'08"0	Abarth Simca 1300
			Sport	1000 cc	T. Spychiger	15'15"8	Fiat Abarth 1000
16 June	Cannes, France	Course de côte de Cannes (hillclimb-7,845 kms)	Touring	850 cc	M. Billard	7'27"6	Fiat Abarth 850 TC
			Touring	up to 1150 cc	J. Reynand	7'31"0	Fiat Abarth 1000 Berlina
			Grand Touring	700 cc	A. Battesti	7'34"9	Fiat Abarth 700 BA
				up to 1000 cc	M. Martin	7'36"1	Fiat Abarth 850 TC
			Grand Touring	1300 cc	J. Clément	7'14"9	Abarth Simca 1300
16 June	Asiago, Italy	Coppa Città di Asiago (hillclimb-14,500 kms)	Touring	850 cc	V. De Giorgio	9'54"6	Fiat Abarth 850 TC
			Touring	1000 cc	M. Polin	9'46"2	Fiat Abarth 1000 Berlina
			Grand Touring	700 cc	R. Ricci	9'55"2	Fiat Abarth 700 BA
			Grand Touring	1000 cc	'Tiger'	9'10"5	Fiat Abarth 1000 BA
			Grand Touring	1300 cc	D. Sigala	9'04"6	Abarth Simca 1300
16 June	Taranto, Italy	Rallye dello Jonio	Touring	850 cc	B. Zavagli	---	Fiat Abarth 850 TC
			Touring	1000 cc	C. Colonna	---	Fiat Abarth 1000 Berlina
16 June	Puerto Canencia, Spagna	Subida Campeonato Castilla de Montana (hillclimb)	Touring	850 cc	A. Alvarez de Leon	---	Fiat Abarth 850 TC
16 June	Portland (Oregon), USA	Rose Cup Race (circuit-2,4 miles x 23 laps)	---	---	R. Householger	---	Fiat Abarth 850 TC
23 June	Hockenheim, Germany	Circuito di Hockenheim (circuit-7,725 kms x 10 laps)	Touring	1000 cc	V. Arena	31'09"9	Fiat Abarth 1000 Berlina
			Grand Touring	1000 cc	E. Mahle	27'05"6	Fiat Abarth 1000 BA
			Grand Touring	1300 cc	T. Pilette	25'18"8	Abarth Simca 1300
23 June	Lignano, Italy	Rallye di Lignano (1 hour)	---	---	G. Cantele	---	Fiat Abarth 850 TC
23 June	Cuneo, Italy	Garessio-Colle San Bernardo (hillclimb-7,100 kms)	Touring	850 cc	A. Marzorati	5'01"7	Fiat Abarth 850 TC
			Touring	up to 1150 cc	P. Airoldi	4'55"0	Fiat Abarth 1000 Berlina
			Grand Touring	1000 cc	P. Marsiaj	4'37"9	Fiat Abarth 1000 BA
			Grand Touring	1300 cc	Acutis	4'34"3	Abarth Simca 1300

DATE	LOCATION	RACE	CATEGORY	CLASS	WINNER	TIME OR DISTANCE	CAR
23 June	Forlì, Italy	Predappio-Rocca delle Caminate (hillclimb-4,000 kms)	Touring	850 cc	'Odar'	3'50"9	Fiat Abarth 850 TC
			Touring	1000 cc	'Donald Duck'	3'52"7	Fiat Abarth 1000 Berlina
			Grand Touring	700 cc	R. Ricci	3'53"8	Fiat Abarth 700 BA
			Grand Touring	1000 cc	A. Zucchi	3'36"4	Fiat Abarth 1000 BA
			Grand Touring	1300 cc	L. Baldini	3'46"9	Abarth Simca 1300
23 June	Avignon, France	Course de côte du Mont Ventoux (hillclimb-21,700 kms)	Touring	1000 cc	H. Herrmann	15'51"0	Fiat Abarth 1000 Berlina
			Grand Touring	1000 cc	R. De la Geneste	14'28"0	Fiat Abarth 1000 BA
			Grand Touring	1300 cc	M. Bianchi	13'36"0	Abarth Simca 1300
			Sport	up to 1150 cc	T. Spychiger	13'13"2	Fiat Abarth 1000
23 June	Weinfelden, Germany	Autoslalom des Thurgauer Automobile Klubs	Touring	up to 1000 cc	K. Boltschauser	---	Fiat Abarth 850 TC
			Grand Touring	1300 cc	O. Model	---	Abarth Simca 1300
23 June	Livorno, Italy	Rallye dei Sette Passi (325,0 kms)	Touring	850 cc	C. Massagrande	---	Fiat Abarth 850 TC
23 June	Vallelunga, Italy	Premio Campagnano di Roma (circuit-3,200 kms x 35 laps)	Touring	850 cc	F. Chiari	1h 05'37"3	Fiat Abarth 850 TC
			Touring	1000 cc	D. Romanello	1h 06'07"9	Fiat Abarth 1000 Berlina
23 June	Metz, France	Course de côte d'Arry (hillclimb-1,200 kms)	Grand Touring	1000 cc	Bourjade	0'45"2	Fiat Abarth 1000 BA
25-29 June	Brescia, Italy	Journalists' Rally	Touring	850 cc	C. Massagrande	---	Fiat Abarth 850 TC
29-30 June	Annecy, France	Rallye International Mont-Blanc (672 kms)	Touring Preparato	850 cc	Krause-Baury	---	Fiat Abarth 850 TC
			Grand Touring	1000 cc	Hanrioud-Sage	---	Fiat Abarth 1000 BA
30 June	Monza, Italy	Gran Premio Grand Touring (circuit-5,750 kms x 3 hours)	Grand Touring	700 cc	G. Capra	kms 407,669	Fiat Abarth 700 BA
			Grand Touring	1000 cc	'Tiger'	kms 481,511	Fiat Abarth 1000 BA
			Grand Touring	1300 cc	G. Rava	kms 532,748	Abarth Simca 1300
30 June	Reggio Emilia, Italy	Vezzano-Casina (hillclimb-12,500 kms)	Touring	850 cc	U. Di Giorgio	8'37"0	Fiat Abarth 850 TC
			Touring	1000 cc	P. Airoldi	8'39"5	Fiat Abarth 1000 Berlina
			Grand Touring	1300 cc	C. Corini	8'06"1	Abarth Simca 1300
			Sport	1000 cc	V. Venturi	7'49"5	Fiat Abarth 1000
30 June	Wasgau, Germany	ADAC Wasgau-Bergprüfung (hillclimb-5,100 kms x 2)	Grand Touring	1300 cc	K, Geiss	6'36"8	Abarth Simca 1300
30 June	Avus, Germany	Grosser Preis von Berlin (circuit-8,300 kms x 10 laps)	Grand Touring	1000 cc	V. Arena	25'39"5	Fiat Abarth 1000 BA
			Grand Touring	1300 cc	T. Pilette	25'51"3	Abarth Simca 1300
30 June	Saint Ursanne, Switzerland	Course de côte Saint Ursanne-Les Rangiers (hillclimb-4,700 kms x 2)	Touring	850 cc	C. Pasche	3'20"5	Fiat Abarth 850 TC
			Touring	1000 cc	S. Hauser	3'04"3	Fiat Abarth 1000 Berlina
			Grand Touring	850 cc	J. R. Kretschi	3'39"2	Fiat Abarth 850 GT
			Grand Touring	1000 cc	P. Ettmüller	2'45"6	Fiat Abarth 1000 BA
			Grand Touring	1300 cc	E. Moll	2'46"6	Abarth Simca 1300
			Sport	1000 cc	T. Spychiger	2'49"1	Fiat Abarth 1000
30 June	Monopoli, Italy	Coppa Città di Monopoli (hillclimb-4,500 kms)	Touring	850 cc	A. Spadaro	2'39"9	Fiat Abarth 850 TC
			Grand Touring	700 cc	A. Rizzo	2'47"2	Fiat Abarth 700 BA
			Grand Touring	1000 cc	M. Moselli	2'27"4	Fiat Abarth 1000 BA
			Grand Touring	1300 cc	'Gordon'	2'24"5	Abarth Simca 1300
3-7 July	Wiesbaden, Germany	Weisbaden Rally (1900 kms)	Grand Touring	1000 cc	P. Ettmüller	---	Fiat Abarth 1000 BA
6 July	Nortena, Spagna	Rally "La Toja" (325 kms)	Touring	850 cc	J. Juncosa	---	Fiat Abarth 850 TC
6-7 July	Siracusa, Italy	3 Ore Notturna (circuit-5,500 kms)	Touring	850 cc	F. Di Stefano	1 hour: 119,438 kms	Fiat Abarth 850 TC
			Grand Touring	1000 cc	F. Patanè	1 hour: 134,763 kms	Fiat Abarth 1000 BA
6-7 July	Zandvoort, Holland	Les Coupes de Benelux (circuit 4,225 kms x 10 laps)	Touring	850 cc	Kernevi	22'04"9	Fiat Abarth 850 TC
6-7 July	Ruselberg, Germany	ADAC Ratisbona Bergrennen (hillclimb-3,8 kms x 2)	Grand Touring	1000 cc	A. Schmalbach	4'40"8	Fiat Abarth 1000 BA
7 July	Lucca, Italy	Rally di Lucca (171 kms)	Touring	850 cc	S. Lipizer	---	Fiat Abarth 850 TC
7 July	Bolzano, Italy	Bolzano-Mendola (hillclimb-14,400 kms)	Touring	850 cc	U. De Giorgio	11'38"5	Fiat Abarth 850 TC
			Touring	1000 cc	P. Villani	11'48"3	Fiat Abarth 1000 Berlina
			Grand Touring	700 cc	M. Pinchetti	11'46"0	Fiat Abarth 700 BA
			Grand Touring	1000 cc	G. Castellina	10'57"7	Fiat Abarth 1000 BA
			Grand Touring	1300 cc	O. Sigala	10'51"2	Abarth Simca 1300
			Sport	1000 cc	T. Spychiger	10'14"9	Fiat Abarth 1000
7 July	Strasbourg, France	Course de côte de Turckeim Trois Epis (hillclimb-6,230 kms)	Touring	850 cc	R. Maggionini	4'31"3	Fiat Abarth 850 TC
			Grand Touring	1000 cc	P. Ettmüller	4'03"1	Fiat Abarth 1000 BA
			Grand Touring	1300 cc	E. Möll	3'54"0	Abarth Simca 1300
7 July	Ehrenhausen, Austria	Platschbergrennen	Touring	850 cc	Müller	2'20"0	Fiat Abarth 850 TC
7 July	Macon, France	Course de côte Macon-Solutre (hillclimb-2,620 kms)	Touring	1000 cc	B. Sechi	1'49"0	Fiat Abarth 1000 Berlina
			Grand Touring	1000 cc	R. Sacchi	1'35"1	Fiat Abarth 1000 BA
			Grand Touring	1300 cc	R. Richard	---	Abarth Simca 1300
			Sport	1000 cc	V. Venturi	1'31"2	Fiat Abarth 1000
7 July	Clermont Ferrand, France	Trophées d'Auvergne (circuit-8,055 kms x 3 hours)	Grand Touring	1300 cc	G. Biscaldi	kms 320,504	Abarth Simca 1300
			Sport	1000 cc	M. Bianchi	kms 344,432	Fiat Abarth 1000
7 July		Rallye de el Cid	---	---	A De Leon	---	Fiat Abarth 850 TC

DATE	LOCATION	RACE	CATEGORY	CLASS	WINNER	TIME OR DISTANCE	CAR
14 July	Trento, Italy	Trento-Bondone (hillclimb-17,300 kms)	Touring	850 cc	A. De Paoli	15'10"5	Fiat Abarth 850 TC
			Touring	1000 cc	M. Crosina	15'7"6	Fiat Abarth 1000 Berlina
			Grand Touring	700 cc	R. Ricci	15'32"1	Fiat Abarth 700 BA
			Grand Touring	1000 cc	Rovetta	14'09"5	Fiat Abarth 1000 BA
			Grand Touring	1300 cc	C. Biscaldi	13'31"8	Abarth Simca 1300
			Sport	1000 cc	T. Spychiger	13'17"5	Fiat Abarth 1000
14 July	Gives, Belgium	Course de côte de Gives (hillclimb-2,400 x 2)	Touring	850 cc	'Thierry'	3'55"4	Fiat Abarth 850 TC
14 July	Adenau, Germany	Nürburgring 12 Stunden (circuit-22,810 kms)	Sport-Prototypes	1300 cc	Moll-Pilette	kms 1481	Abarth Simca 1300
14 July	Madrid, Spagna	Rally Club "600" (246 kms)	---	---	A. Alvarez de Leon	---	Fiat Abarth 850 TC
14 July	Reggio Calabria, Italy	Santo Stefano-Gambarie (hillclimb-7,830 kms)	Touring	850 cc	A. Spadaro	6'12"4	Fiat Abarth 850 TC
			Touring	up to 1150 cc	Spampinato	6'28"0	Fiat Abarth 1000 Berlina
			Grand Touring	up to 1000 cc	Jacono	5'52"3	Fiat Abarth 850 BA
			Grand Touring	1300 cc	D. Scola	5'41"2	Abarth Simca 1300
18-21 July	Moldau, Cecoslovacchia	Rallye Ultava (regolarità)	Touring A/1	up to 1150 cc	Egem	---	Fiat Abarth 850 TC
20 July	Cascais, Portugal	Prova Automovel Club de Portugal (circuit-3,450 kms x 30 laps)	Touring	850 cc	C. Flores	1h01'24"52	Fiat Abarth 850 TC
21 July	Trieste, Italy	Trieste-Opicina (hillclimb-10,150 kms)	Touring	850 cc	U. De Giorgio	5'48"2	Fiat Abarth 850 TC
			Touring	1000 cc	D. Romanello	5'49"5	Fiat Abarth 1000 Berlina
			Grand Touring	700 cc	M. Pinchetti	5'50"5	Fiat Abarth 700 BA
			Grand Touring	1000 cc	K. Zuegg	5'16"9	Fiat Abarth 1000 BA
			Grand Touring	1300 cc	G. Rava	5'21"1	Abarth Simca 1300
21 July	Lyons la Forêt, France	Course de côte Lyons-la Forêt (hillclimb-1,250 kms)	Touring	850 cc	F. Ruata	1'18"0	Fiat Abarth 850 TC
			Grand Touring	1000 cc	Bourjade	1'12"4	Fiat Abarth 1000 BA
			Grand Touring	up to 1600 cc	R. Richard	1'06"4	Abarth Simca 1300
21 July	Brescia, Italy	Scalata al Colle Sant'Eusebio (9,150 kms)	Touring	850 cc	A. Marzorati	5'45"2	Fiat Abarth 850 TC
			Touring	1000 cc	P. Polotti	5'57"4	Fiat Abarth 1000 Berlina
			Grand Touring	700 cc	Zani	5'53"9	Fiat Abarth 700 BA
			Grand Touring	1000 cc	Rovetta	5'10"4	Fiat Abarth 1000 BA
21 July	Nürnberg, Germany	ADAC Norising Rennen (circuit-3,940 kms x 15 laps)	Grand Touring	1300 cc	H. Herrmann	26'58"9	Abarth Simca 1300
21 July	Reisdorf, Luxembourg	Course de côte de Reisdorf (circuit-3,400 kms)	Grand Touring	1300 cc	H. Illert	2'14"2	Abarth Simca 1300
21 July	Long Island (NY), USA	Bridgehampton Race (circuit-2,85 miles x 35 laps)	---	---	H. Kraaz	1h06'3"	Fiat Abarth 1000 BA
28 July	Sestrière, Italy	Cesena-Sestrière (hillclimb-10,400 kms)	Touring	850 cc	M. Crosina	7'08"8	Fiat Abarth 850 TC
			Touring	up to 1150 cc	F. Frescobaldi	6'56"9	Fiat Abarth 1000 Berlina
			Grand Touring	700 cc	G. Varese	7'03"1	Fiat Abarth 700 BA
			Grand Touring	1000 cc	P. Marsiaj	6'36"3	Fiat Abarth 1000 BA
			Grand Touring	1300 cc	G. Biscaldi	6'12"3	Abarth Simca 1300
			Sport	1000 cc	T. Spychiger	6'02"8	Fiat Abarth 1000
28 July	Magny-Cours, France	Circuit Magny-Cours (circuit-2,040 kms x 35 laps)	Grand Touring	up to 1000 cc	R. Richard	35'01"0	Fiat Abarth 850 TC
			Grand Touring	1300 cc	R. Richard	32'13"1	Abarth Simca 1300
28 July	Belluno, Italy	Coppa Agordo-Frassenè (hillclimb-9,000 kms)	Touring	850 cc	U. De Giorgio	6'18"8	Fiat Abarth 850 TC
			Touring	1000 cc	'Clik'	6'20"0	Fiat Abarth 1000 Berlina
			Grand Touring	700 cc	M. Pinchetti	6'15"8	Fiat Abarth 700 BA
			Grand Touring	1000 cc	'Kangaroo'	6'12"8	Fiat Abarth 1000 BA
			Grand Touring	1300 cc	G. Dalla Torre	5'42"2	Abarth Simca 1300
28 July	Stuttgart, Germany	Grand Prix de la Solitude (circuit-11,417 kms x 10 laps)	Grand Touring	1000 cc	L. Bianchi	48'46"5	Fiat Abarth 1000 BA
			Grand Touring	1300 cc	T. Pilette	47'44"2	Abarth Simca 1300
28 July	Mansfield (Ohio), USA	Mid-Ohio S.C.C.A. Divisionals (circuit-2,28 miles x 20 laps)	---	---	J. Sheffield	---	Fiat Abarth 750 GT
3 August	London, United Kingdom	Bank Holiday Sport-Mallory Park (1,35 miles x 10 laps)	Touring	850 cc	G. Gunther	---	Fiat Abarth 850 TC
3-4 August	Mont-Dore, France	Course de côte du Mont-Dore (hillclimb-5,000 kms)	Touring	up to 1150 cc	B. Sechi	4'08"3	Fiat Abarth 1000 Berlina
			Prepared Touring	850 cc	R. Richard	4'07"1	Fiat Abarth 850 TC
			Grand Touring	1300 cc	R. Richard	3'35"4	Abarth Simca 1300
4 August	Nagrada Loke, Jugoslavia	Gara Internazionale di Skofja Loka (circuit-4,500 kms x 8)	---	850 cc	S. Matevz	23'17"3	Fiat Abarth 850 TC
4 August	Adenau, Germany	Rheinland Pfalz Rennen-Nürburgring (circuit-22,810 kms x 7 laps)	Grand Touring	1000 cc	T. Hoffer	1h 18'06"2	Fiat Abarth 1000 BA
4 August	Viterbo, Italy	Corsa lago di Montefiascone (hillclimb-5,000 kms)	Touring	850 cc	B. Faenza	3'01"3	Fiat Abarth 850 TC
			Touring	up to 1150 cc	T. Capuozzo	3'00"8	Fiat Abarth 1000 Berlina
			Grand Touring	1000 cc	L. Mugelli	3'28"1	Fiat Abarth 1000 BA
			Grand Touring	1300 cc	G. Niccolai	2'52"8	Abarth Simca 1300
4 August	Aosta, Italy	Aosta-Pila (hillclimb-11,200 kms)	Touring	850 cc	M. Crosina	8'27"6	Fiat Abarth 850 TC
			Touring	up to 1150 cc	'Jerry'	8'28"8	Fiat Abarth 1000 Berlina
			Grand Touring	700 cc	G. Varese	8'20"3	Fiat Abarth 700 BA
			Grand Touring	1000 cc	A. Zucchi	7'36"0	Fiat Abarth 1000 BA
			Grand Touring	1300 cc	G. Biscaldi	7'20"3	Abarth Simca 1300

DATE	LOCATION	RACE	CATEGORY	CLASS	WINNER	TIME OR DISTANCE	CAR
4 August	Trapani, Italy	Trapani-Monte Erice (hillclimb-6,670 kms)	Touring	850 cc	M. Sancisa	5'26"3	Fiat Abarth 850 TC
			Touring	up to 1150 cc	P. Villani	5'21"1	Fiat Abarth 1000 Berlina
			Grand Touring	700 cc	A. Scimone	5'43"1	Fiat Abarth 700 BA
			Grand Touring	1000 cc	G. Jacono	5'10"3	Fiat Abarth 1000 BA
			Grand Touring	1300 cc	Tagliavia	5'12"3	Abarth Simca 1300
			Sport	1000 cc	V. Venturi	4'44"2	Fiat Abarth 1000
11 August	Freiburg, Germany	ADAC Berpreis Freiburg-Schauinsland (hillclimb-11,200 kms)	Grand Touring	1000 cc	V. Venturi	15'27"73	Fiat Abarth 1000 BA
			Grand Touring	1300 cc	G. Biscaldi	14'55"3	Abarth Simca 1300
			Sport	700 cc	M. Bianchi	15'08"75	Fiat Abarth 700 BA
			Sport	1000 cc	T. Spychiger	14'25"23	Fiat Abarth 1000 SP
11 August	Marlboro (Maryland), USA	6 Hours of Marlboro (circuit 1,800 miles x 184 laps)	---	---	Linton / Heppenstall	---	Abarth Simca 1300
11 August	Cosenza, Italy	Coppa della Sila (hillclimb-14,000 kms)	Touring	850 cc	C. Avventurieri	19'01"1	Fiat Abarth 850 TC
			Touring	up to 1150 cc	F. Pietramala	19'52"3	Fiat Abarth 1000 Berlina
			Grand Touring	1000 cc	C. Giuliani	20'28"2	Fiat Abarth 1000 BA
			Grand Touring	1300 cc	D. Scola	18'26"2	Abarth Simca 1300
18 August	Brindisi, Italy	Coppa-Fasano-Selva (hillclimb-7,400 kms)	Touring	850 cc	N. Persia	4'39"1	Fiat Abarth 850 TC
			Touring	up to 1300 cc	A. Rizzo	4'39"3	Fiat Abarth 1000 Berlina
			Grand Touring	700 cc	'Jamaraschi'	4'59"4	Fiat Abarth 700 BA
			Grand Touring	1000 cc	M. Moselli	4'14"6	Fiat Abarth 1000
			Grand Touring	1300 cc	D. Scola	4'15"4	Abarth Simca 1300
18 August	Herbeumont, Belgium	Course de côte d'Herbeumont (hillclimb-3,600 kms)	Touring	850 cc	'Thierry'	2'37"3	Fiat Abarth 850 TC
			Touring	1000 cc	J. Liegeois	2'20"2	Fiat Abarth Monomille
18 August	Pergusa, Italy	Coppa Città di Enna (circuit-4,797 kms x 65 laps)	Grand Touring	700 cc	G.B. Mainetti	1h 51'25"8	Fiat Abarth 700 BA
			Grand Touring	1000 cc	'Tiger'	1h 50'44"	Fiat Abarth 1000 BA
18 August	Spagna	Rallye Pinares de Alto Duero	Touring	850 cc	J. Juncosa	---	Fiat Abarth 850 TC
25 August	Chieti, Italy	Coppa Città di Chieti (hillclimb-6,200 kms)	Touring	850 cc	F. Chiari	3'48"1	Fiat Abarth 850 TC
			Touring	up to 1150 cc	L. Verrocchio	3'50"1	Fiat Abarth 1000 Berlina
			Grand Touring	700 cc	A. Marino	4'15"2	Fiat Abarth 700 BA
			Grand Touring	up to 1150 cc	A. Antriulli	3'55"4	Fiat Abarth 1000 BA
25 August	Villars, Switzerland	Ollon-Villars (hillclimb-8,000 kms x 2)	Grand Touring	700 cc	H. Affentranger	12'55"0	Fiat Abarth 700 BA
			Grand Touring	850 cc	J. R. Kretschi	12'37"3	Fiat Abarth 850 TC
			Grand Touring	1000 cc	A. Knorr	10'31"6	Fiat Abarth 1000 BA
			Grand Touring	1300 cc	G. Biscaldi	9'59"3	Abarth Simca 1300
			Sport	1000 cc	T. Spychiger	9'50"1	Fiat Abarth 1000 SP
25 August	Zolder, Belgium	Coupes de Terlaemen (circuit-4,184 kms x 16 laps)	Grand Touring	1300 cc	M. Bianchi	31'33"3	Abarth Simca 1300
25 August	Pesaro, Italy	Coppa Città di Pesaro	---	---	M. Caddolo	---	Fiat Abarth 850 TC
27 August	Ipoh, Malesia	Royal Perak Club Meeting (circuit-0,860 kms)	Touring	1000 cc	L. Peng Han	0'30"75	Fiat Abarth 1000 Berlina
			Grand Touring	1000 cc	L. Peng Han	0'30"0	Fiat Abarth 1000 BA
31 August	Genève, Switzerland	Course de côte de Marchairuz (hillclimb-4,800 kms)	Grand Touring	1300 cc	K. Perrot	6'32"2	Abarth Simca 1300
31 August	Waterford (Michigan), USA	Divisional L's Waterford Falls (circuit)	---	---	R. Guerrieri	---	Fiat Abarth 750 GT
1 September	Udine, Italy	Cividale-Castelmonte (hillclimb-7,000 kms)	Touring	850 cc	M. Cintolesi	5'06"5	Fiat Abarth 850 TC
			Touring	up to 1150 cc	D. Romanello	5'06"3	Fiat Abarth 1000 Berlina
			Grand Touring	700 cc	M. Pinchetti	5'13"7	Fiat Abarth 700 BA
			Grand Touring	1000 cc	'Matich'	4'54"5	Fiat Abarth 1000 BA
			Grand Touring	1300 cc	G. Dalla Torre	4'38"6	Abarth Simca 1300
1 September	Grenoble, France	Course de côte Chamrousse (hillclimb-17,200 kms)	Grand Touring	1000 cc	J. P. Hanrioud	14'08"3	Fiat Abarth 1000
			Grand Touring	1300 cc	P. Maublanc	13'44"2	Abarth Simca 1300 GT
1 September	Adenau, Germany	Internationales 500 KMS Nürburgring Rennen (circuit-22,810 kms x 22 laps)	Touring	1000 cc	Biscaldi-Bianchi	4h 12'25"4	Fiat Abarth 1000 Berlina
			Grand Touring	1000 cc	A. Zucchi	4h 08'34"2	Fiat Abarth 1000 BA
			Sport-Prototypes	850 cc	Herrmann-Pilette	4h 03'38"4	Fiat Abarth 850
1 September	Freiburg, Germany	OMK Junioren Pokal (hillclimb-7,200 kms)	Grand Touring	1000 cc	J. Schnitzer	5'38"9	Fiat Abarth 1000 BA
1 September	Avezzano, Italy	Trofeo Micangeli (hillclimb-6,000 kms)	Grand Touring	up to 1300 cc	M. Moselli	3'28"1	Fiat Abarth 1000 BA
8 September	Ascoli Piceno, Italy	Corsa al Colle San Marco (hillclimb-10,200 kms)	Touring	850 cc	M. Crosina	7'07"0	Fiat Abarth 850 TC
			Touring	1000 cc	C. Benelli	7'15"3	Fiat Abarth 1000 Berlina
			Grand Touring	1000 cc	A. Giliberti	6'46"1	Fiat Abarth 1000 BA
			Grand Touring	1300 cc	L. Boldini	6'41"3	Abarth Simca 1300
8 September	Monza, Italy	Coppa Intereuropa (circuit-5,750 kms x 3 hours)	Grand Touring	1300 cc	T. Spychiger	kms 499,550	Abarth Simca 1300
8 September	Hockenheim. Germany	Herbst Pokal Rennen (circuit-6,768 kms x 10 laps)	Grand Touring	1000 cc	F. Kleine	28'11"2	Fiat Abarth 1000 BA
			Grand Touring	1300 cc	K. Geiss	26'32"2	Abarth Simca 1300
8 September	Dijon, France	Course de côte d'Urcy (hillclimb-2,500 kms)	Grand Touring	1300 cc	R. Richard	1'58"2	Abarth Simca 1300
8 September	Napoli, Italy	Targa Vesuvio (hillclimb-9,200 kms)	Touring	850 cc	C. Giuliani	8'20"5	Fiat Abarth 850 TC
			Touring	up to 1150 cc	A. Rizzo	8'06"0	Fiat Abarth 1000 Berlin
			Grand Touring	1300 cc	D. Scola	7'44"8	Abarth Simca 1300
			Sport	1000 cc	R. Bosurgi	8'13"1	Fiat Abarth 1000

DATE	LOCATION	RACE	CATEGORY	CLASS	WINNER	TIME OR DISTANCE	CAR
8 September	Gaisberg, Austria	Gaisberg Bergrpreis von Österreich (hillclimb-8,625 kms)	Grand Touring Sport	1000 cc 1000 cc	K. Zuegg W. Schneider	11'00"9 10'25"2	Fiat Abarth 1000 BA Fiat Abarth 1000
8 September	Barcelona, Spagna	Carrera en cuesta San Feliu Codinas (hillclimb-4,200 kms)	Touring Prepared Touring	850 cc 850 cc	J. Juncosa J. Palau Ribes	3'42"25 3'28"19	Fiat Abarth 850 TC Fiat Abarth 850 TC
14 September	Vaals, Holland	Heuvelklim Vaals (hillclimb-1,200 kms)	Prepared Touring	850 cc	H. Wagner	1'03"4	Fiat Abarth 850 TC
14 September	Long Island (NY), USA	Bridgehampton Double 500 (circuit-2, 84 miles x 110 laps)	---	1300 cc	Fleming / Lynton	---	Abarth Simca 1300
14-15 September	Mont Verdun, France	Course de côte Limonest-Mont Verdun (hillclimb-2,675 kms)	Grand Touring Grand Touring	1000 cc 1300 cc	B. Sechi P. Maublanc	2'16" 2'10"1	Fiat Abarth Monomille Abarth Simca 1300
15 September	Villa Potenza, Italy	Villa Potenza-Macerata (hillclimb-4,700 kms)	Touring Touring	850cc up to 1150 cc	B. Faenza G. Caraceni	2'40"0 2'34"2	Fiat Abarth 850 TC Fiat Abarth 1000 Berlina
15 September	Belluno, Italy	Coppa Alpe del Nevegal (hillclimb-7,500 kms)	Touring Touring Grand Touring Grand Touring Grand Touring	850 cc 1000 cc 700 cc 1000 cc 1300 cc	A De Paoli C. Benelli M. Pinchetti 'Matich' L. Baldini	5'22"6 5'20"2 5'25"6 5'01"2 4'56"0	Fiat Abarth 850 TC Fiat Abarth 1000 Berlina Fiat Abarth 1000 BA Fiat Abarth 1000 BA Abarth Simca 1300
15 September	Innsbruck, Austria	Timmelsjoch Hochalpen Bergrennen	Touring Grand Touring Sport Sport	1000 cc 1000 cc 1000 cc 2000 cc	G. Biscaldi K. Zuegg T. Spychiger H. Herrmann	7'30"9 7'02"1 6'25"7 5'51"8	Fiat Abarth 1000 Berlina Fiat Abarth 1000 BA Fiat Abarth 1000 Fiat Abarth 2000
15 September	Bourscheid, Luxembourg	Course de côte de Burscheid (circuit)	Touring	850 cc	H. Wagner	2'25"1	Fiat Abarth 850 TC
15 September	Madrid, Spagna	Circuito Hoyo de Manzanares (hillclimb)	Touring	850 cc	A. Alvarez De Leon	2'42"8	Fiat Abarth 850 TC
21 September	Cecoslovacchia	Koprivnicer Automobil Wettbewerb (circuit)	Touring	---	Egan	---	Fiat Abarth 850 TC
21-22 September	Mitholz, Switzerland	Mitholz-Kandersteg (hillclimb-2,640 kms x 2)	Touring Touring Prepared Touring Prepared Touring Grand Touring Grand Touring Grand Touring	850 cc 1000 cc 850 cc 1000 cc 700 cc 850 cc 1300 cc	W. Streit R. Müller R. Egli H. Illert P. Ettmüller J. Kretschi A. Hachen	4'26"7 4'10"3 4'05"8 3'55"5 4'06"5 4'19"7 3'31"1	Fiat Abarth 850 TC Fiat Abarth 1000 Berlina Fiat Abarth 850 TC Fiat Abarth 1000 Berlina Fiat Abarth 700 BA Fiat Abarth 850 TC Abarth Simca 1300
22 September	Montlhèry, France	Coupes de Paris (circuit-3,600 kms x 15 laps) (13 laps) (13 laps)	Touring Grand Touring Grand Touring	up to 1000 cc 1000 cc 1300 cc	M. Finquel J. P. Hanrioud R. Richard	30'51"6 23'44"0 23'8"7	Fiat Abarth 850 TC Fiat Abarth 1000 BA Abarth Simca 1300
22 September	Monza, Italy	Coppa d'Autunno (circuit-2,405 kms x 18 laps)	Touring Touring Grand Touring Grand Touring Grand Touring	850 cc 1000 cc 700 cc 1000 cc 1300 cc	M. Cintolesi C. Benelli M. Pinchetti A. Giliberti F. Ghezzi	19'10"9 19'19"9 17'53"6 17'46"0 20'33"2	Fiat Abarth 850 TC Fiat Abarth 1000 Berlina Fiat Abarth 700 BA Fiat Abarth 1000 BA Abarth Simca 1300
22 September	Catania, Italy	Catania-Etna (hillclimb-28,200 kms)	Touring Grand Touring Grand Touring	up to 1150 cc 1000 cc 1300 cc	'Gordon' C. Cannataro A. Monaco	18'00"5 17'10"5 16'10"6	Fiat Abarth 1000 Berlina Fiat Abarth 1000 BA Abarth Simca 1300
22 September	Osimo, Italy	Coppa Luigi Fagioli (hillclimb 5,300 kms)	Touring Touring Grand Touring Grand Touring Grand Touring	850 cc 1000 cc 700 cc 1000 cc 1300 cc	B. Faenza G. Caraceni R. Ricci A. Zucchi S. Ridolfi	3'13"0 3'07"3 3'14"0 2'55"1 2'22"1	Fiat Abarth 850 TC Fiat Abarth 1000 Berlina Fiat Abarth 700 BA Fiat Abarth 1000 BA Abarth Simca 1300
22 September	Houyet, Belgium	Course de côte de Houyet (hillclimb-2,500 kms)	Touring Prepared Touring	850 cc 850 cc	'Thierry' H. Wagner	2'03"4 1'55"0	Fiat Abarth 850 TC Fiat Abarth 850 TC
22 September	Cambridge, United Kingdom	Cambridge Car Club's Driving Test (circuit)	Touring	---	Pinkerton	6'56"8	Fiat Abarth 850 TC
22 September	Kaiserslautern, Germany	Rheinland-Pfalz Bergpreis (hillclimb-3,050 kms)	Touring Grand Touring	850 cc 1000 cc	M. Stephani B. Sechi	4'05"2 3'49"3	Fiat Abarth 850 TC Fiat Abarth 1000 BA
22 September	Novara, Italy	Giro dei Laghi Novaresi (272 kms)	Touring	up to 1000 cc	C. Massagrande	---	Fiat Abarth 850 TC
27-29 September	Cagliari, Italy	Rallye della Sardegna (1425 kms)	Touring	850 cc	Frescobaldi-Martelli	---	Fiat Abarth 850 TC
28-29 September	Washington D.C., USA	Kent's Northwest Grand Prix (circuit)	---	---	D. Snook	---	Fiat Abarth 1000 BA
29 September	Namur, Belgium	Course de côte de Namur (hillclimb-2,000 kms)	Touring Prepared Touring	850 cc 850 cc	'Thierry' H. Wagner	1'55"7 1'46"4	Fiat Abarth 850 TC Fiat Abarth 850 TC
29 September	Freiburg, Germany	Bergrennen am Schauinsland (hillclimb-7,200 kms x 2)	Touring Prepared Touring Grand Touring Grand Touring Grand Touring	1000 cc 1000 cc up to 850 cc 1000 cc 1300 cc	F. Birchler M. Illert P. Ettmüller B. Sechi X. Perrot	11'19"5 10'38"7 10'52"1 10'47"2 9'44"3	Fiat Abarth 1000 Berlina Fiat Abarth 1000 Berlina Fiat Abarth 700 BA Fiat Abarth 1000 BA Abarth Simca 1300
6 October	Siena, Italy	Rallye del Chianti classico (152 kms)	Touring Touring	850 cc 1000 cc	F. De Benedictis Minzoni	--- ---	Fiat Abarth 850 TC Fiat Abarth 1000 Berlina

DATE	LOCATION	RACE	CATEGORY	CLASS	WINNER	TIME OR DISTANCE	CAR
6 October	Montlhéry, France	Coupes du Salon (circuit-3,600 kms)	Touring	1000 cc	M. Finquel	34'12"1	Fiat Abarth 850 TC
			Grand Touring	1300 cc	R. Richard	31'57"0	Abarth Simca 1300
6 October	Monza, Italy	Coppa Leopoldo Carri (circuit-2,405 kms x 20 laps)	Touring	850 cc	A De Paoli	25'07"3	Fiat Abarth 850 TC
			Touring	1000 cc	D. Romanello	25'27"6	Fiat Abarth 1000 Berlina
6 October	Genova, Italy	Pontedecimo-Giovi (hillclimb-9,560 kms)	Touring	850 cc	"Odar"	6'40"9	Fiat Abarth 850 TC
			Grand Touring	700 cc	G. Varese	6'40"9	Fiat Abarth 700 BA
			Grand Touring	1000 cc	G. Rebaudi	6'16"8	Fiat Abarth 1000 BA
			Grand Touring	1300 cc	G. Dalla Torre	5'59"3	Abarth Simca 1300
			Sport	1000 cc	G. Tommasi	6'15"8	Fiat Abarth 1000 Sport
6 October	Palermo, Italy	Alcamo-Monte Bonifato (hillclimb-4,500 kms)	Touring	850 cc	V. Rallo	3'34"4	Fiat Abarth 850 TC
			Touring	up to 1150 cc	G. Virgilio	3'21"2	Fiat Abarth 1000 Berlina
			Grand Touring	700 cc	G. Bonomo	3'40"3	Fiat Abarth 700 BA
			Grand Touring	1000 cc	C. Cannataro	3'17"2	Fiat Abarth 1000 BA
			Grand Touring	1300 cc	D. Scola	3'06"2	Abarth Simca 1300
6 October	Innsbruck, Austria	Preis von Tirol (circuit-2,800 kms x 10 laps)	Touring	1000 cc	H. Herrmann	15'11"57	Fiat Abarth 1000 Berlina
			Grand Touring	1000 cc	T. Hofer	20'04"0	Fiat Abarth 1000 BA
13 October	Cognac, France	Rallye du Cognac (292 kms)	Grand Touring	1300 cc	Pasquier-La Monerie	---	Abarth Simca 1300
13 October	Monza, Italy	Coppa Sport (5,750 kms x 20 laps)	Touring	850 cc	E. Knecht	50'34"2	Fiat Abarth 850 TC
			Touring	1000 cc	F. Birchler	49'33"2	Fiat Abarth 1000 Berlina
			Prepared Touring	up to 1000 cc	R. Egli	50'34"0	Fiat Abarth 850 TC
			Grand Touring	1000 cc	T. Hofer	42'13"2	Fiat Abarth 1000 BA
			Grand Touring	1300 cc	A. Hachen	41'31"8	Abarth Simca 1300
13 October	Pescara, Italy	Cronoscalata Svolte di Popoli (hillclimb-8,000 kms)	Touring	up to 1100 cc	'Jack'	5'30"4	Fiat Abarth 1000 BA
			Grand Touring	up to 1150 cc	R. Ricci	5'07"7	Fiat Abarth 1000 BA
13 October	Varese, Italy	Trofeo Provincia di Varese (150 kms)	Touring	850 cc	A. Reale	---	Fiat Abarth 850 TC
13 October	Trier, Germany	Trierer Flugplatzrennen (circuit-1,900 kms x 27 laps)	Grand Touring	1000 cc	A. Schmalbach	30'26"5	Fiat Abarth 1000 BA
			Grand Touring	1300 cc	K. Geiss	29'42"8	Abarth Simca 1300
13 October	Forlì, Italy	Gara Interregionale (144 kms)	Touring	up to 1150 cc	C. Massagrande	---	Fiat Abarth 850 TC
13 October	Castilla, Spagna	Subida del Campeonato De Montana Castilla (hillclimb)	Touring	850 cc	A. Alvarez De Leon	---	Fiat Abarth 850 TC
13 October	Linz, Austria	Osttiroler Wertungsfahrt (96,000 kms)	Touring	1000 cc	K. Zuegg	---	Fiat Abarth 1000 BA
19 October	Luzern, Switzerland	Hergiswalderbrücke Holder Käppeli (hillclimb-6,740 kms)	Touring	850 cc	E. Wuest	6'21"9	Fiat Abarth 850 TC
20 October	Eberbach, Germany	Bergrennen Eberbach (hillclimb-3,000 kms x 2)	Touring	850 cc	M. Stephani	4'19"0	Fiat Abarth 850 TC
			Grand Touring	1300 cc	K. Geiss	3'58"0	Abarth Simca 1300
20 October	Sanremo, Italy	Trofeo Sanremese della Montagna (hillclimb-3,900 kms x 2)	Touring	850 cc	M. Crosina	5'23"8	Fiat Abarth 850 TC
			Touring	up to 1150 cc	P. Villani	5'35"6	Fiat Abarth 1000 Berlina
			Grand Touring	700 cc	G. Varese	5'40"6	Fiat Abarth 700 BA
			Grand Touring	1000 cc	P. Marsiaj	5'08"4	Fiat Abarth 1000 BA
			Grand Touring	1300 cc	L. Cella	5'00"2	Abarth Simca 1300
			Sport	2000 cc	L. Taramazzo	4'53"1	Fiat Abarth 2000
20 October	Nelson (Ohio), USA	Mahoning Valley Races (circuit-1,07 miles x 18 laps)	---	---	E. Stamm	---	Fiat Abarth 750 BA
20 October	Bern, Switzerland	Slalom	Grand Touring	1300 cc	E. Möll	5'51"8	Abarth Simca 1300
20 October	Madrid, Spagna	Rallye Club "600"-200 kms	---	---	De Leon-Cinto	---	Fiat Abarth 850 TC
26 October	Fribourg, Switzerland	Slalom de la Poya	---	---	F. Haenni	1'47"4	Fiat Abarth 850 TC
26-27 October	Fulda, Germany	Rhön Berggrennen (hillclimb-2,550 kms x 2)	Grand Touring	700 cc	G. Metz	4'31"1	Fiat Abarth 700 BA
3 November	Montlhéry, France	Rallye de l' AGACI (circuit)	Touring	1000 cc	V. Magne	---	Fiat Abarth 1000 Berlina
3 November	Brentford Market, United Kingdom	A.H.C. Guy Fawkes Driving Tests	Touring	---	A. Rogers	5'34"5	Fiat Abarth 850 TC
3 November	Eisenstadt, Austria	Stotzinger Bergrennen (hillclimb-4,000 kms)	Touring	850 cc	Müller	3'27"0	Fiat Abarth 850 TC
17 November	Cagliari, Italy	Corongiu-Campuomu (hillclimb-4,000 kms)	Touring	850 cc	V. Flora-ex aequo S. Cambarau	3'34"7	Fiat Abarth 850 TC
			Grand Touring	up to 1150 cc	M. Rodriguez	3'36"2	Fiat Abarth 1000 BA
18 November	Spagna	Gymkana Escuela de Ingenieros Industriales	Touring	701-850 c c	A. Alvarez De Leon	---	Fiat Abarth 850 TC
24 November	Monza, Italy	Coppa F.I.S.A. (circuit-5,750 kms x 6 laps)	Touring	850 cc	U. De Giorgio	14'21"6	Fiat Abarth 850 TC
			Touring	1000 cc	Zuppini	14'10"2	Fiat Abarth 1000 Berlina
			Grand Touring	700 cc	R. Ricci	13'48"4	Fiat Abarth 700 BA
			Grand Touring	1000 cc	'Tiger'	12'33"1	Fiat Abarth 1000 BA
			Grand Touring	1300 cc	Andreini	12'10"5	Abarth Simca 1300
24 November	Montpellier, France	Criterium des Cevennes (460 kms)	Grand Touring	1000 cc	R. De Lageneste-Du Genestoux	---	Fiat Abarth 1000 BA
8 December	Madrid, Spagna	Rallye Marconi	Touring	850 cc	A. Alvarez De Leon	---	Fiat Abarth 850 TC
15 December	Frascati, Italy	Frascati-Fuscolo (hillclimb-4,200 kms)	Touring	up to 1150 cc	R. Coscioni	---	Fiat Abarth 850 TC

DATE	LOCATION	RACE	CATEGORY	CLASS	WINNER	TIME OR DISTANCE	CAR
1964							
2 February	Igualada, Spagna	Rallye de Invierno (hillclimb-6,000 kms)	Touring	850 cc	'Foca'	---	Fiat Abarth 850 TC
9 February	L'Aquila, Italy	Rallye Autosciatoria Velino-Sirente-Montacagno	Touring	up to 850 cc	R. Coscioni (overall)	---	Fiat Abarth 850 TC
16 February	Roma, Italy	Rallye di Polidoro (152 kms)	Touring	up to 1150 cc	M. Ventriglia	---	Fiat Abarth 1000 Berlina
28 February	Sanremo, Italy	Rallye dei Fiori (1269 kms)	Prepared Touring	850 cc	Bisulli-Andreani	---	Fiat Abarth 850 TC
29 February	Salon de Provence, France	Rallye Mistral (270 kms)	Grand Touring	1300 cc	Maublanc-Dreyfus	---	Abarth Simca 1300
1 March	Roma, Italy	Rallye dell'Artemisio (159 kms)	Touring	700-1150 cc	R. Coscioni	---	Fiat Abarth 850 TC
1 March	Igualada, Spagna	Carrera en cuesta a Las Mayolas (hillclimb-6,000 kms)	Prepared Touring	850 cc	'Foca'	5'01"22	Fiat Abarth 850 TC
1 March	Sable sur Sarthe, France	Rallye du Maine	Grand Touring	up to 2500 cc	R. De la Geneste (overall)	---	Abarth Simca 2000
7-8 March	Padova, Italy	Autosciatoria delle Università Italyne (regolarità-247 kms)	Touring	850 cc	A. Bonomi (overall)	---	Fiat Abarth 850 TC
8 March	Genève, Switzerland	Kilomètre d'Eau Morte	Touring	850 cc	J. Salz	0'35"5	Fiat Abarth 850 TC
			Grand Touring	1000 cc	C. Claudet	0'28"0	Fiat Abarth 1000 BA
15 March	Arezzo, Italy	Rallye Esso di Primavera (regolarità 168 kms)	Touring	1000 cc	F. Naldini	---	Fiat Abarth 1000 Berlina
19 March	Monza, Italy	Trofeo Jolly Club (circuit-57,500 kms)	Touring	600 cc	F. Patria	28'42"2	Fiat Abarth 595
			Touring	850 cc	A. Ratti	24'8"6	Fiat Abarth 850 TC
			Touring	1000 cc	C. Fagioli	24'58"9	Fiat Abarth 1000 Berlina
21 March	Kottingbrunn, Austria	National Flugplatzrennen (circuit-22,000 kms)	Touring	850 cc	H. Kiwish	17'11"4	Fiat Abarth 850 TC
			Grand Touring	1000 cc	K. Zuegg	17'34"0	Fiat Abarth 1000 BA
21 March	El Escorial, Spagna	Subida Cruz Verde-Robledondo (hillclimb-5,100 kms)	Touring	850 cc	A. Alvarez De Leon	4'04"1	Fiat Abarth 850 TC
			Touring	1000 cc	E.D. Moras	4'14"2	Fiat Abarth 1000 Berlina
21 March	Sebring (Florida), USA	12 Hours (circuit-5,205 miles)	Grand Touring	1300 cc	Fleming-Linton-Diaz	---	Abarth Simca 1300
21-22 March	Le Mans, France	Rallye Automobile Nationale de l'Ouest	Grand Touring	up to 2500 cc	R. De la Geneste (overall)	---	Abarth Simca 2000
22 March	Bruxelles, Belgium	Course de côte de Alle sur Semois (hillclimb-4,200 kms)	Touring	1000 cc	Buffin	2'41"8	Fiat Abarth 1000 Berlina
22 March	Erbach, Germany	DMV Krähbergrennen (hillclimb-3,600 kms x 2)	Grand Touring	1000 cc	A. Schmalbach	5'26"1	Fiat Abarth 1000 BA
22 March	Cagliari, Italy	Rallye di Primavera (174 kms)	Grand Touring	1000 cc	M. Rodriguez (overall)	---	Fiat Abarth 1000 BA
22 March	Roma, Italy	Coppa Gallenga (hillclimb-10,400 kms)	Touring	850 cc	M. Cintolesi	5'52"7	Fiat Abarth 850 TC
			Touring	1000 cc	Romanello	5'42"7	Fiat Abarth 1000 Berlina
			Grand Touring	700 cc	Pagano	6'21"7	Fiat Abarth 700 BA
			Grand Touring	1000 cc	M. Moselli	5'31"6	Fiat Abarth 1000 BA
30 March	Bologna, Italy	Coppa Acqua Cerelia (hillclimb-8,000 kms)	Touring	600 cc	A. Lepri	7'13"7	Fiat Abarth 595
			Touring	850 cc	P. Frescobaldi	6'33"	Fiat Abarth 850 TC
			Touring	1000 cc	A. Marzi	6'23"7	Fiat Abarth 1000 Berlina
			Grand Touring	700 cc	D. Morazzoni	6'39"1	Fiat Abarth 700 BA
			Grand Touring	1000 cc	A. Zucchi	6'09"8	Fiat Abarth 1000 BA
			Grand Touring	1300 cc	G. Nicopolous	6'30"1	Abarth Simca 1300
			Sport-Prototypes	1000 cc	V. Venturi	5'48"8	Fiat Abarth 1000
30 March	Vallelunga, Italy	Coppa AGIP (circuit-1,800 kms x 20 laps)	Touring	850 cc	F. Chiari	23'57"4	Fiat Abarth 850 TC
			Touring	1000 cc	F. Chiari	24'00"0	Fiat Abarth 1000 Berlina
5 April	Ursins, Switzerland	Milles Mètres (hillclimb-1,000 kms)	Touring	850 cc	J.P. Brun	0'36"6	Fiat Abarth 850 TC
			Touring	1000 cc	.F. Wipf	0'34"9	Fiat Abarth 1000 Berlina
			Grand Touring	up to 850 cc	S. Hauser	0'35"8	Fiat Abarth 750 BA
			Grand Touring	1000 cc	C. Chaudet	0'30"0	Fiat Abarth 1000 BA
5 April	Verona, Italy	Stallavena-Boscochiesanuova (hillclimb-15,300 kms)	Touring	850 cc	L. Monza	8'57"0	Fiat Abarth 850 TC
			Touring	1000 cc	'Jerry'	8'29"3	Fiat Abarth 1000 Berlina
			Grand Touring	700 cc	'Makye'	8'26"5	Fiat Abarth 700 BA
			Grand Touring	1000 cc	'Matich'	7'49"0	Fiat Abarth 1000 BA
			Grand Touring	1300 cc	G. Dalla Torre	7'46"9	Abarth Simca 1300
			Grand Touring	up to 2500 cc	F. Patria (overall)	7'02"3	Abarth Simca 2000
5 April	Frosinone, Italy	Rallye dei Tre Laghi di Ciociaria (154,850 kms)	Touring	850 cc	A. Greco	---	Fiat Abarth 850 TC
5 April	St. Pölten, Austria	Rallye Voralpenfahrt (190 kms)	Prepared Touring	1000 cc	H. Bennier	---	Fiat Abarth 1000 Berlina
5 April	Innsbruck, Austria	Internationales Weebergrennen (hillclimb-5,000 kms)	Grand Touring	1000 ccc	'Driver'	4'57"4	Fiat Abarth 1000 BA
5 April	Cannes, France	Course de côte de Cannes (hillclimb-7,845 kms)	Grand Touring	1300 cc	A. Raymond	7'31"6	Abarth Simca 1300
			Grand Touring	up to 2500 cc	R. Fraissinet	7'16"4	Abarth Simca 2000
12 April	Aspern, Austria	Preis von Wien (velocità-2,600 kms)	Touring	850 cc	H. Kiwisch	32'25"0	Fiat Abarth 850 TC
			Grand Touring	1000 cc	T. Hofer	27'27"8	Fiat Abarth 1000 BA
			Grand Touring	1300 cc	Hachen	28'05"9	Abarth Simca 1300
			Grand Touring	2000 cc	E. Mahle	26'39"4	Abarth Simca 2000
12 April	Saint Lô, France	Course de côte d'Hébécrevon (hillclimb-1,200 kms)	Grand Touring	1000 cc	M. Martin	0'52"2	Fiat Abarth 1000 BA
			Grand Touring	up to 1600 cc	C. Ballot-Lena	0'51"1	Abarth Simca 1300

DATE	LOCATION	RACE	CATEGORY	CLASS	WINNER	TIME OR DISTANCE	CAR
12 April	Montserrat, Spagna	Carrera en cuesta de Montserrat	Touring	850 cc	A. Alvarez De Leon	6'00"1	Fiat Abarth 850 TC
		(hillclimb-7,000 kms)	Touring	1000 cc	J. Juncosa	5'54"7	Fiat Abarth 1000 Berlina
			Prepared Touring	850 cc	J. Galofre	5'57"7	Fiat Abarth 850 TC
12 April	Eifel, Germany	ADAC Bergpreis Bad Neuenahr	Grand Touring	1000 cc	P. Koepehn	4'35"2	Fiat Abarth 1000 BA
		(hillclimb-3,300 kms x 2)	Grand Touring	1300 cc	K. Geiss	4'29"6	Abarth Simca 1300
12 April	Dakar, Senegal	6 Ore di Dakar (circuit)	Grand Touring	1300 cc	Sarayac-Barthe	kms 686,484	Abarth Simca 1300
18 April	Zandwoort, Holland	Trainingsweekend N.A.V. (circuit-4,225 kms x 10 laps)	Prepared Touring	over 850 cc	P.J. Van Osch	21'37"4	Fiat Abarth 1000 Berlina
19 April	Nancy, France	Course de côte de Maron (hillclimb-1,800 kms)	Grand Touring	1000 cc	J. Hauser	2'24"4	Fiat Abarth 1000 BA
			Grand Touring	1300 cc	C. Ballot-Lena (overall)	2'10"1	Abarth Simca 1300
19 April	Vendrell, Spagna	Carrera en cuesta de Las Ventosas	Touring	1000 cc	J. Juncosa	5'03"3	Fiat Abarth 1000 Berlina
		(hillclimb-5,800 kms)	Prepared Touring	850 cc	J. Galofre	4'58"1	Fiat Abarth 850 TC
19 April	Imola, Italy	Coppa Grand Touring Trofeo Shell	Grand Touring	700 cc	'Makye'	23'16"6	Fiat Abarth 700 BA
		(circuit-5,017 kms x 12 laps)	Grand Touring	1000 cc	A. Zucchi	26'25"9	Fiat Abarth 1000 BA
			Grand Touring	1300 cc	F. Dalla Torre	26'07"1	Abarth Simca 1300
			Grand Touring	up to 2500 cc	F. Patria (overall)	24'07"5	Abarth Simca 2000
19 April	Avola, Italy	Coppa Belmonte (hillclimb-10,200 kms)	Touring	850 cc	E. Di Stefano	5'53"2	Fiat Abarth 850 TC
			Touring	1000 cc	'Gordon'	5'47"0	Fiat Abarth 1000 Berlina
			Grand Touring	700 cc	A. Rizzo	5'12"2	Fiat Abarth 700 BA
			Grand Touring	1000 cc	'Comen'	5'31"5	Fiat Abarth 1000 BA
			Grand Touring	1300 cc	A. Giliberti	5'31"7	Abarth Simca 1300
19 April	Eberbach, Germany	Eberbach Bergrennen (hillclimb-3,000 kms)	Touring	850 cc	M. Stephani	4'11"6	Fiat Abarth 850 TC
			Grand Touring	1000 cc	A. Schmalbach	4'01"7	Fiat Abarth 1000 BA
			Grand Touring	1300 cc	H.D. Dechent	3'47"7	Abarth Simca 1300
19 April	Roma, Italy	Trofeo Edgardo Lazzaroni (rallye-304 kms)	Touring	850 cc	G. Bisulli	---	Fiat Abarth 850 TC
			Touring	1000 cc	M. Ventriglia	---	Fiat Abarth 1000 Berlina
25-26 April	Payerne, Switzerland	Slalom National de Payerne (hillclimb-4,800 kms x 2)	Touring	850 cc	J.P. Brun	3'33"7	Fiat Abarth 850 TC
			Touring	1000 cc	J. Kretschi	3'44"4	Fiat Abarth 1000 Berlina
			Grand Touring	1000 cc	C. Chaudet	3'35"2	Fiat Abarth 1000 BA
			Grand Touring	1300 cc	R. Stier	3'23"5	Abarth Simca 1300
			Sport	1000 cc	P. Ettmüller	3'13"1	Fiat Abarth 1000
26 April	Montseny, Spagna	Carrera en cuesta al Montseny	Touring	850 cc	'Foca'	13'23"3	Fiat Abarth 850 TC
		(hillclimb-15,000 kms)	Touring	1000 cc	J. Juncosa	13'10"7	Fiat Abarth 1000 Berlina
26 April	Gap, France	Course de côte du Col Bayard (hillclimb-5,000 kms)	Touring	850 cc	F. Sapia	3'19"3	Fiat Abarth 850 TC
			Grand Touring	1300 cc	A. Raymond	3'54"2	Abarth Simca 1300
26 April	Palermo, Italy	Targa Florio (circuit-72,000 kms x 10 laps)	Grand Touring	1300 cc	Laureati-Ridolfi	7h36'12"1	Abarth Simca 1300
26 April	Rheingau, Germany	HMSC Taunus Bergrennen Lorch	Grand Touring	1000 cc	A. Schmalbach	4'47"2	Fiat Abarth 1000 BA
		(hillclimb-3,300 kms x 2 laps)	Grand Touring	1300 cc	H.D. Dechent	---	Abarth Simca 1300
26 April	Frosinone, Italy	Magione-Veroli (hillclimb)	Touring	850 cc	F. Chiari	3'35"3	Fiat Abarth 850 TC
			Touring	up to 1150 cc	F. Altieri	3'41"3	Fiat Abarth 1000 Berlina
			Grand Touring	700 cc	F. Palumbo	3'45"2	Fiat Abarth 700 BA
			Grand Touring	1000 cc	F. Amicucci	3'26"1	Fiat Abarth 1000 BA
26 April	Montlhéry, France	Coupes de Vitesse USA (circuit-3,300 kms x 15 laps)	Grand Touring	1000 cc	C. Bobrowski	28'06"7	Fiat Abarth 1000 BA
			Grand Touring	1300 cc	C. Ballot-Lena	34'33"8	Abarth Simca 1300
1 May	Liège, Belgium	Slalom de Liège	Grand Touring	1000 cc	J. Hauser	1'58"2	Fiat Abarth 1000 BA
1-3 May	Madrid, Spagna	Rallye Nacional del RACE (221,7 kms)	Touring	850 cc	'Foca'	---	Fiat Abarth 850 TC
			Touring	1000 cc	J. Juncosa	---	Fiat Abarth 1000 Berlina
1-7 May	Vallelunga, Italy	Coppa Fausto Pesci (circuit-1,800 kms x 40 laps)	Touring	850 cc	F. Chiari	38'58"2	Fiat Abarth 850 TC
			Touring	up to 1300 cc	B. Faenza	38'44"3	Fiat Abarth 1000 Berlina
2 May	Verbois, Switzerland	Course de côte de Verbois (hillclimb-1,150 kms)	Touring	850 cc	P. Transzal	1'05"5	Fiat Abarth 850 TC
			Grand Touring	1000 cc	R. Vaglio	0'58"6	Fiat Abarth 1000 BA
2 May	Grenchen, Switzerland	Slalom Grenchen	Grand Touring	up to 1150 cc	G. Weber	2'09"4	Fiat Abarth 1000 BA
			Grand Touring	1150-1300 cc	F. Della Balda	2'01"8	Abarth Simca 1300
3 May	Frascati, Italy	Frascati-Tuscolo (hillclimb-4,200 kms)	Touring	600 cc	E. Di Benedetto	3'46"5	Fiat Abarth 595
			Touring	850 cc	'Gero'	3'23"2	Fiat Abarth 850 TC
			Touring	1000 cc	B. Faenza	3'23"2	Fiat Abarth 1000 Berlina
			Grand Touring	700 cc	P. Gargano	3'24"6	Fiat Abarth 700 BA
			Grand Touring	1000 cc	E. Buzzetti	3'10"3	Fiat Abarth 1000 BA
			Grand Touring	1300 cc	Brocchi	3'06"2	Abarth Simca 1300
3 May	Monza, Italy	Gran Premio Grand Touring (circuit-5,750 kms x 3 hours)	Grand Touring	1000 cc	A. Zucchi	kms 493,605	Fiat Abarth 1000 BA
			Grand Touring	1300 cc	F. Patria	kms 514,889	Abarth Simca 1300
3 May	Trier, Germany	Trierer Flugplatzrennen (circuit-1,900 kms x 27 laps)	Grand Touring	1000 cc	A. Schmalbach	30'02"8	Fiat Abarth 1000 BA
			Grand Touring	1300 cc	K. Geiss	33'03'5	Abarth Simca 1300
3 May	Rijcka, Jugoslavia	Circuito di Rijcka	Touring	850 cc	S. Matevz	15'05"3	Fiat Abarth 850 TC

DATE	LOCATION	RACE	CATEGORY	CLASS	WINNER	TIME OR DISTANCE	CAR
7 May	Cesena, Italy	Coppa Città di Cesena (167 kms)	Touring Touring	850 cc 1000 cc	G. Bisulli (overall) A. Lambertini	--- ---	Fiat Abarth 850 TC Fiat Abarth 1000 Berlina
7 May	Aix en Provence, France	Course de côte d'Aix-Saint Antonin (hillclimb-2,700 kms)	Grand Touring Grand Touring Grand Touring	700 cc 1000 cc 1300 cc	A. Battesti P. Condriller P. Maublanc	2'14"0 2'01"9 1'54"5	Fiat Abarth 700 BA Fiat Abarth 1000 BA Abarth Simca 1300
8-9 May	Jarvis (Ontario), USA	Harewood Trophy Races (circuit-57,300 miles)	Sport	---	C. Fisher	---	Abarth Simca 1300
9-10 May	Bordeaux, France	Rallye Bordeaux Sud-Ouest (800 kms)	Grand Touring	1300 cc	Sarayac-Jamay	---	Abarth Simca 1300
10 May	Sassari, Italy	Alghero-Scala Picada (hillclimb-8,700 kms)	Touring Grand Touring Grand Touring	850 cc 1000 cc 1300 cc	G. Vacca G. Rebaudi A. Giliberti (overall)	5'29"8 5'19"0 4'57"1	Fiat Abarth 850 TC Fiat Abarth 1000 BA Abarth Simca 1300
10 May	Napoli, Italy	Targa Vesuvio (hillclimb-9,200 kms)	Touring Touring Grand Touring Grand Touring Grand Touring	850 cc 1000 cc 700 cc 1000 cc 1300 cc	L. Culotta T. Capuozzo F. Sorrentino M. Moselli R. Arfè (overall)	8'08"5 7'53"5 9'05"0 7'21"0 7'20"1	Fiat Abarth 850 TC Fiat Abarth 1000 Berlina Fiat Abarth 700 BA Fiat Abarth 1000 BA Abarth Simca 1300
10 May	Ferrara, Italy	Rallye degli Estensi (110 kms)	Touring	500-850 cc	M. Dell'Isola	---	Fiat Abarth 595
10 May	Volterra, Italy	Coppa Città di Volterra (hillclimb-10,060 kms)	Touring Touring Touring Grand Touring Grand Touring Grand Touring	600 cc 850 cc 1000 cc 700 cc 1000 cc 1300 cc	A. Lepri M. Cintolesi G. Rile 'Makye' A. Monselles 'Nicor'	7'48"3 6'59"1 6'58"0 6'56"1 6'44"4 6'43"0	Fiat Abarth 595 Fiat Abarth 850 TC Fiat Abarth 1000 Berlina Fiat Abarth 700 BA Fiat Abarth 1000 BA Abarth Simca 1300
10 May	Marseille, France	Course de côte de la Sainte Baume (hillclimb-8,100 kms)	Grand Touring Grand Touring	700 cc 1300 cc	A. Battesti C. Ballot-Lena	8'49"1 8'05"3	Fiat Abarth 700 BA Abarth Simca 1300
10 May	Wittlich, Germany	Wittlicher ADAC Bergpreis (hillclimb-3,738 kms x 2)	Touring Grand Touring Grand Touring	850 cc 1000 cc 1300 cc	G. Metz A. Schmalbach H.D. Dechent	5'38"2 5'02"6 4'49"8	Fiat Abarth 850 TC Fiat Abarth 1000 BA Abarth Simca 1300
10 May	Longleat, United Kingdom	B.R.S.C.C. Longleat Hill-climb (hillclimb)	Touring	850 cc	J. Anstead	54'34"1	Fiat Abarth 850
10 May	Wien, Austria	Internationale Rallye Martha Goldpokal (circuit-175 kms)	Grand Touring	up to 1300 cc	L. Hlinka	---	Fiat Abarth 1000 BA
10 May	Dudelange, Luxembourg	Course de côte Tele-Dudelange (hillclimb-1,400 kms)	Grand Touring	851-1300 cc	J. Hauser	1'04"1	Fiat Abarth 1000 BA
10 May	Roxilde, Danimarca	Roykilde Ring (circuit-1,400 kms x 8 laps)	Touring	850 cc	A. Larsen	11'31"1	Fiat Abarth 850 TC
10 May	Rio de Janeiro, Brasile	Tres Horas da Barra da Tijuca (circuit-3 hours)	Touring	up to 1300 cc	W. Fittipaldi	---	Fiat Abarth 1000 Berlina
14 May	Digne, France	Course de côte de la Lure (hillclimb-12,700 kms)	Touring Grand Touring Sport	1000 cc 1300 cc 1000 cc	F. Sapia C.Ballot-Lena F.Ruata	9'44"0 8'43"2 9'15"0	Fiat Abarth 1000 Berlina Abarth Simca 1300 Fiat Abarth 1000 SP
16-17 May	Mosport Park (Ontario), USA	Clubmen's Day Races (circuit-20 miles)	---	---	C. Fisher	---	Abarth Simca 1300
17 May	Palermo, Italy	Corsa hillclimb al Monte Pellegrino (8,750 kms)	Touring Touring Touring Grand Touring Grand Touring Grand Touring	600 cc 850 cc 1000 cc 700 cc 1000 cc 1300 cc	S. Calascibetta G. Virgilio V. Arena A. Rizzo I. Capuano D. Scola	6'45"6 6'01"2 5'51"4 6'26"4 5'32"2 5'39"2	Fiat Abarth 595 Fiat Abarth 850 TC Fiat Abarth 1000 Berlina Fiat Abarth 700 BA Fiat Abarth 1000 BA Abarth Simca 1300
17 May	Reggio Emilia, Italy	Giro della Provincia di Reggio Emilia (regolarità-90 kms)	Touring	850 cc	A. Diosy (overall)	---	Fiat Abarth 850 TC
17 May	Lubiana, Jugoslavia	Circuito di Lubiana (14,400 kms)	Touring	850 cc	S. Matevz	1'24"0	Fiat Abarth 850 TC
17 May	Findel, Luxembourg	Slalom du Findel	Touring	850 cc	H. Wagner	1'19"1	Fiat Abarth 850 TC
17 May	Madrid, Spagna	Subida al Puerto de la Morcuera (hillclimb-7,000 kms)	Touring	850 cc	A. Alvarez De Leon	5'09"1	Fiat Abarth 850 TC
23 May	Wallberg, Germany	Wallberg Bergrennen (hillclimb)	Grand Touring	1000 cc	A.Schmalbach	4'35"1	Fiat Abarth 1000 BA
23-24 May	Milano/Venezia, Italy	Rallye di Rialto (336 kms)	Touring	850 cc	V. Tabacchi	---	Fiat Abarth 850 TC
23-24 May	Nancy, France	Rallye de la Lorenne	Touring	850 cc	Erwin-Heissler	---	Fiat Abarth 850 TC
24 May	Firenze, Italy	Coppa della Consuma (hillclimb-12,500 kms)	Touring Touring Touring Grand Touring Grand Touring Grand Touring	600 cc 850 cc 1000 cc 700 cc 1000 cc 1300 cc	R. Ramacciotti G. Raffa A. Marzi (overall) I. Minzoni G. Rebaudi S. Ridolfi	9'09"3 8'07"2 7'58"6 8'43"2 7'49"1 7'25"7	Fiat Abarth 595 Fiat Abarth 850 TC Fiat Abarth 1000 Berlina Fiat Abarth 700 BA Fiat Abarth 1000 BA Abarth Simca 1300
24 May	Foggia, Italy	Valle Bovino-Bovino (hillclimb-6,400 kms)	Touring Touring Grand Touring Grand Touring Sport	850 cc up to 1150 cc 700 cc 1300 cc 1000 cc	B. Massei L. Verrocchio P. Gargano D. Scola (overall) A. Compostella	4'58"2 4'53"2 4'58"4 4'27"1 5'23"7	Fiat Abarth 850 TC Fiat Abarth 1000 Berlina Fiat Abarth 700 BA Abarth Simca 1300 Fiat Abarth 1000 SP
24 May	Zandvoort, Holland	N.A.V. Kampioenschapsraces (circuit-4,225 kms x 15 laps)	Touring	850 cc	E. Swart	17'15"2	Fiat Abarth 850 TC

DATE	LOCATION	RACE	CATEGORY	CLASS	WINNER	TIME OR DISTANCE	CAR
24 May	Barcelona, Spagna	Carrera en cuesta de Vallvidrera (hillclimb-2,450 kms)	Touring	850 cc	'Foca'	5'03"7	Fiat Abarth 850 cc
24 May	Montlhéry, France	Grand Prix de Paris (circuit-3,300 kms x 15 laps)	Touring Touring Grand Touring	850 cc 1000 cc 1000 cc	R. Ruata Rosinski F. Ruata	33'57"0 32'20"6 33'31"4	Fiat Abarth 850 Fiat Abarth 1000 Berlina Fiat Abarth 1000 BA
24 May	Villefranche sur Saône, France	Course de côte de Beaujolais (hillclimb-16,000 kms)	Touring Grand Touring	1000 cc 1300 cc	Fevrier A. Raymond	12'13"4 10'47"0	Fiat Abarth 1000 Berlina Abarth Simca 1300
24 May	Berlin, Germany	Internationale Avus Trophäe für Tourenwagen (circuit-99,600 kms)	Touring	850 cc	H. Garschig	42'40"6	Fiat Abarth 850 TC
24 May	Mitholz, Switzerland	Miltholz-Kandersteg (hillclimb-2,640 kms x 2)	Touring Prepared Touring Prepared Touring Grand Touring Grand Touring Grand Touring Grand Touring Sport	1000 cc 850 cc 1000 cc 700 cc 1000 cc 1300 cc 2000 cc 1000 cc	U.Schütz R.Egli J.Kretschi J.Egli T.Hofer R.Stierli H. Kuhnis P. Ettmüller	4'09"7 4'09"2 4'07"7 4'06"9 3'46"8 3'35"3 3'20"7 3'27"3	Fiat Abarth 1000 Berlina Fiat Abarth 850 TC Fiat Abarth 1000 Berlina Fiat Abarth 700 BA Fiat Abarth 1000 BA Abarth Simca 1300 Abarth Simca 2000 Fiat Abarth 1000 SP
24 May	Castellon de la Plana, Spagna	Subida al Desierto de Las Palmas (hillclimb-4,700 kms)	Touring	1000 cc	J. Juncosa (overall)	4'55'0'	Fiat Abarth 1000 Berlina
29 May	Madrid, Spagna	Subida al Puerto de Galapagar (hillclimb-4,000 kms)	Touring	850 cc	A. Alvarez De Leon	2'32"3	Fiat Abarth 850 TC
30 May	Jarvis (Ontario), USA	Trillium Trophy Race (circuit-15,200 miles)	---	---	C.Fisher	---	Abarth Simca 1300
30-31 May	Limoges, France	Rallye du Limousin (25,800 kms)	Grand Touring	1300 cc	Sarayac-Barthe	---	Abarth Simca 1300
31 May	Herbeumont, Belgium	Course de côte d'Herbeumont (hillclimb-3,600 kms)	Grand Touring	1000 cc	F. Kleine	2'03"3	Fiat Abarth 1000 BA
31 May	Magny Cours, France	Grand Prix de Magny Cours (circuit-2,040 kms x 20 laps)	Touring Grand Touring	850 cc 1000 cc	Lisa F. Ruata	24'7"6 28'22"3	Fiat Abarth 850 TC Fiat Abarth 1000 BA
31 May	Bologna, Italy	Bologna-Passo della Raticosa (hillclimb-32,700 kms)	Touring Touring Grand Touring Grand Touring Grand Touring	850 cc 1000 cc 700 cc 1000 cc 1300cc	A. Stefanelli M. Nesti 'Makye' E. Buzzetti A. Giliberti	23'03"7 22'34"2 21'44"8 22'10"1 20'38"0	Fiat Abarth 850 TC Fiat Abarth 1000 Berlina Fiat Abarth 700 BA Fiat Abarth 1000 BA Abarth Simca 1300
31 May	Adenau, Germany	Nürburgring 1000 Km Rennen (circuit-22,810 kms x 44 laps)	Grand Touring	1300 cc	Herrmann-Jütter	7h14'37"	Abarth Simca 1300
31 May	Belvaux, Luxembourg	Slalom Automobile de Belvaux	Touring Grand Touring	850 cc 851-1150 cc	H. Wagner J. Hauser	1'03"1 1'02"0	Fiat Abarth 850 TC Fiat Abarth 1000 BA
31 May	Monza, Italy	Coppa Automobil Revue (circuit-115 kms) Coppa Dunlop (circuit-115 kms) Coppa Ford (circuit-115 kms) Coppa Kléber-Colombes (circuit-115 kms) Coppa Shell (circuit-115 kms)	Grand Touring Sport Grand Touring Grand Touring Touring Touring	1150-1300 cc 1000 cc 700 cc 850-1150 cc 850 cc 1050 cc	D. Borel P. Ettmüller J. Egli S. Vaglio P. Pichler U. Schütz	42'48"2 38'16"4 45'41"8 43'03"9 51'39"7 50'58"9	Abarth Simca 1300 Fiat Abarth 1000 SP Fiat Abarth 700 BA Fiat Abarth 1000 BA Fiat Abarth 850 TC Fiat Abarth 1000 Berlina
31 May	Ravenna, Italy	Giro di Romagna (156 kms)	Touring	850 cc	C. Massagrande (overall)	---	Fiat Abarth 850 TC
31 May	Kapfenberg, Austria	Kapfenberger Schlossberg Rennen (hillclimb-1,200 kms)	Touring	700-1000 cc	G. Berger	2'49"8	Fiat Abarth 850 TC
2 June	Brescia, Italy	Scalata al Colle di Sant'Eusebio (hillclimb-9,250 kms)	Touring Touring Grand Touring Grand Touring Grand Touring	850 cc 1000 cc 700 cc 1000 cc 1300 cc	G. Gagliardi Gerri M. Pinchetti R. Becchetti R. Bertuzzi	5'56"2 5'34"5 5'44"7 5'26"5 5'15"1	Fiat Abarth 850 TC Fiat Abarth 1000 Berlina Fiat Abarth 700 BA Fiat Abarth 1000 BA Abarth Simca 1300
2 June	Vallelunga, Italy	Gara a Vallelunga (circuit-1,800 kms x 5 laps)	Touring Touring Touring	600 cc 850 cc 1000 cc	E. De Benedetto F. Chiari N. Parenti	6'36"4 6'04"4 6'01"4	Fiat Abarth 595 Fiat Abarth 850 TC Fiat Abarth 1000 Berlina
6-7 June	Pisa, Italy	Trofeo delle Antiche Repubbliche Marinare (regolarità-23 kms)	Touring	850 cc	C. Magnozzi	----	Fiat Abarth 850 TC
6-7 June	Barcelona, Spagna	Rallye de Cataluna (1250 kms)	Touring Touring	850 cc 1000 cc	J. Palau Ribes J. Juncosa	--- ---	Fiat Abarth 850 TC Fiat Abarth 1000 Berlina
7 June	Saint Peray, France	Course de côte du Pin (7,050 kms)	Touring Grand Touring Grand Touring	1000 cc 1000 cc 1300 cc	T. Spychiger R. Vaglio A. Raymond	5'26"2 5'31"8 5'54"4	Fiat Abarth 1000 Berlina Fiat Abarth 1000 BA Abarth Simca 1300
7 June	Berchtesgaden, Germany	Rossfeld Alpen Bergpreis (hillclimb-5,890 kms)	Grand Touring	1300 cc	H. Dechent	7'29"4	Abarth Simca 1300
7 June	Bourges, France	Grand Prix de la Châtre (circuit-1,267 kms x 20 laps)	Grand Touring	1000 cc	F. Ruata	---	Fiat Abarth 1000 BA
7 June	Reisdorf, Luxembourg	Course de côte de Reisdorf (hillclimb-3,400 kms)	Touring Grand Touring	850 cc 850-1300 cc	H. Wagner J. Hauser	2'34"0 2'24"1	Fiat Abarth 850 TC Fiat Abarth 1000 BA

DATE	LOCATION	RACE	CATEGORY	CLASS	WINNER	TIME OR DISTANCE	CAR
7 June	Reggio Emilia, Italy	Vezzano-Casina (hillclimb-12,500 kms)	Touring	850 cc	P. Raffa	8'21"0	Fiat Abarth 850 TC
			Touring	1000 cc	A. Marzi	8'13"2	Fiat Abarth 1000 Berlina
			Grand Touring	700 cc	'Makye'	8'19"5	Fiat Abarth 700 BA
			Grand Touring	1000 cc	G. Rebaudi	7'58"6	Fiat Abarth 1000 BA
			Grand Touring	1300 cc	G. Dalla Torre	7'49"1	Abarth Simca 1300
7 June	Pistoia, Italy	Coppa della Collina (hillclimb-14,700 kms)	Touring	600 cc	G. Giannini	11'48"0	Fiat Abarth 595
			Touring	850 cc	M. Tamburini	10'20"7	Fiat Abarth 850 TC
			Touring	1000 cc	M. Nesti	10'24"4	Fiat Abarth 1000 Berlina
			Grand Touring	1000 cc	A. Zucchi	9'11"3	Fiat Abarth 1000 BA
			Grand Touring	1300 cc	'Nicor'	9'32"6	Abarth Simca 1300
7 June	Torino, Italy	Cuorgné-Alpette (hillclimb-7,000 kms)	Touring	850 cc	M.Acquati	5'06"2	Fiat Abarth 850 TC
			Touring	1000 cc	De Gregorio	5'06"2	Fiat Abarth 1000 Berlina
			Grand Touring	700 cc	G. Priotto	5'07"4	Fiat Abarth 700 BA
			Grand Touring	1000 cc	'Sem'	4'57"1	Fiat Abarth 1000 BA
			Sport	1000 cc	G. Gavelli	4'52"4	Fiat Abarth 1000 SP
13-14 June	Mainz Finthen, Germany	HMSC Flugplatz Rennen (circuit-2,100 kms x 30 laps)	Touring	850 cc	F. Patria	40'17"0	Fiat Abarth 850 TC
			Grand Touring	1000 cc	A. Schmalbach	37'04"2	Fiat Abarth 1000 BA
			Grand Touring	1300 cc	H. Dechent	35'31"8	Abarth Simca 1300
			Grand Touring	2000 cc	F. Patria	35'47"7	Abarth Simca 2000
13-14 June	Cagliari, Italy	Rallye della Sardegna (945 kms)	Touring	850 cc	Costa-Cocco	---	Fiat Abarth 850 TC
			Touring	1000 cc	Pedretti-Nicolini	---	Fiat Abarth 1000 Berlina
14 June	Avignon, France	Course de côte du Mont Ventoux (hillclimb-21,700 kms)	Grand Touring	1000 cc	A. Knorr	13'22"9	Fiat Abarth 1000 BA
			Grand Touring	1300 cc	A. Raymond	13'30"1	Abarth Simca 1300
			Sport	up to 1150 cc	P. Ettmüller	13'14"5	Fiat Abarth 1000 SP
14 June	Palermo, Italy	Alcamo-Monte Bonifato (hillclimb-4,500 kms)	Touring	850 cc	F. Cassarà	3'29"0	Fiat Abarth 850 TC
			Touring	1000 cc	G. Virgilio	3'20"2	Fiat Abarth 1000 Berlina
			Grand Touring	700 cc	P. Bellavia	3'39"0	Fiat Abarth 700 BA
			Grand Touring	1000 cc	A. Catalano	3'12"3	Fiat Abarth 1000 BA
			Grand Touring	1300 cc	A. Gambero	3'03"2	Abarth Simca 1300
14 June	Stans, Switzerland	Slalom A.C. Thurgau (3,000 kms)	Touring	1000 cc	A. Lämmle	3'15"4	Fiat Abarth 1000 Berlina
			Grand Touring	up to 1300 cc	T. Hofer	3'40"8	Fiat Abarth 1000 BA
14 June	Asiago, Italy	Coppa Città di Asiago (hillclimb-14,500 kms)	Touring	850 cc	A. Manavello	9'58"6	Fiat Abarth 850 TC
			Touring	1000 cc	'Ubvar'	9'44'1	Fiat Abarth 1000 Berlina
			Grand Touring	700 cc	G. Varese	9'28"7	Fiat Abarth 700 BA
			Grand Touring	1000 cc	A. Zucchi	8'55"2	Fiat Abarth 1000 BA
14 June	Molins de Rey, Spagna	Subida a Santa Cruz de Olorde (hillclimb-5,700 kms)	Touring	850 cc	J. Vinas	5'06"0	Fiat Abarth 850 TC
			Grand Touring	1000 cc	'Salvador'	4'12"3	Fiat Abarth 1000 BA
14 June	Macerata, Italy	Macerata-Tolentino (274,400 kms)	Touring	850 cc	R. Coscioni (overall)	---	Fiat Abarth 850 TC
14 June	Madrid, Spagna	Gran Premio de Madrid (circuit-56,250 kms)	Touring	up to 1000 cc	A. Alvarez De Leon	38'23"1	Fiat Abarth 850 TC
18-20 June	Mosport Park (Ontario), USA	Annual Sports Car Races (circuit-29,500 miles)	---	---	C. Fisher	---	Abarth Simca 1300
20 June	Genève, Switzerland	Gymcana-Slalom Automobile	Touring	850 cc	J. Brun	---	Fiat Abarth 850 TC
21 June	Cuneo, Italy	Garessio-Colle S.Bernardo (hillclimb-6,200 kms)	Touring	850 cc	G. Raffa	4'50"8	Fiat Abarth 850 TC
			Touring	up to 1150 cc	F. Sapia	4'55"0	Fiat Abarth 1000 Berlina
			Grand Touring	1000 cc	G. Rebaudi	4'29"8	Fiat Abarth 1000 BA
			Grand Touring	up to 1150 cc	L. Pozzo	4'53"7	Abarth Simca 1150
			Grand Touring	1300 cc	G. Giraudi	4'24"8	Abarth Simca 1300
			Grand Touring	over 1300 cc	L. Taramazzo (overall)	4'08'4	Abarth Simca 2000
			Sport	1000 cc	V. Venturi	4'21"2	Fiat Abarth 1000 SP
21 June	Luxembourg	Slalom Luxembourg	Touring	1000 cc	H. Wagner	1'44"3	Fiat Abarth 850 TC
21 June	Scarperia, Italy	Circuito del Mugello (circuit 331 kms)	Touring	1000 cc	G. Virgilio	3h 41'23"3	Fiat Abarth 1000 Berlina
			Grand Touring	1000 cc	E. Buzzetti	3h 44'03"4	Fiat Abarth 1000 BA
			Grand Touring	1300 cc	A. Giliberti	3h 29'59"3	Abarth Simca 1300
21 June	Wolfsberg, Austria	Koralpe Bergrennen (hillclimb-9,500 kms)	Touring	850 cc	H. Hiwisch	7'25"0	Fiat Abarth 850 TC
21 June	San Cugat, Spagna	Carrera en cuesta San Cugat-Tibidabo (hillclimb-4,700 kms)	Touring	850 cc	J. Palau Ribes	3'29"8	Fiat Abarth 850 TC
			Touring	1000 cc	J. Juncosa	3'19"6	Fiat Abarth 1000 Berlina
			Grand Touring	850 cc	J. Vinas	3'34"9	Fiat Abarth 850 TC
			Grand Touring	1000 cc	'Salvador'	2'58"3	Fiat Abarth 1000 BA
21 June	Zandwoort, Holland	N.A.V. Kampioenschapsraces (circuit-4,225 kms x 18 laps)	Touring	1000 cc	J. Van Osch	38'18"3	Fiat Abarth 1000 Berlina
21 June	Arezzo, Italy	Camucia-Cortona (hillclimb-3,500 kms)	Touring	850 cc	M. Tamburini	5'22"5	Fiat Abarth 850 TC
			Touring	1000 cc	G. Sangermano	5'34"9	Fiat Abarth 1000 Berlina
			Grand Touring	1000 cc	'Laki'	4'55"0	Fiat Abarth 1000 BA
28 June	Frankfurt, Germany	DMV Spessart Bergrennen (hillclimb-4,000 kms)	Touring	850 cc	G. Schüler	2'13"2	Fiat Abarth 850 TC
			Grand Touring	1000 cc	W. Bickel	2'14"5	Fiat Abarth 1000 BA
			Grand Touring	1300 cc	K. Geiss	1'57"4	Abarth Simca 1300

DATE	LOCATION	RACE	CATEGORY	CLASS	WINNER	TIME OR DISTANCE	CAR
28 June	Cosenza, Italy	Coppa della Sila (hillclimb-14,000 kms)	Touring	850 cc	N. Persia	11'45"4	Fiat Abarth 850 TC
			Touring	1000 cc	F. Pietramela	11'29"3	Fiat Abarth 1000 Berlina
			Grand Touring	700 cc	'Makye'	11'06"1	Fiat Abarth 700 BA
			Grand Touring	1000 cc	C. Avventurieri	10'29"2	Fiat Abarth 1000 BA
			Grand Touring	1300 cc	D. Scola	10'09"1	Abarth Simca 1300
28 June	Avus, Germany	Grosser Preis von Berlin (circuit-8,300 kms x 10 laps)	Touring	850 cc	K. Ahrens	33'54"0	Fiat Abarth 850 TC
			Grand Touring	1000 cc	E.Kleine	29'04"2	Fiat Abarth 1000 BA
			Grand Touring	1300 cc	H. Dechent	26'34"2	Abarth Simca 1300
			Grand Touring	up to 2500 cc	H. Herrmann (overall)	23'51"4	Abarth Simca 2000
28 June	Gaisberg, Austria	Gaisberg Bergpreis von Österreich (hillclimb-8,625 kms)	Grand Touring	1000 cc	'Driver'	10'38"64	Fiat Abarth 1000 BA
28 June	Forlì, Italy	Predappio-Rocca delle Caminate (hillclimb-4,000 kms)	Touring	850 cc	A. Trenti	3'49"2	Fiat Abarth 850 TC
			Touring	1000 cc	F. Nesti	3'49"7	Fiat Abarth 1000 Berlina
			Grand Touring	700 cc	'Robertino'	3'51"0	Fiat Abarth 700 BA
			Grand Touring	1000 cc	C. Sangiorgi	3'29'7	Fiat Abarth 1000 BA
			Sport	1000 cc	V. Venturi	3'23"1	Fiat Abarth 1000 SP
			F3	---	F. Patria	3'22'3	Fiat Abarth F3
28 June	Kamnik, Jugoslavia	Circuito di Kamnik (20 kms)	Touring	850 cc	S.Matevz	14'14"4	Fiat Abarth 850 TC
29 June	Torino, Italy	Rallye dei Rododendri (179 kms)	Touring	850 cc	R. Coscioni (overall)	---	Fiat Abarth 850 TC
4 July	Siracusa, Italy	Tre Ore Notturna (circuit-5,500 kms x 45')	Touring	850 cc	F. Cassarà	kms 96,779	Fiat Abarth 850 TC
			Touring	1000 cc	G. Virgilio	kms 97,068	Fiat Abarth 1000 Berlina
			Grand Touring	850 cc	S. Danzuso	kms 89,104	Fiat Abarth 850
			Grand Touring	1000 cc	M. Spadafora	kms 101,108	Fiat Abarth 1000 BA
			Grand Touring	1300 cc	A. Gambero	kms 102,148	Abarth Simca 1300
4 July	Banja Luka, Jugoslavia	Corsa hillclimb di Banja Luka (8,870 kms)	Touring	850 cc	S. Matevz	7'13"1	Fiat Abarth 850 TC
4-5 July	Zandvoort, Holland	Les Coupes de Benelux (circuit-4,225 kms x 12 laps)	Touring	850 cc	E. Swart	25'38"4	Fiat Abarth 850 TC
4-5 July	Albi, France	Rallye Automobile des Violettes	Grand Touring	1001-1600 cc	Barthe-'Sarayac'	---	Abarth Simca 1300
5 July	Strasbourg, France	Course de côte Turckeim Trois Epis (hillclimb-6,320 kms)	Touring	850 cc	G. Metz	4'30"3	Fiat Abarth 850 TC
			Touring	1000 cc	F. Ruata	4'03"2	Fiat Abarth 1000 Berlina
			Grand Touring	1150-1600 cc	F. Della Balda	4'08"0	Abarth Simca 1300
			Grand Touring	2000 cc	H. Kuhmis	3'41"4	Abarth Simca 2000
			Sport	up to 1150 cc	P. Ettmüller	3'46"3	Fiat Abarth 1000 SP
5 July	Nürnberg, Germany	ADAC Norisring Rennen (circuit-3,940 kms x 19 laps)	Grand Touring	1000 cc	G. Schlup	35'43"6	Fiat Abarth 1000 BA
			Grand Touring	1300 cc	K. Geiss	34'25"8	Abarth Simca 1300
5 July	Bolzano, Italy	Bolzano-Mendola (hillclimb-14,400 kms)	Touring	850 cc	A. Manavello	12'38"7	Fiat Abarth 850 TC
			Touring	1000 cc	A. Merzario	12'52"4	Fiat Abarth 1000 Berlina
			Grand Touring	700 cc	'Makye'	12'36"8	Fiat Abarth 700 BA
			Grand Touring	1000 cc	E. Buzzetti	12'30"1	Fiat Abarth 1000 BA
			Grand Touring	1300 cc	A. Giliberti	11'45"0	Abarth Simca 1300
			Grand Touring	up to 2500 cc	H. Demetz (overall)	10'54"9	Abarth Simca 2000
			F3	---	T. Spychiger	11'15"2	Fiat Abarth F3
5 July	Monopoli, Italy	Coppa Città di Monopoli (hillclimb-4,500 kms)	Touring	850 cc	Dicadelmo	2'38"8	Fiat Abarth 850 TC
			Touring	1000 cc	T. Capuozzo	2'37"8	Fiat Abarth 1000 Berlina
			Grand Touring	700 cc	P. Gargano	2'43"0	Fiat Abarth 700 BA
			Grand Touring	1000 cc	C. Avventurieri	2'23"0	Fiat Abarth 1000 BA
			Grand Touring	1300 cc	D. Scola	2'18"4	Abarth Simca 1300
5 July	Rosario, Argentina	Premio 15° Anniversario (circuit-1,650 kms x 15 laps)	Touring	850 cc	P. Sancha	16'09"5	Fiat Abarth 850 TC
5 July	Burgos, Spagna	Rallye del Cid (400 kms)	Touring	850 cc	A. Alvarez De Leon	---	Fiat Abarth 850 TC
5 July	Banja Luka, Jugoslavia	Corsa hillclimb di Banja Luka (8,870 kms)	Touring	850 cc	S. Matevz	---	Fiat Abarth 850 TC
10-11 July	Harewood Acres (Ontario), USA	Mid Summer Trophy Races (circuit-19,1 miles)	---	---	C. Fisher	---	Abarth Simca 1300
		(circuit-57,3 miles)	Sport	1300 cc	C. Fisher	---	Abarth Simca 1300
			Sport	1300 cc	C. Fisher	---	Abarth Simca 1300
12 July	Gives, Belgium	Course de côte de Gives (hillclimb, 2,400 kms)	Grand Touring	1000 cc	F. Ruata	1'38"2	Fiat Abarth 1000 BA
12 July	Mamer, Luxembourg	Slalom Mamer	Touring	850 cc	H. Wagner	1'53"2	Fiat Abarth 850 TC
12 July	Barcelona, Spagna	Trofeo Juan Jover-Montjuich (circuit-3,790 kms x 15 laps)	Touring	850 cc	A. Alvarez De Leon	35'02"6	Fiat Abarth 850 TC
			Touring	1000 cc	J. Juncosa	35'04"1	Fiat Abarth 1000 Berlina
12 July	Trento, Italy	Trento-Bondone (hillclimb-17,300 kms)	Touring	600 cc	P. Poggi	16'25"5	Fiat Abarth 595
			Touring	850 cc	A. Trenti	14'56"0	Fiat Abarth 850 TC
			Touring	1000 cc	G. Rile	15'00"7	Fiat Abarth 1000 Berlina
			Grand Touring	700 cc	G. Varese	15'03"0	Fiat Abarth 700 BA
			Grand Touring	1000 cc	'Matich'	14'09"9	Fiat Abarth 1000 BA
			Grand Touring	1300 cc	G. Dalla Torre	13'39"2	Abarth Simca 1300
			Grand Touring	1600-2500 cc	H. Herrmann	12'30"4	Abarth Simca 2000
12 July	Brindisi, Italy	Coppa Selva di Fasano (hillclimb-7,400 kms)	Touring	850 cc	Persia	4'40"7	Fiat Abarth 850 TC
			Touring	1000 cc	T. Capuozzo	4'33"0	Fiat Abarth 1000 Berlina
			Grand Touring	700 cc	'Jacopo'	4'55"2	Fiat Abarth 700 BA
			Grand Touring	1000 cc	C. Avventurieri	4'21"0	Fiat Abarth 1000 BA
			GranTurismo	1300 cc	D. Scola	4'08"0	Abarth Simca 1300

DATE	LOCATION	RACE	CATEGORY	CLASS	WINNER	TIME OR DISTANCE	CAR
12 July	Morges, Switzerland	Slalom Automobile de Morges	Touring	850 cc	J.P. Brun (overall)	1'28"2	Fiat Abarth 850 TC
12 July	Reggio Calabria, Italy	Santo Stefano-Gambarie (hillclimb-7,830 kms)	Touring	850 cc	E. Di Stefano	6'32"1	Fiat Abarth 850 TC
			Touring	1000 cc	Midiri	6'29"8	Fiat Abarth 1000 Berlina
			Grand Touring	600 cc	A. Rizzo	6'48"0	Fiat Abarth 595
			Grand Touring	1000 cc	Spadaro	6'21"3	Fiat Abarth 1000 BA
18-19 July	Crans, Switzerland	Course de côte Lens Crans (hillclimb-3,000 kms)	Touring	850 cc	J.P. Brun	5'20"6	Fiat Abarth 850 TC
			Touring	1000 cc	M. Piller	5'11""3	Fiat Abarth 1000 Berlina
			Grand Touring	1000 cc	R. Vaglio	4'56"0	Fiat Abarth 1000 BA
			Grand Touring	1300 cc	P. Della Balda	4'36"0	Abarth Simca 1300
			Grand Touring	1600-2000 cc	J. Zufferey	4'34"4	Abarth Simca 2000
19 July	Trieste, Italy	Trieste-Opicina (hillclimb-10,150 kms)	Touring	600 cc	S. Calascibetta	6'23"3	Fiat Abarth 595
			Touring	850 cc	G. Raffa	5'47"0	Fiat Abarth 850 TC
			Touring	1000 cc	A. Marzi	5'43"6	Fiat Abarth 1000 Berlina
			Grand Touring	700 cc	M. Acquati	5'56"5	Fiat Abarth 700 BA
			Grand Touring	1000 cc	C. Avventurieri	5'20"5	Fiat Abarth 1000 BA
			Grand Touring	1300 cc	A. Giliberti	5'04"8	Abarth Simca 1300
			Grand Touring	up to 2500 cc	F. Patria (overall)	4'40"2	Abarth Simca 2000
19 July	Viterbo, Italy	Corsa Lago di Montefiascone (hillclimb-5,000 kms)	Touring	850 cc	R. Addamiano	3'01"2	Fiat Abarth 850 TC
			Touring	1000 cc	F. Nesti	2'56"3	Fiat Abarth 1000 Berlina
			Grand Touring	1000 cc	'Laki'	2'48"0	Fiat Abarth 1000 BA
			Grand Touring	1300 cc	S. Ridolfi	2'45"6	Abarth Simca 1300
25 July	Westerham, United Kingdom	Seven Oaks and D.M.C. (hillclimb-700 yarde)	Touring	850 cc	J. Anstead	0'37"3	Fiat Abarth 850 TC
25-26 July	Wolsfeld, Germany	EMSC Bergrennen (hillclimb-1,600 kms x 2)	Grand Touring	1000 cc	F. Kleine	2'50"5	Fiat Abarth 1000 BA
			Grand Touring	1300 cc	H. Dechent	2'43"8	Abarth Simca 1300
26 July	Sestrière, Italy	Cesana-Sestrière (hillclimb-10,400 kms)	Touring	600 cc	P. Poggi	7'50"9	Fiat Abarth 595
			Touring	850 cc	S. Calascibetta	7'03"0	Fiat Abarth 850 TC
			Touring	1000 cc	T. Spychinger	6'53"2	Fiat Abarth 1000 Berlina
			Grand Touring	700 cc	G. Varese	6'50"2	Fiat Abarth 700 BA
			Grand Touring	1000 cc	A. Zucchi	6'18"9	Fiat Abarth 1000 BA
			Grand Touring	1300 cc	G. Dalla Torre	6'13"0	Abarth Simca 1300
			Grand Touring	1600-2500 cc	H. Demetz	5'49"5	Abarth Simca 2000
			Sport	1000 cc	T. Spychiger	5'59"6	Fiat Abarth 1000 SP
26 July	Lignières, Switzerland	Weekend à Lignières (circuit-1,400 kms x 2)	Grand Touring	1300 cc	P. Maron	---	Abarth Simca 1300
26 July	Marlboro, USA	Marlboro 6 Hour Endurance Race (circuit-6 hours)	---	---	J. Cannon	---	Abarth Simca 1300
26 July	Rouen, France	Course de côte des Andelys (hillclimb-1,380 kms)	Grand Touring	1000 cc	F. Ruata	0'58"3	Fiat Abarth 1000 BA
			Grand Touring	1300 cc	G. Ballot-Lena	0'58"2	Abarth Simca 1300
1 August	Adenau, Germany	AVD Rheinland Pfalz-Preis Nürburgring (circuit-22,810 kms x 7 laps)	Grand Touring	1300 cc	G. Bodmer	1h27'18"7	Abarth Simca 1300
1 August	Shrewsbury, United Kingdom	Loton Park Hill-climb (hillclimb)	Grand Touring	over 1600 cc	P. Pycroft	0'44"89	Abarth Simca 2000
2 August	Trapani, Italy	Trapani-Monte Erice (hillclimb-6,670 kms)	Touring	600 cc	N. Barresi	6'07"4	Fiat Abarth 595
			Touring	850 cc	E. Di Stefano	5'23"3	Fiat Abarth 850 TC
			Touring	1000 cc	F. Latteri	5'21"1	Fiat Abarth 1000 Berlina
			Grand Touring	700 cc	G. La Mantia	5'18"1	Fiat Abarth 700 BA
			Grand Touring	1000 cc	I. Capuano	4'56"4	Fiat Abarth 1000 BA
			Grand Touring	1300 cc	A. Giliberti	4'42"2	Abarth Simca 1300
			Sport	1000 cc	V. Venturi	4'37"4	Fiat Abarth 1000 SP
2 August	Belluno, Italy	Agordo-Frassenè (hillclimb-9,000 kms)	Touring	850 cc	G. Piazzi	6'26"8	Fiat Abarth 850 TC
			Touring	1000 cc	P. Cattelan	6'25"8	Fiat Abarth 1000 Berlina
			Grand Touring	700 cc	R. Galasso	6'36"2	Fiat Abarth 700 BA
			Grand Touring	1000 cc	G. Lado	5'57"2	Fiat Abarth 1000 BA
			Grand Touring	1300 cc	G. Dalla Torre	5'35"6	Abarth Simca 1300 cc
			Grand Touring	1600-2500 cc	H. Demetz (overall)	5'16"0	Abarth Simca 2000
2 August	Innsbruck, Austria	Bergrennen zur Axamer Lizum (hillclimb-8,300 kms)	Touring	850 cc	H. Kiwisch	5'01"1	Fiat Abarth 850 TC
			Touring	1000 cc	F. Patria	4'50"3	Fiat Abarth 1000 Berlina
			Grand Touring	1000 cc	'Driver'	4'35"4	Fiat Abarth 1000 BA
			Grand Touring	1300 cc	F. Della Balda	4'32"7	Abarth Simca 1300
			Grand Touring	over 1600 cc	F. Patria (overall)	3'57"0	Abarth Simca 2000
3 August	Brands Hatch, United Kingdom	B.R.S.C.C. International Race Meeting (circuit-2,650 miles)	Grand Touring	1600-2500 cc	B. Burnard	2'08"0	Abarth Simca 1300
8-9 August	Cognac, France	Trophées du Cognac (circuit-4,880 kms x 25 laps)	Grand Touring	1300 cc	Hanrioud	1h0'50"1	Abarth Simca 1300
9 August	Porto d'Ascoli, Italy	Gara Sprint (circuit-0,500 kms)	Touring	600 cc	G. Presti	25"16	Fiat Abarth 595
			Touring	850 cc	B. Scipioni	22"99	Fiat Abarth 850 TC
9 August	Freiburg, Germany	ADAC Bergpreis Freiburg-Schauinsland (hillclimb-11,200 kms)	Grand Touring	1000 cc	F. Ruata	15'37"95	Fiat Abarth 1000 BA
			Grand Touring	1300 cc	H. Dechent	15'16"71	Abarth Simca 1300
			Grand Touring	2000 cc	F. Patria	13'53"58	Abarth Simca 2000

DATE	LOCATION	RACE	CATEGORY	CLASS	WINNER	TIME OR DISTANCE	CAR
9 August	Bomeree, Belgium	Course de côte de Charleroi (hillclimb-2,000 kms)	Touring	850 cc	H. Wagner	1'26"7	Fiat Abarth 850 TC
9 August	Maribor, Jugoslavia	Circuito di Maribor (circuit-2,300 kms x 10 laps)	Touring	850 cc	S. Matevz	18'07"2	Fiat Abarth 850 TC
14-15 August	Harewod Acres (Ontario), USA	Great Lakes Trophy Races (circuit-13,4 miles)	---	---	C. Fisher	---	Abarth Simca 1300
		(circuit-19,1 miles)	---	---	C. Fisher	---	Abarth Simca 1300
		(circuit-57,3 miles)	---	---	C. Fisher	---	Abarth Simca 1300
15 August	Pescara, Italy	Cronoscalata Svolte di Popoli (hillclimb-8,000 kms)	Touring	850 cc	F. Chiari	5'35"0	Fiat Abarth 850 TC
			Touring	1000 cc	N. Parenti	5'38"1	Fiat Abarth 1000 Berlina
			Grand Touring	up to 1150 cc	C. Sangiorgi	5'16"2	Fiat Abarth 1000 BA
			Grand Touring	1300 cc	S. Riboldi (overall)	4'55"6	Abarth Simca 1300
15 August	Nogaro, France	Circuit International de Nogaro (circuit-1,750 kms x 25 laps)	Grand Touring	1000 cc	Faget	---	Fiat Abarth 1000 BA
			Grand Touring	1300 cc	J. Hanriout	---	Abarth Simca 1300
15-16 August	Hannover, Germany	Tausend Kilometer rund um Hannover (regolarità-1000 kms)	Touring	1000 cc	H. Mack	---	Fiat Abarth 850 TC
15-16 August	Mont-Dore, France	Course de côte du Mont-Dore (hillclimb-5,000 kms)	Touring	850 cc	Champeaux	4'18"3	Fiat Abarth 850 TC
			Grand Touring	1000 cc	F. Ruata	3'40"4	Fiat Abarth 1000 BA
			Grand Touring	1300 cc	C. Ballot-Lena	3'37"0	Abarth Simca 1300
			Sport	up to 1150 cc	P. Ettmüller	3'27"2	Fiat Abarth 1000 SP
15-16 August	Avezzano, Italy	Trofeo Micangeli (hillclimb-6,000 kms)	Touring	850 cc	F. Chiari	3'43"6	Fiat Abarth 850 TC
			Touring	1000 cc	A. Merzario	3'45"9	Fiat Abarth 1000 Berlina
			Grand Touring	1000 cc	C. Cannataro	3'33"5	Fiat Abarth 1000 BA
			Grand Touring	1300 cc	S. Ridolfi (overall)	3'19"9	Abarth Simca 1300
16 August	München, Germany	ADAC Flughafen Rennen München (circuit 4,000 kms x 16 laps)	Grand Touring	700 cc	K. Geiss	29'17"5	Fiat Abarth 700 BA
			Grand Touring	1000 cc	A. Schmalbach	27"46"1	Fiat Abarth 1000 BA
			Grand Touring	1300 cc	H. Dechent	27"11'3	Abarth Simca 1300
16 August	Pergusa, Italy	Coppa Città di Enna (circuit-4,797 kms x 70 laps)	Grand Touring	2000 cc	H. Herrmann	1h38'44"2	Abarth Simca 2000
23 August	Ascoli Piceno, Italy	Corsa al Colle San Marco (hillclimb-10,200 kms)	Touring	600 cc	S. Calascibetta	7'30"1	Fiat Abarth 595
			Touring	850 cc	A. Trenti	7'03"1	Fiat Abarth 850 TC
			Touring	1000 cc	A. Marzi	6'54"0	Fiat Abarth 1000 Berlina
			Grand Touring	1000 cc	A. Zucchi	6'35"0	Fiat Abarth 1000 BA
			Grand Touring	1300 cc	A. Giliberti	6'23"3	Abarth Simca 1300
			Sport	1000 cc	V. Venturi	6'27"2	Fiat Abarth 1000 SP
23 August	Brescia, Italy	Trofeo Valle Camonica (hillclimb-8,600 kms)	Touring	850 cc	M. Acquati	6'02"0	Fiat Abarth 850 TC
			Touring	1000 cc	'Jerry'	6'07"0	Fiat Abarth 1000 Berlina
			Grand Touring	700 cc	G. Mainetti	5'56"7	Fiat Abarth 700 BA
			Grand Touring	up to 1150 cc	L. Pozzo	5'59"2	Simca Abarth 1150
			Grand Touring	1300 cc	R. Bertuzzi	5'41"8	Abarth Simca 1300
23 August	Saint-Lary, France	Course de côte de Saint Lary (hillclimb-5,200 kms)	Grand Touring	1300 cc	R. Barone	8'36"1	Abarth Simca 1300
23 August	Wasgau, Germany	Wasgauer Bergprüfung (hillclimb-5,100 kms x 2)	Grand Touring	1300 cc	K. Geiss	5'41"0	Abarth Simca 1300
23 August	Saint Ursanne, Switzerland	Course de côte Saint Ursanne-Les Rangiers (hillclimb-4,700 kms x 2)	Touring	850 cc	J. Brun	6'15"5	Fiat Abarth 850 TC
			Touring	1000 cc	U. Schütz	5'54"4	Fiat Abarth 1000 Berlina
			Grand Touring	700 cc	J. Egli	6'02"7	Fiat Abarth 700 BA
			Grand Touring	1000 cc	A. Knorr	5'22"7	Fiat Abarth 1000 BA
			Grand Touring	1300 cc	P. Maron	5'17"4	Abarth Simca 1300
			Sport	1000 cc	P. Ettmüller	4'55"4	Fiat Abarth 1000 SP
23 August	Zeltweg, Austria	Grosser Preis von Österreich (circuit-3,200 kms x 30 laps)	Grand Touring	1000 cc	H. Bennier	43'50"0	Fiat Abarth 1000 BA
29-30 August	Sierre, Switzerland	Gran Prix d'Europe Sierre-Montana Crans (hillclimb-12,000 kms x 2)	Grand Touring	700 cc	H. Affentranger	17'13"7	Fiat Abarth 700 BA
			Grand Touring	1300 cc	D. Borel	15'08"6	Abarth Simca 1300
			Grand Touring	1600-2000 cc	F. Patria	13'52"7	Abarth Simca 2000
			F2	1000 cc	T. Spychiger	15'33"2	Fiat Abarth F2
29-30 August	Grenoble, France	Course de côte Chamrousse (hillclimb-17,200 kms)	Grand Touring	1000 cc	F. Ruata	12'10"8	Fiat Abarth 1000 BA
			Grand Touring	1300 cc	C. Ballot Lena	12'09"8	Abarth Simca 1300
30 August	Chieti, Italy	Coppa Città di Chieti (hillclimb-6,200 kms)	Touring	850 cc	G. Chiari	3'53"3	Fiat Abarth 850 TC
			Touring	up to 1150 cc	B. Faenza	4'28"1	Fiat Abarth 1000 Berlina
			Grand Touring	up to 1150 cc	C. Cannataro	3'58"0	Fiat Abarth 1000 BA
			Grand Touring	1300 cc	P. Laureati	3'38'4	Abarth Simca 1300
30 August	Udine, Italy	Cividale-Castellamonte (hillclimb 7,000 kms)	Touring	750 cc	A. Lepri	6'36"0	Fiat Abarth 595
			Touring	850 cc	G. Villani	5'33"2	Fiat Abarth 850 TC
			Touring	1000 cc	G. Ranzolin	5'30"3	Fiat Abarth 1000 Berlina
			Grand Touring	700 cc	R. Galasso	5'26"6	Fiat Abarth 700 BA
			Grand Touring	1000 cc	'Matich'	6'52"6	Fiat Abarth 1000 BA
			Grand Touring	1300 cc	G. Dalla Torre	4'46"1	Abarth Simca 1300
30 August	Interlagos, Brasile	Tre Ore di Interlagos (circuit)	Grand Touring	2000 cc	J. Silva (overall)	43 laps	Abarth Simca 2000
30 August	Snetterton, United Kingdom	Motor Racing Meeting Club (circuit)	Grand Touring	1601-2500 cc	B. Burnard	18'34"2	Abarth Simca 2000

DATE	LOCATION	RACE	CATEGORY	CLASS	WINNER	TIME OR DISTANCE	CAR
30 August	Forlì, Italy	Coppa Luigi Arcangeli (196 kms)			V. Balzani	---	Fiat Abarth 850 TC
30 August	Nova Gorica, Jugoslavia	Circuito di Nova Gorica (1,500 kms x 5 laps)	Touring	850 cc	S. Matevz	---	Fiat Abarth 850 TC
30 August	Nürnberg, Germany	ADAC Bergrennen Happurg (hillclimb-3,135 kms x 2)	Grand Touring	1000 cc	A. Schmalbach	3'11"2	Fiat Abarth 1000 BA
5 September	Daytona (Florida), USA	Paul Whiteman Trophy Race (circuit)	---	---	R. Norburn	---	Fiat Abarth 750
5-6 September	Dijon, France	Course de côte d'Urcy (hillclimb-2,500 kms)	Touring	1000 cc	Kretschi	2'12"3	Fiat Abarth 1000 Berlina
			Grand Touring	1000 cc	F. Ruata	2'04"1	Fiat Abarth 1000 BA
			Grand Touring	1300 cc	C. Ballot Lena	2'03"4	Abarth Simca 1300
			Sport	up to 1150 cc	P. Ettmüller	2'01"4	Fiat Abarth 1000 SP
6 September	Parma, Italy	Giro della Provincia di Parma (regolarità-145 kms)	---	---	C. Massagrande	---	Fiat Abarth 850 TC
6 September	Teramo, Italy	Coppa Tortoreto (hillclimb-3,000 kms)	Touring	850 cc	P. Santamato	2'31"4	Fiat Abarth 850 TC
			Touring	up to 1150 cc	B. Faenza	2'29"8	Fiat Abarth 1000 Berlina
			Grand Touring	up to 1150 cc	C. Sangiorgi	2'20"3	Fiat Abarth 1000 BA
			Grand Touring	1300 cc	L. Baldini (overall)	2'14"1	Abarth Simca 1300
			Sport	1000 cc	G. Tommasi	2'14"8	Fiat Abarth 1000 SP
6 September	Adenau, Germany	Internationales 500 Km Nürburgring Rennen (circuit-22,810 kms x 22 laps)	Grand Touring	1300 cc	Dechent-Furtmayr	4h20'01"5	Abarth Simca 1300
			Prototypes	1000 cc	T. Fischhaber	4h29'55"3	Fiat Abarth 1000 Prototipo
			Prototypes	1300 cc	H. Herrmann	4h19'33"8 (overall)	Abarth Simca 1300
6 September	Barcelona, Spagna	Carrera en cuesta San Feliu (hillclimb 4,200 kms)	Touring	850 cc	J. Palau Ribes	3'35"4	Fiat Abarth 850 TC
			Touring	1000 cc	J. Juncosa	3'31"4	Fiat Abarth 1000 Berlina
6 September	Vallelunga, Italy	Trofeo d'Autunno (circuit-3,200 kms x 10 laps)	Touring	850 cc	F. Chiari	19'03"6	Fiat Abarth 850 TC
			Touring	up to 1050 cc	N. Parenti	19'48"4	Fiat Abarth 1000 Berlina
6 September	Freiburg, Germany	ADAC Bergpreis Freiburg-Schauinsland (hillclimb-7,200 kms)	Touring	850 cc	E. Fauser	6'10"9	Fiat Abarth 850 TC
			Grand Touring	up to 1300 cc	A. Schmalbach	6'06"2	Fiat Abarth 1000 BA
6 September	Beograd, Jugoslavia	Circuito di Belgrado (circuit-2,300 kms x 15 laps)	Touring	850 cc	S. Matevz	---	Fiat Abarth 850 TC
7 September	Interlagos, Brasile	500 Km di Interlagos (circuit)	Grand Touring	2000 cc	J.F. Lopes Martins	4h17'58"9	Abarth Simca 2000
11-12 September	Vaals, Holland	Heuvelklim Vaals (hillclimb-1,200 kms)	Touring	850 cc	M. Hooft	1'04"6	Fiat Abarth 850 TC
			Touring	1000 cc	J. Van Osch	1'02"9	Fiat Abarth 1000 Berlina
			Prepared Touring	850 cc	E. Swart	1'01"6	Fiat Abarth 850 TC
			Prepared Touring	1000 cc	E. Swart	0'59"8	Fiat Abarth 1000 Berlina
11-13 September	S.Martino di Castrozza, Italy	Rallye Internazionale di S. Martino di Castrozza (kms 1430)	Touring	700-850 cc	Ferrario-Ranzolin	---	Fiat Abarth 750
12 September	Brighton, United Kingdom	Brighton Speed Trials (race handicap)	Grand Touring	1601-2500 cc	B. Burnard	0'26"31	Abarth Simca 2000
12 September	Augsburg, Germany	Augusta Bergrennen (hillclimb-2,100 kms x 2 laps)	Grand Touring	700 cc	F. Neuser	2'31"6	Fiat Abarth 750
12-13 September	Achum, Germany	Internationales Flugplatzrennen Wunstorf (circuit-3,800 kms x 14 laps)	Touring	850 cc	K. Ahrens	28'38"4	Fiat Abarth 850 TC
			Touring	1000 cc	F. Patria	28'29"6	Fiat Abarth 1000 Berlina
			Grand Touring	1000 cc	A. Schmalbach	27'45"3	Fiat Abarth 1000 BA
			Grand Touring	1300 cc	H. Dechent	27'14"2	Abarth Simca 1300
			Grand Touring	over 1600 cc	F. Patria	23'59"2	Abarth Simca 2000
12-13 September	Mt.Tremblant St.Jovite, Canada	Players Quebec (circuit-99,5 miles)	---	---	C. Fisher	---	Abarth Simca 1300
13 September	Innsbruck, Austria	Timmelsjoch Hochalpen Bergrennen (hillclimb-12,050 kms)	Grand Touring	1000 cc	'Driver'	6'56"7	Fiat Abarth 1000 BA
			Grand Touring	up to 2000 cc	H. Demetz	5'49"8	Abarth Simca 2000
			Grand Touring	up to 2500 cc	L. Taramazzo (overall)	4'17"6	Abarth Simca 2000
13 September	Brescia, Italy	Trofeo Lumezzane (hillclimb-8,500 kms)	Touring	850 cc	G. Raffa	5'08"4	Fiat Abarth 850 TC
			Touring	1000 cc	A. Merzario	5'08"5	Fiat Abarth 1000 Berlina
			Grand Touring	700 cc	M. Pinchetti	5'05"2	Fiat Abarth 700 BA
			Grand Touring	1000 cc	A. Becchetti	4'46"4	Fiat Abarth 1000 BA
			Grand Touring	up to 1150 cc	L. Pozzo	5'16"2	Abarth Simca 1150
			Grand Touring	1300 cc	R. Ramoino	4'42"9	Abarth Simca 1300
13 September	Viterbo, Italy	Coppa del Cimino (hillclimb-10,200 kms)	Touring	850 cc	R. Addamiano	5'32"1	Fiat Abarth 850TC
			Touring	1000 cc	A. Cecchini	5'29"8	Fiat Abarth 1000 Berlina
			Grand Touring	1000 cc	G. Rebaudi	5'05"1	Fiat Abarth 1000 BA
			Grand Touring	1300 cc	S. Ridolfi (overall)	4'58"1	Abarth Simca 1300
			Sport	1000 cc	G. Tommasi	5'11"5	Fiat Abarth 1000 SP
13 September	Schauinsland, Germany	Nationales Bergrennen am Schauinsland (Campionato Montagna Svizzero hillclimb-7,200 kms)	Touring	850 cc	J. Brun	11'24"2	Fiat Abarth 850 TC
			Touring	1000 cc	U. Schütz	11'04"8	Fiat Abarth 1000 Berlina
			Prepared Touring	850-1000 cc	H. Illert	10'10"2	Fiat Abarth 1000 Berlina
			Grand Touring	700 cc	H. Affentranger	11'10"1	Fiat Abarth 700 BA
			Grand Touring	1000 cc	T. Hofer	10'57"6	Fiat Abarth 1000 BA
			Grand Touring	1300 cc	D. Borel	9'50"2	Abarth Simca 1300
			Grand Touring	1600-2000 cc	H. Kuhnis	9'08"6	Abarth Simca 2000
13 September	Civitavecchia, Italy	Rallye Turistico della Tuscia (230 kms)	---	---	R. Coscioni (overall)	---	Fiat Abarth 850 TC

DATE	LOCATION	RACE	CATEGORY	CLASS	WINNER	TIME OR DISTANCE	CAR
13 September	Bourscheid, Luxembourg	Course de côte de Bourscheid (circuit)	Touring	1000 cc	E. Swart	2'18"2	Fiat Abarth 1000 Berlina
			Gran Touring	1000 cc	J. Hauser	2'17"3	Fiat Abarth 1000 BA
15 September	Albacete, Spagna	Gran Premio de Albacete	Touring e Grand Touring	up to 1000 cc	A. Alvarez De Leon (overall)	31'34"3	Fiat Abarth 850 TC
19 September	Thurgau, Switzerland	Surprise Bergrennen (hillclimb-3,000 kms)	Touring	1000 cc	O. Mobel	4'24"3	Fiat Abarth 1000 Berlina
			Grand Touring	1300 cc	T. Hoffer	3'54"9	Fiat Abarth 1000 BA
19-20 September	Mont Verdun, France	Course de côte Limonest-Mont Verdun (hillclimb-2,675 kms)	Grand Touring	1000 cc	S. Vaglio	2'21"9	Fiat Abarth 1000 BA
			Grand Touring	1300 cc	R. Barone	2'10"6	Abarth Simca 1300
			Sport	up to 1150 cc	P. Ettmüller	2'08"6	Fiat Abarth 1000 SP
19-20 September	Kassel, Germany	Bergprüfung Kassel-Langenberge (hillclimb-3,700 kms)	Touring	1000 cc	Rennesland	5'14"9	Fiat Abarth 850 TC
			Grand Touring	1300 cc	F. Kleine	4'19"5	Abarth Simca 1300
19-20 September	Houyet, Belgium	Course de côte de Houyet (hillclimb-2,500 kms)	Touring	850 cc	H. Wagner	1'53"9	Fiat Abarth 850 TC
			Grand Touring	1000 cc	J. Hauser	1'43"6	Fiat Abarth 1000 BA
20 September	Napoli, Italy	Targa Vesuvio (hillclimb-9,200 kms)	Touring	850 cc	B. Montecatini	7'48"0	Fiat Abarth 850 TC
			Touring	1000 cc	F. Altieri	7'40"1	Fiat Abarth 1000 Berlina
			Grand Touring	700 cc	P. Gargano	8'18"3	Fiat Abarth 700 BA
			Grand Touring	1000 cc	V. Cannataro	8'02"7	Fiat Abarth 1000 BA
			Grand Touring	1300 cc	F. Prota	7'27"3	Abarth Simca 1300
20 September	Kaiserslautern, Germany	Rheinland-Pfalz Bergpreis (hillclimb-3,050 kms)	Grand Touring	1000 cc	A. Schmalbach	3'42"9	Fiat Abarth 1000 BA
			Grand Touring	1300 cc	H. Dechent	3'38"9	Abarth Simca 1300
20 September	Siena, Italy	Rallye del Chianti classico (152 kms)	Touring	850 cc	M. Tamburini	---	Fiat Abarth 850 TC
			Touring	1000 cc	A. Cecchini (overall)	---	Fiat Abarth 1000 Berlina
			Grand Touring	1000 cc	'Laki'	---	Fiat Abarth 1000 BA
			Grand Touring	1300 cc	'Nicor'	---	Abarth Simca 1300
20 September	Montlhéry, France	Coupes de Paris (circuit-3,300 kms x 15 laps)	Touring	850 cc	F. Patria	30'08"1	Fiat Abarth 850 TC
			Touring	1150 cc	H. Herrmann	29'10"5	Fiat Abarth 1000 Berlina
			Grand Touring	1000 cc	F. Ruata	27'06"4	Fiat Abarth 1000 BA
			Grand Touring	1300 cc	C. Ballot Lena	26'20"0	Abarth Simca 1300
			Grand Touring	1600-2000 cc	F. Patria	22'40"0	Abarth Simca 2000
20 September	Caltanissetta, Italy	Coppa Nissena (hillclimb-10,000 kms)	Touring	600 cc	S. Calascibetta	5'39"3	Fiat Abarth 595
			Touring	850 cc	A. Trenti	5'15"1	Fiat Abarth 850 TC
			Touring	1000 cc	A. Marzi	5'15"2	Fiat Abarth 1000 Berlina
			Grand Touring	1000 cc	I. Capuano	4'52"1	Fiat Abarth 1000 BA
			Grand Touring	1300 cc	A. Giliberti	4'48"0	Abarth Simca 1300
20 September	Hainburg, Austria	Braunsbergrennen (hillclimb-2,700 kms)	Grand Touring	up to 1300 cc	L. Hlinka	1'53"4	Fiat Abarth 1000 BA
20 September	Warminster, United Kingdom	National British Hillclimb Longleat House	Touring	850 cc	J. Anstead (1ª manche)	0'51"15	Fiat Abarth 850 TC
25-26 September	Mosport Park (Ontario), USA	Canadian Grand Prix (circuit-24,6 miles)	---	---	C. Fisher	---	Abarth Simca 1300
26 September	Lakeville (Connecticut), USA	Lime Rock Park Race (circuit)	---	---	R. Catalano	---	Fiat Abarth 1000 BA
26-27 September	Berlin, Germany	Golden Bär von Berlin (388 kms)	---	---	H. Garschig	---	Fiat Abarth 850 TC
26-27 September	Namur, Belgium	Course de côte de Namur (hillclimb-2,000 kms)	Touring	850 cc	H. Wagner	1'50"6	Fiat Abarth 850 TC
			Touring	1000 cc	E. Swart	1'44"3	Fiat Abarth 1000 Berlina
			Grand Touring	1000 cc	F. Kleine	1'42"2	Fiat Abarth 1000 BA
27 September	Barletta, Italy	Rallye dei Castelli Svevi (298 kms)	---	---	C. Massagrande	---	Fiat Abarth 850 TC
27 September	Trieste, Italy	Rallye San Giusto (196 kms)	---	---	P. Navarra (overall)	---	Fiat Abarth 850 TC
27 September	Annemasse, Switzerland	Circuit Monthoux (circuit-1,000 kms x 20 laps)	Touring	up to 1150 cc	Schaefer	18'12"3	Abarth Simca 1150
27 September	Belluno, Italy	Coppa Alpe del Nevegael (hillclimb-7,500 kms)	Touring	850 cc	G. Piazzi	5'30"2	Fiat Abarth 850 TC
			Touring	1000 cc	E. Schenardi	5'19"6	Fiat Abarth 1000 Berlina
			Grand Touring	700 cc	M. Pinchetti	5'20"4	Fiat Abarth 700 BA
			Grand Touring	1000 cc	'Matich'	5'02"6	Fiat Abarth 1000 BA
			Grand Touring	1300 cc	L. Baldini	4'41"6	Abarth Simca 1300
			Grand Touring	up to 2500 cc	H. Demetz (overall)	4'20"4	Abarth Simca 2000
27 September	Mallorca, Spagna	Carrera en cuesta Puig Major (hillclimb-8,000 kms)	Touring	850 cc	'Jorge Chi'	6'21"83	Fiat Abarth 850 TC
			Touring	1000 cc	J. Juncosa	5'57"62	Fiat Abarth 1000 Berlina
3 October	Castle Combe, United Kingdom	Castle Combe Sprint (circuit)	Grand Touring	2000 cc	P. Pycroft	1'24"2	Abarth Simca 2000
3-4 October	Luzern, Switzerland	Nationales Bergrennen Eigenthal (hillclimb-3,300 kms)	Touring	850 cc	J.P. Brun	6'14"63	Fiat Abarth 850 TC
			Prepared Touring	850-1300 cc	H. Illert	5'39"43	Fiat Abarth 1000 Berlina
			Grand Touring	700 cc	H. Affentranger	6'04"14	Fiat Abarth 700 BA
			Grand Touring	1000 cc	A. Knorr	5'37"76	Fiat Abarth 1000 BA
			Grand Touring	1300 cc	R. Stierli	5'31"15	Abarth Simca 1300
			Grand Touring	2000 cc	H. Kuhnis	5'09"30	Abarth Simca 2000
			Sport	1000 cc	P. Ettmüller	5'27"53	Fiat Abarth 1000 SP

DATE	LOCATION	RACE	CATEGORY	CLASS	WINNER	TIME OR DISTANCE	CAR
3-4 October	Fulda, Germany	Rhön Bergpreis (hillclimb-2,550 kms)	Grand Touring	1000 cc	F. Kleine	3'43"7	Fiat Abarth 1000 BA
4 October	Osimo, Italy	Coppa Luigi Fagioli (hillclimb-5,300 kms)	Grand Touring	1300 cc	P. Laureati	2'51"3	Abarth Simca 1300
4 October	Vianden, Luxembourg	Course de côte de Vianden (hillclimb)	Touring Grand Touring Grand Touring	850 cc 1000 cc 1300 cc	H. Wagner J. Hauser R. Barone	1'54"2 1'44"3 1'38"1	Fiat Abarth 850 TC Fiat Abarth 1000 BA Abarth Simca 1300
4 October	Innsbruck, Austria	Preis von Tirol (circuit-2,800 kms x 20 laps)	Touring Grand Touring Grand Touring Grand Touring Sport	1000 cc 1000 cc 1300 cc 2000 cc 1000 cc	K. Ahrens A. Schmalbach H.D. Dechent K. Ahrens (overall) K. Ahrens	29'15"39 --- --- 25'39"48	Fiat Abarth 1000 Berlina Fiat Abarth 1000 BA Abarth Simca 1300 Abarth Simca 2000 Fiat Abarth 1000 SP
4 October	Montlhéry, France	Coupe du Salon (circuit-3,300 kms x 16 laps)	Touring Touring Grand Touring Grand Touring	850 cc up to 1150 cc 1000 cc 2000 cc	K. Steinmetz F. Patria F. Ruata F. Patria	29'57"5 27'00"5 32'42"2 31'22"7	Fiat Abarth 850 TC Fiat Abarth 1000 Berlina Fiat Abarth 1000 BA Abarth Simca 2000
4 October	Monza, Italy	Coppa Leopoldo Carri (circuit-5,750 kms x 10 laps)	Touring Touring Touring	600 cc 850 cc 1000 cc	R. Ramaciotti G. Raffa C. Zuccoli	31'19"5 28'50"4 28'10"9	Fiat Abarth 595 Fiat Abarth 850 TC Fiat Abarth 1000 Berlina
4 October	Vallelunga, Italy	Trofeo d'Autunno (circuit-32,000 kms)	Touring Touring	850 cc 1000 cc	F. Chiari T. Capuozzo	19'01"2 19'10"3	Fiat Abarth 850 TC Fiat Abarth 1000 Berlina
4 October	Forlì, Italy	Regularity race (144 kms)	Touring Touring Grand Touring	700 cc up to 1150 cc 850 cc	F. Naldini C. Massagrande (overall) G. Ricci	--- --- ---	Fiat Abarth 595 Fiat Abarth 850 TC Fiat Abarth 850 TC
4 October	Massana, Andorra	Costa de la Massana (hillclimb-3,500 kms)	Touring Touring Grand Touring	600 cc 1000 cc up to 1600 cc	J. Juncosa 'Runner' E. Reverter	3'10"70 2'28"09 2'16"0	Fiat Abarth 595 Fiat Abarth 1000 Berlina Fiat Abarth 1000 BA
10 October	Brighton, United Kingdom	New Brighton Sprint (circuit)	Grand Touring	2000 cc	P. Pycroft	14'38"	Abarth Simca 2000
10 October	Castle Combe, United Kingdom	Race Meeting Castle Combe (circuit)	Touring	up to 950 cc	J. Anstead	10'53"8	Fiat Abarth 850 TC
10-11 October	Saint Cloud, France	Rallye des Six-Heures de Saint Cloud (300 kms)	---	---	Henning-Zachariasen (overall)	---	Fiat Abarth 850 TC
11 October	Montlhéry, France	1000 Km de Paris (circuit-7,784 kms x 129 laps)	Grand Touring	1300-1600 cc	Demetz-Fischaber	6h53'21"	Abarth Simca 1300
11 October	Catanzaro, Italy	Coppa Città di Catanzaro (hillclimb-16,000 kms)	Touring Touring Grand Touring Grand Touring Grand Touring	850 cc 1000 cc 1100 cc 1000-1150 cc 1300 cc	M. De Luca P. Ritrovato C. Avventurieri G. Abramo D. Scola (overall)	12'52"1 13'00"8 11'53"6 13'41"7 11'53"6	Fiat Abarth 850 TC Fiat Abarth 1000 Berlina Fiat Abarth 1000 BA Simca Abarth 1150 SS Abarth Simca 1300
11 October	Monza, Italy	Coppa d'Autunno (circuit-5,750 kms x 28 laps)	Touring Touring	850 cc 1000 cc	G. Raffa A. Marzi	1h08'02"2 1h05'09"6	Fiat Abarth 850 TC Fiat Abarth 1000 Berlina
11 October	Lido delle Nazioni, Italy	Autogymcana	---	---	M. Daniele	---	Fiat Abarth 1000 Berlina
17 October	Zandvoort, Holland	Nationale Kampiioenschapsraces (circuit-4,225 kms x 15 laps)	Touring	1000 cc	E. Swart	31'53"6	Fiat Abarth 1000 Berlina
17-18 October	Trembland St.Jovite, Canada	Pepsi Cola's Trophy Race (circuit-32 miles)	---	---	C. Fisher	---	Abarth Simca 1300
18 October	Cagliari, Italy	Corongiu-Campuomu (hillclimb-4,000 kms)	Touring Touring Touring Grand Touring	up to 700 cc 850 cc up to 1150 cc up to 1150 cc	P. Leone V. Flora E. Meloni L. Biggio (overall)	4'01"5 3'03"3 3'39"9 3'23"9	Fiat Abarth 595 Fiat Abarth 850 TC Fiat Abarth 1000 Berlina Fiat Abarth 1000 BA
18 October	Campocatino, Italy	Autoscalata di Campocatino (hillclimb-8,000 kms)	Touring Touring Grand Touring Grand Touring	850 cc 1000 cc 700 cc 1000 cc	M. Terminillo T. Capuozzo 'Jacopo' F. Amicucci	6'50"0 6'36"1 7'06"4 6'32"2	Fiat Abarth 850 TC Fiat Abarth 1000 Berlina Fiat ABarth 700 BA Fiat Abarth 1000 BA
18 October	Angers, France	Course de côte de la Pommeraye (hillclimb-1,700 kms)	Touring Grand Touring Grand Touring	1000 cc 1000 cc 1300 cc	G. Lagier F. Ruata R. Barone	2'11"4 1'57"1 1'56"3	Fiat Abarth 850 TC Fiat Abarth 1000 BA Abarth Simca 1300
18 October	Eberbach, Germany	Bergrennen Eberbach (hillclimb-3,000 kms x 2)	Grand Touring	1300 cc	K. Geiss	4'56"1	Abarth Simca 1300
24 October	Schüpfen, Switzerland	Bergrennen Schüpfen-Gächwill (hillclimb-3,030 kms)	Touring Grand Touring Grand Touring	850 cc 1000 cc 1300 cc	W. Streit G. Schlup P. Maron	4'56"1 4'11"9 3'54"6	Fiat Abarth 850 TC Fiat Abarth 1000 BA Abarth Simca 1300
24 October	Eusserthal, Germany	Tausensuhl Bergrennen (hillclimb-6,030 kms)	Touring (2 manches) Grand Touring	700-1000 cc 1000 cc	D. Eymann T. Reuter	7'57"1 7'36"1	Fiat Abarth 1000 Berlina Fiat Abarth 1000 BA
24-25 October	Barcelona, Spagna	Rallyes de dos Cataluñas (749 kms)	Touring	850 cc 1000 cc	Palau Ribes-Palou 'Runner'-Eche	--- ---	Fiat Abarth 850 TC Fiat Abarth 1000 Berlina

DATE	LOCATION	RACE	CATEGORY	CLASS	WINNER	TIME OR DISTANCE	CAR
24-25 October	Rouen, France	Rallye de l'A.G.A.C.I. (277 kms)	Touring	850 cc	Champeaux-Duchavel	---	Fiat Abarth 850 TC
25 October	Lignières, Switzerland	Course de Lignières (circuit-1,400 kms x 17 laps)	Touring	1000 cc	F. Wipf	16'46"5	Fiat Abarth 1000 Berlina
25 October	Firenze, Italy	Rallye della Toscana (211 kms)	Gruppo B	---	F. Severi (overall)	---	Fiat Abarth 850 TC
26 October	Madrid, Spagna	Circuito Hoyo de Manzanares	Classe D	---	J. de Bragation	2'41"5	Fiat Abarth 850 TC
			Classe C	---	J. de Bragation	2'41"9	Fiat Abarth 850 TC
4 November	Vallelunga, Italy	Conchiglia Shell (circuit-1,800 kms x 15 laps)	Touring	850 cc	R. Addamiano	17'37"0	Fiat Abarth 850 TC
			Touring	1000 cc	T. Capuozzo	17'43"0	Fiat Abarth 1000 Berlina
			Grand Touring	1000 cc	E. Buzzetti	17'27"1	Fiat Abarth 1000 BA
			Grand Touring	1300 cc	S. Ridolfi	17'24"2	Abarth Simca 1300
8 November	Lignières, Switzerland	Course handicap Ecurie Fribourgeoise (circuit-1,400 kms)	---	---	F. Wipf (overall)	21'59"6	Fiat Abarth 1000 Berlina
14-15 November	Vaduz, Liechtenstein	SAR Weekend (hillclimb-2,100 kms)	Grand Touring	1300 cc	H. Cilly	4'30"8	Abarth Simca 1300
29 November	Sanremo, Italy	Trofeo Sanremese della Montagna (hillclimb-3,900 kms x 2)	Touring	850 cc	G. Raffa	5'28"3	Fiat Abarth 850 TC
			Touring	1000 cc	E. Osella	5'23"6	Fiat Abarth 1000 Berlina
			Grand Touring	700 cc	G. Varese	5'22"2	Fiat Abarth 700 BA
			Grand Touring	1000 cc	G. Rebaudi	5'16"2	Fiat Abarth 1000 BA
			Grand Touring	1300 cc	R. Ramoino	5'09"1	Abarth Simca 1300
			Grand Touring	2000 cc	L. Taramazzo (overall)	4'47"8	Abarth Simca 2000
29 November	Chimney Rock (North Carolina), USA	Chimney Rock Hill Climb (hillclimb-1,9 miles)	---	---	P. Feistmann (overall)	---	Fiat Abarth 750 BA
6 December	Madrid, Spagna	Trofeo Federacion Centro Automovilismo			J. de Bragation (overall)	---	Fiat Abarth 850 TC
6 December	Sassari, Italy	Sassari-Santa Teresa (167 kms)	---	1000 cc	F. Locci	---	Fiat Abarth 850 TC
8 December	Monza, Italy	Coppa F.I.S.A. 1964 (circuit-5,750 kms x 6 laps)	Touring	600 cc	G. Danieli	16'02"9	Fiat Abarth 595
			Touring	850 cc	F. Galli	14'36"2/10	Fiat Abarth 850 TC
			Touring	1000 cc	B. Beldi	14'59"4/10	Fiat Abarth 1000 Berlina
			Grand Touring	700 cc	G. Varese	13'42"2	Fiat Abarth 700 BA
			Grand Touring	1000 cc	G. Rebaudi	12'58"6	Fiat Abarth 1000 BA
			Grand Touring	1300 cc	G. Bertuzzi	12'48"7	Abarth Simca 1300
13 December	Roma, Italy	Rallye di Polidoro (193 kms)	---	---	R. Coscioni	---	Fiat Abarth 850 TC
13 December	Vallelunga, Italy	Criterium dell'Ora (circuit-1,800 kms x 1 hour)	Touring	850 cc	B. Montecatino	kms 91,838	Fiat Abarth 850 TC
			Touring	1000 cc	F. Chiari	kms 90,688	Fiat Abarth 1000 Berlina
20 December	Vallelunga, Italy	Criterium dell'Ora (circuit-1,800 kms x 1 hour)	Touring	850 cc	M. Terminiello	kms 90,995	Fiat Abarth 850 TC
			Touring	1000 cc	F. Chiari	kms 89,014	Fiat Abarth 1000 Berlina
			Grand Touring	1300 cc	S. Ridolfi	kms 93,448	Abarth Simca 1300

1965

DATE	LOCATION	RACE	CATEGORY	CLASS	WINNER	TIME OR DISTANCE	CAR
6 January	Vallelunga, Italy	Coppa Epifania (circuit 1,400 kms x 5 laps)	Touring	850 cc	O. Costantini	3'32"6	Fiat Abarth 850 TC
			Touring	1000 cc	F. Chiari	3'32"5	Fiat Abarth 1000 Berlina
31 January	Vallelunga, Italy	Trofeo d'Inverno (circuit-1,800 kms x 20 laps)	Touring	850 cc	O. Costantini	23'32"6	Fiat Abarth 850 TC
			Touring	1000 cc	F. Chiari	23'46"9	Fiat Abarth 1000 Berlina
			Grand Touring	up to1600 cc	S. Ridolfi	21'36"0	Abarth Simca 1300
7 February	Vallelunga, Italy	Trofeo d'Inverno (circuit-1,800 kms x 15 laps)	Touring	850 cc	O. Costantini	17'33"4	Fiat Abarth 850 TC
			Touring	1000 cc	F. Chiari	17'23"0	Fiat Abarth 1000 Berlina
			Grand Touring	up to 1600 cc	E. Buzzetti	15'56"7	Fiat Abarth 1000 BA
8 February	Madrid, Spagna	Subida Puerto de Galapagar (hillclimb-4,000 kms)	---	B	J. de Bagration	2'29"2	Fiat Abarth 850 TC
			---	G	J. de Bagration	2'26"7	Fiat Abarth 850 TC
14 February	Wiesbaden, Germany	HMSC Slalom Finthen	Grand Touring	up to 1300 cc	U. Schütz	1'06"2	Fiat Abarth 1000 BA
20-21 February	Treviso, Italy	Autosciatoria della Marca Trevigiana (rallye-186 kms)	---	---	A. Bonduri (overall)	---	Fiat Abarth 850 TC
21 February	Roma, Italy	Gara Tris-Pista d'Oro (circuit-1,000 kms x 2 laps)	Touring	850 cc	V. Trulli	2'21"7	Fiat Abarth 850 TC
21 February	Nice, France	Course de côte de Saint Cézaire (hillclimb-4,500 kms)	Grand Touring	1300 cc	R. Barone	3'53"4	Abarth Simca 1300
25-28 February	Sanremo, Italy	Rallye dei Fiori (753 kms)	Touring	up to 1000 cc	Rubbieri-Polli	---	Fiat Abarth 850 TC
			Prepared Touring	up to 850 cc	Severi-Venturi	---	Fiat Abarth 850 TC
27-28 February	Salon de Provence, France	Rallye Mistral (270 kms)	---	850 cc	Branche-Melle Brun		Fiat Abarth 850 TC
27-28 February	Le Mans, France	Rallye du Maine (300 kms)	---	1000 cc	Martin-Martin (overall)	---	Fiat Abarth 1000 Berlina
27-28 February	Sintra, Portugal	Rallye das Camelias	---	I	E. Reverter	---	Fiat Abarth 1000 BA
28 February	Salo, Finlandia	Corsa sul ghiaccio (circuit-7,500 kms)	Grand Touring	1000 cc	L. Kinnunen	5'39"0	Fiat Abarth 1000
28 February	Vallelunga, Italy	Coppa Autodromo di Roma (circuit-1,800 kms x 10 laps)	Touring	up to 1000 cc	R. Addamiano	12'04"8	Fiat Abarth 850 TC
			Grand Touring	1300 cc	C. Fabbri	11'02"4	Abarth Simca 1300
7 March	Foligno, Italy	Autosciatoria Foligno-Norcia (regolarità-83,000 kms)	---	---	E. Caporali	---	Fiat Abarth 500
7 March	Erkylä, Finlandia	Corsa sul ghiaccio (7,500 kms)	Grand Touring	1000 cc	L. Kinnunen	6'18"4	Fiat Abarth 1000 Berlina

DATE	LOCATION	RACE	CATEGORY	CLASS	WINNER	TIME OR DISTANCE	CAR
14 March	Vallelunga, Italy	Coppa Autodromo di Roma (circuit-1,800 kms x 10 laps)	Touring	850 cc	R. Addamiano	11'45"2	Fiat Abarth 850 TC
			Touring (I handicap)	1000 cc	P. Bruscolotti	13'07"4	Fiat Abarth 1000 Berlina
			Touring (II handicap)	850 cc	R. Addamiano	12'04"5	Fiat Abarth 850 TC
			Touring	1000 cc	G. Saudelli	12'11"5	Fiat Abarth 1000 Berlina
18-19 March	Monza, Italy	4 Ore del Jolly Club (circuit-10,000 kms)	Touring	850 cc	G. Raffa	kms 619,774	Fiat Abarth 850 TC
			Touring	1000 cc	Demetz-Herrmann	kms 626,458	Fiat Abarth 1000 Berlina
19 March	Vallelunga, Italy	Meeting di apertura (circuit-1,800 kms x 15 laps)	Touring	850 cc	R. Addamiano	18'18"3	Fiat Abarth 850 TC
			Touring	1000 cc	B. Faenza	17'49"8	Fiat Abarth 1000 Berlina
			Grand Touring	1300 cc	C. Fabbri	17'03"4	Abarth Simca 1300
21 March	Monza, Italy	Coppa Automobile Club di Genova (circuit-2,405 kms x 20 laps)	Touring	600 cc	S. del Cason	26'55"5	Fiat Abarth 595
			Touring	850 cc	G. Pianta	26'00"1'	Fiat Abarth 850 TC
			Touring	up to 1150 cc	G. Bardini	25'01"4	Fiat Abarth 1000 Berlina
			Grand Touring	1000 cc	G. Rebaudi	23'15"9	Fiat Abarth 1000 BA
21 March	Wien, Austria	Dopplerhüttewertungfahrt (hillclimb-2,700 kms)	Touring	1000 cc	W. Pöltinger	2'36"8	Fiat Abarth 1000 Berlina
21 March	Montseny, Spagna	Carrera en cuesta al Montseny (hillclimb-15,000 kms)	Touring	600 cc	M. Juncosa	14'9"5	Fiat Abarth 595
			Touring	850 cc	'Foca'	13'19"5	Fiat Abarth 850 TC
			Touring	1000 cc	J. Juncosa	12'58"0	Fiat Abarth 1000 Berlina
			Prepared Touring	850 cc	'Jorge Chi'	13'34"8	Fiat Abarth 850 TC
21 March	Magdalena Mexhuca, Messico	Primer Premio Benito Juarez (circuit)	Touring	---	F. Van Beuren	40'46"0	Fiat Abarth 750
27-28 March	Padova, Italy	Coppa Mario Della Favera (regolarità-247 kms)	---	---	M. Rigo	---	Fiat Abarth 850 TC
27-28 March	Grasse, France	Criterium Fleurs et Parfums (528 kms)	Touring	up to 850 cc	Branche-Jullien	---	Fiat Abarth 850 TC
28 March	Eisenstadt, Austria	Eisenstadter Bergwertungsfahrt (hillclimb-4,000 kms)	Touring	1000 cc	W. Pöltinger	3'25"2	Fiat Abarth 1000 Berlina
28 March	Kautenbach, Luxembourg	Course de côte de Kautenbach (hillclimb)	Grand Touring	1000 cc	J. Hauser	2'14"4	Fiat Abarth 1000 BA
			Grand Touring	from 1601 to 2000 cc	J. Rudaz	2'03"4	Abarth Simca 2000
28 March	Roma, Italy	Coppa Gallenga (hillclimb-10,400 kms)	Touring	600 cc	L. Durst	6'15"7	Fiat Abarth 595
			Touring	850 cc	M. Cintolesi	5'49"4	Fiat Abarth 850 TC
			Touring	1000 cc	P. Lado	5'32"9	Fiat Abarth 1000 Berlina
			Grand Touring	1000 cc	E. Buzzetti	5'10"4	Fiat Abarth 1000 BA
			Grand Touring	1300 cc	B. Carotenuto	5'17"7	Abarth Simca 1300
28 March	Wittlich, Germany	Wittlicher ADAC Bergpreis (hillclimb-3,738 kms x 2)	Grand Touring	1000 cc	H. Krause	5'09"0	Fiat Abarth 1000 BA
			Grand Touring	1300 cc	K. Federhofer	4'56"4	Abarth Simca 1300
28 March	Lignières, Switzerland	Course de côte de Lignières (circuit-1,400 kms x 10 laps)	Touring	850 cc	R. Piccard	26'03"36	Fiat Abarth 850 TC
1-4 April	Madrid, Spagna	Rallye del R.A.C.E. (2,217 kms)	Touring	up to 850 cc	E. Doncel	---	Fiat Abarth 850 TC
			Touring	from 850 to 1000 cc	'Runner'	---	Fiat Abarth 1000 Berlina
			Grand Touring	1000 cc	E. Reverter	---	Fiat Abarth 1000 Bialbero
4 April	Verona, Italy	Stallavena-Boscochiesanuova (hillclimb-15,300 kms)	Touring	850 cc	B. Paglieno	8'34"0	Fiat Abarth 850 TC
			Touring	1000 cc	P. Lado	8'04"4	Fiat Abarth 1000 Berlina
			Grand Touring	700 cc	G. Varese	8'19"9	Fiat Abarth 700 BA
			Grand Touring	1000 cc	G. Avventurieri	7'39"5	Fiat Abarth 1000 BA
			Grand Touring	1300 cc	R. Bertuzzi	7'39"2	Abarth Simca 1300
4 April	Saint-Lô, France	Course de côte d'Hébécrevon (hillclimb-1,200 kms)	Touring	850 cc	B. Lagier	0'51"1	Fiat Abarth 850 TC
			Grand Touring	1000 cc	J. Mazzanti	0'47"1	Fiat Abarth 1000 BA
			Grand Touring	from 1601 to 2000 cc	B. Lagier	0'44"0	Abarth Simca 2000
4 April	Cagliari, Italy	Monte Acuto-Campuomu (hillclimb-14,000 kms)	Touring	600 cc	B. Massidda	10'56"3	Fiat Abarth 595
			Touring	850 cc	G. Vacca	9'51"9	Fiat Abarth 850 TC
			Touring	up to 1300 cc	N. Parenti	10'03"7	Fiat Abarth 1000 Berlina
			Grand Touring	up to 1150 cc	F. Amicucci	10'23"2	Fiat Abarth 1000 BA
			Grand Touring	1300 cc	P. Gargano	10'12"7	Abarth Simca 1300
4 April	Wien, Austria	Slalom Wertungsfahrt (43 kms)	Touring	up to 1150 cc	W. Pöltinger	---	Fiat Abarth 1000 Berlina
4 April	Jämsankoski, Finlandia	Corsa sul ghiaccio (5,000 kms)	Grand Touring	1000 cc	L. Kinnunen	4'19"	Fiat Abarth 1000 Berlina
10-11 April	Berlin, Germany	ADAC Nachtprüfungsfahrt (regolarità-210 kms)	Touring	850 cc	H. Mack	---	Fiat Abarth 850 TC
11 April	Aspern, Austria	Preis von Wien (circuit-2,600 kms)	Touring (1 hour)	1000 cc	K. Steinmetz	kms 111,8	Fiat Abarth 1000 Berlina
			Grand Touring (20 laps)	up to 1300 cc	A. Schmalbach	24'40"5	Fiat Abarth 1000 BA
			Grand Touring	up to 2000 cc	J. Rindt	24'18"8	Abarth Simca 2000
			Sport	up to 2000 cc	H. Herrmann	24'01"1	Fiat Abarth 2000

DATE	LOCATION	RACE	CATEGORY	CLASS	WINNER	TIME OR DISTANCE	CAR
11 April	Messina, Italy	Coppa Colle S. Rizzo (hillclimb-9,000 kms)	Touring	1000 cc	I. Capuano	6'46"3	Fiat Abarth 1000 Berlina
			Grand Touring	700 cc	R. Gervasi	7'29"1	Fiat Abarth 700 BA
			Grand Touring	1000 cc	M. Spadafora (overall)	6'16"3	Fiat Abarth 1000 BA
			Grand Touring	1300 cc	F. Piccione	6'54"5	Abarth Simca 1300
			Sport	up to 2000 cc	A. Riolo	6'30"9	Fiat Abarth 1000
11 April	Draguignan, France	Course de côte d'Ampus (hillclimb-6,800 kms)	Touring	850 cc	J. Raquin	6'03"8	Fiat Abarth 850 TC
			Grand Touring	1300 cc	R. Barone	5'39"0	Abarth Simca 1300
11 April	Eifel, Germany	ADAC Bergpreis Bad Neuenahr (hillclimb-3,300 kms x 2)	Grand Touring	1000 cc	K. Geiss	4'34"8	Fiat Abarth 1000 BA
			Grand Touring	1300 cc	K. Federhofer	4'21"9	Abarth Simca 1300
11 April	Imola, Italy	Gran Premio Shell (circuit-5,017 kms x 3 hours)	Grand Touring	1000 cc	B. Bonini	kms 400,547	Fiat Abarth 1000 BA
			Grand Touring	1300 cc	H. Demetz	kms 422,860	Abarth Simca 1300
11 April	Livorno, Italy	Rosa della Montagna (regolarità-131 kms)	---	---	E. Caporali	---	Fiat Abarth 595
11 April	Liège, Belgium	Course de côte La Roche Samree (hillclimb-2,700 kms)	Touring	850 cc	P. Henning	4'03"3	Fiat Abarth 850 TC
			Touring	1000 cc	C. Ballot Lena	3'48"9	Fiat Abarth 1000 Berlina
			Grand Touring	1000 cc	J. Hauser	3'58"5	Fiat Abarth 1000 BA
11 April	Trieste, Italy	Coppa del Biancospino (regolarità-144 kms)	---	---	P. Navarra	---	Fiat Abarth 850 TC
11 April	Zandvoort, Holland	Trainingsdag (circuit-4,225 kms)	Grand Touring	up to 1300 cc	E. Swart	22'46"4	Fiat Abarth 1000 Berlina
19 April	Innsbruck, Austria	Internationales Weerbergrennen (hillclimb-5,000 kms)	Grand Touring	1000 cc	H. Affentranger	4'39"6	Fiat Abarth 1000 BA
			Grand Touring	1300 cc	K. Federhofer	4'17"7	Abarth Simca 1300
19 April	Vallelunga, Italy	Conchiglia Shell (circuit-1,800 kms x 56 laps)	Touring	850 cc	S. Morando	1h05'08"8	Fiat Abarth 850 TC
			Touring	1000 cc	A. Marzi	1h04'44"5	Fiat Abarth 1000 Berlina
19 April	Snetterton, United Kingdom	B.R.S.G.C. Cup Race (circuit-2,700 miles)	Touring	850 cc	G. Günther	17'29"3	Fiat Abarth 850 TC
19 April	Zandvoort, Holland	RAC West Autosprint (circuit-4,225 kms)	Touring	850 cc	Hooft	---	Fiat Abarth 850 TC
24 April	Verbois, Switzerland	Course de côte de Verbois (hillclimb-1,150 kms)	Touring	850 cc	F. Aeschimann	1'04"3	Fiat Abarth 850 TC
			Grand Touring	1000 cc	S. Vaglio	0'57"3	Fiat Abarth 1000 BA
25 April	Gap, France	Course de côte du Col Bayard (hillclimb-5,000 kms)	Touring	850 cc	A. Dole	4'07"0	Fiat Abarth 850 TC
			Touring	1000 cc	I. Giunti	3'52"3	Fiat Abarth 1000 Berlina
			Grand Touring	up to 850 cc	J. Branche	4'49"2	Fiat Abarth 700 BA
			Grand Touring	1000 cc	Y. Barone	4'06"0	Fiat Abarth 1000 BA
			Grand Touring	1300 cc	R. Barone	3'47"2	Abarth Simca 1300
			Grand Touring	from 1600 to 2000 cc	P. Maublanc	3'29"3	Abarth Simca 2000
25 April	Monza, Italy	Coppa Intereuropa (circuit-5,750 kms x 3 hours)	Grand Touring	1000 cc	C. Sangiorgi	kms 473,947	Fiat Abarth 1000 BA
			Grand Touring	1300 cc	K. Steinmetz	kms 502,678	Abarth Simca 1300
25 April	Amalfi, Italy	Coppa della Primavera (hillclimb-8,750 kms)	Touring	850 cc	R. Bosurgi	7'05"9	Fiat Abarth 850 TC
			Touring	1000 cc	V. Ferro	7'19"9	Fiat Abarth 1000 Berlina
			Grand Touring	700 cc	'Jacopo'	7'15"5	Fiat Abarth 700 BA
			Grand Touring	1000 cc	'Ellevu'	7'24"4	Fiat Abarth 1000 BA
			Grand Touring	1300 cc	R. Arfé	6'59"0	Abarth Simca 1300
25 April	Avola, Italy	Coppa Belmonte (hillclimb-10,200 kms)	Touring	from 700 to 1000 cc	N. Barbagallo	7'17"9	Fiat Abarth 1000 Berlina
			Grand Touring	700 cc	L. Alce	7'44"7	Fiat Abarth 700 BA
			Grand Touring	1000 cc	C. Avventurieri	7'00"8	Fiat Abarth 1000 BA
25 April	Erbach, Germany	DMW Kräbergrennen (hillclimb-3,600 kms x 2)	Grand Touring	700 cc	K. Geiss	5'13"6	Fiat Abarth 700 BA
25 April	Monthléry, France	Coupes de Vitesse (circuit-3,340 kms x 15 laps)	Touring	850 cc	P. Henning	30'16"5	Fiat Abarth 850 TC
			Touring	1000 cc	C. Ballot Lena	30'07"0	Fiat Abarth 1000 Berlina
			Grand Touring (19 laps)	1000 cc	J. Mazzanti	33'00"5	Fiat Abarth 1000 BA
25 April	Roma, Italy	Competizioni Pista d'Oro (circuit-1,000 kms x 2)	Touring	up to 1000 cc	V. Trulli	2'19"1	Fiat Abarth 850 TC
			Grand Touring	up to 1300 cc	M. Tommasini	2'25"5	Fiat Abarth 1000 BA
1 May	Mannheim, Germany	Internationales Automobil Turnier (circuit-1,800 kms)	Grand Touring	1000 cc	K. Jacoby	---	Fiat Abarth 1000 BA
1 May	Oberuhldingen, Germany	A.C.S. Auto Slalom (circuit-0,450 kms x 5 laps)	Touring	1000 cc	K. Köstli	2'51"6	Fiat Abarth 1000 Berlina
			Grand Touring	up to 1300 cc	T. Hofer	2'47"4	Fiat Abarth 1000 BA
1-2 May	Luxembourg	Slalom-Coupe d'Europe des Ecuries (1,000 kms x 2 laps)	Touring	850 cc	H. Wagner	1'17"5	Fiat Abarth 850 TC
			Touring	1000 cc	E. Swart	1'16"5	Fiat Abarth 1000 Berlina
			Grand Touring	1000 cc	J. Hauser	1'16"7	Fiat Abarth 1000 BA
			Grand Touring	from 1600 to 2000 cc	H. Kuhnis	1'15"7	Abarth Simca 2000
			Sport	up to 1600 cc	W. Rufenacht	1'16"3	Fiat Abarth 1000 Sport
1-2 May	Dachau, Germany	ADAC Peissenberg Rennen (hillclimb-3,100 kms x 2)	Grand Touring	700 cc	K. Geiss	4'07"8	Fiat Abarth 700 BA
			Grand Touring	1000 cc	H. Krause	4'09"5	Fiat Abarth 1000 BA
			Grand Touring	1300 cc	K. Federhofer	3'51"0	Abarth Simca 1300

DATE	LOCATION	RACE	CATEGORY	CLASS	WINNER	TIME OR DISTANCE	CAR
2 May	Nevers, France	Circuit Jean Behra (2,000 kms)	Touring (20 laps) Grand Touring (30 laps)	1000 cc 1000 cc	C. Ballot Lena J. Mazzanti	22'38"7 35'21"2	Fiat Abarth 1000 Berlina Fiat Abarth 1000 BA
2 May	Cannes, France	Course de côte de Cannes (hillclimb-7,845 kms)	Prepared Touring Prepared Touring Grand Touring	850 cc 1000 cc up to 850 cc	M. Cintolesi A. Dole R. Jullien	7'14"1 7'04"8 7'30"3	Fiat Abarth 850 TC Fiat Abarth 1000 Berlina Fiat Abarth 700 BA
2 May	Montelimar, France	Course de côte du Colombier (hillclimb-3,300 kms)	Touring Grand Touring Grand Touring Grand Touring	850 cc 1000 cc 1300 cc from 1300 to 1600 cc	J. Raquin Y. Barone R. Barone P. Maublanc	3'20"1 3'11"3 2'55"9 2'47"7	Fiat Abarth 850 TC Fiat Abarth 1000 BA Abarth Simca 1300 Abarth Simca 2000
2 May	Udine, Italy	Cividale-Castelmonte (hillclimb-7,000 kms)	Touring Touring Grand Touring Grand Touring Grand Touring Grand Touring	850 cc 1000 cc 700 cc 1000 cc 1300 cc over 1600 cc	S. Uberti P. Lado G. Mainetti M. Baldo G. Dalla Torre A. Giliberti	5'10"7 4'53"6 5'17"6 4'47"8 4'38"6 4'16"5	Fiat Abarth 850 TC Fiat Abarth 1000 Berlina Fiat Abarth 700 BA Fiat Abarth 1000 BA Abarth Simca 1300 Abarth Simca 2000
2 May	Las Palmas, Spagna	Circuito Ciudad de Las Palmas (circuit-2,580 kms)	Prepared Touring	up to 1150 cc	R. Spencer	---	Fiat Abarth 850 TC
2 May	Kobenhavn, Danimarca	Gara d'apertura circuit di Roskilde (8 laps x 2)	Touring	850 cc	A. Larsen	14'22"5	Fiat Abarth 850 TC
7 May	Harewood Acres (Ontario), USA	Great Lakes Trophy Races (circuit-1,18 miles x 10 laps)	--- Sport	--- up to 1600 cc	C. Fisher C. Fisher	--- ---	Abarth Simca 1300 Abarth Simca 1300
8 May	Zandvoort, Holland	N.A.V. Kampioenschapsraces (circuit-4,225 kms x 15 laps)	Touring	850 cc	H. Veder	34'26"2	Fiat Abarth 850 TC
8 May	Adenau, Germany	ADAC Hansa Nürburgring Pokalrennen (circuit-7,700 kms x 8 laps)	Touring	1000 cc	E. Bitter	31'46"2	Fiat Abarth 1000 Berlina
8-9 May	Abano Terme, Italy	Rallye delle Prealpi Venete e dei Colli Euganei (660,500 kms)	Touring	850 cc	R. Zanini	---	Fiat Abarth 850 TC
8-9 May	Bordeaux, France	Rallye Bordeaux Sud Ouest (564 kms)	Grand Touring	from 1150 to 1600 cc	Sarayac-Chevallier	---	Abarth Simca 1300
9 May	Lorch am Rhein, Germany	Taunus Bergprüfung (hillclimb-3,300 kms x 2)	Grand Touring	1300 cc	T. Reuter	2'24"9	Abarth Simca 1300
9 May	Adenau, Germany	Nürburgring Handicap-Rennen (circuit-22,810 kms x 5 laps)	Touring	1000 cc	E. Bitter	---	Fiat Abarth 1000 Berlina
9 May	Marseille, France	Course de côte de la Sainte Baume (hillclimb-8,100 kms)	Touring Touring Grand Touring Grand Touring Grand Touring	850 cc 1000 cc 850 cc 1000 cc from 1600 to 2000 cc	I. Giunti S. Stefani J. Raquin C. Ballot Lena P. Maublanc	7'38"1 7'48"0 8'21"0 7'47"1 7'10"1	Fiat Abarth 850 TC Fiat Abarth 1000 Berlina Fiat Abarth 850 TC Fiat Abarth 1000 BA Abarth Simca 2000
9 May	Payerne, Switzerland	Slalom Aérodrome de Payerne (circuit-4,600 kms x 2)	Touring Prepared Touring Grand Touring Grand Touring Grand Touring Grand Touring Sport	850 cc up to 1150 cc 700 cc 1000 cc 1300 cc from 1600 to 2000 cc 750 cc	E. Scheiter P. Grisel H. Affentranger J. Kretschi D. Borel H. Kuhnis F. Marti	4'14"2 3'59"0 4'00"6 3'41"6 3'30"5 3'23"2 5'20"6	Fiat Abarth 850 TC Fiat Abarth 1000 Berlina Fiat Abarth 700 BA Fiat Abarth 1000 BA Abarth Simca 1300 Abarth Simca 2000 Fiat Abarth 750
9 May	Anvers, Belgium	Journée des Records Belges	Grand Touring	1000 cc	J. Hauser	19"8	Fiat Abarth 1000 BA
9 May	Barcelona, Spagna	Trofeo Juan Jover-Montjuich (circuit-3,790 kms)	Touring (15 laps) Grand Touring (20 laps)	850 cc up to 1300 cc	J. de Bragation J. Piera	35'16"2 40'57"7	Fiat Abarth 850 TC Fiat Abarth 1000 BA
9 May	Palermo, Italy	Targa Florio (circuit-72,000 kms x 10 laps)	Grand Touring Sport-Prototypes	1300 cc from 1000 to 1600 cc	Calascibetta-Virgilio Herrmann-Cella	7h47'12"4 7h17'23"0	Abarth Simca 1300 Fiat Abarth OT 1600
9 May	Volterra, Italy	Coppa Città di Volterra (hillclimb-10,060 kms)	Touring Touring Touring Grand Touring Grand Touring Sport	600 cc 850 cc 1000 cc 700 cc 1300 cc 1000 cc	L. Durst S. Morando A. Marzi B. Mainetti 'Nicor' A. Rusconi	7'32"6 6'49"7 6'37"8 7'00" 6'30"4 6'48"4	Fiat Abarth 595 Fiat Abarth 850 TC Fiat Abarth 1000 Berlina Fiat Abarth 700 BA Abarth Simca 1300 Fiat Abarth 1000
9 May	Piacenza, Italy	Castell'Arquato-Vernasca (hillclimb-9,775 kms)	Touring Touring Grand Touring Grand Touring Grand Touring	850 cc up to 1150 cc 1000 cc up to 1150 cc 1300 cc	'Jerry' A. Grillo P. Scantimburgo L. Pozzo R. Bertuzzi	6'17"7 6'05"2 6'28"3 6'11"2 5'27"4	Fiat Abarth 850 TC Fiat Abarth 1000 Berlina Fiat Abarth 1000 BA Simca Abarth 1150 Abarth Simca 1300

DATE	LOCATION	RACE	CATEGORY	CLASS	WINNER	TIME OR DISTANCE	CAR
16 May	Reisdorf, Luxembourg	Course de côte de Reisdorf (hillclimb-3,400 kms)	Touring	850 cc	H. Wagner	2'28"1	Fiat Abarth 850 TC
			Grand Touring	1000 cc	J. Hauser	2'17"3	Fiat Abarth 1000 BA
			Grand Touring	from 1600 to 2000 cc	H. Kuhnis	2'02"1	Abarth Simca 2000
16 May	Francorchamps, Belgium	Coupes de Spa (circuit-14,100 kms x 11 laps)	Touring	1000 cc	E. Swart	59'12"9	Fiat Abarth 1000 Berlina
16 May	Eberbach, Germany	Eberbach Bergrennen (hillclimb-3,000 kms)	Grand Touring	1300 cc	T. Reuter	3'53"8	Abarth Simca 1300
16 May	Livorno, Italy	Antignano-Monteburrone (hillclimb-3,500 kms)	Touring	600 cc	G. Caroli	5'25"1	Fiat Abarth 595
			Touring	850 cc	A. Monselles	4'53"2	Fiat Abarth 850 TC
			Touring	1000 cc	M. Nesti	4'39"6	Fiat Abarth 1000 Berlina
			Grand Touring	700 cc	G.B. Mainetti	4'55"6	Fiat Abarth 700 BA
			Grand Touring	1000 cc	A. Becchetti	4'40"5	Fiat Abarth 1000 BA
			Grand Touring	1300 cc	R. Arfé	4'34"9	Abarth Simca 1300
16 May	Ferrara, Italy	Trofeo Armando François (105 kms)	---	---	'Picci'	---	Fiat Abarth 850 TC
16 May	Sassari, Italy	Alghero-Scala Picada (hillclimb-8,700 kms)	Touring	600 cc	B. Massidda	6'13"0	Fiat Abarth 595
			Touring	850 cc	S. Gambarau	5'25"2	Fiat Abarth 850 TC
			Touring	1000 cc	F. Locci	5'24"8	Fiat Abarth 1000 Berlina
			Grand Touring	1000 cc	G. Vacca	5'05"3	Fiat Abarth 1000 BA
16 May	Beograd, Jugoslavia	Derby a Belgrado (circuit-2,300 kms)	---	850 cc	S. Matevz	13'11"1	Fiat Abarth 850 TC
20-23 May	Saint Raphaël, France	Saint Raphaël Feminin Criterium (1342 kms)	Grand Touring	1000 cc	Y. Barone	---	Fiat Abarth 1000 BA
22 May	Goodwood, United Kingdom	Goodwood Races (circuit-2,400 miles)	Grand Touring	from 1600 to 2000 cc	R. Burnard	17'03"6	Abarth Simca 2000
22 May	Frutigen, Switzerland	A.C.S. "Surprise" (circuit)	Touring	850 cc	E. Schnetter	4'48"9	Fiat Abarth 850 TC
			Prepared Touring	up to 850 cc	R. Egli	5'02"7	Fiat Abarth 750
			Grand Touring	up to 1150 cc	H. Affentranger	4'37"2	Fiat Abarth 700 BA
			Grand Touring	1300 cc	F. Della Balda	4'15"8	Abarth Simca 1300
22-23 May	Amiens, France	Rallye de Picardie (400 kms)	Grand Touring	1000 cc	Lemaitre-Legrand	---	Fiat Abarth 1000 BA
22-23 May	Nice, France	Criterium Jean Behra (424 kms)	Grand Touring	up to 850 cc	Branche-Jullien	---	Fiat Abarth 700 BA
23 May	Adenau, Germany	ADAC Nürburgring 1000 kms Rennen (circuit-22,810 kms x 44 laps)	Grand Touring	1300 cc	Virgilio-Calascibetta	6h53'49"9	Abarth Simca 1300
23 May	Voralberg, Austria	Bergrennen Schwarzenberg Bödele (hillclimb-3,850 kms x 2)	Touring	1000 cc	K. Husel	2'44"9	Fiat Abarth 1000 Berlina
			Grand Touring	up to 1150 cc	J. Kretschi	2'26"7	Fiat Abarth 1000 BA
			Grand Touring	up to 2000 cc	H. Kuhnis	2'15"8	Abarth Simca 2000
23 May	Madrid, Spagna	Subida al Puerto de la Morentera (hillclimb-7,000 kms)	Touring	850 cc	J. de Bagration	4'56"6	Fiat Abarth 850 TC
23 May	Montlhéry, France	Grand Prix de Paris (circuit-3,360 kms x 15 laps)	Touring	850 cc	P. Henning	32'05"6	Fiat Abarth 850 TC
			Touring	1000 cc	C. Ballot Lena	30'42"1	Fiat Abarth 1000 Berlina
			Grand Touring	1300 cc	D. Borel	24'52"9	Abarth Simca 1300
			Grand Touring (20 laps)	from 1600 to 2000 cc	F. Ruata	31'55"0	Abarth Simca 2000
			Sport	1000 cc	C. Ballot Lena	25'08"7	Fiat Abarth 1000
23 May	Reggio Emilia, Italy	Vezzano-Casina (hillclimb-12,500 kms)	Touring	600 cc	L. Durst	8'56"5	Fiat Abarth 595
			Touring	850 cc	S. Morando	8'07"9	Fiat Abarth 850 TC
			Touring	1000 cc	P. Lado	7'53"9	Fiat Abarth 1000 Berlina
			Grand Touring	700 cc	G.B. Mainetti	8'23"5	Fiat Abarth 700 BA
			Grand Touring	1000 cc	G. Rebaudi	7'43"2	Fiat Abarth 1000 BA
			Grand Touring	1300 cc	G. Dalla Torre	7'33"3	Abarth Simca 1300
23 May	Sorrento, Italy	Sorrento-Sant'Agata (hillclimb-10,500 kms)	Touring	1000 cc	M. Terminiello	7'03"5	Fiat Abarth 850 TC
			Grand Touring	700 cc	'Jacopo'	7'24"3	Fiat Abarth 700 BA
			Grand Touring	1000 cc	M. Moselli	6'31"9	Fiat Abarth 1000 BA
			Grand Touring	1300 cc	R. Arfé	6'47"2	Abarth Simca 1300
23 May	Berlin, Germany	Goldner Bär von Berlin (325 kms)	Touring	850 cc	H. Garschig	3h01'40"	Fiat Abarth 850 T
23 May	Vaasa, Finlandia	Corsa in circuit (15,000 kms)	Grand Touring	1000 cc	L. Kinnunen	8'49"9	Fiat Abarth 1000 Berlina
27 May	Barcelona, Spagna	Carrera en cuesta de Vallvidrera	Grand Touring	from 850 to 1300 cc	'Salvador'	4'08"5	Fiat Abarth 1000 BA
27 May	Aix en Provence, France	Course de côte d'Aix-Saint Antonin (hillclimb-2,700 kms)	Touring	1000 cc	A. Raymond	2'02"7	Fiat Abarth 1000 Berlina
			Grand Touring	1300 cc	J.P. Raquin	2'14"6	Fiat Abarth 850 TC
			Grand Touring	1300 cc	R. Barone	1'57"6	Abarth Simca 1300
			Grand Touring	from 1600 to 2500 cc	P. Maublanc	1'48"8	Abarth Simca 2000
27 May	Pistoia, Italy	Coppa della Collina (hillclimb-14,700 kms)	Touring	up to 600 cc	C. Bilenghi	10'56"9	Fiat Abarth 595
			Touring	850 cc	R. Basurgi	9'47"4	Fiat Abarth 850 TC
			Touring	1000 cc	M. Nesti	9'24"4	Fiat Abarth 1000 Berlina
			Grand Touring	up to 1000 cc	'Jamaraschi'	10'24"2	Fiat Abarth 700 BA
			Grand Touring	1300 cc	'Nicor'	10'09"6	Abarth Simca 1300

DATE	LOCATION	RACE	CATEGORY	CLASS	WINNER	TIME OR DISTANCE	CAR
27 May	Salerno, Italy	Cava dei Tirreni-Badia (hillclimb-2,300 kms)	Touring	850 cc	M. Terminiello	1'52"0	Fiat Abarth 850 TC
			Touring	1000 cc	V. Ferro	1'49"7	Fiat Abarth 1000 Berlina
			Grand Touring	700 cc	'Cicci'	1'54"5	Fiat Abarth 700 BA
			Grand Touring	1000 cc	V. Cannataro	1'56"0	Fiat Abarth 1000 BA
			Grand Touring	1300 cc	R. Arfé	1'43"3	Abarth Simca 1300
			Grand Touring	up to 2500 cc	D. Scola	1'46"1	Abarth Simca 2000
27 May	Villars, Switzerland	Course de côte Fontaines Villars-Bourquin (hillclimb-1,800 kms)	Touring	850 cc	J.P. Brun	1'33"7	Fiat Abarth 850 TC
			Touring	1000 cc	P. Grisel	1'29"5	Fiat Abarth 1000 Berlina
			Grand Touring	1000 cc	D. Pache	1'26"6	Fiat Abarth 1000 BA
			Grand Touring	1300 cc	D. Borel	1'18"7	Abarth Simca 1300
30 May	Bologna, Italy	Bologna-Passo della Raticosa (hillclimb-32,700 kms)	Touring	850 cc	D. Martino	22'51"6	Fiat Abarth 850 TC
			Touring	up to 1150 cc	G. Galimberti	22'14"3	Fiat Abarth 1000 Berlina
			Grand Touring	700 cc	G. Varese	21'53"3	Fiat Abarth 700 BA
			Grand Touring	1000 cc	G. Rebaudi	20'59"0	Fiat Abarth 1000 BA
			Grand Touring	1300 cc	G. Dalla Torre	20'30"0	Abarth Simca 1300
			Grand Touring	from 1600 to 2500 cc	L. Cella	19'31"0	Abarth Simca 2000
			Sport	up to 2000 cc	H. Herrmann	19'14"4	Fiat Abarth OT 2000
30 May	Herbeumont, Belgium	Course de côte d'Herbeumont (hillclimb-3,600 kms)	Touring	from 850 to 1000 cc	T. Pilette	2'03"5	Fiat Abarth 1000 Berlina
30 May	Dudelange, Luxembourg	Course de côte Tele-Dudelange (hillclimb-1,400 kms)	Touring	from 850 to 1000 cc	E. Swart	1'03"4	Fiat Abarth 1000 Berlina
			Grand Touring	up to 1600 cc	J. Hauser	1'03"0	Fiat Abarth 1000 BA
30 May	Cesena, Italy	Coppa Città di Cesena (rallye-192 kms)	Touring	up to 1000 cc	F. Merelli	---	Fiat Abarth 850 TC
30 May	Väinölönniemi, Finlandia	Corsa in circuit (9,000 kms)	Touring	up to 1000 cc	L. Kinnunen	6'11"0	Fiat Abarth 1000 Berlina
30 May	Lignières, Switzerland	Circuit de Lignières (circuit-1,400 kms)	---	---	F. Wipf	---	Fiat Abarth 1000 Berlina
30 May	Bayerischzell, Germany	ADAC Sudelfeld Bergrennen (hillclimb-3,500 kms x 2)	Grand Touring	1000 cc	A. Schmalbach	4'28"0	Fiat Abarth 1000 BA
			Grand Touring	1300 cc	R. Stierli	4'19"3	Abarth Simca 1300
			Sport	1000 cc	W. Rufenacht	4'18"2	Fiat Abarth 1000
31 May	Thompson (Connecticut), USA	Thompson Raceway (circuit)	---	---	A. Cosentino	---	Fiat Abarth 850 TC
2 June	Treviso, Italy	Vittorio Veneto-Cansiglio (hillclimb-11,000 kms)	Touring	600 cc	'Napoleone'	8'34"9	Fiat Abarth 595
			Touring	850 cc	G. Gallo	7'51"9	Fiat Abarth 850 TC
			Touring	1000 cc	A. Marcon	7'50"9	Fiat Abarth 1000 Berlina
			Grand Touring	700 cc	L. Galli	8'33"8	Fiat Abarth 700 BA
			Grand Touring	up to 1150 cc	T. Bindoni	7'54"8	Fiat Abarth 1000 BA
6 June	Avignon, France	Course de côte du Mont Ventoux (hillclimb-21,700 kms)	Prepared Touring	850 cc	S. Calascibetta	14'12"1	Fiat Abarth 850 TC
			Prepared Touring	1000 cc	A. Marzi	13'34"2	Fiat Abarth 1000 Berlina
			Sport	up to 1150 cc	W. Rufenacht	13'24"3	Fiat Abarth 1000
			Sport	from 1151 to 2000 cc	H. Herrmann	11'16"5	Fiat Abarth OT 2000
6 June	Frankfurt, Germany	Fast Spessart Bergrennen (hillclimb-4,000 kms)	Grand Touring	1300 cc	K. Geiss	2'07"3	Abarth Simca 1300
6 June	Trapani, Italy	Alcamo-Monte Bonifato (hillclimb-4,500 kms)	Touring	600 cc	De Francisci	4'00"7	Fiat Abarth 595
			Touring	850 cc	'G.A.'	3'26"9	Fiat Abarth 850 TC
			Touring	up to 1150 cc	G. Gianno	3'23"0	Fiat Abarth 1000 Berlina
			Grand Touring	700 cc	D. Giambianco	4'01"0	Fiat Abarth 700 BA
			Grand Touring	1000 cc	I. Capuano	3'30"7	Fiat Abarth 1000 BA
			Grand Touring	1300 cc	A. Gambero	3'00"1	Abarth Simca 1300
6 June	Torino, Italy	Cuorgné-Alpette (hillclimb-7,200 kms)	Touring	up to 700 cc	G. Dezutti	7'17"5	Fiat Abarth 595
			Touring	1000 cc	G. Vivarelli	6'13"3	Fiat Abarth 1000 Berlina
			Prepared Touring	up to 600 cc	A. Poggio	6'12"6	Fiat Abarth 595
			Prepared Touring	850 cc	B. Pagliero	5'34"9	Fiat Abarth 850 TC
			Prepared Touring	up to 1150 cc	A. Grillo	5'18"3	Fiat Abarth 1000 Berlina
			Grand Touring	1000 cc	P. Scantimburgo	5'28"4	Fiat Abarth 1000 BA
			Grand Touring	up to 1150 cc	F. Pozzo	5'30"5	Simca Abarth 1150
			Grand Touring	1300 cc	G. Giraudi	5'13"6	Abarth Simca 1300
			Sport	1000 cc	E. Gonella	5'18"2	Fiat Abarth 1000
6 June	Vallelunga, Italy	Coppa Fausto Pesci (circuit-1,800 kms x 20 laps)	Touring	1000 cc	A. Saltari	23'26"9	Fiat Abarth 1000 Berlina
			Grand Touring	up to 1600 cc	'Ellevu'	2"18"7	Fiat Abarth 1000 BA
6 June	Jazersko, Jugoslavia	Corsa hillclimb	Touring	850 cc	R. Fikfak	---	Fiat Abarth 850
7 June	Zandvoort, Holland	N.A.V. Kampioenschapsraces en Sportwagenraces (circuit-4,250 kms x 16 laps)	Touring	1000 cc	E. Swart	33'44"8	Fiat Abarth 1000 Berlina
7 June	Snetterton, United Kingdom	Nottingham S.C.C. at Silverstone (circuit-1,608 miles x 10 laps)	Grand Touring	from 1601 to 2500 cc	B. Burnard	11'36"0	Abarth Simca 2000
12 June	Salamanca, Spagna	Premio de Salamanca (circuit-2,000 kms x 20 laps)	Touring	850 cc	J. de Bagration	28'02"8	Fiat Abarth 850 TC

DATE	LOCATION	RACE	CATEGORY	CLASS	WINNER	TIME OR DISTANCE	CAR
12-13 June	Le Touquet, France	Rallye du Touquet (410 kms)	Touring	850 cc	Ruata-De Poleon	---	Fiat Abarth 850 TC
13 June	Palermo, Italy	Corsa al Monte Pellegrino (hillclimb-8,750 kms)	Touring	600 cc	A. Laluce	6'48"7	Fiat Abarth 595
			Touring	850 cc	S. Sancisa	5'53"0	Fiat Abarth 850 TC
			Touring	up to 1150 cc	A. Romano	5'45"8	Fiat Abarth 1000 Berlina
			Grand Touring	700 cc	F. Lisitano	6'35"4	Fiat Abarth 700 BA
			Grand Touring	1000 cc	I. Capuano	5'22"0	Fiat Abarth 1000 BA
			Grand Touring	1300 cc	A. Gambero	5'47"0	Abarth Simca 1300
13 June	Adenau, Germany	Nürburgring 6 Stunden (circuit-22,810 kms)	Touring	1000 cc	Ahrens-Steinmetz	kms 729,600	Fiat Abarth 1000 Berlina
13 June	Lodève, France	Course de côte de Lodève (hillclimb-5,400 kms)	Prepared Touring	850 cc	J. Raquin	4'35"1	Fiat Abarth 850 TC
			Grand Touring	up to 1150 cc	R. Barone	4'05"1	Fiat Abarth 1000 BA
13 June	Belvaux, Luxembourg	Slalom Automobile de Belvaux	Touring	from 851 to 1000 cc	M. Müller	1'56"1	Fiat Abarth 1000 Berlina
13 June	Berchtesgaden, Germany	Rossfeld Alpen Bergpreis (hillclimb-5,890 kms)	Touring	from 700 to 1000 cc	A. Marzi	7'54"4	Fiat Abarth 1000 Berlina
			Grand Touring	1000 cc	J. Kretschi	7'39"8	Fiat Abarth 1000 BA
13 June	Frascati, Italy	Frascati-Tuscolo (hillclimb-4,200 kms)	Touring	600 cc	E. Di Benedetto	3'39"6	Fiat Abarth 595
			Touring	850 cc	O. Costantini	3'18"7	Fiat Abarth 850 TC
			Touring	1000 cc	A. Saltari	3'15"6	Fiat Abarth 1000 Berlina
			Grand Touring	700 cc	'Jacopo'	3'16"7	Fiat Abarth 700 BA
			Grand Touring	1000 cc	N. Pica	3'25"8	Fiat Abarth 1000 BA
13 June	Asiago, Italy	Coppa Città di Asiago (hillclimb-14,500 kms)	Touring	up to 700 cc	'Napoleone'	10'47"6	Fiat Abarth 595
			Touring	850 cc	G. Gallo	9'48"3	Fiat Abarth 850 TC
			Touring	1000 cc	P. Lado	9'20"5	Fiat Abarth 1000 Berlina
			Grand Touring	700 cc	G. Varese	9'24"3	Fiat Abarth 700 BA
			Grand Touring	1000 cc	G. Rebaudi	8'46"0	Fiat Abarth 1000 BA
			Grand Touring	1300 cc	G. Dalla Torre	8'34"0	Abarth Simca 1300
13 June	Seinäjoki, Finlandia	Corsa in circuit (5,000 kms)	Grand Touring	1000 cc	L. Kinnunen	3'01"9	Fiat Abarth 1000 Berlina
13 June	Snetterton, United Kingdom	Whitmore Trophy Race Meeting (circuit-2,700 miles x 8 laps)	Touring	850 cc	G. Anstead	---	Fiat Abarth 850 TC
13 June	Vineland (NJ), USA	Vineland Raceway (circuit)	---	---	A. Cosentino	---	Fiat Abarth 850 TC
17 June	Vallelunga, Italy	Coppa AGIP (circuit-1,800 kms x 56 laps)	Touring	600 cc	L. Durst	1h09'26"7	Fiat Abarth 595
			Touring	1000 cc	G. Rebaudi	1h04'26"4	Fiat Abarth 1000 Berlina
			Grand Touring	1300 cc	R. Bertuzzi	1h04'18"2	Abarth Simca 1300
17 June	Brescia, Italy	Scalata al Colle di Sant'Eusebio (hillclimb-9,500 kms)	Touring	850 cc	S. Uberti	5'32"9	Fiat Abarth 850 TC
			Touring	1000 cc	C. Galimberti	5'31"6	Fiat Abarth 1000 Berlina
			Grand Touring	700 cc	G. Mainetti	5'26"3	Fiat Abarth 700 BA
			Grand Touring	1000 cc	A. Becchetti	5'09"7	Fiat Abarth 1000 BA
			Grand Touring	1300 cc	F. Rovetta	5'06"8	Abarth Simca 1300
			Grand Touring	up to 2500 cc	M. Pasotti	4'34"8	Abarth Simca 2000
17 June	Bologna, Italy	Coppa Acqua Cerelia (hillclimb-8,000 kms)	Touring	850 cc	G. Villani	6'07"4	Fiat Abarth 850 TC
			Touring	1000 cc	A. Stefanelli	6'06"8	Fiat Abarth 1000 Berlina
			Grand Touring	700 cc	'Jamaraschi'	6'10"5	Fiat Abarth 700 BA
			Grand Touring	1000 cc	E. Guarnieri	6'48"3	Fiat Abarth 1000 BA
			Grand Touring	1300 cc	'Micor'	5'59"1	Abarth Simca 1300
19 June	Roma, Italy	Competizione Tris-Pista d'Oro (circuit-1,000 kms)	Touring	1000 cc	M. Tommasini	2'20"5	Fiat Abarth 1000 Berlina
			Touring Special	up to 1000 cc	V. Trulli	2'18"9	Fiat Abarth 850 TC
19 June	Lime Rock (Connecticut), USA	Lime Rock National Race (circuit-10 laps)	---	---	A. Cosentino	---	Fiat Abarth 850 TC
19-20 June	Wasgau, Germany	ADAC Wasgau Bergprüfung (hillclimb-5,100 kms x 2)	Grand Touring	1000 cc	K. Geiss	5'41"4	Fiat Abarth 1000 BA
			Grand Touring	1300 cc	K. Federhofer	5'41"2	Abarth Simca 1300
20 June	Cuneo, Italy	Garessio-Colle San Bernardo (hillclimb-6,200 kms)	Touring	850 cc	S. Morando	4'41"2	Fiat Abarth 850 TC
			Touring	1000 cc	A. Grillo	4'36"9	Fiat Abarth 1000 Berlina
			Grand Touring	up to 1000 cc	G. Varese	4'37"7	Fiat Abarth 700 BA
			Grand Touring	1300 cc	G. Ramoino	4'37"3	Abarth Simca 1300
			Sport	1000 cc	G. Virgilio	4'18"8	Fiat Abarth 1000
20 June	Arezzo, Italy	Camuccia-Cortona (hillclimb-3,500 Km x 2)	Touring	600 cc	C. Bilenghi	5'47"2	Fiat Abarth 595
			Touring	850 cc	M. Tamburini	5'12"1	Fiat Abarth 850 TC
			Touring	up to 1150 cc	A. Stefanelli	5'05"7	Fiat Abarth 1000 Berlina
			Grand Touring	1000 cc	'Jamaraschi'	5'00"9	Fiat Abarth 1000 BA
			Grand Touring	1300 cc	'Micor'	4'49"2	Abarth Simca 1300
20 June	Parma, Italy	Castione Baratti-Neviano degli Arduini (hillclimb-6,000 kms)	Touring	850 cc	G. Baruffaldi	4'12"0	Fiat Abarth 850 TC
			Touring	up to 1150 cc	M. Nesti	3'49"4	Fiat Abarth 1000 Berlina
			Grand Touring	700 cc	B. Bocconi	4'18"3	Fiat Abarth 595
			Grand Touring	up to 2500 cc	C. Corini	3'30"3	Abarth Simca 2000
20 June	Metz, France	Course de côte d'Arry (hillclimb-1,200 kms)	Grand Touring	1000 cc	J. Kretschi	0'45"3	Fiat Abarth 1000 BA
			Grand Touring	1300 cc	R. Stierli	0'44"1	Abarth Simca 1300

DATE	LOCATION	RACE	CATEGORY	CLASS	WINNER	TIME OR DISTANCE	CAR
20 June	Molins de Rey, Spagna	Subida Santa Cruz de Olorde (hillclimb-5,700 kms)	Grand Touring	1000 cc	'Salvador'	4'08"6	Fiat Abarth 1000 BA
20 June	Valence, France	Course de côte du Pin (hillclimb-7,050 kms)	Touring Touring Grand Touring	850 cc 1000 cc 1000 cc	M. Cintolesi I. Giunti R. Barone	5'31"9 5'09"2 5'19"1	Fiat Abarth 850 TC Fiat Abarth 1000 Berlina Fiat Abarth 1000 BA
20 June	Lignières, Switzerland	Circuit de Lignières (circuit-1,400 kms x 20 laps)	Gran Touring	up to 1300 cc	H. Kennel	22'09"9	Fiat Abarth 1000 BA
26 June	Mühlehorn-Stocken, Switzerland	Bergprüfungsfahrt "Surprise" (hillclimb-2,000 kms)	Prepared Touring Grand Touring	850 cc 1000 cc	G. Naef K. Hauser	3'59"6 3'43"4	Fiat Abarth 850 TC Fiat Abarth 1000 Berlina Corsa
26 June	Zolder, Belgium	Coupes de Terlaemen (circuit-4,200 kms x 25 laps)	Touring Touring	850 cc 1000 cc	E. Swart L. Cella (overall)	52'33"8 51'24"1	Fiat Abarth 850 TC Fiat Abarth 1000 Berlina
26-27 June	Avus, Germany	Grosser Preis von Berlin (circuit-8,300 kms x 10 laps)	Touring Grand Touring	1000 cc 1300 cc	E. Bitter H. Dechent	31'43"0 27'23"4	Fiat Abarth 1000 Berlina Corsa Abarth Simca 1300
26-27 June	Marlboro (Maryland), USA	June Sprints Regional Championship Races (circuit-1,7 miles x 5 laps)	---	---	A. Cosentino	---	Fiat Abarth 850 TC
27 June	Vianden, Luxembourg	Course de côte Vianden (hillclimb-1,800 kms)	Touring Grand Touring Grand Touring	1000 cc 1000 cc 1300 cc	T. Pilette H. Affentranger P. Maron	1'46"1 1'45"0 1'39"6	Fiat Abarth 1000 Berlina Fiat Abarth 1000 BA Abarth Simca 1300
27 June	Strasbourg, France	Course de côte Turckheim Trois Epis (hillclimb-6,320 kms)	Touring Touring Grand Touring	850 cc 1000 cc up to 1150 cc	S. Stefani I. Giunti J. Kretschi	4'17"3 4'03"4 4'02"0	Fiat Abarth 850 TC Fiat Abarth 1000 Berlina Fiat Abarth 1000 BA
27 June	Latina, Italy	Valvisciolo-Bassiano (hillclimb-6,200 kms)	Touring Touring Touring Grand Touring	600 cc 850 cc 1000 cc 1000 cc	E. Di Benedetto E. Palma V. Trulli F. Amicucci	4'45"9 4'25"3 4'03"7 4'05"7	Fiat Abarth 595 Fiat Abarth 850 TC Fiat Abarth 1000 Berlina Fiat Abarth 1000 BA
27 June	Thun, Switzerland	Autoslalom Thun	Touring Prepared Touring Grand Touring	850 cc up to 850 cc 1000 cc	K. Schneiter R. Egli H. Kennel	6'43"1 6'45"4 6'15"0	Fiat Abarth 850 TC Fiat Abarth 750 Fiat Abarth 1000 BA
27 June	Luzern, Switzerland	Autoslalom Emmen	Touring	1000 cc	J. Tschopp	3'46"4	Fiat Abarth 1000 Berlina
27 June	San Cugat, Spagna	Carrera en cuesta San Cugat-Tibidabo (hillclimb-4,700 kms)	Grand Touring	1000 cc	'Salvador'	2'58"6	Fiat Abarth 1000 BA
27 June	Nîmes, France	Ronde Cevenole (rallye-400 kms)	Touring	1000 cc	P. Raymond	---	Fiat Abarth 1000 Berlina
27 June	Forlì, Italy	Predappio-Rocca delle Caminate	Touring Touring Touring Grand Touring Grand Touring Grand Touring	up to 700 cc 850 cc 1000 cc 1000 cc 1300 cc over 1600 cc	C. Bilenghi G. Villani M. Nesti 'Ellevu' C. Sangiorgi A. Brini	4'07"8 3'46"2 3'43"1 3'37"0 3'39"9 3'22"2	Fiat Abarth 595 Fiat Abarth 850 TC Fiat Abarth 1000 Berlina Fiat Abarth 1000 BA Abarth Simca 1300 Abarth Simca 2000
27 June	Kamnik, Jugoslavia	Circuito di Kamnik (2,000 kms)	Touring	850 cc	P. Knezevic	---	Fiat Abarth 850 TC
29 June	Cosenza, Italy	Coppa della Sila (hillclimb-14,000 kms)	Touring Touring Touring Grand Touring Grand Touring Grand Touring Grand Touring	600 cc 850 cc up to 1150 cc 700 cc 1000 cc 1300 cc up to 2500 cc	M. De Luca V. Bilotti 'Tre Teste' 'Jacopo' V. Cannataro G. Dalla Torre D. Scola	12'31"5 11'20"6 11'17"6 11'37"7 10'57"5 10'25"9 10'58"8	Fiat Abarth 595 Fiat Abarth 850 TC Fiat Abarth 1000 Berlina Fiat Abarth 700 BA Fiat Abarth 1000 BA Abarth Simca 1300 Abarth Simca 2000
3 July	Siracusa, Italy	3 Hours Night Race (circuit-5,500 kms)	Prepared Touring Prepared Touring Grand Touring	850 cc 1000 cc up to 2500 cc	E. Di Stefano A. Marendino A. Gambero	kms 120,772 kms 129,994 kms 135,569	Fiat Abarth 850 TC Fiat Abarth 1000 Berlina Abarth Simca 1300
3 July	Goodwood, United Kingdom	Brooklands Memorial Trophy (circuit-2,274 miles x 10 laps)	Grand Touring	from 1601 to 2500 cc	R. Burnard	16'15"8	Abarth Simca 2000
4 July	Woburn Park, United Kingdom	Woburn Park Hill Club (hillclimb-800 yards x 2)	Touring	from 901 to 1300 cc	M. Healey	0'21"3	Fiat Abarth 1000 Berlina
4 July	Banja Luka, Jugoslavia	Corsa hillclimb di Banja Luka (8,870 kms)	Touring	850 cc	S. Matevz	---	Fiat Abarth 850 TC
4 July	Rochefort la Tourne, Switzerland	Course de côte de la Tourne (hillclimb-5,000 kms)	Touring Grand Touring Grand Touring Grand Touring Sport	1000 cc 1000 cc 1300 cc from 1600 to 2000 cc 1000 cc	U. Schutz J. Kretschi D. Borel H. Kuhnis W. Rufenacht	3'50"5 3'30"6 3'08"4 2'55"2 3'06"1	Fiat Abarth 1000 Berlina Fiat Abarth 1000 BA Abarth Simca 1300 Abarth Simca 2000 Fiat Abarth 1000

DATE	LOCATION	RACE	CATEGORY	CLASS	WINNER	TIME OR DISTANCE	CAR
4 July	Zandvoort, Holland	Les Coupes de Benelux (circuit-4,225 kms x 12 laps)	Touring	1000 cc	E. Swart	23'15"1	Fiat Abarth 1000 Berlina
4 July	Nürnberg, Germany	ADAC Norisring Rennen (circuit-3,940 kms x 19 laps)	Grand Touring	1300 cc	H.D. Dechent	34'02"1	Abarth Simca 1300
4 July	Bolzano, Italy	Bolzano-Mendola (hillclimb-14,400 kms)	Touring	850 cc	G. Villani	12'26"4	Fiat Abarth 850 TC
			Touring	1000 cc	'Matich'	12'29"3	Fiat Abarth 1000 Berlina
			Grand Touring	700 cc	S. La Valle	12'49"4	Fiat Abarth 700 BA
			Grand Touring	1000 cc	G. Rebaudi	12'09"4	Fiat Abarth 1000 BA
			Grand Touring	1300 cc	G. Dalla Torre	11'38"7	Abarth Simca 1300
			Grand Touring	up to 2500 cc	H. Demetz (overall)	11'01"2	Abarth Simca 2000
4 July	Fiuggi, Italy	Targa Fiuggi (hillclimb-5,000 kms)	Touring	850 cc	G. Casati	3'37"5	Fiat Abarth 850 TC
			Touring	1000 cc	F. Palumbo	3'45"6	Fiat Abarth 1000 Berlina
			Grand Touring	1000 cc	F. Amicucci	3'37"3	Fiat Abarth 1000 BA
			Grand Touring	up to 1150 cc	R. Chianta	4'15"4	Simca Abarth 1150
			Grand Touring	up to 2500 cc	L. Marignani	3'30"3	Abarth Simca 2000
4 July	Graz, Austria	Stubenbergrennen (hillclimb-6,500 kms)	Touring	1000 cc	W. Pöltinger	4'39"7	Fiat Abarth 1000 Berlina
4 July	Pori, Finlandia	Corsa in circuit (5,000 kms)	Grand Touring	1000 cc	L. Kinnunen	3'12"6	Fiat Abarth 1000 Berlina
10-11 July	Mainz Finthen, Germany	HMSC Flugplatz Rennen (circuit-2,100 kms x 30 laps)	Touring	1000 cc	E. Swart	38'11"7	Fiat Abarth 1000 Berlin
			Grand Touring	1000 cc	W. Nolte	34'08"6	Fiat Abarth 1000 BA
			Grand Touring	1300 cc	H. Dechent	33'28"7	Abarth Simca 1300
10-11 July	Adenau, Germany	ADAC Girling Bergpreis Nürburgring (hillclimb-3,000 kms)	Touring	1000 cc	A. Ostermann	---	Fiat Abarth 1000 Berlina
			Grand Touring	from 700 to 1300 cc	H. Krause	---	Fiat Abarth 1000 BA
10-11 July	Long Island (New York), USA	Bridgehampton Circuit Inter-Club Championship Event (circuit-2,85 miles)	---	---	A. Cosentino	---	Fiat Abarth 850 TC
			---	---	R. Catalano	---	Fiat Abarth 1000 Berlina
10-11 July	Santiago, Spagna	Rallye Camino de Santiago (824,425 kms)	Grand Touring	1000 cc	'Runner'	---	Fiat Abarth 1000 Berlina
10-11 July	Gives, Belgium	Course de côte de Gives (hillclimb-2,400 kms)	Touring	1000 cc	T. Pilette	1'36"9	Fiat Abarth 1000 Berlina
			Grand Touring	1000 cc	J. Hauser	1'37"3	Fiat Abarth 1000 BA
11 July	Trento, Italy	Trento-Bondone (hillclimb-17,300 kms)	Touring	850 cc	G. Villani	14'27"7	Fiat Abarth 850 TC
			Touring	1000 cc	A. Marzi	14'00"8	Fiat Abarth1000 Berlina
			Grand Touring	1000 cc	G. Rebaudi	13'57"5	Fiat Abarth 1000 BA
			Grand Touring	1300 cc	G. Dalla Torre	13'39"5	Abarth Simca 1300
11 July	Teramo, Italy	Giro del Gran Sasso (regolarità-183,300 kms)	---	---	F. Severi (overall)	---	Fiat Abarth 850 TC
11 July	Ascoli Piceno, Italy	S. Benedetto del Tronto-Acquaviva Picena (hillclimb-7,000 kms)	Touring	850 cc	S. Fratticioli	5'11"0	Fiat Abarth 850 TC
			Touring	1000 cc	M. Tommasini	5'18"4	Fiat Abarth 1000 Berlina
			Grand Touring	1000 cc	N. Pica	5'19"2	Fiat Abarth 1000 Berlina
			Grand Touring	1300 cc	A. Brocchi	5'16"4	Abarth Simca 1300
			Grand Touring	over 1300 cc	L. Marignani	4'49"8	Abarth Simca 2000
11 July	Biella, Italy	Coppa Carlo Felice Trossi (hillclimb-6,500 kms)	Touring	850 cc	G. Bona	4'43"1	Fiat Abarth 850 TC
			Touring	up to 1150 cc	'Gibi'	4'35"1	Fiat Abarth 1000 Berlina
			Grand Touring	700 cc	P. Rolle	5'12"8	Fiat Abarth 595
			Grand Touring	1000 cc	P. Giacomini	4'29"3	Fiat Abarth 1000 Berlina
			Grand Touring	1300 cc	R. Ramoino	4'27"3	Abarth Simca 1300
			Grand Touring	over 1300 cc	A. Brini	4'15"6	Abarth Simca 2000
11 July	Reggio Calabria, Italy	Santo Stefano-Gambarie (hillclimb-7,830 kms)	Touring	up to 1150 cc	A. Marendino	5'59"0	Fiat Abarth 1000 Berlina
			Grand Touring	700 cc	G. Calabrese	6'41"6	Fiat Abarth 700 BA
			Grand Touring	up to 1150 cc	C. Avventurieri	5'53"5	Fiat Abarth 1000 BA
11 July	Graz, Austria	Bergrennen Reinischkogel (hillclimb-4,200 kms)	Touring	1000 cc	W. Pöltinger	3'11"6	Fiat Abarth 1000 Berlina
11 July	Tuttlingen, Germany	Gara di abilità di Tuttlingen	Touring	850 cc	K. Kaspryk	---	Fiat Abarth 850 TC
11 July	Lignières, Switzerland	Circuit de Lignières (circuit-1,400 kms x 10 laps)	Touring	850 cc	A. Wittenbach	11'04"9	Fiat Abarth 850 TC
			Touring	1000 cc	F. Wipf	11'02"3	Fiat Abarth 1000 Berlina
			Grand Touring	1000 cc	W. Friedrich	10'22"6	Fiat Abarth 1000 BA
13 July	Maribor, Jugoslavia	Circuito di Maribor (circuit-2,300 kms x 10 laps)	Touring	850 cc	S. Matevz	---	Fiat Abarth 850 TC
18 July	Zandvoort, Holland	Nationale Toerwagenraces (circuit-4,225 kms x 12 laps)	Touring	1000 cc	E. Swart	24'19"2	Fiat Abarth 1000 Berlina
18 July	Hockenheim, Germany	Nationales Rundstrecken (circuit-2,690 kms x 15 laps)	Grand Touring	1000 cc	J. Kretschi	23'55"7	Fiat Abarth 1000 BA
			Grand Touring	1300 cc	R. Stierli	24'34"6	Abarth Simca 1300
18 July	Termini Imerese, Italy	Termini-Caccamo (hillclimb-10,000 kms)	Grand Touring	1300 cc	V. Tipa	6'43"6	Abarth Simca 1300
18 July	Trieste, Italy	Trieste-Opicina (hillclimb-10,150 kms)	Touring	600 cc	L. Durst	6'17"0	Fiat Abarth 595
			Touring	850 cc	S. Morando	5'40"6	Fiat Abarth 850 TC
			Touring	1000 cc	P. Lado	5'20"8	Fiat Abarth 1000 Berlina
			Grand Touring	700 cc	S. La Valle	5'39"7	Fiat Abarth 700 BA
			Grand Touring	1000 cc	M. Cintolesi	5'23"1	Fiat Abarth 1000 BA
			Grand Touring	1300 cc	M. Nardari	5'16"8	Abarth Simca 1300
			Grand Touring	up to 2500 cc	'Matich'	4'52"3	Abarth Simca 2000
18 July	Bomeree, Belgium	Course de côte de Charleroi (hillclimb-2,000 kms)	Touring	1000 cc	T. Pilette	1'17"0	Fiat Abarth 1000 Berlina
			Grand Touring	1000 cc	J. Hauser	1'19"7	Fiat Abarth 1000 BA

DATE	LOCATION	RACE	CATEGORY	CLASS	WINNER	TIME OR DISTANCE	CAR
18 July	Magny Cours, France	Criterium de Vitesse (circuit-2,040 kms x 20 laps)	Touring	850 cc	F. Ruata	23'02"4	Fiat Abarth 850 TC
			Touring	up to 1150 cc	C. Ballot Lena	22'02"4	Fiat Abarth 1000 Berlina
			Grand Touring (30 laps)	up to 2000 cc	B. Lagier	31'14"5	Abarth Simca 2000
24-25 July	Bamberg, Germany	ADAC Ellerberg Rennen (hillclimb-6,080 kms)	Grand Touring	1000 cc	H. Krause	3'58"3	Fiat Abarth 1000 BA
			Grand Touring	1300 cc	K. Federhofer	3'56"0	Abarth Simca 1300
24-25 July	Trier, Germany	Trierer Flugplatzrennen (circuit-1,900 kms x 27 laps)	Touring	1000 cc	E. Bitter	30'38"5	Fiat Abarth 1000 Berlina
			Grand Touring	1000 cc	A. Schmalbach	29'34"2	Fiat Abarth 1000 BA
			Grand Touring	1300 cc	H. Dechent	29'43"8	Abarth Simca 1300
			Sport	1300 cc	H. Dechent	28'59"9	Abarth Simca 1300
25 July	Sestrière, Italy	Cesana-Sestrière (hillclimb- 10,400 kms)	Touring	600 cc	L. Durst	7'45"4	Fiat Abarth 595
			Touring	850 cc	S. Dal Cason	7'51"0	Fiat Abarth 850 TC
			Touring	1000 cc	G. Gavelli	7'42"1	Fiat Abarth 1000 Berlina
			Grand Touring	700 cc	G. Varese	6'51"1	Fiat Abarth 700 BA
			Grand Touring	1000 cc	G. Rebaudi	6'06"1	Fiat Abarth 1000 BA
			Grand Touring	1300 cc	G. Dalla Torre	6'11"1	Abarth Simca 1300
			Sport	1000 cc	L. Cella	6'06"1	Fiat Abarth OT 1000
25 July	Brindisi, Italy	Coppa Selva di Fasano (hillclimb-7,400 kms)	Touring	850 cc	V. Bilotti	4'34"4	Fiat Abarth 850 TC
			Touring	1000 cc	'Tre Teste'	4'26"7	Fiat Abarth 1000 Berlina
			Grand Touring	700 cc	G.B. Mainetti	4'30"6	Fiat Abarth 700 BA
			Grand Touring	1000 cc	G. Benini	4'20"0	Fiat Abarth 1000 BA
			Grand Touring	1300 cc	R. Arfé	4'20""4	Abarth Simca 1300
			Grand Touring	over 1300 cc	D. Scola	3'59"6	Abarth Simca 2000
25 July	Rouen, France	Course de côte des Andelys (hillclimb-1,380 kms)	Touring	from 401 to 1000 cc	C. Ballot Lena	0'47"1	Fiat Abarth 1000 Berlina
			Grand Touring	from 401 to 1000 cc	R. Barone	0'49"2	Fiat Abarth 1000 BA
			Grand Touring	1300 cc	D. Borel	0'43"2	Abarth Simca 1300
			Grand Touring	from 1301 to 2500 cc	B. Lagier	0'46"3	Abarth Simca 2000
25 July	Pergusa, Italy	Trofeo A.C. Siciliani (circuit-4,797 kms x 10 laps)	Touring	850 cc	S. Danzuso	19'29"3	Fiat Abarth 850 TC
			Touring (15 laps)	up to 1150 cc	'Jenko'	27'03"1	Fiat Abarth 1000 Berlina
25 July	Cascais, Portugal	Prova Automovel Club de Portugal (circuit-3,450 kms x 22 laps)	Touring	850 cc	C. Flores	47'07"3	Fiat Abarth 850 TC
31 July	Adenau, Germany	ADAC Nürburgring Trophäe (circuit-22,810 kms x 5 laps)	Touring	1000 cc	E. Bitter	55'12"3	Fiat Abarth 1000 Berlina
1 August	Innsbruck, Austria	Bergrennen zur Axamer Lizum (hillclimb-8,300 kms)	Touring	850 cc	L. Cella	9'20"7	Fiat Abarth 850 TC
			Touring	1000 cc	H. Herrmann	9'02"5	Fiat Abarth 1000 Berlina
			Grand Touring	1000 cc	H. Affentranger	8'50"5	Fiat Abarth 1000 BA
			Grand Touring	1300 cc	P. Maron	8'24"9	Abarth Simca 1300
1 August	Grenoble, France	Course de côte Chamrousse (hillclimb-17,200 kms)	Touring	850 cc	S. Stefani	12'50"3	Fiat Abarth 850 TC
			Touring	1000 cc	A. Raymond	12'39"0	Fiat Abarth 1000 Berlina
			Grand Touring	1300 cc	D. Borel	13'27"5	Abarth Simca 1300
			Grand Touring	from 1600 to 2000 cc	F. Ruata	13'32"0	Abarth Simca 2000
1 August	Belluno, Italy	Agordo-Frassenè (hillclimb-9,000 kms)	Touring	up to 700 cc	A. Triboldi	7'55"2	Fiat Abarth 595
			Touring	850 cc	G. Villani	6'55"1	Fiat Abarth 850 TC
			Touring	1000 cc	P. Lado	6'49"6	Fiat Abarth 1000 Berlina
			Grand Touring	700 cc	L. Galli	7'22"0	Fiat Abarth 700 BA
			Grand Touring	1000 cc	A. Becchetti	7'05"0	Fiat Abarth 1000 BA
			Grand Touring	1300 cc	G. Dalla Torre	6'26"1	Abarth Simca 1300
			Grand Touring	from 1600 to 2000 cc	'Matich'	6'30"5	Abarth Simca 2000
1 August	Trapani, Italy	Trapani-Monte Erice (hillclimb-6,670 kms)	Touring	600 cc	D. Stabile	5'53"5	Fiat Abarth 595
			Touring	850 cc	G. Adragna	5'11"0	Fiat Abarth 850 TC
			Touring	1000 cc	A. Romano	5'07"9	Fiat Abarth 1000 Berlina
			Grand Touring	700 cc	F. Lisitano	5'41"3	Fiat Abarth 700 BA
			Grand Touring	1000 cc	A. Rizzo	5'10"7	Fiat Abarth 1000 BA
			Sport	1000 cc	P. Ciraolo	5'52"6	Fiat Abarth 1000
1 August	Adenau, Germany	AVD Rheinland Pfalz-Preis (Nürburgring) (circuit-22,810 kms x 7 laps)	Grand Touring	1300 cc	H. Dechent	1h15'30"4	Abarth Simca 1300
1 August	Viterbo, Italy	Corsa Lago di Montefiascone (hillclimb-5,000 kms)	Touring	850 cc	M. Tamburini	2'57"1	Fiat Abarth 850 TC
			Touring	1000 cc	M. Tommasini	2'58"8	Fiat Abarth 1000 Berlina
			Grand Touring	1000 cc	G. Spicciani	2'46"8	Fiat Abarth 1000 BA
			Sport	1000 cc	'Mister Brandy'	2'45"5	Fiat Abarth 1000
7-8 August	Karlskoga, Sweden	Canonloppet (circuit-3,000 kms x 35 laps)	Touring	850 cc	H. Koster	1h01'56"4	Fiat Abarth 850 TC
			Touring	1000 cc	E. Swart	1h00'16"9	Fiat Abarth 1000 Berlina

DATE	LOCATION	RACE	CATEGORY	CLASS	WINNER	TIME OR DISTANCE	CAR
8 August	Avezzano, Italy	Trofeo Micangeli (hillclimb-6,000 kms)	Touring	850 cc	S. Fraticcioli	3'45"3	Fiat Abarth 850 TC
			Touring	up to 1150 cc	B. Faenza	3'46"6	Fiat Abarth 1000 Berlina
			Grand Touring	from 700 to 1150 cc	'Ellevu'	3'31"4	Fiat Abarth 1000 BA
			Grand Touring	1300 cc	S. Ridolfi	3'28"3	Abarth Simca 1300
8 August	Cagliari, Italy	Corongiu-Campuomu (hillclimb-4,000 kms)	Prepared Touring	600 cc	B. Massida	3'43"2	Fiat Abarth 595
			Prepared Touring	850 cc	G. Vacca	3'18"2	Fiat Abarth 850 TC
			Prepared Touring	1000 cc	F. Locci	3'31"2	Fiat Abarth 1000 Berlina
			Grand Touring	1000 cc	'Lubi' (overall)	3'12"3	Fiat Abarth 1000 BA
			Grand Touring	over 1000 cc	G. Ambu	3'29"4	Abarth Simca 1300
8 August	Monopoli, Italy	Coppa Città di Monopoli (hillclimb-4,500 kms)	Touring	850 cc	'Gordon'	2'34"4	Fiat Abarth 850 TC
			Touring	1000 cc	'Tre Teste'	2'30"8	Fiat Abarth 1000 Berlina
			Grand Touring	1000 cc	G. Benini	2'27"2	Fiat Abarth 1000 BA
			Grand Touring	1300 cc	P. Gargano	2'22"3	Abarth Simca 1300
			Grand Touring	2000 cc	D. Scola (overall)	2'14"7	Abarth Simca 2000
8 August	Freiburg, Germany	ADAC Bergpreis Freiburg-Schauinsland (hillclimb-11,200 kms)	Grand Touring	1000 cc	S. Spiess	15'22"4	Fiat Abarth 1000 BA
			Grand Touring	1300 cc	R. Stierli	15'11"4	Abarth Simca 1300
			Sport	1000 cc	L. Cella	14'19"1	Fiat Abarth OT 1000
8 August	Porto d'Ascoli, Italy	Gara Sprint (circuit-0,500 kms)	Touring	700 cc	E. Marconi	25"42	Fiat Abarth 595
			Touring	1000 cc	R. Rubini	22"09	Fiat Abarth 1000 Berlina
			Prepared Touring	850 cc	F. Massetti	21"14	Fiat Abarth 850 TC
			Prepared Touring	1000 cc	A. Liberati	23'44"0	Fiat Abarth 1000 Berlina
8 August	Murska Sobota, Jugoslavia	Corsa di velocità su strada	Touring	850 cc	J. Malezic	---	Fiat Abarth 850 TC
8 August	London (New Hampshire), USA	Bryar Motorsport Park (circuit-1,600 miles x 30')	---	---	A. Cosentino	---	Fiat Abarth 850 TC
8 August	Castro, Spagna	Rallye Las Rias Bajas	---	---	P. De Leonibus	12'55"4	Abarth Simca 1300
14 August	Roskilde, Danimarca	G. P. Kobenhavn (circuit-1,400 kms x 8 laps x 2)	Touring	1000 cc	T. Pilette	20'27"7	Fiat Abarth 1000 Berlina
14 August	Los Molinos, Spagna	Rallye Los Molinos	---	---	J. de Bagration (overall)	---	Fiat Abarth 850 TC
15 August	Mont-Dore, France	Course de côte du Mont-Dore (hillclimb-5,000 kms)	Touring	850 cc	J. Brun	4'14"0	Fiat Abarth 850 TC
			Touring	1000 cc	C. Ballot Lena	3'46"2	Fiat Abarth 1000 Berlina
			Grand Touring	2000 cc	F. Ruata	3'25"3	Abarth Simca 2000
15 August	Snetterton, United Kingdom	500 Km di Snetterton (circuit-2,700 miles)	Touring	1000 cc	J. Anstead	3h57'34"8	Fiat Abarth 1000 Berlina
15 August	Nürnberg, Germany	ADAC Bergrennen Happurg (hillclimb-3,130 kms x 2)	Grand Touring	1000 cc	H. Krause	3'18"8	Fiat Abarth 1000 BA
15 August	Soria, Spagna	Rallye a Pinares de Alto Duero	---	---	de Bagration-Barcenas (overall)	---	Fiat Abarth 850 TC
15 August	Pescara, Italy	Cronoscalata Svolte di Popoli (hillclimb-8,000 kms)	Touring	850 cc	S. Del Cason	5'37"4	Fiat Abarth 850 TC
			Touring	1000 cc	S. Genovesi	5'34"2	Fiat Abarth 1000 Berlina
			Grand Touring	700 cc	A. D'Antinone	6'03"6	Fiat Abarth 700 BA
			Grand Touring	1000 cc	'Ellevu'	5'16"8	Fiat Abarth 1000 BA
			Grand Touring	1300 cc	'Nicor'	5'06"9	Abarth Simca 1300
			Grand Touring	over 1600 cc	'Gim'	5'08"2	Abarth Simca 2000
			Sport	1000 cc	'M. Brandy'	5'07"7	Fiat Abarth 1000
22 August	Alcalà de Henares, Spagna	Subida al Gurugu (hillclimb)	Touring	850 cc	J. de Bagration	1'52"0	Fiat Abarth 850 TC
22 August	Saint Lary, France	Course de côte de Saint Lary (hillclimb-5,200 kms)	Touring	1000 cc	C. Ballot Lena	3'08"8	Fiat Abarth 1000 Berlina
			Grand Touring	1000 cc	Y. Barone	3'22"3	Fiat Abarth 1000 BA
			Grand Touring	1300 cc	R. Barone	3'01"7	Abarth Simca 1300
			Grand Touring	2000 cc	F. Ruata	2'46"2	Abarth Simca 2000
22 August	Regensburg, Germany	Regensburg Bergrennen (hillclimb-3,700 kms x 2)	Touring	850 cc	J. Abt	2'21"6	Fiat Abarth 850 TC
			Grand Touring	1000 cc	K. Geiss	4'11"8	Fiat Abarth 1000 BA
22 August	Le Marquisard, Belgium	Course de côte du Marquisard (hillclimb-2,850 kms)	Touring	850 cc	G. Harris	2'13"1	Fiat Abarth 850 TC
			Touring	1000 cc	T. Pilette	2'07"4	Fiat Abarth 1000 Berlina

DATE	LOCATION	RACE	CATEGORY	CLASS	WINNER	TIME OR DISTANCE	CAR
22 August	Saint Ursanne, Switzerland	Course de côte Saint Ursanne-Les Rangiers (hillclimb-4,700 kms x 2)	Prepared Touring	850 cc	I. Giunti	6'14"1	Fiat Abarth 850 TC
			Prepared Touring	1000 cc	E. Swart	5'48"9	Fiat Abarth 1000 Berlina
			Grand Touring	700 cc	H. Affentranger	6'32"7	Fiat Abarth 700 BA
			Grand Touring	1000 cc	J. Kretschi	5'40"5	Fiat Abarth 1000 BA
			Grand Touring	1300 cc	D. Borel	5'50"9	Abarth Simca 1300
			Sport	1000 cc	W. Rufenacht	5'22"6	Fiat Abarth OT 1000
			Sport	1300 cc	H. Zweifel	5'39"6	Fiat Abarth 1300
22 August	Brescia, Italy	Trofeo Valle Camonica (hillclimb-8,600 kms)	Touring	850 cc	G. Vedovello	6'36"6	Fiat Abarth 850 TC
			Touring	1000 cc	B. Guarneri	6'10"7	Fiat Abarth 1000 Berlina
			Grand Touring	700 cc	G. Mainetti	6'45"3	Fiat Abarth 700 BA
			Grand Touring	1000 cc	M. Baldo	6'32"7	Fiat Abarth 1000 BA
			Grand Touring	1300 cc	R. Bertuzzi	6'37"6	Abarth Simca 1300
			Grand Touring	up to 2500 cc	G.F. Rovetta	6'28"0	Abarth Simca 2000
22 August	Ascoli Piceno, Italy	Corsa al Colle San Marco (hillclimb-10,200 kms)	Touring	600 cc	G.P. Caroli	7'44"0	Fiat Abarth 595
			Touring	850 cc	G. Villani	6'47"4	Fiat Abarth 850 TC
			Touring	1000 cc	A. Marzi	6'31"3	Fiat Abarth 1000 Berlina
			Grand Touring	1000 cc	G. Rebaudi	6'29"1	Fiat Abarth 1000 BA
			Grand Touring	1300 cc	R. Arfé	6'36"3	Abarth Simca 1300
			Grand Touring	up to 2500 cc	'Nann'	6'42"1	Abarth Simca 2000
22 August	Mamer, Luxembourg	Slalom Mamer	Grand Touring	up to 1600 cc	J. Hauser	2'13"2	Fiat Abarth 1000 BA
22 August	Snetterton, United Kingdom	National Benzole Trophy (circuit-2,700 kms x 10 laps)	Grand Touring	over 1600 cc	R. Burnard	19'23"0	Abarth Simca 2000
29 August	Zandvoort, Holland	Zandvoort Trophy (circuit-4,225 kms x 24 laps)	Touring	850 cc	K. Ahrens	48'54"8	Fiat Abarth 850 TC
			Touring	1000 cc	E. Swart	48'02"8	Fiat Abarth 1000 Berlina Corsa
29 August	Villars, Switzerland	Course de côte Ollon-Villars (hillclimb-8,000 kms x 2)	Grand Touring	1000 cc	J. Kretschi	10'24"6	Fiat Abarth 1000 BA
			Grand Touring	1300 cc	D. Borel	9'38"7	Abarth Simca 1300
			Sport-Prototypes	1000 cc	W. Rufenacht	9'45"2	Fiat Abarth 1000
			Sport-Prototypes	1300 cc	H. Zweifel	9'39"4	Fiat Abarth 1300
			Sport-Prototypes	1600 cc	H. Herrmann	9'04"9	Fiat Abarth OT 1600
			Corsa	from 1500 to 2000 cc	H. Demetz	8'44"4	Fiat Abarth OT 2000
29 August	Villefranche sur Saône, France	Course de côte de Beaujolais (hillclimb-16,000 kms)	Touring	850 cc	M. Cintolesi	11'00"4	Fiat Abarth 850 TC
			Grand Touring	1300 cc	R. Barone	10"29"3	Abarth Simca 1300
			Grand Touring	2000 cc	P. Maublanc	9'49"3	Abarth Simca 2000
29 August	Chieti, Italy	Coppa Città di Chieti (hillclimb-6,200 kms)	Touring	850 cc	G. Gentile	3'53"6	Fiat Abarth 850 TC
			Touring	1000 cc	A. Saltari	3'37"3	Fiat Abarth 1000 Berlina
			Grand Touring	1000 cc	A. D'Antinone	4'06"4	Fiat Abarth 1000 Ba
			Grand Touring	1300 cc	P. Laureati	3'32"1	Abarth Simca 1300
			Sport	1000 cc	G. Tommasi	3'23"4	Fiat Abarth 1000
29 August	Veldenz, Germany	ADAC Hunsrück Bergrennen (hillclimb-3,600 kms x 2)	Grand Touring	1300 cc	H. Iffert	5'00"9	Abarth Simca 1300
30 August	Brands Hatch, United Kingdom	The Redex Trophy (circuit-4,625 miles x 20 laps)	Grand Touring	from 1600 to 2500 cc	R. Burnard	37'01"6	Abarth Simca 2000
4-5 September	Le Chable Bruson, Switzerland	Course de côte Le Chable-Bruson (hillclimb-2,400 kms x 2)	Touring	850 cc	J.P. Brun	5'33"8	Fiat Abarth 850 TC
			Prepared Touring	1000 cc	M. Piller	5'39"9	Fiat Abarth 1000 Berlina
			Grand Touring	over 1600 cc	H. Kuhnis	4'48"7	Abarth Simca 2000
5 September	Rouen, France	Course de côte de Belbeuf (hillclimb-2,800 kms)	Touring	850 cc	P. Henning	2'05"2	Fiat Abarth 850 TC
			Touring	up to 1150 cc	C. Ballot Lena	1,55"4	Fiat Abarth 1000 Berlina
			Grand Touring	1000 cc	J. Mazzanti	1'56"3	Fiat Abarth 1000 BA
			Grand Touring	from 1600 to 2500 cc	B. Lagier	1'49"4	Abarth Simca 2000
5 September	San Lorenzo de Savall, Spagna	Carrera en cuesta San Feliu de Codinas (hillclimb-4,200 kms)	Touring	850 cc	'Jorge Chi'	3'33"6	Fiat Abarth 850 TC
			Prepared Touring	up to 1150 cc	'Runner'	3'20"4	Fiat Abarth 1000 Berlina
5 September	Augsburg, Germany	Augusta Bergrennen (hillclimb- 2,100 kms x 2)	Touring	1000 cc	M. Endress	2'23"2	Fiat Abarth 1000 Berlina
			Grand Touring	1000 cc	H. Krause	2'25"3	Fiat Abarth 1000 BA
			Grand Touring	1300 cc	K. Federhofer	2'17"5	Abarth Simca 1300
5 September	Ceyreste, France	Course de côte de Ceyreste (hillclimb-5,200 kms)	Touring	600 cc	J. Chocquel	4'44"8	Fiat Abarth 595
			Touring	850 cc	A. Rizzo	4'10"5	Fiat Abarth 850 TC
			Touring	1000 cc	A. Romano	3'56"2	Fiat Abarth 1000 Berlina
			Grand Touring	up to 850 cc	J. Raquin	4'12"4	Fiat Abarth 700 BA
			Grand Touring	2000 cc	P. Maublanc	3'28"3	Abarth Simca 2000
5 September	Nancy, France	Course de côte de Belleau (hillclimb-2,000 kms)	Grand Touring	1000 cc	J. Hauser	1'38"4	Fiat Abarth 1000 BA
			Grand Touring	1300 cc	R. Stierli	1'33"1	Abarth Simca 1300

DATE	LOCATION	RACE	CATEGORY	CLASS	WINNER	TIME OR DISTANCE	CAR
5 September	Adenau, Germany	Internationales 500 Km Nürburgring Rennen (circuit-22,810 kms x 22 laps)	Touring	850 cc	E. Swart	3h50'43"7	Fiat Abarth 850 TC
			Touring	1000 cc	J. Jüttner	3h46'31"2	Fiat Abarth 1000 Berlina
			Grand Touring	1000 cc	J. Neerpasch	3h48'56"4	Fiat Abarth 1000 BA
			Grand Touring	1300 cc	K. Ahrens	3h46'13"0	Abarth Simca 1300
			Sport-Prototypes	850 cc	E. Furtmayr	3h44'10"7	Fiat Abarth 850
			Sport-Prototypes	1000 cc	L. Cella	3h44'20"7	Fiat Abarth OT 1000
5 September	Brescia, Italy	Trofeo Lumezzane (hillclimb-8,500 kms)	Touring	700 cc	V. Gatti	5'25"2	Fiat Abarth 595
			Touring	850 cc	P. Cazzullo	5'05"5	Fiat Abarth 850 TC
			Touring	1000 cc	C. Galimberti	5'00"6	Fiat Abarth 1000 Berlina
			Grand Touring	700 cc	G. Mainetti	5'00"3	Fiat Abarth 700 BA
			Grand Touring	1000 cc	G. Giacomini	4'48"1	Fiat Abarth 1000 BA
			Grand Touring	1300 cc	R. Bertuzzi	4'30"6	Abarth Simca 1300
5 September	Savona, Italy	Giro dei Monti Savonesi (255 kms)	---	---	V. Pené (overall)	---	Fiat Abarth 1000 Berlina
5 September	Macerata, Italy	Coppa Città di Tolentino (hillclimb-4,100 kms)	Touring	600 cc	M. Longarino	3'25"3	Fiat Abarth 595
			Touring	700 cc	R. Borgiani	3'03"2	Fiat Abarth 695
			Touring	850 cc	P. Bianchi	2'50"3	Fiat Abarth 850 TC
			Touring	1000 cc	V. Trulli	2'43"2	Fiat Abarth 1000 Berlina
			Grand Touring	1300 cc	P. Laureati	2'30"0	Abarth Simca 1300
5 September	Arezzo, Italy	Pieve S. Stefano-Passo dello Spino (hillclimb-12,700 kms)	Touring	850 cc	G. Casati	8'07"9	Fiat Abarth 850 TC
			Touring	1000 cc	F. Ricci	8'24"4	Fiat Abarth 1000 Berlina
			Grand Touring	up to 1000 cc	C. Bilenghi	8'54"6	Fiat Abarth 595
			Grand Touring	1300 cc	L. Baldini	7'32"0	Abarth Simca 1300
			Grand Touring	over 1600 cc	'Luca Fabri'	7'51"3	Abarth Simca 2000
5 September	Elmore Plain, United Kingdom	Elmore Plain Sport (circuit)	Touring	850 cc	W. Pratt	1'24"2	Fiat Abarth 850 TC
5 September	Lignières, Switzerland	Circuit de Lignières (1,400 kms x 2)	Touring	1000 cc	R. Egli	---	Fiat Abarth 750
5 September	Beograd, Jugoslavia	Circuito di Belgrado (2,300 kms x 15 laps)	Touring	850 cc	S. Matevz	---	Fiat Abarth 850 TC
10-11 September	Vaals, Holland	Heuvelklim Vaals (hillclimb-1,200 kms)	Prepared Touring	1000 cc	E. Swart	0'57"9	Fiat Abarth 1000 Berlina
			Grand Touring	1000 cc	J. Hauser	1'00"5	Fiat Abarth 1000 BA
11 September	Brighton, United Kingdom	Brighton Speed Trials	Grand Touring	from 1600 to 2000 cc	R. Burnard	25'27"0	Abarth Simca 2000
11-12 September	Miltholz, Switzerland	Bergrennen Miltholz-Kandersteg (hillclimb-2,640 kms x 2)	Prepared Touring	850 cc	M. Lanz	4'24"0	Fiat Abarth 850 TC
			Grand Touring	700 cc	H. Affentranger	4'12"2	Fiat Abarth 700 BA
			Grand Touring	1000 cc	J. Kretschi	3'57"7	Fiat Abarth 1000 BA
			Grand Touring	1300 cc	D. Borel	3'41"4	Abarth Simca 1300
			Grand Touring	2000 cc	E. Kuhnis	3'30"5	Abarth Simca 2000
			Sport	1000 cc	W. Rufenacht	3'38"4	Fiat Abarth 1000
12 September	Dijon, France	Course de côte d'Urcy (hillclimb-2,500 kms)	Touring	1000 cc	C. Ballot Lena	2'02"3	Fiat Abarth 1000 Berlina
			Grand Touring	1300 cc	'Celestin'	2'03"4	Abarth Simca 1300
			Grand Touring	1000 cc	J. Mazzanti	1'57"0	Fiat Abarth 1000 Bialbero
			Sport	over 1600 cc	J. Guichet	1'46"3	Fiat Abarth 2000
12 September	Zwieselstein, Austria	Timmelsjoch Hochalpen Bergrennen (hillclimb- 12,100 kms)	Touring	850 cc	J. Abt	10'23"5	Fiat Abarth 850 TC
			Touring	up to 1000 cc	M. Endress	10'02"6	Fiat Abarth 850 TC
12 September	Kaiserslautern, Germany	Rheinland Pfalz Bergpreis (hillclimb-3,050 kms)	Grand Touring	1000 cc	H. Krause	3'56"3	Fiat Abarth 1000 BA
12 September	Born, Luxembourg	Course de côte de Born (hillclimb-1,885 kms)	Touring	850 cc	E. Fauser	1'38"4	Fiat Abarth 850 TC
			Grand Touring	1000 cc	J. Hauser	1'34"0	Fiat Abarth 1000 BA
12 September	Frosinone, Italy	Guarcino-Campocatino (hillclimb-16,800 kms)	Touring	850 cc	M. Terminiello	12'38"1	Fiat Abarth 850 TC
			Touring	1000 cc	P. Finzi	12'25"4	Fiat Abarth 1000 Berlina
			Grand Touring	1000 cc	F. Amicucci	12'26"2	Fiat Abarth 1000 BA
			Grand Touring	1300 cc	P. Laureati	12'07"3	Abarth Simca 1300
12 September	Brands Hatch, United Kingdom	British Automobile Racing Club (circuit-4,265 miles x 10 laps)	Grand Touring	from 1600 to 2500 cc	R. Burnard	10'01"8	Abarth Simca 2000
18-19 September	Mont Verdun, France	Course de côte Limonest-Mt. Verdun (hillclimb-2,675 kms)	Touring	850 cc	J. Raquin	2'32"0	Fiat Abarth 850 TC
			Touring	1000 cc	A. Raymond	2'17"0	Fiat Abarth 1000 Berlina
			Grand Touring	2000 cc	P. Maublanc	2'05"0	Abarth Simca 2000
19 September	Houyet, Belgium	Course de côte de la Houyet (hillclimb-2,500 kms)	Touring	850 cc	G. Harris	1'53"3	Fiat Abarth 850 TC
			Touring	1000 cc	T. Pilette	1'46"0	Fiat Abarth 1000 Berlina
19 September	Gaisberg, Austria	Grosser Bergpreis von Österreich (hillclimb-8,625 kms)	Touring	1000 cc	M. Endress	12'02"8	Fiat Abarth 1000 Berlina
			Grand Touring	1000 c	J. Kretschi	11'53"5	Fiat Abarth 1000 BA
			Grand Touring	1300 cc	K. Federhofer	11'24"2	Abarth Simca 1300
19 September	Aosta, Italy	Aosta-Pila (hillclimb-11,200 kms)	Touring	700 cc	L. Durst	9'03"3	Fiat Abarth 595
			Touring	850 cc	A. Trenti	7'57"5	Fiat Abarth 850 TC
			Touring	up to 1150 cc	A. Marzi	7'39"5	Fiat Abarth 1000 Berlina
			Grand Touring	700 cc	G. Varese	8'05"1	Fiat Abarth 700 BA
			Grand Touring	1000 cc	G. Rebaudi	7'24"1	Fiat Abarth 1000 BA
			Grand Touring	1300 cc	G. Dalla Torre	7'22"5	Abarth Simca 1300
			Sport	1000 cc	K. Steinmetz	7'17"5	Fiat Abarth OT 1000
			Sport	2000 cc	H. Herrmann	6'53"0	Fiat Abarth OT 2000

DATE	LOCATION	RACE	CATEGORY	CLASS	WINNER	TIME OR DISTANCE	CAR
19 September	Osimo, Italy	Coppa Luigi Fagioli (hillclimb-5,270 kms)	Touring	850 cc	F. Cardinali	3'36"2	Fiat Abarth 850 TC
			Touring	1000 cc	G. Caraceni	3'30"4	Fiat Abarth 1000 Berlina
			Grand Touring	1000 cc	G. Spicciani	3'41"2	Fiat Abarth 1000 BA
			Grand Touring	1300 cc	P. Laureati	3'25"2	Abarth Simca 1300
			Grand Touring	up to 2500 cc	A. Brini	3'16"3	Abarth Simca 2000
			Sport	1000 cc	G. Tommasi	3'22"1	Fiat Abarth 1000
19 September	Montlhéry, France	Coupes de Paris (circuit-3,360 kms x 15 laps)	Touring	850 cc	P. Henning	28'49"5	Fiat Abarth 850 TC
			Touring	1000 cc	C. Ballot Lena	28'27"3	Fiat Abarth 1000 Berlina
			Grand Touring	1000 cc	F. Bichat	27'58"1	Fiat Abarth 1000 BA
19 September	Sondrio, Italy	Coppa Valtellina (166 kms)	---	---	A. Villiger (overall)	---	Fiat Abarth 850 TC
19 September	Caltanissetta, Italy	Coppa Nissena (hillclimb-10,000 kms)	Touring	600 cc	D. Stabile	6'43"4	Fiat Abarth 595
			Touring	700 cc	R. Parodi	6'29"3	Fiat Abarth 695
			Touring	850 cc	G. Villani	5'51"4	Fiat Abarth 850 TC
			Touring	1000 cc	P. Lado	5'41"7	Fiat Abarth 1000 Berlina
			Grand Touring	1000 cc	A. Rizzo	6'06"0	Fiat Abarth 1000 BA
			Grand Touring	1300 cc	A. Gambero	5'37"8	Abarth Simca 1300
19 September	Carlow, Irlanda	Carlow Autocross (circuit-3 laps)	Touring	850 cc	P. Curran	2'32"8	Fiat Abarth 850 TC
25-26 September	Hockenheim, Germany	Rundstreckenrennen Preis von Hockenheim (circuit-6,768 kms x 10 laps)	Touring	850 cc	M. Piller	34'07"9	Fiat Abarth 850 TC
			Sport	1000 cc	W. Rufenacht	30'03"4	Fiat Abarth 1000
26 September	Bruxelles, Belgium	Championnat de Belgique (circuit-2,000 kms x 2)	Touring	1000 cc	G. Harris	3'07"9	Fiat Abarth 1000 Berlina
			Grand Touring	up to 1300 cc	J. Hauser	3'39"0	Fiat Abarth 1000 BA
26 September	Neuffen, Germany	Bergpreis Schwäbische Alb (hillclimb-4,200 kms x 2)	Touring	850 cc	J. Abt	6'07"6	Fiat Abarth 850 T
			Grand Touring	1300 cc	H. Iffert	5'44"0	Abarth Simca 1300
26 September	Sewen, France	Course de côte de Sewen-Lac d'Alfeld (hillclimb-6,000 kms)	Touring	850 cc	J. Raquin	5'11"2	Fiat Abarth 850 TC
26 September	Mallorca, Spagna	Carrera en cuesta Puig Major (hillclimb-8,000 kms)	Touring	700 cc	M. Juncosa	6'22"5	Fiat Abarth 695 SS
			Touring	850 cc	'Jorge Chi'	6'00"1	Fiat Abarth 850 TC
			Prepared Touring	1000 cc	J. Juncosa	5'52"4	Fiat Abarth 1000 Berlina
			Grand Touring	850 cc	R. Marsans	5'35"3	Fiat Abarth 850 TC corsa
			Grand Touring	1000 cc	'Runner'	5'24"8	Fiat Abarth 1000 BA
26 September	Siena, Italy	Coppa del Chianti classico (hillclimb-5,800 kms)	Touring	600 cc	G. Caroli	4'24"9	Fiat Abarth 595
			Touring	850 cc	A. Monselles	4'04"9	Fiat Abarth 850 TC
			Grand Touring	1000 cc	'Jamarashi'	4'09"7	Fiat Abarth 1000 BA
			Grand Touring	1300 cc	L. Baldini	3'55"6	Abarth Simca 1300
			Grand Touring	over 1600 cc	A. Brini	3'54"8	Abarth Simca 2000
26 September	Belluno, Italy	Coppa Alpe del Nevegal (hillclimb-7,500 kms)	Touring	up to 700 cc	A. Minen	6'50"3	Fiat Abarth 595
			Touring	850 cc	C. Margoni	6'15"3	Fiat Abarth 850 TC
			Touring	1000 cc	G. Nart	6'15"1	Fiat Abarth 1000 Berlina
			Grand Touring	700 cc	G. Mainetti	6'16"2	Fiat Abarth 700 BA
			Grand Touring	1000 cc	M. Baldo	6'00"1	Fiat Abarth 1000 BA
			Grand Touring	1300cc	G. Dalla Torre	5'28"2	Abarth Simca 1300
			Sport	2000 cc	'Matich'	5'26"3	Abarth Simca 2000
26 September	Genova, Italy	Mignanego-Passo dei Giovi (hillclimb-4,000 kms)	Touring	600 cc	L. Durst	3'09"9	Fiat Abarth 595
			Touring	700 cc	R. Parodi	3'12"8	Fiat Abarth 695
			Touring	850 cc	P. Cazzullo	3'05"6	Fiat Abarth 850 TC
			Touring	1000 cc	A. Grillo	3'00"8	Fiat Abarth 1000 Berlina
			Grand Touring	1000 cc	G. Giacomini	2'52"2	Fiat Abarth 1000 BA
			Grand Touring	up to 1600 cc	F. Genta	2'52"4	Abarth Simca 1300
			Grand Touring	over 1600 cc	F. Rovetta (overall)	2'43"1	Abarth Simca 2000
26 September	Roskilde, Danimarca	Roskilde Ring (circuit-1,400 kms)	Touring	1000 cc	F. Möller	14'06"1	Fiat Abarth 1000 Berlina
2 October	Silverstone, United Kingdom	Fourth Clubmen Championship (circuit-2,920 miles x 10 laps)	Grand Touring	2000 cc	R. Burnard	18'19"2	Abarth Simca 2000
2-3 October	Namur, Belgium	Course de côte de Namur (hillclimb-2,000 kms)	Touring	850 cc	G. Harris	1'44"0	Fiat Abarth 850 TC
			Touring	1000 cc	T. Pilette	1'41"4	Fiat Abarth 1000 Berlina
3 October	Montlhéry, France	Coupes du Salon (circuit-3,600 kms x 16 laps)	Touring	850 cc	De Pollon	31'03"5	Fiat Abarth 850 TC
			Touring	1000 cc	C. Ballot Lena	29'44"7	Fiat Abarth 1000 Berlina
			Grand Touring (20 laps)	1000 cc	J. Mazzanti	---	Fiat Abarth 1000 BA
			Grand Touring (20 laps)	2000 cc	F. Ruata	31'1"5	Abarth Simca 2000
3 October	Wunstorf, Germany	Flugplatzrennen Wunstorf (circuit-5,100 kms x 10 laps)	Touring	1000 cc	E. Bitter	25'20"6	Fiat Abarth 1000 Berlina
			Grand Touring	1300 cc	H. Dechent	23'10"0	Abarth Simca 1300
3 October	Genève, Switzerland	Course de côte de Marchairuz (hillclimb-4,800 kms x 2)	Touring	850 cc	O. Model	6'26"9	Fiat Abarth 850 TC
			Touring	1000 cc	K. Huser	6'13"5	Fiat Abarth 1000 Berlina
			Prepared Touring	850 cc	M. Cintolesi	6'13"3	Fiat Abarth 850 TC
			Prepared Touring	1000 cc	S. Stefani	6'06"7	Fiat Abarth 1000 Berlina
			Grand Touring	1300 cc	D. Borel	5'30"9	Abarth Simca 1300
			Grand Touring	2000 cc	H. Kuhnis	5'19"0	Abarth Simca 2000

DATE	LOCATION	RACE	CATEGORY	CLASS	WINNER	TIME OR DISTANCE	CAR
3 October	Dortmund, Germany	ADAC Sauerland Bergpreis (hillclimb-3,400 kms x 2)	Grand Touring	1000 cc	A. Hessel	4'58"6	Fiat Abarth 1000 BA
			Grand Touring	up to 1600	K. Federhofer	4'30"4	Abarth Simca 1300
3 October	Angers, France	Course de côte de la Pommeraye (hillclimb-1,700 kms)	Touring	850 cc	J. Raquin	1'16"0	Fiat Abarth 850 TC
			Grand Touring	1000 cc	Y. Barone	1'10"3	Fiat Abarth 1000 BA
			Grand Touring	1300 cc	R. Barone	1'07"1	Abarth Simca 1300
3 October	Wien, Austria	Internationales Höhenstrassenrennen (hillclimb-2,400 kms)	Touring	850 cc	K. Steinmetz	1'48"6	Fiat Abarth 850 TC
			Touring	1000 cc	H. Herrmann	1'47"7	Fiat Abarth 1000 Berlina
			Grand Touring	1000 cc	E. Fabian	1'48"7	Fiat Abarth 1000 BA
			Sport	up to 1600 cc	K. Steinmetz	1'34"4	Fiat Abarth OT 1600
			Sport	up to 2000 cc	H. Herrmann	1'32"8	Fiat Abarth OT 2000
3 October	Zandvoort, Holland	Nationaal Kampioenschap Toerwagens (circuit-4,225 kms x 12 laps)	Touring	1000 cc	J. van Osch	24'42"8	Fiat Abarth 1000 Berlina
3 October	Monza, Italy	Coppa Leopoldo Carri (circuit-5,750 kms x 10 laps)	Touring	600 cc	L. Durst	48'59"3	Fiat Abarth 595
			Touring	850 cc	S. Morando	41'28"4	Fiat Abarth 850 TC
			Touring	1000 cc	A. Marzi	39'39"9	Fiat Abarth 1000 Berlina
3 October	Sao Paolo, Brasile	Premio Rodovia do Café (circuit)	Grand Touring	2000 cc	J. Silva (overall)	4h34'01"3	Abarth Simca 2000
3 October	Vallelunga, Italy	Coppa Settecolli (circuit-1,800 kms x 12 laps)	Touring (12 laps)	850 cc	F. Chiari	14'08"4	Fiat Abarth 850 TC
			Grand Touring (15 laps)	1000 cc	M. Moselli	16'39"5	Fiat Abarth 1000 BA
9 October	Thöringen, Germany	Bergrennen Thöringen (hillclimb-2,400 kms x 2)	Touring	850 cc	A. Wyttenbach	3'10"4	Fiat Abarth 850 TC
			Touring	1000 cc	P. Engler	3'01"6	Fiat Abarth 1000 Berlina
9 October	Silverstone, United Kingdom	Club Meeting B.A.R.C. (circuit-2,900 miles x 7 laps)	Grand Touring	from 1601 to 2500 cc	R. Burnard	9'27"4	Abarth Simca 2000
9-10 October	Santhià, Italy	Autoraduno S. Vincent (94,700 kms)	---	---	'Villi' (overall)	---	Fiat Abarth 850 TC
10 October	Monza, Italy	Coppa Kléber Colombes (circuit-5,750 kms x 20 laps)	Prepared Touring	850 cc	M. Piller	49'20"5	Fiat Abarth 850 TC
			Grand Touring	up to 850 cc	H. Affentranger	47'33"8	Fiat Abarth 700 BA
		Coppa Dunlop (circuit-5,750 kms x 20 laps)	Grand Touring	1000 cc	J. Kretschi	42'25"6	Fiat Abarth 1000 BA
			Grand Touring	1300 cc	D. Borel	40'54"2	Abarth Simca 1300
10 October	Massana, Andorra	Costa de la Massana (hillclimb-3,500 kms)	Touring	700 cc	J. Juncosa	3'13"0	Fiat Abarth 695 SS
			Touring	850 cc	G. Moragas	2'47"6	Fiat Abarth 850 TC
			Prepared Touring	850 cc	'Jorge Chi'	2'54"7	Fiat Abarth 850 TC
			Grand Touring	from 1000 to 1600 cc	R. Barone	2'28"3	Abarth Simca 1300
10 October	Innsbruck, Austria	Preis von Tirol (circuit-2,800 kms x 42 laps)	Touring (42 laps)	1000 cc	J. Abt	---	Fiat Abarth 1000 Berlina
			Grand Touring (22 laps)	1000 cc	A. Schmalbach	30'59"1	Fiat Abarth 1000 BA
			Grand Touring (25 laps)	2000 cc	J. Rindt	30'40"7	Abarth Simca 2000
			Sport (20 laps)	2000 cc	J. Rindt	24'07"2	Fiat Abarth OT 2000
10 October	Dudelange, Luxembourg	Slalom Dudelange	Grand Touring	1000 cc	J. Hauser	1'13"1	Fiat Abarth 1000 BA
10 October	Cagliari, Italy	Coppa Città di Iglesias (hillclimb-6,700 kms)	Touring	600 cc	F. Collu	6'05"3	Fiat Abarth 595
			Touring	850 cc	F. Melis	5'36"4	Fiat Abarth 850 TC
			Touring	1000 cc	F. Locci	5'40"7	Fiat Abarth 1000 Berlina
			Grand Touring	1000 cc	G. Vacca	5'36"4	Fiat Abarth 1000 BA
			Grand Touring	1300 cc	G. Ambu	5'39"2	Abarth Simca 1300
10 October	Como, Italy	Trofeo Val d'Intelvi (hillclimb-8,700 kms)	Touring	850 cc	S. Uberti	6'40"5	Fiat Abarth 850 TC
			Touring	up to 1150 cc	G. Galimberti	6'24"4	Fiat Abarth 1000 Berlina
			Grand Touring	700 cc	S. La Valle	6'32"8	Fiat Abarth 700 BA
			Grand Touring	1000 cc	G. Rebaudi	6'04"5	Fiat Abarth 1000 BA
			Grand Touring	1300 cc	F. Genta	6'06"9	Abarth Simca 1300
10 October	Sonntag, Austria	Sonntag Bergrennen (hillclimb-3,000 kms)	Grand Touring	1000 cc	E. Fabian	1'50"8	Fiat Abarth 1000 BA
10 October	Lignières, Switzerland	Circuit de Lignières (1,400 kms x 20)	Touring	850 cc	H. Wengert	---	Fiat Abarth 850 TC
16 October	Ober Uhldingen, Switzerland	Slalom Ober Uhldingen	Touring	850 cc	O. Model	3'39"7	Fiat Abarth 850 TC
			Touring	1000 cc	E. Stäheli	3'35"2	Fiat Abarth 1000 Berlina
16 October	Eigenthal, Switzerland	Nationales Bergrennen Eigenthal (hillclimb-3,300 kms)	Touring	1000 cc	M. Wuest	5'14"4	Fiat Abarth 1000 Berlina
			Grand Touring	1300 cc	R. Stierli	4'45"4	Abarth Simca 1300
			Grand Touring	2000 cc	H. Kuhnis	4'44"8	Abarth Simca 2000
			Sport	from 851 to 1600 cc	H. Zweifel	4'20"4	Fiat Abarth 1300 Sport
17 October	Kassel, Germany	ADAC Bergprüfung Kassel-Langenberge	Touring	850 cc	E. Fauser	4'39"5	Fiat Abarth 850 TC
			Grand Touring	1000 cc	H. Krause	4'32"5	Fiat Abarth 1000 BA
			Grand Touring	1300 cc	K. Federhofer	4'12"9	Abarth Simca 1300
17 October	Aspern, Austria	Donaupokal Flugplatz-Rennen (circuit-2,600 kms x 22 laps)	Touring (1 hour)	1000 cc	J. Abt	kms 116,423	Fiat Abarth 1000 Berlina
			Grand Touring	1000 cc	A. Schmalbach	28'40"8	Fiat Abarth 1000 BA
			Sport	1600 cc	K. Steinmetz	23'14"3	Fiat Abarth OT 1600
			Sport	2000 cc	J. Rindt	22'34"6	Fiat Abarth OT 2000

DATE	LOCATION	RACE	CATEGORY	CLASS	WINNER	TIME OR DISTANCE	CAR
17 October	Bergamo, Italy	Castione-Passo della Presolana (hillclimb-6,400 kms)	Touring Touring Touring Grand Touring Grand Touring	600 cc 850 cc 1000 cc 1000 cc 1300 cc	G. Danieli S. Uberti G. Galimberti A. Becchetti R. Ramoino	4'33"5 4'00"2 3'52"1 3'45"9 3'34"3	Fiat Abarth 595 Fiat Abarth 850 TC Fiat Abarth 1000 Berlina Fiat Abarth 1000 BA Abarth Simca 1300
17 October	Vallelunga, Italy	Premio Campagnano (circuit-3,200 kms x 32 laps)	Touring Touring Touring	600 cc 850 cc 1000 cc	R. Pinto M. Terminiello P. Lado	1h01'58"5 57'57"4 57'31"1	Fiat Abarth 595 Fiat Abarth 850 TC Fiat Abarth 1000 Berlina
17 October	Forlì, Italy	Gara Nazionale di Regolarità (120 kms)	Touring	1000 cc	F. Severi	---	Fiat Abarth 850 TC
23 October	Oberhallau, Switzerland	Bergrennen Oberhallauberg (hillclimb-3,000 kms)	Touring Touring Grand Touring	850 cc 1000 cc up to 1300 cc	O. Model K. Köstli T. Hofer	2'10"7 2'09"6 1'51"8	Fiat Abarth 850 TC Fiat Abarth 1000 Berlina Fiat Abarth 1000 BA
23-24 October	Barcelona, Spagna	Rallye de dos Catalunas (749 kms)	Touring Grand Touring	850 cc 850 cc	'Pele'-Sensat Moragas-Marsans	--- ---	Fiat Abarth 695 Fiat Abarth 850 TC
24 October	Trieste, Italy	Coppa San Giusto (208 kms)	---	---	P. Navarra (overall)	---	Fiat Abarth 850 TC
24 October	Le Mans, France	Criterium de Vitesse de l'A.S. de l'A.C.O. (circuit-4,422 kms x 15 laps)	Grand Touring	1300 cc	R. Barone	35'14"9	Abarth Simca 1300
31 October	Vallelunga, Italy	Coppa API (circuit-1,800 kms x 20 laps)	Touring	1000 cc	A. Stefanelli	23'13"2	Fiat Abarth 1000 Berlina
4 November	Vallelunga, Italy	Trofeo delle Cinque Conchiglie (circuit-1,800 kms x 20 laps)	Touring	1000 cc	A. Saltari	23'00"3	Fiat Abarth 1000 Berlina
6 November	Lignières, Switzerland	Circuit de Lignières (1,400 kms x 17 laps)	Sport	1300 cc	W. Rufenacht	17'09"1	Fiat Abarth 1000
7 November	Vallelunga, Italy	Gara Velocità Tris (circuit-1,800 kms)	Touring Touring Special	1000 cc 1000 cc	R. Tirabassi V. Trulli	2'25"1 2'23"1	Fiat Abarth 1000 Berlina Fiat Abarth 1000 Berlina
7 November	Manresa, Spagna	Rallye de los 2000 Virajes (580 kms)	---	---	'Paco' (overall)	---	Fiat Abarth 1000 Berlina
14 November	Viterbo, Italy	Trofeo Marathon Viterbo-Poggio Nibbio (hillclimb-10,200 kms)	Touring Touring Touring Grand Touring Grand Touring	600 cc 850 cc 1000 cc 1000 cc over 1600 cc	B. Bollati F. Domini B. Faenza G. Spicciani D. Scola	6'32"0 5'39"6 5'33"1 5'05"2 4'45"6	Fiat Abarth 595 Fiat Abarth 850 TC Fiat Abarth 1000 Berlina Fiat Abarth 1000 BA Abarth Simca 2000
27-28 November	Daytona Beach (Florida), USA	International Speedway (circuit-1,630 miles x 32 laps)	---	---	P. Feistmann	---	Fiat Abarth 750
27-28 November	Montpellier, France	Criterium de Cevennes (rallye-570,228 kms)	Touring	1000 cc	Maublanc-Roussin	---	Fiat Abarth 1000 Berlina
28 November	Vallelunga, Italy	Criterium dell'Ora (circuit-1,800 kms x 1 hour)	Touring Touring	850 cc 1000 cc	M. Terminiello V. Finzi	kms 89,762 kms 92,315	Fiat Abarth 850 TC Fiat Abarth 1000 Berlina
19 December	Vallelunga, Italy	Criterium dell'Ora (circuit-1,800 kms x 1 hour)	Touring Touring	850 cc 1000 cc	M. Terminiello V. Finzi	kms 92,725 kms 92,315	Fiat Abarth 850 TC Fiat Abarth 1000 Berlina

1966

DATE	LOCATION	RACE	CATEGORY	CLASS	WINNER	TIME OR DISTANCE	CAR
6 January	Vallelunga, Italy	Coppa Epifania (circuit-1,800 kms)	Touring Touring	850 cc 1000 cc	F. Chiari A. Saltari	1'13"3 1'12"6	Fiat Abarth 850 TC Fiat Abarth 1000 Berlina
16 January	Otsolahti, Finlandia	Prepared (1,500 kms)	Grand Touring	1000 cc	L. Kinnunen	5'45"1	Fiat Abarth 1000 Berlina
23 January	Vammala, Finlandia	Ice race (1,100 kms)	Grand Touring	1000 cc	L. Kinnunen	4'03"2	Fiat Abarth 1000 Berlina
6 February	Acireale, Italy	Autogiro Etneo-Trofeo Paoletti (Ice race 137,600 kms)	---	---	D. Cali (overall)	---	Fiat Abarth 850 TC
13 February	Konstanz, Germany	Winter Rallye (Ice race 400 kms)	Touring	850 cc	S. Urban	---	Fiat Abarth 850 TC
13 February	Kokkola, Finlandia	Ice race (16,000 kms) (8,000 kms)	--- Touring	--- 1000 cc	L. Kinnunen L. Kinnunen	12'31"0 5'30"4	Fiat Abarth 1000 Berlina Fiat Abarth 1000 Berlina
20 February	Joenkoeping, Sweden	Ice race	Touring	1000 cc	B. Johansson	11'46"0	Fiat Abarth 1000 Berlina
20 February	Novara, Italy	Trofeo 250 minuti	---	---	C. Brustia (overall)	---	Fiat Abarth 1000 Berlina
20 February	Westend, Finlandia	Ice race (14,100 kms)	---	1000 cc	L. Kinnunen	11'55"6	Fiat Abarth 1000 Berlina
24-27 February	Alicante, Spagna	Rallye Alicante Internacional (Ice race 2604 kms)	Gruppo 2 Gruppo 2 ---	6 7 over 2000 cc	J. Juncosa A. Baturone J. Juncosa	--- --- ---	Fiat Abarth 850 TC Fiat Abarth 1000 Berlina Fiat Abarth 1000 BA
27 February	Vallelunga, Italy	Coppa Autodromo di Roma (circuit-1,800 kms x 40 laps)	Sport	1000 cc	G. Rebaudi	43'44"8	Fiat Abarth 1000 BA
5-6 March	Gerona, Spagna	Rallye de Gerona (516 kms)	Touring Prepared Touring	up to 1150 cc 850 cc	'Paco' M. Juncosa	--- ---	Fiat Abarth 1000 Berlina Fiat Abarth 850 TC
5-6 March	Reggio Emilia, Italy	Autodromo Sciatorio Cerreto-Longo (regolarità-90 kms)	---	---	C. Massagrande	---	Fiat Abarth 850 TC
6 March	Mänttä, Finlandia	Ice race (13,500 kms)	---	1000 cc	L. Kinnunen	9'38"8	Fiat Abarth 1000 Berlina

DATE	LOCATION	RACE	CATEGORY	CLASS	WINNER	TIME OR DISTANCE	CAR
12-13 March	Monza, Italy	Coppa FISA (circuit-5,750 kms x 10 laps)	Touring	600 cc	R. Pinto	24'11"9	Fiat Abarth 595
			Touring	850 cc	F. Galli	23'21"5	Fiat Abarth 850 TC
			Touring	1000 cc	'Matich'	22'33"1	Fiat Abarth 1000 Berlina
			Grand Touring	1300 cc	G. Dalla Torre	20'16"7	Abarth Simca 1300
			Sport	1000 cc	M. Pinchetti	21'47"0	Fiat Abarth 1000 BA
			Sport-Prototypes	1300 cc	K. Steinmetz	18'54"1	Fiat Abarth OT 1300
13 March	Fleron, Belgium	Course de côte de Fleron (hillclimb-2,000 kms x 2)	Sport	1000 cc	J. Goossens	1'44"7	Fiat Abarth 1000
13 March	Savona, Italy	Giro dei Monti Savonesi (255 kms)	Touring	850 cc	C. Massagrande	---	Fiat Abarth 850 TC
			Touring	up to 1150 cc	V. Pené	---	Fiat Abarth 1000 Berlina
13 March	Lisalmi, Finlandia	Ice race at Porcveden (14,000 kms)	---	1000 cc	L. Kinnunen	10'53"4	Fiat Abarth 1000 Berlina
19-20 March	Monza, Italy	Coppa 4 Ore del Jolly Club (circuit-10,000 kms x 4 hours)	Touring	850 cc	J. Ortner	kms 638,603	Fiat Abarth 850 TC
			Touring	1000 cc	G. Baghetti (overall)	kms 666,615	Fiat Abarth 1000 Berlina
20 March	Albi, France	Coupes d'Albi	Grand Touring	2000 cc	F. Ruata	32'49"4	Abarth Simca 2000
20 March	Liège, Belgium	Course de côte La Roche-Samree	Touring	850 cc	P. Henning	4'01"1	Fiat Abarth 850 TC
			Sport	1000 cc	H. Affentranger	4'08"4	Fiat Abarth 1000 B
20 March	Königsstetten, Austria	Dopplerhütten Wertungsfahrt (hillclimb-2,700 kms)	Touring	1000 cc	W. Pöltinger	2'26"8	Fiat Abarth 1000 Berlina
20 March	Montserrat, Spagna	Carrera en cuesta de Monserrat (hillclimb-7,000 kms)	Touring	850 cc	'Jorge Chi'	5'56"2	Fiat Abarth 850 TC
			Touring	up to 1150 cc	'Paco'	5'30"5	Fiat Abarth 1000 Berlina
			Touring Special	1000 cc	'Pelé'	5'46"7	Fiat Abarth 1000 Berlina
20 March	Ceyreste, France	Course de côte de Ceyreste (hillclimb-4,800 kms)	Touring	850 cc	L. Gaglio	3'18"1	Fiat Abarth 850 TC
			Touring	1000 cc	A. Romano	3'22"5	Fiat Abarth 1000 Berlina
			Sport	1000 cc	A. Battesti	3'24"0	Fiat Abarth 1000 SP
20 March	Kemi, Finlandia	Ice race	---	1000 cc	L. Kinnunen	7'09"7	Fiat Abarth 1000 Berlina
25 March	Madrid, Spagna	Subida Puerto de Galapagar (hillclimb-4,000 kms)	Touring	1000 cc	F. Lencina	2'22"1	Fiat Abarth 1000 Berlina
25 March	Sebring (Florida), USA	4 Hours (circuit-5,212 miles)	---	---	A. Cosentino	---	Fiat Abarth 1000 Berlina
27 March	Eisenstadt, Austria	Eisenstadter Bergwertungsfahrt (hillclimb-4,000 kms)	Touring	1000 cc	W. Pöltinger	3'23"2	Fiat Abarth 1000 Berlina
27 March	Roma, Italy	Coppa Gallenga (hillclimb-10,800 kms)	Touring	600 cc	R. Pinto	6'36"6	Fiat Abarth 595
			Touring	850 cc	S. Uberti	5'55"3	Fiat Abarth 850 TC
			Touring	1000 cc	G.B. Guarnieri	5'55"0	Fiat Abarth 1000 Berlina
			Grand Touring	2000 cc	D. Scola	5'38"9	Abarth Simca 2000
			Sport	1000 cc	G. Tommasi	6'20"4	Fiat Abarth 1000 SP
			Sport	up to 1300 cc	M. Spadafora	5'57"0	Fiat Abarth 1000 BA
			Sport-Prototypes	up to 2000 cc	'Nicor'	5'57"7	Fiat Abarth OT 1300
27 March	Pitea, Sweden	Ice race	Touring	1000 cc	B. Johansson	10'39"0	Fiat Abarth 1000 Berlina
27 March	Ylitornio, Finlandia	Ice race at Aavasaksa	Touring	1000 cc	L. Kinnunen	5'29"1	Fiat Abarth 1000 Berlina
3 April	Detmold, Germany	ADAC Bergrennen Teutoburger Wald (hillclimb-2,250 kms x 2)	Touring	1000 cc	E. Bitter	2'29"7	Fiat Abarth 1000 Berlina
3 April	Draguignan, France	Course de côte d'Ampus (hillclimb-6,800 kms)	Grand Touring	2000 cc	Benedetto	6'04"0	Abarth Simca 2000
3 April	Sanremo, Italy	Trofeo Sanremo della Montagna (hillclimb-5,900 kms x 2)	Touring	600 cc	G. Recordati	9'52"5	Fiat Abarth 595
			Touring	700 cc	G. Barillaro	9'50"5	Fiat Abarth 695
			Touring	850 cc	S. Uberti	9'10"8	Fiat Abarth 850 TC
			Touring	1000 cc	G. Guarneri	9'03"9	Fiat Abarth 1000 Berlina
			Grand Touring	1300 cc	G. Dalla Torre	8'34"3	Abarth Simca 1300
			Sport-Prototypes	2000 cc	L. Taramazzo (overall)	8'22"2	Fiat Abarth OT 2000
3 April	Amalfi, Italy	Coppa della Primavera (hillclimb-8,750 kms)	Touring	850 cc	B. Mazzei	7'16"0	Fiat Abarth 850 TC
			Touring	1000 cc	V. Ferro	7'05"5	Fiat Abarth 1000 Berlina
			Grand Touring	1300 cc	G. Barella	7'11"2	Abarth Simca 1300
3 April	Washington (D.C.), USA	Marlboro Speedway (circuit)	---	---	A. Cosentino	---	Fiat Abarth 1000 Berlina
10 April	Hockenheim, Germany	Trainingstag am Hockenheimring (circuit-2,690 kms x 35 laps)	Touring	850 cc	R. Stierli	61'28"4	Fiat Abarth 850 TC
			Touring	1000 cc	G. Brun	61'21"5	Fiat Abarth 1000 Berlina
10 April	Imola, Italy	Coppa Touring A.C. Bologna (circuit-5,017 kms x 10 laps)	Touring	600 cc	R. Pinto	26'44"2	Fiat Abarth 595
			Touring	700 cc	U. Barilaro	26'45"7	Fiat Abarth 695
11 April	Vallelunga, Italy	Coppa Marathon (circuit-1,800 kms x 40 laps)	Sport	1300 cc	B. Faenza	44'49"5	Fiat Abarth 1000 Sport
11 April	Hockenheim, Germany	Rundstreckenrennen Hockenheim (circuit-2,690 kms x 10 laps)	Touring	850 cc	J. Schneiter	16'18"0	Fiat Abarth 850 TC
			Touring	1000 cc	P. Engler	17'19"9	Fiat Abarth 1000 Berlina
			Sport	from 700 to 1300 cc	P. Maron	14'52"7	Fiat Abarth OT 1300
11 April	Lorentzweiler, Luxembourg	Course de côte de Lorentzweiler (hillclimb-1,800 kms)	Touring	850 cc	R. Stierli	2'42"0	Fiat Abarth 850 TC
			Touring	1000 cc	E. Welti	2'41"9	Fiat Abarth 1000 Berlina
			Sport	1000 cc	J. Hauser	2'43"6	Fiat Abarth 1000 BA
11 April	Goodwood, United Kingdom	Sussex Trophy Race (circuit-2,400 miles x 15 laps)	Grand Touring	2000 cc	R. Burnard	24'25"0	Abarth Simca 2000

DATE	LOCATION	RACE	CATEGORY	CLASS	WINNER	TIME OR DISTANCE	CAR
16 April	Hockenheim, Germany	Rundstrekenrennen am Hockenheim (circuit-2,690 kms x 15 laps)	Touring	850 cc	R. Stierli	24'03"3	Fiat Abarth 850 TC
			Touring	1000 cc	E. Welti	22'38"1	Fiat Abarth 1000 Berlina
			Grand Touring	up to 1600 cc	H. Kennel	23'21"6	Fiat Abarth 1000 BA
			Sport-Prototypes	---	J. Kretschi	22'34"1	Fiat Abarth OT 1300
16-17 April	Vineland (N.J.), USA	Vineland Raceway (circuit)	---	---	A. Cosentino	---	Fiat Abarth 1000 Berlina
16-17 April	Berlin, Germany	ADAC Nachtprüfungsfahrt (220 kms)	Touring	850 cc	H.J. Mack	---	Fiat Abarth 850 TC
17 April	Verona, Italy	Stallavena-Boscochiesanuova (hillclimb-15,300 kms)	Touring	600 cc	A. Triboldi	10'11"7	Fiat Abarth 595
			Touring	700 cc	R. Rasia dal Polo	9'15"8	Fiat Abarth 695
			Touring	850 cc	M. Favaretto	8'35"1	Fiat Abarth 850 TC
			Touring	1000 cc	'Matich'	8'12"0	Fiat Abarth 1000 Berlina
			Grand Touring	1300 cc	M. Nardari	7'56"9	Abarth Simca 1300
			Sport	1000 cc	M. Moselli	7'34"0	Fiat Abarth 1000 BA
			Sport-Prototypes	1300 cc	S. Morando	7'35"2	Fiat Abarth OT 1300
16-17 April	Grasse, France	Rallye Fleurs et Parfums (528 kms) (gruppo 2)	Prepared Touring	850 cc	Armando-Pronesti	---	Fiat Abarth 850 TC
16-17 April	Grasse, France	Criterium International Alpin (528 kms) (gruppi 2 e 5)	Prepared Touring	850 cc	Branche-Jullien	---	Fiat Abarth 850 TC
17 April	Herbeumont, Belgium	Course de côte d'Herbeumont-Ste Cecile (hillclimb-3,600 kms)	Sport	1000 cc	H. Affentranger	2'14"3	Fiat Abarth 1000 BA
17 April	Saint-Lô, France	Course de côte d'Hébécrevon (hillclimb-1,200 kms)	Touring	850 cc	P. Henning (overall)	49"3	Fiat Abarth 850 TC
17 April	Avola, Italy	Coppa Belmonte (hillclimb-10,200 kms)	Touring	600 cc	G. Quintieri	8'10"6	Fiat Abarth 595
			Touring	850 cc	P. Amato	7'16"3	Fiat Abarth 850 TC
			Touring	1000 cc	A. Romano	6'55"5	Fiat Abarth 1000 Berlina
17 April	Belvaux, Luxembourg	Slalom Automobile de Belvaux	Touring	850 cc	R. Stierli	49"3	Fiat Abarth 850 TC
			Touring	1000 cc	U.P. Dietrich	47"7	Fiat Abarth 1000 Berlina
			Sport-Prototypes	1300 cc	J. Kretschi	48"0	Fiat Abarth OT 1300
17 April	Manresa, Spagna	Subida en cuesta a San Mateo de Bages (hillclimb-7,000 kms)	Touring	up to 1150 cc	'Paco'	4'32"0	Fiat Abarth 1000 Berlina
			Touring	850 cc	M. Juncosa	4'41"1	Fiat Abarth 850 TC
			Special Sport	1000 cc	J. Juncosa (overall)	4'24"0	Fiat Abarth 1000 BA
17 April	Vallelunga, Italy	Coppa Primavera (circuit-1,800 kms x 15 laps)	Touring	700 cc	'Marino'	18'52"5	Fiat Abarth 695
			Touring	850 cc	F. Chiari	17'41"5	Fiat Abarth 850 TC
			Touring	1000 cc	V. Finzi	17'43"2	Fiat Abarth 1000 Berlina
17 April	Trieste, Italy	Coppa del Biancospino (67 kms x 2)	Touring	850 cc	G. Pavanello	---	Fiat Abarth 850 TC
17 April	Eifel, Germany	ADAC Bergpreis Bad Neuenahr (hillclimb-300 kms x 2)	Touring	1000 cc	W.S. Kauhsen	4'21"0	Fiat Abarth 1000 Berlina
			Grand Touring	1300 cc	H. Dechent	4'16"0	Abarth Simca 1300
17 April	Wien, Austria	Preis von Wien (circuit-2,600 kms x 1 hour)	Touring	850 cc	J. Ortner	kms 113,669	Fiat Abarth 850 TC
			Touring	1000 cc	G. Baghetti	kms 115,734	Fiat Abarth 1000 Berlina
23 April	Lime Rock (Connecticut), USA	Area I Championship Races (circuit)	---	---	A. Cosentino	---	Fiat Abarth 1000 Berlina
23-24 April	Gap, France	Course de côte du Col Bayard (hillclimb-6,300 kms)	Touring	850 cc	S. Piana	4'58"8	Fiat Abarth 850 TC
			Touring	1000 cc	P. Maublanc	4'29"0	Fiat Abarth 1000 Berlina
23-24 April	Zandvoort, Holland	Trainingsweekend N.A.V. (circuit-4,225 kms x 10 laps)	Touring	850 cc (1st race)	A. Hezemans	22'58"8	Fiat Abarth 850 TC
			Touring	850 cc (2nd race)	M. Jonker	26'20"6	Fiat Abarth 850 T
24 April	Rieka, Jugoslavia	Circuito di Rieka	Touring	850 cc	R. Jalovean	---	Fiat Abarth 850 TC
24 April	Innsbruck, Austria	Internationales Weerbergrennen (hillclimb-5,000 kms)	Touring	1000 cc	J. Abt	4'28"9	Fiat Abarth 1000 Berlina
			Sport	up to 1600 cc	H. Tschiemer	4'09"3	Abarth Simca 1600
24 April	Snetterton, United Kingdom	Clubman's Car Races (circuit-2,700 miles x 8 laps)	Touring	850 cc	J. Anstead	17'25"8	Fiat Abarth 850 TC
24 April	Astorp, Sweden	Ring Knutstorp (circuit-1,100 kms x 20 laps)	Touring	1000 cc	B. Johansson	16'40"2	Fiat Abarth 1000 Berlina
24 April	Montlhéry, France	Coupes de Vitesse (circuit-3,340 kms x 16 laps)	Touring	850 cc	P. Maury	29'26"5	Fiat Abarth 850 TC
			Touring	1000 cc	C. Ballot Lena	28'05"9	Fiat Abarth 1000 Berlina
			Sport	1000 cc	J. Mazzanti	35'20"1	Fiat Abarth 1000 BA
24 April	Kolind, Danimarca	Ring Djursland-Abningslob (hillclimb)	Touring	1000 cc	S. Borch	---	Fiat Abarth 1000 Berlina
24 April	Erbach, Germany	DMV Krähbergrennen (hillclimb-3,600 kms)	Sport	1300 cc	B. Hofmann	4'46"1	Abarth Simca 1300
25 April	Monza, Italy	1000 Km di Monza (circuit-10,000 kms x 100 laps)	Sport-Prototypes	1300 cc	Baghetti-Furtmayr	6h8'5"8	Fiat Abarth OT 1300
25 April	Vallelunga, Italy	Coppa Fiat (circuit-1,800 kms x 40 laps)	Touring	850 cc	S. Uberti	44'27"6	Fiat Abarth 850 TC
			Touring	1000 cc	M. Nesti (overall)	44'51"7	Fiat Abarth 1000 Berlina
30 April	Grenchen, Switzerland	Automobilslalom Grenchen	Touring	1000 cc	R. Egli	2'48"7	Fiat Abarth 1000 Berlina
30 April	Verbois, Switzerland	Course de côte de Verbois (hillclimb-1,150 kms x 2)	Touring	1000 cc	J.P. Brun	0'57"9	Fiat Abarth 1000 Berlina
			Sport	up to 1150 cc	R. Ghezzi	0'57"8	Fiat Abarth 1000 BA
30 April	Danville (Virginia), USA	National Championship No. Carolina (circuit-3,200 miles x 5 laps)	---	---	A. Cosentino	---	Fiat Abarth 1000 Berlina

DATE	LOCATION	RACE	CATEGORY	CLASS	WINNER	TIME OR DISTANCE	CAR
30 April	Valencia, Spagna	Rallye Fallas-Trofeo Barreira (531 kms)	Touring	850 cc	J. Juncosa	---	Fiat Abarth 850 TC
			Touring	up to 1150 cc	E. Baturone	---	Fiat Abarth 1000 Berlina
30 April-1 May	Nice, France	Criterium Automobile Jean Behra (regolarità-374,350 kms)	Grand Touring	up to 850 cc	Branche-Jullien	---	Fiat Abarth 695
1 May	Emmen, Switzerland	Autoslalom Emmen	Touring	1000 cc	J. Gloggner	3'42"0	Fiat Abarth 1000 Berlina
			Grand Touring	1000 cc	M. Wuest	3'55"5	Fiat Abarth 1000 OTS
1 May	Pistoia, Italy	Coppa della Collina (hillclimb-14,700 kms)	Touring	700 cc	R. Rasia dal Polo	10'30"7	Fiat Abarth 695
			Touring	850 cc	G. Casati	9'49"6	Fiat Abarth 850 TC
			Touring	1150 cc	M. Nesti	9'15"2	Fiat Abarth 1000 Berlina
			Sport	1000 cc	M. Pinchetti	9'02"1	Fiat Abarth 1000 BA
			Sport	1300 cc	G. Rebaudi	8'49"7	Fiat Abarth OT 1300
			Sport	2000 cc	M. Baldi	9'05"2	Abarth Simca 2000
			Sport-Prototypes	1000 cc	M. Brandi	9'13"6	Fiat Abarth 1000
1 May	Dachau, Germany	ADAC Peissenberg Rennen (hillclimb-3,100 kms x 2)	Touring	850 cc	R. Stierli	4'04"0	Fiat Abarth 850 TC
			Touring	1000 cc	J. Abt	3'53"2	Fiat Abarth 1000 Berlina
			Sport-Prototypes	1150 cc	H. Zweifel	3'54"0	Fiat Abarth 1150
1 May	Igualada, Spagna	Subida en cuesta a Las Mayolas (hillclimb-6,100 kms)	Touring	850 cc	'Jorge Chi'	4'59"4	Fiat Abarth 850 TC
			Sport	1000 cc	J. Juncosa (overall)	4'23"3	Fiat Abarth 1000 BA
1 May	Hockenheim, Germany	A.C.S. Hockenheimring (circuit-2,690 kms x 10 laps)	Sport	1000 cc	T. Hofer	15'33"7	Fiat Abarth 1000 BA
1 May	Cannes, France	Course de côte de Cannes (hillclimb-7,845 kms)	Touring	1000 cc	P. Maublanc	6'51"6	Fiat Abarth 1000 Berlina
			Grand Touring	1300 cc	A. Battesti	7'01"0	Abarth Simca 1300
1 May	Stadtsteinach, Germany	ADAC Frankenwald Bergrennen (hillclimb-3,545 kms x 2)	Touring	1000 cc	P. Lotz	3'36"0	Fiat Abarth 1000 Berlina
1 May	Trier, Germany	Trierer Flugplatzrennen (circuit-1,900 kms x 27 laps)	Touring	850 cc	J. Van Osch	31'12"4	Fiat Abarth 850 TC
			Touring	1000 cc	E. Bitter	29'38"0	Fiat Abarth 1000 Berlina
			Grand Touring	1000 cc	E. Bitter	35'41"4	Fiat Abarth OT 1000
			Sport	1300 cc	H. Dechent	29'40"2	Fiat Abarth OT 1300
1 May	Magny Cours, France	Circuit de Magny Cours (circuit-2,040 kms x 20 laps)	Touring	1000 cc	Schleter	23'58"2	Fiat Abarth 1000 Berlina
1 May	Brasilia, Brasile	Mille Km di Brasilia (circuit)	Touring	up to 1300 cc	Gancia-Albertine	---	Fiat Abarth 1000 Berlina
1 May	Snetterton, United Kingdom	Thetford Trophy Car Races (2,700 miles x 8 laps)	Touring	850 cc	J.A. Anstead	17'35"0	Fiat Abarth 850 TC
1 May	Danville (Virginia), USA	National Championship No. Carolina (circuit-3,2 miles x 15 laps)	---	---	A. Cosentino	---	Fiat Abarth 1000 Berlina
1 May	Koper, Jugoslavia	Gara di regolarità	---	---	R. Krasna (overall)	---	Fiat Abarth 850 TC
1 May	Mannheim, Germany	Mannheimer Automobil Turnier Slalom (45,000 kms x 2)	Touring	1000 cc	P. Engler	---	Fiat Abarth 1000 Berlina
1 May	Bodanrück, Germany	Bergprüfung auf dem Bodanrück (hillclimb-1,600 kms)	Touring	850 cc	Kaspuzvk	---	Fiat Abarth 850 TC
7 May	Dübendorf, Switzerland	Automobil-Slalom Wangen	Touring	850 cc	R. Stierli	9'51"6	Fiat Abarth 850 TC
			Touring	1000 cc	J. Brun	9'38"5	Fiat Abarth 1000 Berlina
			Sport	1000 cc	T. Hofer	9'04"7	Fiat Abarth 1000 BA
			Sport	1150 cc	H. Zweifel	9'04'9	Fiat Abarth 1150
			Sport	1300 cc	J. Kretschi	9'25"7	Fiat Abarth OT 1300
			Sport	1600 cc	W. Rufenacht	8'45"5	Fiat Abarth OT 1600
			Sport	2000 cc	W. Biedermann	9'01"6	Abarth Simca 2000
7-8 May	Arezzo, Italy	Pieve S. Stefano-Passo dello Spino (hillclimb-12,700 kms)	Touring	600 cc	A. Magnoni	8'46"4	Fiat Abarth 595
			Touring	700 cc	R. Rasia dal Polo	8'25"9	Fiat Abarth 695
			Touring	850 cc	D. Martino	7'45"7	Fiat Abarth 850 TC
			Touring	1000 cc	M. Nesti	7'32"6	Fiat Abarth 1000 Berlina
			Sport	up to 1300 cc	G. Spicciani	7'32"5	Fiat Abarth 1000 BA
8 May	Zolder, Belgium	Grand Prix du Limbourg (circuit-4,200 kms x 32 laps)	Touring	850 cc	M. Joncker	1h04'42"7	Fiat Abarth 850 TC
			Touring	1000 cc	J. Ortner (overall)	1h 04'18"9	Fiat Abarth 1000 Berlina
8 May	Reisdorf, Luxembourg	Course de côte de Reisdorf (hillclimb-3,400 kms)	Touring	850 cc	R. Stierli	2'24"2	Fiat Abarth 850 TC
			Touring	1000 cc	J. Brun	2'19"3	Fiat Abarth 1000 Berlina
			Sport	up to 1600 cc	J. Kretschi	2'10"0	Fiat AbarthOT 1300
8 May	Sorrento, Italy	Sorrento-Sant'Agata (hillclimb-10,500 kms)	Touring	700 cc	G. Lopez	7'42"4	Fiat Abarth 695
			Touring	850 cc	M. Terminiello	6'47"6	Fiat Abarth 850 TC
			Touring	1000 cc	V. Ferro	6'56"7	Fiat Abarth 1000 Berlina
			Sport	1000 cc	Lanza	7'04"8	Fiat Abarth 1000 BA
			Sport	1300 cc	R. Arfé	6'41"0	Abarth Simca 1300
			Sport	2000 cc	D. Scola (overall)	6'28"1	Abarth Simca 2000
8 May	Beograd, Jugoslavia	Criterium Européen (circuit-2,520 kms)	Touring	850 cc	R. Jelovean	---	Fiat Abarth 850 TC
8 May	Marseille, France	Course de côte de la Sainte Baume (hillclimb-8,100 kms)	Touring	600 cc	J. Choquel	9'08"0	Fiat Abarth 595
			Sport-Prototypes	from 1600 to 2000 cc	P. Maublanc (overall)	6'56"1	Fiat Abarth OT 2000
8 May	Adenau, Germany	ADAC Hansa Nürburgring (circuit-7,700 kms x 8 laps)	Touring	1000 cc	E. Britter	29'45"0	Fiat Abarth 1000 Berlina
			Grand Touring	1300 cc	H. Dechent	28'54"7	Fiat Abarth OT 1300

DATE	LOCATION	RACE	CATEGORY	CLASS	WINNER	TIME OR DISTANCE	CAR
8 May	Empoli, Italy	Coppa Città di Empoli (160,100 kms)	---	---	G. Padredii (overall)	---	Fiat Abarth 850 TC
8 May	Roxilde, Danimarca	Roxilde Ring (circuit-1,400 kms x 8 laps)	Touring	1000 cc	B. Christensen	13'46"2	Fiat Abarth 1000 Berlina
14-15 May	Bordeaux, France	Rallye Bordeaux-Sud Ouest (774 kms)	Sport	700 cc	Branche-Jullien	---	Fiat Abarth 700 BA
15 May	Snetterton, United Kingdom	Forest Trophy (circuit-2,700 miles x 7 laps)	Touring	850 cc	G. Anstead	15'29"0	Fiat Abarth 850 TC
15 May	Eberbach, Germany	Bergrennen Eberbach (hillclimb-3,000 kms)	Touring	1000 cc	M. Endress	4'03"1	Fiat Abarth 1000 Berlina
			Sport	1300 cc	J. Kretschi	3'51"6	Fiat Abarth OT 1300
15 May	Aarhus, Danimarca	Jyllands Ringen (circuit-1,600 kms x 3)	Touring	1000 cc	O. Wijk	22'48"5	Fiat Abarth 1000 Berlina
15 May	Montseny, Spagna	Carrera en cuesta al Montseny (hillclimb-15,000 kms)	Touring	850 cc	M. Juncosa	12'48"4	Fiat Abarth 850 TC
			Touring	1000 cc	'Paco'	12'28"4	Fiat Abarth 1000 Berlina
			Touring Gr. 5	1000 cc	J. Juncadella	13'23"4	Fiat Abarth 1000 Berlina
			Sport	up to 1300 cc	J. Juncosa	13'15"8	Fiat Abarth 1000 BA
15 May	Montlhéry, France	Grand Prix de Paris (circuit-3,360 kms x 15 laps)	Touring	1000 cc	C. Ballot Lena (overall)	27'44"9	Fiat Abarth 1000 Berlina
			Sport (20 laps)	1000 cc	J. Mazzanti	28'42"3	Fiat Abarth 1000 BA
15 May	Bruxelles, Belgium	Course de côte de Alle sur Semois (hillclimb-4,200 kms)	Touring	700 cc	H. Veder	2'34"3	Fiat Abarth 695
			Touring	850 cc	R. Stierli	2'27"3	Fiat Abarth 850 TC
			Touring	1000 cc	A. Hezemans	2'17"4	Fiat Abarth 1000 Berlina
			Grand Touring	from 1150 to 1600 cc	M. Abels	2'18"6	Fiat Abarth OT 1300
			Sport	1000 cc	J. Hauser	2'28"8	Fiat Abarth 1000 BA
			Sport	1600 cc	W. Rufenacht	2'09"1	Fiat Abarth OT 1600
15 May	Sassari, Italy	Alghero-Scala Picada (hillclimb-8,700 kms)	Touring	850 cc	F. Melis	5'29"8	Fiat Abarth 850 TC
			Touring	up to 1150 cc	F. Locci	5'19"5	Fiat Abarth 1000 Berlina
			Sport	1000 cc	G. Vacca	4'59"0	Fiat Abarth 1000 BA
15 May	Udine, Italy	Cividale-Castelmonte (hillclimb-7,000 kms)	Touring	600 cc	G. Bassi	5'45"3	Fiat Abarth 595
			Touring	850 cc	V. Tabucchi	5'05"0	Fiat Abarth 850 TC
			Touring	1000 cc	P. Cattelan	4'54"2	Fiat Abarth 1000 Berlina
			Sport	1000 cc	M. Pinchetti	4'42"3	Fiat Abarth 1000 BA
			Sport	1300 cc	G. Dalla Torre	4'33"7	Abarth Simca 1300
			Sport	1600 cc	W. Rufenacht	2'09"1	Fiat Abarth OT 1600
15 May	Parma, Italy	Castione Baratti-Neviano Arduini (hillclimb-5,500 kms)	Touring	700 cc	M. Zanetti	4'13"0	Fiat Abarth 695
			Touring	850 cc	G. Pasolini	4'01"8	Fiat Abarth 850 TC
			Touring	up to 1300 cc	G. Parecchini	3'55"8	Fiat Abarth 1000 Berlina
			Sport	1000 cc	A. Zucchi	3'41"2	Fiat Abarth 1000 BA
			Sport	1300 cc	G. Rebaudi	3'26"2	Fiat Abarth OT 1300
			Sport	over 1600 cc	A. Brini	3'25"9	Abarth Simca 2000
			Sport-Prototypes	2000 cc	V. Venturi (overall)	3'24"9	Fiat Abarth OT 2000
15 May	Zagreb, Jugoslavia	Circuito di Stribicke Toplice (29,000 kms)	Touring	850 cc	R. Jelovcan	19'54"3	Fiat Abarth 850 TC
15 May	Berlin, Germany	Goldner Bär von Berlin (325 kms)	Touring	up to 1000 cc	H. Garschig	2h50'51"0	Fiat Abarth 850 TC
15 May	Cumberland, USA	Cumberland National Races (circuit-1,6 miles x 30')	---	---	A. Cosentino	---	Fiat Abarth 1000 Berlina
15 May	Nîmes, France	Course de côte du Minier (hillclimb-21,500 kms)	Touring	600 cc	J. Choquel	19'03"9	Fiat Abarth 595 SS
			Touring	1000 cc	Fortmann	15'37"4	Fiat Abarth 1000 Berlina
17 May	Roulers, Belgium	Slalom de Roulers	Touring	850 cc	J. Jones	2'57"6	Fiat Abarth 850 TC
18-19 May	Monza, Italy	Mobil Economy Run (731 kms)	Touring	1000 cc	Massagrande-Vicini	5h48'12"4	Fiat Abarth OT 1000
19 May	Volterra, Italy	Coppa Città di Volterra (hillclimb-10,060 kms)	Touring	600 cc	A. Magnoni	7'20"3	Fiat Abarth 595
			Touring	700 cc	M. Zanetti	7'08"6	Fiat Abarth 695
			Touring	850 cc	D. Martino	6'37"4	Fiat Abarth 850 TC
			Touring	1000 cc	M. Nesti	6'24"7	Fiat Abarth 1000 Berlina
			Sport	1000 cc	E. Ostini	6'30"2	Fiat Abarth 1000 BA
			Sport	1300 cc	G. Bona	6'05"5	Fiat Abarth OT 1300
			Sport-Prototypes	2000 cc	L. Taramazzo	6'05"6	Fiat Abarth OT 2000
			Sport-Prototypes	2200 cc	M. Baldi	6'17"3	Abarth Simca 2200
19 May	Aix en Provence, France	Course de côte d'Aix-Saint Antonin (hillclimb-2,700 kms)	Touring	1000 cc	C. Ballot Lena	1'51"1	Fiat Abarth 1000 Berlina
			Sport-Prototypes	2000 cc	P. Maublanc	1'46"1	Fiat Abarth OT 2000
19 May	Dudelange, Luxembourg	Course de côte Tele-Dudelange (hillclimb-1,400 kms)	Touring	1000 cc	W. Kauhsen	1'04"1	Fiat Abarth 1000 Berlina
			Sport	1000 cc	J. Hauser	1'04"0	Fiat Abarth 1000 BA
19 May	Cesena, Italy	Coppa Città di Cesena (150 kms)	Touring	700 cc	C. Sangiorgi	---	Fiat Abarth 695
			Touring	up to 1000 cc	S. Foti	---	Fiat Abarth 850 TC
19 May	Ausburg, Germany	Dasca Autoslalom	Sport	1300 cc	K. Federhofer	---	Fiat Abarth OT 1300
21 May	Bern, Switzerland	Surprise Bergrennen (hillclimb)	Touring	850 cc	K. Schneiter	4'28"4	Fiat Abarth 850 TC
			Sport	850 cc	R. Egli	4'39"3	Fiat Abarth 850
21-22 May	Wiesbaden, Germany	Taunus Bergprüfung (hillclimb-3,300 kms x 2)	Touring	1000 cc	J. Abt	2'19"1	Fiat Abarth 1000 Berlina
			Sport	1300 cc	K. Federhofer	2'13"1	Fiat Abarth OT 1300

DATE	LOCATION	RACE	CATEGORY	CLASS	WINNER	TIME OR DISTANCE	CAR
21-22 May	Stuttgart, Germany	ADAC Rennen Slalom	---	up to 1300 cc	P. Lotz	---	Fiat Abarth 1000 BA
22 May	Cagliari, Italy	Monte Acuto-Campuomu (hillclimb-14,000 kms)	Touring	600 cc	F. Collu	11'19"3	Fiat Abarth 595
			Touring	700 cc	S. Farris	10'46"5	Fiat Abarth 695
			Touring	850 cc	G. Ambu	9'44"6	Fiat Abarth 850 TC
			Touring	1000 cc	F. Locci	9'52"0	Fiat Abarth 1000 Berlina
			Sport	1000 cc	S. Gambarau	9'24"5	Fiat Abarth 1000 BA
22 May	Bologna, Italy	Bologna-Passo della Raticosa (hillclimb-32,700 kms)	Touring	850 cc	D. Martino	21'32"5	Fiat Abarth 850 TC
			Touring	up to 1150 cc	M. Nesti	21'08"8	Fiat Abarth 1000 Berlina
			Grand Touring	1000 cc	D. Banchelli	23'38"8	Fiat Abarth 1000 OTS
			Sport	1000 cc	V. Cannataro	24'10"1	Fiat Abarth 1000 BA
			Sport	1300 cc	G. Rebaudi	19'56"3	Fiat Abarth OT 1300
			Sport	2000 cc	V. Venturi	19'37"2	Fiat Abarth OT 2000
			Sport-Prototypes	2200 cc	M. Baldi	20'49"5	Abarth Simca 2200
22 May	Piacenza, Italy	Castell'Arquato-Vernasca (hillclimb-9,775 kms)	Touring	700 cc	G. Caffi	6'30"0	Fiat Abarth 695
			Touring	850 cc	G. Vedovello	6'13"2	Fiat Abarth 850 TC
			Touring	up to 1150 cc	F. Zanardelli	5'49"8	Fiat Abarth 1000 Berlina
22 May	Wittlich, Germany	Wittlicher ADAC Bergrpreis (hillclimb-3,738 kms x 2)	Touring	1000 cc	E. Bitter	5'06"8	Fiat Abarth 1000 Berlina
			Sport	1300 cc	M. Abels	5'00"8	Fiat Abarth OT 1300
22 May	Francorchamps, Belgium	Coupes de Spa (circuit-14,100 kms x 12 laps)	Touring	from 850 to 1150 cc	A. Hezemans	58'37"1	Fiat Abarth 1000 Berlina
22 May	Digne, France	Course de côte de la Lure (hillclimb-12,700 kms)	Prepared Touring	600 cc	J. Choquel	11'59"2	Fiat Abarth 595 SS
			Prepared Touring	850 cc	M. Cintolesi	9'21"8	Fiat Abarth 850 TC
			Sport	from 1600 to 2500 cc	P. Maublanc (overall)	7'58"2	Fiat Abarth OT 2000
22 May	Monza, Italy	GP di Monza (circuit-5,750 kms x 20 laps)	Touring	850 cc	R. Blatter	46'38"9	Fiat Abarth 850 TC
			Touring	1000 cc	E. Welti	45'40"0	Fiat Abarth 1000 Berlina
			Sport	700 cc	H. Affentranger	42'01"8	Fiat Abarth 700 BA
			Sport	1000 cc	J. Hauser	40'45"3	Fiat Abarth 1000 BA
			Sport	1300 cc	J. Kretschi	40'31"4	Fiat Abarth OT 1300
22 May	Abreschviller, France	Course de côte d'Abreschviller (hillclimb-2,000 kms)	Touring Special	850 cc	Maury	1'29"4	Fiat Abarth 850 TC
22 May	Barcelona, Spagna	Carrera en cuesta de Vallvidrera (hillclimb-2,450 kms)	Prepared Touring	850 cc	M. Juncosa	2'13"0	Fiat Abarth 850 TC
			Prepared Touring	1000 cc	'Paco'	2'13"2	Fiat Abarth 1000 Berlina
			Sport	1000 cc	J. Juncosa	2'08"9	Fiat Abarth 1000 BA
28 May	Palermo-Trieste, Italy	Trofeo Jolly Hotels (3370 kms)	Touring	850-1000 cc	Calascibetta-Bilotti	---	Fiat Abarth 1000 Berlina
29 May	Savona, Italy	Savona-Cadibona (hillclimb-7,800 kms)	Touring	600 cc	R. Rasia dal Polo	5'05"5	Fiat Abarth 595 SS
			Touring	850 cc	G. Carena	4'58"4	Fiat Abarth 850 TC
			Touring	1000 cc	V. Pené	5'06"0	Fiat Abarth 1000 Berlina
			Sport	1000 cc	P. Scantimburgo	4'26"1	Fiat Abarth 1000 BA
			Sport	1300 cc	R. Ramoino	4'40"0	Fiat Abarth OT 1300
			Sport-Prototypes	2200 cc	M. Baldi	4'39"4	Abarth Simca 2200
29 May	New England, USA	Thompson New England Regional Race (circuit-1,6 miles x 15 laps)	---	---	A. Cosentino	---	Fiat Abarth 1000 Berlina
29 May	Autun, France	Course de côte Autun-Montjeu (hillclimb-4,100 kms)	Touring	850-1150 cc	P. Maublanc (overall)	5'10"3	Fiat Abarth 1000 Berlina
			Sport	850-1150 cc	J. Mazzanti	5'08"5	Fiat Abarth 1000 BA
			Sport	2000 cc	P. Maublanc	4'34"7	Fiat Abarth OT 2000
29 May	Castellon, Spagna	Subida en cuesta al Desierto de Las Palmas (hillclimb-7,200 kms)	Touring	850 cc	M. Bertolin	4'55"0	Fiat Abarth 850 TC
			Touring	over 850 cc	J. Juncosa	4'44"0	Fiat Abarth 1000 Berlina
29 May	Hockenheim, Germany	100 Meilen von Hockenheim (circuit-6,768 kms x 24 laps)	Touring	1000 cc	D. Mantzel	1h06'35"0	Fiat Abarth 1000 Berlina
			Grand Touring	1000 cc	E. Bitter	53'25"0	Fiat Abarth 1000 OTS
			Sport	1300 cc	M. Abels	54'51"0	Fiat Abarth OT 1300
29-30 May	Lerida, Spagna	Rallye Valle de Aran (838 kms)	Touring	850 cc	'Jorge Chi'	---	Fiat Abarth 850 TC
30 May	Innsbruck, Austria	Bergpreis von Tirol (hillclimb-4,600 kms x 2)	Touring	1000 cc	J. Abt	4'17"7	Fiat Abarth 1000 Berlina
			Sport	1300 cc	P. Maron	3'39"4	Fiat Abarth OT 1300
			Sport	1600 cc	W. Rufenacht	3'50"0	Fiat Abarth OT 1600
			Sport-Prototypes	1000 cc	J. Ortner	3'53"5	Fiat Abarth 1000
30 May	New England, USA	Thompson Raceway (circuit-1,6 miles x 15 laps)	---	---	A. Cosentino	---	Fiat Abarth 1000 Berlina
30 May	Snetterton, United Kingdom	Closed Race Meeting (circuit-2,700 miles x 8 laps)	Touring	850 cc	J. Anstead	17'17"0	Fiat Abarth 850 TC
30 May	Zandvoort, Holland	N.A.V. National Kampioenschapsraces (circuit-4,225 kms x 16 laps)	Touring	850 cc	J. van Osch	35'53"2	Fiat Abarth 850 TC
			Touring	1000 cc	A. Hezemans	31'08"5	Fiat Abarth 1000 Berlina
2 June	Brescia, Italy	Scalata al Colle Sant'Eusebio (hillclimb-9,150 kms)	Touring	700 cc	G. Caffi	5'47"9	Fiat Abarth 695
			Touring	850 cc	S. Uberti	5'14"9	Fiat Abarth 850 TC
			Touring	1000 cc	F. Zanardelli	5'10"3	Fiat Abarth 1000 Berlina
			Sport	1000 cc	G.B. Mainetti	5'02"8	Fiat Abarth 1000 BA
			Sport	1300 cc	G. Dalla Torre	4'52"4	Abarth Simca 1300

DATE	LOCATION	RACE	CATEGORY	CLASS	WINNER	TIME OR DISTANCE	CAR
4-5 June	Vineland (New Jersey), USA	Garden State Trophy (circuit)	---	---	A. Cosentino	---	Fiat Abarth 1000 Berlina
4-5 June	Zandvoort, Holland	Stunden Rennen Zandvoort (circuit-4,225 kms x 1 hour)	Touring	850-1000 cc	J. Menneken	53 laps	Fiat Abarth 1000 Berlina
4-5 June	Heidelberg, Germany	Heidelberg Preis (circuit-2,600 kms x 20 laps)	Sport Sport	1000 cc 1300 cc	F. Neuser D. Dau	--- ---	Fiat Abarth 1000 BA Fiat Abarth OT 1300
4-5 June	Antibes, France	Rallye des Roses (395,400 kms)	Touring Grand Touring	600-850 cc 500-850 cc	Armando-Camatte Branche-Jullien	--- ---	Fiat Abarth 850 TC Fiat Abarth 700 BA
5 June	Treviso, Italy	Vittorio Veneto-Cansiglio (10,000 kms)	Touring Touring Touring Sport Sport	700 cc 850 cc 1000 cc 1000 cc 1300 cc	G. Caffi S. Uberti 'Matich' M. Pinchetti A. Becchetti	8'07"2 7'22"8 7'17"3 7'00"3 7'14"2	Fiat Abarth 695 Fiat Abarth 850 TC Fiat Abarth 1000 Berlina Fiat Abarth 1000 BA Fiat Abarth OT 1300
5 June	Villach, Austria	Dobratsch Bergrennen (hillclimb-15,800 kms)	Touring	up to 1300 cc	M. Endress	9'03"1	Fiat Abarth 1000 Berlina
5 June	Adenau, Germany	ADAC Nürburgring 1000 Km Rennen (circuit-22,810 kms x 44 laps)	Sport Sport-Prototypes	1300 cc 1300 cc	Ahrens-Furtmayr Abels-Bitter	7h04'12"2 7h10'17"1	Fiat Abarth OT 1300 Fiat Abarth OT 1300
5 June	Rothenburg, Germany	ADAC Bergrennen Rothenburg (hillclimb-3,000 kms)	Touring	1000 cc	H.P. Klein	3'58"8	Fiat Abarth 1000 Berlina
5 June	Payerne, Switzerland	Slalom National de Payerne (5,200 kms x 2)	Touring Touring Sport Sport	850 cc 1000 cc 1000 cc 1300 cc	R. Stierli A. Wittenbach K. Bolzhauser J. Kretschi	3'19"4 3'17"5 3'17"6 3'07"2	Fiat Abarth 850 TC Fiat Abarth 1000 Berlina Fiat Abarth 1000 BA Fiat Abarth OT 1300
5 June	Gubbio, Italy	Coppa Mengara (hillclimb-5,200 kms)	Touring Touring Touring Touring Sport	600 cc 700 cc 850 cc 1000 cc 1300 cc	G. Caroli R. Borgiani 'Cristi' R. Puccinelli G. Villani	3'20"3 3'39"5 2'53"5 2'56"9 2'43"9	Fiat Abarth 595 Fiat Abarth 695 SS Fiat Abarth 850 TC Fiat Abarth 1000 Berlina Fiat Abarth OT 1300
5 June	Torino, Italy	Cuorgné-Alpette (hillclimb-7,000 kms)	Touring Touring Touring Sport Sport Sport	600 cc 850 cc up to 1150 cc 1000 cc 1300 cc 2000 cc	A. Poggio B. Pagliero G. Gavelli F. Pilone G. Rebaudi A. Brini (overall)	5'49"5 5'15"2 5'14"8 4'55"7 4'44"0 4'43"8	Fiat Abarth 595 Fiat Abarth 850 TC Fiat Abarth 1000 Berlina Fiat Abarth 1000 BA Fiat Abarth OT 1300 Abarth Simca 2000
5 June	Bourges, France	Grand Prix de la Châtre (circuit-1,267 kms x 20 laps)	Touring	1000 cc	C. Ballot Lena (overall)	15'50"5	Fiat Abarth 1000 Berlina
5 June	Kolind, Danimarca	Ring Kolind Djursland (circuit)	Touring	1000 cc	H. Jurgensen	---	Fiat Abarth 1000 Berlina
5 June	Taranto, Italy	Rallye di Puglia e Lucania (475 kms)	Touring	1000 cc	G. Tommasi	---	Fiat Abarth 1000 Berlina
9 June	Palermo, Italy	Alcamo-Monte Bonifato (hillclimb-4,500 kms)	Touring Touring Touring Sport Sport Sport	600 cc 850 cc up to 1150 cc 1000 cc 1300 cc 2000 cc	I. Serse F. Mangiapane S. Calascibetta M. Spadafora S. Alongi S. Andolina	3'49"8 3'30"6 3'12"6 3'12"6 2'59"1 2'56"3	Fiat Abarth 595 Fiat Abarth 850 TC Fiat Abarth 1000 Berlina Fiat Abarth 1000 BA Abarth Simca 1300 Abarth Simca 2000
9 June	Asti, Italy	Circuito di S. Secondo (157 kms)	---	---	'Villi' (overall)	---	Fiat Abarth 850 TC
11 June	Noordwijk, Holland	G.P. Autoslalom	Touring Touring	850 cc 1000 cc	J. van Osch A. Hezemans	1'12"1 1'04"4	Fiat Abarth 850 TC Fiat Abarth 1000 Berlina
11 June	Goodwood, United Kingdom	Race Meeting Goodwood (circuit-2,400 miles x 10 laps)	Touring	850 cc	J. Anstead	18'43"6	Fiat Abarth 850 TC
11 June	Lime Rock (Connecticut), USA	Championship Races Lime Rock (circuit-15 laps)	---	---	A. Cosentino	---	Fiat Abarth 1000 Berlina
11 June	Zagreb, Jugoslavia	Premio di Hrvatskeg Zagorja (circuit)	Touring	850 cc	R. Jelovcan	19'55"0	Fiat Abarth 850 TC
12 June	Livorno, Italy	Antignano-Monteburrone (hillclimb-3,500 kms x 2)	Touring Touring Touring	600 cc 850 cc 1000 cc	'Antarcs' B. Pagliero F. Zanardelli	5'36"7 4'45"7 4'44"1	Fiat Abarth 595 Fiat Abarth 850 TC Fiat Abarth 1000 Berlina
12 June	Palermo, Italy	Corsa hillclimb al Monte Pellegrino (8,750 kms)	Touring Touring Touring Sport Sport	600 cc 850 cc 1000 cc 1000 cc 1300 cc	I. Serse F. Mangiapane S. Calascibetta F. Patané S. Alongi	6'30"1 5'57"2 5'38"2 5'50"8 5'27"5	Fiat Abarth 595 Fiat Abarth 850 TC Fiat Abarth 1000 Berlina Fiat Abarth 1000 BA Abarth Simca 1300
12 June	Bologna, Italy	Coppa Acqua Cerelia (hillclimb-8,000 kms)	Touring Touring Touring Grand Touring Sport Sport Sport Sport-Prototypes	600 cc 850 cc 1000 cc 1000 cc 1000 cc 1300 cc 2000 cc 1000 cc	R. Pinto S. Uberti M. Nesti F. Salomoni E. Guerrieri 'Yama-Arashi' A. Brini 'Mister Brandy'	9'35"7 8'42"5 8'32"0 10'21"0 10'00"1 11'20"3 8'13"8 9"19"0	Fiat Abarth 595 S Fiat Abarth 850 TC Fiat Abarth 1000 Berlina Fiat Abarth 1000 OTS Fiat Abarth 1000 BA Fiat Abarth OT 1300 Abarth Simca 2000 Fiat Abarth 1000

DATE	LOCATION	RACE	CATEGORY	CLASS	WINNER	TIME OR DISTANCE	CAR
12 June	San Cugat, Spagna	Carrera en cuesta San Cugat-Tibidabo (hillclimb-4,700 kms)	Touring	700 cc	J. Juncosa	3'51"4	Fiat Abarth 695
			Touring	850 cc	M. Juncosa	3'15"6	Fiat Abarth 850 TC
			Touring	851-1000 cc	J. Juncosa	3'05"6	Fiat Abarth 1000 Berlina
			Special				
			Sport	851-1000 cc	'Juanco'	3'08"1	Fiat Abarth 1000 BA
12 June	Berchtesgaden, Germany	Rossfeld Alpen Bergpreis (hillclimb-5,890 kms)	Touring	1000 cc	J. Abt	7'52"01	Fiat Abarth 1000 Berlina
			Sport	1000 cc	K. Huser	7'31"3	Fiat Abarth 1000 BA
			Sport	1300 cc	M. Maron	7'20"2	Fiat Abarth OT 1300
12 June	Valence, France	Course de côte du Pin (hillclimb-7,050 kms)	Touring	850 cc	A. Tedeschi	5'33"9	Fiat Abarth 850 TC
			Touring	1000 cc	C. Ballot Lena	5'04"1	Fiat Abarth 1000 Berlina
			Sport	1601-2000 cc	P. Maublanc (overall)	4'41"1	Fiat Abarth OT 2000
12 June	Salamanca, Spagna	Premio Automovilistico de Salamanca (circuit-1,400 kms x 25 laps)	Touring	850 cc	E. Villancieros	27'51"9	Fiat Abarth 850 TC
			Touring	851-1150 cc	F. Lencina	26'49"3	Fiat Abarth 1000 Berlina
			Sport	850 cc	E. Villancieros	27'00"0'	Fiat Abarth 850
12 June	Helsinki, Finlandia	Corsa Gr. 2	Touring	1000 cc	H. Liedes	18'23"8	Fiat Abarth 1000 Berlina
12 June	Snetterton, United Kingdom	N.G.K. Trophy Race (circuit-2,700 miles x 8 laps)	Touring	up to 875 cc	J. Anstead	18'21"0	Fiat Abarth 695
			Touring	up to 900 cc	J. Anstead	17'45"0	Fiat Abarth 850 TC
18 June	Silverstone, United Kingdom	B.A.R.C. Race Meeting (circuit-2,920 miles x 10 laps)	Touring	850 cc	J. Anstead	12'59"6	Fiat Abarth 850 TC
18 June	Long Island (New York), USA	Bridgehampton Circuit (2,85 miles x 30')	---	---	A. Cosentino	---	Fiat Abarth 1000 Berlina
19 June	Arezzo, Italy	Camucia-Cortona (hillclimb-3,500 kms x 2)	Touring	600 cc	A. Andreini	5'47"9	Fiat Abarth 595
			Touring	850 cc	P. Falorni	5'06"3	Fiat Abarth 850 TC
			Touring	1000 cc	R. Puccinelli	5'20"4	Fiat Abarth 1000 Berlina
			Sport	1300 cc	'Nicor'	4'49"7	Fiat Abarth OT 1300
19 June	Cosenza, Italy	Coppa della Sila (hillclimb-14,000 kms)	Touring	600 cc	G. Lopez	12'13"3	Fiat Abarth 595
			Touring	700 cc	U. Barilaro	11'44"1	Fiat Abarth 695
			Touring	850 cc	M. Terminiello	11'22"1	Fiat Abarth 850 TC
			Touring	up to 1150 cc	S. Calascibetta	10'38"4	Fiat Abarth 1000 Berlina
			Grand Touring	1000 cc	R. Lo Celso	12'12"4	Fiat Abarth 1000 OTS
			Sport	1000 cc	M. Spadafora	10'48"0	Fiat Abarth 1000 BA
			Sport	1300 cc	C. Avventurieri	10'04"1	Fiat Abarth OT 1300
			Sport-Prototypes	1000 cc	G. Tommasi	10'59"3	Fiat Abarth 1000
19 June	Mainz Finthen, Germany	HMSC Flugplatzrennen (circuit-2,100 kms x 30 laps)	Touring	1000 cc	A. Hezemans	38'10"6	Fiat Abarth 1000 Berlina
			Grand Touring	1000 cc	E. Bitter	39'04"8	Fiat Abarth 1000 OTS
			Sport	up to 1150 cc	F. Neusser	34'02"0	Fiat Abarth 1000 BA
			Sport	1300 cc	S. Dau	33'53"2	Fiat Abarth OT 1300
19 June	Hockenheim, Germany	A.C.S. Rundstreckerennen Hockenheim (circuit-2,690 kms x 20 laps)	Touring	850 cc	R. Blatter	31'53"0	Fiat Abarth 850 TC
			Touring	1000 cc	J. Brun	31'54"2	Fiat Abarth 1000 Berlina
			Sport (28 laps)	up to 850 cc	H. Affentranger	47'23"3	Fiat Abarth 700 BA
			Sport (30 laps)	1000 cc	N. Gody	47'31"5	Fiat Abarth1000 BA
			Sport (30 laps)	1600 cc	W. Rufenacht	44'50"4	Fiat Abarth OT 1600
19 June	Deggendorf, Germany	ADAC Rusel Bergrennen (hillclimb-5,100 kms)	Touring	1000 cc	J. Abt	2'50"6	Fiat Abarth 1000 Berlina
			Sport (2 manches)	1300 cc	E. Furtmayr	5'02"9	Fiat Abarth OT 1300
19 June	Cuneo, Italy	Garessio-Colle S. Bernardo (hillclimb-6,200 kms)	Touring	700 cc	R. Rasia dal Polo	4'55"1	Fiat Abarth 695
			Touring	850 cc	B. Pagliero	4'39"8	Fiat Abarth 850 TC
			Touring	up to 1150 cc	G. Gavelli	5'08"2	Fiat Abarth 1000 Berlina
			Sport	1000 cc	D. Lo Coco	5'10"8	Fiat Abarth 1000 BA
			Sport-Prototypes	1000 cc	G. Varese	4'45"7	Fiat Abarth 1000
			Sport-Prototypes	2000 cc	K. Steinmetz (overall)	4'08"9	Fiat Abarth OT 2000
19 June	Vallelunga, Italy	Coppa Vallelunga (circuit-1,800 kms x 20 laps)	Touring	up to 700 cc	C. Francisci	24'49"3	Fiat Abarth 595
			Touring	850 cc	P. Amato	22'49"6	Fiat Abarth 850 TC
			Touring	1000 cc	A. Saltari	22'27"0	Fiat Abarth 1000 Berlina
			Sport	1000 cc	G. Spicciani	22'29"0	Fiat Abarth 1000 BA
19 June	Macerata, Italy	Macerata-Tolentino (274,400 kms)	Touring	700 cc	A. Mozzarecchia	---	Fiat Abarth 695
25 June	Nîmes, France	Ronde Cevenole (450 kms)	Touring	1000 cc	Maublanc-Roussin	---	Fiat Abarth 1000 Berlina
25 June	Zürich, Switzerland	Slalom	Grand Touring	1000 cc	K. Meyer	4'44"1	Fiat Abarth 1000 BA
25-26 June	Albi, France	Coupe des violettes (circuit)	Sport	1300 cc	I. Barthe	1'40"5	Fiat Abarth OT 1300
26 June	Avüs, Germany	Grosser Preis von Berlin (circuit-8,300 kms x 10 laps)	Touring	1000 cc	W. Mantzel	31'06"5	Fiat Abarth 1000 Berlina
			Sport	up to 1600 cc	H. Dechent	23'04"8	Fiat Abarth OT 1300
26 June	Avignon, France	Course de côte du Mont Ventoux (hillclimb-21,700 kms)	Touring	850 cc	M. Cintolesi	14'13"0	Fiat Abarth 850 TC
			Touring	1000 cc	J. Ortner	13'09"8	Fiat Abarth 1000 Berlina
			Sport	1000 cc	J. Mazzanti	12'56"3	Fiat Abarth 1000 BA
			Sport	1600 cc	W. Rufenacht	12'17"6	Fiat Abarth OT 1600
26 June	Gau, Germany	ADAC Taubensuhl Bergrennen (hillclimb-6,000 kms x 2)	Touring	1000 cc	E. Bitter	7'55"3	Fiat Abarth 1000 Berlina
			Sport	1300 cc	S. Dau	7'52"7	Fiat Abarth OT 1300

DATE	LOCATION	RACE	CATEGORY	CLASS	WINNER	TIME OR DISTANCE	CAR
26 June	Strasbourg, France	Course de côte Turckheim Trois Epis (hillclimb-6,320 kms)	Touring Special	up to 1150 cc	P. Engler	4'13"4	Fiat Abarth 1000 Berlina
26 June	Villars, Switzerland	Course de côte de Villars Burquin (hillclimb-1,800 kms)	Sport	2000 cc	H. Muller	1'13"6	Fiat Abarth OT 2000
26 June	Innsbruck, Austria	Bergrennen zur Axamer Lizum (hillclimb-8,300 kms)	Touring	1000 cc	M. Endress	8'59"5	Fiat Abarth 1000 Berlina
			Sport	1300 cc	E. Furtmayr	7'43"5	Fiat Abarth OT 1300
26 June	Thun, Switzerland	Slalom Thun	Touring	850 cc	K. Schneiter	8'10"8	Fiat Abarth 850 TC
			Sport	1000 cc	H. Herren	8'14"4	Fiat Abarth 1000 BA
26 June	Snetterton, United Kingdom	Cambridge Cup Car Races (circuit-2,700 miles x 8 laps)	Touring	850 cc	G. Anstead	17'15"2	Fiat Abarth 850 TC
			Touring	1000 cc	A. Mylius (overall)	16'38"2	Fiat Abarth 1000 Berlina
26 June	Zürich, Switzerland	Zürich Surprise	Grand Touring	1000 cc	W. Friedrich	5'06"8	Fiat Abarth Monomille
26 June	Modena, Italy	Coppa Appennino Modenese (212 kms)	Touring	850 cc	A. Dalan	---	Fiat Abarth 850 TC
26 June	Frascati, Italy	Frascati-Tuscolo (hillclimb-4,200 kms)	Touring	600 cc	C. Francisci	3'27"0	Fiat Abarth 595
			Touring	700 cc	R. Causo	3'26"4	Fiat Abarth 695
			Touring	850 cc	A. Del Giovane	3'14"0	Fiat Abarth 850 TC
			Touring	1000 cc	F. Acasbo	3'21"6	Fiat Abarth 1000 Berlina
26 June	Salerno, Italy	Cava dei Tirreni-Badia (hillclimb-2,400 kms x 2)	Touring	850 cc	G. Esposito	3'44"3	Fiat Abarth 850 TC
			Touring	1000 cc	L. Rizzo	3'39"2	Fiat Abarth 1000 Berlina
			Sport	1300 cc	C. Avventurieri	3'28"3	Fiat Abarth OT 1300
			Sport	2000 cc	D. Scola (overall)	3'18"9	Abarth Simca 2000
26 June	Parma, Italy	Parma-Poggio di Berceto (hillclimb-13,000 kms)	Touring	700 cc	A. Polli	10'12"9	Fiat Abarth 695
			Touring	850 cc	E. Crivellari	9'18"8	Fiat Abarth 850 TC
			Touring	1000 cc	F. Zanardelli	8'55"5	Fiat Abarth 1000 Berlina
			Grand Touring	1000 cc	'Pedrin'	10'06"8	Fiat Abarth 1000 OTS
			Sport	1000 cc	P. Polin	8'54"3	Fiat Abarth 1000 BA
			Sport	1300 cc	E. Tondelli	8'29"3	Fiat Abarth OT 1300
			Sport-Prototypes	1000 cc	M. Nesti	8'29"1	Fiat Abarth 1000
29 June	Forli, Italy	Predappio-Rocca delle Caminate (hillclimb-4,000 kms)	Touring	700 cc	A. Polli	4'04"0	Fiat Abarth 695
			Touring	850 cc	B. Pagliero	3'43"3	Fiat Abarth 850 TC
			Touring	1000 cc	M. Nesti	3'35"8	Fiat Abarth 1000 Berlina
			Sport	1000 cc	D. Lo Coco	3'26"8	Fiat Abarth 1000 BA
29 June	Latina, Italy	Valviscolo-Bassiano (hillclimb-6,200 kms)	Touring	600 cc	C. Francisci	4'32"2	Fiat Abarth 595
			Touring	700 cc	R. Causo	4'30"2	Fiat Abarth 695
			Touring	850 cc	M. Tommasini	4'15"8	Fiat Abarth 850 TC
			Touring	1000 cc	V. Trulli	4'10"6	Fiat Abarth 1000 Berlina
			Sport	1300 cc	B. Faenza	4'05"3	Fiat Abarth 1000 BA
30 June	Principato d'Andorra	Pujada fra Miquel (hillclimb-10,900 kms)	Touring	1000 cc	E. Baturone	7'51"2	Fiat Abarth 1000 Berlina
			Touring Special	850 cc	J. Juncosa	8'05"7	Fiat Abarth 850 TC
2 July	Siracusa, Italy	3 Hours Night Race (circuit-5,500 kms)	Touring	850 cc	G. Nicoletti	53'14"4	Fiat Abarth 850 TC
			Touring	1000 cc	'Gordon'	52'26"5	Fiat Abarth 1000 Berlina
			Sport	up to 3000 cc	F. Patanè	47'39"2	Fiat Abarth 1000 BA
2 July	Goodwood, United Kingdom	Race Meeting Goodwood (circuit-2,274 miles x 10 laps)	Touring	850 cc	J. Anstead	18'18"4	Fiat Abarth 850 TC
			Touring	1000 cc	A. Mylius (overall)	17'43"6	Fiat Abarth 1000 Berlina
2 July	Villeneuve, Switzerland	Course de côte Villeneuve Surprise (hillclimb-1,850 kms)	Touring	1000 cc	F. Wipf	1'29"3	Fiat Abarth 1000 Berlina
2-3 July	Oberhallau, Switzerland	Bergpreis de Ostschweiz in Oberhallau (hillclimb-3,000 kms x 2)	Touring	850 cc	R. Stierli	4'00"9	Fiat Abarth 850 TC
			Touring	1000 cc	J. Brun	3'46"6	Fiat Abarth 1000 Berlina
			Grand Touring	1000 cc	J. Gloggner	4'13"5	Fiat Abarth 1000 OTS
			Sport	1000 cc	C. Ballot Lena	3'38"5	Fiat Abarth 1000 BA
3 July	Graz, Austria	Stubenbergrennen (hillclimb-6,500 kms)	Touring	1000 cc	J. Abt	4'08"5	Fiat Abarth 1000 Berlina
3 July	Dalsland, Sweden	Dalsland Ring (circuit-12 laps)	Touring	1000 cc	B. Johansson	11'33"8	Fiat Abarth 1000 Berlina
3 July	Vianden, Luxembourg	Course de côte Vianden (hillclimb-1,800 kms)	Touring	850 cc	J. van Osch	1'53"3	Fiat Abarth 850 TC
			Touring	1000 cc	P. Engler	1'52"3	Fiat Abarth 1000 Berlina
3 July	Adenau, Germany	Nürburgring Grosser Preis der Tourenwagen (circuit-22,810 kms x 6 hours)	Touring	1000 cc	Mantzel-Baghetti	6h03'57"9	Fiat Abarth 1000 Berlina
3 July	Bolzano, Italy	Bolzano-Mendola (hillclimb-14,400 kms)	Touring	850 cc	B. Pagliero	11'0"66	Fiat Abarth 850 TC
			Touring	1000 cc	P. Cattelan	11'05"5	Fiat Abarth 1000 Berlina
			Sport	1000 cc	M. Pinchetti	10'27"3	Fiat Abarth 1000 BA
			Sport	1300 cc	A. Becchetti	10'53"0	Fiat Abarth OT 1300
3 July	Monopoli, Italy	Coppa Città di Monopoli (hillclimb-4,500 kms)	Touring	600 cc	A. Angheloni	3'01"6	Fiat Abarth 595
			Touring	700 cc	S. Pezzolla	2'47"7	Fiat Abarth 695
			Touring	850 cc	G. Della Mura	2'35"3	Fiat Abarth 850 TC
			Touring	1000 cc	S. Calascibetta	2'26"3	Fiat Abarth 1000 Berlina
			Sport	1000 cc	M. Spadafora	2'20"7	Fiat Abarth 1000 BA
			Sport-Prototypes	1000 cc	G. Tommasi	2'20"7	Fiat Abarth 1000

DATE	LOCATION	RACE	CATEGORY	CLASS	WINNER	TIME OR DISTANCE	CAR
3 July	Frosinone, Italy	10 Miglia, Guarcino-Campocatino (hillclimb-16,950 kms)	Touring	600 cc	A. Magnoni	13'03"4	Fiat Abarth 595
			Touring	700 cc	R. Rasia Dal Polo	12'50"8	Fiat Abarth 695
			Touring	850 cc	M. Terminiello	12'31"7	Fiat Abarth 850 TC
			Touring	1000 cc	S. Mingotti	12'12"3	Fiat Abarth 1000 Berlina
4 July	Lime Rock (Connecticut), USA	Championship Races (circuit-24 laps)	---	---	A. Cosentino	---	Fiat Abarth 1000 Berlina
10 July	Trento, Italy	Trento-Bondone (hillclimb-17,300 kms)	Touring	850 cc	S. Uberti	14'36"8	Fiat Abarth 850 TC
			Touring	1000 cc	M. Nesti	14'07"3	Fiat Abarth 1000 Berlina
			Grand Touring	1000 cc	'Pedrin'	16'01"4	Fiat Abarth 1000 OTS
			Sport	1000 cc	M. Pinchetti	14'09"9	Fiat Abarth 1000 BA
			Sport	1300 cc	G. Rebaudi	13'35"8	Fiat Abarth OT 1300
10 July	Brindisi, Italy	Coppa Selva di Fasano (hillclimb-6,500 kms)	Touring	600 cc	G. Lopez	4'18"4	Fiat Abarth 595
			Touring	700 cc	R. Causo	4'03"3	Fiat Abarth 695
			Touring	850 cc	B. Pagliero	3'44"8	Fiat Abarth 850 TC
			Touring	1000 cc	S. Calascibetta	3'33"1	Fiat Abarth 1000 Berlina
			Grand Touring	1000 cc	V. Carrieri	4'11"2	Fiat Abarth 1000 OTS
			Sport	1000 cc	M. Spadafora	3'27"5	Fiat Abarth 1000 BA
			Sport	up to 1600 cc	C. Avventurieri	3'22"9	Fiat Abarth OT 1300
			Sport-Prototypes	1000 cc	G. Tommasi (overall)	3'28"9	Fiat Abarth 1000
10 July	Gives, Belgium	Course de côte de Gives (hillclimb-2,400 kms)	Touring	850 cc	J. van Osch	1'43"8	Fiat Abarth 850 TC
10 July	Neuss, Germany	ADAC Autoslalom (1,600 kms x 2 laps)	Grand Touring	up to 1300 cc	F. Kalkuhl	---	Fiat Abarth 1000 OTS
10 July	Corsavy, France	Course de côte de Corsavy (circuit-5,200 kms)	Touring	850 cc	S. Stefani	3'59"4	Fiat Abarth 850 TC
			Touring	1000 cc	E. Baturone	3'59"4	Fiat Abarth 1000 Berlina
			Sport	up to 1150 cc	C. Ballot Lena	3'57"0	Fiat Abarth 1000 BA
16 July	Monthoux, Switzerland	Rundkurs Monthoux (circuit-10 laps x 2)	Touring	1000 cc	E. Welti	15'18"9	Fiat Abarth 1000 Berlina
			Sport	1600 cc	W. Rufenacht	14'34"1	Fiat Abarth OT 1600
16-17 July	Hockenheim, Germany	Firestone Pokal (circuit-2,690 kms x 25 laps)	Touring	1000 cc	J. Brun	39'45"8	Fiat Abarth 1000 Berlina
17 July	Magny Cours, France	Criterium de Vitesse (circuit-2,040 kms x 25 laps)	Sport	up to 1150 cc	C. Ballot Lena	25'39"5	Fiat Abarth 1000 BA
17 July	Scarperia, Italy	Circuito del Mugello (circuit-66,200 kms x 8 laps)	Sport	1000 cc	Zurchi-Moselli	5h23'40"0	Fiat Abarth 1000 BA
			Sport	1300 cc	Taramazzo-Bona	5h13'55"6	Fiat Abarth OT 1300
			Sport-Prototypes	1000 cc	M. Nesti	4h59'31"2	Fiat Abarth 1000
17 July	Wasgau, Germany	ADAC Wasgau Bergprüfung (hillclimb-5,100 kms x 2)	Touring	1000 cc	J. Abt	5'48"9	Fiat Abarth 1000 Berlina
			Sport	1300 cc	S. Dau	5'25"9	Fiat Abarth OT 1300
17 July	Belluno, Italy	Agordo-Frassenè (hillclimb-10,020 kms)	Touring	600 cc	G. Bassi	8'26"4	Fiat Abarth 595
			Touring	700 cc	M. Zanetti	8'08"8	Fiat Abarth 695
			Touring	850 cc	'Girasole'	7'38"6	Fiat Abarth 850 TC
			Touring	1000 cc	P. Lado	7'26"0	Fiat Abarth 1000 Berlina
			Sport	1000 cc	L. Caprioli	7'17"0	Fiat Abarth 1000 BA
			Sport	1300 cc	G. Dalla Torre	7'25"1	Abarth Simca 1300
			Sport	2000 cc	P. Lado	7'03"6	Abarth Simca 2000
17 July	Biella, Italy	Coppa Carlo Felice Trossi (hillclimb-6,500 kms)	Touring	600 cc	A. Poggio	5'22"2	Fiat Abarth 595
			Touring	850 cc	S. Piana	4'59"4	Fiat Abarth 850 TC
			Touring	1000 cc	G. Gavelli	5'14"7	Fiat Abarth 1000 Berlina
			Sport	1000 cc	F. Pilone	4'55"3	Fiat Abarth 1000 BA
			Sport	1300 cc	E. Tondelli	4'52"7	Fiat Abarth OT 1300
			Sport	2000 cc	A. Caffi (overall)	4'48"1	Abarth Simca 2000
17 July	Verona, Italy	Gymkana Auto Zevio	---	up to 750 cc	Broggi-Moiola	---	Fiat Abarth 595
22-23 July	Zandvoort, Holland	Grand Prix Zandvoort (circuit-4,225 kms x 16 laps)	Touring	1000 cc	E. Swart	32'13"2	Fiat Abarth 1000 Berlina
23-24 July	St. Germain Bini, Switzerland	Course de côte de Saviese (hillclimb-2,700 kms x 2)	Touring	1000 cc	J. Brun	3'20"6	Fiat Abarth 1000 Berlina
			Sport	1300 cc	D. Borel	3'08"6	Abarth Simca 1300
24 July	Sestrière, Italy	Cesana-Sestrière (hillclimb-10,400 kms)	Touring	600 cc	A. Poggio	7'59"5	Fiat Abarth 595
			Touring	850 cc	B. Pagliero	6'53"5	Fiat Abarth 850 TC
			Touring	1000 cc	F. Zanardelli	7'00"8	Fiat Abarth 1000 Berlina
			Grand Touring	1000 cc	G. Silecchia	7'43"8	Fiat Abarth 1000 OTS
			Sport	1000 cc	M. Pinchetti	6'56"2	Fiat Abarth 1000 BA
			Sport	1300 cc	G. Bona	6'32"7	Fiat Abarth OT 1300
			Sport-Prototypes	1000 cc	G. Varese	6'45"6	Fiat Abarth 1000
24 July	Hockenheim, Germany	Grosser Preis für Tourenwagen (circuit-2,700 kms x 15 laps)	Touring	1000 cc	W. Mantzel	41'35"0	Fiat Abarth 1000 Berlina
			Touring Special	701-1000 cc	L. Cirincione	46'35"5	Fiat Abarth 850 TC
			Sport	up to 1300 cc	W. Dau	37'16"3	Fiat Abarth OT 1300
24 July	Bamberg, Germany	ADAC Ellerberg Rennen (hillclimb-6,080 kms)	Touring	1000 cc	M. Endress	4'09"6	Fiat Abarth 1000 Berlina
			Sport	1000 cc	F. Neuser	4'07"3	Fiat Abarth 1000 BA
24 July	Rouen, France	Course de côte des Andelys (hillclimb-1,380 kms)	Sport	1000 cc	J. Mazzanti	0'43"0	Fiat Abarth 1000 BA
24 July	Connesville Airport, USA	Connesville National Races (circuit)	---	---	A. Cosentino	---	Fiat Abarth 1000 Berlina

DATE	LOCATION	RACE	CATEGORY	CLASS	WINNER	TIME OR DISTANCE	CAR
30-31 July	Grenoble, France	Course de côte Chamrousse (hillclimb-11,850 kms)	Touring Touring	850 cc 1000 cc	A. Monselles A. Raymond	9'13"9 9'26"1	Fiat Abarth 850 TC Fiat Abarth 1000 Berlina
31 July	Snetterton, United Kingdom	500 kms of Snetterton (circuit-2,710 miles x 105 laps)	Touring Touring	850 cc 1000 cc	Anstead-Anstead J. Fitzpatrick	4h13'19"8 4h13'49"4	Fiat Abarth 850 TC Fiat Abarth 1000 Berlina
31 July	Freiburg, Germany	ADAC Bergpreis Freiburg-Schauinsland (hillclimb-11,200 kms)	Sport Sport	1000 cc 1300 cc	R. Stierli K. Federhofer	14'45"3 14'59"9	Fiat Abarth 1000 BA Fiat Abarth OT 1300
31 July	Termini Imerese, Italy	Termini-Caccamo (hillclimb-10,000 kms)	Touring	600 cc	I. Serse	7'03"3	Fiat Abarth 595
			Touring	700 cc	A. La Luce	6'52"7	Fiat Abarth 695
			Touring	850 cc	F. Mangiapane	6'42"0	Fiat Abarth 850 TC
			Touring Sport	1000 cc 1300 cc	S. Calascibetta V. Ferlito	6'19"1 6'12"5	Fiat Abarth 1000 Berlina Abarth Simca 1300
31 July	Trieste, Italy	Trieste-Opicina (hillclimb-10,150 kms)	Touring	600 cc	A. Magnoni	6'11"2	Fiat Abarth 595
			Touring	850 cc	S. Uberti	5'36"0	Fiat Abarth 850 TC
			Touring Sport	1000 cc 1000 cc	G. Franzolin M. Pinchetti	5'34"6 5'07"6	Fiat Abarth 1000 Berlina Fiat Abarth 1000 BA
			Sport	1300 cc	C. Avventurieri	5'03"8	Fiat Abarth OT 1300
31 July	Oberpfalz, Germany	Bergpreis der Oberpfalz (hillclimb-5,800 kms)	Sport Sport	1000 cc 1300 cc	F. Neuser H. Krause	--- ---	Fiat Abarth 1000 BA Fiat Abarth OT 1300
31 July	Kolind, Danimarca	Kanonlobet Ring Djursland (circuit)	Touring	1000 cc	H. Jurgensen	23'18"7	Fiat Abarth 1000 Berlina
31 July	Viterbo, Italy	Corsa Lago di Montefiascone (hillclimb-5,000 kms)	Touring Touring Touring Sport Sport	600 cc 850 cc 1000 cc 1000 cc 1300 cc	R. Mazzeschi A. Del Giovane A. Saltari S. Pandolfo L. Mugelli	3'20"9 2'55"8 2'47"1 2'47"8 2'42"8	Fiat Abarth 595 Fiat Abarth 850 TC Fiat Abarth 1000 Berlina Fiat Abarth 1000 BA Abarth Simca 1300
6 August	Adenau, Germany	ADAC Nürburgring Trophäe (circuit-22,810 kms x 5 laps)	Touring	1000 cc	W. Kauhsen	54'19"4	Fiat Abarth 1000 Berlina
7 August	Udine, Italy	Tolmezzo-Verzeignis (hillclimb-4,260 kms x 2)	Touring Touring Touring Touring Sport	600 cc 700 cc 850 cc 1000 cc 1300 cc	A. Lepori R. Galasso V. Tabucchi P. Cattelan G. Vettor	6'26"4 6'30"5 5'31"6 5'24"4 6'33"4	Fiat Abarth 595 Fiat Abarth 695 Fiat Abarth 850 TC Fiat Abarth 1000 Berlina Abarth Simca 1300
7 August	Adenau, Germany	AVD Rheinland Pfalz Preis (Nürburgring) (circuit-22,810 kms x 7 laps)	Grand Touring Sport	1000 cc 1300 cc	F. Kleine M. Abels	1h17'01"5 1h19'51"7	Fiat Abarth 1000 OTS Fiat Abarth OT 1300
7 August	Stainz, Austria	Stainzer Bergrennen (hillclimb-5,500 kms)	Touring Touring	850 cc 1000 cc	R. Stierli J. Abt	4'03"1 3'52"07	Fiat Abarth 850 TC Fiat Abarth 1000 Berlina
7 August	Pergusa, Italy	Coppa Città di Enna (circuit-4,797 kms x 70 laps)	Sport Sport	1000 cc 1300 cc	F. Patanè A. Gambero	1h41'52"6 1h42'19"2	Fiat Abarth 1000 BA Fiat Abarth OT 1300
7 August	Siena, Italy	Abbadia S. Salvatore-Vetta Amiata (hillclimb-12,700 kms)	Touring Touring Touring Touring Sport	600 cc 700 cc 850 cc 1000 cc 1300 cc	G. Caroli G. Brancadori B. Pagliero M. Nesti C. Avventurieri	9'53"0 9'27"9 8'45"9 8'28"5 7'54"2	Fiat Abarth 595 Fiat Abarth 695 Fiat Abarth 850 TC Fiat Abarth 1000 Berlina Fiat Abarth OT 1300
7 August	Ulm, Germany	ADAC Bergpreis Ulm (hillclimb)	Touring Sport Sport	850-1150cc 1000 cc 1300 cc	M. Endress F. Neuser K. Federhofer	3'02"9 3'07"9 3'00"6	Fiat Abarth 1000 Berlina Fiat Abarth 1000 BA Abarth Simca 1300
7 August	London (New Hampshire), USA	Bryar Motor Sport Park (circuit-1,600 miles)	---	---	A. Cosentino	---	Fiat Abarth 1000 Berlina
7 August	Avezzano, Italy	Trofeo Micangeli (hillclimb-6,000 kms)	Touring Touring	up to 700 cc 850 cc	P. Valeri M. Tommasini	4'01"5 3'39"9	Fiat Abarth 595 Fiat Abarth 850 TC
13-14 August	Wunstorf, Germany	Flugplatzrennen Wunstorf (circuit-5,100 kms x 10 laps)	Touring Grand Touring	1000 cc 1000 cc	E. Bitter E. Bitter	25'26"5 27'33"9	Fiat Abarth 1000 Berlina Fiat Abarth 1000 OTS
13-14 August	Zandvoort, Holland	Grossen Girling Bremsen Preises (circuit-4,225 kms x 25 laps)	Sport	1300 cc	M. Abels	---	Fiat Abarth OT 1300
13-14 August	Beck, Holland	Vliegveld Beck (slalom)	Touring	up to 1000 cc	J. van Osch	1'27"7	Fiat Abarth 850 cc
14 August	Hockenheim, Germany	Grosser Preis von Hockenheim (circuit-6,768 kms x 57 laps)	Sport	1300 cc	H. Krause	2h42'07"6	Fiat Abarth OT 1300
14 August	Marlboro (Maryland), USA	Marlboro 12 Hours (circuit-1,700 miles)	---	---	Cuomo-Fisher	---	Fiat Abarth 1000 Berlina
14 August	Nogaro, France	Circuit International de Nogaro (circuit-1,750 kms x 20 laps)	Sport	up to 1150 cc	J. Mazzanti	22'56"0	Fiat Abarth 1000 BA
14 August	Lydden, United Kingdom	Race Meeting Lydden Hill (circuit x 12 laps)	Touring	850 cc	J. Anstead	8'51"6	Fiat Abarth 850 TC
15 August	Bockholtz-Goesdorf, Luxembourg	Course de côte Goesdorf-Ardennes (hillclimb-2,800 kms)	Touring Touring Sport	850 cc 851-1300 cc 1000 cc	R. Blatter W. Kauhsen J. Hauser	1'46"2 1'44"3 1'43"4	Fiat Abarth 850 TC Fiat Abarth 1000 Berlina Fiat Abarth 1000 BA

DATE	LOCATION	RACE	CATEGORY	CLASS	WINNER	TIME OR DISTANCE	CAR
15 August	Lucca, Italy	Coppa Fornaci di Barca (hillclimb-3,200 kms x 2)	Touring	600 cc	A. Andreini	5'29"5	Fiat Abarth 595
			Touring	700 cc	A. Polli	5'26"3	Fiat Abarth 695
			Touring	850 cc	A. Mostardini	5'13"5	Fiat Abarth 850 TC
			Touring	1000 cc	M. Nesti	4'46"3	Fiat Abarth 1000 Berlina
			Grand Touring	1000 cc	'Pedrin'	5'33"9	Fiat Abarth 1000 OTS
			Sport	1000 cc	G. Sangermano	5'40"9	Fiat Abarth 1000 BA
15 August	Mont-Dore, France	Course de côte du Mont-Dore (hillclimb-5,000 kms)	Touring	850 cc	Maury	4'16"8	Fiat Abarth 850 TC
			Sport	1000-1150 cc	C. Ballot Lena	3'29"8	Fiat Abarth 1000 BA
15 August	Pescara, Italy	Cronoscalata Svolte di Popoli (hillclimb-8,000 kms)	Touring	700 cc	M. Zanetti	5'39"4	Fiat Abarth 695
			Touring	850 cc	E. Gatti	5'30"7	Fiat Abarth 850 TC
			Touring	1000 cc	S. Calascibetta	5'07"4	Fiat Abarth 1000 Berlina
			Sport	1300 cc	C. Avventurieri	4'51"8	Fiat Abarth OT 1300
20 August	Watkins Glen (New York), USA	Watkins Touring Trophy (circuit-2,300 miles x 20 laps)	---	---	A. Cosentino	---	Fiat Abarth 1000 Berlina
21 August	Saint Ursanne, Switzerland	Course de côte Saint Ursanne-Les Rangiers (hillclimb-4,700 kms x 2)	Touring	850 cc	A. Monselles	5'44"8	Fiat Abarth 850 TC
			Touring	1000 cc	J.P. Brun	5'37"6	Fiat Abarth 1000 Berlina
			Sport	1000 cc	H. Kennel	5'19"4	Fiat Abarth 1000 BA
			Sport	1150-1300 cc	J. Kretschi	5'13"5	Fiat Abarth OT 1300
			Sport	1301-1600 cc	W. Rufenacht	4'59"5	Fiat Abarth OT 1600
21 August	Ascoli Piceno, Italy	Corsa al Colle San Marco (hillclimb-10,200 kms)	Touring	600 cc	A. Magnoni	7'27"1	Fiat Abarth 595
			Touring	700 cc	M. Zanetti	7'08"2	Fiat Abarth 695
			Touring	850 cc	F. Galli	6'46"3	Fiat Abarth 850 TC
			Touring	1000 cc	S. Calascibetta	6'35"0	Fiat Abarth 1000 Berlina
			Sport	1300 cc	C. Avventurieri	6'18"1	Fiat Abarth OT 1300
			Sport-Prototypes	1000 cc	G. Tommasi	6'29"3	Fiat Abarth 1000
21 August	Pergusa, Italy	Trofeo A.C. Siciliani (circuit-4,797 kms x 10 laps)	Touring	600 cc	A. Guagliardo	23'19"0	Fiat Abarth 595
			Touring	850 cc	'Jimmy'	19'308	Fiat Abarth 850 TC
			Touring	1000 cc	'Gordon'	27'25"3	Fiat Abarth 1000 Berlina
21 August	Brescia, Italy	Trofeo Valle Camonica (hillclimb-8,600 kms)	Touring	700 cc	G. Caffi	6'08"3	Fiat Abarth 695
			Touring	850 cc	V. Tabucchi	5'45"7	Fiat Abarth 850 TC
			Touring	1000 cc	F. Zanardelli	5'30"9	Fiat Abarth 1000 Berlina
			Grand Touring	1000 cc	G. Dolci	6'24"8	Fiat Abarth 1000 OTS
			Sport	1000 cc	D. Lo Coco	5'15"4	Fiat Abarth 1000 BA
			Sport	1300 cc	F. Genta	5'11"5	Fiat Abarth OT 1300
27 August	Lime Rock (Connecticut), USA	Regional Race-Lime Rock (circuit-10 laps)	---	---	A. Cosentino	---	Fiat Abarth 1000 Berlina
27 August	Rotterdam, Holland	Vlicoveld Zestienhoven (circuit)	Touring	850 cc	J. van Osch	1'10"0	Fiat Abarth 850 TC
			Touring	1000 cc	A. Hezemans	1'04"7	Fiat Abarth 1000 Berlina
28 August	Montana Crans, Switzerland	Gran Prix d'Europe de la Montagne Sierre-Montana Crans (hillclimb-11,000 kms x 2)	Touring	850 cc	R. Blatter	17'05"0	Fiat Abarth 850 TC
			Touring	1000 cc	J.P. Brun	15'58"5	Fiat Abarth 1000 Berlina
			Grand Touring	1000 cc	J. Gloggner	17'05"9	Fiat Abarth 1000 OTS
			Sport	up to 1000 cc	J. Kretschi	15'30"6	Fiat Abarth 1000 BA
			Sport	up to 1150 cc	E. Ostini	15'50"8	Fiat Abarth 1000 BA
			Sport	1300 cc	J. Grandjean	14'45"9	Abarth Simca 1300
			Sport-Prototypes	850 cc	G. Carena	17'35"5	Fiat Abarth 850
			Sport-Prototypes	1300-1600 cc	D. Borel	14'22"4	Fiat Abarth OT 1600
28 August	Chieti, Italy	Coppa Città di Chieti (hillclimb-6,200 kms)	Touring	700 cc	L. Finelli	4'12"9	Fiat Abarth 695
			Touring	850 cc	'Cristi'	3'48"2	Fiat Abarth 850 TC
			Touring	up to 1150 cc	A. Saltari	3'36"1	Fiat Abarth 1000 Berlina
			Sport	1000 cc	G. Barrella	3'48"0	Fiat Abarth 1000 BA
			Sport	1300 cc	P. Laureati	3'34"4	Abarth Simca 1300
			Sport	2000 cc	P. Gargano	3'38"8	Abarth Simca 2000
28 August	Mantova, Italy	Caprino-Spiazzi (hillclimb-10,000 kms)	Touring	600 cc	G. Poggi	6'24"2	Fiat Abarth 595
			Touring	700 cc	G. Caffi	6'20"2	Fiat Abarth 695
			Touring	850 cc	'Opicina'	5'46"0	Fiat Abarth 850 TC
			Touring	1000 cc	F. Zanardelli	5'34"3	Fiat Abarth 1000 Berlina
			Sport	1000 cc	M. Pinchetti	5'13"3	Fiat Abarth 1000 BA
28 August	Trapani, Italy	Trapani-Monte Erice (hillclimb-6,670 kms)	Touring	600 cc	I. Serse	5'39"0	Fiat Abarth 595
			Touring	700 cc	A. La Luce	5'35"7	Fiat Abarth 695
			Touring	850 cc	F. Mangiapane	5'18"0	Fiat Abarth 850 TC
			Touring	up to 1150 cc	'Zerimar'	5'14"8	Fiat Abarth 1000 Berlina
			Sport	1000 cc	F. Patanè	5'15"5	Fiat Abarth 1000 BA
			Sport	1300 cc	C. Avventurieri	4'46"5	Fiat Abarth OT 1300
28 August	Villefranche sur Saône, France	Course de côte du Beaujolais (hillclimb-16,000 kms)	Touring	850 cc	M. Cintolesi	10'52"3	Fiat Abarth 850 TC
			Touring	851-1050 cc	P. Maublanc (overall)	10'06"2	Fiat Abarth 1000 Berlina
			Sport	1000 cc	C. Ballot Lena	10'21"0	Fiat Abarth 1000 BA
			Sport	1300 cc	J. Mazzanti	10'26"0	Fiat Abarth OT 1300
28 August	Ixelles, Belgium	Slalom du Marathon (3,000 kms)	Touring	850 cc	Jones	3'17"1	Fiat Abarth 850 TC

DATE	LOCATION	RACE	CATEGORY	CLASS	WINNER	TIME OR DISTANCE	CAR
28 August	Augsburg, Germany	Augusta Bergrennen (hillclimb-2,100 kms x 2)	Sport	1300 cc	H. Krause	2'19"0	Fiat Abarth OT 1300
28 August	Kägeröd, Sweden	Ring Knutstorp (circuit-3,000 kms x 12)	Touring	1000 cc	L. Englund (overall)	9'51"6	Fiat Abarth 1000 Berlina
28 August	Veldenz, Germany	ADAC Hunsrück Bergrennen (hillclimb-3,600 kms x 2)	Touring	1000 cc	J. Abt	5'11"7	Fiat Abarth 1000 Berlina
			Sport	1300 cc	S. Dau	4'48"2	Fiat Abarth OT 1300
28 August	Bilbao, Spagna	Rallye Exemo	Touring	700 cc	M. Juncosa	---	Fiat Abarth 695
			Sport	1000 cc	J. Juncosa	---	Fiat Abarth 1000 BA
29 August	Snetterton, United Kingdom	Meeting Bank Holiday (circuit-2,700 miles x 8 laps)	Touring	850 cc	J. Anstead	17'42"6	Fiat Abarth 850 TC
1-3 September	S. Martino di Castrozza	Rallye di San Martino (1304 kms)	Touring	1000 cc	Rubbieri-Bini	---	Fiat Abarth 850 TC
4 September	Vallelunga, Italy	Trofeo Ettore Bettoja (circuit-3,200 kms x 35 laps)	Sport	1000 cc	M. Moselli	58'56"4	Fiat Abarth 1000 BA
			Sport	1300 cc	G. Bona	57'16"2	Fiat Abarth OT 1300
			Sport-Prototypes	1000 cc	G. Tommasi	1h00'14"5	Fiat Abarth 1000
4 September	Zandvoort, Holland	International Touring Car Races (circuit-4,225 kms x 35 laps)	Touring	850 cc	M. Jonker	1h09'01"2	Fiat Abarth 850 TC
			Touring	1000 cc	W. Mantzel	1h07'07"1	Fiat Abarth 1000 Berlina
4 September	Rouen, France	Course de côte de Belbeuf (hillclimb-2,800 kms)	Touring	1000 cc	C. Ballot Lena	1'58"2	Fiat Abarth 1000 Berlina
4 September	Gaisberg, Austria	Grosser Bergpreis von Österreich (hillclimb-8,625 kms)	Touring	1000 cc	M. Endress	5'07"3	Fiat Abarth 1000 Berlina
			Sport	1000 cc	J. Kretschi	5'21"8	Fiat Abarth 1000 BA
			Sport	1300 cc	K. Federhofer	5'01"4	Abarth Simca 1300
4 September	Adenau, Germany	Internationales 500 Km Nürburgring Rennen (circuit-22,810 kms)	Touring	1000 cc	W. Kauhsen	3h42'40"7	Fiat Abarth 1000 Berlina
			Sport-Prototypes	1000 cc	H. Müller	3h39'20"3	Fiat Abarth 1000 SP
			Sport-Prototypes	1300 cc	E. Furtmayr (overall)	3h39'17"3	Fiat Abarth OT 1300
4 September	Nürnberg, Germany	ADAC Bergrennen Happurg (hillclimb-3,130 kms x 2)	Touring	1000 cc	J. Abt	3'13"1	Fiat Abarth 1000 Berlina
			Sport	1000 cc	F. Neuser	3'12"1	Fiat Abarth 1000 BA
			Sport	1300 cc	H. Krause	3'12"0	Fiat Abarth OT 1300
4 September	Gex France	Course de côte de la Faucille (hillclimb-10,396 kms)	Touring	up to 1150 cc	P. Maublanc	6'32"2	Fiat Abarth 1000 Berlina
			Sport	up to 1150 cc	S. Vaglio	6'45"1	Fiat Abarth 1000 BA
			Sport	1151-1600 cc	W. Rufenacht	6'16"1	Fiat Abarth OT 1600
4 September	Bourscheid, Luxembourg	Course de côte de Bourscheid (circuit)	Touring	701-1000 cc	R. Stierli	2'13"3	Fiat Abarth 850 TC
			Sport	1000 cc	J. Hauser	2'12"3	Fiat Abarth 1000 BA
4 September	Bologna, Italy	Trofeo dell'Appennino Bolognese (210,300 kms)	Touring	850 cc	C. Massagrande (overall)	---	Fiat Abarth 850 TC
4 September	Lendinara, Italy	Gymkana Automobilistica	---	---	M. Rizzi	---	Fiat Abarth 595
4 September	Domodossola, Italy	Trofeo Valli Ossolane (176,100 kms)	---	---	A. Dalan (overall)	---	Fiat Abarth 850 TC
4 September	Macerata, Italy	Coppa Città di Tolentino (hillclimb-4,700 kms)	Touring	600 cc	M. Longarino	6'17"0	Fiat Abarth 595
			Touring	up to 750 cc	P. Valeri	5'52"1	Fiat Abarth 695
			Touring	850 cc	M. Tommasini	5'31"7	Fiat Abarth 850 TC
			Touring	1000 cc	G. Ancarani	5'34"8	Fiat Abarth 1000 Berlina
4 September	San Lorenzo de Savall, Spagna	Carrera en cuesta San Feliu (hillclimb-4,200 kms)	Touring	700 cc	J. Juncosa	3'45"4	Fiat Abarth 695
			Touring	850 cc	'Jorge Chi'	3'28"2	Fiat Abarth 850 TC
			Touring	1000 cc	M. Juncosa	3'14"8	Fiat Abarth 1000 Berlina
			Sport	1000 cc	J. Juncosa	3'10"0	Fiat Abarth 1000 BA
			Sport	1300 cc	J. de Vilar	3'07"2	Fiat Abarth OT 1300
4 September	Montlhéry, France	Epreuve Nationale de Vitesse (circuit-3,341 kms 18 laps)	Sport	1000 cc	J. Mazzanti	---	Fiat Abarth 1000 BA
9-10 September	Vaals, Holland	Heuvelklin Vaals (hillclimb-1,200 kms)	Touring	850 cc	M. Joker	1'00"8	Fiat Abarth 850 TC
			Touring	1000 cc	E. Swart	0'56"7	Fiat Abarth 1000 Berlina
			Sport	1300 cc	J. Hauser	0'57"6	Fiat Abarth OT 1300
10-11 September	Mitholz, Switzerland	Bergrennen Mitholz-Kandersteg (hillclimb-2,640 kms x 2)	Touring	850 cc	R. Stierli	3'57"3	Fiat Abarth 850 TC
			Touring	1000 cc	A. Wittenbach	3'58"8	Fiat Abarth 1000 Berlina
			Grand Touring	1000 cc	J. Gloggner	4'09"0	Fiat Abarth 1000 OTS
			Sport	1000 cc	R. Vaglio	3'51"7	Fiat Abarth 1000 BA
			Sport	1300 cc	W. Egger	3'42"1	Abarth Simca 1300
			Sport	1600 cc	D. Borel	3'22"6	Fiat Abarth OT 1600
10-11 September	Dijon, France	Course de côte d'Urcy (hillclimb-2,500 kms)	Touring	850 cc	S. Stefani	4'06"6	Fiat Abarth 850 TC
			Touring	1000 cc	P. Maublanc	3'55"8	Fiat Abarth 1000 Berlina
			Sport	1000 cc	J. Mazzanti	3'53"5	Fiat Abarth 1000 BA
			Sport	1300 cc	D. Grandjean	3'48"4	Abarth Simca 1300
11 September	Kaiserslautern, Germany	Rheinland-Pfalz Bergpreis (hillclimb-3,050 kms)	Touring	1000 cc	J. Abt	3'45"7	Fiat Abarth 1000 Berlina
			Sport	up to 1600 cc	S. Dau	3'33"1	Fiat Abarth OT 1300
11 September	Snetterton, United Kingdom	Closed Race Meeting (circuit-2,700 miles x 8 laps)	Touring	850 cc	G. Anstead	17'54"4	Fiat Abarth 850 TC
			Touring	1000 cc	A. Mylius	16'58"8	Fiat Abarth 1000 Berlina
11 September	Guadalupe, Spagna	Gran Premio Circuito de Guadalupe (circuit-3,900 kms x 1 hour)	---	---	J. Juncosa	kms 95,668	Fiat Abarth 1000 Berlina
11 September	Zeltweg, Austria	Grosser Preis von Österreich (circuit-3,200 kms x 157 laps)	Sport	1300 cc	J. Ortner	3h07'55"0	Fiat Abarth OT 1300

DATE	LOCATION	RACE	CATEGORY	CLASS	WINNER	TIME OR DISTANCE	CAR
11 September	Viterbo, Italy	Coppa del Cimino (hillclimb-10,200 kms)	Touring	600 cc	R. Mazzeschi	6'34"0	Fiat Abarth 595
			Touring	700 cc	G. Brancadori	6'00"5	Fiat Abarth 695
			Touring	850 cc	S. Pandolfo	5'25"6	Fiat Abarth 850 TC
			Touring	1000 cc	A. Saltari	5'23"7	Fiat Abarth 1000 Berlina
			Sport	1000 cc	B. Faenza	5'02"5	Fiat Abarth 1000 BA
			Sport	1300 cc	P. Laureati	4'59"7	Abarth Simca 1300
11 September	Bergamo, Italy	Castione-Passo della Presolana (hillclimb-6,400 kms)	Touring	700 cc	G. Caffi	4'17"5	Fiat Abarth 695
			Touring	850 cc	F. Galli	4'01"2	Fiat Abarth 850 TC
			Touring	1000 cc	G. Guarnieri	3'52"6	Fiat Abarth 1000 Berlina
			Sport	1000 cc	F. Pilone	3'41"5	Fiat Abarth 1000 BA
			Sport	1300 cc	C. Avventurieri	3'31"1	Fiat Abarth OT 1300
17 September	Frauenfeld, Switzerland	Autoslalom Thurgau (1,200 kms)	Touring	up to 1300 cc	O. Model	3'02"5	Fiat Abarth 850 TC
			Grand Touring	up to 1150 cc	T. Hofer	3'05"4	Simca Abarth 1150
17-18 September	Bayerischzell, Germany	ADAC Sudelfeld Bergrennen (hillclimb-3,500 kms x 2)	Touring	850 cc	R. Stierli	4'32"4	Fiat Abarth 850 TC
			Touring	1000 cc	J. Abt	4'21"2	Fiat Abarth 1000 Berlina
			Sport	1300 cc	K. Federhofer	4'03"4	Abarth Simca 1300
18 September	Mont Verdun, France	Course de côte Limonest-Mont Verdun (hillclimb-2,675 kms)	Touring	850 cc	A. Nesi	2'17"2	Fiat Abarth 850 TC
			Touring	1000 cc	P. Maublanc (overall)	2'06"4	Fiat Abarth 1000 Berlina
			Sport	1000 cc	S. Vaglio	2'11"2	Fiat Abarth 1000 BA
			Sport	1151-1600 cc	D. Borel	2'01"1	Fiat Abarth OT 1600
18 September	Gargellen, Austria	Internationales Steinbock Bergrennen Gargellen	Touring	up to 1600 cc	R. Blatter	4'19'51	Fiat Abarth 850 TC
			Sport	up to 1600 cc	W. Rufenacht	7'35"0	Fiat Abarth OT 1600
18 September	S. Sebastian, Spagna	Carrera en cuesta a El Farrel de Caldas de Montbuy (hillclimb-7,750 kms)	Touring	700 cc	J. Juncosa	8'18"0	Fiat Abarth 695
			Sport	1000 cc	J. Juncosa	6'53"2	Fiat Abarth 1000 BA
			Sport	1300 c	J. de Vilar (overall)	6'52"4	Fiat Abarth OT 1300
18 September	Brescia, Italy	Trofeo Lumezzane (hillclimb-8,500 kms)	Touring	700 cc	G. Caffi	5'25"7	Fiat Abarth 695
			Touring	850 cc	'Lentel'	5'10"6	Fiat Abarth 850 TC
			Touring	1000 cc	P. Lado	4'49"9	Fiat Abarth 1000 Berlina
			Grand Touring	1000 cc	G. Dolci	5'59"8	Fiat Abarth 1000 OTS
			Sport	1000 cc	G. Mainetti	4'42"3	Fiat Abarth 1000 BA
			Sport	1300 cc	G. Villani	4'38"2	Fiat Abarth OT 1300
			Sport	2000 cc	'Matich'	4'31"5	Abarth Simca 2000
18 September	Genova, Italy	Pontedecimo-Passo dei Giovi (hillclimb-4,000 kms)	Touring	600 cc	L. Durst	3'31"4	Fiat Abarth 595
			Touring	700 cc	A. Poggio	3'29"6	Fiat Abarth 695
			Touring	850 cc	G. Cazzullo	3'29"7	Fiat Abarth 850 TC
			Touring	1000 cc	I. Grillo	3'32"2	Fiat Abarth 1000 Berlina
			Sport	1000 cc	P. Scantimburgo	3'24"4	Fiat Abarth 1000 BA
18 September	Aosta, Italy	Aosta-Pila (hillclimb-11,200 kms)	Touring	up to 700 cc	R. Pinto	8'54"6	Fiat Abarth 595
			Touring	850 cc	M. Chiantelassa	8'03"9	Fiat Abarth 850 TC
			Touring	1000 cc	A. D'Onofrio	8'27"1	Fiat Abarth 1000 Berlina
			Sport	1300 cc	G. Rebaudi	7'10"0	Fiat Abarth OT 1300
			Sport-Prototypes	1000 cc	L. Cella	7'16"1	Fiat Abarth 1000 SP
			Sport-Prototypes	2000 cc	J. Ortner	6'40"9	Fiat Abarth OT 2000
18 September	Nürnberg, Germany	Tourenwagen Rennen Nürnberg (circuit-1,700 kms x 30 laps)	Touring	1000 cc	W. Mantzel	30'02"0	Fiat Abarth 1000 Berlina
			Sport	1300 cc	H. Dechent (overall)	28'47"9	Fiat Abarth OT 1300
18 September	Vallelunga, Italy	Trofeo Coca Cola (circuit-1,800 kms x 56 laps)	Touring	850 cc	D. Martino	1h03'11"9	Fiat Abarth 850 TC
18 September	Wilmot, USA	Midwest Council (circuit-30 laps)	---	---	D. Parent	---	Fiat Abarth 1000 Berlina
18 September	Houyet, Belgium	Course de côte de Houyet (hillclimb-2,500 kms)	Touring	1000 cc	W. Kauhsen	1'44"8	Fiat Abarth 1000 Berlina
18 September	Berghem, Holland	Groot Berghem (circuit)	Touring	850 cc	J. van Osch	---	Fiat Abarth 850 TC
24-25 September	Eigenthal, Switzerland	Nationales Bergrennen Eigenthal	Touring	850 cc	R. Stierli	5'33"8	Fiat Abarth 850 TC
			Touring	1000 cc	J. Brun	5'27"7	Fiat Abarth 1000 Berlina
			Grand Touring	1000 cc	J. Gloggner	5'44"9	Fiat Abarth 1000 OTS
			Sport	1000 cc	G. Naef	5'19"0	Fiat Abarth 1000 BA
			Sport	1300 cc	W. Egger	5'36"1	Abarth Simca 1300
25 September	Montlhéry, France	Coupes de Paris (circuit-3,600 kms x 15 laps)	Touring	1000 cc	J. Ortner	26'53"3	Fiat Abarth 1000 Berlina
			Sport	1000 cc	J. Ortner	25'44"8	Fiat Abarth 1000
			Sport	1300 cc	C. Ballot Lena	33'26"9	Fiat Abarth OT 1300
			Sport	1600 cc	J. Ortner	30'10"2	Fiat Abarth OT 1600
25 September	Belluno, Italy	Coppa Alpe del Nevegal (hillclimb-7,500 kms)	Touring	600 cc	P.G. Poggio	5'58"8	Fiat Abarth 595
			Touring	700 cc	G. Caffi	5'43"5	Fiat Abarth 695
			Touring	850 cc	'Opicina'	5'14"4	Fiat Abarth 850 TC
			Touring	1000 cc	P. Lado	5'04"6	Fiat Abarth 1000 Berlina
			Sport	1000 cc	M. Pinchetti	4'52"3	Fiat Abarth 1000 BA
			Sport	1300 cc	C. Avventurieri	4'46"9	Fiat Abarth OT 1300

DATE	LOCATION	RACE	CATEGORY	CLASS	WINNER	TIME OR DISTANCE	CAR
25 September	Osimo, Italy	Coppa Luigi Fagidi (hillclimb-5,300 kms)	Touring	700 cc	L. Cestra	3'19"1	Fiat Abarth 695
			Touring	up to 1000 cc	C. Catena	3'09"0	Fiat Abarth 850 TC
			Sport	1000 cc	E. Ostini	2'50"3	Fiat Abarth 1000 BA
			Sport	1300 cc	P. Laureati	2'47"4	Abarth Simca 1300
			Sport-Prototypes	1000 cc	G. Tommasi	2'46"1	Fiat Abarth 1000
			Sport-Prototypes	2000 cc	L. Taramazzo	2'36"0	Fiat Abarth 2000
25 September	Caltanissetta, Italy	Coppa Nissena (hillclimb-10,000 kms)	Touring	600 cc	I. Serse	6'46"2	Fiat Abarth 595
			Touring	700 cc	D. Stabile	6'34"4	Fiat Abarth 695
			Touring	850 cc	'Jimmy'	6'23"3	Fiat Abarth 850 TC
			Touring	1000 cc	S. Calascibetta	5'59"0	Fiat Abarth 1000 Berlina
			Sport	1300 cc	G. Rebaudi	5'35"8	Fiat Abarth OT 1300
25 September	Dortmund, Germany	ADAC Sauerland Bergpreis (hillclimb-3,400 kms x 2)	Touring	1000 cc	H. Hessel	4'34"8	Fiat Abarth 1000 Berlina
			Grand Touring	1000 cc	F. Kleine	4'51"6	Fiat Abarth 1000 OTS
25 September	Mallorca, Spagna	Carrera en cuesta Puig Major (hillclimb-8,000 kms)	Touring	701-1150 cc	'Jorge Chi'	5'50"9	Fiat Abarth 850 TC
25 September	Dalsland, Sweden	Dalsland Ring (circuit-16 laps)	Touring	1000 cc	L. Englund	14'50"4	Fiat Abarth 1000 Berlina
25 September	Adenau, Germany	ADAC Nürburgring Eifelpokalrennen (circuit-7,700 kms x 7 laps)	Touring	up to 1000 cc	D. Mantzel	---	Fiat Abarth 1000 Berlina
25 September	Marlboro, USA	Marlboro Cup (circuit-1,7 miles x 15 laps)	---	---	A. Cosentino	30'09"4	Fiat Abarth 1000 Berlina
25 September	Bruxelles, Belgium	Championnat de Belgique (circuit-2,000 kms x 2)	Sport	1000 cc	J. Hauser	2'41"6	Fiat Abarth 1000 BA
25 September	Cremona, Italy	Coppa Luigi Faraboli (circuit-2,000 kms)	Touring	1000 cc	G.C. Guidetti	0'51"9	Fiat Abarth 1000 Berlina
25 September	Siena, Italy	Coppa del Chianti classico (hillclimb-8,200 kms)	Prepared Touring	600 cc	G. Caroli	5'40"6	Fiat Abarth 595
			Prepared Touring	700 cc	G. Brancadori	5'17"3	Fiat Abarth 695
			Prepared Touring	850 cc	P. Falorni	5'08"2	Fiat Abarth 850 TC
			Prepared Touring	1000 cc	M. Cintolesi	5'06"1	Fiat Abarth 1000 Berlina
			Sport	1300 cc	A. Trenti	4'43"8	Fiat Abarth OT 1300
1-2 October	Aschaffenburg, Germany	ADAC Spessart-Zuverlässigkeitsfahrt (hillclimb)	Touring	850 cc	O. Pfister	---	Fiat Abarth 850 TC
2 October	Massana, Andorra	Costa de la Massana (hillclimb-3,500 kms)	Touring	700 cc	J. Juncosa	3'03"5	Fiat Abarth 695
			Touring	850 cc	M. Juncosa	2'43"3	Fiat Abarth 850 TC
			Touring	1000 cc	G. Moragas	2'34"5	Fiat Abarth 1000 Berlina
			Sport	1000 cc	R. Barone	2'28"9	Fiat Abarth 1000 BA
			Sport	1300 cc	J. de Vilar	2'30"8	Fiat Abarth OT 1300
2 October	Kassel, Germany	Bergprüfung Kassel Langenberge (hillclimb-3,700 kms)	Touring	1000 cc	K. Göttig	4'26"4	Fiat Abarth 1000 Berlina
			Sport	up to 1300 cc	P. Lotz	4'29"8	Fiat Abarth 1000 BA
2 October	Rieti, Italy	Rieti-Terminillo (hillclimb-15,100 kms)	Touring	600 cc	M. Urbani	11'44"2	Fiat Abarth 595
			Touring	700 cc	A. Poli	11'06"0	Fiat Abarth 695
			Touring	850 cc	S. Pandolfo	9'48"2	Fiat Abarth 850 TC
			Touring	1000 cc	V. Trulli	9'57"1	Fiat Abarth 1000 Berlina
			Sport	1000 cc	G. Benini	9'10"4	Fiat Abarth 1000 BA
			Sport	2000 cc	L. Marignani	8'13"0	Abarth Simca 2000
2 October	Angers, France	Course de côte de la Pommeraye (hillclimb-1,700 kms)	Sport	1300 cc	M. Martin	1'06"2	Fiat Abarth OT 1300
2 October	Zandvoort, Holland	Trophy of the Dunes (circuit-4,225 kms x 12)	Touring	850 cc	M. Joker	25'29"1	Fiat Abarth 850 TC
			Touring	1000 cc	E. Swart	26'14"3	Fiat Abarth 1000 Berlina
2 October	Messina, Italy	Capo d'Orlando-Naso (hillclimb-8,200 kms)	Touring	600 cc	'Harca'	7'36"1	Fiat Abarth 595
			Touring	700 cc	D. Stabile	7'13"0	Fiat Abarth 695
			Touring	850 cc	G. Asciutto	7'15"0	Fiat Abarth 850 TC
			Touring	1000 cc	A. Romano	6'40"4	Fiat Abarth 1000 Berlina
			Sport	1300 cc	A. Gambero	6'43"1	Fiat Abarth OT 1300
2 October	Neuffen, Germany	Bergpreis Schwäbische Alb (hillclimb-4,200 kms x 2)	Sport	1300 cc	K. Federhofer	5'12"1	Fiat Abarth OT 1300
2 October	Monza, Italy	Coppa Leopoldo Carri (circuit-5,750 kms x 18 laps)	Touring	600 cc	A. Magnoni	47'58"6	Fiat Abarth 595
			Touring	850 cc	D. Martino	41'38"1	Fiat Abarth 850 TC
			Touring	1000 cc	M. Nesti	40'10"9	Fiat Abarth 1000 Berlina
8-9 October	Genève, Switzerland	Course de côte du Marchairuz (hillclimb-4,800 kms x 2)	Touring	850 cc	R. Stierli	7'06"2	Fiat Abarth 850 TC
			Touring	1000 cc	J. Brun	7'07"4	Fiat Abarth 1000 Berlina
			Sport	1300 cc	J. Grandjean	6'34"7	Abarth Simca 1300
8-9 October	Fulda, Germany	Rhön Bergpreis (hillclimb-2,550 kms)	Touring	1000 cc	H. Kleine	4'32"8	Fiat Abarth 1000 Berlina
			Sport	1000 cc	F. Neuser	4'16"7	Fiat Abarth 1000 BA
			Sport	1300 cc	K. Federhofer	4'05"3	Fiat Abarth OT 1300
9 October	Innsbruck, Austria	Preis von Tirol (circuit-2,800 kms x 25 laps)	Touring	1000 cc	D. Mantzel	28'19"6	Fiat Abarth 1000 Berlina
			Sport	1300	H. Dechent	---	Fiat Abarth OT 1300
9 October	Catanzaro, Italy	Coppa Città di Catanzaro (hillclimb-8,400 kms)	Touring	600 cc	G. Angheloni	7'39"4	Fiat Abarth 595
			Touring	700 cc	D. Calendino	7'32"2	Fiat Abarth 695
			Touring	850 cc	L. Ferragina	6'42"0	Fiat Abarth 850 TC
			Touring	1000 cc	A. Romano	6'49"4	Fiat Abarth 1000 Berlina
			Grand Touring	1000 cc	O. Ferragino	6'58"9	Fiat Abarth 1000 OTS
			Sport	1000 cc	G. Benini	6'43"4	Fiat Abarth 1000 BA
			Sport	1300 cc	C. Avventurieri (overall)	6'09"0	Fiat Abarth OT 1300

DATE	LOCATION	RACE	CATEGORY	CLASS	WINNER	TIME OR DISTANCE	CAR
9 October	Sewen, France	Course de côte de Sewen-Lac d'Alfeld (hillclimb-6,000 kms)	Touring	1000 cc	P. Maublanc (overall)	4'45"3	Fiat Abarth 1000 Berlina
			Sport	1000 cc	C. Ballot Lena	4'50"4	Fiat Abarth 1000 BA
9 October	Forlì, Italy	Gara Nazionale Automobilistica (regolarità-126 kms)	Touring	700 cc	F. Rubbieri	---	Fiat Abarth 695
9 October	Manresa, Spagna	Subida en cuesta Pont de Vilumara (hillclimb-5,000 kms)	Touring	850 cc	'Jorge Chi'	2'26"1	Fiat Abarth 850 TC
			Touring Special	850 cc	M. Juncosa	2'18"1	Fiat Abarth 850 TC
			Sport	up to 1600 cc	J. de Vilar	2'06"4	Fiat Abarth OT 1300
9 October	Namur, Belgium	Course de côte de Namur (hillclimb-2,000 kms)	Touring	1000 cc	H. Hessel	1'52"4	Fiat Abarth 1000 Berlina
			Sport	1000 cc	J. Hauser	1'56"7	Fiat Abarth 1000 BA
9 October	Vallelunga, Italy	Coppa Settecolli (circuit-1,800 kms x 12 laps)	Touring	700 cc	G. Brancadori	14'25"2	Fiat Abarth 695
			Touring	up to 100 cc	'Cristi'	13'45"6	iat Abarth 850 TC
			Prototypes (35 laps)	1000 cc	G. Tommasi	39'23"0	Fiat Abarth 1000
9 October	Nice, France	Course de côte de Sainte Cezaire (hillclimb-4,500 kms)	Touring	850 cc	M. Chiantelassa (overall)	3'50"2	Fiat Abarth 850 TC
9 October	Nice, France	Circuit de Longvic	Sport	1300 cc	B. Belin	---	Fiat Abarth OT 1300
12 October	Manresa, Spagna	Rallye de los 2000 Virajes (580 kms)	Sport	1300 cc	J. de Vilar	---	Fiat Abarth OT 1300
15 October	Kerenzer, Switzerland	Kerenzer Bergrennen (hillclimb-2,000 kms)	Touring	850 cc	R. Blatter	1'50"1	Fiat Abarth 850 TC
			Touring	1000 cc	E. Welti	1'50"0	Fiat Abarth 1000 Berlina
			Sport	up to 1100 cc	H. Affentranger	1'48"7	Fiat Abarth 700 BA
			Sport	1300 cc	W. Egger	1'49"5	Abarth Simca 1300
			Sport	1600 cc	W. Rufenacht	1'47"2	Fiat Abarth OT 1600
16 October	Aspern, Austria	Donau Pokal Flugplatzrennen (circuit-2,600 kms x 25 laps)	Touring	700 cc	M. Zanetti	32'01"3	Fiat Abarth 695
			Touring	1000 cc	D. Mantzel	31'56"3	Fiat Abarth 1000 Berlina
			Sport	1300 cc	H.D. Dechent	27'25"7	Fiat Abarth OT 1300
16 October	Cagliari, Italy	Coppa Città di Iglesias (hillclimb-6,700 kms)	Touring	600 cc	P. Leone	5'48"5	Fiat Abarth 595
			Touring	700 cc	S. Farris	5'57"7	Fiat Abarth 695
			Touring	850 cc	A. Capitta	5'20"7	Fiat Abarth 850 TC
			Touring	1000 cc	F. Locci	5'19"3	Fiat Abarth 1000 Berlina
16 October	Malmedy, Belgium	Course de côte des Trois Marets (hillclimb-4,800 kms)	Touring	1000 cc	F. Hessel	3'06"4	Fiat Abarth 1000 Berlina
			Sport	1000 cc	J. Hauser	3'17"7	Fiat Abarth 1000 BA
16 October	Draguignan, France	Rallye du Var (regolarità)	Touring	1000 cc	Maublanc-Roussin (overall)	1h25'21"3	Fiat Abarth 1000 Berlina
16 October	Karlskoga, Sweden	Kanonloppet (circuit-3,000 kms x 10 laps)	Touring	1000 cc	L. Englund	18'09"6	Fiat Abarth 1000 Berlina
16 October	Monthléry, France	1000 Km de Paris (circuit-3,600 kms)	Sport	850-1300 cc	J. Mazzanti	---	Fiat Abarth 1000 BA
22 October	Bern, Switzerland	Surprise Bergrennen (hillclimb)	Touring	850 cc	K. Schneiter	3'15"9	Fiat Abarth 850 TC
			Touring	1000 cc	A. Wittenbach	2'59"3	Fiat Abarth 1000 Berlina
			Sport	1000 cc	H. Affentranger	2'52"7	Fiat Abarth 1000 BA
			Sport	1300 cc	P. Maron	2'47"0	Fiat Abarth OT 1300
22 October	Hockenheim, Germany	ACS Hockenheim Ring (circuit-6,768 kms x 10 laps)	Touring	up to 1300 cc	R. Blatter	14'44"1	Fiat Abarth 850 TC
			Sport	---	W. Rufenacht (overall)	14'40"0	Fiat Abarth OT 1600
23 October	Barcelona, Spagna	Trofeo Montjuich (circuit)	Touring	850 cc	'Jorge Chi'	43'47"0	Fiat Abarth 850 TC
			Touring	1000 cc	J. Juncosa	42'48"0	Fiat Abarth 1000 Berlina
23 October	Nancy, France	Course de côte de Belleau (hillclimb-2,000 kms)	Touring	850 cc	G. Caroli	1'41"2	Fiat Abarth 850 TC
			Touring	1000 cc	H. Hessel	1'33"3	Fiat Abarth 1000 Berlina
			Sport	1000 cc	P. Henning	1'28"4	Fiat Abarth 1000 BA
			Sport	1300 cc	M. Martin	1'30"2	Fiat Abarth OT 1300
			Sport	1600 cc	D. Borel	1'23"3	Fiat Abarth OT 1600
23 October	Tucson (Arizona), USA	National Race (circuit-25 laps)	---	---	F. Pletkin	33'04"8	Fiat Abarth
23 October	Innsbruck, Austria	Bergrennen Wörgl Wildschönau (hillclimb-2,800 kms x 2)	Touring	1000 cc	J. Abt	1'10"0	Fiat Abarth 1000 Berlina
			Sport	1300 cc	K. Federhofer	1'03"1	Abarth Simca 1300
23 October	Siracusa, Italy	Coppa Val d'Anapo-Sortino (hillclimb-6,700 kms)	Touring	600 cc	G. Vassallo	5'47"6	Fiat Abarth 595
			Touring	700 cc	P. Marino	5'56"7	Fiat Abarth 695
			Touring	850 cc	Laut	5'36"7	Fiat Abarth 850 TC
			Touring	1000 cc	S. Calascibetta	5'08"0	Fiat Abarth 1000 Berlina
			Sport	1300 cc	G. Virgilio	4'57"9	Abarth Simca 1300
			Sport	2000 cc	V. Ferlito	5'08"9	Abarth Simca 2000
			Sport-Prototypes	2000 cc	Raveto (overall)	4'50"2	Abarth Simca 2000
23 October	Verona, Italy	Gymkana Caprino Veronese	---	up to 750 cc	D. Toffoli	0'46"2	Fiat Abarth 595
23 October	Firenze, Italy	Giro della Toscana (regolarità-171,400 kms)	Touring	700 cc	A. Polli	---	Fiat Abarth 695
			Touring	700-1000 cc	C. Massagrande	---	Fiat Abarth 850 TC
23 October	Fribourg, Switzerland	Slalom de la Poya	---	---	M. Piller	1'50"0	Fiat Abarth 1000 Berlina
29 October	Verona, Italy	Gymkana Automobilistica	---	up to 750 cc	Toffoli-Rizzo	1'37"9	Fiat Abarth 595
29-30 October	Rouen, France	Rallye Automobile Jeanne d'Arc (287,900 kms)	Grand Touring	0-1150 cc	Joneaux-Homo (overall)	---	Fiat Abarth 1000 BA

DATE	LOCATION	RACE	CATEGORY	CLASS	WINNER	TIME OR DISTANCE	CAR
30 October	Barcelona, Spagna	Carrera en cuesta de la Rabassada (hillclimb)	Touring Touring Sport Sport	850 cc 1000 cc 1000 cc 1300 cc	'Jorge Chi' E. Baturone J. Juncosa J. de Vilar	3'15"5 3'02"4 3'01"5 2'55"0	Fiat Abarth 850 TC Fiat Abarth 1000 Berlina Fiat Abarth 1000 BA Fiat Abarth OT 1300
30 October	Treviso, Italy	Coppa del Piave (213,400 kms)	Touring	700 cc	L. Roberto	---	Fiat Abarth 595
30 October	Orvieto, Italy	Trofeo Città di Orvieto (hillclimb-7,600 kms)	Touring Touring Touring Touring Grand Touring Sport	600 cc 700 cc 850 cc 1000 cc 1000 cc 1300 cc	P. Poggi G. Brancadori G. Casati M. Cintolesi E. Ostini G. Villani	5'57"9 5'58"1 5'30"0 5'27"9 5'06"5 4'49"8	Fiat Abarth 595 Fiat Abarth 695 Fiat Abarth 850 TC Fiat Abarth 1000 Berlina Fiat Abarth 1000 BA Fiat Abarth OT 1300
30 October	Augsburg, Germany	ASC Rennslalom	Sport	up to 1300 cc	F. Neuser	---	Fiat Abarth 1000 BA
1 November	Alassio, Italy	Coppa del Cesio (61,500 kms)	--- ---	up to 650 cc up to 1000 cc	G. Tilli (overall) V. Pené	--- ---	Fiat Abarth 595 Fiat Abarth 1000 Berlina
4 November	Vallelunga, Italy	Trofeo delle Cinque Conchiglie (circuit-1,800 kms x 20 laps)	Touring Touring	up to 700 cc 1000 cc	M. Zanetti A. Saltari	18'49"5 17'48"0	Fiat Abarth 695 Fiat Abarth 1000 Berlina
5 November	Hockenheim, Germany	ACS 3 Stunden Rennen (circuit-6,768 kms)	Sport	up to 1600 cc	Maron-Kretschi	3h02'44"0	Fiat Abarth OT 1300
6 November	Barcelona, Spagna	Rallye 600 (555 kms)	Touring	850 cc	'Jorge Chi'	---	Fiat Abarth 850 TC
6 November	Zandvoort, Holland	Nationale Toerwagenraces (circuit-4,225 kms x 10 laps)	Touring Touring	850 cc 1000 cc	M. Joker A. Hezemans	16'53"3 15'28"4	Fiat Abarth 850 TC Fiat Abarth 1000 Berlina
6 November	Buochs, Switzerland	Autoslalom (1,840 kms)	Touring	1000 cc	M. Wüest	2'09"4	Fiat Abarth 1000 Berlina
11 November	Nice, France	Course de côte de L'Orme (hillclimb-10,000 kms)	Touring Touring Sport Sport-Prototypes	850 cc 1000 cc 1000 cc 2000 cc	Reano Silvy P. Henning L. Taramazzo (overall)	9'38"4 10'17"2 10'19"3 9'28"5	Fiat Abarth 850 TC Fiat Abarth 1000 Berlina Fiat Abarth 1000 BA Fiat Abarth 2000
13 November	Vallelunga, Italy	Coppa Fausto Pesci (circuit-1,800 kms x 56 laps)	Touring Touring Sport Sport	600 cc 700 cc 1000 cc 1300 cc	R. Pinto M. Zanetti M. Moselli G. Bona (overall)	1h08'06"0 1h06'01"0 59'51"7 58'03"2	Fiat Abarth 595 Fiat Abarth 695 Fiat Abarth 1000 BA Fiat Abarth OT 1300
13 November	Berlin, Germany	DMV Herbstfahrt Rallye	Touring	850 cc	H.J. Mack	---	Fiat Abarth 850 TC
19-20 November	Barcelona, Spagna	Rallye Barcelona-Andorra (603 kms)	Touring	850 cc	'Jorge Chi'	---	Fiat Abarth 850 TC
20 November	Roma, Italy	Campionato Laziale Tris (circuit-1,000 kms x 2 laps)	Touring	850 cc	Marras	2'32"9	Fiat Abarth 850 TC
23-27 November	Riverside, USA	Riverside International Raceway (2,100 miles x 24 laps)	--- ---	--- ---	D. Parkinson D. Parkinson	--- ---	Fiat Abarth 1000 Berlina Fiat Abarth 1000
27 November	Robledondo, Spagna	Trofeo Cruz Verde (hillclimb-5,100 kms)	Touring	850-1150 cc	J. Juncosa	3'42"7	Fiat Abarth 1000 Berlina
27 November	Vallelunga, Italy	Criterium dell'Ora (circuit-1,800 kms x 1 hour)	Touring Touring Touring	700 cc 850 cc 1000 cc	'Marino' F. Altieri A. Saltari	kms 87,832 kms 92,386 kms 92,007	Fiat Abarth 695 Fiat Abarth 850 TC Fiat Abarth 1000 Berlina
3-4 December	Cannes, France	Rallye de Cannes (440 kms)	Touring	850 cc	Branche-Jullien	1h02'54"6	Fiat Abarth 850 TC
4 December	Vallelunga, Italy	Criterium dell'Ora (circuit-1,800 kms x 1 hour)	Touring Touring Touring	700 cc 850 cc 1000 cc	M. Zanetti V. Cristi A. Saltari	kms 91,549 kms 92,549 kms 88,346	Fiat Abarth 695 Fiat Abarth 850 TC Fiat Abarth 1000 Berlina
4 December	Luanda, Angola	Taca de Giro de Angola (circuit-1,832 kms x 80 laps)	Touring Grand Touring	1000 cc 1000 cc	J. Silvera Machado J. Silvera Machado	1h32'10"0 1h26'09"8	Fiat Abarth 1000 Berlina Fiat Abarth 1000 BA
8 December	Rabasada, Spagna	Rallye Les Tres Cortes (201,000 kms)	Touring	850 cc	M. Juncosa	---	Fiat Abarth 850 TC
11 December	Manresa, Spagna	Mini Rallye Biela Club (215,200 kms)	Touring	850 cc	'Jorge Chi'	---	Fiat Abarth 850 TC
18 December	Jarama, Spagna	Circuito del Jarama (3,432 kms x 14 laps)	Touring Touring	850 cc 1000 cc	M. Juncosa J. Juncosa	--- ---	Fiat Abarth 850 TC Fiat Abarth 1000 Berlina
1967							
8 January	Washington (D.C.), USA	Refrigerator Bowl Races (circuit)	---	---	R. Cuomo	---	Fiat Abarth 1000 Berlina
15 January	Roma, Italy	Competizione Tris-Pista d'Oro (circuit-1,000 kms x 2 laps)	Touring	850 cc	L. Marras	2'16"0	Fiat Abarth 850 TC
21-22 January	Luanda, Angola	Rallye 1° Anniversario do Banco Comercial de Angola	Touring	1000 cc	J. Silveira Machado	---	Fiat Abarth 1000 Berlina
29 January	Sestrière, Italy	Coppa Sestrière (183 kms)	---	1000 cc	A.P. Borghesio	---	Fiat Abarth 1000 OTS
29 January	Vallelunga, Italy	Gara Velocità Tris (circuit-1,800 kms x 3 laps)	Touring	850 cc	V. Cristi	3'35"5	Fiat Abarth 850 TC
29 January	Luanda, Angola	Circuito Automobilistico de Cidade de Cabinda (circuit-2,400 kms x 70 laps)	Touring	1000 cc	C. Pereira	1h47'38"1	Fiat Abarth 1000 Berlina

DATE	LOCATION	RACE	CATEGORY	CLASS	WINNER	TIME OR DISTANCE	CAR
5 February	Kotka, Finlandia	Teilrennen (ice race-16,000 kms)	Touring	850 cc	S. Martti	12'29"8	Fiat Abarth 850 TC
5 February	Vallelunga, Italy	Gara Velocità Tris (circuit-1,800 kms x 3 laps)	Touring	850 cc	S. Silvano	3'56"0	Fiat Abarth 850 TC
10-12 February	Lille, France	Rallye des Routes du Nord (1048 kms)	Grand Touring	701-1000 cc	Mulard-Callens	---	Fiat Abarth 1000 OTS
12 February	Vallelunga, Italy	Gara Velocità Tris (circuit-1,800 kms x 3 laps)	Touring	850 cc	V. Cristi	3'30"8	Fiat Abarth 850 TC
			Touring	1000 cc	F. Acasbo	3'23"4	Fiat Abarth 1000 Berlina
12 February	Kungsgärden, Sweden	SMK Storviks Hastighetstävling (ice race)	Touring	1000 cc	L. Englund	---	Fiat Abarth 1000 Berlina
12 February	Piikkio, Finlandia	Ice race (7,000 kms)	Touring	850 cc	A. Vaara	5'47"2	Fiat Abarth 850 TC
			Touring	1000 cc	H. Liedes	5'33"9	Fiat Abarth 1000 Berlina
			Touring Special	850 cc	A. Vaara	5'40"8	Fiat Abarth 850 TC
26 February	Helsinki, Finlandia	Ice race (8,000 kms)	Touring	1000 cc	H. Liedes	6'47"8	Fiat Abarth 1000 Berlina
			Touring Special	1000 cc	H. Liedes	---	-
26 February	Valpineda, Spagna	Rallye de Invierno	Touring	850 cc	M. Juncosa	---	Fiat Abarth 850 TC
			Touring	1000 cc	J. Juncosa	---	Fiat Abarth 1000 Berlina
			Sport	up to 1600 cc	E. Baturone	---	Fiat Abarth 1000 BA
26 February	Helsinki, Finlandia	Ice race (8,000 kms)	Touring	1000 cc	A. Vaara	6'34"0	Fiat Abarth 1000 Berlina
4-5 March	Gerona, Spagna	Rallye de Gerona (502 kms)	Touring	851-1600 cc	M. Juncosa	---	Fiat Abarth 1000 Berlina
5 March	Yverdon, Switzerland	Mille mètres d'Ursins (acceleration race)	Touring	1000 cc	F. Wipf	16'47"	Fiat Abarth 850 TCS
5 March	Stockholm, Sweden	Flatenloppet (ice race-7,000 kms)	Touring	1000 cc	L. Englund	9'27"8	Fiat Abarth 1000 Berlina
5 March	Novara, Italy	Coppa Monterosso (297 kms)	Touring	600 cc	M. Ranzini	---	Fiat Abarth 595
5 March	Vallelunga, Italy	Gara Velocità Tris (circuit-1,800 kms x 3 laps)	Touring	850 cc	V. Cristi	3'28"2	Fiat Abarth 850 TC
			---	1000 cc	F. Acasbo	6'48"5	Fiat Abarth 100 Berlina
12 March	Savona, Italy	Giro dei Monti Savonesi (300 kms)	Touring	850 cc	P. Bollati	-	Fiat Abarth 850 TC
12 March	Vallelunga, Italy	Trofeo Sambuca Molinari (circuit-1,800 kms x 30 laps x 2)	Touring	700 cc	M. Zanetti	1h12'10"7	Fiat Abarth 695
12 March	Bellevue, Belgium	Course de côte de Fleron (hillclimb-2,000 kms)	Touring	1000 cc	F. Neri	1'28"5	Fiat Abarth 1000 Berlina
12 March	Montserrat, Spagna	Carrera en cuesta de Montserrat (hillclimb-7,000 kms)	Touring	850 cc	J. Juncosa	5'31"8	Fiat Abarth 850 TC
			Touring	851-1150 cc	G. Moragas	5'43"3	Fiat Abarth 1000 Berlina
			Sport	1000 cc	E. Baturone	5'25"0	Fiat Abarth 1000 BA
18 March	Mocamades, Angola	Rallye Automovel das VI Festas do Mar (196 kms)	Touring	1000 cc	C. Pereira	2h17'48"8	Fiat Abarth 1000 Berlina
18-19 March	Le Mans, France	Grand Rallye de l'Ouest (864 kms)	Grand Touring	1000 cc	'Bianca'-Gandre	-	Fiat Abarth 1000 OTS
19 March	Monza, Italy	4 Ore del Jolly Club (circuit-10,000 kms)	Touring	850 cc	'Opicina'-Marcon	kms 544,717	Fiat Abarth 850 TC
			Touring	1000 cc	M. Zanetti	kms 583,262	Fiat Abarth 1000 Berlina
19 March	Draguignan, France	Course de côte d'Ampus (hillclimb-6,800 kms)	Touring Special	1000 cc	A. Tarasco	5'15"1	Fiat Abarth 1000 Berlina
			Sport	up to 1600 cc	F. Pilone	4'35"0	Fiat Abarth OT 1300
			Corsa	2000 cc	J. Ortner	4'14"1	Fiat Abarth 2000
19 March	Liège, Belgium	Course de côte La Roche Samree (hillclimb-8,000 kms)	Touring	1000 cc	F. Neri	4'26"0	Fiat Abarth 1000 Berlina
27 March	Vallelunga, Italy	Coppa Fiat (circuit-3,120 kms x 38 laps)	Touring	600 cc	M. Maeri	1h16'41"3	Fiat Abarth 595
			Sport (48 laps)	1000 cc	M. Polin	1h16'34"1	Fiat Abarth 1000 BA
			Sport (48 laps)	up to 1300 cc	G. Rebaudi	1h16'11'3	Fiat Abarth OT 1300
27 March	Lorentzweiler, Luxembourg	Course de côte de Lorentzweiler (hillclimb-1,800 kms)	Sport	1000 cc	R. Stierli	1'10"4	Fiat Abarth 1000 SP
			Sport	1001-1600 cc	K. Federhofer	1'13"1	Fiat Abarth OT 1300
31 March	Sebring (Florida), USA	4 Hours Championship Race (circuit-5,212 miles)	Touring	up to 1000 cc	R. Cuomo	---	Fiat Abarth 1000 Berlina
2 April	Aspern, Austria	Preis von Wien (circuit-2,600 kms x 1 hour)	Touring	850 cc	F. Galli	kms 114,824	Fiat Abarth 850 TC
			Touring	1000 cc	W. Kauhsen	kms 120,278	Fiat Abarth 1000 Berlina
2 April	Tucson (Arizona), USA	National Races Tucson	---	---	D. Parkinson	---	Fiat Abarth 1000 Berlina
2 April	Erbach, Germany	DMV Krähberg Rennen (hillclimb-3,600 kms)	Touring	up to 1150 cc	J. Abt	4'51"1	Fiat Abarth 1000 Berlina
			Sport-Prototypes	1000 cc	R. Stierli	4'29"0	Fiat Abarth 1000 SP
2 April	Vallelunga, Italy	Coppa Gallenga (circuit-3,120 kms x 1h10')	Touring	850 cc	V. Cristi	1h11'49"2	Fiat Abarth 850 TC
			Touring	1000 cc	P. Botalla	1h10'48"7	Fiat Abarth 1000 Berlina
			Grand Touring	1000 cc	P. Brambilla	1h11'24"3	Fiat Abarth 1000 OTS
			Sport (1h30')	1000 cc	B. Faenza	1h32'01"0	Fiat Abarth 1000 BA
				1300 cc	G. Danieli	1h31'05"6	Fiat Abarth OT 1300
8-9 April	Wiesbaden, Germany	ADAC Heilbronner Bergpreis (hillclimb-2,500 kms x 2)	Touring	850 cc	K. Auer	3'24"4	Fiat Abarth 850 TC
			Sport	up to 1300 cc	G. Bredow	3'20"2	Fiat Abarth OT 1300
9 April	Pressberg, Germany	Taunus Bergprüfung (hillclimb-3,300 kms x 2)	Touring	851-1150 cc	J. Abt	2'17"5	Fiat Abarth 1000 Berlina
			Sport	1300 cc	M. Krause	2'14"5	Fiat Abarth OT 1300
9 April	Marlboro, USA	Marlboro Park Speedway (circuit-1,7 miles x 18 laps)	---	---	A. Cosentino	---	Fiat Abarth 1000 Berlina
9 April	Avola, Italy	Coppa Belmonte (hillclimb-10,200 kms)	Touring	600 cc	V. Traina	8'15"7	Fiat Abarth 595
			Touring	700 cc	D. Stabile	7'49"5	Fiat Abarth 695
			Touring	850 cc	G. Amodeo	7'11"6	Fiat Abarth 850 TC
			Touring	1000 cc	S. Calascibetta	6'52"7	Fiat Abarth 1000 Berlina

DATE	LOCATION	RACE	CATEGORY	CLASS	WINNER	TIME OR DISTANCE	CAR
15 April	Grenchen, Switzerland	Automobilslalom Grenchen	Touring	850-1300 cc	P. Engler	2'44"2	Fiat Abarth 1000 Berlina
15 April	Verbois, Switzerland	Course de côte de Verbois (hillclimb-1,200 kms)	Touring	850 cc	H. Herren	1'04"0	Fiat Abarth 850 TC
			Sport	1000 cc	R. Vaglio	0'53"5	Fiat Abarth 1000 BA
15-16 April	Zandvoort, Holland	Trainingsweekend N.A.V. (circuit-4,225 kms x 10 laps)	Touring	1000 cc	A.B. Goedemans	19'53"2	Fiat Abarth 1000 Berlina
15-16 April	Saint-Lô, France	Course de côte de Hébécrevon (hillclimb-1,200 kms)	Sport	1000 cc	J. Lemaitre	0'48'4	Fiat Abarth 1000 BA
			Sport	1301-2000 cc	P. Maublanc (overall)	0'41"0	Fiat Abarth 2000
16 April	Vallelunga, Italy	Trofeo Marathon (circuit-1,800 kms x 66 laps)	Touring	850 cc	F. Galli	1h15'25"8	Fiat Abarth 850 TC
			Touring	1000 cc	N. Parenti	1h15'03"0	Fiat Abarth 1000 Berlina
16 April	Cagliari, Italy	Monte Acuto-Campuomu (hillclimb-14,000 kms)	Touring	600 cc	A. Palmas	6'13"0	Fiat Abarth 595
			Touring	700 cc	S. Farris	6'14"4	Fiat Abarth 695
			Touring	850 cc	A. Capitta	5'40"4	Fiat Abarth 850 TC
			Sport	1000 cc	G. Vacca	5'27"0	Fiat Abarth 1000 BA
16 April	Eifel, Germany	ADAC Bergpreis Bad Neuenahr (hillclimb-3,300 kms x 2)	Touring	850-1150 cc	K. Göttig	4'15"7	Fiat Abarth 1000 Berlina
			Sport	1300 cc	K. Federhofer	3'58"5	Fiat Abarth OT 1300
16 April	Offenburg, Germany	ADAC Slalom Offenburg	Touring	850 cc	E. Fauser	---	Fiat Abarth 850 TC
16 April	Tarrasa, Spagna	Carrera en cuesta Col d'Estenalles (hillclimb-5,600 kms)	Touring	850 cc	M. Juncosa	4'15"6	Fiat Abarth 850 TC
			Touring	1000 cc	'Paco'	4'41"4	Fiat Abarth 1000 Berlina
			Grand Touring	1000 cc	'Pelè'	4'12"8	Fiat Abarth 1000 OTS
			Sport	1000 cc	J.F. Cerveto	4'18"5	Fiat Abarth 1000 BA
16 April	Innsbruck, Austria	Weerbergrennen (hillclimb-5,000 kms)	Touring	1000 cc	J. Abt	4'22"5	Fiat Abarth 1000 Berlina
			Sport	1000 cc	R. Stierli	4'09"6	Fiat Abarth 1000 BA
16 April	Lodrino, Switzerland	Slalom di Lodrino (2,350 kms)	---	---	H. Regusci	2'18"6	Fiat Abarth OT 2000
16 April	Verona, Italy	Stallavena-Boscochiesanuova (hillclimb-15,300 kms)	Touring	600 cc	L. Pesce	9'20"5	Fiat Abarth 595
			Touring	700 cc	R. Galasso	9'25"6	Fiat Abarth 695
			Touring	850 cc	V. Tabucchi	8'18"0	Fiat Abarth 850 TC
			Touring	1000 cc	A. De Paoli	8'04"6	Fiat Abarth 1000 Berlina
			Sport	1000 cc	P. Lado	7'24"1	Fiat Abarth 1000 BA
			Sport	1300 cc	'Matich'	7'20"1	Fiat Abarth OT 1300
			Sport-Prototypes	1000 cc	A. Hezemans	7'10"3	Fiat Abarth 1000 SP
			Sport-Prototypes	2000 cc	J. Ortner	6'40"3	Fiat Abarth 2000
22 April	Madrid, Spagna	Circuito Hoyo de Manzanares (hillclimb)	---	---	J.A. Guerrea	---	Fiat Abarth 1000 Berlina
22 April	Lime Rock (Connecticut), USA	Area 1 Championship Races (circuit)	---	---	A. Cosentino	---	Fiat Abarth 1000 Berlina
22-23 April	Elbeuf, France	Rallye Automobile d'Elbeuf (300 kms)	Sport	1000 cc	Joneaux-Homo	---	Fiat Abarth 1000 BA
23 April	Montlhéry, France	Coupes de Vitesse (circuit-3,340 kms x 15 laps)	Touring	1000 cc	A. Hezemans	27'28"8	Fiat Abarth 1000 Berlina
			Grand Touring	1000 cc	A. Biancarelli	25'59"8	Fiat Abarth 1000 OTS
			Sport-Prototypes	1000 cc	G. Gandre	25'12"5	Fiat Abarth 1000
			Sport-Prototypes	2000 cc	J. Ortner	24'13"9	Fiat Abarth 2000
23 April	Eberbach, Germany	Bergrennen Eberbach (hillclimb-3,000 kms)	Touring	up to 1150 cc	J. Abt	4'26"6	Fiat Abarth 1000 Berlina
			Sport	1300 cc	B. Hofmann	4'19"9	Abarth Simca 1300
23 April	Gap, France	Course de côte du Col Bayard (hillclimb-5,000 kms)	Touring	1000 cc	F. Tomas	4'36"9	Fiat Abarth 1000 Berlina
			Sport	1300 cc	F. Canin	4'04"4	Fiat Abarth OT 1300
23 April	Payerne, Switzerland	Slalom National de Payerne (4,800 kms x 2)	Touring	850 cc	R. Blatter	3'41"5	Fiat Abarth 850 TC
			Touring	up to 1150 cc	M. Wuest	3'34"2	Fiat Abarth 1000 Berlina
23 April	Snetterton, United Kingdom	Race Meeting at Snetterton (circuit-2,700 miles x 8 laps)	Sport	up to 1150 cc	M. Walton	13'37"8	Fiat Abarth 1000 BA
23 April	Turku, Finlandia	Rennen für 100 Wagen (17,000 kms)	Touring	850 cc	A. Vaara	15'08"5	Fiat Abarth 850 TC
			Touring	1000 cc	H. Liedes	14'12"0	Fiat Abarth 1000 Berlina
23 April	Detmold, Germany	ADAC Bergrennen Teutoburger Wald (hillclimb-2,250 kms x 2)	Touring	up to 1150 cc	H. Klein	2'22"9	Fiat Abarth 1000 Berlina
23 April	Genova, Italy	Raduno Autoviatorio (gimkana-5 laps)	Prepared Touring	850 cc	U. Barilaro	1'53"7	Fiat Abarth 850 TC
23 April	Waiblingen, Germany	ADAC Automobil Slalom (1,800 kms)	Touring	1000 cc	C. Kubon	1'16"9	Fiat Abarth 1000 Berlina
29 April	Bern, Switzerland	Surprise Bergrennen (hillclimb)	Touring	850 cc	H. Herren	4'02"5	Fiat Abarth 850 TC
			Sport	850 cc	P. Rohrer	3'54"5	Fiat Abarth 850
			Sport	1000 cc	H. Affentranger	3'39"8	Fiat Abarth 1000 BA
			Sport	1300 cc	W. Egger	3'37"2	Fiat Abarth OT 1300
29-30 April	Fribourg, Switzerland	Course de côte de Fribourg (hillclimb-1,400 kms x 2)	Touring	850 cc	R. Neumann	---	Fiat Abarth 850 TC
			Touring	1000 cc	P. Engler	---	Fiat Abarth 1000 Berlina
			Sport	1000 cc	J.P. Brun	---	Fiat Abarth 1000 BA
30 April	Fassberg, Germany	AD Rundstreckenrennen (hillclimb-4,500 kms)	Touring	up to 1150 cc	H.P. Koepchen	30'46"4	Fiat Abarth 1000 Berlina
			Sport	1300 cc	H. Krause	25'46"2	Fiat Abarth OT 1300
30 April	Beograd, Jugoslavia	Grand Prix de Belgrade (circuit-4,350 kms x 2 hours)	Touring	up to 1000 cc	A. Merzario	---	Fiat Abarth 850 TC

DATE	LOCATION	RACE	CATEGORY	CLASS	WINNER	TIME OR DISTANCE	CAR
30 April	Arezzo, Italy	Pieve S. Stefano-Passo dello Spino (hillclimb-12,700 kms)	Touring	600 cc	M. Magri	8'18"6	Fiat Abarth 595
			Touring	700 cc	G. Brancadori	8'24"2	Fiat Abarth 695
			Touring	850 cc	A. Mostardini	7'46"1	Fiat Abarth 850 TC
			Touring	1000 cc	N. Parenti	7'23"9	Fiat Abarth 1000 Berlina
			Sport	1300 cc	G. Villani	7'11"8	Fiat Abarth OT 1300
30 April	Freiburg, Germany	ADAC Freiburg Autoslalom (2,000 kms x 2)	Touring	850 cc	E. Fauser	---	Fiat Abarth 850 TC
30 April	Cannes, France	Course de côte de Cannes (hillclimb-7,845 kms)	Touring	1000 cc	F. Tomas	6'55"8	Fiat Abarth 1000 Berlina
			Sport	up to 1600 cc	F. Canin	6'30"6	Fiat Abarth OT 1300
			Sport	up to 2500 cc	P. Maublanc	6'13"3	Fiat Abarth 2000
30 April	Villach, Austria	Dobratsch Bergrennen (hillclimb-15,800 kms)	Touring	1000 cc	G. König	8'51"2	Fiat Abarth 1000 Berlina
			Prototypes	2000 cc	A. Hezemans	7'19"8	Fiat Abarth 2000
30 April	Montelimar, France	Course de côte du Colombier (hillclimb-3,300 kms)	Sport	1000 cc	S. Vaglio	2'51"7	Fiat Abarth 1000 BA
30 April	Danville (Virginia), USA	National Races Virginia (circuit-176 miles)	---	---	S. Perry	---	Fiat Abarth 1000 Berlina
30 April	Savona, Italy	Savona-Cadibona (hillclimb-7,800 kms)	Touring	600 cc	L. Pesce	5'00"9	Fiat Abarth 595
			Touring	700 cc	R. Origlia	5'21"8	Fiat Abarth 695
			Touring	850 cc	P.L. De Zordo	4'44"9	Fiat Abarth 850 TC
			Touring	up to 1150 cc	D. Scardi	4'40"5	Fiat Abarth 1000 Berlina
			Sport	1300 cc	E. Tondelli	4'24"8	Fiat Abarth OT 1300
30 April-1 May	Magny Cours, France	Circuit de Magny Cours (2,040 kms x 20 laps)	Touring Special	up to 1150 cc	Bourguignon	23'14"9	Fiat Abarth 850 TC
30 April-1 May	Barcelona, Spagna	Rallye Costa Brava (396 kms)	Touring	850 cc	M. Juncosa	---	Fiat Abarth 850 TC
			Touring	1000 cc	'Pelè'	---	Fiat Abarth 1000 Berlina
30 April-1 May	Zandvoort, Holland	Stunden-Rennen Zandvoort (circuit-4,225 kms x 48 laps)	Touring	1000 cc	M. Kranefuss	---	Fiat Abarth 1000 Berlina
30 April-1 May	Roskilde, Danimarca	Borch-Christensen Abnings-Lobet (circuit-1,400 kms x 12 laps x 2)	Touring	850 cc	A. Larsen	21'17"5	Fiat Abarth 850 TC
1 May	Roma, Italy	Gara Tris-Pista d'Oro (circuit-1,000 kms x 2 laps)	Touring	850 cc	F. Murra	2'15"5	Fiat Abarth 850 TC
1 May	Zweibrücken, Germany	Westrich Bergpreis (hillclimb-1,500 kms)	Sport	1300 cc	K. Federhofer	2'51"8	Fiat Abarth OT 1300
4 May	Volterra, Italy	Coppa Città di Volterra (hillclimb-10,060 kms)	Touring	600 cc	F. Gragnoli	7'25"5	Fiat Abarth 595
			Touring	700 cc	G. Brancadori	7'07"2	Fiat Abarth 695
			Touring	850 cc	A. Mostardini	6'45"4	Fiat Abarth 850 TC
			Touring	1000 cc	N. Parenti	6'25"3	Fiat Abarth 1000 Berlina
			Grand Touring	1000 cc	G. Trombetta	7'30"0	Fiat Abarth 1000 OTS
			Sport	1000 cc	E. Ostini	6'19"7	Fiat Abarth 1000 BA
			Sport	1300 cc	'Nicor'	6'03"9	Fiat Abarth OT 1300
			Sport-Prototypes	1000 cc	P. Lado	6'06"2	Fiat Abarth 1000 SP
4 May	Helsinki, Finlandia	Kelsingin Vauhtikisat (circuit-3,300 kms x 10 laps)	Touring	850 cc	A. Vaara	18'23"8	Fiat Abarth 850 TC
4 May	Vicenza, Italy	Bassano-Montegrappa (hillclimb-11,750 kms)	Touring	600 cc	L. Pellanda	10'32"7	Fiat Abarth 595
			Touring	700 cc	R. Galasso	10'04"0	Fiat Abarth 695
			Touring	850 cc	'Opicina'	9'00"7	Fiat Abarth 850 TC
			Touring	1000 cc	G. Danieli	8'40"1	Fiat Abarth 1000 Berlina
4 May	Aix en Provence, France	Course de côte d'Aix-Saint Antonin (hillclimb-2,700 kms)	Touring	1000 cc	F. Tomas	1'57"5	Fiat Abarth 1000 Berlina
			Sport	1150-1600 cc	F. Canin	1'49"4	Fiat Abarth OT 1300
6-7 May	Stadtsteinach, Germany	ADAC Frankenwald Bergrennen (hillclimb-3,545 kms x 2)	Touring	up to 1150 cc	J. Abt	3'44"4	Fiat Abarth 1000 Berlina
			Sport	1300 cc	K. Federhofer	3'27"6	Fiat Abarth OT 1300
6 May	Zürich, Switzerland	Automobilslalom Zürich (4,000-5,000 kms)	Touring	850 cc	F. Wipf	9'52"8	Fiat Abarth 850 TC
			Touring	1000 cc	M. Wuest	9'30"6	Fiat Abarth 1000 Berlina
			Sport	850 cc	P. Maron	9'40"9	Fiat Abarth 850
			Sport	1000 cc	G. Naef	8'51"0	Fiat Abarth 1000 BA
7 May	Bruxelles, Belgium	Course de côte de Alle sur Semois (hillclimb-4,200 kms)	Touring	1000 cc	F. Neri	2'23"0	Fiat Abarth 1000 Berlina
			Sport	1000 cc	H. Affentranger	2'30"0	Fiat Abarth 1000 BA
7 May	Manresa, Spagna	Carrera en cuesta Can Masana (hillclimb-8,650 kms)	Touring	1000 cc	"Pelè"	4'22"3	Fiat Abarth 1000 Berlina
7 May	Imola, Italy	Coppa Bruno Deserti (circuit-5,017 kms x 30 laps)	Touring	600 cc	M. Magri	1h09'09"4	Fiat Abarth 595
			Touring	700 cc	U. Barilaro	1h08'05"8	Fiat Abarth 695
			Touring	850 cc	D. Martino	1h07'21"6	Fiat Abarth 850 TC
			Touring	1000 cc	M. Zanetti	1h06'41"5	Fiat Abarth 1000 Berlina
7 May	Krieglach, Austria	Internationales Alpen Bergrennen (hillclimb-6,500 kms)	Touring	1000 cc	G. König	3'48"2	Fiat Abarth 1000 Berlina
7 May	Bourscheid, Luxembourg	Course de côte de Bourscheid (circuit)	Touring	700 cc	H. Veder	2'23"2	Fiat Abarth 695 SS
7 May	Messina, Italy	Coppa Colle S. Rizzo (hillclimb-9,000 kms)	Touring	600 cc	"Valman"	7'24"8	Fiat Abarth 595
			Touring	700 cc	V. Virgilio	7'20"4	Fiat Abarth 695
			Touring	850 cc	G. Amodeo	6'48"7	Fiat Abarth 850 TC
			Touring	1000 cc	T. D'Alia	6'58"7	Fiat Abarth 1000 Berlina
			Grand Touring	up to 1300 cc	G. Passalacqua	7'12"3	Fiat Abarth 1000 OTS
			Sport	1000 cc	M. Calabrò	7'03"2	Fiat Abarth 1000 BA

DATE	LOCATION	RACE	CATEGORY	CLASS	WINNER	TIME OR DISTANCE	CAR
7 May	Udine, Italy	Cividale-Castelmonte (hillclimb, 7,000 kms)	Touring	600 cc	'Arch'	5'53"1	Fiat Abarth 595
			Touring	850 cc	B. Pagliaro	4'56"5	Fiat Abarth 850 TC
			Touring	1000 cc	P. Cattelan	4'48"8	Fiat Abarth 1000 Berlina
			Grand Touring	1000 cc	S. Piana	5'30"4	Fiat Abarth 1000 OTS
			Sport	1000 cc	S. Caldart	5'06"6	Fiat Abarth 1000 BA
			Sport-Prototypes	1000 cc	'Matich'	4'29"7	Fiat Abarth 1000 SP
7 May	Frascati, Italy	Frascati-Tuscolo (hillclimb-4,200 kms)	Touring	600 cc	F. Chiari	3'27"7	Fiat Abarth 595
			Touring	850 cc	A. Del Giovane	3'06"6	Fiat Abarth 850 TC
			Touring	1000 cc	F. Acasbo	3'06"1	Fiat Abarth 1000 Berlina
			Sport	1000 cc	G. Barbetti	3'19"6	Fiat Abarth 1000 BA
7 May	Adenau, Germany	ADAC Hansa Nürburgring Pokalrennen (circuit-7,700 kms x 8 laps)	Touring	up to 1150 cc	A.H. Hessel	---	Fiat Abarth 1000 Berlina
14 May	Innsbruck, Austria	Preis des Olympiastadt Innsbruck (circuit-2,800 kms x 10 laps)	Touring	1000 cc	J. Abt	14'59"9	Fiat Abarth 1000 Berlina
			Sport (13 laps)	up to 1300 cc	K. Miedaner	18'58"3	Fiat Abarth 1000 BA
14 May	Sassari, Italy	Alghero-Scala Picada (hillclimb-8,700 kms)	Touring	600 cc	F. Collu	5'54"8	Fiat Abarth 595
			Touring	850 cc	L. Capitta	5'12"0	Fiat Abarth 850 TC
			Sport	up to 1300 cc	S. Gambarau	4'49"1	Fiat Abarth 1000 BA
14 May	Frankfurt, Germany	DMV Spessart Bergrennen (hillclimb-3,400 kms)	Touring	1000 cc	K. Göttig	4'20"2	Fiat Abarth 1000 Berlina
			Grand Touring	1000 cc	H. Kämpf	4'34"9	Fiat Abarth 1000 OTS
			Sport	1300 cc	H. Krause	3'57"3	Fiat Abarth OT 1300
14 May	Vallelunga, Italy	Trofeo Coca Cola (circuit-1,800 kms x 62 laps)	Touring	600 cc	F. Mariani	1h17'33"9	Fiat Abarth 595
			Touring (66 laps)	850 cc	A. Del Giovane	1h14'35"9	Fiat Abarth 850 TC
			Touring (66 laps)	1000 cc	A. Merzario	1h14'35"3	Fiat Abarth 1000 Berlina
14 May	Amalfi, Italy	Coppa della Primavera (hillclimb-8,750 kms)	Touring	700 cc	S. Grillo	8'09"8	Fiat Abarth 695
			Touring	850 cc	F. Palumbo	7'01"0	Fiat Abarth 850 TC
			Touring	1000 cc	C. Palumbo	7'00"8	Fiat Abarth 1000 Berlina
			Sport	1300 cc	M. Terminiello	6'30"8	Fiat Abarth OT 1300
			Sport-Prototypes	2000 cc	P. Gargano	6'28"1	Fiat Abarth 2000
14 May	Parma, Italy	Castione Baratti-Neviano Arduini (hillclimb-5,500 kms)	Touring	850 cc	E. Gatti	3'47"7	Fiat Abarth 850 TC
			Touring	1000 cc	'Raffy'	3'45"5	Fiat Abarth 1000 Berlina
			Sport	1300 cc	A. Trenti	3'23"7	Fiat Abarth OT 1300
14 May	Palermo, Italy	Targa Florio (circuit-72,000 kms x 10 laps)	Sport	1300 cc	Garufi-Ferlito	7h35'04"2	Fiat Abarth OT 1300
15 May	Zandvoort, Holland	N.A.V. Nationale Pinksterraces (circuit-4,225 kms x 12 laps)	Touring	850 cc	H. Veder	27'07"5	Fiat Abarth 850 TC
			Touring	1000 cc	E.H. Swart	26'47"6	Fiat Abarth 1000 Berlina
18 May	Long Island (N.Y.), USA	Bridgehampton National Championship Race (circuit-2,850 miles)	---	---	A. Cosentino	---	Fiat Abarth 1000 Berlina
19-21 May	Montseny, Spagna	Carrera en cuesta al Montseny (hillclimb-15,000 kms)	Touring	850 cc	J. Juncosa	13'06"7	Fiat Abarth 850 TC
			Touring	up to 1150 cc	M. Juncosa	12'18"5	Fiat Abarth 1000 Berlina
			Sport	1000 cc	J.F. Cerveto	13'23"2	Fiat Abarth 1000 BA
			Sport-Prototypes	1300 cc	P. Maron	11'55"4	Fiat Abarth OT 1300
20-21 May	Zandvoort, Holland	ADAC Juniores Trophy (circuit-4,225 kms x 25 laps)	Touring	1000 cc	M. Kranefuss	---	Fiat Abarth 1000 Berlina
20-21 May	Abano Terme, Italy	Rallye delle Prealpi Venete (616 kms)	Touring	850 cc	O. Sfreddo	---	Fiat Abarth 850 TC
20-21 May	Bödele, Switzerland	Schwarzenberg Bödele Bergpreis (hillclimb-3,850 kms)	Touring	1000 cc	A. Benzinger	2'33"4	Fiat Abarth 1000 Berlina
21 May	Wissembourg, France	Course de côte Wissembourg-Col du Pigeonnier (hillclimb-4,700 kms)	Sport	up to 1150 cc	R. Stierli	2'32"2	Fiat Abarth 1000 BA
			Sport	up to 1600 cc	H. Krause	2'27"4	Fiat Abarth OT 1300
			Sport	over 1600 cc	Tritenne	2'25"1	Fiat Abarth 2000
21 May	Zolder, Belgium	Grand Prix du Limbourg (circuit-4,200 kms x 28 laps)	Touring	850 cc	M. Reano	1h01'36"3	Fiat Abarth 850 TC
			Touring	1000 cc	W. Kauhsen	1h00'58"9	Fiat Abarth 1000 Berlina
21 May	Bologna, Italy	Bologna-Passo della Raticosa (hillclimb-32,700 kms)	Touring	600 cc	R. Serpieri	23'43"0	Fiat Abarth 595
			Touring	850 cc	D. Martino	21'27"5	Fiat Abarth 850 TC
			Touring	1000 cc	G. Larini	21'51"7	Fiat Abarth 1000 Berlina
			Sport	1000 cc	E. Ostini	21'46"9	Fiat Abarth 1000 BA
			Sport	1300 cc	G. Villani	19'50"0	Fiat Abarth OT 1300
			Sport-Prototypes	1000 cc	P. Schetty	19'12"4	Fiat Abarth 1000 SP
			Sport-Prototypes	2000 cc	J. Ortner	18'47"8	Fiat Abarth 2000
21 May	Sorrento, Italy	Sorrento-Sant'Agata (hillclimb-10,500 kms)	Touring	700 cc	G. D'Ippolito	7'24"3	Fiat Abarth 695
			Touring	850 cc	F. Palumbo	6'35"6	Fiat Abarth 850 TC
			Touring	1000 cc	C. Palumbo	6'39"0	Fiat Abarth 1000 Berlina
			Sport	1000 cc	M. Terminiello	6'08"8	Fiat Abarth 1000 BA
			Sport-Prototypes	1000 cc	G. Tommasi	6'15"8	Fiat Abarth 1000 SP
			Sport-Prototypes	2000 cc	D. Scola	6'23"6	Fiat Abarth 2000
21 May	Jyllands, Danimarca	Jyllands Ringen (circuit)	Touring	850 cc	A. Larsen	1h14'4"	Fiat Abarth 850 TC
21 May	Obernzell, Germany	ADAC Nibelungen Bergrennen (hillclimb-3,100 kms)	Touring	850 cc	E. Schuder	4'13"5	Fiat Abarth 850 TC
21 May	Wittlich, Germany	Wittlicher ADAC Bergpreis (hillclimb-3,738 kms)	Touring	up to 1150 cc	J. Abt	5'01"0	Fiat Abarth 1000 Berlina
25 May	San Mateo de Bages, Spagna	Carrera en cuesta a San Mateo (hillclimb-5,500 kms)	Touring	up to 1150 cc	M. Juncosa	4'15"2	Fiat Abarth 1000 Berlina
			Sport	1000 cc	P. Basols	4'35"0	Fiat Abarth 1000 BA

DATE	LOCATION	RACE	CATEGORY	CLASS	WINNER	TIME OR DISTANCE	CAR
25 May	Cremona, Italy	Coppa Luigi Feraboli (acceleration stage)	Touring Touring	850 cc 1000 cc	A. Picchi F. Zanardelli	0'44"5 0'39"5	Fiat Abarth 850 TC Fiat Abarth 1000 Berlina
25 May	Augsburg, Germany	ASC-Rennslalom	Touring	850 cc	C. Kubon	---	Fiat Abarth 850 TC
27 May	Montlhéry, France	Grand Prix de Paris (circuit-3,360 kms x 20 laps)	Sport Sport	1000 cc 1300-1600 cc	G. Deforge Saint Quintin	36'16"6 35'58"6	Fiat Abarth 1000 BA Fiat Abarth OT 1600
27-28 May	Thompson (Connecticut), USA	Area 1 Championship Race (circuit-1,6 miles)	---	---	A. Cosentino	---	Fiat Abarth 1000 Berlina
27-28 May	Monza, Italy	Gran Premio di Monza (circuit-5,750 kms x 15 laps)	Touring Sport	1000 cc 1000 cc	M. Wuest R. Stierli	35'42"3 31'05"5	Fiat Abarth 1000 Berlina Fiat Abarth 1000 BA
28 May	Roskilde, Danimarca	Kobenhavan Cup (circuit-1,400 kms x 12 laps)	Touring	850 cc	A. Larsen	20'53"8	Fiat Abarth 850 TC
28 May	Abreschviller, France	Course de côte d'Abreschviller (hillclimb-2,000 kms)	Touring	up to 1000 cc	R. Neumann	1'23"3	Fiat Abarth 850 TC
28 May	Trieste, Italy	Trieste-Opicina (hillclimb-10,150 kms)	Touring Touring Touring Grand Touring Sport Sport Sport-Prototypes	600 cc 850 cc 1000 cc 1000 cc 1000 cc 1300 cc 1000 cc	R. Serpieri A. Mostardini M. Zanetti P. Brambilla M. Polin S. Morando P. Schetty	6'31"3 5'35"1 5'24"8 6'14"8 5'07"5 4'54"6 4'51"2	Fiat Abarth 595 Fiat Abarth 850 TC Fiat Abarth 1000 Berlina Fiat Abarth 1000 OTS Fiat Abarth 1000 BA Fiat Abarth OT 1300 Fiat Abarth 1000 SP
28 May	Piacenza, Italy	Castell'Arquato-Vernasca (hillclimb-9,775 kms)	Touring Touring Prototypes	850 cc up to 1150 cc 2000 cc	E. Gatti F. Zanardelli R. Bertuzzi (overall)	6'08"2 5'56"1 5'07"2	Fiat Abarth 850 TC Fiat Abarth 1000 Berlina Fiat Abarth OT 2000
28 May	Hockenheim, Germany	A.C.S. Rundstreckenrennen Hockenheim (circuit-2,690 kms x 12 laps)	Touring Touring	850 cc 1000 cc	H. Herren P. Engler	20'45"4 19'47"4	Fiat Abarth 850 TC Fiat Abarth 1000 Berlina
28 May	Kuopio, Finlandia	Väinolänniemen (circuit-19,000 kms)	Touring	850 cc	H. Liedes	11'32"2	Fiat Abarth 1000 Berlina
2 June	Gubbio, Italy	Coppa Mengara (hillclimb-5,200 kms)	Touring Touring Touring Touring Sport	600 cc 700 cc 850 cc 1000 cc 1000 cc	M. Cangelli L. Cestra V. Cristi L. Laringi N. Camilli	3'18"9 3'13"6 2'56"3 2'54"5 3'30"3	Fiat Abarth 595 Fiat Abarth 695 Fiat Abarth 850 TC Fiat Abarth 1000 Berlina Fiat Abarth 1000 BA
3-4 June	Hockenheim, Germany	A.C.S. Nationales Rundstreckenrennen Hockenheim (circuit-2,690 kms)	Touring Sport (30 laps) Sport (30 laps)	850 cc 850 cc 1000 cc	R. Blatter H. Affentranger R. Stierli	31'00"1 45'30"4 44'08"7	Fiat Abarth 850 TC Fiat Abarth 850 Fiat Abarth 1000 BA
4 June	Guadalupe, Spagna	Gran Premio Ciudad de Alcaniz (circuit-3,900 kms x 3)	Touring	1000 cc	M. Juncosa	---	Fiat Abarth 1000 Berlina
4 June	Zandvoort, Holland	Grand Prix Holland (circuit-4,225 kms x 14 laps)	Touring	1000 cc	A. Goedemans	27'03"3	Fiat Abarth 1000 Berlina
4 June	Berchtesgaden, Germany	Rossfeld Alpen Bergpreis (hillclimb-5,890 kms)	Sport	1300 cc	B. Hofmann	7'18"5	Abarth Simca 1300
4 June	Biella, Italy	Coppa Carlo Felice Trossi (hillclimb-6,500 kms)	Touring Touring Sport	850 cc 1000 cc 1300 cc	P. Tali A. Merzario R. Ramoino	4'42"4 4'16"1 4'04"1	Fiat Abarth 850 TC Fiat Abarth 1000 Berlina Fiat Abarth OT 1300
4 June	Rieti, Italy	Coppa Bruno Carotti (hillclimb-15,100 kms)	Touring Touring Touring Sport Sport-Prototypes	600 cc 700 cc 850 cc 1000 cc 1000 cc	M. Longarini L. Cestra A. Del Giovane E. Ostini G. Tommasi	11'40"2 10'17"1 9'24"0 9'00"5 8'31"1	Fiat Abarth 595 Fiat Abarth 695 Fiat Abarth 850 TC Fiat Abarth 1000 BA Fiat Abarth 1000 SP
4 June	Treviso, Italy	Vittorio Veneto-Cansiglio (hillclimb-10,000 kms)	Touring Touring Grand Touring Sport Sport	850 cc 1000 cc 1000 cc 1000 cc 1300 cc	V. Tabacchi A. De Paoli S. Piana L. Rassega S. Morando	7'32"4 7'18"6 8'49"1 7'36"2 7'07"5	Fiat Abarth 850 TC Fiat Abarth 1000 Berlina Fiat Abarth 1000 OTS Fiat Abarth 1000 BA Fiat Abarth OT 1300
4 June	Curborough, United Kingdom	Curborough Sprint (circuit)	Sport	up to 1600 cc	R. Eccles	39'9"0	Fiat Abarth OT 1600
4 June	Digne, France	Course de côte de la Lure (hillclimb-12,700 kms)	Touring Sport	1000 cc 1300 cc	F. Tomas F. Canin	9'16"4 8'17"0	Fiat Abarth 1000 Berlina Fiat Abarth OT 1300
4 June	Dijon, France	Course de côte Autun-Montjeu (hillclimb-4,100 kms)	Sport	2000 cc	Tritenne	4'41"9	Fiat Abarth 2000
10-11 June	Antibes, France	Rallye des Roses (395 kms)	Touring	850 cc	Lanteri-Fiquet	48'53"5	Fiat Abarth 850 TC
10-11 June	Le Mans, France	24 Heures du Mans (circuit-13,461 kms x 24 hours)	Sport	1300 cc	Martin-Mesange	kms 3531,530	Fiat Abarth OT 1300
10-11 June	Rothenburg, Germany	ADAC Bergprüfung in Rothenburg (hillclimb-6,05 kms x 2)	Touring Sport	up to 1150 cc 1300 cc	P. Klein H. Krause	4'18"2 4'24"4	Fiat Abarth 1000 Berlina Fiat Abarth OT 1300
11 June	Perigueux, France	Course de côte Perigueux-Coursac (hillclimb-4,100 kms)	Touring	850 cc	Y. Barone	1'56"0	Fiat Abarth 850 TC
11 June	Herbeumont, Belgium	Course de côte Herbeumont-Sainte Cecile (hillclimb-3,600 kms)	Touring Sport	1000 cc 1000 cc	F. Neri H. Affentranger	2'06"8 2'04"3	Fiat Abarth 1000 Berlina Fiat Abarth 1000 BA
11 June	Catanzaro, Italy	Nicastro-Passo di Acquavona (hillclimb-14,550 kms)	Touring Touring Sport Sport	600 cc 850 cc 1000 cc 2000 cc	G. D'Ippolito L. Ferragina V. Cannataro D. Scola (overall)	11'39"7 10'35"7 10'35"9 9'26"7	Fiat Abarth 595 Fiat Abarth 850 TC Fiat Abarth 1000 BA Fiat Abarth 2000

DATE	LOCATION	RACE	CATEGORY	CLASS	WINNER	TIME OR DISTANCE	CAR
11 June	Bichlbach, Austria	Bichlbach-Berwang Bergrennen (hillclimb-2,600 kms x 2)	Touring	1000 cc	J. Abt	3'27"9	Fiat Abarth 1000 Berlina
			Sport	1300 cc	G. Bredow	3'27"3	Fiat Abarth OT 1300
11 June	Budapest, Hungary	Grand Prix de Budapest (circuit-5,200 kms)	Touring	1000 cc	M. Zanetti	---	Fiat Abarth 1000 Berlina
11 June	Bologna, Italy	Coppa Acqua Cerelia (hillclimb-12,200 kms)	Touring	600 cc	M. Magri	9'31"2	Fiat Abarth 595
			Touring	700 cc	P. Valeri	10'18"5	Fiat Abarth 695
			Touring	850 cc	D. Martino	8'45"3	Fiat Abarth 850 TC
			Touring	1000 cc	S. Mingotti	8'41"3	Fiat Abarth 1000 Berlina
			Sport	1300 cc	A. Trenti	8'14"3	Fiat Abarth OT 1300
11 June	Asiago, Italy	Coppa Città di Asiago (hillclimb-14,500 kms)	Touring	600 cc	P. Cozzarolo	12'00"5	Fiat Abarth 595
			Touring	700 cc	R. Galasso	10'46"8	Fiat Abarth 695
			Touring	850 cc	V. Tabucchi	9'29"6	Fiat Abarth 850 TC
			Touring	1000 cc	A. De Paoli	9'05"7	Fiat Abarth 1000 Berlina
			Grand Touring	1000 cc	G. Trombetta	10'27"4	Fiat Abarth 1000 OTS
			Sport	1000 cc	M. Polin	8'48"9	Fiat Abarth 1000 BA
11 June	Livorno, Italy	Antignano-Monteburrone (hillclimb-3,400 kms x 2)	Touring	850 cc	B. Pagliaro	4'40"3	Fiat Abarth 850 TC
			Touring	1000 cc	R. Santini	4'30"7	Fiat Abarth 1000 Berlina
			Sport	up to 1600 cc	B. Guarneri	4'21"8	Fiat Abarth OT 1300
11 June	Valence, France	Course de côte du Pin (hillclimb-7,050 kms)	Touring	1000 cc	F. Tomas	5'22"4	Fiat Abarth 1000 Berlina
			Sport	1000 cc	Thian	5'11"6	Fiat Abarth 1000 BA
			Sport	up to 1600 cc	F. Canin	5'10"7	Fiat Abarth OT 1300
11 June	Vallelunga, Italy	Trofeo Mario Angiolini (circuit-1,800 kms x 20 laps)	Touring	700 cc	L. Cestra	24'17"9	Fiat Abarth 695
			Touring	1000 cc	F. Acasbo	25'40"2	Fiat Abarth 1000 Berlina
			Sport	up to 2000 cc	G. Peroni	25'01"1	Fiat Abarth OT 1300
11 June	Villars, Switzerland	Course de côte Villars-Burquin (hillclimb-1,800 kms)	Touring	850 cc	R. Neumann	1'26"8	Fiat Abarth 850 TC
			Touring	1000 cc	P. Engler	1'27"9	Fiat Abarth 1000 Berlina
11 June	San Cugat, Spagna	Carrera en cuesta San Cugat-Tibidabo (hillclimb-5,900 kms)	Touring	850 cc	'Jorge Chi'	4'03"4	Fiat Abarth 850 TC
			Touring	up to 2500 cc	M. Juncosa	3'49"3	Fiat Abarth 1000 Berlina
			Sport	1000 cc	J.F. Cerveto	3'50"2	Fiat Abarth 1000 BA
17 June	Reitnau, Switzerland	Bergslalom Reitnau	Sport	up to 1300 cc	H. Affentranger (overall)	---	Fiat Abarth 1000 BA
17-18 June	Wiesbaden, Germany	HMSC Flugplatz Rennen (circuit-2,100 kms x 30 laps)	Touring	850 cc	K.H. Scholl	38'36"9	Fiat Abarth 850 TC
			Touring	up to 1150 cc	K. Hessel	38'01"1	Fiat Abarth 1000 Berlina
			Sport	1300 cc	E. Bitter	33'26"1	Fiat Abarth OT 1300
			Prototypes	1300 cc	H. Krause	33'28"7	Fiat Abarth OT 1300
18 June	Novara, Italy	Gimkana di Cerano	---	---	G.P. Pedretti	---	Fiat Abarth 1000 Berlina
18 June	Bremen, Germany	ADAC Bremer Slalom	---	up to 1150 cc	H.P. Klein	---	Fiat Abarth 1000 Berlina
18 June	Metz, France	Course de côte d'Arry (hillclimb-1,200 kms)	Touring	1000 cc	Bourguignon	0'52"1	Fiat Abarth 1000 Berlina
			Sport	1300 cc	R. Stierli	0'45"3	Fiat Abarth 1000 BA
			---	---	P. Maublanc	0'39"2	Fiat Abarth 2000
18 June	Cuneo, Italy	Garessio-Colle S. Bernardo (hillclimb-7,100 kms)	Touring	850 cc	B. Pagliero	4'36"9	Fiat Abarth 850 TC
			Touring	up to 1150 cc	S. Dal Cason	4'32"5	Fiat Abarth 1000 Berlina
			Sport	over 1300 cc	L. Taramazzo	3'47"2	Fiat Abarth 1000 BA
18 June	Arezzo, Italy	Camucia-Cortona (hillclimb-3,500 kms x 2)	Touring	600 cc	F. Gragnoli	5'35"2	Fiat Abarth 595
			Touring	700 cc	R. Canestrelli	5'46"1	Fiat Abarth 695
			Touring	850 cc	P. Falani	5'07"5	Fiat Abarth 850 TC
			Touring	1000 cc	R. Santini	5'02"4	Fiat Abarth 1000 Berlina
18 June	Cosenza, Italy	Coppa della Sila (hillclimb-14,000 kms)	Touring	600 cc	G. Lopez	12'01"4	Fiat Abarth 595
			Touring	700 cc	P. De Luca	11'59"7	Fiat Abarth 695
			Touring	850 cc	L. Ferragina	11'03"2	Fiat Abarth 850 TC
			Touring	up to 1150 cc	F. Sorrentino	11'15"4	Fiat Abarth 1000 Berlina
			Grand Touring	1000 cc	A. Torricelli	11'38"5	Fiat Abarth 1000 OTS
			Sport	1000 cc	M. Terminiello	10'27"2	Fiat Abarth 1000 BA
			Sport-Prototypes	up to 2000 cc	D. Scola (overall)	9'29"3	Fiat Abarth 2000
18 June	Reggio Emilia, Italy	Vezzano-Casina (hillclimb-12,500 kms)	Touring	600 cc	M. Magri	8'30"7	Fiat Abarth 595
			Touring	700 cc	G.F. Loda	8'43"5	Fiat Abarth 695
			Touring	850 cc	D. Martino	7'49"6	Fiat Abarth 850 TC
			Touring	1000 cc	A. Merzario	7'43"0	Fiat Abarth 1000 Berlina
			Sport	1000-1600 cc	S. Morando (overall)	7'11"5	Fiat Abarth OT 130
18 June	Zandvoort, Holland	N.A.V. Trophy of the Dunes (circuit-4,225 kms)	Touring	850 cc	H. Veder	31'21"9	Fiat Abarth 850 TC
			Touring	1000 cc	A. Goedemans	28'58"5	Fiat Abarth 1000 Berlina
23 June	Rovaniemi, Filandia	Ounasvaaran Vauhtikisat (circuit-1,950 kms x 10 laps)	Touring	850 cc	A. Vaara	11'40"8	Fiat Abarth 850 TC
24-25 June	Valencia, Spagna	Rallye Fallas (425 kms)	---	---	M. Juncosa	---	Fiat Abarth 1000 Berlina
24-25 June	Wasgau, Germany	ADAC Wasgau Bergprüfung (hillclimb-5,100 kms x 2)	Touring	850 cc	R. Fischer	6'01"1	Fiat Abarth 850 TC
			Touring	up to 1150 cc	J. Abt	5'45"2	Fiat Abarth 1000 Berlina
			Touring Special	1000 cc	H. Schöne	5'55"6	Fiat Abarth 1000 Berlina

DATE	LOCATION	RACE	CATEGORY	CLASS	WINNER	TIME OR DISTANCE	CAR
24-25 June	Sierre, Switzerland	Lens-Crans sur Sierre (hillclimb-3,100 kms x 2)	Touring	850 cc	R. Neumann	4'43"8	Fiat Abarth 850 TC
			Touring	1000 cc	M. Wuest	4'38"1	Fiat Abarth 1000 Berlina
			Sport	1000 cc	G. Naef	4'16"6	Fiat Abarth 1000 BA
24-25 June	Huambo, Angola	Pequeno Circuito Automovel do Huambo (circuit-1,200 kms x 90 laps)	---	---	C. Pereira	18'48"8	Fiat Abarth 1000 Berlina
25 June	Torino, Italy	Cuorgné-Alpette (hillclimb-7,000 kms)	Touring	700 cc	L. Pesce	5'57"2	Fiat Abarth 695
			Touring	850 cc	P. Tali	5'30"2	Fiat Abarth 850 TC
			Touring	up to 1150 cc	S. Dal Cason	5'10"2	Fiat Abarth 1000 Berlina
			Sport	1300 cc	S. Morando	4'39"8	Fiat Abarth OT 1300
25 June	Strasbourg, France	Course de côte de Turckheim Trois Epis (hillclimb-6,230 kms)	Touring	up to 1150 cc	P. Engler	4'04"1	Fiat Abarth 1000 Berlina
25 June	Avus, Germany	Grosser Preis von Berlin (circuit-8,300 kms x 10 laps)	Touring	850 cc	J. Ortner	31'39"7	Fiat Abarth 850 TC
			Touring	up to 1150 cc	K. Ahrens	30'09"3	Fiat Abarth 1000 Berlina
			Sport	1300 cc	K. Ahrens	25'26"2	Fiat Abarth OT 1300
25 June	Parma, Italy	Fornovo-Monte Cassio (hillclimb-13,000 kms)	Touring	600 cc	P.G. Poggi	9'58"8	Fiat Abarth 595
			Touring	700 cc	U. Barilaro	9'53"1	Fiat Abarth 695
			Touring	850 cc	F. Galli	9'07"4	Fiat Abarth 850 TC
			Touring	1000 cc	M. Campanini	8'56"6	Fiat Abarth 1000 Berlina
			Grand Touring	1000 cc	P. Brambilla	10'01"5	Fiat Abarth 1000 OTS
			Sport	1300 cc	A. Trenti	8'23"8	Fiat Abarth OT 1300
			Sport-Prototypes	2000 cc	D. Scola	8'11"6	Fiat Abarth 2000
25 June	Palermo, Italy	Corsa al Monte Pellegrino (hillclimb-8,750 kms)	Touring	600 cc	'Harca'	6'25"7	Fiat Abarth 595
			Touring	700 cc	P. De Luca	6'16"7	Fiat Abarth 695
			Touring	850 cc	'Gam'	5'46"0	Fiat Abarth 850 TC
			Touring	1000 cc	S. Scigliano	5'57"2	Fiat Abarth 1000 Berlina
			Sport	1300 cc	A. Gambero	5'23"0	Fiat Abarth OT 1300
			Prototypes	1000 cc	S. Calascibetta	5'40"9	Fiat Abarth 1000 SP
25 June	Nîmes, France	Course de côte du Minier (hillclimb-21,500 kms)	Touring	600 cc	J. Choquel	18'51"7	Fiat Abarth 595
			Prototypes	up to 2500 cc	P. Maublanc	13'36"0	Fiat Abarth 2000
29 June	Brescia, Italy	Scalata al Colle Sant'Eusebio (hillclimb-9,150 kms)	Touring	600 cc	R. Amighini	5'48"2	Fiat Abarth 595
			Touring	700 cc	L. Lazzeri	6'09"3	Fiat Abarth 695
			Touring	850 cc	L. Benini	5'47"9	Fiat Abarth 850 TC
			Touring	1000 cc	F. Zanardelli	5'08"7	Fiat Abarth 1000 Berlina
			Grand Touring	1000 cc	P. Brambilla	5'50"6	Fiat Abarth 1000 OTS
			Sport	1000 cc	L. Rassega	5'11"3	Fiat Abarth 1000 BA
			Sport	1300 cc	A. Becchetti	4'53"3	Fiat Abarth OT 1300
29 June	Forlì, Italy	Predappio-Rocca delle Caminate (hillclimb-4,000 kms)	Touring	700 cc	G. Gatta	3'57"1	Fiat Abarth 695
			Touring	850 cc	P. Falorni	3'42"1	Fiat Abarth 850 TC
			Touring	1000 cc	G. Parri	3'48"6	Fiat Abarth 1000 Berlina
1 July	Woburn, United Kingdom	Woburn Hillclimb (hillclimb-800 yarde)	Touring	850 cc	J. Pratt	---	Fiat Abarth 850 TC
1-2 July	Orense, Spagna	Rallye Orense (425 kms)	Touring	up to 1150 cc	Juncosa-'Artemi	---	Fiat Abarth 1000 Berlina
2 July	Marsannay, France	Course de côte Marsannay (hillclimb)	Grand Touring	1000 cc	J. Chabroud	3'22"3	Fiat Abarth 1000 OTS
			Prototypes	1000 cc	'Thiani'	2'58"9	Fiat Abarth 1000 SP
			Prototypes	2000 cc	P. Maublanc	2'36"2	Fiat Abarth 2000
2 July	Adenau, Germany	Nürburgring Grosser Preis der Tourenwagen (circuit-22,810 kms x 6 hours)	Touring	850 cc	Galli-Raffa	6h02'53"3	Fiat Abarth 850 TC
			Touring	1000 cc	Uberti-'Pam'	6h02'23"1	Fiat Abarth 1000 Berlina
2 July	Bolzano, Italy	Bolzano-Mendola (hillclimb-14,400 kms)	Touring	850 cc	V. Tabucchi	10'55"7	Fiat Abarth 850 TC
			Touring	1000 cc	P. Cattelan	10'56"1	Fiat Abarth 1000 Berlina
			Sport	1000 cc	M. Polin	12'03"8	Fiat Abarth 1000 BA
			Sport	1300 cc	A. Becchetti	12'22"6	Fiat Abarth OT 1300
2 July	Vallelunga, Italy	Premio Campagnano (circuit-3,200 kms x 35 laps)	Touring	600 cc	R. Mazzeschi	1h10'08"6	Fiat Abarth 595
			Touring	700 cc	L. Finelli	1h11'05"1	Fiat Abarth 695 SS
			Touring	850 cc (40 laps)	D. Martino	1h11'13"8	Fiat Abarth 850 TC
2 July	Dalsand, Danimarca	Dalsand Ring (circuit-2,000 kms x 16 laps)	Touring	1000 cc	L. Englund	11'02"1	Fiat Abarth 1000 Berlina
2 July	Courtland (Alabama), USA	National Courtland Alabama (circuit)	---	---	G. Wood	---	Fiat Abarth 1000 Berlina
2 July	Lodève, France	Course de côte de Lodève (hillclimb-5,400 kms)	Sport-Prototypes	over 1300 cc	L. Gayraud	3'51"0	Fiat Abarth 2000
2 July	Zandvoort, Holland	Les Coupes Benelux (circuit-4,225 kms x 12 laps)	Touring	850 cc	H. Veder	25'39"5	Fiat Abarth 850 TC
			Touring	1000 cc	H. Lenz	24'49"5	Fiat Abarth 1000 Berlina
			Sport	1000 cc	F. Neri	23'21"3	Fiat Abarth 1000 BA
4 July	Lime Rock (Connecticut), USA	National Races-Lime Rock Park (circuit-1,5 miles x 15 laps)	---	---	A. Cosentino	---	Fiat Abarth 1000 Berlina
8-9 July	Höxter, Germany	ADAC Weser Bergpreis Höxter (hillclimb-3,400 kms x 3)	Touring	850 cc	K.H. Trettan	6'18"5	Fiat Abarth 850 TC
			Touring	up to 1150 cc	K. Göttig	6'02"6	Fiat Abarth 1000 Berlina

DATE	LOCATION	RACE	CATEGORY	CLASS	WINNER	TIME OR DISTANCE	CAR
9 July	Lucca, Italy	Coppa Fornaci di Barga-Barga (hillclimb-3,500 kms x 2)	Touring Touring Touring Touring Sport	600 cc 700 cc 850 cc up to 1150 cc 1000 cc	R. Serpieri G. Brancadori F. Gatti G. Ranzolin A. Cittadini	2'34"8 2'39"6 2'23"7 2'18"9 2'26"0	Fiat Abarth 595 Fiat Abarth 695 Fiat Abarth 850 TC Fiat Abarth 1000 Berlina Fiat Abarth 1000 BA
9 July	Palermo, Italy	Alcamo-Monte Bonifato (hillclimb-4,500 kms)	Touring Touring Touring Touring Sport Sport	600 cc 700 cc 850 cc up to 1150 cc 1000 cc 1300 cc	'Procopio' A. La Luce G. Amodeo T. D'Alia M. Calabrò A. Gambero	3'51"6 3'36"9 3'20"0 3'19"0 3'15"9 2'59"6	Fiat Abarth 595 SS Fiat Abarth 695 Fiat Abarth 850 TC Fiat Abarth 1000 Berlina Fiat Abarth 1000 BA Fiat Abarth OT 1300
9 July	Gives, Belgium	Course de côte de Gives (hillclimb-2,400 kms x 2)	Touring Sport	1000 cc 1000 cc	F. Neri F. Neri	1'36"7 1'27"9	Fiat Abarth 1000 Berlina Fiat Abarth 1000 BA
9 July	Monopoli, Italy	Coppa Città di Monopoli (hillclimb-4,500 kms)	Prepared Touring Prepared Touring Sport Sport-Prototypes Sport-Prototypes	850 cc 1000 cc 1000 cc 1000 cc 2000 cc	F. Altieri F. Sorrentino G. Benini G. Tommasi D. Scola (overall)	2'34"2 2'33"5 2'31"1 2'17"3 2'09"3	Fiat Abarth 850 TC Fiat Abarth 1000 Berlina Fiat Abarth 1000 BA Fiat Abarth 1000 SP Fiat Abarth 2000
9 July	Trento, Italy	Trento-Bondone (hillclimb-17,300 kms)	Touring Touring Grand Touring Sport Sport Sport-Prototypes	850 cc 1000 cc 1000 cc 1000 cc 1300 cc 1000 cc	C. Margoni J. Abt P. Brambilla E. Ostini F. Pilone E. Buzzetti	14'26"7 14'11"7 16'52"9 14'15"9 13'30"8 13'49"4	Fiat Abarth 850 TC Fiat Abarth 1000 Berlina Fiat Abarth 1000 OTS Fiat Abarth 1000 BA Fiat Abarth OT 1300 Fiat Abarth 1000 SP
9 July	Hockenheim, Germany	Preis von Baden (circuit-2,768 kms x 45 laps)	Sport	1300 cc	A. Hezemans	1h47'50"4–	Fiat Abarth OT 1300
9 July	Blackhall Rock, United Kingdom	Autocross Meeting (750 yards)	---	up to 1200 cc	P. Irvine	---	Fiat Abarth 1000 Berlina
9 July	Lima, Perù	Las Seis Horas Peruanas (circuit-154 laps)	Grand Touring	1000 cc	H. Grimm	6h00'03"3	Fiat Abarth 1000 OTS
9 July	Wilmot Hills (Illinois), USA	Wilmot Hills Race (circuit-15 laps)	---	---	N.E. Schley	---	Fiat Abarth 1000 Berlina
15 July	Long Island (N.Y.), USA	Bridgehampton National Championship Races (circuit-2,85 miles)	---	---	A. Cosentino	---	Fiat Abarth 1000 Berlina
15-16 July	Taubensuhl, Germany	ADAC Taubensuhl Bergrennen (hillclimb-6,000 kms)	Touring Touring Sport	850 cc up to 1150 cc 1300 cc	K.H. Trettan H. Schöne R. Jöst	7'38"8 7'35"3 7'39"9	Fiat Abarth 850 TC Fiat Abarth 1000 Berlina Fiat Abarth OT 1300
15-16 July	Magny Cours, France	Criterium de Vitesse (circuit-2,040 kms x 20 laps)	Touring	up to 1150 cc	A. Maury	22'56"2	Fiat Abarth 1000 Berlina
16 July	Hämeenlinnan, Finlandia	Hämeenlinnan Ajot (circuit-2,900 kms x 10 laps)	Touring	850 cc	A. Vaara	18'29"8	Fiat Abarth 850 TC
16 July	Landebahn, Austria	Flugplatzrennen Tulln-Langenlebarn (circuit-2,700 kms x 25 laps)	Touring	1000 cc	G. König	34'48"1	Fiat Abarth 1000 Berlina
16 July	Sestrière, Italy	Cesana-Sestrière (hillclimb-10,400 kms)	Touring Touring Touring Grand Touring Sport Sport	600 cc 850 cc 1000 cc 1000 cc 1000 cc 1300 cc	P.G. Poggi F. Galli S. Dal Cason G.F. Silecchia G. Varese J. Ortner	7'38"5 6'55"4 6'42"2 7'25"2 6'17"3 5'50"6	Fiat Abarth 595 Fiat Abarth 850 TC Fiat Abarth 1000 Berlina Fiat Abarth 1000 OTS Fiat Abarth 1000 BA Fiat Abarth OT 1300
16 July	Corsavy, France	Course de côte Corsavy (hillclimb-5,200 kms)	Touring	1000 cc	J. Stivin	4'15"0	Fiat Abarth 1000 Berlina
21 July	Thöringen, Switzerland	Bergrennen Thöringen-Linden (hillclimb-2,500 kms)	Sport	1000 cc	H. Affentranger	1'14"6	Fiat Abarth 1000 BA
22-23 July	Thompson (Connecticut), USA	Thompson National Races (circuit-1,6 miles)	---	---	A. Cosentino	---	Fiat Abarth 1000 Berlina
23 July	Ellerberg, Germany	Ellerberg-rennen (hillclimb-2,690 kms)	Sport	1300 cc	G. Dioler	---	Fiat Abarth OT 1300
23 July	Staplefield, United Kingdom	Staplefield Autocross (circuit-700 yarde)	Touring	1000 cc	M. Knights	---	Fiat Abarth 1000 Berlina
23 July	Monthoux, Switzerland	Rundkurs Monthoux (circuit-10 laps x 2)	Touring Sport	1000 cc 1300 cc	P. Walther P. Toher	16'16"1 15'51"9	Fiat Abarth 1000 Berlina Fiat Abarth OT 1300
23 July	Reggio Calabria, Italy	Santo Stefano-Gambarie (hillclimb-7,800 kms)	Touring Touring Touring Sport Sport Prototypes	600 cc 850 cc 1000 cc up to 1300 cc over 1300 cc 2000 cc	Angheloni Laganà A. Bonaccorsi A. Gambero P. Gargano D. Scola	6'46"1 5'58"4 6'19"4 5'40"4 5'14"7 5'07"4	Fiat Abarth 595 Fiat Abarth 850 TC Fiat Abarth 1000 Berlina Fiat Abarth OT 1300 Fiat Abarth 2000 Fiat Abarth 2000
23 July	Scarperia, Italy	Gran Premio del Mugello (circuit-66,200 kms x 8 laps)	Sport Sport Sport-Prototypes	1000 cc 1300 cc 1000 cc	Baldo-Margoni Nesti-Furtmayr E. Buzzetti	4h22'38"8 4h24'36"6 4h30'56"9	Fiat Abarth 1000 BA Fiat Abarth OT 1300 Fiat Abarth 1000 SP
29 July	Lime Rock (Connecticut), USA	Area 1 Championship Races (circuit-1,5 miles x 15 laps)	---	---	A. Cosentino	---	Fiat Abarth 1000 Berlina

DATE	LOCATION	RACE	CATEGORY	CLASS	WINNER	TIME OR DISTANCE	CAR
29 July	Adenau, Germany	Nürburgring ADAC Rheinghauer Zuverlassigkeitsfahrt (circuit-22,810 kms x 18 laps)	Grand Touring	1000 cc	G. Binzenbach	---	Fiat Abarth 1000 OTS
29-30 July	Bomeree, Belgium	Course de côte du Charleroi (hillclimb-2,000 kms)	Touring Sport	1000 cc 1000 cc	S. Tucci H. Affentranger	1'20"2 1'10"2	Fiat Abarth 1000 Berlina Fiat Abarth 1000 BA
30 July	Barberust, France	Course de côte d'Argèles Gazost (hillclimb-3,000 kms)	Touring	850 cc	H. Stivin	3'08"1	Fiat Abarth 850 TC
30 July	Zandvoort, Holland	Grosser Preis von Zandvoort (circuit-4,225 kms x 12 laps)	Touring Touring	850 cc 1000 cc	H. Veder A. Goedemans	24'22"5 23'57"8	Fiat Abarth 850 TC Fiat Abarth 1000 Berlina
30 July	Freiburg, Germany	ADAC Bergpreis Freiburg-Schauinsland (hillclimb-11,200 kms)	Touring Touring Sport	850 cc up to 1150 cc 1300 cc	R. Fischer J. Abt E. Bitter	16'31"1 15'04"6 13'55"5	Fiat Abarth 850 TC Fiat Abarth 1000 Berlina Fiat Abarth OT 1300
30 July	Siena, Italy	Abbadia S. Salvatore-Vetta Amiata (hillclimb-12,700 kms)	Touring Touring Touring Touring Sport Sport	600 cc 700 cc 850 cc 1000 cc 1000 cc 1300 cc	R. Mazzeschi G. Brancadori V. Cristi N. Parenti S. Pellegrino 'Nicor'	9'21"3 9'16"8 8'44"1 8'16"0 9'40"2 8'08"9	Fiat Abarth 595 Fiat Abarth 695 Fiat Abarth 850 TC Fiat Abarth 1000 Berlina Fiat Abarth 1000 BA Fiat Abarth OT 1300
30 July	Cuneo, Italy	Rallye Ruota d'Oro (368 kms)	Touring	850 cc	P. Bollati	---	Fiat Abarth 850 TC
5 August	Västkustoppet, Sweden	Vastkustoppet (circuit)	Touring	850 cc	L. Englund	---	Fiat Abarth 850 TC
6 August	Pergusa, Italy	Coppa Città di Enna (circuit-4,797 kms x 63 laps)	Sport Sport	1000 cc 1300 cc	F. Patané A. Gambero	1h43'41"0 1h42'38"4	Fiat Abarth 1000 BA Fiat Abarth OT 1300
6 August	Adenau, Germany	Nürburgring Grosser Preis von Deutschland (circuit-22,810 kms x 6 laps)	Touring Touring	850 cc up to 1150 cc	K.H. Trettan E. Swart	1h08'20"3 1h05'29"3	Fiat Abarth 850 TC Fiat Abarth 1000 Berlina
6 August	Mittendorf, Austria	Tauplitzalm-Bergrennen	Touring	1000 cc	G. König	6'35"2	Fiat Abarth 1000 Berlina
6 August	La Coruna, Spagna	Gran Premio de La Coruna (circuit-1,800 kms x 30 laps)	A	850 cc	F. Serra	33'36"0	Fiat Abarth 850 TC
6 August	Richterswil, Switzerland	Autoslalom	---	1000 cc	R. Stierli	4'30"2	Fiat Abarth 1000 SP
6 August	Saint Lary, France	Course de côte de Saint Lary (hillclimb-4,200 kms)	Sport	over 1600 cc	P. Maublanc	2'40"8	Fiat Abarth 2000
12 August	Marlboro (Maryland), USA	Marlboro Double 300 Trans American Race (circuit-1,8 miles)	---	---	Cuomo-Fisher	---	Fiat Abarth 1000 Berlina
12-13 August	Wunstorf, Germany	Flugplatzrennen Wunstorf (circuit-5,100 kms x 10 laps)	Touring Sport (20 laps)	850 cc 1300 cc	K. Scholl E. Bitter	24'34"0 46'37"8	Fiat Abarth 850 TC Fiat Abarth OT 1300
13 August	Prats del Rey, Spagna	Carrera en cuesta a Las Mayolas (hillclimb-6,100 kms)	Grand Touring	1000 cc	F. Serra	4'27"1	Fiat Abarth 1000 Berlina
13 August	Friedenfels, Germany	Bergpreis von Oberpfalz (hillclimb-3,100 kms)	Touring	up to 1150 cc	H. Schöne	3'54"6	Fiat Abarth 1000 Berlina
13 August	Termini Imerese, Italy	Termini-Caccamo (hillclimb-10,000 kms)	Touring Touring Touring Touring Sport	600 cc 700 cc 850 cc 1000 cc 1300 cc	G. Mercadante V. Virgilio D. Stabile T. D'Alia G. Gambero	7'45"8 6'54"8 6'21"6 6'25"4 5'58"4	Fiat Abarth 595 Fiat Abarth 695 Fiat Abarth 850 TC Fiat Abarth 1000 Berlina Fiat Abarth OT 1300
13 August	Stainz, Austria	Stainzer Bergrennen (hillclimb-5,500 kms)	Touring	1000 cc	G. König	4'00"4	Fiat Abarth 1000 Berlina
15 August	Dudelange, Luxembourg	Course de côte Tele-Dudelange (hillclimb-1,400 kms)	Touring Sport	1000 cc 1000 cc	P. Engler H. Affentranger	1'04"2 0'57"2	Fiat Abarth 1000 Berlina Fiat Abarth 1000 BA
15 August	Pescara, Italy	Cronoscalata Svolte di Popoli (hillclimb-8,000 kms)	Touring Touring Prototypes	850 cc 1000 cc 1000 cc	F. Galli N. Parenti G. Tommasi	5'24"4 5'09"8 4'53"0	Fiat Abarth 850 TC Fiat Abarth 1000 Berlina Fiat Abarth 1000 SP
15 August	Mont-Dore, France	Course de côte du Mont-Dore (hillclimb-5,000 kms)	Touring Prototypes	850 cc 2000 cc	P. Champean P. Maublanc	4'08"9 3'02"0	Fiat Abarth 850 TC Fiat Abarth 2000
15 August	Walding, Austria	Bergrennen Walding (hillclimb-3,750 kms)	Touring	1000 cc	G. König	4'06"5	Fiat Abarth 1000 Berlina
18-20 August	Vigo, Spagna	Rallye a Las Rias Bajas (620 kms)	Touring	1000 cc	M. Juncosa	---	Fiat Abarth 1000 Berlina
19 August	Watkins Glen (N.Y.), USA	The Glen 500 Race (circuit-2,3 miles x 20 laps)	---	---	A. Cosentino	---	Fiat Abarth 1000 Berlina
19-20 August	Regensburg, Germany	Regensburg Bergrennen (hillclimb-3,700 kms)	Touring Sport	1000 cc 1300 cc	J. Abt K. Federhofer	2'09"4 3'59"5	Fiat Abarth 1000 Berlina Fiat Abarth OT 1300
20 August	Zeltweg, Austria	Grosser Preis von Österreich (circuit-3,200 kms x 157 laps)	Sport	1300 cc	J. Ortner	3h17'11"9	Fiat Abarth OT 1300
20 August	Saint-Ursanne, Switzerland	Course de côte Saint-Ursanne-Les Rangiers (hillclimb-4,700 kms x 2)	Touring Touring	850 cc 1000 cc	R. Blatter W. Kauhsen	5'33"8 5'19"5	Fiat Abarth 850 TC Fiat Abarth 1000 Berlina

DATE	LOCATION	RACE	CATEGORY	CLASS	WINNER	TIME OR DISTANCE	CAR
20 August	Ascoli Piceno, Italy	Corsa al Colle San Marco (hillclimb-10,200 kms)	Touring	850 cc	A. Mostardini	6'58"8	Fiat Abarth 850 TC
			Touring	1000 cc	N. Parenti	6'35"8	Fiat Abarth 1000 Berlina
			Sport	1000 cc	M. Terminiello	6'31"2	Fiat Abarth 1000 BA
			Sport	1300 cc	C. Zuccoli	6'14"8	Fiat Abarth OT 1300
			Sport-Prototypes	1000 cc	G. Tommasi	6'18"6	Fiat Abarth 1000 SP
20 August	Brescia, Italy	Trofeo Valle Camonica (hillclimb-8,600 kms)	Touring	600 cc	P.A. Poggi	6'17"0	Fiat Abarth 595
			Touring	700 cc	"Byron"	6'12"5	Fiat Abarth 695
			Touring	850 cc	V. Tabucchi	5'43"7	Fiat Abarth 850 TC
			Touring	1000 cc	P. Botella	5'40"4	Fiat Abarth 1000 Berlina
			Grand Touring	1000 cc	P. Brambilla	6'26"9	Fiat Abarth 1000 OTS
			Sport	1000 cc	D. Lo Cocco	5'24"2	Fiat Abarth 1000 BA
			Sport	1300 cc	A. Becchetti	5'27"6	Fiat Abarth OT 1300
			Prototypes	1000 cc	L. Caprioli	5'19"1	Fiat Abarth 1000 SP
20 August	Brindisi, Italy	Ostuni-Coppa Apruzzi (hillclimb-2,500 kms)	Touring	up to 700 cc	A. D'Alessio	1'27"1	Fiat Abarth 595
			Touring	850 cc	N. Leogrande	1'16"0	Fiat Abarth 850 TC
			Touring	up to 1150 cc	D. Giannotte	1'21"3	Fiat Abarth 1000 Berlina
			Sport	1000 cc	P. Gargano	1'05"3	Fiat Abarth 1000 BA
20 August	Adenau, Germany	ADAC Nürburgring Mittel Rhein-Rennen (circuit-22,810 kms x 5 laps)	Touring	850 cc	K.H. Scholl	56'27"3	Fiat Abarth 850 TC
			Touring	up to 1150 cc	H. Ranke	55'17"7	Fiat Abarth 1000 Berlina
20 August	Villefranche sur Saône, France	Course de côte du Beaujolais (hillclimb-16,000 kms)	Sport-Prototypes	up to 1000 cc	Rochet	10'22"6	Fiat Abarth 1000 SP
			Sport-Prototypes	over 1600 cc	P. Maublanc	8'53"1	Fiat Abarth 2000
26 August	Lime Rock (Connecticut), USA	Area 1 Championship Race (circuit-1,5 miles x 15 laps)	---	---	A. Cosentino	---	Fiat Abarth 1000 Berlina
26-27 August	Namur, Belgium	Course de côte de Namur (hillclimb-2,000 kms)	Touring	1000 cc	F. Neri	1'44"6	Fiat Abarth 1000 Berlina
			Sport	1000 cc	F. Neri	1'37"1	Fiat Abarth 1000 BA
27 August	Udine, Italy	Tolmezzo-Verzegnis (hillclimb-4,260 kms x 2)	Touring	600 cc	"Arch"	6'06"9	Fiat Abarth 595
			Touring	700 cc	P.L. Fuganti	6'00"5	Fiat Abarth 695
			Touring	850 cc	V. Tabucchi	5'25"6	Fiat Abarth 850 TC
			Touring	1000 cc	P. Cattelan	5'14"8	Fiat Abarth 1000 Berlina
			Sport	1000 cc	M. Polin	5'19"3	Fiat Abarth 1000 BA
27 August	Pori, Finlandia	Porin Maarata Ajot (circuit-5,000 kms)	Touring	850 cc	P. Saikkonen	3'25"7	Fiat Abarth 850 TC
27 August	Mainz Finthen, Germany	Flugplatzrennen (circuit-2,100 kms x 30 laps)	Touring	up to 1150 cc	F. Kerckerinck	37'37"2	Fiat Abarth 1000 Berlina
27 August	Zandvoort, Holland	Zandvoort Trophy (circuit-4,225 kms x 35 laps)	Touring	850 cc	H. Veder	1h08'00"2	Fiat Abarth 850 TC
			Touring	1000 cc	E. Swart	1h07'12"8	Fiat Abarth 1000 Berlina
27 August	Villars, Switzerland	Course de côte Ollon-Villars (hillclimb-8,000 kms x 2)	Touring	up to 1150 cc	G. König	10'24"5	Fiat Abarth 1000 Berlina
			Sport	1300 cc	A. Hezemans	9'20"8	Fiat Abarth OT 1300
			Sport-Prototypes	up to 1150 cc	R. Stierli	9'49"9	Fiat Abarth 1000 SP
27 August	Grenoble, France	Course de côte Chamrousse (hillclimb-17,200 kms)	Touring	850 cc	J. Stivin	12'11"9	Fiat Abarth 850 TC
			Touring	up to 1150 cc	F. Tomas	11'16"4	Fiat Abarth 1000 Berlina
			Sport	up to 1600 cc	F. Canin	9'57"8	Fiat Abarth OT 1300
27 August	Trapani, Italy	Trapani-Monte Erice (hillclimb-6,670 kms)	Touring	600 cc	'Harca'	5'42"0	Fiat Abarth 595
			Touring	700 cc	A. La Luce	5'33"9	Fiat Abarth 695
			Touring	850 cc	G. Amodeo	5'03"9	Fiat Abarth 850 TC
			Touring	1000 cc	S. Scigliano	5'14"0	Fiat Abarth 1000 Berlina
			Sport	1300 cc	A. Gambero	4'41"6	Fiat Abarth OT 1300
27 August	Augsburg, Germany	Augusta Bergrennen (hillclimb-2,100 kms x 2)	Touring	850 cc	K.H. Trettan	2'26"9	Fiat Abarth 850 TC
			Touring	1000 cc	J. Abt	3'19"1	Fiat Abarth 1000 Berlina
			Sport	1300 cc	F. Neuser	2'17"7	Fiat Abarth OT 1300
27 August	Kägeröd, Sweden	Ring Knutstorp (circuit-16 laps)	Touring	1000 cc	L. Englund	12'52"6	Fiat Abarth 1000 Berlina
2-3 September	Gex, France	Course de côte de la Faucille (hillclimb-10,396 kms)	Sport	up to 1150 cc	Thian	6'22"0	Fiat Abarth 1000 BA
3 September	Bergamo, Italy	Castione-Passo della Presolana (hillclimb-6,100 kms)	Touring	700 cc	U. Barilaro	4'03"8	Fiat Abarth 695
			Touring	850 cc	B. Pagliero	3'46"0	Fiat Abarth 850 TC
			Touring	1000 cc	F. Zanardelli	3'39"8	Fiat Abarth 1000 Berlina
			Sport	1300 cc	C. Zuccoli	3'19"8	Fiat Abarth OT 1300
			Sport-Prototypes	1000 cc	L. Caprioli	3'22"7	Fiat Abarth 1000 SP
3 September	San Feliu de Codinas, Spagna	Carrera en cuesta a San Feliu de Codinas (hillclimb-4,200 kms)	Touring	850 cc	F. Serra	3'14"7	Fiat Abarth 850 TC
			Touring	up to 1150 cc	'Pelè'	3'18"8	Fiat Abarth 1000 Berlina
			Sport	up to 2000 cc	M. Juncosa	3'05"1	Fiat Abarth 2000
3 September	Rouen, France	Course de côte de Belbeuf (hillclimb-2,800 kms)	Sport	1300 cc	D. Joneaux	1'56"0	Fiat Abarth OT 1300
3 September	Gaisberg, Austria	Grosser Bergpreis von Österreich (hillclimb-8,625 kms)	Touring	1000 cc	J. Abt	10'06"8	Fiat Abarth 1000 Berlina
			Sport	1300 cc	A. Hezemans	8'45"4	Fiat Abarth OT 1300
3 September	Adenau, Germany	Internationales 500 Km Nürburgring Rennen (circuit-22,815 kms x 22 laps)	Touring	1000 cc	G.D. Edelhoff	3h48'15"6	Fiat Abarth 1000 Berlina
			Sport	1300 cc	E. Furtmayr	3h40'32"2	Fiat Abarth OT 1300
3 September	Macerata, Italy	Coppa Città di Tolentino (hillclimb-4,100 kms)	Touring	600 cc	F. Ferroni	2'54"9	Fiat Abarth 595
			Touring	700 cc	L. Finelli	2'48"4	Fiat Abarth 695
			Touring	850 cc	F. Cardinali	2'44"3	Fiat Abarth 850 TC

DATE	LOCATION	RACE	CATEGORY	CLASS	WINNER	TIME OR DISTANCE	CAR
3 September	Monbazillac, France	Course de côte du Bergeracois (hillclimb-3,150 kms)	Sport	1300 cc	F. Canin	1'53"2	Fiat Abarth OT 1300
9 September	Vaals, Holland	Heuvelklim Vaals (hillclimb-1,200 kms)	Touring	1000 cc	A. Goedemans	---	Fiat Abarth 1000 Berlina
9-10 September	Ulm, Germany	ADAC Flugplatzrennen Ulm-Laupheim (hillclimb-4,200 kms x 20 laps)	Touring	850 cc	K.H. Trettan	32'53"5	Fiat Abarth 850 TC
9-10 September	Avus, Germany	Internationales Avus Rennen (circuit-8,300 kms x 10 laps)	Sport	1300 cc	R. Jöst	---	Fiat Abarth OT 1300
9-10 September	Dijon, France	Course de côte d'Urcy (hillclimb-2,500 kms)	Touring	850 cc	Champeaux	2'09"3	Fiat Abarth 850 TC
			Sport	up to 1150 cc	Thiam	1'56"2	Fiat Abarth 1000 BA
			Sport-Prototypes	2000 cc	P. Maublanc	1'43"1	Fiat Abarth 2000
10 September	Viterbo, Italy	Coppa del Cimino (hillclimb-10,200 kms)	Touring	600 cc	P.V. Vannini	6'22"0	Fiat Abarth 595
			Touring	700 cc	A. Fracassini	6'05"4	Fiat Abarth 695
			Touring	850 cc	V. Cristi	5'27"9	Fiat Abarth 850 TC
			Touring	1000 cc	N. Parenti	5'12"0	Fiat Abarth 1000 Berlina
			Sport	1000 cc	G. Cucchiarelli	5'10"0	Fiat Abarth 1000 BA
10 September	Engelhartszells, Austria	Bergrennen Engelhartszell (hillclimb-4,500 kms x 2)	Touring	1000 cc	G. König	6'22"0	Fiat Abarth 1000 Berlina
10 September	Caltanissetta, Italy	Coppa Nissena (hillclimb-10,000 kms)	Touring	600 cc	'Harca'	6'39"2	Fiat Abarth 595
			Touring	700 cc	A. La Luce	6'39"3	Fiat Abarth 695
			Touring	850 cc	S. Campione	6'12"4	Fiat Abarth 850 TC
			Touring	1000 cc	P. De Francisci	6'12"4	Fiat Abarth 1000 Berlina
			Sport	1300 cc	A. Gambero	5'41"5	Fiat Abarth OT 1300
			Sport-Prototypes	1000 cc	S. Calascibetta	5'51"5	Fiat Abarth 1000 SP
10 September	Turku, Finlandia	Itäharjun Ajo (circuit-20,000 kms)	Touring	850 cc	P. Saikkonen	10'28"8	Fiat Abarth 850 TC
10 September	Hockenheim, Germany	100 Meilen von Hockenheim (circuit-2,768 kms x 24 laps)	Touring	850 cc	K.H. Scholl	---	Fiat Abarth 850 TC
			Touring	up to 1150 cc	F. Kercherinck	---	Fiat Abarth 1000 Berlina
			Sport	up to 1300 cc	E. Bitter	---	Fiat Abarth OT 1300
10 September	Montlhéry, France	Coupes de Paris (circuit-6,320 kms x 10 laps)	Touring	1000 cc	A. Maury	31'52"3	Fiat Abarth 1000 Berlina
			Sport	1300 cc	Brel	27'05"1	Fiat Abarth OT 1300
16-17 September	Nice, France	Criterium Jean Behra (763 kms)	Touring	850 cc	Lanteri-Fiquet	---	Fiat Abarth 850 TC
17 September	Siena, Italy	Coppa Caltex (gimkana)	---	850 cc	C. Moretti	---	Fiat Abarth 850 TC
17 September	Kaufbeuren, Germany	ADAC Auerbergrennen (hillclimb-2,900 kms x 2)	Touring	850 cc	K.H. Trettan	4'08"3	Fiat Abarth 850 TC
			Sport	1300 cc	G. Disfler	3'38"7	Fiat Abarth OT 1300
17 September	Brescia, Italy	Trofeo Lumezzane (hillclimb-8,500 kms)	Touring	600 cc	R. Amighini	5'30"2	Fiat Abarth 595
			Touring	700 cc	E. Pasolini	5'24"0	Fiat Abarth 695
			Touring	850 cc	D. Comini	4'59"5	Fiat Abarth 850 TC
			Touring	1000 cc	A. Campanini	4'51"5	Fiat Abarth 1000 Berlina
			Grand Touring	1000 cc	G. Galmozzi	5'16"5	Fiat Abarth 1000 OTS
			Sport	1000 cc	L. Rassega	4'39"0	Fiat Abarth 1000 BA
			Sport	1300 cc	S. Morando	4'31"3	Fiat Abarth OT 1300
			Sport-Prototypes	1000 cc	L. Caprioli	4'32"0	Fiat Abarth 1000 SP
			Sport-Prototypes	2000 cc	P. Schetty (overall)	3'55"5	Fiat Abarth 2000
17 September	Kaiserslautern, Germany	Rheinland-Pfalz-Bergpreis (hillclimb-3,050 kms)	Touring	850 cc	R. Neumann	4'01"9	Fiat Abarth 850 TC
			Touring	up to 1150 cc	J. Abt	3'45"7	Fiat Abarth 1000 Berlina
17 September	Belluno, Italy	Coppa Alpe del Nevegal (hillclimb-11,500 kms)	Touring	600 cc	C. Bellin	9'41"0	Fiat Abarth 595
			Touring	700 cc	G. Zanchetti	9'23"3	Fiat Abarth 695
			Touring	850 cc	V. Tabucchi	8'20"8	Fiat Abarth 850 TC
			Touring	1000 cc	A. De Paoli	8'06"5	Fiat Abarth 1000 Berlina
			Grand Touring	1000 cc	P. Brambilla	9'04"8	Fiat Abarth 1000 OTS
			Sport	1000 cc	M. Polin	7'49"0	Fiat Abarth 1000 BA
			Sport	1300 cc	G. Danieli	9'13"7	Fiat Abarth OT 1300
17 September	Mont Verdun, France	Course de côte Limonest-Mont Verdun (hillclimb-2,675 kms)	Touring	850 cc	R. Barone	2'18"4	Fiat Abarth 850 TC
			Touring	1000 cc	F. Tomas	2'11"6	Fiat Abarth 1000 Berlina
			Sport	up to 1150 cc	S. Vaglio	2'04"2	Fiat Abarth 1000 BA
			Sport	over 1600 cc	P. Maublanc	1'51"2	Fiat Abarth 2000
17 September	Snetterton, United Kingdom	Snetterton 500 Km (circuit)	Touring	850 cc	Galli-Botella	3h44'49"6	Fiat Abarth 850 TC
			Touring	1000 cc	Hezemans-Kauhsen	3h44'40"6	Fiat Abarth 1000 Berlina
17 September	Gargellen, Austria	Steinbockrennen Gargellen (hillclimb-7,260 kms)	Touring	1000 cc	G. König	4'03"4	Fiat Abarth 1000 Berlina
			Sport	1000 cc	G. Naef	3'58"4	Fiat Abarth 1000 BA
17 September	Catania, Italy	Catania-Etna (hillclimb-33,000 kms)	Touring	600 cc	G. Mercadante	22'33"3	Fiat Abarth 595
			Touring	850 cc	A. Bonaccorsi	19'51"8	Fiat Abarth 850 TC
			Touring	up to 1150 cc	Scigliano	20'12"9	Fiat Abarth 1000 Berlina
			Grand Touring	1000 cc	W. Manthe	23'36"2	Fiat Abarth 1000 OTS
17 September	Caldas de Montbuy, Spagna	Carrera en cuesta al Farrell de Caldas de Montbuy (hillclimb-7,800 kms)	Touring	850 cc	F. Serra	7'02"0	Fiat Abarth 850 T
			Touring	up to 1150 cc	M. Juncosa	7'05"8	Fiat Abarth 1000 Berlina
			Sport	1000 cc	J. Juncosa	6'58"4	Fiat Abarth 1000 BA
17 September	Roskilde, Danimarca	Roskilde Ring (circuit-1,400 kms x 12 laps)	Touring	1000 cc	P. Petersen	19'50"0	Fiat Abarth 1000 Berlina

DATE	LOCATION	RACE	CATEGORY	CLASS	WINNER	TIME OR DISTANCE	CAR
21-24 September	Madrid, Spagna	Rallye Nacional R.A.C.E. (625 kms)	Touring	1000 cc	M. Juncosa	---	Fiat Abarth 1000 Berlina
23-24 September	Houyet, Belgium	Course de côte de Houyet (hillclimb-2,500 kms)	Touring	1000 cc	F. Neri	1'48"0	Fiat Abarth 1000 Berlina
			Sport	1000 cc	H. Affentranger	1'38"2	Fiat Abarth 1000 BA
23-24 September	Zolder, Belgium	ADAC Westfalen Pokal (circuit-4,184 kms x 24 laps)	Touring	850 cc	K.H. Scholl	52'49"1	Fiat Abarth 850 TC
			Touring	1000 cc	"H. Lenz"	49'58"8	Fiat Abarth 1000 Berlina
			Grand Touring Special	1000 cc	K.H. Scholl	48'17"9	Fiat Abarth 1000 Berlina
			Sport	1000 cc	W. Gras	49'40"1	Fiat Abarth 1000 BA
23-24 September	Rouen, France	Rallye Automobile Jeanne d'Arc (288 kms)	Sport	1000 cc	Joneaux-Homo	---	Fiat Abarth 1000 BA
24 September	Bad Oldesloe, Gemania	ADAC Travering-Rennen Bad Oldesloe (circuit-6,500 kms x 10 laps)	Touring	1000 cc	H.P. Klein	6'11"0	Fiat Abarth 1000 Berlina
24 September	Kassel, Germany	Bergprüfung Kassel Langenberge (hillclimb-3,700 kms)	Sport	1300 cc	F. Molino	4'51"0	Fiat Abarth OT 1300
24 September	Ajore, Switzerland	Surprise Bergrennen (hillclimb-2,950 kms)	Touring	850 cc	K. Schwarzgruner	---	Fiat Abarth 850 TC
			Sport	1000 cc	E. Wyss	---	Fiat Abarth 1000 BA
24 September	Aosta, Italy	Aosta-Pila (hillclimb-11,200 kms)	Touring	700 cc	U. Barilaro	8'35"7	Fiat Abarth 695
			Touring	850 cc	B. Pagliero	7'59"2	Fiat Abarth 850 TC
			Touring	1000 cc	S. Dal Cason	7'54"5	Fiat Abarth 1000 Berlina
			Grand Touring	1000 cc	P. Brambilla	8'36"3	Fiat Abarth 1000 OTS
			Sport	1000 cc	F. Marino	8'25"9	Fiat Abarth 1000 BA
			Sport	1300 cc	S. Morando	7'10"2	Fiat Abarth OT 1300
			Prototypes	1000 cc	G. Varese	7'12"0	Fiat Abarth 1000 SP
			Prototypes	2000 cc	P. Schetty	6'12"1	Fiat Abarth 2000
24 September	Neuffen, Germany	Bergpreis Schwäbische Alb (hillclimb-4,200 kms x 2)	Touring	850 cc	R. Neumann	5'40"4	Fiat Abarth 850 TC
			Touring	up to 1150 cc	J. Abt	5'26"7	Fiat Abarth 1000 Berlina
			Sport-Prototypes	up to 1150 cc	R. Stierli	5'12"2	Fiat Abarth 1000 SP
24 September	Ales, France	Course de côte d'Ales (hillclimb-1,500 kms)	Touring	600 cc	J. Choquel	1'50"6	Fiat Abarth 595
1 October	Dortmund, Germany	ADAC Sauerland Bergpreis (hillclimb-3,400 kms x 2)	Touring	1000 cc	J. Abt	4'26"8	Fiat Abarth 1000 Berlina
1 October	Angers, France	Course de côte de la Pommeraye (hillclimb-1,700 kms)	Sport	1000 cc	P. Champeaux	1'13"0	Fiat Abarth 1000 BA
1 October	Aspern, Austria	Donau Pokal Flugplatzrennen (circuit-2,600 kms)	Touring	1000 cc	W. Kauhsen	24'37"9	Fiat Abarth 1000 Berlina
			Sport	1300 cc	J. Ortner	24'43"8	Fiat Abarth OT 1300
			Sport-Prototypes	1300 cc	G. König	24'13"1	Fiat Abarth 1300
1 October	La Roche, Belgium	Course de côte de La Roche Samree (hillclimb-2,700 kms)	Touring	1000 cc	F. Neri	1'33"6	Fiat Abarth 1000 Berlina
			Sport	1000 cc	F. Neri	1'26"1	Fiat Abarth 1000 BA
1 October	Massana, Andorra	Costa de la Massana (hillclimb-3,500 kms)	Touring	850 cc	F. Serra	2'43"9	Fiat Abarth 850 TC
			Touring	1000 cc	J. Juncosa	2'34"5	Fiat Abarth 1000 Berlina
1 October	Hockenheim, Germany	Preis von Hockenheim (circuit-6,768 kms x 10 laps)	Touring	850 cc	K.H. Scholl	29'19"3	Fiat Abarth 850 TC
			Touring	1000 cc	H. Lenz	28'41"7	Fiat Abarth 1000 Berlina
			Touring Special	1000 cc	W. Zanders	25'29"4	Fiat Abarth 1000 Berlina
1 October	Nancy, France	Course de côte de Belleau (hillclimb-2,000 kms)	Sport	1000 cc	M. Presumey	1'33"3	Fiat Abarth 1000 BA
1 October	Monza, Italy	Coppa Leopoldo Carri (circuit-5,750 kms x 27 laps)	Touring	850 cc	D. Martino	1h00'54"2	Fiat Abarth 850 TC
			Touring	1000 cc	N. Parenti	58'52"8	Fiat Abarth 1000 Berlina
7-8 October	Bryars, USA	Bryar National Race	---	---	A. Cosentino	---	Fiat Abarth 1000 Berlina
7-8 October	Genève, Switzerland	Course de côte du Marchairuz (hillclimb-4,800 kms x 2)	Touring	850 cc	R. Blatter	6'35"8	Fiat Abarth 850 TC
			Sport	1000 cc	R. Stierli	6'02"9	Fiat Abarth 1000 BA
8 October	Nice, France	Course de côte de Sainte Cezaire (hillclimb-4,500 kms)	Touring	1000 cc	F. Tomas	3'37"3	Fiat Abarth 1000 Berlina
			Sport	1300 cc	F. Pilone	3'20"0	Fiat Abarth OT 1300
			Sport	2000 cc	P. Maublanc	3'04"4	Fiat Abarth 2000
8 October	Fulda, Germany	ADAC Rhön Bergpreis (hillclimb-2,550 kms x 2)	Touring	up to 1150 cc	H.P. Klein	2'16"3	Fiat Abarth 1000 Berlina
8 October	Ferrier, France	Course de côte du Ferrier (hillclimb-4,100 kms)	Touring	1000 cc	F. Tomas	2'43"9	Fiat Abarth 1000 Berlina
			Sport	1300 cc	F. Pilone	2'23"5	Fiat Abarth OT 1300
8 October	Innsbruck, Austria	Preis von Tirol (circuit-2,800 kms x 15 laps)	Touring	1000 cc	G. König	21'16"23	Fiat Abarth 1000 Berlina
			Sport (20 laps)	1000 cc	L. Rassega	25'53"0	Fiat Abarth 1000 BA
			Sport (20 laps)	1300 cc	R. Jöst	25'35"1	Fiat Abarth OT 1300
8 October	Montlhéry, France	Coupes du Salon (circuit-3,600 kms x 15 laps)	Touring	1000 cc	A. Hezemans	26'18"8	Fiat Abarth 1000 Berlina
			Grand Touring	1000 cc	Biancarelli	---	Fiat Abarth 1000 OTS
			Sport	1300 cc	G. Brel	---	Fiat Abarth OT 1300
			Sport	2000 cc	P. Schetty	22'26"0	Fiat Abarth 2000
8 October	Ragusa, Italy	Coppa dei Monti Iblei (hillclimb-5,200 kms)	Touring	600 cc	G. Marcadante	4'35"5	Fiat Abarth 595
			Touring	700 cc	P. De Luca	4'25"0	Fiat Abarth 695
			Touring	850 cc	G. Nicoletti	4'18"5	Fiat Abarth 850 TC
			Sport	1300 cc	A. Gambero (overall)	3'50"5	Fiat Abarth OT 1300
			Sport-Prototypes	1000 cc	S. Calascibetta	4'00"2	Fiat Abarth 1000 SP
11-14 October	Madrid, Spagna	Trofeo R.A.C.E. Internacional Rallye (742 kms)	Touring	850 cc	Serra-Tugues	---	Fiat Abarth 850 TC
			Touring	1000 cc	Juncosa-Oliva	---	Fiat Abarth 1000 Berlina

DATE	LOCATION	RACE	CATEGORY	CLASS	WINNER	TIME OR DISTANCE	CAR
12 October	Montserrat, Spagna	Rallye Nacional de los 2000 Virages (580 kms)	Touring	1000 cc	M. Juncosa	---	Fiat Abarth 1000 Berlina
12 October	Barcelona, Spagna	Carrera en cuesta a la Rabassada (hillclimb-4,500 kms)	Touring	851-1600 cc	'Pelé'	3'12"4	Fiat Abarth 1000 Berlina
14 October	Denver (Colorado), USA	Gimmick Car Rally (circuit-71,9 miles)	---	----	Miller-Ziering	---	Fiat Abarth
14 October	Vallelunga, Italy	Trofeo Ettore Bettoja (circuit-3,200 kms x 1h10')	Touring	850 cc	D. Martino	1h11'28"5	Fiat Abarth 850 TC
			Touring	1000 cc	N. Parenti	1h10'25"6	Fiat Abarth 1000 Berlina
			Grand Touring	1000 cc	P. Brambilla	1h10'57"9	Fiat Abarth 1000 OTS
			Sport (1h30')	1000 cc	L. Caprioli	1h30'37"0	Fiat Abarth 1000 BA
14-15 October	Draguignan, France	Rallye National du Var	Touring	850 cc	Lanteri-Duhaze	1h34'14"	Fiat Abarth 850 TC
14-15 October	Malval, France	Course de côte du Col de Malval (hillclimb-4,200 kms)	Touring	1000 cc	B. Mauris	3'00"0	Fiat Abarth 1000 Berlina
			Touring Special	850 cc	G. Stivin	2'51"0	Fiat Abarth 850 TC
			Sport	850 cc	Vernet	3'03"1	Fiat Abarth 850 TC
14-15 October	Mollis, Switzerland	Kerenzen Bergrennen (hillclimb-2,200 kms)	Touring	1000 cc	M. Wuest	3'09"9	Fiat Abarth 1000 Berlina
			Sport	1000 cc	R. Stierli	2'49"1	Fiat Abarth 1000 BA
15 October	Wolfsburger, Germany	ADAC Autoslalom Wolfsbuger	Touring	up to 1150 cc	H.P. Klein	---	Fiat Abarth 1000 Berlina
15 October	Siracusa, Italy	Coppa Val d'Anapo-Sortino (hillclimb-6,000 kms)	Touring	600 cc	G. Mercadante	5'17"7	Fiat Abarth 595
			Touring	700 cc	A. La Luce	5'12"3	Fiat Abarth 695
			Touring	850 cc	S. Campione	4'50"9	Fiat Abarth 850 TC
			Touring	1000 cc	G. Calabrese	5'05"0	Fiat Abarth 1000 Berlina
			Sport	1000 cc	S. Sutera	4'48"5	Fiat Abarth 1000 BA
			Sport	1300 cc	A. Gambero (overall)	4'25"7	Fiat Abarth OT 1300
15 October	Genova, Italy	Pontedecimo-Giovi (hillclimb-4,000 kms)	Touring	700 cc	G. Barbasio	3'04"9	Fiat Abarth 695
			Touring	850 cc	P.L. De Zordo	3'05"4	Fiat Abarth 850 TC
			Touring	1000 cc	P.G. Cazzulo	2'57"4	Fiat Abarth 1000 Berlina
			Sport	1000 cc	R. Grasso	3'32"8	Fiat Abarth 1000 BA
			Sport	1300 cc	A. Dutto	3'12"7	Fiat Abarth OT 1300
			Prototypes	2000 cc	P. Schetty (overall)	2'27"2	Fiat Abarth 2000
21 October	Solothurn, Switzerland	Bergrennen Solothurn (hillclimb)	Sport	850 cc	P. Roher	2'46"3	Fiat Abarth 850 TC
			Sport	1000 cc	H. Affentranger	2'30"7	Fiat Abarth 1000 BA
21-22 October	Montlhéry, France	Rallye National A.G.A.C.I. (363 kms)	Sport	up to 1150 cc	Gandre-Beaulay	---	Fiat Abarth 1000 Berlina
22 October	Zolder, Belgium	ADAC Rhein Ruhr Pokal Rennen (circuit-4,184 kms x 25 laps)	Touring	850 cc	K.H. Scholl	50'52"5	Fiat Abarth 850 TC
			Grand Touring	up to 1300 cc	G. Artelli	52'24"2	Fiat Abarth 1000 OTS
22 October	Lerida, Spagna	Rallye Valle de Aran (569 kms)	Touring	850 cc	F. Serra	---	Fiat Abarth 850 TC
			Touring	up to 1150 cc	M. Juncosa	---	Fiat Abarth 1000 Berlina
22 October	Cagliari, Italy	Coppa Città di Iglesias (hillclimb-6,700 kms)	Touring	600 cc	A. Palmas	5'46"4	Fiat Abarth 595
			Touring	700 cc	G. Martis	5'36"0	Fiat Abarth 695
			Touring	850 cc	P.L. Mele	5'35"3	Fiat Abarth 850 TC
			Touring	1000 cc	F. Locci	5'04"2	Fiat Abarth 1000 Berlina
			Sport	1000 cc	G. Vacca	4'44"9	Fiat Abarth 1000 BA
			Sport	1300 cc	A. Graziano	5'01"2	Fiat Abarth OT 1300
22 October	Bevence à Mont, Belgium	Course de côte des Trois Marets (hillclimb-4,800 kms)	Sport	1000 cc	F. Neri	3'01"4	Fiat Abarth 1000 BA
22 October	Wörgl, Austria	Bergrennen Wörgl-Wildschönau (hillclimb-2,000 kms)	Touring	1000 cc	G. König	3'29"5	Fiat Abarth 1000 Berlina
22 October	Perpignan, France	Rallye International du Roussillon (391,100 kms)	---	1000 cc	Brel-Paulet	---	Fiat Abarth 1000 Berlina
22 October	Roma, Italy	Campionato Laziale Tris (circuit-1,000 kms x 2 laps)	Touring	850 cc	E. Salvatori	2'20"6	Fiat Abarth 850 TC
22 October	Orvieto, Italy	Coppa Città di Orvieto (hillclimb-7,600 kms)	Touring	600 cc	P.G. Poggi	5'40"7	Fiat Abarth 595
			Touring	700 cc	A. Fracassini	5'46"3	Fiat Abarth 695
			Touring	850 cc	V. Cristi	5'22"5	Fiat Abarth 850 TC
			Touring	1000 cc	N. Parenti	5'19"2	Fiat Abarth 1000 Berlina
			Grand Touring	1000 cc	P. Brambilla	5'48"3	Fiat Abarth 1000 OTS
			Sport	1000 cc	M. Terminiello	5'04"3	Fiat Abarth 1000 BA
22 October	Catanzaro, Italy	Coppa Città di Catanzaro (hillclimb-11,300 kms)	Touring	600 cc	C. Aloe	9'12"1	Fiat Abarth 595
			Touring	850 cc	L. Ferragina	8'20"2	Fiat Abarth 850 TC
			Touring	1000 cc	G. Calabrese	8'45"0	Fiat Abarth 1000 Berlina
			Grand Touring	1000 cc	F. Di Benedetto	9'12"1	Fiat Abarth 1000 OTS
			Sport	1000 cc	G. Benini	8'27"6	Fiat Abarth 1000 BA
			Sport-Prototypes	2000 cc	D. Scola (overall)	7'37"9	Fiat Abarth 2000
28-29 October	Vallelunga, Italy	Coppa Fausto Pesci (circuit-1,800 kms x 56 laps)	Touring	850 cc	V. Cristi	1h10'26"3	Fiat Abarth 850 TC
			Touring	1000 cc	N. Parenti	1h11'11"1	Fiat Abarth 1000 Berlina
			Sport (1h30')	1000-1300 cc	G. Rebaudi	1h30'23"3	Fiat Abarth OT 1300
29 October	Nürnberg, Germany	Automobil-Slalom Nürnberg (1,400 kms x 2)	Sport	up to 1300 cc	F. Neuser	1'44"8	Fiat Abarth 1000 BA
29 October	Köln, Germany	ADAC Autoslalom Köln (1,600 kms x 2)	Touring	850 cc	K.H. Scholl	---	Fiat Abarth 850 TC
29 October	Marseille, France	Course de côte de la Sainte Baume (hillclimb-8,100 kms)	Touring	850 cc	J. Choquel	8'55"0	Fiat Abarth 850 TC
			Touring	1000 cc	F. Tomas	6'66"8	Fiat Abarth 1000 Berlina
			Sport Prototypes	2000 cc	P. Maublanc	6'18"2	Fiat Abarth 2000

DATE	LOCATION	RACE	CATEGORY	CLASS	WINNER	TIME OR DISTANCE	CAR
4-5 November	Celle-Wietzenbruch, Germany	ADAC Hansa Automobilprüfung (circuit-4,810 kms x 12 laps)	Touring	850 cc	H.K. Scholl	---	Fiat Abarth 850 TC
			Touring	up to 1150 cc	H.P. Klein	---	Fiat Abarth 1000 Berlina
			Touring Special	up to 1300 cc	K.H. Scholl	---	Fiat Abarth 1000 Berlina
			Sport	1300 cc	H. Krause	---	Fiat Abarth OT 1300
5 November	La Zapatera, Spagna	Subida en cuesta a La Zapatera (hillclimb-4,300 kms)	Touring	850 cc	E. Boujoch	2'50"0	Fiat Abarth 850 TC
11 November	Luceram, France	Course de côte du Col de l'Orme (hillclimb-10,000 kms)	Touring	1000 cc	F. Tomas	9'16"8	Fiat Abarth 1000 Berlina
			Sport	up to 1150 cc	F. Marino	9'26"3	Fiat Abarth 1000 BA
			Sport	up to 1600 cc	S. Morero	8'57"5	Fiat Abarth OT 1300
			Sport	up to 2500 cc	P. Maublanc	8'19"0	Fiat Abarth 2000
12 November	Barcelona, Spagna	Rallye Club 600	Touring	up to 1150 cc	M. Juncosa	---	Fiat Abarth 1000 Berlina
18-19 November	Barcelona, Spagna	Rallye Barcelona-Andorra (kms 505)	Touring	up to 1150 cc	M. Juncosa	---	Fiat Abarth 1000 Berlina
			Sport	1000 cc	J. Juncosa	---	Fiat Abarth 1000 BA
19 November	Frenkendorf, Switzerland	Autoslalom Frenkendorf, Switzerland (0,600 kms)	Touring	1000 cc	J.P. Läubin	---	Fiat Abarth 1000 Berlina
19 November	Vallelunga, Italy	Coppa Supercortemaggiore (circuit-3,120 kms x 12 laps)	Touring	850 cc	V. Cristi	20'58"2	Fiat Abarth 850 TC
			Touring	1000 cc	L. Rinaldi	21'01"8	Fiat Abarth 1000 Berlina
			Sport-Prototypes	1000 cc	L. Caprioli	---	Fiat Abarth 1000 SP
25-26 November	Daytona (Florida), USA	Daytona International Race (circuit-1,63 miles x 33 laps)	-	-	D. Parkinson	---	Fiat Abarth 1000 Berlina
26 November	Vilumara, Spagna	Carrera en cuesta Pont de Vilumara (hillclimb-3,000 kms)	Touring	850 cc	F. Serra	2'04"0	Fiat Abarth 850 TC
			Touring	1000 cc	M. Juncosa	2'17"3	Fiat Abarth 1000 Berlina
			Sport	1000 cc	J. Juncosa	2'14"0	Fiat Abarth 1000 BA
3 December	Vallelunga, Italy	Trofeo Rocca di Mezzo (circuit-3,200 kms)	Touring	850 cc	Salvatori	---	Fiat Abarth 850 TC
8 December	Fontpineda, Spagna	Rallye las tres Costes (223 kms)	Touring	850 cc	F. Serra	---	Fiat Abarth 850 TC
10 December	Witten, Germany	Wittener Autoslalom	Sport	1300 cc	E. Günter	---	Fiat Abarth OT 1300
30-31 December	Luanda, Angola	Rallye do fim do Ano (533 kms)	Sport	1300	Vieria-Silva	---	Fiat Abarth OT 1300

1968

DATE	LOCATION	RACE	CATEGORY	CLASS	WINNER	TIME OR DISTANCE	CAR
24 March	Monza, Italy	4 Ore del Jolly Club (circuit 10,000 kms)	Touring	850 cc	Stabile-Galli	kms 561,816	Fiat Abarth 850 TC
			Touring	1000 cc	Schetty-Goedemans	kms 595,735	Fiat Abarth 1000 Berlina
24 March	Ceyreste, France	Course de côte de Ceyreste (hillclimb-4,800 kms)	Sport-Prototypes	2000 cc	P. Maublanc	---	Fiat Abarth 2000
31 March	Vallelunga, Italy	Coppa Autodromo di Roma (circuit-1,800 kms x 40 laps)	Touring	600cc	C. Bettazzi	1h 10'26"8	Fiat Abarth 595
			Touring	700 cc	R. Tampolli	1h 10'46"6	Fiat Abarth 695
			Touring	850 cc	'Gibi'	1h 10'09"6	Fiat Abarth 850 TC
			Touring	1000 cc	N. Parenti	1h 10'24"9	Fiat Abarth 1000 Berlina
31 March	Avola, Italy	Coppa Belmonte (hillclimb-10,200 kms)	Touring	600 cc	G. D'Ippolito	7'45"5	Fiat Abarth 595
			Touring	700 cc	A. Guagliardo	7'50"3	Fiat Abarth 695
			Touring	850 cc	S. Sansica	7'09"0	Fiat Abarth 850 TC
			Touring	1000 cc	V. Virgilio	7'02"0	Fiat Abarth 1000 Berlina
			Grand Touring	1000 cc	A. Torricelli	7'36"3	Fiat Abarth 1000 OTS
			Sport	1000 cc	S. Calascibetta	6'45"3	Fiat Abarth 1000 BA
			Sport	1300 cc	A. Gambero	6'33"0	Fiat Abarth OT 1300
			Sport	2000 cc	G. Savona	8'23"7	Abarth Simca 2000
			Sport-Prototypes	2000 cc	D. Scola	6'21"6	Fiat Abarth 2000
7 April	Draguignan, France	Course de côte D'Ampus (hillclimb-6,800 kms)	Sport-Prototypes	2000 cc	P. Schetty	4'00"	Fiat Abarth 2000
7 April	Aspern, Austria	Preis Von Wien (circuit-2,600 kms x1 hour)	Touring	1000 cc	A. Goedemans	---	Fiat Abarth 1000 Berlina
12 April	Snetterton, United Kingdom	B.R.S.C.E. Guards 500 kms (circuit-4,365 kms)	Touring	1000 cc	Hezemans - Goedemans	3h 48'21"	Fiat Abarth 1000 Berlina
21 April	Verona, Italy	Stallavena-Boscochiesanuova (hillclimb-15,300 kms)	Touring	600 cc	'Arch'	9'22"5	Fiat Abarth 595
			Touring	700 cc	E. Pasolini	9'06"3	Fiat Abarth 695
			Touring	850 cc	G. Gatti	8'42"2	Fiat Abarth 850 TC
			Touring	1000 cc	N. Parenti	7'55"1	Fiat Abarth 1000 Berlina
			Grand Touring	1000 cc	G. C. Galmozzi	8'55"5	Fiat Abarth 1000 OTS
			Sport	1000 cc	P. Lando	7'15"0	Fiat Abarth 1000 SP
			Sport	1300 cc	F. Pilone	7'18"1	Fiat Abarth OT 1300
			Sport-Prototypes	1000 cc	M. Baldo	8'04"8	Fiat Abarth 1000
			Sport-Prototypes	2000 cc	P. Schetty	6'21"8	Fiat Abarth 2000
21 April	Beograd, Jugoslavia	Grand Prix de Belgrade (circuit-4,350 kms x 2 hours)	Touring	1000 cc	M. Pasotti	---	Fiat Abarth 1000 Berlina
21 April	Cagliari, Italy	S. Gregorio-Campuomu (hillclimb-10,000 kms)	Touring	600 cc	E. Di Rienzo	8'30"6	Fiat Abarth 595
			Touring	700 cc	S. Fabbris	8'15"6	Fiat Abarth 695
			Touring	850 cc	G. Querzoli	8'06"0	Fiat Abarth 850 TC
			Touring	1000 cc	G. Papa	7'49"0	Fiat Abarth 1000 Berlina
			Sport	1000 cc	G. Vacca	7'11"1	Fiat Abarth 1000
25 April	Monza, Italy	1000 Km di Monza (circuit-10,000 kms x 100 laps)	Sport	1000 cc	'Pal Joe'-Botalla	82 laps	Fiat Abarth 1000 SP

DATE	LOCATION	RACE	CATEGORY	CLASS	WINNER	TIME OR DISTANCE	CAR
28 April	Arezzo, Italy	Pieve S. Stefano-Passo dello Spino (hillclimb-12,700 kms)	Touring	600 cc	A. Bianchi	8'58"3	Fiat Abarth 595
			Touring	700 cc	A Sacchetti	8'53"1	Fiat Abarth 695
			Touring	850 cc	A. Mostardini	8'04"9	Fiat Abarth 850 TC
			Touring	1000 cc	N. Parenti	7'53"6	Fiat Abarth 1000 Berlina
			Grand Touring	1000 cc	L. Carmignani	9'25"1	Fiat Abarth 1000 OTS
			Sport	1300 cc	'Drason'	7'55"0	Fiat Abarth OT 1300
			Sport-Prototypes	2000 cc	P. Schetty	6'48"8	Fiat Abarth 2000
28 April	Amalfi, Italy	Coppa della Primavera (hillclimb-8,750 kms)	Touring	600 cc	G. D'Ippolito	7'16"0	Fiat Abarth 595
			Touring	700 cc	I. Micelli	7'22"7	Fiat Abarth 695
			Touring	850 cc	G. Spadafora	6'45"8	Fiat Abarth 850 TC
			Touring	1000 cc	N. Barbagallo	6'53"4	Fiat Abarth 1000 Berlina
			Grand Touring	1000 cc	A. Torricelli	7'07"4	Fiat Abarth 1000 OTS
			Sport	1000 cc	N. Terminiello	6'37"7	Fiat Abarth 1000 BA
			Sport	over 1000 cc	P. Gargano	6'15"1	Abarth Simca 2000
			Sport-Prototypes	2000	D. Scola	6'09"2	Abarth Simca 2000
28 April	Brescia, Italy	Scalata al Colle Sant'Eusebio (hillclimb-9,500 kms)	Touring	600 cc	Verbotto	6'29"3	Fiat Abarth 595
			Touring	850 cc	'Moxee'	5'24"2	Fiat Abarth 850 TC
			Touring	1000 cc	F. Zanardelli	5'02"2	Fiat Abarth 1000 Berlina
			Grand Touring	1000 cc	G. C. Galmozzi	5'33"3	Fiat Abarth 1000 OTS
			Sport	1000 cc	L. Caprioli	4'49"5	Fiat Abarth 1000 ST
			Sport	1300 cc	F. Pilone	5'05"5	Fiat Abarth OT 1300
			Sport	2000 cc	T. Brescianini	4'51"1	Abarth Simca 2000
			Sport-Prototypes	1000 cc	Donati	5'05"9	Fiat Abarth 1000 SP
28 April	Gap, France	Course de côte du Col Bayard (hillclimb-5,000 kms)	Sport-Prototypes	2000 cc	P. Maublanc	---	Fiat Abarth 2000
5 May	Vallelunga, Italy	Trofeo Marathon (circuit-1,800 kms x 66 laps)	Touring	600 cc	R. Serpieri	1h 11'49"9	Fiat Abarth 595
			Touring	1000 cc	N. Parenti	1h 10'57"4	Fiat Abarth 1000 Berlina
5 May	Cannes, France	Course de côte de Cannes (hillclimb-7,845 kms)	Touring	1000 cc	A. Merzario	6'24"0	Fiat Abarth 1000 Berlina
			Sport-Prototypes	1000 cc	J. Ortner	6'15"4	Fiat Abarth 1000 SP
			Sport-Prototypes	2000 cc	P. Maublanc	6'05"8	Fiat Abarth 2000
1 May	Vallelunga, Italy	Coppa G. Smargiassi (circuit-1,800 kms)	Touring	700 cc	L. Cestra	23'37"5	Fiat Abarth 695
5 May	Palermo, Italy	Targa Florio (circuit-72,000 kms, x 10 laps)	Sport	1000 cc	'Don Pedrito' - Bruno di Belmonte	7h 30'14"4	Fiat Abarth 1000 SP
			Sport	1300 cc	Rizzo-Alongi	6h 57'53"1	Abarth Simca 1300
			Sport-Prototypes	1000 cc	Patanè-'Oras'	7h 12'11"9	Fiat Abarth 1000 SP
5 May	Savona, Italy	Savona-Cadibona (hillclimb-7,800 kms)	Touring	600 cc	L. Grisolia	5'10"4	Fiat Abarth 595
			Touring	850 cc	P. L. De Zordo	4'46"0	Fiat Abarth 850 TC
			Touring	up to1300 cc	A. Soria	4'38"2	Fiat Abarth 1000 Berlina
			Sport	1300 cc	M. Roasio	4'21"2	Fiat Abarth 1000 SP
			Sport	over 1300 cc	T. Brescianini	4'35"0	Abarth Simca 2000
			Corsa	1000 cc	A. Soria	4'38"2	Fiat Abarth 1000
5 May	Zolder, Belgium	Grand Prix du Limbourg (circuit-4,200 kms)	Touring	850 cc	E. Swart	---	Fiat Abarth 850 TC
			Touring	1000 cc	A. Goedemans	---	Fiat Abarth 1000 Berlina
12 May	Parma, Italy	Castione Baratti-Neviano Arduini (hillclimb-5,500 kms)	Touring	600 cc	P. G. Poggi	4'17"5	Fiat Abarth 595
			Touring	700 cc	P. L. Fuganti	4'24"0	Fiat Abarth 695
			Touring	850 cc	E. Gatti	3'58"6	Fiat Abarth 850 TC
			Touring	1000 cc	M. Campanini	3'47"8	Fiat Abarth 1000 Berlina
			Sport	1300 cc	G. Boeris	3'49"4	Fiat Abarth OT 1300
			Sport-Prototypes	1000 cc	A. Merzario	3'30"4	Fiat Abarth SP
			Sport-Prototypes	2000 cc	J. Ortner	3'17"3	Fiat Abarth 2000
12 May	Zandvoort, Holland	Trophy of the Dunes (circuit-4,225 kms)	Touring	850 cc	E. Swart	---	Fiat Abarth 850 TC
			Touring	1000 cc	A. Hezemans	---	Fiat Abarth 1000 Berlina
12 May	Montelimar, France	Course de côte de Colombier (hillclimb-3,300 kms)	Prototypes	2000	P. Maublanc	---	Fiat Abarth 2000
23 May	Volterra	Coppa Città di Volterra (hillclimb-10,060 kms)	Touring	600 cc	G. Brancadori	7'08"6	Fiat Abarth 595
			Touring	700 cc	A. Polli	7'26"5	Fiat Abarth 695
			Touring	850 cc	P. Falorni	6'36"8	Fiat Abarth 850 TC
			Touring	1000 cc	M. Campanini	6'22"8	Fiat Abarth 1000 Berlina
			Gran turismo	1000 cc	I. Galmossi	6'52"3	Fiat Abarth OT 1000
			Sport	1000 cc	P. Lado	6'01"8	Fiat Abarth 1000 SP
			Sport	1300 cc	C. Zuccoli	5'39"0	Fiat Abarth OT 1300
			Prototypes	1000 cc	L. Caprioli	6'01"2	Fiat Abarth 1000
			Prototypes	2000 cc	J. Ortner	5'17"8	Fiat Abarth 2000
23 May	Vallelunga, Italy	Coppa Canaletti Gaudenti (circuit-1,800 kms x 20 laps)	Touring	850 cc	A. Savarese	23'25"1	Fiat Abarth 850 TC
26 May	Monza, Italy	Coppa Automobile Club di Genova (circuit-2,405 kms x 15 giorni)	Touring	850 cc	E. Crivellari	18'30"2	Fiat Abarth 850 TC
			Touring	up to 1150 cc	'Pal Joe'	17'52"5	Fiat Abarth 1000 Berlina

DATE	LOCATION	RACE	CATEGORY	CLASS	WINNER	TIME OR DISTANCE	CAR
26 May	Piacenza, Italy	Castell'Arquato-Vernasca (hillclimb-9,775 kms)	Touring	700 cc	R. Amighini	6'59"0	Fiat Abarth 695
			Touring	850 cc	E. Gatti	5'58"0	Fiat Abarth 850 TC
			Touring	up to 1150 cc	M. Campanini	5'41"8	Fiat Abarth 1000 Berlina
			Grand Touring	1000 cc	P. Brambilla	6'23"0	Fiat Abarth 1000 OTS
			Sport	1300 cc	G. Pesenti	5'35"3	Fiat Abarth OT 1300
2 June	Vallelunga, Italy	Gran Premio della Repubblica (circuit 1,800 kms x 68 laps)	Sport	1000 cc	P. Lado	1h 31'39"0	Fiat Abarth 1000 SP
			Sport	1300 cc	C. Zuccoli	1h 32'45"7	Fiat Abarth OT 1300
			Sport-Prototypes	1000 cc	A. Merzario	1h 30'19"9	Fiat Abarth 1000 SP
2 June	Biella, Italy	Coppa Carlo Felice Trossi (hillclimb-6,500 kms)	Touring	600 cc	P. Bertotto	5'13"6	Fiat Abarth 595
			Touring	850 cc	P. Tali	4'29"1	Fiat Abarth 850 TC
			Touring	1000 cc	A. Soria	4'24"0	Fiat Abarth 1000 Berlina
			Sport	up to 1300 cc	L. Pasotto	4'04"4	Fiat Abarth 1000 SP
9 June	Bologna, Italy	Coppa Acqua Cerelia (hillclimb-kms 12,200)	Touring	600 cc	R. Amighini	9'38"1	Fiat Abarth 595
			Touring	700 cc	A. Santulli	10'08"4	Fiat Abarth 695 SS
			Touring	850 cc	P. Falorni	9'44"8	Fiat Abarth 850 TC
			Touring	1000 cc	G. C. Casati	9'15"5	Fiat Abarth 1000 Berlina
			Sport	1000 cc	L. Caprioli	8'55"1	Fiat Abarth 1000 SP
			Sport	1300 cc	'Onid'	10'23"1	Abarth Simca 1300
			Sport-Prototypes	1000 cc	A. Merzario	8'40"1	Fiat Abarth 1000 SP
			Sport-Prototypes	2000 cc	J. Ortner	8'07"1	Fiat Abarth 2000
9 June	Rieti, Italy	Coppa Bruno Carotti (hillclimb-15,100 kms)	Touring	600 cc	P. G. Poggi	10'07"1	Fiat Abarth 595
			Touring	700 cc	E. Pasolini	9'55"7	Fiat Abarth 695
			Touring	850 cc	G. Cucchiarelli	9'41"5	Fiat Abarth 850 TC
			Touring	1000 cc	N. Parenti	8'40"3	Fiat Abarth 1000 Berlina
			Grand Touring	1000 cc	I. Cardinali	11'20"6	Fiat Abarth 1000 OTS
			Sport	1000 cc	S. Calascibetta	8'29"0	Fiat Abarth 1000 SP
			Sport	1300 cc	R. Ramoino	9'53"9	Fiat Abarth OT 1300
			Sport-Prototypes	1000 cc	G. Tommasi	8'41"9	Fiat Abarth 1000 SP
9 June	Frascati, Italy	Frascati-Tuscolo (hillclimb-4,200 kms)	Touring	600 cc	B. Carotenuto	3'37"4	Fiat Abarth 595
			Touring	700 cc	G. Miranda	3'44"8	Fiat Abarth 695
			Touring	850 cc	D. Busilacchio	3'25"1	Fiat Abarth 850 TC
			Sport	1000 cc	S. Sebastiani	3'14"1	Fiat Abarth 1000 SP
			Sport	2000 cc	P. Gargano	2'55"8	Fiat Abarth 2000
13 June	Bassano del Grappa, Italy	Bassano-Montegrappa (hillclimb-11,750 kms)	Touring	600 cc	R. Amighini	10'52"0	Fiat Abarth 595
			Touring	700 cc	R. Galasso	10'47"9	Fiat Abarth 695 SS
			Touring	850 cc	F. Stefanelli	10'14"0	Fiat Abarth 850 TC
			Touring	1000 cc	U. Grano	10'00"2	Fiat Abarth 1000 Berlina
			Sport	1000 cc	'Match'	9'31"9	Fiat Abarth 1000 SP
			Sport	1300 cc	A. Becchetti	10'19"1	Fiat Abarth OT 1300
			Sport-Prototypes	1000 cc	L. Pasotto	9'02"5	Fiat Abarth 1000 SP
			Sport-Prototypes	2000 cc	E. Tondelli	9'32"6	Fiat Abarth 2000
13 June	Livorno, Italy	Antignano-Monteburrone (hillclimb-3,400 kms x 2)	Touring	600 cc	C. Tabacco	5'10"8	Fiat Abarth 595
			Touring	700 cc	R. Tampoldi	5'02"6	Fiat Abarth 695
			Touring	850 cc	P. Falorni	4'38"9	Fiat Abarth 850 TC
			Touring	1000 cc	A. Stefanelli	4'31"1	Fiat Abarth 1000 Berlina
			Grand Touring	1000 cc	P. Brambilla	4'55"0	Fiat Abarth 1000 OTS
			Sport	1000 cc	Filippi	5'03"9	Fiat Abarth 1000 SP
			Sport	1300 cc	L. Ungelli	4'27"8	Fiat Abarth OT 1300
16 June	Treviso, Italy	Vittorio Veneto-Cansiglio (hillclimb-10,800 kms)	Touring	850 cc	'Gibi'	7'17"9	Fiat Abarth 850 TC
			Touring	1000 cc	D. Gargan	7'02"3	Fiat Abarth 1000 Berlina
			Grand Touring	1000 cc	P. Brambilla	7'47"7	Fiat Abarth 1000 OTS
			Sport	1000 cc	M. Nardini	6'34"3	Fiat Abarth 1000 SP
			Prototypes	1000 cc	A. Merzario	6'06"7	Fiat Abarth 1000 SP
			Prototypes	2000 cc	P. Schetty	5'45"2	Fiat Abarth 2000
16 June	Reggio Calabria, Italy	Santo Stefano-Gambarie (hillclimb-7,830 kms)	Touring	600 cc	G. D'Ippolito	6'27"9	Fiat Abarth 595
			Touring	700 cc	A. Bonaccorsi	6'10"7	Fiat Abarth 695
			Touring	850 cc	L. Ferragina	5'51"6	Fiat Abarth 850 TC
			Touring	up to 1150 cc	N. Barbagallo	5'51"6	Fiat Abarth 1000 Berlina
			Sport	up to 1300 cc	Benini	5'48"8	Fiat Abarth OT 1300
			Sport	over 1300 cc	P. Gargano	5'10"3	Fiat Abarth 2000
			Sport-Prototypes	up to 2000 cc	D. Scola	4'57"1	Fiat Abarth 2000
16 June	Cortona, Italy	Coppa Città di Cortona (hillclimb-3,500 kms)	Touring	700 cc	R. Tampolli	5'31"1	Fiat Abarth 695
			Touring	850 cc	P. Falorni	5'03"8	Fiat Abarth 850 TC
			Sport	1300 cc	L. Mugelli	4'46"5	Fiat Abarth OT 1300
16 June	Cremona, Italy	Coppa Feraboli (1 flying kilometre)	Touring	1000 cc	F. Zanardelli	0'16"4	Fiat Abarth 1000 Berlina
23 June	Bologna, Italy	Bologna-Passo della Raticosa (hillclimb-32,700 kms)	Touring	600 cc	L. Fontanesi	23'24"3	Fiat Abarth 595
			Touring	700 cc	G. Gatta	23'08"1	Fiat Abarth 695
			Touring	850 cc	E. Gatti	21'59"8	Fiat Abarth 850 TC
			Touring	1000 cc	M. Campanini	21'08"8	Fiat Abarth 1000 Berlina
			Sport	1300 cc	'Onid'	23'08"9	Abarth Simca 1300
			Prototypes	1000 cc	V. Venturi	19'22"7	Fiat Abarth 1000 SP
			Prototypes	2000 cc	J. Ortner	17'38"5	Fiat Abarth 2000

DATE	LOCATION	RACE	CATEGORY	CLASS	WINNER	TIME OR DISTANCE	CAR
23 June	Vallelunga, Italy	Coppa Gallenga (circuit-3,120 kms)	Sport	1300 cc	R. Ramoino	1h 33'11"7	Fiat Abarth OT 1300
			Prototypes	1000 cc	P. Lado	1h 32'04"8	Fiat Abarth 1000 SP
23 June	Torino, Italy	Cuorgné-Alpette (hillclimb-7,000 kms)	Touring	600 cc	C. Tabacco	6'03"0	Fiat Abarth 595
			Touring	850 cc	P. Tali	5'42"5	Fiat Abarth 850 TC
			Touring	1000 cc	A. Soria	5'22"0	Fiat Abarth 1000 Berlina
			Grand Touring	1000 cc	"Amo"	5'50"1	Fiat Abarth 1000 OTS
			Sport	1000 cc	M. Roasio	5'06"6	Fiat Abarth 1000 SP
30 June	Monza, Italy	Trofeo Mario Angiolini (circuit-2,405 kms x 63 laps)	Touring	600 cc	R. Amighini	1h 25'52"2	Fiat Abarth 595
			Touring	700 cc	L. Lazzari	1h 26'01"8	Fiat Abarth 695
			Touring	850 cc	"Opicina"	1h 14'00"0	Fiat Abarth 850 TC
			Touring	1000 cc	M. Campanini	1h 13'56"2	Fiat Abarth 1000 Berlina
30 June	Cosenza, Italy	Coppa della Sila (hillclimb-14,000 kms)	Touring	600 cc	G. D'Ippolito	12'04"3	Fiat Abarth 595
			Touring	700 cc	G. Marotta	11'51"4	Fiata Abarth 695
			Touring	850 cc	L. Ferragina	11'19"2	Fiat Abarth 850 TC
			Touring	1000 cc	V. Virgilio	11'14"3	Fiat Abarth 1000 Berlina
			Grand Touring	1000 cc	F. Di Benedetto	12'33"3	Fiat Abarth 1000 OTS
			Sport	1000 cc	S. Calascibetta	10'14"3	Fiat Abarth 1000 SP
			Sport	1300 cc	A. Gambero	10'41"2	Fiat Abarth 1300 OT
			Sport-Prototypes	1000 cc	G. Tommasi	10'37"0	Fiat Abarth 1000 SP
			Sport-Prototypes	2000	P. Schetty	8'58"3	Fiat Abarth 2000
7 July	Adenau, Germany	Nürburgring 6 Stunden (circuit-22,810 kms)	Touring	850 cc	Swart-Zanetti	---	Fiat Abarth 850 TC
			Touring	1000 cc	Goedemans-Hezemans	---	Fiat Abarth 1000 Berlina
7 July	Bolzano, Italy	Bolzano-Mendola (hillclimb-14,400 kms)	Touring	600 cc	P. G. Poggi	11'57"3	Fiat Abarth 595
			Touring	850 cc	'Gibi'	11'03"2	Fiat Abarth 850 TC
			Touring	1000 cc	U. Grano	10'42"0	Fiat Abarth 1000 Berlina
			Sport	1000 cc	L. Pasotto	9'47"3	Fiat Abarth 1000 SP
			Sport-Prototypes	1000 cc	A. Merzario	9'23"2	Fiat Abarth 1000 SP
			Sport-Prototypes	2000 cc	P. Schetty	8'27"0	Fiat Abarth 2000
14 July	Trento, Italy	Trento-Bondone (hillclimb-17,300 kms)	Touring	850 cc	E. Gatti	16'28"1	Fiat Abarth 850 TC
			Touring	1000 cc	M. Campanini	16'06"1	Fiat Abarth 1000 Berlina
			Grand Touring	1000 cc	G. Galmozzi	17'41"5	Fiat Abarth 1000 OTS
			Sport	1000 cc	G. Varese	13'17"0	Fiat Abarth 1000 SP
			Sport	1300 cc	K. Federhofer	13'06"6	Fiat Abarth OT 1300
21 July	Sestrière, Italy	Cesana-Sestrière (hillclimb-10,400 kms)	Touring	600 cc	C. Bettazzi	7'22"4	Fiat Abarth 595
			Touring	850 cc	'Gibi'	6'43"6	Fiat Abarth 850 TC
			Touring	1000 cc	G. C. Casati	6'29"0	Fiat Abarth 1000 Berlina
			Grand Touring	1000cc	P. Brambilla	7'18"1	Fiat Abarth 1000 OTS
			Sport	1000 cc	G. Varese	6'02"5	Fiat Abarth 1000 SP
			Sport	1300 cc	K. Federthofer	6'11"4	Fiat Abarth OT 1300
28 July	Freiburg, Germany	ADAC Bergpreis Freiburg-Schauinsland (hillclimb-11.500 kms)	Sport	1300 cc	K. Federhofer	13'15"19	Fiat Abarth OT 1300
28 July	Scarperia, Italy	Gran Premio del Mugello (circuit-66,200 kms x 8 laps)	Sport	1000 cc	Calascibetta-Ferlaino	4h31'50"5	Fiat Abarth 1000 SP
4 August	Trapani, Italy	Trapani-Monte Erice (hillclimb-6,670 kms)	Touring	600 cc	C. Aloè	5'37"5	Fiat Abarth 595
			Touring	700 cc	G. Marotta	5'27"2	Fiat Abarth 695
			Touring	850 cc	S. Sansica	5'04"2	Fiat Abarth 850 TC
			Touring	1000 cc	V. Virgilio	4'59"1	Fiat Abarth 1000 Berlina
			Sport	1000 cc	'Figi'	5'18"9	Fiat Abarth 1000 SP
			Sport	1300 cc	G. Valenzia	5'36"0	Abarth Simca 1300
			Sport-Prototypes	1000 cc	S. Calascibetta	4'38"4	Fiat Abarth 1000 SP
4 August	Avezzano, Italy	Trofeo Micangeli (hillclimb-6,000 kms)	Touring	850 cc	'Tato'	3'12"5	Fiat Abarth 850 TC
			Grand Touring	1000 cc	L. Carmignani	3'29"7	Fiat Abarth 1000 OTS
			Sport	1000 cc	P. Gargano	2'56"5	Fiat Abarth 1000 BA
			Sport-Prototipi	1000 cc	G. Tommasi	2'51"0	Fiat Abarth 1000 SP
11 August	Palermo, Italy	Alcamo-Monte Bonifato (hillclimb-4,500 kms)	Touring	600 cc	G. Caci	3'41"3	Fiat Abarth 595
			Touring	700 cc	G. Maratta	3'30"1	Fiat Abarth 695
			Touring	850 cc	S. Sansica	3'17"3	Fiat Abarth 850 TC
			Touring	1000 cc	V. Virgilio	3'11"2	Fiat Abarth 1000 Berlina
			Sport	1000 cc	S. Calascibetta	2'51"4	Fiat Abarth 1000 SP
			Sport	1300 cc	G. Valenza	3'23"0	Abarth Simca 1300
15 August	Pescara, Italy	Cronoscalata Svolte di Popoli (hillclimb-8,000 kms)	Touring	600 cc	C. Aloe	5'54"4	Fiat Abarth 595
			Touring	700 cc	L. Cestra	5'32"4	Fiat Abarth 695
			Touring	850 cc	E. Gatti	5'20"2	Fiat Abarth 850 TC
			Touring	1000 cc	A. Soria	5'10"7	Fiat Abarth 1000 Berlina
			Grand Touring	1000 cc	A. Torricelli	5'31"4	Fiat Abarth 1000 OTS
			Sport	1300 cc	'Mici'	5'08"2	Fiat Abarth OT 1300
			Sport-Prototypes	1000 cc	G. Tommasi	4'43"0	Fiat Abarth 1000 SP
15 August	Pergusa, Italy	Coppa Città di Enna (circuit-4,797 kms x 40 laps)	Prototypes	1000 cc	S. Calascibetta	1h 09'53"3	Fiat Abarth 1000 SP

DATE	LOCATION	RACE	CATEGORY	CLASS	WINNER	TIME OR DISTANCE	CAR
18 August	Viterbo, Italy	Corsa lago di Montefiascone (hillclimb-5,000 kms)	Touring	600 cc	A. Bianchi	3'15"3	Fiat Abarth 595
			Touring	950 cc	E. Gatti	2'53"5	Fiat Abarth 850 TC
			Touring	1000 cc	T. Acasbo	2'46"3	Fiat Abarth 1000 Berlina
			Sport	1000 cc	P. Gargano	2'41"0	Fiat Abarth 1000 SP
			Sport-Prototypes	1000 cc	G. Spicciani	2'42"9	Fiat Abarth 1000 SP
25 August	Brescia, Italy	Trofeo Valle Camonica (hillclimb-8,600 kms)	Touring	600 cc	R. Amighini	6'01"6	Fiat Abarth 595
			Touring	850 cc	'Tato'	5'36"2	Fiat Abarth 850 TC
			Touring	1000 cc	E. Pasolini	5'26"4	Fiat Abarth 1000 Berlina
			Grand Touring	1000 cc	G. Galmozzi	6'28"9	Fiat Abarth 1000 OTS
			Sport	1000 cc	'Maurizio'	6'34"9	Fiat Abarth 1000 SP
			Sport	1300 cc	G. A. Pesenti	5'55"7	Fiat Abarth OT 1300
			Sport-Prototypes	1000 cc	M. Baldo	6'21"9	Fiat Abarth 1000 SP
			Sport-Prototypes	2000 cc	E. Tondelli	5'51"6	Fiat Abarth 2000
25 August	Ascoli Piceno, Italy	Ascoli-Colle San Marco (hillclimb-10,200 kms)	Touring	600 cc	P. Poggi	7'57"0	Fiat Abarth 595
			Touring	700 cc	L. Finelli	7'50"0	Fiat Abarth 695
			Touring	850 cc	'Gibi'	7'15"3	Fiat Abart 850 TC
			Touring	1000 cc	G. Casati	7'04"2	Fiat Abarth 1000 Berlina
			Sport	1000 cc	V. Bernasconi	6'57"1	Fiat Abarth 1000 SP
			Sport	1300 cc	R. Ramoino	6'58"7	Fiat Abarth OT 1300
			Prototypes	1000 cc	P. Gargano	6'45"5	Fiat Abarth 1000 SP
1 September	Adenau, Germany	Internationales 500 Km Nürburgring Rennen (circuit-22,810 kms x 22 laps)	Touring	1000 cc	'Pam'	3h 32'35"	Fiat Abarth 1000 Berlina
			Sport	1300 cc	J. Walton	3h 39'54"6	Fiat Abarth OT 1300
			Sport-Prototypes	1600 cc	P. Schetty	3h 29'7"	Fiat Abarth 1600
1 September	Verona, Italy	Caprino Veronese-Spiazzi di Monte Baldo (hillclimb-10,400 kms)	Touring	600 cc	R. Amighini	6'16"0	Fiat Abarth 595
			Touring	700 cc	L. Moreschi	6'17"5	Fiat Abarth 695
			Touring	850 cc	'Tato'	5'49"4	Fiat Abarth 850 TC
			Touring	1000 cc	U. Grano	5'40"4	Fiat Abarth 1000 Berlina
			Grand Touring	1000 cc	G. Galmozzi	6'07"5	Fiat Abarth 1000 OTS
			Sport	1000 cc	P. Lado	5'12"3	Fiat Abarth 1000 SP
			Sport	1300 cc	F. Pilone	5'13"3	Fiat Abarth OT 1300
			Sport-Prototypes	1000 cc	M. Baldo	5'42"8	Fiat Abarth 1000 SP
1 September	Macerata, Italy	Coppa Città di Tolentino (hillclimb-4,100 kms)	Touring	600 cc	V. Gattafoni	2'47"0	Fiat Abarth 595
			Touring	700 cc	L. Finelli	2'41"4	Fiat Abarth 695
			Touring	850 cc	P. Falorni	2'39"3	Fiat Abarth 850 TC
			Sport	1000 cc	G. Benini	2'31"3	Fiat Abarth 1000 SP
			Sport-Prototypes	1000 cc	P. Gargano	2'24"	Fiat Abarth 1000 SP
8 September	Caltanissetta, Italy	Coppa Nissena (hillclimb-10,000 kms)	Touring	600 cc	G. D'Ippolito	6'44"2	Fiat Abarth 595
			Touring	700 cc	A. Bonaccorsi	6'32"7	Fiat Abarth 695
			Touring	850 cc	S. Campione	6'15"5	Fiat Abarth 850 TC
			Touring	1000 cc	V. Virgilio	6'06"1	Fiat Abarth 1000 Berlina
			Grand Touring	1000 cc	A. Torricelli	6'26"4	Fiat Abarth 1000 OTS
			Sport	1000 cc	S. Calascibetta	5'39"6	Fiat Abarth 1000 SP
			Sport	1300 cc	"Vick"	6'14"3	Fiat Abarth OT 1300
8 September	Brescia, Italy	Autoscalata della Maddalena (hillclimb-5,200 kms)	Touring	600 cc	R. Amighini	4'17"3	Fiat Abarth 595
			Touring	850 cc	"Tato"	-	Fiat Abarth 850 TC
			Touring	1000 cc	E. Pasolini	3'51"7	Fiat Abarth 1000 Berlina
			Grand Touring	1000 cc	G. C. Galmozzi	4'10"2	Fiat Abarth 1000 OTS
			Sport	1000 cc	'Matich'	3'43"8	Fiat Abarth 1000 SP
			Sport	1300 cc	A. Becchetti	3'55"3	Fiat Abarth OT 1300
			Sport-Prototypes	2000 cc	P. Schetty	3'08"2	Fiat Abarth 2000
8 September	Imola, Italy	500 kms di Imola (circuit 5,017 kms x 100 laps)	Sport-Prototypes	1000 cc	Lado-Donà	3h 06'23"7	Fiat Abarth 1000 SP
15 September	Catania, Italy	Catania-Etna (hillclimb-33,000 kms)	Touring	600 cc	C. Leone	22'25"0	Fiat Abarth 595
			Touring	700 cc	G. Marotta	21'30"1	Fiat Abarth 695
			Touring	850 cc	S. Campione	19'48"3	Fiat Abarth 850 TC
			Touring	up to 1150 cc	C. Bonaccorsi	20'02"2	Fiat Abarth 1000 Berlina
			Grand Touring	1000 cc	A. Torricelli	20'49"7	Fiat Abarth 1000 OTS
			Sport	1000 cc	S. Calascibetta	17'48"6	Fiat Abarth 1000 SP
15 September	Belluno, Italy	Coppa Alpe del Nevegal (hillclimb-11,500 kms)	Touring	600 cc	P. Cozzarolo	10'59"4	Fiat Abarth 595
			Touring	700 cc	'Antares'	10'40"4	Fiat Abarth 695
			Touring	850 cc	'Gibi'	9'10"4	Fiat Abarth 850 TC
			Touring	1000 cc	M. Campanini	8'56"7	Fiat Abarth 1000 Berlina
			Grand Touring	1000 cc	P. Brambilla	9'59"9	Fiat Abarth 1000 OTS
			Sport	1000 cc	M. Nardari	8'48"9	Fiat Abarth 1000 SP
			Sport	1300 cc	G. Uliana	10'06"1	Abarth Simca 1300
			Sport-Prototypes	1000 cc	M. Baldo	9'28"3	Fiat Abarth 1000 SP
			Sport-Prototypes	2000 cc	P. Schetty	7'45"1	Fiat Abarth 2000

DATE	LOCATION	RACE	CATEGORY	CLASS	WINNER	TIME OR DISTANCE	CAR
15 September	Bergamo, Italy	Castione-Passo della Presolana (hillclimb-6,400 kms)	Touring	600 cc	R. Amighini	4'25"3	Fiat Abarth 595
			Touring	700 cc	P. Poggi	4'15"3	Fiat Abarth 695
			Touring	850 cc	G. Carena	4'12"1	Fiat Abarth 850 TC
			Touring	1000 cc	A. Soria	3'55"8	Fiat Abarth 1000 Berlina
			Grand Touring	1000 cc	"Amo"	4'14"8	Fiat Abarth 1000 OTS
			Sport	1000 cc	G. Boeris	3'46"4	Fiat Abarth 1000 SP
			Sport	1300 cc	G.Pesenti	3'49"7	Fiat Abarth OT 1300
			Sport-Prototypes	1000 cc	E. Velardita	4'26"7	Fiat Abarth 1000 SP
22 September	Vallelunga, Italy	Premio Campagnano (circuit-3,200 kms x 35 laps)	Touring	600 cc	P. Poggi	1h 10'24"0	Fiat Abarth 595
			Touring	700 cc	R. Tampolli	1h 12'11"6	Fiat Abarth 695
			Touring	850 cc	S. Uberti	1h 11'14"2	Fiat Abarth 850 TC
			Touring	1000 cc	F. Acasbo	1h 10'37"1	Fiat Abarth 1000 Berlina
22 September	Ragusa, Italy	Coppa dei Monti Iblei (hillclimb-5,200 kms)	Touring	600 cc	V. Virgilio	4'37"5	Fiat Abarth 595
			Touring	700 cc	G. Marotta	4'24"6	Fiat Abarth 695
			Touring	850 cc	'Jimmy'	4'20"9	Fiat Abarth 850 TC
			Touring	1000 cc	I. Serse	4'21"1	Fiat Abarth 1000 Berlina
			Sport	1300 cc	A. Gambero	4'12"7	Fiat Abarth OT 1300
			Sport-Prototypes	1000 cc	S. Calascibetta	4'12"6	Fiat Abarth 1000 SP
			Sport-Prototypes	2000 cc	D. Scola	3'56"0	Fiat Abarth 2000
22 September	Udine, Italy	Tolmezzo-Verzegnis (hillclimb-4,260 kms x 2)	Touring	600 cc	'Arch'	4'59"8	Fiat Abarth 595
			Touring	700 cc	R. Galasso	4'56"6	Fiat Abarth 695
			Touring	850 cc	M. Favaretto	4'31"7	Fiat Abarth 850 TC
			Touring	1000 cc	U. Grano	4'16"2	Fiat Abarth 1000 Berlina
			Grand Touring	1000 cc	P. Brambilla	4'46"2	Fiat Abarth 1000 OTS
			Sport	1000 cc	M. Nardari	4'08"6	Fiat Abarth 1000 SP
			Sport	up to 1600 cc	G. Uliana	5'00"3	Abarth Simca 1300
22 September	Aosta, Italy	Aosta-Pila (hillclimb-11,200 kms)	Touring	600 cc	A. Ori	9'09"9	Fiat Abarth 595
			Touring	700 cc	G. Dalla Chiesa	8'33"2	Fiat Abarth 695
			Touring	850 cc	G. Carena	8'08"6	Fiat Abarth 850 TC
			Touring	1000 cc	A. Soria	7'55"2	Fiat Abarth 1000 Berlina
			Grand Touring	1000 cc	G. Galmozzi	8'19"9	Fiat Abarth 1000 OTS
			Sport	1000 cc	G. Varese	7'08"1	Fiat Abarth 1000 SP
			Sport	1300 cc	R. Ramoino	7'16"2	Fiat Abarth OT 1300
			Sport-Prototypes	2000 cc	P. Schetty	6'12"1	Fiat Abarth 2000
29 September	Siena, Italy	Coppa del Chianti classico (hillclimb-8,200 kms)	Touring	600 cc	G. Brancadori	4'37"8	Fiat Abarth 595
			Touring	700 cc	F. Gragnoli	4'50"1	Fiat Abarth 695
			Touring	850 cc	A. Mostardini	4'17"4	Fiat Abarth 850 TC
			Touring	1000 cc	A. Stefanelli	4'10"9	Fiat Abarth 1000 Berlina
			Grand Touring	1000 cc	P. Brambilla	4'37"2	Fiat Abarth 1000 OTS
			Sport	1000 cc	P. Botalla	4'01"5	Fiat Abarth 1000 SP
29 September	Palermo, Italy	Cefalù-Gibilmanna (hillclimb-9,500 kms)	Touring	600 cc	C. Aloc	7'31"0	Fiat Abarth 595
			Touring	700 cc	E. Salerno	7'43"8	Fiat Abarth 695
			Touring	850 cc	G. Spadafora	7'00"4	Fiat Abarth 850 TC
			Touring	1000 cc	V. Virgilio	6'56"0	Fiat Abarth 1000 Berlina
			Sport	1000 cc	'Fifi'	7'12"4	Fiat Abarth 1000 SP
			Sport	1300 cc	G. Marino	8'04"7	Abarth Simca 1300
			Sport	over 1300 cc	V. Ferlito	6'25"9	Abarth Simca 2000
29 September	Madrid, Spagna	3 Ore di Jarama (circuit-3,404 kms)	Touring	850 cc	E. Swart	kms 286,398	Fiat Abarth 850 TC
			Touring	1000 cc	'Pam'	kms 305,000	Fiat Abarth 1000 Berlina
6 October	Gubbio, Italy	Coppa Mengara (hillclimb-5,200 kms)	Touring	600 cc	R. Mazzeschi	2'59"2	Fiat Abarth 595
			Touring	700 cc	'Antares'	3'03"0	Fiat Abarth 695
			Touring	850 cc	A. Mostardini	2'38"6	Fiat Abarth 850 TC
			Touring	1000 cc	A. Stefanelli	2'36"8	Fiat Abarth 1000 Berlina
			Grand Touring	1000 cc	P. Brambilla	2'57"1	Fiat Abarth 1000 OTS
			Sport	1000 cc	P. Botalla	2'29"6	Fiat Abarth 1000 SP
			Sport	1300 cc	A. Fabroni	2'41"8	Fiat Abarth OT 1300
5 October	Vallelunga, Italy	Trofeo Luigi Musso (circuit-3,120 kms)	Touring	600 cc	P. Rianna	22'57"8	Fiat Abarth 595
			Touring	700 cc	L. Finelli	22'22"9	Fiat Abarth 695
			Touring	850 cc	L. Moreschi	22'04"1	Fiat Abarth 850 TC
			Touring	1000 cc	F. Acasbo	23'18"1	Fiat Abarth 1000 Berlina
6 October	Aspern, Austria	Donau Pokal Flugplatzrennen (circuit-2,600 kms)	Touring	1000 cc	A. Hezemans	24'50"4	Fiat Abarth 1000 Berlina
			Sport	1000 cc	W. Donà	---	Fiat Abarth 1000 SP
			Sport-Prototypes	3000 cc	P. Schetty	25'54"38	Fiat Abarth 3000
6 October	Siracusa, Italy	Coppa Val d'Anapo-Sortino (hillclimb-6,000 kms)	Touring	600 cc	G. D'Ippolito	5'06"9	Fiat Abarth 595
			Touring	700 cc	A. Bonaccorsi	4'53"8	Fiat Abarth 695
			Touring	850 cc	G. Spadafora	4'44"4	Fiat Abarth 850 TC
			Touring	1000 cc	I. Serse	4'41"9	Fiat Abarth 1000 Berlina
			Sport	1000 cc	S. Calascibetta	4'16"4	Fiat Abarth 1000 SP
			Sport	1300 cc	A. Gambero	4'17"0	Fiat Abarth OT 1300

DATE	LOCATION	RACE	CATEGORY	CLASS	WINNER	TIME OR DISTANCE	CAR
13 October	Catanzaro, Italy	Ponte Coriace-Tiriolo (hillclimb-10,417 kms)	Touring	600 cc	C. Aloe	8'56"0	Fiat Abarth 595
			Touring	700 cc	L. Magdalone	8'56"9	Fiat Abarth 695
			Touring	850 cc	L. Ferragina	8'25"4	Fiat Abarth 850 TC
			Touring	1000 cc	A. Bonaccorsi	8'20"4	Fiat Abarth 1000 Berlina
			Grand Touring	1000 cc	A. Crispo	8'49"	Fiat Abarth 1000 OTS
			Sport	1000 cc	V. Bernasconi	7'56"6	Fiat Abarth 1000 SP
			Sport	1300 cc	A. Gambaro	7'55"0	Abarth Simca 1300
			Prototypes	1000 cc	R. Ricciardi	8'01"6	Fiat Abarth 1000 SP
13 October	Monza, Italy	Coppa Leopoldo Carri (circuit-5,750 kms x 27 laps)	Touring	850 cc	M. Favaretto	1h 1'33"6	Fiat Abarth 850 TC
			Touring	1000 cc	G. Bassani	1h 0'54"6	Fiat Abarth 1000 Berlina
20 October	Innsbruck, Austria	Preis Von Tirol (circuit-2,800 kms x 20 laps)	Touring	1000 cc	J. Ortner	---	Fiat Abarth 1000 Berlina
			Sport-Prototypes	2000 cc	A. Hezemans	22'50"25	Fiat Abarth 2000
			Sport-Prototypes	3000 cc	P. Schetty	22'09"29	Fiat Abarth 3000
20 October	Cagliari, Italy	Coppa Città d'Iglesias (hillclimb-6,700 kms)	Touring	600 cc	E. Di Rienzo	5'02"2	Fiat Abarth 595
			Touring	700 cc	G. Martis	4'47"2	Fiat Abarth 695
			Touring	850 cc	G. Bertagnolli	4'52"6	Fiat Abarth 850 TC
			Touring	1000 cc	G. Fadda	4'40"2	Fiat Abarth 1000 Berlina
			Grand Touring	1300 cc	A. Giliberti	4'03"8	Abarth Simca 1300
20 October	Orvieto, Italy	Trofeo Città di Orvieto (hillclimb-7,600 kms)	Touring	600 cc	R. Mazzeschi	4'46"0	Fiat Abarth 595
			Touring	700 cc	L. Cestra	4'33"4	Fiat Abarth 695
			Touring	850 cc	R. Mostardini	4'19"0	Fiat Abarth 850 TC
			Touring	1000 cc	G. Casati	4'11"3	Fiat Abarth 1000 Berlina
			Grand Touring	1000 cc	P. Brambilla	4'44"2	Fiat Abarth 1000 OTS
			Sport	1000 cc	P. Botalla	3'38"0	Fiat Abarth 1000 SP
			Sport	1300 cc	A. Fabroni	4'14"0	Fiat Abarth OT 1300
27 October	Marseille, France	Course de côte de la Sainte Baume (hillclimb-8,100 kms)	Sport-Prototypes	2000	Rouverayn	---	Fiat Abarth 2000
1969							
9 March	Bellevue, Belgium	Course de côte de Fleron (hillclimb-2,000 kms)	Sport-Prototypes	2000 cc	E. Tondelli	---	Fiat Abarth 2000
16 March	Vallelunga, Italy	Trofeo Dalla Vecchia (circuit-1,800 kms)	Touring	600 cc	R. De Santis	14'22"0	Fiat Abarth 595
			Touring	700 cc	A. Mola	14'23"0	Fiat Abarth 695
			Touring	850 cc	P. Falorni	14'38"2	Fiat Abarth 850 TC
			Touring	1000 cc	R. Mostardini	13'22"1	Fiat Abarth 1000 Berlina
			Grand Touring	1000 cc	P. Brambilla	14'17"4	Fiat Abarth 1000 OTS
			Sport	1000 cc	P. Monticone	13'31"8	Fiat Abarth 1000 SP
			Sport	1300 cc	A. Fabroni	14'44"1	Fiat Abarth OT 1300
			Sport-Prototypes	1000 cc	S. Sebastiani	13'53"6	Fiat Abarth 1000 SP
			Sport-Prototypes	2000 cc	G.P. Biscoldi	11'20"8	Fiat Abarth 2000
23 March	Avola, Italy	Coppa Belmonte (hillclimb-10,200 kms)	Touring	600 cc	C. Alce	7'36"9	Fiat Abarth 595
			Touring	700 cc	G.Platania	7'35"6	Fiat Abarth 695
			Touring	850 cc	G. Spatafora	7'01"2	Fiat Abarth 850 TC
			Touring	1000 cc	E. Fichera	7'04"1	Fiat Abarth 1000 Berlina
			Grand Touring	1000 cc	A. Torricelli	7'15"0	Fiat Abarth 1000 OTS
			Sport	1000 cc	S. Calascibetta	6'22"1	Fiat Abarth 1000 SP
			Sport	1300 cc	A. Gambero	6'40"5	Fiat Abarth OT 1300
			Sport-Prototypes	2000 cc	P. Bellavia	6'50"8	Abarth Simca 2000
			Sport-Prototypes	3000 cc	D. Scola	6'03"8	Fiat Abarth 3000
23 March	Monza, Italy	4 Ore del Jolly Club (circuit-10,000 kms)	Touring	1000 cc	S. Swart	kms 548,678	Fiat Abarth 1000 Berlina
30 March	Monza, Italy	Coppa Automobile Club di Genova (circuit-2,405 kms x 15 laps)	Touring	850 cc	'Opicina'	18'09"6	Fiat Abarth 850 TC
			Touring	1000 cc	F. Acasbo	17'39"6	Fiat Abarth 1000 Berlina
6 April	Monza, Italy	Trofeo Mario Angiolini (circuit-2,405 kms x 63 laps)	Touring	700 cc	L. Cestra	1h31'53"6	Fiat Abarth 695
			Touring	850 cc	M. Barone	1h22'26"3	Fiat Abarth 850 TC
			Touring	1000 cc	F. Acasbo	1h21'58"8	Fiat Abarth 1000 Berlina
13 April	Gubbio, Italy	Coppa Mengara (hillclimb-5,200 kms)	Touring	700 cc	R. Scoccia	3'07"4	Fiat Abarth 695
			Touring	850 cc	A. Gunnella	2'54"5	Fiat Abarth 850 TC
			Touring	1000 cc	A. Stefanelli	2'39"0	Fiat Abarth 1000 Berlina
			Grand Touring	1000 cc	G. Brambilla	3'15"5	Fiat Abarth 1000 OTS
			Sport	1000 cc	G. Bonella	3'13"7	Fiat Abarth 1000 SP
20 April	Amalfi, Italy	Coppa Primavera (hillclimb-8,750 kms)	Touring	600 cc	V. Nigro	---	Fiat Abarth 595
			Touring	700 cc	'Zio Po'	---	Fiat Abarth 695
			Touring	850 cc	M. De Furia	---	Fiat Abarth 850 TC
			Touring	1000 cc	'Matteo'	---	Fiat Abarth 1000 Berlina
			Grand Touring	1000 cc	A. Torricelli	---	Fiat Abarth 1000 OTS
			Sport	1000 cc	M. Terminiello	---	Fiat Abarth 1000 SP
			Sport-Prototypes	1000 cc	G. Ricciardi	---	Fiat Abarth 1000 SP
			Sport-Prototypes	2000 cc	D. Scola	---	Fiat Abarth 2000

DATE	LOCATION	RACE	CATEGORY	CLASS	WINNER	TIME OR DISTANCE	CAR
20 April	Beograd, Jugoslavia	Grand Prix de Belgrade (circuit-4,350 kms x 2 hours)	Touring	1000 cc	M. Pasotti	---	Fiat Abarth 1000 Berlina
20 April	Villach, Austria	Dobrascht Bergrennen (hillclimb-15,800 kms)	Sport Prototypes	3000 cc	J. Ortner	6'09"9	Fiat Abarth 3000
25 April	Monza, Italy	1000 Km di Monza (circuit-10,000 kms x 100 laps)	Sport	1000 cc	'White'-Grano	---	Fiat Abarth 1000 SP
27 April	Brescia, Italy	Scalata al Colle Sant'Eusebio (hillclimb-9,500 kms)	Touring	600 cc	C. Tabacco	5'53"5	Fiat Abarth 595
			Touring	700 cc	D. Alasia	5'54"3	Fiat Abarth 695
			Touring	850 cc	C. Margoni	5'32"3	Fiat Abarth 850 TC
			Touring	1000 cc	'Pooky'	6'37"5	Fiat Abarth 1000 Berlina
			Grand Touring	1000 cc	G. Galmozzi	5'50"5	Fiat Abarth 1000 OTS
			Sport	1000 cc	G. Pesenti	5'05"9	Fiat Abarth OT 1300
			Sport-Prototypes	1000 cc	'Pam'	4'59"4	Fiat Abarth 1000 SP
			Sport-Prototypes	2000 cc	F. Pilone	4'21"4	Fiat Abarth 2000
27 April	Arezzo, Italy	Pieve S. Stefano-Passo dello Spino (hillclimb-12,700 kms)	Touring	600 cc	P. Vannini	8'31"3	Fiat Abarth 595
			Touring	700 cc	G. Gatta	8'22"2	Fiat Abarth 695
			Touring	850 cc	'Tito'	8'07"1	Fiat Abarth 850 TC
			Touring	1000 cc	A. Mostardini	7'28"6	Fiat Abarth 1000 Berlina
			Grand Touring	1000 cc	P. Brambilla	8'19"0	Fiat Abarth 1000 OTS
			Sport	1000 cc	A. Fracastoro	7'51"9	Fiat Abarth 1000 SP
			Sport	1300 cc	'Nicor'	7'16"3	Fiat Abarth OT 1300
			Sport	over 1300 cc	D. Scola	6'37"7	Fiat Abarth 2000
			Sport-Prototypes	1000 cc	P. Gargano	7'38"3	Fiat Abarth 1000 SP
27 April	Cagliari, Italy	San Gregorio-Campuomu (hillclimb-10,000 kms)	Touring	700 cc	A. Fina	8'08"6	Fiat Abarth 695
			Touring	850 cc	D. Stabile	7'27"9	Fiat Abarth 850 TC
			Sport	---	V. Virgilio	7'03"1	Fiat Abarth 1000 SP
1 May	Gap, France	Course de côte du Col Bayard (hillclimb-6,500 kms)	Sport-Prototypes	2000 cc	A. Finkel	---	Fiat Abarth 2000
1 May	Piacenza, Italy	Castell'Arquato-Vernasca (hillclimb-9,775 kms)	Touring	700 cc	F. Piroli	6'27"12	Fiat Abarth 695
			Touring	850 cc	M. Barone	5'54"4	Fiat Abarth 850 TC
			Touring	up to 1150 cc	E. Gatti	5'46"4	Fiat Abarth 1000 Berlina
			Grand Touring	1000 cc	P. Brambilla	6'24"3	Fiat Abarth 1000 OTS
			Sport	1300 cc	P. Botalla	5'17"1	Fiat Abarth OT 1300
			Sport	over 1300 cc	E. Lualdi	4'53"7	Fiat Abarth 2000
4 May	Palermo, Italy	Targa Florio (circuit-72,000 kms x 10 laps)	Sport	up to 1300 cc	Calascibetta-Ferlito	6h34'19"2	Fiat Abarth 1000 SP
			Sport	up to 5000 cc	Bitter-Kelleners	6h17'50"7	Fiat Abarth 2000
			Sport-Prototypes	1000 cc	Patanè-Oras	4h55'12"	Fiat Abarth 1000 SP
4 May	Savona, Italy	Savona-Cadibona (hillclimb-7,800 kms)	Touring	600 cc	A. Corazza	5'21"1	Fiat Abarth 595
			Touring	850 cc	G. Carena	4'43"5	Fiat Abarth 850 TC
			Touring	1000 cc	A. Soria	4'35"0	Fiat Abarth 1000 Berlina
			Grand Touring	1000 cc	E. Rossi	4'58"0	Fiat Abarth 1000 OTS
			Sport	1000 cc	G. Varese	4'23"8	Fiat Abarth 1000 SP
			Sport	2000 cc	F. Pilone	4'09"1	Fiat Abarth 2000
4 May	Vallelunga, Italy	Coppa Agip (circuit-3,120 kms)	Touring	600 cc	P. Poggi	1h11'34"7	Fiat Abarth 595
			Touring	700 cc	L. Cestra	1h12'06"9	Fiat Abarth 695
			Touring	850 cc	Tandoi	1h11'17"8	Fiat Abarth 850 TC
11 May	Volterra, Italy	Coppa Città di Volterra (hillclimb-10,060 kms)	Touring	600 cc	P. Poggi	7'09"7	Fiat Abarth 595
			Touring	700 cc	G. Del Panta	7'12"1	Fiat Abarth 695
			Touring	850 cc	L. Ferragina	6'40"2	Fiat Abarth 850 TC
			Touring	1000 cc	A. Stefanelli	6'22"2	Fiat Abarth 1000 Berlina
			Grand Touring	1000 cc	G. Galmozzi	6'50"7	Fiat Abarth 1000 OTS
			Sport	1000 cc	G. Bernasconi	6'18"6	Fiat Abarth 1000 SP
			Sport	1300 cc	'Nicor'	6'13"4	Fiat Abarth OT 1300
			Sport	over 1600 cc	D. Scola	5'32"3	Fiat Abarth 2000
15 May	Cremona, Italy	Coppa Feraboli (1 flying kilometre)	Touring	850 cc	F. Tango	---	Fiat Abarth 850 TC
			Touring	1000 cc	L. Codeluppi	---	Fiat Abarth 1000 Berlina
18 May	Parma, Italy	Castione Baratti-Neviano Arduini (hillclimb-5,500 kms)	Touring	600 cc	D. Martino	4'04"3	Fiat Abarth 595
			Touring	700 cc	'Renzo'	4'09"7	Fiat Abarth 695
			Touring	850 cc	M. Barone	3'44"7	Fiat Abarth 850 TC
			Touring	1000 cc	A. Soria	3'37"7	Fiat Abarth 1000 Berlina
			Sport	1000 cc	P. Botalla	3'19"2	Fiat Abarth 1000 SP
			Sport-Prototypes	1000 cc	M. Roasio	3'20"4	Fiat Abarth 1000 SP
			Sport-Prototypes	2000 cc	E. Tondelli	3'10"5	Fiat Abarth 2000
25 May	Trieste, Italy	Trieste-Opicina (hillclimb-10,150 kms)	Touring	600 cc	C. Battezzi	6'06"8	Fiat Abarth 595
			Touring	850 cc	'Gibi'	5'34"3	Fiat Abarth 850 TC
			Touring	1000 cc	A. Soria	5'27"4	Fiat Abarth 1000 Berlina
			Grand Touring	1000 cc	E. Rossi	5'49"0	Fiat Abarth 1000 OTS
			Sport	1000 cc	P. Botalla	4'55"7	Fiat Abarth 1000 SP
			Sport	1300 cc	A. Fabroni	5'21"2	Fiat Abarth OT 1300
			Sport	2000 cc	F. Pilone	4'27"3	Fiat Abarth 2000
			Sport-Prototypes	1000 cc	M. Zanetti	4'52"4	Fiat Abarth 1000 SP
25 May	Biella, Italy	Coppa Carlo Felice Trossi (hillclimb-6,500 kms)	Touring	850 cc	G. Carena	4'44"5	Fiat Abarth 850 TC
			Touring	1000 cc	P. Giachino	4'57"0	Fiat Abarth 1000 Berlina
			Sport	1000 cc	M. Roasio	4'16"5	Fiat Abarth 1000 SP

DATE	LOCATION	RACE	CATEGORY	CLASS	WINNER	TIME OR DISTANCE	CAR
1 June	Treviso, Italy	Vittorio Veneto-Cansiglio (hillclimb-10,800 kms)	Touring	850 cc	M. Favaretto	---	Fiat Abarth 850 TC
			Touring	1000 cc	D. Gargan	---	Fiat Abarth 1000 Berlina
			Grand Touring	1000 cc	G. Galmozzi	---	Fiat Abarth 1000 OTS
			Sport	1000 cc	'White'	---	Fiat Abarth 1000 SP
			Sport	2000 cc	F. Pilone	---	Fiat Abarth 2000
5 June	Catanzaro, Italy	Nicastro-Acquavona (hillclimb-14,450 kms)	Touring	600 cc	G. D'Ippolito	11'2"8	Fiat Abarth 595
			Touring	700 cc	G. Crescenti	11'00"0	Fiat Abarth 695
			Touring	850 cc	A. Vitali	10'13"2	Fiat Abarth 850 TC
			Touring	1000 cc	V. Varese	10'01"4	Fiat Abarth 1000 Berlina
			Grand Touring	1000 cc	A. Torricelli	10'22"1	Fiat Abarth 1000 OTS
			Sport	1000 cc	M. Terminiello	9'48"0	Fiat Abarth 1000 SP
			Sport	1300 cc	M. Di Gioia	11'04"8	Abarth Simca 1300
			Sport Prototypes	1000 cc	P. Gargano	9'23"8	Fiat Abarth 1000 SP
			Sport Prototypes	2000 cc	D. Scola	8'32"0	Fiat Abarth 2000
8 June	Bassano del Grappa, Italy	Bassano-Montegrappa (hillclimb-11,750 kms)	Touring	600 cc	R. Amighini	9'28"5	Fiat Abarth 595
			Touring	700 cc	G. Gatta	9'27"5	Fiat Abarth 695
			Touring	850 cc	C. Margoni	9'06"2	Fiat Abarth 850 TC
			Touring	1000 cc	D. Gargan	8'29"8	Fiat Abarth 1000 Berlina
			Grand Touring	1000 cc	G. Galmozzi	9'30"9	Fiat Abarth 1000 OTS
			Sport	1000 cc	M. Nardari	8'33"2	Fiat Abarth 1000 SP
8 June	Berchtesgaden, Germany	Rossfeld Alpen Bergpreis (hillclimb-5,890 kms)	Touring	850 cc	K.H. Trettau	---	Fiat Abarth 595
			Touring	1000 cc	J. Abt	---	Fiat Abarth 1000 Berlina
			Sport	1300 cc	H. Wengat	---	Fiat Abarth OT 1300
15 June	Vallelunga, Italy	Coppa Gallenga (circuit-3,120 kms)	Touring	600 cc	R. De Sanctis	1h16'06"8	Fiat Abarth 595
			Touring	700 cc	L. Finelli	1h14'05"3	Fiat Abarth 695
			Touring	850 cc	'Gibi'	1h11'08"3	Fiat Abarth 850 TC
			Touring	1000 cc	F. Acasbo	1h13'01"3	Fiat Abarth 1000 Berlina
15 June	Aosta, Italy	Aosta-Pila (hillclimb-11,200 kms)	Touring	600 cc	E. Bettini	9'04"8	Fiat Abarth 595
			Touring	850 cc	P. Tali	7'52"5	Fiat Abarth 850 TC
			Touring	1000 cc	S. Pellegrino	7'54"4	Fiat Abarth 1000 Berlina
			Grand Touring	1000 cc	C. Spica	8'30"3	Fiat Abarth 1000 OTS
			Sport	1000 cc	P. Monticone	8'04"8	Fiat Abarth 1000 SP
			Sport	over 1300 cc	A. Fabroni	8'25"0	Fiat Abarth 1600
			Sport	2000 cc	J. Ortner	7'26"0	Fiat Abarth 2000
15 June	S. Benedetto del Tronto, Italy	S. Benedetto-Acquaviva (hillclimb-6,500 kms)	Touring	600 cc	'Handi-Cap'	5'16"0	Fiat Abarth 595
			Touring	850 cc	S. Sani	5'01"6	Fiat Abarth 850 TC
			Touring	1000 cc	A. Guaiano	5'00"2	Fiat Abarth 1000 Berlina
			Grand Touring	1000 cc	P. Brambilla	4'57"8	Fiat Abarth 1000 OTS
			Sport	1300 cc	'Onid'	4'54"0	Abarth Simca 1300
15 June	Reggio Calabria, Italy	S. Stefano-Gambarie (hillclimb-7,830 kms)	Touring	850 cc	Laganà	5'51"8	Fiat Abarth 850 TC
			Touring	up to 1150 cc	Cavarra	5'50"7	Fiat Abarth 1000 Berlina
			Sport	1000 cc	P. Zoccoli	6'28"4	Fiat Abarth 1000 SP
			Sport	1300 cc	A. Gambero	5'24"1	Fiat Abarth OT 1300
			Sport-Prototypes	1000 cc	Paladino	6'29"5	Fiat Abarth 1000 SP
			Sport-Prototypes	2000 cc	'Bitter'	4'46"6	Fiat Abarth 2000
22 June	Palermo, Italy	Corsa al Monte Pellegrino (hillclimb-8,750 kms)	Touring	600 cc	C. Aloe	6'15"7	Fiat Abarth 595
			Touring	700 cc	G. Musumeci	6'07"4	Fiat Abarth 695
			Touring	850 cc	G. Spadafora	5'42"1	Fiat Abarth 850 TC
			Touring	1000 cc	E. Fichera	5'42"5	Fiat Abarth 1000 Berlina
			Sport	1000 cc	V. Virgilio	5'06"2	Fiat Abarth 1000 SP
			Sport-Prototypes	1000 cc	P. Gargano	5'30"3	Fiat Abarth 1000 SP
22 June	Bologna, Italy	Bologna-Passo della Raticosa (hillclimb-32,700 kms)	Touring	600 cc	P. Poggi	23'05"4	Fiat Abarth 595
			Touring	700 cc	A. Mola	23'05"3	Fiat Abarth 695
			Touring	850 cc	'Gibi'	21'31"4	Fiat Abarth 850 TC
			Touring	1000 cc	A. Stefanelli	21'51"1	Fiat Abarth 1000 OTS
			Sport	1000 cc	E. Buzzetti	21'19"2	Fiat Abarth 1000 SP
			Sport	1300 cc	A. Fabroni	21'32"5	Fiat Abarth OT 1300
			Sport	2000 cc	E. Tondelli	19'05"9	Fiat Abarth 2000
29 June	Cuneo, Italy	Garessio-San Bernardo (hillclimb-6,200 kms)	Touring	600 cc	C. Tabacco	5'04"5	Fiat Abarth 595
			Touring	700 cc	A. Mola	4'55"2	Fiat Abarth 695
			Touring	850 cc	G. Carena	4'38"7	Fiat Abarth 850 TC
			Touring	1000 cc	P. Giachino	4'34"7	Fiat Abarth 1000 Berlina
			Grand Touring	1000 cc	G. Galmozzi	4'46"0	Fiat Abarth 1000 OTS
			Sport	1000 cc	P. Monticone	4'02"7	Fiat Abarth 1000 SP
			Sport	1300 cc	U. Cinto	4'20"7	Fiat Abarth OT 1300
			Sport	2000 cc	L. Taramazzo	3'47"2	Fiat Abarth 2000
			Sport-Prototypes	1000 cc	R. Origlia	4'30"1	Fiat Abarth 1000 SP

DATE	LOCATION	RACE	CATEGORY	CLASS	WINNER	TIME OR DISTANCE	CAR
29 June	Cosenza, Italy	Coppa della Sila (hillclimb-14,000 kms)	Touring	600 cc	'Frank Mc Boden'	11'45"3	Fiat Abarth 595
			Touring	700 cc	L. Cestra	11'28"1	Fiat Abarth 695
			Touring	850 cc	M. Terminiello	10'55"4	Fiat Abarth 850 TC
			Touring	1000 cc	V. Varese	10'46"4	Fiat Abarth 1000 Berlina
			Grand Touring	1000 cc	A. Torricelli	11'24"2	Fiat Abarth 1000 OTS
			Sport	1000 cc	V. Virgilio	11'05"1	Fiat Abarth 1000 SP
			Sport-Prototypes	1000 cc	G. Ricciardi	10'31"4	Fiat Abarth 1000 SP
			Sport-Prototypes	2000 cc	D. Scola	8'53"2	Fiat Abarth 2000
6 July	Adenau, Germany	Nürburgring 6 Stunden (circuit-22,810 kms)	Touring	1000 cc	'Pam'-Uberti	kms 752,730	Fiat Abarth 1000 Berlina
6 July	Bolzano, Italy	Bolzano-Mendola (hillclimb-14,400 kms)	Touring	1000 cc	K. Wendlinger	10'39"9	Fiat Abarth 1000 Berlina
			Sport	1000 cc	W. Donà	11'35"0	Fiat Abarth 1000 SP
			Sport	2000 cc	A. Merzario	10'10"5	Fiat Abarth 2000
			Sport-Prototypes	1000 cc	M. Baldo	12'02"7	Fiat Abarth 1000 SP
			Sport-Prototypes	2000 cc	J. Ortner	10'23"5	Fiat Abarth 2000
13 July	Trento, Italy	Trento-Bondone (hillclimb-17,300 kms)	Touring	850 cc	C. Margoni	14'08"31	Fiat Abarth 850 TC
			Touring	1000 cc	P. Lado	13'59"1	Fiat Abarth 1000 Berlina
			Grand Touring	1000 cc	C. Spica	15'29"0	Fiat Abarth 1000 OTS
			Sport	1000 cc	'White'	13'02"0	Fiat Abarth 1000 SP
			Sport	1300 cc	K. Federhofer	12'50"4	Fiat Abarth OT 1300
			Sport	2000 cc	A. Merzario	11'33"6	Fiat Abarth 2000
			Sport-Prototypes	1000 cc	V. Locatelli	12'35"7	Fiat Abarth 1000 SP
13 July	Pistoia, Italy	Coppa della Collina (hillclimb-14,700 kms)	Touring	700 cc	R. Mazzeschi	10'12"3	Fiat Abarth 695
			Touring	850 cc	Bresci	9'42"7	Fiat Abarth 850 TC
			Grand Touring	1000 cc	P. Brambilla	10'01"2	Fiat Abarth 1000 OTS
			Sport	1300 cc	P. Falorni	9'02"5	Fiat Abarth OT 1300
			Sport	over 1300 cc	'Bitter'	7'39"9	Fiat Abarth 2000
			Sport-Prototypes	2000 cc	D. Scola	7'45"2	Fiat Abarth 2000
13 July	Messina, Italy	Premio Messina (hillclimb-3,400 kms)	Touring	1000 cc	A. Bonaccorsi	30'01"4	Fiat Abarth 1000 Berlina
13 July	Gives, Belgium	Course de côte de Gives (hillclimb-2,400 kms)	Sport	2000 cc	E. Tondelli	1'17"8	Fiat Abarth 2000
20 July	Scarperia, Italy	Gran Premio del Mugello (circuit-66,200 kms x 8 laps)	Sport	1300 cc	Buzzetti-Donà	4h33'33"6	Fiat Abarth 1000 SP
			Sport	2000 cc	Ortner-Van Lennep	4h18'54"7	Fiat Abarth 2000
			Sport-Prototypes	2000 cc	A. Merzario	4h15'55"8	Fiat Abarth 2000
20 July	Macerata, Italy	Sarnano-Sassotetto (hillclimb-12,400 kms)	Touring	600 cc	P.G. Poggi	9'20"6	Fiat Abarth 595
			Touring	700 cc	G. Gatta	9'10"8	Fiat Abarth 695
			Touring	850 cc	E. Pasolini	8'22"4	Fiat Abarth 850 TC
			Sport	1300 cc	U. Curto	8'17"2	Fiat Abarth OT 1300
			Sport-Prototypes	1000 cc	G. Ricciardi	8'05"8	Fiat Abarth 1000 SP
			Sport-Prototypes	2000 cc	D. Scola	6'55"1	Fiat Abarth 2000
20 July	Ragusa, Italy	Coppa dei Monti Iblei (hillclimb-5,200 kms)	Touring	600 cc	A. Fina	4'28"1	Fiat Abarth 595
			Touring	700 cc	G. Musumeci	4'17"8	Fiat Abarth 695
			Touring	850 cc	A. Gambero	4'11"5	Fiat Abarth 850 TC
			Touring	1000 cc	A. Cernigliaro	4'06"1	Fiat Abarth 1000 Berlina
			Sport	1300 cc	A. Gambero	3'54"9	Fiat Abarth OT 1300
			Sport-Prototypes	2000 cc	M. Spadafora	3'36"7	Fiat Abarth 2000
27 July	Freiburg, Germany	ADAC Bergpreis Freiburg-Schauinsland (hillclimb-11,500 kms)	Touring	850 cc	K.H. Trettau	15'00"7	Fiat Abarth 850 TC
			Touring	1000 cc	J. Abt	14'21"4	Fiat Abarth 1000 Berlina
			Sport	1300 cc	K. Federhofer	13'11"2	Fiat Abarth OT 1300
			Sport	2000 cc	A. Merzario	11'28"0	Fiat Abarth 2000
27 July	Sansepolcro, Italy	Coppa S. Giustino-Trabaria (hillclimb-11,900 kms)	Touring	600 cc	R. Mazzeschi	9'17"7	Fiat Abarth 595
			Touring	700 cc	R. Scoccia	9'25"5	Fiat Abarth 695
			Touring	850 cc	S. Scalabrini	8'59"7	Fiat Abarth 850 TC
			Touring	1000 cc	P. Falorni	8'30"8	Fiat Abarth 1000 SP
			Sport	1300 cc	'Onid'	9'08"3	Abarth Simca 1300
			Sport	over 1300 cc	D. Scola	6'58"1	Fiat Abarth 2000
3 August	Sestrière, Italy	Cesana-Sestrière (hillclimb-10,400 kms)	Touring	600 cc	P.G. Poggi	7'12"7	Fiat Abarth 595
			Touring	700 cc	A. Mola	7'05"8	Fiat Abarth 695
			Touring	850 cc	M. Barone	6'36"3	Fiat Abarth 850 TC
			Touring	1000 cc	A. Soria	6'35"9	Fiat Abarth 1000 Berlina
			Grand Touring	1000 cc	C. Spica	7'00"9	Fiat Abarth 1000 OTS
			Sport	1300 cc	L. Moreschi	6'12"9	Fiat Abarth OT 1300
			Sport	2000 cc	A. Merzario	5'01"0	Fiat Abarth 2000
			Sport-Prototypes	1000 cc	M. Zanetti	5'48"6	Fiat Abarth 1000 SP
3 August	Viterbo, Italy	Corsa lago di Montefiascone (hillclimb-5,000 kms)	Touring	600 cc	R. De Sanctis	3'06"7	Fiat Abarth 595
			Touring	700 cc	A. Fracassini	3'21"2	Fiat Abarth 695
			Touring	850 cc	L. Russo	3'23"4	Fiat Abarth 850 TC
			Grand Touring	1000 cc	P. Brambilla	3'36"3	Fiat Abarth 1000 OTS
			Sport	up to 1300 cc	F. Palumbo	3'18"6	Fiat Abarth 1000 SP

DATE	LOCATION	RACE	CATEGORY	CLASS	WINNER	TIME OR DISTANCE	CAR
10 August	Rieti, Italy	Coppa Carotti (hillclimb-15,100 kms)	Touring	700 cc	G. Del Panta	10'09"5	Fiat Abarth 695
			Touring	850 cc	L. Russo	9'09"6	Fiat Abarth 850 TC
			Grand Touring	1000 cc	P. Brambilla	10'03"4	Fiat Abarth 1000 OTS
			Sport	1000 cc	E. Buzzetti	8'09"5	Fiat Abarth 1000 SP
			Sport	1300 cc	A. Fabroni	9'02"6	Fiat Abarth OT 1300
			Sport	2000 cc	F. Pilone	7'12"1	Fiat Abarth 2000
			Sport-Prototypes	1000 cc	G. Fasano	8'55"0	Fiat Abarth 1000 SP
			Sport-Prototypes	2000 cc	D. Scola	7'15"0	Fiat Abarth 2000
10 August	Trapani, Italy	Trapani-Monte Erice (hillclimb-6,670 kms)	Touring	600 cc	'Frank Mc Boden'	5'31"5	Fiat Abarth 595
			Touring	700 cc	G. Musumeci	5'33"4	Fiat Abarth 695
			Touring	850 cc	G. Spatafora	5'05"6	Fiat Abarth 850 TC
			Touring	1000 cc	A. Cernigliaro	5'01"6	Fiat Abarth 1000 Berlina
			Sport	1000 cc	V. Vincenzo	4'34"7	Fiat Abarth 1000 SP
			Sport	1300 cc	G. Garufi	5'18"1	Fiat Abarth OT 1300
			Sport	over 1300 cc	'Bitter'	4'11"9	Fiat Abarth 2000
			Sport-Prototypes	1000 cc	G. Barba	4'42"0	Fiat Abarth 1000 SP

1970

DATE	LOCATION	RACE	CATEGORY	CLASS	WINNER	TIME OR DISTANCE	CAR
15 February	Barcelona, Spagna	Rallye 600 (441 kms)	Sport	over 1600 cc	M. Juncosa	---	Fiat Abarth 2000
15 March	Monza, Italy	4 Ore del Jolly Club (circuit-10,000 kms)	Touring	600 cc	Magri-Gatta	kms 505,445	Fiat Abarth 595
			Touring	850 cc	F. Tandoj	kms 542,987	Fiat Abarth 850 TC
			Touring	1000 cc	Leso-Faraoni	kms 571,410	Fiat Abarth 1000 Berlina
15 March	Bellevue, Belgium	Course de côte de Fleron (hillclimb-1,810 kms)	Sport-Prototypes	over 1000 cc	E. Tondelli	1'08"0	Fiat Abarth 2000
22 March	Draguignan, France	Course de côte d'Ampus (hillclimb-6,800 kms)	Sport-Prototypes	up to 1000 cc	G. Brancato	5'27"3	Fiat Abarth 1000
			Sport-Prototypes	up to 2000 cc	M. Casoni	3'50"2	Fiat Abarth 2000
22 March	Avola, Italy	Coppa Belmonte (hillclimb-10,200 kms)	Touring	850 cc	'Jimmy'	7'06"8	Fiat Abarth 850 TC
			Touring	1000 cc	'Bernardo'	6'52"2	Fiat Abarth 1000 Berlina
			Sport	1000 cc	G. Sapienza	7'55"9	Fiat Abarth 1000 BA
			Sport	up to 1300 cc	S. Figuera	7'38"2	Abarth Simca 1300
			Sport	2000 cc	M. Spadafora	5'55"7	Fiat Abarth 2000
			Sport-Prototypes	1000 cc	V. Virgilio	6'34"2	Fiat Abarth 1000 SP
22 March	Kautenbach, Luxembourg	Course de côte de Kautenbach (hillclimb-2,400 kms)	Touring	up to 700 cc	E. Barbanti	2'37"2	Fiat Abarth 695
22 March	Zotzenbach, Germany	Bergpreis Zotzenbach (hillclimb-3,790 kms x 2)	Touring	850 cc	K. Weiss	5'02"0	Fiat Abarth 850 TC
			Touring	1000 cc	R. Schmidt	5'08"5	Fiat Abarth 1000 Berlina
30 March	Barisart, Belgium	Course de côte de Spa-Barisart (hillclimb-3,200 kms)	Touring	850 cc	Greco	2'23"5	Fiat Abarth 850 TC
			Sport-Prototypes	over 1000 cc	F. Neri	1'53"5	Fiat Abarth 3000
30 March	Zandvoort, Holland	Internationale Paasraces (circuit-4,225 kms x 15 laps)	Touring	850 cc	H. Kiviet	34'18"4	Fiat Abarth 850 TC
			Touring	1000 cc	D. van Staalduissen	32'04"1	Fiat Abarth 1000 Berlina
5 April	Monza, Italy	Coppa A. C. Genova (circuit-2,385 kms x 15 laps)	Touring	up to 600 cc	A. Corazza	20'08"6	Fiat Abarth 595
			Touring	1000 cc	'Pal Joe'	16'59"1	Fiat Abarth 1000 Berlina
5 April	Krähberg, Germany	Internationales Krähberg Rennen (hillclimb-3,600 kms x 2)	Touring	850 cc	K. Weiss	5'01"3	Fiat Abarth 850 TC
			Sport	1300 cc	K. Federhofer	4'54"4	Fiat Abarth OT 1300
12 April	Salzburg, Austria	Austria Trophäe (circuit-4,238 kms x 75 laps)	Touring	1000 cc	J. Abt	2h01'53"5	Fiat Abarth 1000 Berlina
12 April	Monza, Italy	Coppa A. C. Verona (circuit-5,750 kms x 18 laps)	Touring	1000 cc	D. Gargan	42'43"6	Fiat Abarth 1000 Berlina
			Grand Touring (27 laps)	1000 cc	G. Galmuzzi	55'10"7	Fiat Abarth 1000 OTS
			Sport	1000 cc	G. Varese	53'57"1	Fiat Abarth 1000 SP
12 April	Eifel, Germany	ADAC Bergpreis Bad Neunahr (hillclimb-3,100 kms x 2)	Touring	850 cc	K. Weiss	4'26"3	Fiat Abarth 650 TC
			Touring	1000 cc	K. Trettan	4'09"3	Fiat Abarth 1000 Berlina
			Sport	up to 1300 cc	K. Federhofer	3'52"7	Fiat Abarth OT 1300
18 April	Payerne, Switzerland	Autoslalom de Payerne	Touring	up to 1000 cc	A. Benzinger	1'45"6	Fiat Abarth 1000 Berlina
19 April	Presberg, Germany	Taunus Bergpreis (hillclimb-3,300 kms x 2)	Touring	850 cc	G. Weber	2'20"5	Fiat Abarth 850 TC
			Touring	1000 cc	K. Trettan	2'07"2	Fiat Abarth 1000 Berlina
19 April	Bruckberg, Germany	Bruckberg Bergrennen (hillclimb-3,000 kms x 2)	Touring	850 cc	H. Monath	2'37"4	Fiat Abarth 850 TC
			Touring	1000 cc	J. Abt	2'18"4	Fiat Abarth 1000 Berlina
			Sport	1300 cc	J. Federhofer	2'07"4	Fiat Abarth OT 1300
			Sport	up to 2000 cc	P. Wiechert	2'07"1	Fiat Abarth 2000
19 April	Oberhallau, Switzerland	Bergrennen in Oberhallau (hillclimb-3,000 kms)	Touring	1000 cc	F. Benzinger	1'53"5	Fiat Abarth 1000 Berlina
			Sport-Prototypes	1000 cc	R. Blatter	1'54"2	Fiat Abarth 1000 SP
19 April	Adenau, Germany	Nürburgring 300 Km Rennen (circuit-22,810 kms)	Touring	850 cc	R. Kreuzer	2h33'14"6	Fiat Abarth 850 TC
19 April	Toulon, France	Trophée International Paul Ricard (circuit-3,263 kms x 100 laps)	Sport	up to 2000 cc	M. Casoni	2h20'23"5	Fiat Abarth 2000
25 April	Monza, Italy	1000 Km di Monza (circuit-5,750 kms x 174 laps)	Sport	up to 1600 cc	Vaglio-Vaglio	4h19'21"6	Fiat Abarth 1000
26 April	Bockholtz, Luxembourg	Course de côte Goesdorf-Ardennes (hillclimb-2,800 kms)	Sport-Prototypes	over 1600 cc	W. Scheeren	1'33"9	Fiat Abarth 2000
26 April	Detmold, Germany	Bergrennen Teutoburger Wald (hillclimb-2,250 kms x 2)	Touring	850 cc	A. Joswig	2'32"55	Fiat Abarth 850 TC

DATE	LOCATION	RACE	CATEGORY	CLASS	WINNER	TIME OR DISTANCE	CAR
26 April	Eberbach, Germany	Bergrennen Eberbach (hillclimb-3,000 kms x 2)	Touring	850 cc	K. Weiss	2'59"6	Fiat Abarth 850 TC
			Touring	1000 cc	A. Benzinger	2'50"8	Fiat Abarth 1000 Berlina
			Sport	up to 2000 cc	P. Wiechert	2'25"1	Fiat Abarth 2000
26 April	Riedenburg, Germany	Bergrennen Riedenburg (hillclimb-2,200 kms x 2)	Touring	850 cc	H. Monath	3'26"1	Fiat Abarth 850 TC
			Touring	1000 cc	J. Abt	3'00"9	Fiat Abarth 1000 Berlina
			Sport	1300 cc	K. Federhofer	2'57"6	Fiat Abarth OT 1300
26 April	Amalfi, Italy	Coppa Primavera (hillclimb-8,750 kms)	Touring	600 cc	P. Vena	7'30"1	Fiat Abarth 595
			Touring	700 cc	M. De Luca	7'01"5	Fiat Abarth 695
			Touring	850 cc	G. D'Ippolito	6'13"9	Fiat Abarth 850 TC
			Touring	1000 cc	V. Varese	6'41"1	Fiat Abarth 1000 Berlina
			Sport	up to 1300 cc	G. Barrella	7'02"9	Fiat Abarth OT 1300
			Sport	over 1300 cc	"Riccardone" (overall)	5'41"0	Fiat Abarth 2000
			Sport-Prototypes	1000 cc	N. Amoroso	7'55"8	Fiat Abarth 1000
26 April	Cagliari, Italy	Coppa Mario Costa (hillclimb-10,250 kms)	Touring	700 cc	S. Tornatore	8'20"1	Fiat Abarth 695
			Touring	850 cc	E. Fichera	7'25"9	Fiat Abarth 850 TC
			Touring	up to 1150 cc	G. Bonaccorso	7'38"4	Fiat Abarth 1000 Berlina
			Sport	up to 1300 cc	M. Codeluppi	7'35"6	Fiat Abarth 1000 SP
			Sport	up to 2000 cc	M. Spatafora	6'35"6	Fiat Abarth 2000
26 April	Fribourg, Switzerland	Course de côte de St. Antoni-Obermouten (hillclimb-1,400 kms x 2)	Touring	850 cc	R. Piccard	2'02"1	Fiat Abarth 850 TC
2 May	Wagen, Switzerland	Autoslalom Wagen (5,000 kms x 2)	Touring	1000 cc	F. Spicher	8'40"2	Fiat Abarth 1000 Berlina
3 May	Aspern, Austria	Preis von Wien (circuit-2,600 kms x 38 laps)	Touring	1000 cc	K. Wendlinger	---	Fiat Abarth 1000 Berlina
3 May	Kaufbeuren, Germany	Flugplatzrennen in Kaufbeuren (circuit-4,500 kms x 12 laps)	Touring	850 cc	K. Weiss	29'25"1	Fiat Abarth 850 TC
			Touring	1000 cc	J. Abt	28'29"8	Fiat Abarth 1000 Berlina
			Sport	up to 1300 cc	R. Walter	27'19"2	Fiat Abarth OT 1300
3 May	Palermo, Italy	Targa Florio (circuit-72,000 kms x 11 laps)	Sport	1300 cc	Garufi-'Black and White'	6h44'58"8	Fiat Abarth OT 1300
			Sport	2000 cc	'Pam'-'Gibi'	6h48'16"3	Fiat Abarth 2000
			Sport-Prototypes	1000 cc	Ruspa-Pellegrin	6h43'40"8	Fiat Abarth 1000
3 May	Arezzo, Italy	Pieve S. Stefano-Passo dello Spino (hillclimb-12,500 kms)	Touring	600 cc	L. Galluzzo	8'05"9	Fiat Abarth 595
			Touring	700 cc	P. Barberini	8'20"1	Fiat Abarth 695
			Touring	850 cc	M. Barone	7'51"1	Fiat Abarth 850 TC
			Touring	1000 cc	R. Becchetti	7'39"7	Fiat Abarth 1000 Berlina
			Sport	up to 1300 cc	P. Falorni	7'34"2	Fiat Abarth 1000 BA
			Sport	2000 cc	"Riccardone"	6'16"5	Fiat Abarth 2000
3 May	Pocono, USA	Pocono International Raceway (circuit-25 laps)	---	---	A. Cosentino	---	Fiat Abarth 1000 Berlina
7 May	Holtz, Luxembourg	Course de côte de Luxembourg à Holtz (hillclimb)	Touring	1000 cc	A. Benzinger	1'39"7	Fiat Abarth 1000 Berlina
			Prototypes	over 1600 cc	W. Scheeren	1'28"5	Fiat Abarth 2000
9-10 May	Fassberg, Germany	Rundstreekenrennen Fassberg (circuit-4,500 kms x 12 laps)	Touring	850 cc	K. Weiss	27'19"6	Fiat Abarth 850 TC
			Sport	1000 cc	R. Zwolsmann	24'52"8	Fiat Abarth 1000
			Sport-Prototypes (40 laps)	2000 cc	E. Swart	26'56"6	Fiat Abarth 2000
9-10 May	Zandvoort, Holland	ADAC 2 Stunden Rennen (circuit-4,225 kms x 24 laps)	Touring	1000 cc	E. Stegemann	---	Fiat Abarth 1000 Berlina
			Grand Touring	1000 cc	R. Schmidt	---	Fiat Abarth 1000 OTS
10 May	Budapest, Hungary	Grand Prix de Budapest (circuit-5,300 kms x 2 hours)	Touring	1000 cc	J. Abt	---	Fiat Abarth 1000 Berlina
10 May	Monza, Italy	Coppa Piemonte (circuit-2,405 kms x 20 laps)	Touring	850 cc	A. Tarasco	29'04"4	Fiat Abarth 850 TC
			Touring	up to 1150 cc	A. Goria	28'12"1	Fiat Abarth 1000 Berlina
			Sport	1000 cc	G. Varese	25'32"2	Fiat Abarth 1000
			Sport	2000 cc	'Pogo'	24'48"1	Fiat Abarth 2000
10 May	Eggberg, Germany	ADAC Bergrennen (hillclimb-3,800 kms)	Touring	850 cc	G. Weber	3'12"2	Fiat Abarth 850 TC
			Touring	1000 cc	K. Trettan	2'53"5	Fiat Abarth 1000 Berlina
			Sport	up to 1600 cc	K. Federhofer	3'03"4	Fiat Abarth OT 1300
			Sport	over 1600 cc	P. Wiechert	2'57"4	Fiat Abarth 2000
10 May	Montelimar, France	Course de côte de Colombier (hillclimb-3,300 kms)	Sport	up to 1600 cc	'Cyprien'	2'59"0	Fiat Abarth 1600
			Sport-Prototypes	2000 cc	A. Finkel	2'35"0	Fiat Abarth 2000
16 May	Cumberland, USA	National Sedan Races (circuit-1,6 miles)	---	---	A. Cosentino	---	Fiat Abarth 1000 Berlina
17 May	Passau, Germany	Nibelungen Bergrennen (hillclimb-3,600 kms x 2)	Touring	850 cc	G. Weber	4'26"1	Fiat Abarth 850 TC
			Touring	1000 cc	A. Benzinger	4'00"4	Fiat Abarth 1000 Berlina
			Sport	up to 1300 cc	K. Federhofer	3'45"6	Fiat Abarth OT 1300
17 May	San Marino	6 Ore del Titano (3,500 kms x 6 laps)	---	up to 700 cc	A. Zannoni	---	Fiat Abarth 695
17 May	Volterra, Italy	Coppa Città di Volterra (hillclimb-10,150 kms)	Touring	850 cc	S. Scalabrini	6'30"8	Fiat Abarth 850 TC
			Sport	1000 cc	P. Falorni	6'20"8	Fiat Abarth 1000
			Sport	up to 1300 cc	L. Moreschi	6'04"2	Fiat Abarth OT 1300
			Sport	2000 cc	M. Casoni	5'10"3	Fiat Abarth 2000
			Sport-Prototypes	up to 1000 cc	'Riccardone'	5'36"8	Fiat Abarth 1000 SP
			Sport-Prototypes	up to 2000 cc	J. Ortner	5'06"4	Fiat Abarth 2000
17 May	Zolder, Belgium	Rhein Ruhr Pokalrennen (circuit-4,184 kms x 25 laps)	Touring	850 cc	K. Weiss	51'54"0	Fiat Abarth 850 TC
			Touring	1000 cc	E. Casteel	49'18"4	Fiat Abarth 1000 Berlina

DATE	LOCATION	RACE	CATEGORY	CLASS	WINNER	TIME OR DISTANCE	CAR
24 May	Wasgau, Germany	Wasgau Bergpreis (hillclimb-5,100 kms x 2)	Touring	850 cc	K. Weiss	5'46"8	Fiat Abarth 850 TC
			Touring	1000 cc	K. Trettan	5'25"7	Fiat Abarth 1000 Berlina
			Sport	up to 1300 cc	K. Federhofer	5'19"8	Fiat Abarth OT 1300
			Sport	over 1600 cc	P. Wiechert	5'09"0	Fiat Abarth 2000
24 May	Montseny, Spagna	Carrera en cuesta al Montseny (hillclimb-16,300 kms)	Touring	up to 1150 cc	'Jorge Chi'	13'04"5	Fiat Abarth 850 TC
			Sport	up to 1300 cc	M. Juncosa	10'39"8	Fiat Abarth OT 1300
			Sport	2000 cc	M. Casoni	9'30"3	Fiat Abarth 2000
24 May	Salzburg, Austria	Tauernpokal (circuit-4,238 kms x 80 laps)	Sport	2000 cc	J. Williams	1h45'18"0	Fiat Abarth 2000
24 May	Cagliari, Italy	Monte Acuto-Campuomu (hillclimb-3,930 kms)	Touring	700 cc	G. Barbarossa	10'07"1	Fiat Abarth 695
			Touring	850 cc	E. Fichera	9'26"7	Fiat Abarth 850 TC
			Touring	1000 cc	C. Bonaccorso	9'33"5	Fiat Abarth 1000 Berlina
			Sport	1000 cc	G. Fadda	9'23"9	Fiat Abarth 1000
			Sport	up to 1300 cc	A. Bellavia	10'29"5	Abarth Simca 1300
			Sport	2000 cc	G. Piras	9'30"3	Fiat Abarth 2000
24 May	Trieste, Italy	Trieste-Opicina (hillclimb-10,150 kms)	Touring	700 cc	C. Garrone	4'54"3	Fiat Abarth 695
			Touring	850 cc	S. Scalabrini	4'26"0	Fiat Abarth 850 TC
			Touring	1000 cc	M. Campanini	4'02"3	Fiat Abarth 1000 Berlina
			Grand Touring	1000 cc	G. Galmozzi	4'33"4	Fiat Abarth 1000 OTS
			Sport	1000 cc	'White'	3'53"6	Fiat Abarth 1000
			Sport	1300 cc	L. Moreschi	4'00"9	Fiat Abarth OT 1300
			Sport	2000 cc	'Gibi'	3'25"4	Fiat Abarth 2000
			Sport-Prototypes	2000 cc	P. Botalla	3'25"5	Fiat Abarth 2000
28 May	Udine, Italy	Tolmezzo-Verzegnis (hillclimb-6,200 kms)	Touring	600 cc	A. Pieraccini	4'52"3	Fiat Abarth 595
			Touring	700 cc	G. Lise	4'57"7	Fiat Abarth 695
			Touring	850 cc	G. Olper	4'49"9	Fiat Abarth 850 TC
			Touring	1000 cc	F. Stefanelli	4'22"1	Fiat Abarth 1000 Berlina
			Grand Touring	1000 cc	G. Ronchi	4'57"0	Fiat Abarth 1000 OTS
			Sport	1000 cc	M. Nardari	3'58"2	Fiat Abarth 1000
			Sport	1300 cc	L. De Nicola	4'22"5	Abarth Simca 1300
			Sport	2000 cc	White	3'35"1	Fiat Abarth 2000
			Sport-Prototypes	2000 cc	A. Zadra	3'29"0	Fiat Abarth 2000
31 May	Hohegeiss, Germany	PMSC Bergpreis (hillclimb-3,550 kms x 2)	Touring	850 cc	K. Weiss	4'59"2	Fiat Abarth 850 TC
31 May	Livorno, Italy	Ardenza-Gabbro (hillclimb-7,500 kms)	Touring	700 cc	E. Bogani	4'45"5	Fiat Abarth 695
			Touring	850 cc	'Tito'	4'22"5	Fiat Abarth 850 TC
			Touring	1000 cc	A. Stefanelli	4'16"2	Fiat Abarth 1000 Berlina
			Sport	1300 cc	P. Falorni	4'06"8	Fiat Abarth 1000
			Sport	over 1300 cc	'Riccardone'	3'38"3	Fiat Abarth 2000
31 May	Autun, France	Course de côte d'Autun Montjeu (hillclimb-4,100 kms x 2)	Sport-Prototypes	up to 2000 cc	Belin	2'06"9	Fiat Abarth 2000
31 May	Charnizay, France	Course de côte de Charnizay (hillclimb-1,900 kms x 2)	Sport-Prototypes	up to 2000 cc	A. Finkel	0'56"7	Fiat Abarth 2000
31 May	Parma, Italy	Castione Baratti-Neviano Arduini (hillclimb-5,500 kms)	Touring	600 cc	F. Piroli	3'57"6	Fiat Abarth 595
			Touring	700 cc	A. Audisio	3'55"7	Fiat Abarth 695
			Touring	850 cc	M. Barone	3'45"1	Fiat Abarth 850 TC
			Sport-Prototypes	1000 cc	J. Ortner	3'08"5	Fiat Abarth 1000
			Sport-Prototypes	up to 2000 cc	M. Casoni	2'59"1	Fiat Abarth 2000
31 May	Sondrio, Italy	Bormio-Stelvio (hillclimb-5,000 kms)	Touring	1000 cc	R. Trevisan	3'44"0	Fiat Abarth 1000 Berlina
			Grand Touring	1000 cc	P. Brambilla	3'56"8	Fiat Abarth 1000 OTS
			Sport	1000 cc	A. Bonini	4'03"0	Fiat Abarth 1000
			Sport	up to 1300 cc	F. Marenco	3'35"7	Fiat Abarth OT 1300
31 May	Macerata, Italy	Sarnano-Sassotetto (hillclimb-12,400 kms)	Touring	700 cc	G. Del Panta	9'29"3	Fiat Abarth 695
			Touring	850 cc	G. D'Ippolito	8'48"4	Fiat Abarth 850 TC
			Touring	1000 cc	E. Gatti	8'20"1	Fiat Abarth 1000 Berlina
			Sport	1000 cc	A. Bersano	8'47"4	Fiat Abarth 1000
			Sport	up to 1300 cc	L. Moreschi	8'11"2	Fiat Abarth OT 1300
			Sport	2000 cc	'Gibi'	7'06"5	Fiat Abarth 2000
			Sport-Prototypes	1000 cc	P. Pellegrini	8'08"8	Fiat Abarth 1000
7 June	Berchtesgaden, Germany	Rossfeld Alpen Bergpreis (hillclimb-5,890 kms)	Touring	850 cc	K. Weiss	7'55"9	Fiat Abarth 850 TC
			Touring	1000 cc	J. Abt	7'09"0	Fiat Abarth 1000 Berlina
			Sport	up to 1300 cc	K. Federhofer	6'50"4	Fiat Abarth OT 1300
7 June	Anderstorp, Sweden	Race Meeting at Scandinavian Raceway (circuit-4,018 kms x 80 laps)	Sport	2000 cc	J. Ortner	2h12'02"8	Fiat Abarth 2000
14 June	Hockenheim, Germany	Rhein Pokal Rennen (circuit-6,768 kms x 15 laps)	Touring	850 cc	K. Weiss	46'38"0	Fiat Abarth 850 TC
			Touring	1000 cc	E. Casteel	46'04"8	Fiat Abarth 1000 Berlina
21 June	Reggio Calabria, Italy	S. Stefano-Gambarie (hillclimb-7,830 kms)	Touring	600 cc	S. Sturniolo	7'01"2	Fiat Abarth 595
			Touring	700 cc	A. Testa	6'55"3	Fiat Abarth 695
			Touring	850 cc	M. De Furia	6'19"2	Fiat Abarth 850 TC
			Sport	1000 cc	R. Caratozzolo	6'18"6	Fiat Abarth 1000
			Sport-Prototypes	up to 2000 cc	D. Scola	5'06"7	Fiat Abarth 2000

DATE	LOCATION	RACE	CATEGORY	CLASS	WINNER	TIME OR DISTANCE	CAR
21 June	Avignon, France	Course de côte du Mont Ventoux (hillclimb-21,600 kms)	Sport	1300 cc	'Cyprien'	14'40"8	Fiat Abarth OT 1300
			Sport	2000 cc	J. Ortner	12'00"8	Fiat Abarth 2000
			Sport-Prototypes	3000 cc	J. Ortner (overall)	10'04"8	Fiat Abarth 3000
21 June	Mainz Finthen, Germany	HMSC Flugplatz Rennen (circuit-2,100 kms x 30 laps)	Touring	850 cc	K. Weiss	35'41"5	Fiat Abarth 850 TC
			Touring	1000 cc	D. Machtius	34'09"7	Fiat Abarth 1000 Berlina
			Sport (50 laps)	2000 cc	H. Gillen	45'48"8	Fiat Abarth 2000
21 June	Pistoia, Italy	Coppa della Collina (hillclimb-5,200 kms)	Touring	700 cc	G. Del Panta	3'51"6	Fiat Abarth 695
			Touring	850 cc	Scalabrini	3'43"5	Fiat Abarth 850 TC
			Sport	1000 cc	P. Falorni	6'54"0	Fiat Abarth 1000
			Sport	1300 cc	L. Moreschi	6'37"2	Fiat Abarth OT 1300
			Sport-Prototypes	up to 1000 cc	M. Nesti	6'21"3	Fiat Abarth 1000
			Sport-Prototypes	up to 2000 cc	'Riccardone'	5'52"5	Fiat Abarth 2000
28 June	Cosenza, Italy	Coppa della Sila (hillclimb-15,100 kms)	Touring	600 cc	R. Mazzeschi	11'20"0	Fiat Abarth 595
			Touring	700 cc	S. Famoso	11'15"1	Fiat Abarth 695
			Touring	850 cc	G. D'Ippolito	10'51"0	Fiat Abarth 850 TC
			Touring	1000 cc	M. Campanini	10'27"1	Fiat Abarth 1000 Berlina
			Grand Touring	1000 cc	Di Benedetto	11'32"4	Fiat Abarth 1000 OTS
			Sport	1000 cc	'White'	10'00"0	Fiat Abarth 1000
			Sport	1300 cc	L. Moreschi	9'52"1	Fiat Abarth OT 1300
			Sport-Prototypes	up to 1000 cc	V. Bernasconi	10'14"3	Fiat Abarth 1000
			Sport-Prototypes	up to 2000 cc	D. Scola	8'44"1	Fiat Abarth 2000
29 June	Bolzano, Italy	Bolzano-Mendola (hillclimb-14,400 kms)	Touring	1000 cc	F. Stefanelli	10'45"9	Fiat Abarth 1000 Berlina
			Sport	1000 cc	M. Nardari	10'12"9	Fiat Abarth 1000
			Sport	1300 cc	R. Terigi	10'43"0	Fiat Abarth OT 1300
			Sport-Prototypes	up to 1000 cc	M. Nesti	9'19"5	Fiat Abarth 1000
			Sport-Prototypes	up to 2000 cc	'Pam'	8'26"9	Fiat Abarth 2000
29 June	Cuneo, Italy	Garessio-Colle San Bernardo (hillclimb-6,200 kms)	Touring	600 cc	E. Bettini	5'01"9	Fiat Abarth 595
			Touring	700 cc	A. Audisio	4'58"3	Fiat Abarth 695
			Touring	850 cc	A. Tarasco	4'45"0	Fiat Abarth 850 TC
			Touring	1000 cc	C. Bonaccorso	4'24"1	Fiat Abarth 1000 Berlina
			Grand Touring	1000 cc	P. Brambilla	4'55"8	Fiat Abarth 1000 OTS
			Sport	1000 cc	G. Varese	3'57"8	Fiat Abarth 1000
			Sport	1300 cc	F. Marenco	4'26"8	Fiat Abarth OT 1300
			Sport	over 1300 cc	F. Pilone	3'40"1	Fiat Abarth 2000
5 July	Zandvoort, Holland	Coupes Benelux (circuit-4,225 kms)	Touring	1000 cc	R. Dijkstra	22'42"6	Fiat Abarth 1000 Berlina
			Sport	1300 cc	R. Zwolsman	21'36"1	Fiat Abarth 1000
5 July	Ascoli Piceno, Italy	S. Benedetto del Tronto-Acquaviva (hillclimb-6,500 kms)	Touring	600 cc	V. Gattafoni	4'54"4	Fiat Abarth 595
			Touring	700 cc	A. Fantoni	5'44"6	Fiat Abarth 695
			Touring	1000 cc	M. Cevasco	4'38"2	Fiat Abarth 1000 Berlina
			Grand Touring	1000 cc	S. Fortini	4'48"2	Fiat Abarth 1000 OTS
			Sport	over 1600 cc	P. Laureati	3'35"0	Fiat Abarth 2000
5 July	Vuillafans, France	Course de côte Vuillafans-Echevannes (hillclimb-4,800 kms x 2)	Touring	1000 cc	A. Benzinger	2'52"5	Fiat Abarth 1000 Berlina
5 July	Trento, Italy	Trento-Bondone (hillclimb-17,300 kms)	Touring	850 cc	C. Margoni	14'26"4	Fiat Abarth 850 TC
			Touring	1000 cc	A. Soria	13'51"8	Fiat Abarth 1000
			Grand Touring	1000 cc	P. Friso	15'44"1	Fiat Abarth 1000 OTS
			Sport	1000 cc	A. Bersano	13'31"1	Fiat Abarth 1000
			Sport	1300 cc	K. Federhofer	12'55"5	Fiat Abarth OT 1300
			Sport	2000 cc	F. Pilone	11'48"8	Fiat Abarth 2000
			Sport-Prototypes	up to 1000 cc	M. Nesti	12'35"3	Fiat Abarth 1000
			Sport-Prototypes	up to 2000 cc	A. Zadra	11'43"5	Fiat Abarth 2000
12 July	Adenau, Germany	Nürburgring 6 Stunden Rennen (circuit-22,810 kms x 6 hours)	Touring	850 cc	R. Krenzer	kms 663,741	Fiat Abarth 850 TC
			Touring	1000 cc	J. Abt	kms 776,312	Fiat Abarth 1000 Berlina
12 July	Sestrière, Italy	Cesana-Sestrière (hillclimb-10,400 kms)	Touring	600 cc	R. Mazzeschi	7'15"5	Fiat Abarth 595
			Touring	700 cc	S. Famoso	7'06"9	Fiat Abarth 695
			Touring	850 cc	A. Gunnella	6'41"0	Fiat Abarth 850 TC
			Touring	1000 cc	M. Campanini	6'11"7	Fiat Abarth 1000 Berlina
			Grand Touring	1000 cc	G. Galmozzi	6'58"1	Fiat Abarth 1000 OTS
			Sport	1000 cc	G. Varese	5'46"0	Fiat Abarth 1000
			Sport	1300 cc	L. Moreschi	6'00"0	Fiat Abarth OT 1300
			Sport	2000 cc	J. Ortner	5'07"4	Fiat Abarth 2000
			Sport-Prototypes	up to 1000 cc	M. Nesti	5'45"3	Fiat Abarth 1000
			Sport-Prototypes	up to 2000 cc	A. Merzario	5'03"6	Fiat Abarth 2000
12 July	Trapani, Italy	Trapani-Monte Erice (hillclimb-6,670 kms)	Touring	600 cc	A. Fina	5'30"0	Fiat Abarth 595
			Touring	700 cc	Frank Mc Boden	5'27"5	Fiat Abarth 695
			Touring	850 cc	'Bernardo'	5'09"5	Fiat Abarth 850 TC
			Touring	1000 cc	C. Bonaccorso	5'02"5	Fiat Abarth 1000 Berlina
			Grand Touring	1000 cc	G. La Barbera	5'33"1	Fiat Abarth 1000 OTS
			Sport	up to 1300 cc	R. Caratozzolo	4'49"5	Fiat Abarth 1000
			Sport-Prototypes	up to 2000 cc	'Amphicar'	4'17"2	Fiat Abarth 2000

DATE	LOCATION	RACE	CATEGORY	CLASS	WINNER	TIME OR DISTANCE	CAR
19 July	Scarperia, Italy	Gran Premio del Mugello (66,200 kms x 5 laps)	Sport	up to 1300 cc	'White'	2h24'13"5	Fiat Abarth 1000
			Sport	2000 cc	'Pogo'	2h38'13"6	Fiat Abarth 2000
			Sport-Prototypes	up to 1000 cc	M. Nesti	2h49'07"7	Fiat Abarth 1000 SP
			Sport-Prototypes	up to 2000 cc	A. Merzario	2h30'28"8	Fiat Abarth 2000
19 July	Diepholz, Germany	Flugplatzrennen Diepholz (circuit-2,600 kms x 20 laps)	Touring	850 cc	D. Schulz	28'25"4	Fiat Abarth 850 TC
			Touring	1000 cc	D. Machatius	26'44"6	Fiat Abarth 1000 Berlina
			Sport	1300 cc	W. Buls	---	Fiat Abarth OT 1300
19 July	Ulm, Germany	Flugplatzrennen Ulm Laupheim (circuit-4,620 kms)	Touring	850 cc	K. Weiss	29'51"9	Fiat Abarth 850 TC
			Touring	1000 cc	A. Benzinger	28'32"4	Fiat Abarth 1000 Berlina
25-26 July	Francorchamps, Belgium	24 Heures de Francorchamps (circuit-14,120 kms x 24 hours)	Touring	1000 cc	Abt-Grano	kms 3505,313	Fiat Abarth 1000 Berlina
26 July	Freiburg, Germany	ADAC Bergpreis Freiburg-Schauinsland (hillclimb-11,200 kms)	Touring	up to 1150 cc	R. Neumann	14'50"1	Fiat Abarth 1000 Berlina
			Sport	1300 cc	K. Federhofer	13'13"4	Fiat Abarth OT 1300
			Sport	2000 cc	J. Ortner	11'38"9	Fiat Abarth 2000
26 July	Ascoli Piceno, Italy	Ascoli-Colle San Marco (hillclimb-10,200 kms)	Touring	700 cc	S. Famoso	7'13"6	Fiat Abarth 695
			Touring	850 cc	M. Barone	6'40"8	Fiat Abarth 850 TC
			Touring	1000 cc	M. Campanini	6'15"0	Fiat Abarth 1000 Berlina
			Sport	1000 cc	'White'	6'12"9	Fiat Abarth 1000
			Sport	1300 cc	L. Moreschi	6'10"5	Fiat Abarth OT 1300
			Sport	2000 cc	'Amphicar'	5'27"8	Fiat Abarth 2000
			Sport-Prototypes	up to 1000 cc	M. Nesti	5'54"9	Fiat Abarth 1000
			Sport-Prototypes	up to 2000 cc	'Riccardone'	5'20"1	Fiat Abarth 2000
9 August	Rieti, Italy	Rieti-Terminillo (hillclimb-15,100 kms)	Touring	700 cc	S. Famoso	9'55"9	Fiat Abarth 695
			Touring	850 cc	A. Gunnella	9'12"2	Fiat Abarth 850 TC
			Sport	1000 cc	'White'	11'59"0	Fiat Abarth 1000 Berlina
			Sport	2000 cc	P. Laureati	11'57"6	Fiat Abarth 2000
			Sport-Prototypes	up to 1000 cc	P. Pellegrini	12'17"2	Fiat Abarth 1000
9 August	Mont-Dore, France	Course de côte du Mont-Dore (hillclimb-5,000 kms)	Sport	2000 cc	J. Ortner	6'14"4	Fiat Abarth 2000
14 August	Watkins Glen (N.Y.), USA	Watkins Touring Trophy (circuit-2,300 miles x 19 laps)	---	---	C. Fisher	31'05"6	Fiat Abarth 1000 Berlina
15 August	Pescara, Italy	Cronoscalata Svolte di Popoli (hillclimb-8,000 kms)	Touring	600 cc	V. Gattafoni	5'48"2	Fiat Abarth 595
			Touring	700 cc	D. Ministrini	5'52"0	Fiat Abarth 695
			Touring	850 cc	A. Soria	5'00"5	Fiat Abarth 1000 Berlina
			Sport	2000 cc	E. Lualdi	4'13"0	Fiat Abarth 2000
			Sport-Prototypes	up to 1000 cc	V. Bernasconi	4'45"9	Fiat Abarth 1000
			Sport-Prototypes	up to 2000 cc	'Riccardone'	4'08"5	Fiat Abarth 2000
23 August	Jyllands, Danimarca	Hella Grand Prix (circuit-1,475 kms x 15 laps x 3)	Sport	2000 cc	E. Swart	22'57"3	Fiat Abarth 2000
6 September	Monza, Italy	Coppa Intereuropa (circuit-5,750 kms x 40 laps)	Grand Touring	1000 cc	G. Galmozzi	1h17'00"3	Fiat Abarth 1000 OTS
6 September	Adenau, Germany	Internationales Nürburgring 500 Km Rennen (circuit-22,810 kms x 22 laps)	Sport-Prototypes	up to 2000 cc	A. Merzario	3h10'05"9	Fiat Abarth 2000
20 September	Bergamo, Italy	Clusone-Passo Presolana (hillclimb-6,100 kms)	Touring	700 cc	C. Garrone	3'58"9	Fiat Abarth 695
			Touring	850 cc	A. Tarasco	3'50"2	Fiat Abarth 850 TC
			Touring	1000 cc	A. Soria	3'29"9	Fiat Abarth 1000 Berlina
			Grand Touring	1000 cc	C. Galmozzi	3'51"7	Fiat Abarth 1000 OTS
			Sport	1000 cc	G. Varese	3'09"7	Fiat Abarth 1000
			Sport	1300 cc	'John John'	3'32"1	Fiat Abarth OT 1300
			Sport	2000 cc	L. Moreschi	2'52"8	Fiat Abarth 2000
			Sport-Prototypes	3000 cc	A. Merzario	2'39"8	Fiat Abarth 3000
27 September	Villach, Austria	Grosser Bergpreis von Österreich (hillclimb-6,393 kms x 2)	Touring	1000 cc	K. Wendlingler	7'24"7	Fiat Abarth 1000 Berlina
			Sport	1300 cc	K. Federhofer	7'03"5	Fiat Abarth OT 1300
			Sport	2000 cc	J. Ortner	6'02"8	Fiat Abarth 2000
4 October	Siracusa, Italy	Val d'Anapo-Sortino (hillclimb-6,000 kms)	Touring	600 cc	'Otello'	5'08"6	Fiat Abarth 595
			Touring	700 cc	G. Musumeci	5'00"8	Fiat Abarth 695
			Touring	850 cc	'Jymmy'	4'47"8	Fiat Abarth 850 TC
			Touring	1000 cc	C. Bonaccorso	4'36"6	Fiat Abarth 1000 Berlina
			Grand Touring	1000 cc	A. Rubino	5'15"9	Fiat Abarth 1000 OTS
			Sport	1000 cc	R. Caratozzolo	4'36"5	Fiat Abarth 1000
			Sport	1300 cc	'Frank Mc Boden'	4'35"7	Fiat Abarth OT 1300
			Sport	2000 cc	D. Scola	4'07"0	Fiat Abarth 2000
4 October	Auerber, Germany	Auerbergrennen (hillclimb-3,100 kms x 2)	Touring	850 cc	G. Weber	4'13"1	Fiat Abarth 850 TC
			Touring	1000 cc	K. Wendlinger	3'45"7	Fiat Abarth 1000 Berlina
			Sport	1300 cc	K. Brand	3'41"8	Fiat Abarth OT 1300
			Sport	2000 cc	H. Wengert	3'29"8	Fiat Abarth 2000
11 October	Steibis, Germany	Bergpreis Steibis (hillclimb-2,000 kms x 2)	Touring	850 cc	K. Weiss	2'56"8	Fiat Abarth 850 TC
			Touring	1000 cc	J. Abt	2'42"9	Fiat Abarth 1000 Berlina
			Sport	1300 cc	K. Brand	2'36"7	Fiat Abarth OT 1300
			Sport	2000 cc	J. Wengert	2'24"7	Fiat Abarth 2000
25 October	München, Germany	Flugplatzrennen München-Neubiberg (circuit-5,750 kms x 10 laps)	Touring	1000 cc	J. Abt	26'49"3	Fiat Abarth 1000 Berlina
			Sport	up to 1300 cc	G. Serblin	27'12"5	Fiat Abarth 1000

DATE	LOCATION	RACE	CATEGORY	CLASS	WINNER	TIME OR DISTANCE	CAR
1971							
6-7 February	Barcelona, Spagna	Rallye 600 (383 kms)	Sport	over 1600 cc	M. Juncosa	---	Fiat Abarth 2000
7 March	Madrid, Spagna	Copa Alfil en el Jarama (circuit-3,404 kms x 21 laps)	Touring	up to 1000 cc	'Targa I'	47'59"9	Fiat Abarth 850 TC
14 March	Monza, Italy	Trofeo Mario Angiolini (circuit-5,750 kms x 12 laps)	Touring	700 cc	G. Giudici	30'17"8	Fiat Abarth 695
			Touring	850 cc	A. Gunnella	30'30"5	Fiat Abarth 850 TC
			Touring	1000 cc	G. Ravinale	30'13"1	Fiat Abarth 1000 Berlina
14 March	Bellevue, Belgium	Course de côte de Fleron (hillclimb-2,200 kms x 2)	Touring	850 cc	M. Greco	1'37"7	Fiat Abarth 850 TC
			Sport-Prototypes	over 1000 cc	F. Neri	1'18"1	Fiat Abarth 3000
21 March	Liège, Belgium	Course de côte du Condroz (hillclimb-2,500 kms x 2)	Touring	850 cc	M. Greco	1'51"3	Fiat Abarth 850 TC
21 March	Madrid, Spagna	Trofeo de Primavera en el Jarama (circuit-3,404 kms x 22 laps)	Touring	up to 1000 cc	'Targa I'	48'50"1	Fiat Abarth 850 TC
27-28 March	Zolder, Belgium	ADAC Bergischer Löwe Streckenrennen (circuit-4,184 kms x 24 laps)	Touring	850 cc	K. Schmidt	51'20"0	Fiat Abarth 850 TC
			Touring	1000 cc	P. Ernst	47'34"5	Fiat Abarth 1000 Berlina
28 March	Zotzenbach, Germany	Bergpreis Zotzenbach (hillclimb-3,790 kms x 2)	Touring	850 cc	G. Weber	5'03"9	Fiat Abarth 850 TC
			Sport	up to 2000 cc	P. Stürtz	4'36"4	Fiat Abarth OT 1300
28 March	Draguignan, France	Course de côte d'Ampus (hillclimb-6,800 kms)	Sport-Prototypes	2000 cc	L. Taramazzo	3'53"1	Fiat Abarth 2000
			Sport-Prototypes	up to 3000 cc	F. Pilone	3'41"8	Fiat Abarth 3000
4 April	Roquette, France	Course de côte de la Roquette (hillclimb-6,100 kms)	Sport-Prototypes	2000 cc	L. Taramazzo	4'31"6	Fiat Abarth 2000
4 April	Montjuic, Spagna	Carrera en cuesta al Montjuic (hillclimb-3,400 kms x 2)	Sport-Prototypes	over 1600 cc	M. Juncosa	2'10"5	Fiat Abarth 2000
4 April	Kassel, Germany	Autoslalom Kassel (1,500 kms x 2)	Touring	1000 cc	B. Ziegler	---	Fiat Abarth 1000 Berlina
11 April	Murs, France	Course de côte du Col de Murs (hillclimb-5,950 kms)	Sport-Prototypes	up to 1000 cc	G. Boeris	4'38"6	Fiat Abarth 1000
			Sport-Prototypes	2000 cc	E. Ostini	4'21"0	Fiat Abarth 2000
12 April	Barisart, Belgium	Course de côte de Spa-Barisart (hillclimb-3,200 kms)	Touring	850 cc	M. Greco	2'14"8	Fiat Abarth 850 TC
			Sport-Prototypes	over 1300 cc	W. Scheeren	1'39"8	Fiat Abarth 2000
12 April	Adenau, Germany	Nürburgring 300 kms Rennen (circuit-7,747 kms x 13 laps)	Touring	1000 cc	H. Hessel	43'58"4	Fiat Abarth 1000 Berlina
17 April	Monthoux, Switzerland	Rundrennen in Monthoux (circuit-1,000 kms x 25 laps)	Touring	1000 cc	C. Waser	20'37"1	Fiat Abarth 1000
18 April	Liège, Belgium	Course de côte de la Roche Samree (hillclimb-8,000 kms)	Touring	850 cc	M. Greco	4'10"7	Fiat Abarth 850 TC
			Sport-Prototypes	over 1300 cc	F. Neri	2'58"6	Fiat Abarth 3000
18 April	Presberg, Germany	Taunus Bergpreis (hillclimb-3,300 kms x 2)	Touring	850 cc	G. Weber	2'15"8	Fiat Abarth 850 TC
			Touring	1000 cc	P. Ernst	2'01"7	Fiat Abarth 1000 Berlina
			Sport	up to 1300 cc	P. Stürtz	1'58"6	Fiat Abarth OT 1300
24 April	Villach, Austria	Grosser Bergpreis von Österreich (hillclimb-6,392 kms x 2)	Sport	over 2000 cc	J. Ortner	5'56"3	Fiat Abarth 3000
25 April	Sembach, Germany	Flugplatzrennen Sembach (circuit-5,610 kms x 10 laps)	Touring	850 cc	R. Rach	---	Fiat Abarth 850 TC
			Touring	1000 cc	R. Caprez	---	Fiat Abarth 1000 Berlina
			Sport	2000 cc	D. Machatius	---	Fiat Abarth 2000
25 April	Krähberg, Germany	Internationales Krähberg Rennen (hillclimb-3,600 kms x 2)	Touring	850 cc	G. Weber	5'11"2	Fiat Abarth 850 TC
			Touring	1000 cc	P. Ernst	4'56"3	Fiat Abarth 1000 Berlina
25 April	Gap, France	Course de côte du Col Bayard (hillclimb-5,000 kms)	Sport-Prototypes	up to 3000 cc	P. Maublanc	3'08"8	Fiat Abarth 3000
25 April	Collias, France	Course de côte de Nîmes, Collias (hillclimb-4,100 kms)	Sport-Prototypes	2000 cc	J. Danmet	2'54"2	Fiat Abarth 2000
25 April	Heyd, Belgium	Course de côte du Val d'Aisne	Touring	850 cc	M. Greco	2'54"0	Fiat Abarth 850 TC
25 April	Monza, Italy	1000 Km di Monza (circuit-5,750 kms x 174 laps)	Sport	2000 cc	'Pal Joe'-Becchetti	4h15'42"1	Fiat Abarth 2000
1 May	Mannheim, Germany	Autoslalom von Mannheim (1,200 kms x 2)	Touring	1000 cc	L. Dammers	---	Fiat Abarth 1000 Berlina
2 May	Monza, Italy	Coppa A.C. Verona (circuit-5,750 kms x 18 laps)	Touring	850 cc	E. Gatti	44'06"8	Fiat Abarth 850 TC
			Touring	1000 cc	A. Stefanelli	42'31"0	Fiat Abarth 1000 Berlina
			Grand Touring	1000 cc	C. Spica	51'27"5	Fiat Abarth 1000 OTS
2 May	Heilbronn, Germany	ADAC Heilbronner Bergpreis (hillclimb-6,030 kms x 2)	Touring	1000 cc	G. Volle	---	Fiat Abarth 1000 Berlina
			Sport	up to 1300 cc	E. Ell	---	Fiat Abarth OT 1300
2 May	Amalfi, Italy	Coppa Primavera (hillclimb-8,750 kms)	Touring	600 cc	M. De Luca	7'29"4	Fiat Abarth 595
			Touring	700 cc	M. De Luca	7'08"2	Fiat Abarth 695
			Touring	850 cc	L. Russo	6'39"3	Fiat Abarth 850 TC
			Touring	1000 cc	C. Bonaccorso	6'32"2	Fiat Abarth 1000 Berlina
			Grand Touring	1000 cc	A. Torricelli	6'56"3	Fiat Abarth 1000 OTS
			Sport	1300 cc	R. Caratozzolo	6'22"7	Fiat Abarth OT 1300
			Sport	2000 cc	'Amphicar'	5'33"3	Fiat Abarth 2000
2 May	Passau, Germany	Nibelungen Bergrennen (hillclimb-3,600 kms x 2)	Touring	850 cc	G. Weber	5'06"0	Fiat Abarth 850 TC
			Touring	1000 cc	H. Schenzl	4'56"1	Fiat Abarth 1000 Berlina
			Sport	1300 cc	G. Lehmann	4'37"6	Fiat Abarth OT 1300
			Sport-Prototypes	up to 1150 cc	J. Abt	4'34"9	Fiat Abarth 1000
2 May	Jura, France	Course de côte des Monts du Jura (hillclimb)	Sport-Prototypes	1000 cc	'Skip'	2'11"3	Fiat Abarth 1000
			Sport-Prototypes	up to 3000 cc	P. Maublanc	2'08"2	Fiat Abarth 3000

DATE	LOCATION	RACE	CATEGORY	CLASS	WINNER	TIME OR DISTANCE	CAR
2 May	Arezzo, Italy	Pieve S. Stefano-Passo dello Spino (hillclimb-12,500 kms)	Touring	600 cc	R. Mazzeschi	8'31"0	Fiat Abarth 595
			Touring	700 cc	M. Magri	8'20"9	Fiat Abarth 695
			Touring	850 cc	'Ragastas'	8'06"7	Fiat Abarth 850 TC
			Touring	1000 cc	G. Giovannelli	8'32"7	Fiat Abarth 1000 Berlina
			Grand Touring	1000 cc	F. Indignozzi	8'31"9	Fiat Abarth 1000 OTS
			Sport	1000 cc	V. Venturi	6'49"9	Fiat Abarth 1000
			Sport	1300 cc	R. Terigi	7'33"2	Fiat Abarth OT 1300
			Sport-Prototypes	2000 cc	J. Ortner	6'00"5	Fiat Abarth 2000
2 May	Pforzheim, Germany	Slalom den Goldstadtpokal Pforzheim (1,100 kms x 2)	Touring	1000 cc	L. Dammers	2'13"9	Fiat Abarth 1000
8 May	Wallberg, Germany	Wallberg Rennen (hillclimb-3,100 kms x 2)	Touring	850 cc	G. Weber	4'33"6	Fiat Abarth 850 TC
			Touring	1000 cc	P. Ernst	4'08"2	Fiat Abarth 1000 Berlina
			Sport	1300 cc	P. Stürtz	3'52"5	Fiat Abarth OT 1300
			Sport	2000 cc	H. Wengert	3'37"8	Fiat Abarth 2000
			Sport-Prototypes	1000 cc	J. Abt	3'52"8	Fiat Abarth 1000
9 May	Monza, Italy	Coppa Piemonte e Valle d'Aosta (circuit-2,405 kms x 15 laps)	Touring	700 cc	G. Boggiatto	2é'11"4	Fiat Abarth 695
			Grand Touring	1000 cc	A. Mercadante	16'51"3	Fiat Abarth 1000 OTS
			Sport-Prototypes	2000 cc	F. Pilone	14'23"9	Fiat Abarth 2000
9 May	Duren, Germany	Durener Autoslalom (1,300 kms x 2)	Touring	1000 cc	L. Dammers	2'36"3	Fiat Abarth 1000 Berlina
15 May	Freiburg, Germany	ADAC Bergpreis Freiburg-Schauinsland (hillclimb-5,600 kms x 2)	Touring	1000 cc	G. Volla	6'55"7	Fiat Abarth 1000 Berlina
16 May	Bruxelles, Belgium	Course de côte de Alles sur Samois (hillclimb-4,200 kms)	Touring	850 cc	M. Greco	2'36"1	Fiat Abarth 850 TC
			Sport-Prototypes	over 1300 cc	F. Neri	1'58"1	Fiat Abarth 3000
16 May	Obermässing, Germany	Eurohill Bergrennen (hillclimb-3,080 kms x 2)	Sport	1300 cc	S. Siebenbürger	3'20"2	Fiat Abarth OT 1300
16 May	Piacenza, Italy	Castell'Arquato-Vernasca (hillclimb-9,775 kms)	Touring	700 cc	A. Ruffoli	6'21"8	Fiat Abarth 695
			Touring	850 cc	G.F. Del Bono	5'58"1	Fiat Abarth 850 TC
			Sport	1000 cc	F. Marenco	5'30"6	Fiat Abarth 1000
			Sport	1300 cc	V. Motta	5'57"9	Fiat Abarth OT 1300
16 May	Vicenza, Italy	Bassano-Monte Grappa (hillclimb-11,750 kms)	Touring	600 cc	'Tocai'	9'50"0	Fiat Abarth 595
			Touring	700 cc	F. Muratori	10'11"7	Fiat Abarth 695
			Touring	850 cc	'Tino'	9'16"7	Fiat Abarth 850 TC
			Touring	1000 cc	D. Gargan	8'30"8	Fiat Abarth 1000 Berlina
			Grand Touring	1000 cc	G. Ronchi	9'06"0	Fiat Abarth 1000 OTS
			Sport	1000 cc	G. Rossi	7'57"4	Fiat Abarth 1000
			Sport	up to 1600 cc	C. Zaldini	9'25"8	Fiat Abarth OT 1300
16 May	Sainte Anne, France	Course de côte de Sainte Anne (hillclimb-5,800 kms)	Sport-Prototypes	2000 cc	J. Porcier	3'28"2	Fiat Abarth 2000
16 May	Palermo, Italy	Targa Florio (circuit-72,000 kms x 11 laps)	Sport-Prototypes	1000 cc	Barone-Campanini	6h44'49"3	Fiat Abarth 1000 SP
16 May	Kaufbeuren, Germany	Flugplatzrennen in Kaufbeuren (circuit-4,500 kms x 12 laps)	Touring	850 cc	F. Lampe	29'03"3	Fiat Abarth 850 TC
			Touring	1000 cc	K. Trettan	26'54"4	Fiat Abarth 1000 Berlina
			Sport	1300 cc	K. Brand	23'55"1	Fiat Abarth OT 1300
			Sport-Prototypes	up to 1150 cc	J. Abt	24'26"2	Fiat Abarth 1000
16 May	Alp, Austria	Alp Bergrennen (hillclimb-6,500 kms)	Touring	1000 cc	H. Schenzel	4'02"2	Fiat Abarth 1000 Berlina
			Sport-Prototypes	2000 cc	J. Ortner	2'55"7	Fiat Abarth 2000
16 May	Harz, Germany	ADAC Harz Bergpreis (hillclimb-3,550 kms x 2)	Touring	850 cc	G. Weber	4'39"9	Fiat Abarth 850 TC
16 May	Abreschviller, France	Course de côte d'Abreschviller (hillclimb-2,100 kms)	Sport-Prototypes	up to 3000 cc	P. Maublanc	1'00"9	Fiat Abarth 3000
20 May	Neuchatel en Bray, France	Course de côte de Neuchatel en Bray (hillclimb-2,200 kms)	Sport	up to 1300 cc	J. Lemerle	1'11"9	Fiat Abarth 1000 SP
20 May	Aix-en-Provence, France	Course de côte d'Aix-Saint Antonin (hillclimb-2,700 kms)	Sport-Prototypes	2000 cc	'Cyprien'	1'57"0	Fiat Abarth 2000
23 May	Sassari, Italy	Alghero-Scala Picada (hillclimb-8,700 kms)	Touring	600 cc	I. Gasola	5'40"7	Fiat Abarth 595
			Touring	850 cc	C. Ferretti	5'16"3	Fiat Abarth 850 TC
			Sport	1000 cc	A. Tola	4'51"7	Fiat Abarth 1000
			Sport	2000 cc	'Amphicar'	3'49"9	Fiat Abarth 2000
23 May	Trieste, Italy	Trieste-Opicina (hillclimb-10,150 kms)	Touring	600 cc	J. Strenghetto	5'41"0	Fiat Abarth 595
			Touring	700 cc	M. Ruoso	5'50"6	Fiat Abarth 695
			Touring	850 cc	U. Smittarello	5'16"6	Fiat Abarth 850 TC
			Touring	1000 cc	L. Pellanda	5'06"8	Fiat Abarth 1000 Berlina
			Grand Touring	1000 cc	G. Galmozzi	5'10"8	Fiat Abarth 1000 OTS
			Sport	1000 cc	L. Rassega	4'43"1	Fiat Abarth 1000
			Sport	1300 cc	G. Savoldi	5'04"6	Fiat Abarth OT 1300
23 May	Parma, Italy	Castione Baratti-Neviano Arduini (hillclimb-5,500 kms)	Touring	850 cc	G. Del Bono	3'44"50	Fiat Abarth 850 TC
			Touring	1000 cc	M. Campanini	3'36"3	Fiat Abarth 1000 Berlina
			Sport	1300 cc	C. Turizio	3'32"8	Fiat Abarth OT 1300
23 May	Wasgau, Germany	Wasgau Bergpreis (hillclimb-5,100 kms x 2)	Touring	1000 cc	B. Pfeiffer	5'31"9	Fiat Abarth 1000 Berlina
			Sport	1300 cc	P. Stürtz	5'30"7	Fiat Abarth OT 1300
			Sport	over 1600 cc	K. Federhofer	6'00"7	Fiat Abarth 2000
			Sport-Prototypes	1300 cc	J. Fuchs	5'53"4	Fiat Abarth 1300
23 May	Hessen, Germany	Autoslalom Hessen (1,400 kms x 2)	Touring	1000 cc	B. Ziegler	---	Fiat Abarth 1000 Berlina

DATE	LOCATION	RACE	CATEGORY	CLASS	WINNER	TIME OR DISTANCE	CAR
23 May	Bremgarten, Germany	Flugplatzrennen Bremgarten (circuit-3,700 kms x 12 laps)	Touring	850 cc	K. Schneiter	30'19"0	Fiat Abarth 850 TC
			Touring	1000 cc	R. Caprez	23'41"5	Fiat Abarth 1000 Berlina
23 May	North Carolina, USA	3 Hour Piedmont (circuit-1,75 miles)	---	---	A. Cosentino	---	Fiat Abarth 1000 Berlina
23 May	Wittlich, Germany	Wittlicher Bergpreis (hillclimb-3,720 kms x 2)	Touring	850 cc	Streuber	5'51"0	Fiat Abarth 850 TC
23 May	Zandvoort, Holland	ADAC Junioren Trophy Zandvoort (circuit-4,225 kms x 25 laps)	Touring	850 cc	F. Lampe	---	Fiat Abarth 850 TC
23 May	Digne, France	Course de côte de la Lure (hillclimb-12,700 kms)	Sport-Prototypes	2000 cc	J. Porcier	7'24"9	Fiat Abarth 2000
			Sport-Prototypes	3000 cc	P. Maublanc	6'57"5	Fiat Abarth 3000
30 May	Neuhausen, Germany	Flugplatzrennen Neuhausen (circuit-x 17 laps)	Touring	850 cc	K. Schmidt	25'33"6	Fiat Abarth 850 TC
30 May	Mont Revard, France	Course de côte du Mont Revard (hillclimb)	Sport-Prototypes	up to 3000 cc	P. Maublanc	5'16"5	Fiat Abarth 3000
30 May	Pistoia, Italy	Coppa della Collina (hillclimb-5,200 kms)	Touring	850 cc	'Tito'	3'43"7	Fiat Abarth 850 TC
			Touring	1000 cc	M. Banchieri	3'46"2	Fiat Abarth 1000 Berlina
			Sport	1000 cc	E. Tinghi	6'27"5	Fiat Abarth 1000
			Sport	up to 1600 cc	R. Terigi	6'47"4	Fiat Abarth OT 1300
			Sport-Prototypes	2000 cc	P. Laureati	6'08"2	Fiat Abarth 2000
30 May	Treviso, Italy	Vittorio Veneto-Cansiglio (hillclimb-10,800 kms)	Touring	850 cc	U. Smittarello	8'22"3	Fiat Abarth 850 TC
			Touring	1000 cc	D. Gargan	7'43"8	Fiat Abarth 1000 Berlina
			Grand Touring	1000 cc	G. Galmozzi	8'39"5	Fiat Abarth 1000 OTS
			Sport-Prototypes	2000 cc	A. Merzario (overall)	7'00"5	Fiat Abarth 2000
30 May	Salzburg, Austria	Silberschildrennen Salzburgring (circuit-4,238 kms x 15 laps)	Touring	1000 cc	H. Schenzel	29'30"8	Fiat Abarth 1000 Berlina
			Sport	1300 cc	S. Siebenbürger	25'10"2	Fiat Abarth OT 1300
30 May	Sondrio, Italy	Bormio-Stelvio (hillclimb-6,000 kms)	Touring	600 cc	P. Poggi	4'47"6	Fiat Abarth 595
			Touring	up to 850 cc	G. Boggiatto	5'16"8	Fiat Abarth 695
			Touring	1000 cc	P. Gasparini	4'53"7	Fiat Abarth 1000 Berlina
			Grand Touring	1000 cc	P. Brambilla	5'25"4	Fiat Abarth 1000 OTS
			Sport	1000 cc	A. La Masson	4'43"2	Fiat Abarth 1000
			Sport	1300 cc	B. Rebai	4'13"5	Abarth Simca 1300
			Sport-Prototypes	1000 cc	F. Marenco	4'43"3	Fiat Abarth 1000
30 May	Zolder, Belgium	Rhein Ruhr Pokalrennen (circuit-4,184 kms x 25 laps)	Touring	1000 cc	E. Casteel	---	Fiat Abarth 1000 Berlina
2 June	Vallelunga, Italy	Premio della Repubblica (circuit-3,200 kms x 40')	Sport	1000 cc	E. Tinghi	kms 80,000	Fiat Abarth 1000
5 June	Dübendorf, Switzerland	Automobilslalom Wangen (5,000 kms)	Touring	1000 cc	R. Caprez	4'03"2	Fiat Abarth 1000 Berlina
6 June	Autun, France	Course de côte d'Autun Montjeu (hillclimb-4,100 kms x 2)	Sport-Prototypes	up to 3000 cc	P. Maublanc	1'53"4	Fiat Abarth 3000
6 June	Cosenza, Italy	Coppa della Sila (hillclimb-15,100 kms)	Touring	600 cc	E. Santucci	11'34"3	Fiat Abarth 595
			Touring	700 cc	M. Casciaro	11'15"0	Fiat Abarth 695
			Touring	850 cc	G. D'Ippolito	10'45"3	Fiat Abarth 850 TC
			Touring	1000 cc	L. Ferragina	10'18"2	Fiat Abarth 1000 Berlina
			Grand Touring	1000 cc	A. Torricelli	11'09"2	Fiat Abarth 1000 OTS
			Sport	1000 cc	R. Caratozzolo	10'10"4	Fiat Abarth 1000
			Sport	1300 cc	'Frank Mc Boden'	10'50"4	Fiat Abarth OT 1300
			Prototypes	up to 3000 cc	D. Scola	8'51"0	Fiat Abarth 3000
6 June	Berchtesgaden, Germany	Rossfeld Alpen Bergpreis (hillclimb-5,890 kms)	Touring	850 cc	G. Weber	8'26"5	Fiat Abarth 850 TC
			Touring	1000 cc	K. Weiss	7'18"3	Fiat Abarth 1000 Berlina
			Sport	1300 cc	P. Stürtz	7'00"3	Fiat Abarth OT 1300
			Sport	2000 cc	H. Wengert	6'51"7	Fiat Abarth 2000
			Sport-Prototypes	up to 3000 cc	J. Ortner	6'08"2	Fiat Abarth 3000
6 June	Udine, Italy	Tolmezzo-Verzegnis (hillclimb-6,200 kms)	Touring	600 cc	A. Pieraccini	4'46"5	Fiat Abarth 595
			Touring	700 cc	M. Ruoso	4'59"0	Fiat Abarth 695
			Touring	850 cc	'Tino'	4'26"6	Fiat Abarth 850 TC
			Touring	1000 cc	D. Gargan	4'02"3	Fiat Abarth 1000 Berlina
			Grand Touring	1000 cc	G. Ronchi	4'59"0	Fiat Abarth 1000 OTS
			Sport	1000 cc	G. Rossi	4'51"1	Fiat Abarth 1000
			Sport	1300 cc	A. Rossetto	4'45"7	Fiat Abarth OT 1300
6 June	Alzey, Germany	Rheinhessen Bergrennen Teufelsrutsch (hillclimb-2,250 kms x 3)	Touring	850 cc	B. Rack	4'16"9	Fiat Abarth 850 TC
10 June	Imola, Italy	Trofeo Autosprint (circuit-5,018 kms x 12 laps)	Touring	600 cc	E. Santucci	30'07"6	Fiat Abarth 595
			Touring	700 cc	C. Garrone	30'10"5	Fiat Abarth 695
			Touring (14 laps)	850 cc	E. Gatti	33'09"8	Fiat Abarth 850 TC
			Touring (15 laps)	1000 cc	M. Campanini	32'45"3	Fiat Abarth 1000 Berlina
13 June	Lodrino, Switzerland	Slalom Nazionale (2,000 kms)	Touring	1000 cc	R. Caprez	1'50"3	Fiat Abarth 1000 Berlina
			Touring Special	1000 cc	F. Gilardoni	2'02"9	Fiat Abarth 1000 Berlina
13 June	Asiago, Italy	Coppa Città di Asiago (hillclimb-14,500 kms)	Touring	600 cc	'Tocai'	9'30"4	Fiat Abarth 595
			Touring	700 cc	G.A. Ongaro	11'08"5	Fiat Abarth 695
			Touring	850 cc	'Tino'	8'51"9	Fiat Abarth 850 TC
			Touring	1000 cc	R. Ferretti	8'58"3	Fiat Abarth 1000 Berlina
			Grand Touring	1000 cc	P. Friso	9'20"6	Fiat Abarth 1000 OTS
			Sport	1000 cc	G. Rossi	7'55"9	Fiat Abarth 1000
			Sport	up to 1600 cc	L. De Nicola	9'12"2	Abarth Simca 1300
			Sport-Prototypes	1000 cc	G. Serblin	7'46"6	Fiat Abarth 1000

DATE	LOCATION	RACE	CATEGORY	CLASS	WINNER	TIME OR DISTANCE	CAR
13 June	Brescia, Italy	Scalata al Colle della Maddalena (hillclimb-5,590 kms)	Touring	600 cc	A. Pieraccini	4'16"6	Fiat Abarth 595
			Touring	700 cc	G. Boggiato	4'45"9	Fiat Abarth 695
			Touring	850 cc	S. Valerio	4'06"6	Fiat Abarth 850 TC
			Touring	1000 cc	G. Gasparini	4'01"6	Fiat Abarth 1000 Berlina
			Grand Touring	1000 cc	G. Brambilla	4'26"0	Fiat Abarth 1000 OTS
			Sport	1000 cc	L. Rassega	3'45"9	Fiat Abarth 1000
			Sport	1300 cc	G. Savoldi	3'52"4	Fiat Abarth OT 1300
			Sport	2000 cc	'Pal Joe'	3'36"1	Fiat Abarth 2000
			Prototypes	up to 3000 cc	'Pam'	3'07"9	Fiat Abarth 3000
13 June	Neumarkt, Germany	Jura Bergrennen Neumarkt (hillclimb-2,490 kms x 3)	Touring	850 cc	G. Weber	4'29"4	Fiat Abarth 850 TC
			Touring	1000 cc	K. Weiss	4'00"4	Fiat Abarth 1000 Berlina
			Sport	1300 cc	P. Stürtz	3'46"6	Fiat Abarth OT 1300
			Sport	over 1300 cc	K. Federhofer	3'30"3	Fiat Abarth 2000
			Sport-Prototypes	up to 1150 cc	J. Abt	3'43"4	Fiat Abarth 1000
13 June	Cruseilles, France	Cruseilles-Le Salève (hillclimb-7,800 kms)	Sport-Prototypes	up to 3000 cc	P. Maublanc	4'32"2	Fiat Abarth 3000
20 June	Avignon, France	Course de côte du Mont Ventoux (hillclimb-21,600 kms)	Sport-Prototypes	2000 cc	J. Belin	11'33"8	Fiat Abarth 2000
			Sport-Prototypes	up to 3000 cc	P. Maublanc	10'32"9	Fiat Abarth 3000
20 June	Pont de la Berwinne, Belgium	Course de côte du Val Dieu (hillclimb-2,200 kms)	Touring	850 cc	T. Ernould	1'42"2	Fiat Abarth 850 TC
			Sport-Prototypes	over 1300 cc	F. Neri	1'18"6	Fiat Abarth 3000
20 June	Eggberg, Germany	ADAC Bergrennen (hillclimb-5,100 kms x 2)	Touring	850 cc	G. Weber	6'14"3	Fiat Abarth 850 TC
			Touring	1000 cc	G. Volle	5'53"6	Fiat Abarth 1000 Berlina
			Sport	over 1600 cc	H. Wengert	5'03"0	Fiat Abarth 2000
20 June	Macerata, Italy	Sarnano-Sassotetto (hillclimb-12,400 kms)	Touring	600 cc	E. Santucci	8'57"2	Fiat Abarth 595
			Touring	700 cc	P. Meratti	9'11"8	Fiat Abarth 695
			Touring	850 cc	G. D'Ippolito	9'20"8	Fiat Abarth 850 TC
			Touring	1000 cc	M. Campanini	7'44"3	Fiat Abarth 1000
			Sport	2000 cc	P. Laureati	6'51"1	Fiat Abarth 2000
			Sport-Prototypes	2000 cc	'Bitter'	6'44"1	Fiat Abarth 2000
20 June	Parma, Italy	Salsomaggiore-Monte S. Antonio (hillclimb)	Touring	600 cc	P. Poggi	4'20"9	Fiat Abarth 595
			Touring	700 cc	M. Tramonti	4'40"0	Fiat Abarth 695
			Touring	850 cc	P. Biselli	4'14"3	Fiat Abarth 850 TC
			Touring	1000 cc	P. Gasparini	4'14"1	Fiat Abarth 1000 Berlina
			Sport	up to 1300 cc	'Renzo'	3'51"9	Fiat Abarth 1000
			Sport	2000 cc	'Nicor'	3'30"7	Fiat Abarth 2000
20 June	Rotenburg, Germany	Bergrennen Rotenburg (hillclimb-3,000 kms)	Touring	850 cc	A. Joswig	4'17"6	Fiat Abarth 850 TC
			Sport-Prototypes	2000 cc	P. Wiechert	3'59"3	Fiat Abarth 2000
20 June	Mainz Finthen, Germany	Flugplatz Rennen in Mainz Finthen (circuit-2,100 kms x 30 laps)	Touring	850 cc	B. Rack	37'40"5	Fiat Abarth 850 TC
			Touring	1000 cc	K. Weiss	34'09"3	Fiat Abarth 1000 Berlina
			Sport (50 laps)	up to 1300 cc	K. Brand	36'17"1	Fiat Abarth OT 1300
			Sport-Prototypes (50 laps)	2000 cc	D. Machatius	49'56"7	Fiat Abarth 2000
27 June	Orvieto, Italy	Corsa della Castellana (hillclimb-7,600 kms)	Touring	700 cc	D. Minestrini	5'38"4	Fiat Abarth 695
			Touring	850 cc	A. Gunnella	5'14"5	Fiat Abarth 850 TC
			Touring	1000 cc	A. Stefanelli	4'52"9	Fiat Abarth 1000 Berlina
			Grand Touring	1000 cc	F. Indignozzi	5'25"4	Fiat Abarth 1000 OTS
			Sport-Prototypes	1000 cc	M. Barone	4'33"7	Fiat Abarth 1000
27 June	Palermo, Italy	Corsa al Monte Pellegrino (hillclimb-8,750 kms)	Touring	600 cc	'Alfa'	5'55"3	Fiat Abarth 595
			Touring	700 cc	'Cucciolo'	6'09"0	Fiat Abarth 695
			Touring	850 cc	'Jimmy'	5'49"3	Fiat Abarth 850 TC
			Touring	1000 cc	S. Famoso	5'30"1	Fiat Abarth 1000 Berlina
			Sport	2000 cc	'Amphicar'	4'40"9	Fiat Abarth 2000
			Sport-Prototypes	1000 cc	F. Patanè	5'30"5	Fiat Abarth 1000
27 June	Sion, Switzerland	Automobilslalom de Sion	Touring	1000 cc	S. Plastina	2'02"6	Fiat Abarth 1000 Berlina
27 June	Piesting, Austria	Bergpreis Piesting (hillclimb-2,700 kms)	Touring	1000 cc	H. Schenzel	2'17"8	Fiat Abarth 1000 Berlina
27 June	Schotten, Germany	Bergpreis Schottenring (hillclimb-6,500 kms x 2)	Touring	850 cc	B. Rack	4'48"4	Fiat Abarth 850 TC
			Sport	1300 cc	N. Knatterton	4'44"1	Fiat Abarth OT 1300
			Sport	2000 cc	W. Scheeren	3'51"6	Fiat Abarth 2000
27 June	Bremen, Germany	Slalom Silbernen Schlüssen von Bremen (2,000 kms)	Touring	1000 cc	B. Ziegler	2'22"3	Fiat Abarth 1000 Berlina
27 June	Bolzano, Italy	Bolzano-Mendola (hillclimb-14,400 kms)	Touring	600 cc	A. Pieraccini	11'33"1	Fiat Abarth 595
			Touring	700 cc	B. Caramaschi	11'53"6	Fiat Abarth 695
			Touring	850 cc	I. Pain	10'57"5	Fiat Abarth 850 TC
			Touring	1000 cc	R. Amighini	10'13"0	Fiat Abarth 1000 Berlina
			Sport	1000 cc	G. Rossi	9'50"7	Fiat Abarth 1000
			Sport	1300 cc	'Frank Mc Boden'	10'06"5	Fiat Abarth OT 1300
27 June	Clouneaux, France	Course de côte de Clouneaux (hillclimb-2,500 kms)	Sport-Prototypes	up to 3000 cc	P. Maublanc	1'29"8	Fiat Abarth 3000
27 June	Nîmes, France	Course de côte du Minier (hillclimb-21,000 kms)	Sport-Prototypes	2000 cc	J. Daumet	12'33"2	Fiat Abarth 2000
4 July	Echelette, France	Course de côte de l'Echelette (hillclimb-3,000 kms)	Sport-Prototypes	up to 3000 cc	J. Daumet	1'55"6	Fiat Abarth 2000

DATE	LOCATION	RACE	CATEGORY	CLASS	WINNER	TIME OR DISTANCE	CAR
4 July	Vuillifans, France	Course de côte Vuillifans-Echevannes (hillclimb-4,800 kms x 2)	Sport-Prototypes	2000 cc	J. Belin	2'36"1	Fiat Abarth 2000
			Sport-Prototypes	up to 3000 cc	P. Maublanc	2'24"7	Fiat Abarth 3000
4 July	Trento, Italy	Trento-Bondone (hillclimb-17,300 kms)	Touring	850 cc	E. Gatti	14'36"0	Fiat Abarth 850 TC
			Touring	1000 cc	L. Ferragina	13'51"0	Fiat Abarth 1000 Berlina
			Sport	1000 cc	E. Tinghi	12'53"1	Fiat Abarth 1000
			Sport	1300 cc	P. Stürtz	12'50"5	Fiat Abarth OT 1300
			Sport	over 2000 cc	J. Ortner	11'16"8	Fiat Abarth 3000
4 July	Ascoli Piceno, Italy	S. Benedetto del Tronto-Acquaviva (hillclimb-6,500 kms)	Touring	600 cc	F. Ferroni	4'28"1	Fiat Abarth 595
			Touring	700 cc	V. Gattafoni	4'27"5	Fiat Abarth 695
			Touring	850 cc	'Svig'	4'12"7	Fiat Abarth 850 TC
			Sport	2000 cc	P. Laureati	3'22"8	Fiat Abarth 2000
11 July	Gives, Belgium	Course de côte de Gives (hillclimb-2,400 kms x 2)	Touring	1300 cc	A. Person	1'43"7	Fiat Abarth 1300
11 July	Hockenheim, Germany	Rundstreckenrennen (circuit-2,630 kms x 18 laps)	Touring	850 cc	F. Lampe	---	Fiat Abarth 850 TC
			Touring (20 laps)	1000 cc	K. Trettan	---	Fiat Abarth 1000 Berlina
11 July	Potenza, Italy	Trofeo Città di Potenza (hillclimb-7,100 kms)	Touring	600 cc	M. De Luca	6'04"1	Fiat Abarth 595
			Touring	700 cc	M. Casciaro	6'11"6	Fiat Abarth 695
			Touring	850 cc	G. D'Ippolito	5'46"9	Fiat Abarth 850 TC
			Touring	1000 cc	M. De Luca	5'29"9	Fiat Abarth 1000 Berlina
			Grand Touring	1000 cc	R. Pezzella	7'06"5	Fiat Abarth 1000 OTS
			Sport	1000 cc	'Sergio'	5'20"2	Fiat Abarth 1000
			Sport	1300 cc	C. Turizio	5'18"2	Abarth Simca 1300
			Sport	2000 cc	'Bitter'	4'40"7	Fiat Abarth 2000
11 July	Adenau, Germany	Nürburgring 6 Stunden Rennen (circuit-22,810 kms x 6 hours)	Touring	1000 cc	Freiherr-Ranke	kms 752,734	Fiat Abarth 1000 Berlina
11 July	Sestrière, Italy	Cesana-Sestrière (hillclimb-10,400 kms)	Touring	600 cc	R. Mazzeschi	7'10"7	Fiat Abarth 595
			Touring	700 cc	P. Bernardi	7'11"2	Fiat Abarth 695
			Touring	850 cc	E. Gatti	6'40"6	Fiat Abarth 850 TC
			Touring	1000 cc	M. Campanini	6'16"9	Fiat Abarth 1000 Berlina
			Grand Touring	1000 cc	C. Spica	6'59"7	Fiat Abarth 1000 OTS
			Sport	1000 cc	F. Marenco	6'12"6	Fiat Abarth 1000
			Sport	1300 cc	G. Savoldi	6'29"5	Fiat Abarth OT 1300
			Sport	up to 3000 cc	J. Ortner	4'59"7	Fiat Abarth 3000
			Sport-Prototypes	2000 cc	A. Merzario	5'02"1	Fiat Abarth 2000
			Sport-Prototypes	up to 3000 cc	L. Taramazzo	5'20"2	Fiat Abarth 3000
11 July	Belluno, Italy	Pedavena-Croce d'Anne (hillclimb-8,000 kms)	Touring	600 cc	J. Strenghetto	6'12"6	Fiat Abarth 595
			Touring	700 cc	M. Ruoso	6'13"4	Fiat Abarth 695
			Touring	850 cc	I. Pain	5'45"5	Fiat Abarth 850 TC
			Touring	1000 cc	D. Gargan	5'17"9	Fiat Abarth 1000 Berlina
			Grand Touring	1000 cc	G. Ronchi	5'57"9	Fiat Abarth 1000 OTS
			Sport	1000 cc	R. De Gan	5'15"6	Fiat Abarth 1000
			Sport	1300 cc	G. Trebbia	5'49"8	Fiat Abarth OT 1300
			Sport	2000 cc	'Amphicar'	4'25"0	Fiat Abarth 2000
			Sport-Prototypes	1000 cc	G. Serblin	4'49"0	Fiat Abarth 1000
18 July	Taubensuhl, Germany	Taubensuhl Bergrennen (hillclimb-6,030 kms x 2)	Touring	1000 cc	G. Volle	7'11"3	Fiat Abarth 1000 Berlina
			Sport	1300 cc	K. Brand	7'03"0	Fiat Abarth OT 1300
			Sport-Prototypes	up to 1300 cc	J. Fuchs	7'08"3	Fiat Abarth 1300
			Sport-Prototypes	over 1600 cc	P. Wiechert	6'58"2	Fiat Abarth 2000
18 July	Oberpfalz, Germany	Bergpreis der Oberpfalz (hillclimb-3,140 kms x 2)	Touring	850 cc	F. Neitzel	3'49"3	Fiat Abarth 850 TC
			Sport	2000 cc	S. Stebenbürger	3'26"5	Fiat Abarth 2000
			Prototypes	up to 1600 cc	E. Ele	3'17"4	Fiat Abarth 1300
18 July	Cacharat, France	Course de côte de Cacharat (hillclimb-2,200 kms)	Sport	2000 cc	'Cyprien'	1'44"2	Fiat Abarth 2000
25 July	Goesdorf, Luxembourg	Course de côte de Goesdorf (hillclimb-2,800 kms)	Sport	2000 cc	W. Scheeren	1'24"6	Fiat Abarth 2000
25 July	Hockenheim, Germany	Rundstreckenrennen Hessen (circuit-2,630 kms x 20 laps)	Touring	850 cc	K. Schmidt	29'18"8	Fiat Abarth 850 TC
			Touring	1000 cc	K. Weiss	27'55"8	Fiat Abarth 1000 Berlina
25 July	Villars, Switzerland	Course de côte Villars-Burquin (hillclimb)	Touring	850 cc	K. Schneider	1'24"4	Fiat Abarth 850 TC
25 July	Potzberg, Germany	Rheinland Pfalz Bergpreis (hillclimb)	Touring	850 cc	F. Neitzel	6'50"3	Fiat Abarth 850 TC
			Touring	1000 cc	G. Volle	6'11"2	Fiat Abarth 1000 Berlina
			Sport	up to 1600 cc	K. Brand	6'11"2	Fiat Abarth OT 1300
			Prototypes	up to 1300 cc	J. Fuchs	5'50"1	Fiat Abarth 1300
25 July	Ascoli Piceno, Italy	Ascoli-Colle San Marco (hillclimb-10,200 kms)	Touring	600 cc	E. Santucci	6'52"1	Fiat Abarth 595
			Touring	700 cc	M. Magri	6'59"7	Fiat Abarth 695
			Touring	850 cc	S. Valerio	6'41"4	Fiat Abarth 850 TC
			Touring	1000 cc	L. Ferragina	6'13"6	Fiat Abarth 1000 Berlina
			Grand Touring	1000 cc	C. Spica	6'50"4	Fiat Abarth 1000 OTS
			Sport	1000 cc	F. Marenco	6'12"2	Fiat Abarth 1000
			Sport	1300 cc	C. Turizio	6'15"1	Fiat Abarth OT 1300
			Prototypes	up to 1300 cc	G. Boeris	5'42"9	Fiat Abarth 1300

Abarth, the Records

From June 1956 to October 1966, special Abarth cars established 113 world and international records on the full Monza circuit, including the high speed banked section over a total of 10,000 kms. With the exception of the Alfa Abarth 1100 in May 1965, all the other cars were Fiat Abarths.

DATE	CAR	CLASS	DRIVERS	DISTANCE OR TIME	TIME OR DISTANCE COVERED	AVERAGE (kms/h)
17-18 giungo 1956	750 Bertone	H	R. Cattini, U. Maglioli, M. Poltronieri, A. Thiele	24 hours	kms 3.743,642	155,985
27-29 June 1956	750 Bertone	H	B. Cahier, R. Cattini, P. Frère	5000 Kms	34h 26' 27"	145,175
			W. Honneger, G. Lurani,	5000 miles	55h 26' 53"	145,121
			M. Poltronieri, A. Thiele,	10.000 Kms	71h 26' 53"	140,658
			H.U. Wieselmann, G. Wilkins	48 hours	kms 6.978,931	145,394
				72 hours	kms 10.125,560	140,632
27 July 1956	500 Bertone	I	R. Cattini, A. Della Beffa,	2000 Kms	14h 17'16"	139,978
			G. Fraccari, M. Poltronieri, A. Thiele	2000 miles	23h 20' 55"	137,771
				24 hours	kms 3.310,769	137,948
9 August 1956	800 Bertone	G	R. Cattini, M. Poltronieri, A. Thiele	1000 miles	9h 32, 31"	168,659
				2000 Kms	11h 44' 47"	170,264
				2000 miles	19h 13' 59"	167,350
				5000 Kms	29h 48' 58"	167,694
				12 hours	kms 2.044,924	170,410
				24 hours	kms 4.025,333	167,722
12 October 1956	500 Bertone	I	R. Cattini, M. Poltronieri, A. Thiele	500 mile	5h 21' 39"	150,098
				1000 Kms	6h 44' 20"	148,387
				1000 miles	10h 53' 55"	147,664
				6 hours	kms 901,102	150,183
				12 hours	kms 1.175,788	147,982
18 May 1957	Alfa-Abarth 1100	G	P. Frère, A. Thiele	500 Kms	2h 31' 02"	198,618
				500 miles	4h 10' 05"	193,046
				1000 Kms	5h 12' 20"	192,098
				3 hours	kms 580,116	193,372
				6 hours	kms 1.153,338	192,223
24-27 July 1957	750 Pininfarina	H	G. Cabianca, R. Cattini, M. Guarnieri,	500 Kms	2h 43' 15"	183,752
			C. Manfredini, M. Poltronieri	3 hours	kms 551,719	183,906
				500 miles	4h 25' 09"	182,079
				1000 Kms	5h 28' 29"	182,652
				6 hours	kms 1.086,022	181,003
				1000 miles	8h 59' 25"	179,007
				2000 Kms	11h 07' 01"	179,904
				12 hours	kms 2.161,498	180,124
				2000 miles	18h 09' 43"	177,220
				24 hours	kms 4.258,909	177,454
				5000 Kms	28h 43' 29"	174,065
				5000 miles	47h 14'00"	170,345
				48 hours	kms 8.174,476	170,301
				10.000 Kms	59h 03' 02"	169,346
				72 hours	kms 11.906,473	165,367
25 October 1957	750 Pininfarina	H	P. Frère, M. Guarnieri	200 miles	1h 35' 25"	202,347
				500 Kms	2h 30' 36"	199,189
				3 hours	kms 592,746	197,582
13-20 February 1958	500 Berlina	I	R. Cattini, A. Giuberti,	4 days	kms 10.457,874	108,936
			M. Guarnieri, C. Manfredini,	5 days	kms 12.993,716	108,280
			M. Poltronieri, E. Zagato	15.000 Kms	139h 16' 33"	107,699
				6 days	kms 15.531,957	107,860
				10.000 miles	149h 9' 29"	107,894
				7 days	kms 18.186,440	108,252
27 September- 7 October 1958	500 Pininfarina	I	G. Baghetti, G. Bassi, R. Cattini,	5000 Kms	38h 13' 32"	130,804
			A. Giuberti, E. Crivellari, M. Guarnieri,	48 hours	6.,985334	131,978
			C. Manfredini, M. Poltronieri, M. Toselli	5000 miles	61h 31' 52"	130,752
				72 hours	kms 9.450,7787	131,286
				10.000 Kms	76h 10' 01"	131,290
				96 hours	kms 12.079,130	125,824
				15.000 Kms	118h 50' 35"	126,216

DATE	CAR	CLASS	DRIVERS	DISTANCE OR TIME	TIME OR DISTANCE COVERED	AVERAGE (kms/h)
				120 hours	kms 15.154,958	126,291
				10.000 miles	135h 11' 28"	119,041
				6 days	kms 17.258,447	119,850
				20.000 Kms	165h 39' 39"	120,728
				7 days	kms 20.303,508	120,854
				8 days	kms 23.151,847	120,580
				15.000 miles	199h 45' 37"	120,845
				25.000 Kms	208h 45' 40"	119,754
				9 days	kms 25.336,969	117,300
				10 days	kms 27.931,439	116,380
21 October 1958	500 Pininfarina	I	G. Baghetti, R. Cattini, M. Poltronieri	500 miles	5h 31' 21"	154,485
				6 hours	kms 930,196	155,032
				1000 Kms	6h 26' 21"	155,296
				1000 miles	10h 21' 45"	155,303
				12 hours	kms 1.852,093	154,341
				2000 Kms	13h 02' 46"	153,319
25 October 1958	750 Pininfarina	H	R. Cattini	100 miles	47' 07"	204,924
				200 Kms	58' 21"	205,655
				1 Ora	kms 205,760	205,760
				200 miles	1h 33' 22"	206,811
				500 Kms	2h 25' 04"	206,796
1-3 August 1959	500 Pininfarina	I	G. Baghetti, M. Poltronieri	500 miles	4h 51' 18"	165,740
				6 hours	kms 993,953	165,658
				1000 Kms	6h 02' 12"	165,652
				1000 miles	6h 02' 12"	159,677
				12 hours	kms 1.923,330	160,277
22 September 1960	750 Pininfarina	H	U. Maglioli	100 miles	45' 24"1	212,672
				200 Kms	56' 13"2	213,434
				2000 miles	1h 29' 54" 1	214,809
				1 hour	kms 213,643	213,643
28 September - 1 October 1960	1000 Pininfarina	G	G. Baghetti, G. Bassi, G. Castellina, R. Cattini, M. Leto di Priolo, U. Maglioli, C. Manfredini, M. Poltronieri, G. Rigamonti, A. Thiele	12 hours	kms 2.443,875	203,656
				2000 miles	16h 00'14" 4	201,115
				24 hours	kms 4.771,082	198,795
				5.000 Kms	25h 05' 43" 4	199,238
				5.000 miles	41h 43' 08" 2	192,878
				48 hours	kms 9.132,672	190,264
				10.000 Kms	52h 16' 21" 2	191,376
				72 hours	kms 13.441,498 (world record)	186,68
1-5 March 1963	2300 S Coupé	D	E. Arena, G. Balzarini, M. Bianchi, R. Cattini, G. Demetz, W. Gausch, P. Ickx, T. Pilette, M. Poltronieri, K. Steinmetz	5000 miles	44h 47' 44" 4	179,631
				10.000 Kms	55h 15' 28" 3	180,778
				48 hours	kms 8.650,013	180,208
				72 hours	kms 12.824,545	178,11
20 October 1965	1000 monoposto F2	G	C. Abarth	1/4 mile	13" 62	106,344
				500 metres	15" 38	117,035
21 October 1965	2000 monoposto F2	E	C. Abarth	1/4 mile	11" 55 (world record)	125,348
				500 metres	13" 37 (world record)	134,62
27-28 October 1965	750 monoposto F2	H	M. Poltronieri	1/4 mile	15" 125	95,762
				500 metres	17" 15	104,95
27-28 October 1965	1450 OT Sport	F	K. Steinmetz	1/4 mile	12" 87	112,541
				500 metres	14" 915	124,86
28 October 1965	1000 monoposto F2	G	K. Steinmetz	1 Km	21" 25	143,884
11 December 1965	2000 OT Sport	E	K. Steinmetz	1 Km	21" 535	167,253
27-31 October 1966	750 monoposto F2	H	J. Ortner	1/4 mile	14" 585	99,307
				500 metres	16" 955	106,163
31 October 1966	1000 monoposto F2	G	J. Ortner	1/4 mile	13" 035	111,116
27-31 October 1966	2000 monoposto	E	J. Ortner	1/4 mile	11" 095 (world record)	130,545
				500 metres	12" 855 (world record)	140,023

Bibliography

L'Auto Italyna, 1930-1969
Autosprint, 1961-1971
Balestra Nino, De Agostini Cesare, *Cisitalia*, Automobilia, Milano 1991
Braden Pat, Schmidt Greg, *Abarth*, Osprey, London 1983
Caballo Ernesto, *Pininfarina-Nato con l'automobile*, Palazzi, Milano 1968
Conradt Dirk Michael, *Porsche 356*, Motorbuch Verlag, Stuttgart 1989 (edizione italiana Giorgio Nada Editore, Vimodrone 1991)
Cosentino Alfred S. *Abarth Guide*, Nigensha, Tokyo 1984
Deganello Elvio, Donati Renato, *Abarth 595/695*, Giorgio Nada Editore, Vimodrone 1998
Deganello Aligi, Deganello Elvio, *Abarth 850 TC e 1000*, Giorgio Nada Editore, Vimodrone 2001
Fagiuoli Gianfranco, Gerosa Guido, *Carlo Abarth*, L'Editrice dell'Automobile, Roma 1967
Giacosa Dante, *I miei 40 anni di progettazione alla Fiat*, Automobilia, Milano 1979
Greggio Luciano, *Bertone*, Giorgio Nada Editore, Vimodrone 1992
Orsini Luigi, *Mille miles*, Edizioni Abiemme, Milano 1990
Prunet Antoine, *Pininfarina-Arte e Industria 1930-2000*, Giorgio Nada Editore, Vimodrone 2000
Wimpffen János L. *Time and two seats*, Motorsport Research Group, Redmond (USA) 1999
Zampini Salazar Carlo Felice, *Abarth-Catalogue Raisonné*, Automobilia, Milano 1986

Index

Index of technical specifications

The victories-Index

Printed by Grafiche D'Auria,
Ascoli Piceno, Italy April 2002.